Advances in Intelligent Systems and Computing

Volume 1370

The series "Advances in Intelligent Systems and Computing" contains publications on theory, applications, and design methods of Intelligent Systems and Intelligent Computing. Virtually all disciplines such as engineering, natural sciences, computer and information science, ICT, economics, business, e-commerce, environment, healthcare, life science are covered. The list of topics spans all the areas of modern intelligent systems and computing such as: computational intelligence, soft computing including neural networks, fuzzy systems, evolutionary computing and the fusion of these paradigms, social intelligence, ambient intelligence, computational neuroscience, artificial life, virtual worlds and society, cognitive science and systems, Perception and Vision, DNA and immune based systems, self-organizing and adaptive systems, e-Learning and teaching, human-centered and human-centric computing, recommender systems, intelligent control, robotics and mechatronics including human-machine teaming, knowledge-based paradigms, learning paradigms, machine ethics, intelligent data analysis, knowledge management, intelligent agents, intelligent decision making and support, intelligent network security, trust management, interactive entertainment, Web intelligence and multimedia.

The publications within "Advances in Intelligent Systems and Computing" are primarily proceedings of important conferences, symposia and congresses. They cover significant recent developments in the field, both of a foundational and applicable character. An important characteristic feature of the series is the short publication time and world-wide distribution. This permits a rapid and broad dissemination of research results.

Indexed by DBLP, INSPEC, WTI Frankfurt eG, zbMATH, Japanese Science and Technology Agency (JST).

All books published in the series are submitted for consideration in Web of Science.

More information about this series at http://www.springer.com/series/11156

Falgun Thakkar · Geetali Saha · Celia Shahnaz · Yu-Chen Hu
Editors

Proceedings of the International e-Conference on Intelligent Systems and Signal Processing

e-ISSP 2020

Springer

Editors
Falgun Thakkar
Department of Electronics
and Communication
G H Patel College of Engineering
and Technology
Vallabh Vidyanagar, Anand, Gujarat, India

Geetali Saha
Department of Electronics
and Communication
G H Patel College of Engineering
and Technology
Vallabh Vidyanagar, Anand, Gujarat, India

Celia Shahnaz
Department of Electrical and Electronics
Engineering
Bangladesh University of Engineering
and Technology
Dhaka, Bangladesh

Yu-Chen Hu
Department of Computer Science
and Information Management
Providence University
Taichung, Taiwan

ISSN 2194-5357 ISSN 2194-5365 (electronic)
Advances in Intelligent Systems and Computing
ISBN 978-981-16-2122-2 ISBN 978-981-16-2123-9 (eBook)
https://doi.org/10.1007/978-981-16-2123-9

This Springer imprint is published by the registered company Springer Nature Singapore Pte Ltd.
The registered company address is: 152 Beach Road, #21-01/04 Gateway East, Singapore 189721, Singapore

Preface

International e-Conference on Intelligent Systems and Signal Processing (e-ISSP 2020) is the First Virtual Conference organized by the Department of Electronics and Communication, G H Patel College of Engineering and Technology, V. V. Nagar, Gujarat.

This was a three-day event, with the first day—December 28, 2020—allocated for Pre-Conference workshops. The Keynote talks were organized on the 29th and 30th of December, 2020. The Dignitaries invited were Dr. Ami Wiesel, University of Jerusalem, Dr. Latif Ladid, University of Luxembourg, Dr. Rangaraj Rangayan, University of Calagry, and Dr. Ram Bilas Pachori, IIT Indore. All of them enlightened the attendees with their inspiring talks.

The Conference provided scope for traditional topics of Research and the tracks were Track 1—Signal Processing and Communication, Track 2—Applied Electronics and Emerging Technology, Track 3—Computer Vision and Artificial Intelligence, Track 4—Big Data IoT and Cloud Computing, and Track 5—Industrial Automation and Robotics. Keeping in mind the diverse developments, an additional track dedicatedly for Interdisciplinary studies was also introduced.

A total of 128 papers were received and after a rigorous peer-review process, 61 were shortlisted and presented, that is, amounting to 53.12% of Acceptance rate.

The Valedictory was organized on the concluding day, December 30, 2020, where many attendees shared their wonderful experiences of participation and learning.

Vallabh Vidyanagar, Anand, India — Dr. Falgun Thakkar
Vallabh Vidyanagar, Anand, India — Prof. Geetali Saha
Dhaka, Bangladesh — Dr. Celia Shahnaz
Taichung, Taiwan — Prof. Yu-Chen Hu

Contents

About the Editors

Dr. Falgun Thakkar obtained Ph.D. from National Institute of Technology, Allahabad, in February 2018. He graduated from Birla Vishvakarma Mahavidyalaya (BVM) in year 2004 and completed his Masters of Engineering in Communication from GCET, Sardar Patel University (SPU), V. V. Nagar, in year 2010. Dr. Falgun has published more than 30 research articles in various international and national journals and conferences. He has served as Reviewer in many international journals and conferences of repute. His areas of interests include multimedia security, evolutionary algorithms, medical imaging, machine and deep learning, RF and microwave antenna design. He has also received a grant to organize one-week faculty development program on topic 'Recent trends in medical image processing' by ISTE, GTU and GUJCOST. At present, he is working on cardiovascular imaging, diabetic retinopathy, medical image security and DRA design for various applications.

Prof. Geetali Saha has submitted her doctoral thesis to Gujarat Technological University (GTU), Gujarat, India, and is awaiting her final defense. She graduated from G H Patel College of Engineering and Technology, Sardar Patel University (SPU), in the year 2001 and completed her Master's in 2011. She is presently working as Associate Professor at G H Patel College of Engineering and Technology, Department of Electronics and Communication. She is invited as Technical Program Committee Member and Reviewer for the 7th International Conference on Time Series and Forecasting (ITISE-2021), due from April 7–9, 2021, at Gran Canaria, Spain. Apart from that, she has served as Technical Program Committee Member and Reviewer for International Conference on Computational Performance Evaluation (ComPE-2020) during July 2–4, 2020, at North-Eastern Hill University (NEHU), Shillong, India, and was associated as IEEE Technical Session Chair for International Conference on Innovations in Power and Advanced Computing Technologies: iPact'17 and iPact'19 at Vellore Institute of Technology, VIT, Vellore, India. Department of Science and Technology (DST) and Texas Instruments (TI), as a part of INDIA Innovation Challenge Design Contest 2018, have awarded her recognition in appreciation for fostering an ecosystem bridging government, industry and academia. She has a few Springer book chapters and IEEE papers published in her domains of research interest that include artificial neural networks, deep neural networks, time series forecasting, IoT applications, real-time object tracking and encryption/decryption methods.

Dr. Celia Shahnaz has completed her Ph.D. from Electrical and Computer Engineering Department of Concordia University Montreal, Quebec, Canada, in 2009. At present, she is Professor in Electrical and Electronic Engineering Department of Bangladesh University of Engineering and Technology (BUET). She is also serving as Chair, IEEE Bangladesh section, and also Candidate for 2021–2022 IEEE Region 10 Director-Elect/Delegate-Elect. She is also having many other portfolios in IEEE Bangladesh section like Member of IEEE New Initiative Committee, Member of IEEE Signal Processing Society (SPS) Women in Signal Processing Committee, Member of IEEE WIE WePOWER Subcommittee and many others. In addition to that, she is Founder & Chair of IEEE SPS, IAS Bangladesh Chapter. Her research interest includes the area of deep learning, pattern recognition and machine learning for audio, video, biomedical and power signals. Also, she is working in signal processing for speech analysis and enhancement, multimodal emotion recognition, multimedia communication and many others.

Dr. Yu-Chen Hu received his Ph.D. degree in Computer Science and Information Engineering from the Department of Computer Science and Information Engineering, National Chung Cheng University, Chiayi, Taiwan, in 1999. Currently, Dr. Hu is Professor in the Department of Computer Science and Information Management, Providence University, Sha-Lu, Taiwan. He is Senior Member of IEEE. He is also Member of Computer Vision, Graphics, and Image Processing (CVGIP), Chinese Cryptology and Information Security Association (CCISA), Computer Science and Information Management (CSIM) and Phi Tau Phi Society of the Republic of China. He joints the editorial boards of several other journals. His research interests include digital forensics, information hiding, image and signal processing, data compression, information security and data engineering.

Signal Processing and Communication Engineering

Design and Analysis of Modified Split Ring Resonator Structured Multiband Antenna for WCDMA and WiMAX Applications

Upesh Patel, Trushit Upadhyaya, Rajat Pandey, Arpan Desai, and Killol Pandya

Abstract A miniaturized slotted multiband antenna is presented. The antenna consists of a square Split Ring Resonator-shaped slot in the radiating patch. The split ring resonator structure-based antenna presents operational capability at 2.18, 2.56 and 3.52 GHz, respectively. The achieved return loss at the resonance frequencies are 21.38, 12.00 and 25.46 dB and the bandwidth of the proposed antenna are 2.51, 2.01 and 3.27% for respective frequencies. The standard low-loss cost-effective FR-4 laminates of 1.6 mm thickness substrate were used for the design. The proposed prototype antenna was fabricated and measured; the results show reliable agreement with the simulated results. The antenna possesses adequate gain values at all resonating frequencies. The reflection coefficient, radiation pattern and other antenna parameters are exhibiting quite satisfactory results, fulfilling the requirement for WCDMA and WiMAX applications.

Keywords Metamaterial antenna · Multiband planar antenna · Split ring resonator · WCDMA · WiMAX

1 Introduction

With the increasing development in wireless technologies, there is higher demand for portable antennas which operate on multiple applications at multiple frequencies. Due to the smaller size of mobile terminals, multiple antennas cannot be accommodated within the available space for realizing various applications. One of the promising solutions is to use the multiband antenna that resonates at a multi-frequency band thus avoid the use of numerous antennas on a single device [1]. Designing and analyzing a conventional monopole or dual band microstrip patch antennas have various methods such as cavity model or transmission line model but

U. Patel (✉) · T. Upadhyaya · R. Pandey · A. Desai · K. Pandya
Department of Electronics & Communication Engineering, Chandubhai S Patel Institute of Technology, Charotar University of Science and Technology, Changa, India
e-mail: upeshpatel.ec@charusat.ac.in

F. Thakkar et al. (eds.), *Proceedings of the International e-Conference on Intelligent Systems and Signal Processing*, Advances in Intelligent Systems and Computing 1370,
https://doi.org/10.1007/978-981-16-2123-9_1

for the multiband antenna, there is not any specific method available in the literature for any specific applications [2]. Despite significant usage and importance of microstrip antennas, miniaturization of the antenna is the biggest challenge, although there are multiple methods available in the literature for the same like introducing slots in patch [3], use of metamaterials [4], defected ground structures (DGS) [4–6], split ring resonators (SRR) [4, 5], etc. In the last decade, a lot of noticeable research has been done in the development of multiband antennas. A compact F-shaped triple-band antenna consists of radiator and defected ground structure was proposed in [6]. A pair of rectangular slits along with the inverted U-shaped slot on patch and a hexagon-shaped slot cut on the ground plane are presented in [7, 8]. In [8], a multiband antenna radiating at 2.4, 3.52, and 5.68 GHz frequencies is proposed that comprises a U slot inverted slot, circular-shaped radiating patch, and rectangular slits. The ground plane consists of a slot having a hexagonal shape that leads to defected ground structure. A multiband compact antenna fed using a meandered coplanar waveguide is reported in 9] that is formed using complementary split ring-shaped slots. There are also a few other techniques available in the literature to obtain multiband nature of antennas such as one open-ended circular ring along with the two circular rings [10], PIN diode switches [10] used for reconfigure the multiband antenna and cylindrical dielectric resonator reported in [11, 12].

Negative refractive index materials are also known as left-handed materials due to a few properties they exhibit. There are two primary elements of left-handed materials. The first element is split ring resonator (SRR) which is a compact LC resonator in which inductance L is generated because of the metallic loops and capacitance C is generated due to dielectric gaps. SRRs are responsible for induction of magnetic dipole moment. The second element is thin rods which induce electric dipole moment. Split ring resonators are the important phenomena for the miniaturization and gain improvement. The conventional split ring resonator unit cell is composed by two circular coplanar metallic rings, either circular or square, each with a split displaced of 180°. These rings were patterned on a low-loss dielectric substrate having the same center and alienated from each other by a short gap distance. SRRs are of various shapes and size such as rectangular, square and triangular [12]. A wide range of applications can be covered by split ring resonators such as frequency selective surfaces (FSSs) [13, 14], absorbers [15], sensors [16], antenna structures, high-frequency modulators [17–19] and composite materials [20–23]. Negative refractive materials are also used for the miniaturization in multiple methods reported in [24–28].

In this paper, the proposed multiband antenna is able to support Wideband Code Division Multiple Access (WCDMA) and Worldwide Interoperability for Microwave Access (WiMAX) applications.

2 Antenna Geometry

The proposed antenna configuration is illustrated in Fig. 1. As a substrate, the cost-effective FR4 of dielectric constant 4.4 and thickness of 1.6 mm is used. The substrate

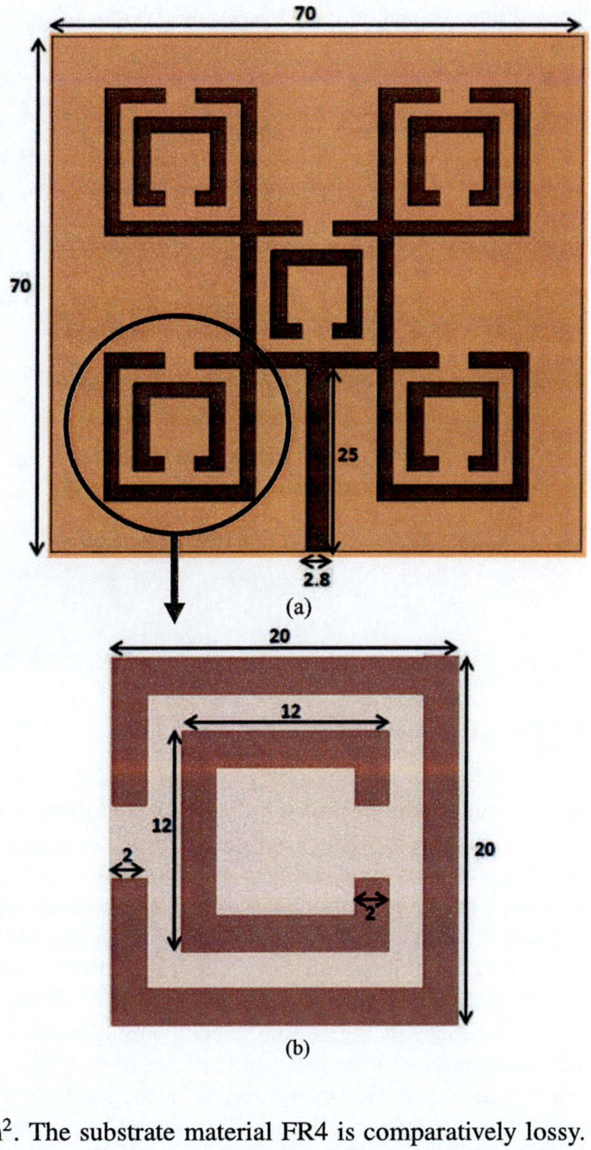

Fig. 1 Antenna geometry **a** microstrip patch **b** the unit cell

dimensions are 70 × 70 mm². The substrate material FR4 is comparatively lossy. However, due to higher value of loss tangent, this material supports mass production of devices. Optimal and appropriate feed location have been selected for the feeding matching at 50 Hz for excitation. Front view of the antenna is shown in Fig. 1a with the detailed dimension, and the unit cell SRR dimensions are illustrated in Fig. 1b.

Figure 2 illustrated the fabricated prototype of the proposed antenna using PCB fabrication machine.

Fig. 2 Fabricated prototype

3 Results and Discussion

The various antenna parameters are analyzed using a finite element method-based High-Frequency Structure Simulator. Reflection coefficient measurement of the antenna was measured using Keysight Vector Network Analyzer N9912A and radiation pattern measurement was carried out in an anechoic chamber.

Figure 3 Illustrated simulated and measured reflection coefficient values at different resonating frequencies like 21.38 dB at 2.18 GHz, 12 dB at 2.56 GHz and 25.46 dB at 3.52 GHz, respectively.

As illustrated in Fig. 4, the anechoic chamber is used to measure the antenna radiation patterns for both E and H planes. The typical size of anechoic chamber is 3 m × 3 m × 3 m. However, because of the electrical compactness of the proposed antenna, this chamber is suitable for radiation pattern measurement.

The software simulated and measured antenna radiation patterns at the target resonant frequencies are shown in Fig. 3. The electrical radiation characteristics of the proposed antenna were studied through simulator, and all the results were compared with the fabricated antenna. The simulated radiation patterns at the frequencies 2.18, 2.56 and 3.52 GHz are shown in Fig. 5. The simulated and measured results show good similar radiation patterns in the E and H Planes.

Figure 6 illustrates the current distribution at the modified SRR patch of the antenna. Higher amount of current density is visible because of the lossy nature of the split ring resonators.

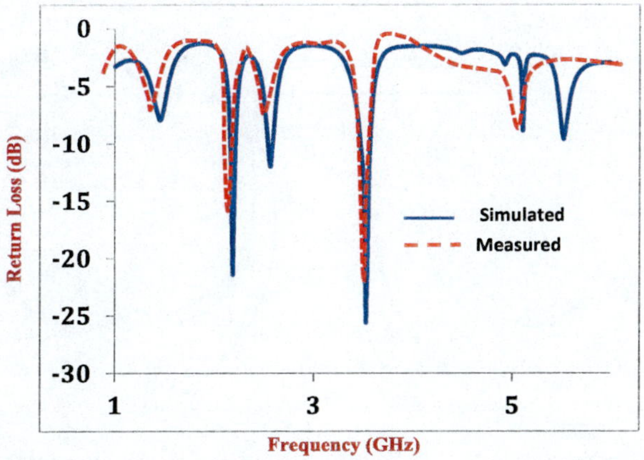

Fig. 3 Reflection coefficient (simulated and measured)

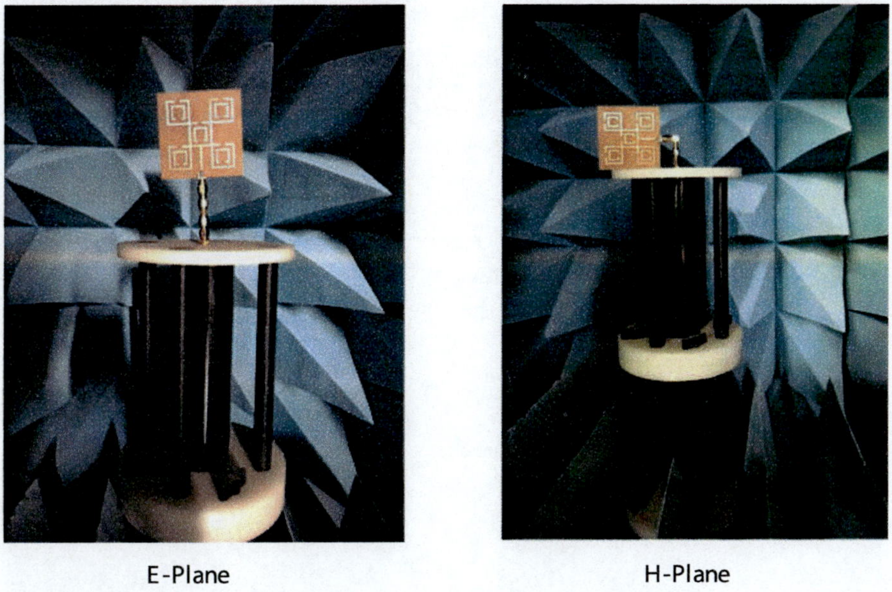

Fig. 4 Radiation pattern measurement

Table 1 shows the comparisons of various antenna parameters at targeted frequency for simulated and measured results. Due to complex mechanical structure of the antenna, slight difference in the measured results occurs.

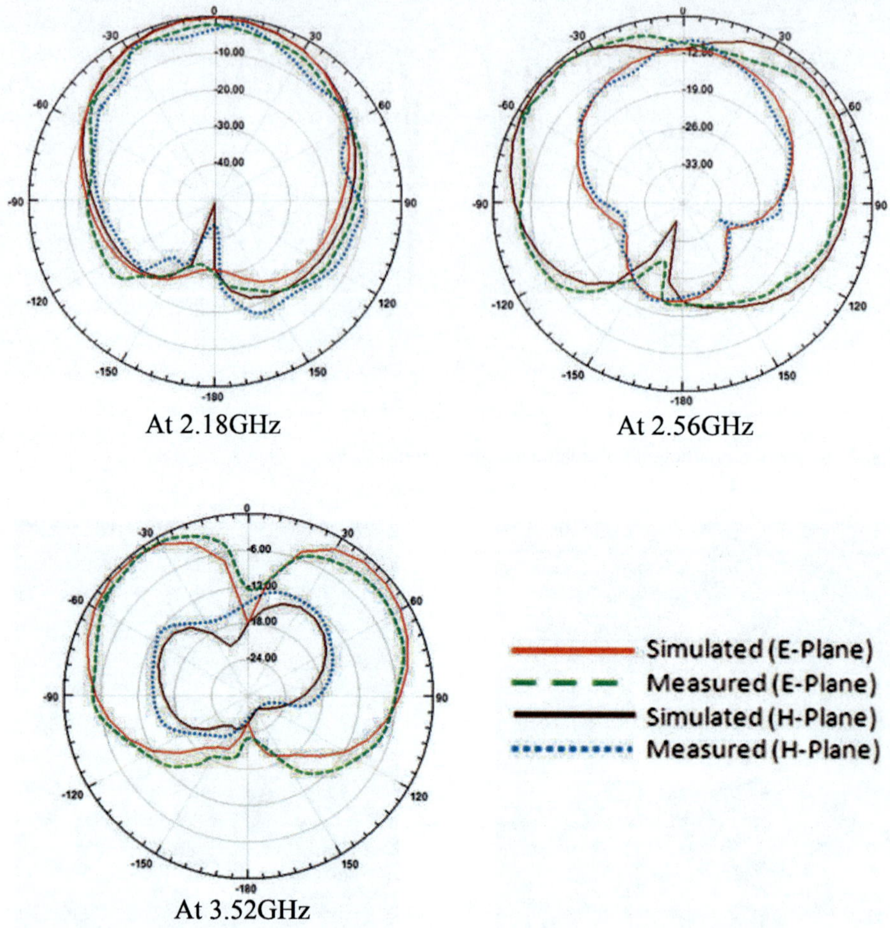

Fig. 5 Radiation patterns at resonance frequencies

4 Conclusion

The proposed multiband antenna possesses adequate applications for a few wireless applications like WCDMA at 2.18 GHz and for WiMAX at 2.56 and 3.52 GHz, respectively, at target resonance frequencies. This antenna has also adequate satisfactory gain and bandwidth for optimal performance of wireless communication devices. Satisfactory performance has been achieved in terms of other antenna electrical parameters like return loss, radiation pattern and current distribution. In addition, the gain and directivity also show good agreement for WCDMA and WiMAX applications. Negative refractive index materials and split ring resonators play a major role in higher bandwidth achievement. Additionally, modified Split Ring Resonator (SRR)-based microstrip patch offers good current density in the patch region.

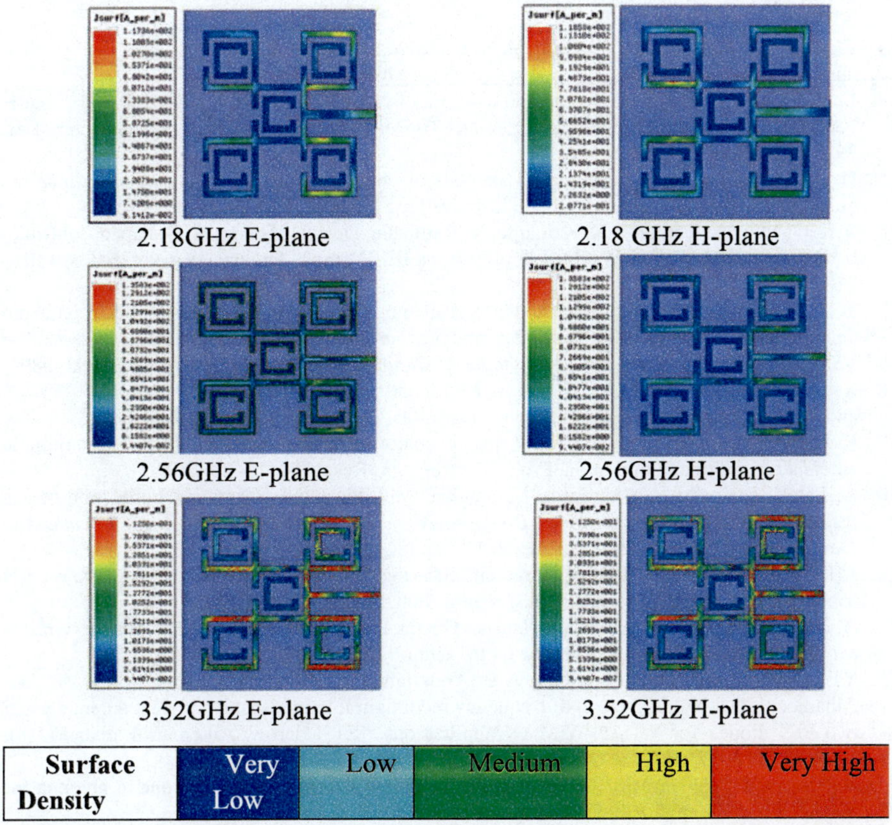

Fig. 6 Current distribution at resonance frequencies

Table 1 Comparisons of antenna parameters at resonance frequencies

Antenna parameters	First resonance	Second resonance	Third resonance
Frequency (GHz)	2.18	2.56	3.52
VSWR	1.18	1.67	1.10
Bandwidth (%)	2.51	2.01	3.27
Peak gain (dBi)	0.50	0.88	0.69

References

1. R. Zhi, M. Han, J. Bai, W. Wu, G. Liu, Miniature multiband antenna for WLAN and X-band satellite communication applications. Prog. Electromagn. Res. **75**, 13–18 (2018)
2. A. Mahesh, K.S. Shushrutha, R.K. Shukla, Design of multi-band antennas for wireless communication applications, in *2017 IEEE Applied Electromagnetics Conference (AEMC)*, (IEEE,

2017), pp. 1–2
3. T. Ali, K.D. Prasad, R.C. Biradar, A miniaturized slotted multiband antenna for wireless applications. J. Comput. Electron. **17**(3), 1056–1070 (2018)
4. U. Patel, T.K. Upadhyaya, Design and analysis of compact μ-negative material loaded wideband electrically compact antenna for WLAN/WiMAX applications. Prog. Electromagn. Res. **79**, 11–22 (2019)
5. U. Patel, T. Upadhyaya, Design and analysis of periodic split ring resonator array inspired defected ground plane antenna. Int. J. Microw. Opt. Technol. **13**(5), 401–407 (2018)
6. A.K. Gautam, L. Kumar, B.K. Kanaujia, K. Rambabu, Design of compact F-shaped slot triple-band antenna for WLAN/WiMAX applications. IEEE Trans. Antennas Propag. **64**(3), 1101–1105 (2016)
7. T. Upadhyaya, R. Patel, A. Desai, U. Patel, K. Pandya, K.P. Kaur, Electrically tilted broadband antenna using negative refractive index material, in *2019 Third International conference on I-SMAC (IoT in Social, Mobile, Analytics and Cloud)(I-SMAC)*, (IEEE, 2019), pp. 685–689
8. X. Dong, Z. Liao, J. Xu, Q. Cai, G. Liu, Multiband and wideband planar antenna for WLAN and WiMAX applications. Prog. Electromagn. Res. Lett. **46**, 101–106 (2014)
9. R. Pandeeswari, Complimentary split ring resonator inspired meandered CPW-fed monopole antenna for multiband operation. Prog. Electromagn. Res. C **80**, 13–20 (2018)
10. A. Patel, Y. Kosta, N. Chhasatia, K. Pandya, Multiple band waveguide based microwave resonator, in *IEEE—International Conference On Advances In Engineering, Science And Management (ICAESM-2012)* (IEEE, 2012), pp. 84–87
11. J.H. Yoon, Y.C. Rhee, Modified three-circular-ring monopole antenna for WiMAX/WLAN triple-band operations. Microw. Opt. Technol. Lett. **56**(3), 625–631 (2014)
12. A. Desai, T. Upadhyaya, M. Palandoken, C. Gocen, Dual band transparent antenna for wireless MIMO system applications. Microw. Opt. Technol. Lett. **61**(7), 1845–1856 (2019)
13. Y.I. Abdulraheem, G.A. Oguntala, A.S. Abdullah, H.J. Mohammed, R.A. Ali, R.A. Abd Alhameed, J.M. Noras, Design of frequency reconfigurable multiband compact antenna using two PIN diodes for WLAN/WiMAX applications. IET Microw. Antennas Propag. **11**(8), 1098–1105 (2017)
14. A. Sharma, P. Ranjan, R.K. Gangwar, Multiband cylindrical dielectric resonator antenna for WLAN/WiMAX application. Electron. Lett. **53**(3), 132–134 (2017)
15. Y. Zhang, W. Hong, C. Yu, Z.Q. Kuai, Y.D. Don, J.Y. Zhou, Planar ultrawideband antennas with multiple notched bands based on etched slots on the patch and/or split ring resonators on the feed line. IEEE Trans. Antennas Propag. **56**(9), 3063–3068 (2008)
16. T. Upadhyaya, S. Kosta, R. Jyoti, M. Palandöken, Novel stacked μ-negative material-loaded antenna for satellite applications. Int. J. Microw. Wirel. Technol. **8**(2), 229–235 (2016). https://doi.org/10.1017/S175907871400138X
17. L. Sun, H. Cheng, Y. Zhou, J. Wang, Broadband metamaterial absorber based on coupling resistive frequency selective surface. Opt. Express **20**(4), 4675–4680 (2012)
18. N.I. Landy, S. Sajuyigbe, J.J. Mock, D.R. Smith, W.J. Padilla, Perfect metamaterial absorber. Phys. Rev. Lett. **100**(20), 207402 (2008)
19. Y. Cheng, H. Yang, Z. Cheng, N. Wu, Perfect metamaterial absorber based on a split-ring-cross resonator. Appl. Phys. A **102**(1), 99–103 (2011)
20. R. Patel, T. Upadhyaya, An electrically small antenna for nearfield biomedical applications. Microw. Opt. Technol. Lett. **60**(3), 556–561 (2018)
21. E. Cubukcu, S. Zhang, Y.S. Park, G. Bartal, X. Zhang, Split ring resonator sensors for infrared detection of single molecular monolayers. Appl. Phys. Lett. **95**(4), 043113 (2009)
22. M. Gil, J. Bonache, J. Garcia-Garcia, F. Martin, New left handed microstrip lines with complementary split rings resonators (CSRRs) etched in the signal strip, in *2007 IEEE/MTT-S International Microwave Symposium* (IEEE, 2007), pp. 1419–1422
23. M. Gil, J. Bonache, J. Garcia-Garcia, J. Martel, F. Martin, Composite right/left-handed metamaterial transmission lines based on complementary split-rings resonators and their applications to very wideband and compact filter design. IEEE Trans. Microw. Theory Tech. **55**(6), 1296–1304 (2007)

24. Z. Han, K. Kohno, H. Fujita, K. Hirakawa, H. Toshiyoshi, Tunable terahertz filter and modulator based on electrostatic MEMS reconfigurable SRR array. IEEE J. Sel. Top. Quantum Electron. **21**(4), 114–122 (2014)
25. U. Patel, T.K. Upadhyaya, Dual band planar antenna for GSM and WiMAX applications with inclusion of modified split ring resonator structure. Prog. Electromagn. Res. **91**, 1–7 (2020)
26. H. Mosallaei, Y. Rahmat-Samii, Composite materials with negative permittivity and permeability properties: concept, analysis, and characterization, in *IEEE Antennas and Propagation Society International Symposium. 2001 Digest. Held in conjunction with: USNC/URSI National Radio Science Meeting (Cat. No. 01CH37229)*, vol. 4 (IEEE, 2001), pp. 378–381
27. A.F. Starr, P.M. Rye, D.R. Smith, S. Nemat, Fabrication and characterization of a negative-refractive-index composite metamaterial. Phys. Rev. B **70**(11), 113102 (2004)
28. T.K. Upadhyaya, V.V. Dwivedi, S.P. Kosta, Y.P. Kosta, Miniaturization of tri band patch antenna using metamaterials, in *Fourth International Conference on Computational Intelligence and Communication Networks*, Mathura (2012), pp. 45–48. https://doi.org/10.1109/CICN.2012.147

A Wearable Finger Exoskeleton for Motor Rehabilitation Using Mobile Application

U. Snekhalatha, Satya Harsha Vardhan Muppidi, and Lakshmi Swathi Akkiraju

Abstract Major stroke-affected persons suffer with a poorly operating hemiparetic broad palm (metacarpus), and require treatments that include physical therapy directed with intense activity and movement. This paper presents a novel finger exoskeleton device to facilitate therapy exercises while keeping it lightweight, compact, and affordable. The self-operated exoskeleton is designed to assist finger flexion and extension motions for the restoration of voluntary function, which is driven using servos and string mechanism, and controlled using the mobile application. Testing results show that this wearable finger exoskeleton can produce required torque to flex each finger toward the broad palm for rehabilitation with 90% accuracy.

Keywords Finger rehabilitation · Hand exoskeleton · Hemiparetic hand · Mobile application

1 Introduction

The number of stroke patients is increasing drastically, and a majority of them suffer from motor impairment or muscle tightness or immobility. In India, approximately 9 million people, or 0.8% of the population suffer from paralysis or immobility of one or both upper or lower extremities [1]. The potential causes of paralysis are stroke and spinal cord injuries (29% and 23%, respectively) [2]. A stroke can result in decreased motor function which disrupts the signaling activity to a muscle or group of muscles, leading to paralysis of the affected area which directly influences the daily activities of a person. The symptoms include decreased sensations, spasms, loss of muscle tone, and curling of the fingers.

Inactivity or immobilization of the limbs after being affected by paralysis can lead to loss of muscle mass and an increase in fat, also known as Sarcopenia [3]. This in

U. Snekhalatha (✉) · S. H. V. Muppidi · L. S. Akkiraju
Department of Biomedical Engineering, College of Engineering and Technology, SRM Institute of Science and Technology, Kattankulathur, Tamil Nadu, India
e-mail: snehalau@srmist.edu.in

F. Thakkar et al. (eds.), *Proceedings of the International e-Conference on Intelligent Systems and Signal Processing*, Advances in Intelligent Systems and Computing 1370,
https://doi.org/10.1007/978-981-16-2123-9_2

turn reduces muscle strength. The decrease in muscle mass reduces the number of contractile proteins (actin and myosin) and eventually decreases the overall force.

Hand has 27 Degrees of Freedom (DOFs) [4]. The human broad palm consists of carpals, metacarpals, and the phalanges (proximal, medial, and distal). These are connected by the carpometacarpal joints, the inter-metacarpal joints, the meta-carpophalangeal joints, the proximal and distal inter-phalangeal joints (PIP and DIP, respectively), and the inter-phalangeal joint at the thumb. The distal inter-phalangeal joints and meta-carpophalangeal joints can bend up to 90° while PIP can bend up to 110°, thus making the hand rehabilitation complex. However, research studies have reported that human neurons exhibit plasticity, which forms the basis for learning and memory [5]. Motor cortex functions can be re-mapped through repeated motor exercises.

The treatment of hand paralysis is usually done with the help of physiotherapists, who bend fingers manually to maximum range of joint angle. This procedure is time-consuming and expensive. Advanced technology has contributed significantly to physical rehabilitation. The use of rehabilitation devices has shown great benefits such as precise and repeatable therapeutic exercises, reduction of dependence on physiotherapists, and collection of quantitative data, and also providing feedback mechanisms [6].

Standard stroke rehabilitation types of equipment are labor intensive, complex, and require frequent interactions with therapists. These systems are highly expensive and are meant for clinical use. This project aids the patients in acquiring rehabilitation at a feasible cost. Here, we describe a novel method for training the distal upper limb (forearm and wrist), a model-driven by SG90 servos. Controlling work was fulfilled by Arduino UNO. The uniqueness of this project is that the whole setup can be controlled remotely using a mobile application with any smartphone. Since all the components used are light and smaller in size, it is portable and easily affordable. Any patient can easily operate the model using a smartphone.

2 Literature Review

Stroke is one of the leading causes of permanent disability and impairment [7]. Just like any other health issues, impairment has both therapeutic and surgical treatment methods. However, surgical methods haven't shown any superiority [8]. And the most common treatment option for any disability is rehabilitation. In the past two decades, there has been a lot of research in post-stroke rehabilitation. Different types of techniques have been tried to evaluate how effective and efficient they are in rehabilitating stroke affected extremities. All the different types of rehab methodologies fall under two therapies [9], namely Conventional and Enhanced. While conventional therapy is all about simple physiotherapy which is described by Bobath [10], enhanced is the one with advanced and sophisticated machinery.

Among both the therapies, conventional physiotherapy stands out ahead of enhanced therapy in terms of patient comfort, ease of use, and most importantly affordability. According to the theory inspired by Carr and Shepherd, post-stroke rehabilitation must focus on functional tasks, regardless of several motor strategies used [11] and physiotherapy is one such method of rehabilitation. It is also proven that 2 hours of physiotherapy is sufficient [12] to improve the function of the paretic hand in chronic stroke patients if done in the first 3 months after being affected by stroke.

There are various devices available to assist in physiotherapy, and robotics is the most recent advancement that is being used mostly these days for post-stroke rehabilitation. Robotic rehabilitation devices have many advantages over conventional physiotherapy equipment such as automation, precise control over a different range of motions (ROM), and most importantly increased degrees of freedom (DOF) [13]. However, there are different types of robotic rehabilitation devices in the market, and most of them have their own downsides. For instance, Armeospring, an orthotic exoskeleton with an integrated spring system to compensate gravitation [14], has a very large setup such that another base is required to support the whole weight of the system. While fingers are very important in performing various functional tasks, MIT Manus (m power 10,004) [15] is an elbow orthotic arm with only one DOF. This system only concentrates on elbow support and rehabilitation that might help patient in performing various complex tasks but is not efficient in rehabilitation.

Alongside with hefty, complicated, and sophisticated robotic-orthotic systems, there are a few compact and portable-wearable haptic systems [16, 17] for use in virtual environment for emulating rehabilitation and sports. But these systems are again restricted to the elbow and wrist, which isn't sufficient for simulating all the functional tasks to assist in faster motor recovery. Considering the systems discussed in earlier sections, it can be said that there are very few devices that are simple to use, affordable, and could efficiently assist in physiotherapy. Having the theory proposed by Carr and Shepherd considered, we have come up with a simple prototype design that can be easily controlled by the user with a mobile app and most importantly, it focuses mainly on finger movements and assists in basic functional tasks to emulate everyday activities of the patient for efficient stimulation of forearm muscles that eventually aids in faster recovery.

3 Architecture

The mobile application was designed using MIT app inventor 2 which provided a simplified user interface. The app lets the patient control all the 5 servos with ease. The app can be used in any android phone that has access to APK installation. HC-05 was used as a Bluetooth module that offers high signal strength, long-range communication, and more importantly less power consumption. The Bluetooth module converts the digital signals into encoded Bluetooth signals that carry the signal to the receiver Bluetooth module. Bluetooth receiver module receives the encoded signals from the mobile and converts the signals into a digital language which can be under-stood by the Microcontroller board. Arduino UNO, an open-source microcontroller was used. 12 volts DC adapter as a power source and voltage regulator for differential power supply to servos were used. Servo motors were used for driving the strings and performing finger actuation (Fig. 1).

3.1 Exoskeleton Assembly

As seen in Fig. 2, an elastic band was used to place the Arduino, servomotors, and other components around the arm. Servos are fastened on the inner side of the hand and other components like battery and circuit board on the outer side. Nylon strings were used to connect servos to fingers. The frame which holds the string in place was 3D printed and made of PLA. PLA material has great strength (36–55 Mpa) and is easy to customize [18]. This was designed using SOLIDWORKS software. The overall weight of the exoskeleton is less than 0.3 kg.

3.2 Circuit Design

The whole circuit consists of Arduino, HC-05 Bluetooth module, Power regulating module, and Servos. All the servos are connected to the Arduino's digital pins (pin2–pin6). +5volt and GND pins of the servos are connected to the Arduino, respectively. Power from Arduino is regulated before going to the servos. The power regulating module has capacitors, resistors, and transistors. TXD and RXD pins of the Bluetooth module are connected to Arduino, respectively. Power and GND pins are connected to the Arduino pins, respectively. When the Bluetooth module receives a signal from the mobile phone, it transmits that signals to Arduino via digital pins. The microcontroller in the Arduino will control the servos according to the signal received and the program written in the Arduino (Fig. 3).

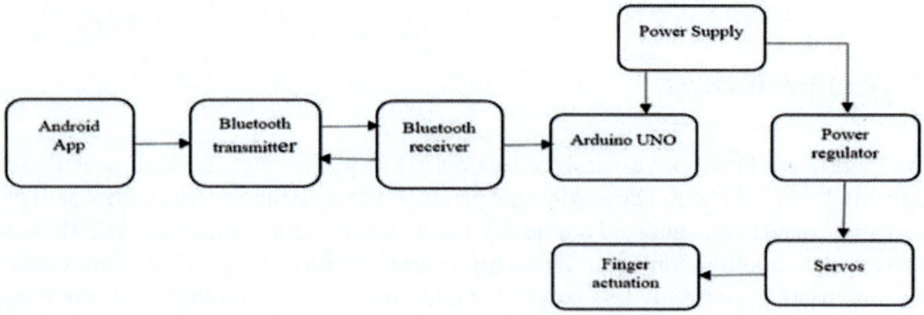

Fig. 1 Overall system architecture of finger exoskeleton that is controlled with mobile application using Bluetooth module. System contains smartphone with a customized mobile app that allows patient to seamlessly control orthotic arm

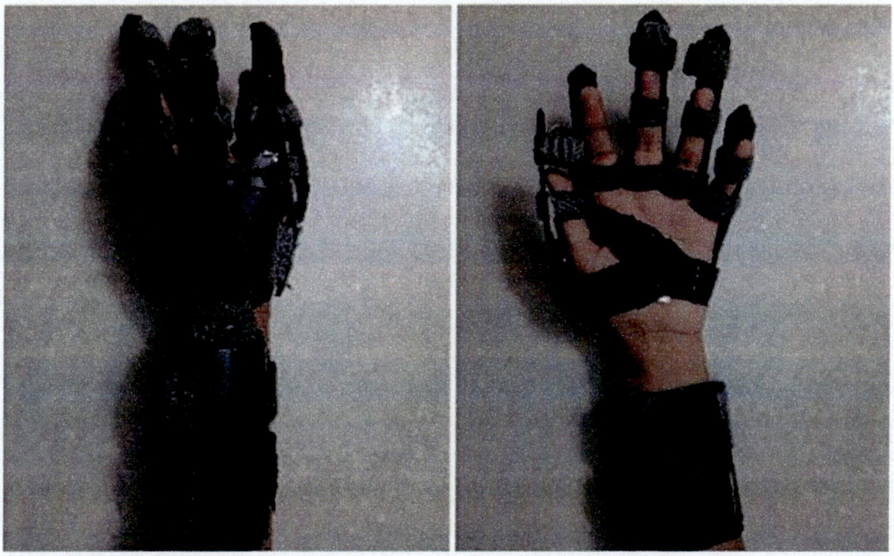

Fig. 2 Prototype of finger exoskeleton

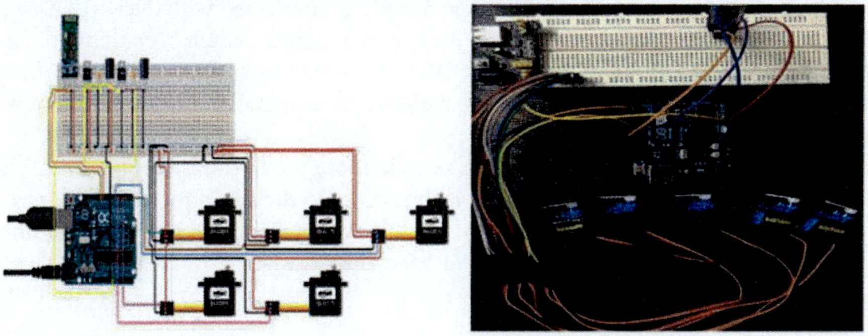

Fig. 3 Circuit

3.3 Arduino Programming

The Arduino board was programmed in the Arduino software. Five digital pins of Arduino were linked to five servos each using the Attach () function. The angular position of servos was set between 0 and 180° using the write () function. The position of the servo determines the bending angle of the finger. The 0° leaves the finger relaxed and 180° bends the finger toward the palm (Flexion).

3.4 Mobile Application

The mobile application was developed by MIT App Inventor-2, an open-source android application designing website with a block-based coding method. It is a virtual app developing an online platform where the app appears step-by-step as the blocks are added to it so that the work can be tested and when it is done, we can package the app and produce a stand-alone application to install. The app developed for controlling this exoskeleton was "Finger Exoskeleton-1". Just like any other app developed using app inventor, Finger Exoskeleton-1 also has two important parts, App or User interface and background block code.

Application Specifications

- Size: 3.59 Mb;
- Operating platform: Android (8.0 and above);
- Build Version: 1.0.1.

User interface

Finger Exoskeleton-1 has a very simple and straightforward user interface with one toggle switch, a Bluetooth connection status indicator, and five sliding bars. The "Not connected" indicator on the user interface is the Bluetooth connection status indicator that turns to "Connected" when the mobile is connected with the exoskeleton's Bluetooth module. Right below the indicator is a toggle switch "Available device" that can be used to see all the available Bluetooth devices around the mobile phone. It helps users to check whether the exoskeleton is available and to connect it with a mobile phone.

Apart from the indicator and toggle switch, there are five sliding bars for all five fingers. These sliding bars are named with respect to the servo motor that control a specific finger's flexion and extension. When sliding along the bar, these switches control the angle of rotation of the servo head (Fig. 4).

Background Blocks

The most important reason for developing a mobile app using MIT App Inventor-2 is block-based programming. Though it has its own limitations when compared to classical app development methodologies, the one side that makes App Inventor a better option for beginners is its intuitive error-free programming technique. The background of this app has a few important blocks. The first one is the "Initializing block" which initiates the application and turns ON the Bluetooth module. It is also responsible for the Bluetooth connection status indicator.

The second and third blocks are responsible for the list picking activity and Bluetooth initiation, respectively. The list picker block has a sub-block "List picker element" that is responsible for collecting Bluetooth data from the mobile's Bluetooth module and displaying it over the toggling interface. The next five important blocks of the code are called "position changing" blocks and are used for controlling sliding bars and their interfacing with the servos.

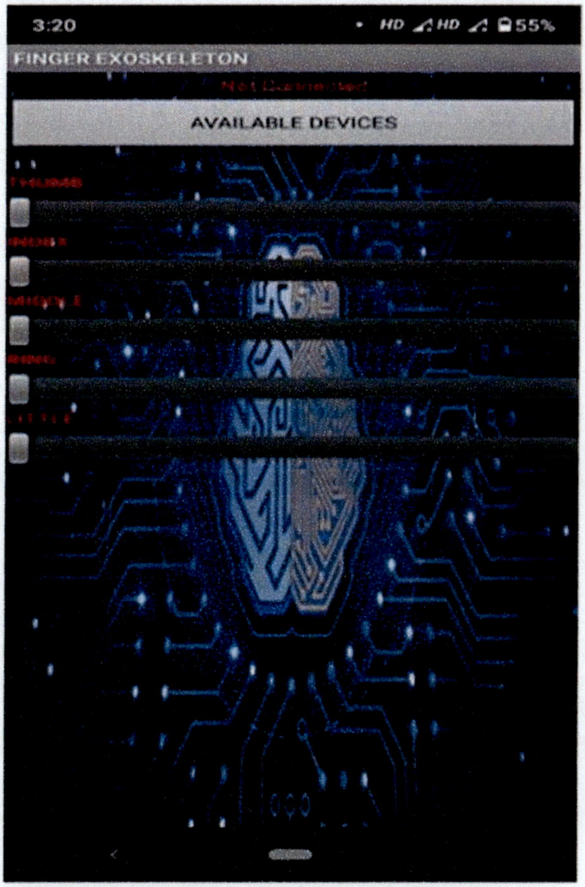

Fig. 4 Mobile application interface

These blocks along with various sub-blocks are used to control the digital data that in turn controls the servo's action. The overall sliding length of the sliding bars will be divided into 1000 equal parts that in turn translates as 1000 bytes. These 1000 bytes of information consecutively will be translated to the digital data transferred to the embedded board of servos (Fig. 5).

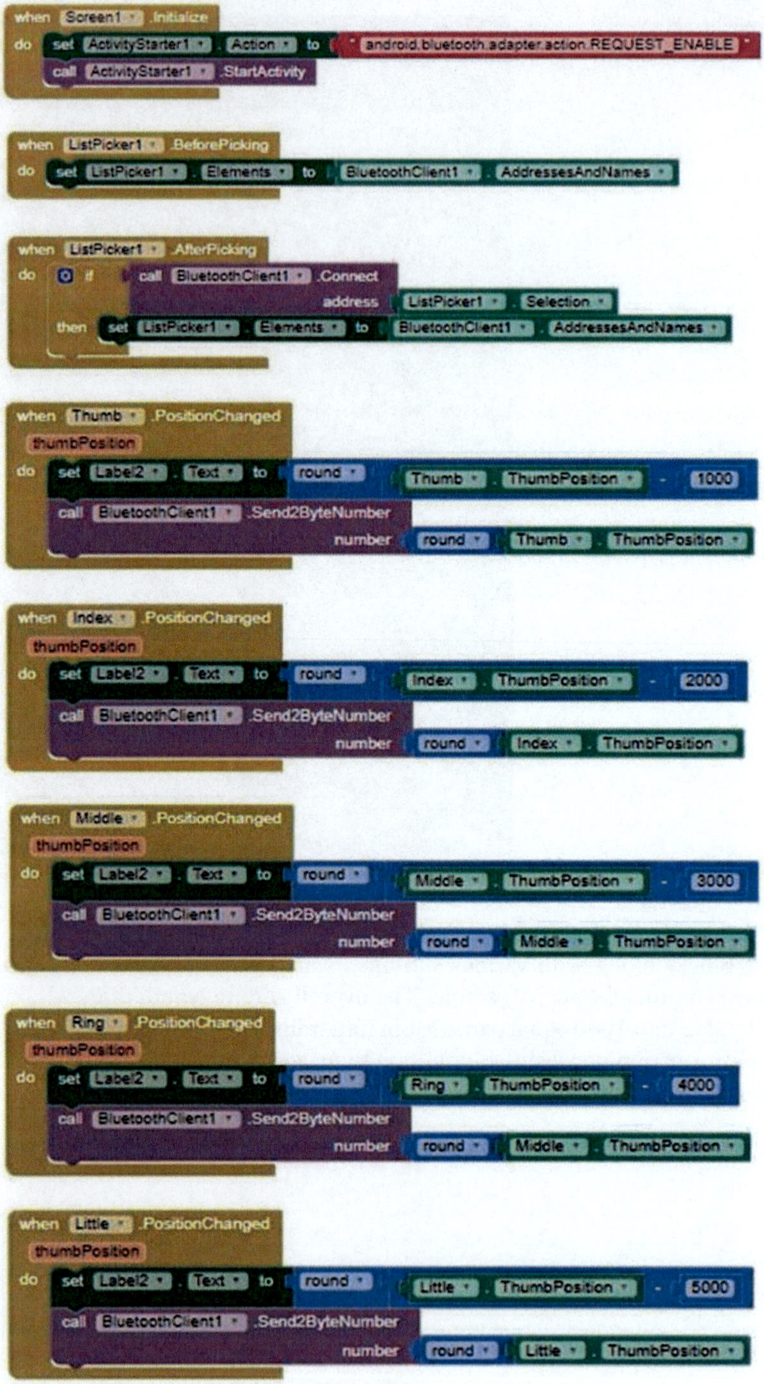

Fig. 5 MIT App inventor block code

4 Experiments and Results

It is very important to test the system with various subjects in order to evaluate its efficiency. A few parameters were recorded from various subjects to determine the efficiency and durability. For finger rehabilitation, the Range of motion (ROM) of various joints of the fingers that is achieved by the system is the primary parameter that has to be tested to evaluate the efficiency of the prototype. Various finger angles that are measured in this experiment are Metacarpophalangeal (MCP), Proximal (PIP), and Distal (DIP) joints. Attitude sensor (MPU 6050) was used to achieve this task (Fig. 6).

In the experiment, the attitude sensor was placed on top of metacarpals of each finger, and range of motion (ROM) of proximal, distal, and metacarpophalangeal joints were measured from healthy and paralyzed subjects. Note that the angle of various joints of fingers isn't zero in the relaxed position. To calculate the accurate flexion and extension angle of joints of respective fingers, the fully bent angle of the finger is subtracted from relaxed finger angle. The relaxed angle was calculated every time for each individual finger to calibrate attitude sensor. Data was collected five times for each and every individual and the average values of bending angles and their ranges were tabulated (Tables 1 and 2).

The next parameter tested was torque stability. Since actuation was achieved by string mechanism, testing the consistency of torque translated from servos to the fingertips is very necessary. This was done using a torque meter which gives continuous torque measurement in real time. Torque meter was placed in various

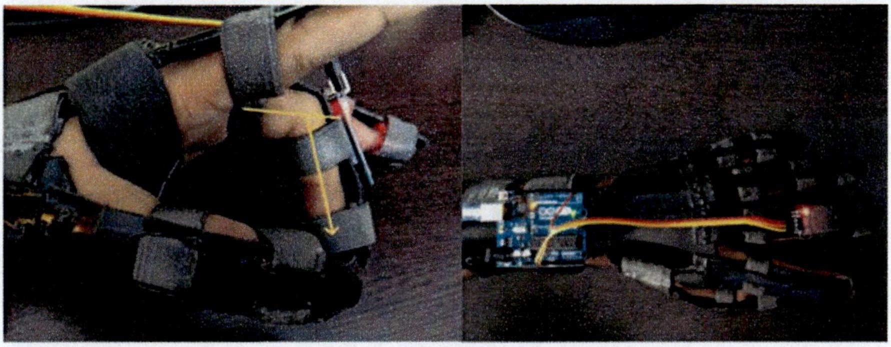

Fig. 6 Finger bending angle measurement setup

Table 1 Range of bending angles

Joint	Range of angle
DIP	**0°–42°**
PIP	**0°–59°**
MCP	**0°–54°**

Table 2 Averaged bending angles of each finger achieved after servo actuation

Finger	MCP	PIP	DIP
Thumb	**52°**	**55°**	
Index	**49°**	**54°**	**38°**
Middle	**49°**	**53°**	**39°**
Ring	**47°**	**53.1°**	**41°**
Little	**48.2°**	**51°**	**38°**

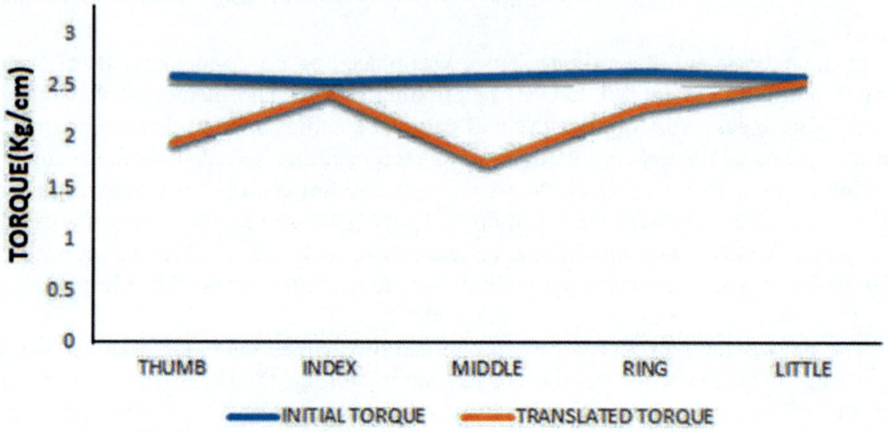

Fig. 7 Translated torque measurement

joints and positions where string bends, and the average of various positions for five fingers is given by the following graph. Generally, the torque output of SG-90 servos is 2.7 kg/cm.

The data mentioned in Fig. 7 is the averaged output of data from 10 subjects. Torque was calculated from the proximal and distal ends of the strings between servos and the fingertips. The significant deviation from the ideal amount of torque in the case of thumb and middle fingers is because of bending in the string. Overall, 85% of torque translation efficiency was achieved.

5 Conclusions and Future Works

This paper proposes a simple and cost-effective alternative for finger exoskeleton in motor rehabilitation based on a drive-by-string mechanism and mobile application. Various experiments on multiple subjects have shown that this design is easy to use, and controlling it using mobile applications adds up to the comfort in handling the system. The output evaluated from the torque meters and attitude sensors also proves that the system is efficient almost up to 90%. This one might not be as complicated and highly sophisticated with various sensors and modules as conventional rehabilitation equipment used for physiotherapy, but the efficiency achieved with the simplest arrangement possible is pretty much enough to stimulate forearm muscles that could eventually regain the motor abilities of the paralyzed arm.

Independent movements of all the five fingers assist not only in the rehabilitation of the finger muscles but also in maintaining the posture while performing various simple tasks that enhance the quality of life if not similar to the one with a healthy hand but better than how it was with a paralyzed hand.

Coming to the future work, we intend to extend the exoskeleton up to the elbow with 8 new degrees of freedom (DOF). This could stimulate various new muscles for efficient and faster regain of motor abilities. To know whether the muscles are being stimulated or not, there are various methods and EMG is one such important modality to detect any impulses in the muscles. We also intend to develop a system to quantify the change in the motor abilities of the paretic hand by monitoring and comparing the peak amplitude of the H-wave in hand muscle's EMG after every physiotherapy session.

References

1. CureResearch. http://www.cureresearch.com/p/paralysis/stats-country_printer.htm. Accessed 2 Sept 2020
2. Sci-Info-Pages, Paralysis and spinal cord injury statistics. https://www.sci-info-pages.com Accessed 2 Sept 2020
3. V. Gray, C.L. Rice, S.J. Garland, Factors that influence muscle weakness following stroke and their clinical implications: a critical review. Physiotherapy Canada. Physiotherapie Canada **64**(4), 415–426 (2012). https://doi.org/10.3138/ptc.2011-03
4. A. Gustus, G. Stillfried, J. Visser et al., Human hand modelling: kinematics, dynamics, applications. Biol. Cybern. **106**, 741–755 (2012). https://doi.org/10.1007/s00422-012-0532-4
5. Z. Yue, X. Zhang, J. Wang, Hand rehabilitation robotics on poststroke motor recovery. Behav. Neurol. **2017**, 3908135 (2017). https://doi.org/10.1155/2017/3908135
6. C.Y. Chu, R.M. Patterson, Soft robotic devices for hand rehabilitation and assistance: a narrative review. J. Neuroeng. Rehabilit. **15**(1), 9 (2018). https://doi.org/10.1186/s12984-018-0350-6
7. About Stroke, Internet Stroke Center. http://www.strokecenter.org/pat/about.htm. Accessed 9 Sept 2020
8. L. Masia, H.I. Krebs, P. Cappa, N. Hogan, Whole-arm rehabilitation following stroke: hand module, in *The First IEEE/RAS-EMBS International Conference on Biomedical Robotics and Biomechatronics, 2006*. BioRob 2006, Pisa, pp. 1085–1089 (2006). https://doi.org/10.1109/BIOROB.2006.1639236
9. A. Sunderland, D.J. Tinson, E.L. Bradley, D. Fletcher, R. Langton Hewer, D.T. Wade, Enhanced physical therapy improves recovery of arm function after stroke. A randomised controlled trial. J. Neurol. Neurosur. Psychia. **55**(7), 530–535 (1992). https://doi.org/10.1136/jnnp.55.7.530
10. B. Bobath, *Adult Hemiplegia: Evaluation and Treatment* (William and Heinemann, London, 1978)
11. J.H. Carr, R.B. Shepherd, Stroke Rehabilitation: Guidelines for Exercises and Training to Optimize Motor Skill (Butterworth-Heinemann Oxford, 2002)
12. J.H. Cauraugh, S.B. Kim, Chronic stroke motor recovery: duration of active neuromuscular stimulation. J. Neurol. Sci. **215**(1–2), 13–19 (2003). https://doi.org/10.1016/s0022-510x(03)00169-2

13. J.J. Daly, N. Hogan, E.M. Perepezko, H.I. Krebs, J.M. Rogers, K.S. Goyal, M.E. Dohring, E. Fredrickson, J. Nethery, R.L. Ruff, Response to upper-limb robotics and functional neuromuscular stimulation following stroke. J. Rehabil. Res. Dev. **42**(6), 723–736 (2005). https://doi.org/10.1682/jrrd.2005.02.0048
14. S.J. Housman, V. Le, T. Rahman, R.J. Sanchez, D.J. Reinkensmeyer, Arm-training with T-WREX after chronic stroke: preliminary results of a randomized controlled trial, in *IEEE 10th International Conference on Rehabilitation Robotics, 2007*. ICORR 2007, pp. 562–568 (2007)
15. H.I. Krebs, B.T. Volpe, D. Williams, J. Celestino, S.K. Charles, D. Lynch, N. Hogan, Robot-aided neurorehabilitation: a robot for wrist rehabilitation. IEEE Trans. Neural Syst. Rehabilit. Eng. Publicat. IEEE Eng. Med. Biol. Soc. **15**(3), 327–335 (2007). https://doi.org/10.1109/TNSRE.2007.903899
16. M. Folgheraiter, E.A. Kirchner, A. Seeland, et al., A multimodal brain-arm interface for operation of complex robotic systems and upper limb motor recovery, in *BIODEVICES 2011—Proceedings of the International Conference on Biomedical Electronics and Devices*, pp. 150–162
17. D. Tsetserukou, K. Sato, S. Tachi, ExoInterfaces: novel exosceleton haptic interfaces for virtual reality, augmented sport and rehabilitation. AH '10 (2010)
18. A. Roshdy, S.A. Kork, S. Said, T. Beyrouthy, A wearable exoskeleton rehabilitation device for paralysis—a comprehensive study. Adv. Sci. Technol. Eng. Syst. J. **4**, 17–26 (2019)

Game Theoretical Approach for Cluster-Based Routing Protocol in Wireless Sensor Network

Namrata Sahayam, Anjana Jain, and Shekhar Sharma

Abstract One of the emerging optimization techniques formulating on the interest of conflict or cooperation for achieving maximum outcome on the bases of prediction is the Game Theory. From the last two decades, it has gradually been used in wireless communication and slowly stepped into wireless sensor networks. Wireless sensor networks are made up of wireless nodes working on batteries and so the life of the networks depends on the life of the battery. Applying the best cluster-based routing protocol with appropriate game has increased network lifetime, and enhanced its energy efficiencies. Present paper summarizes the literature and the research carried out for the cluster-based routing protocol in wireless sensor network using game theory.

Keywords Game theory · Wireless sensor network · Sensor nodes · Cluster-based routing protocol · Energy efficiencies · Cluster-head

1 Introduction

In a few decades of time, Wireless Sensor Network (WSN) is enormously gaining popularity in research community, military applications, and environmental monitoring [1] as well as day-to-day commercial applications. It consists of many sensor nodes having the ability to sense, communicate, and do some signal processing. These sensor nodes are battery operated with limited energy and computational processing capability. If the sensor nodes exhaust their energy, network partition occurs terminating the network so energy-efficient mechanisms are required to increase network lifetime. There are many ways of increasing the network lifetime in which routing plays an important role; routing means sending data between sensor nodes and sink to establish communication.

N. Sahayam (✉) · A. Jain · S. Sharma
Department of Electronics & Telecommunication Engineering, SGSITS, Indore, MP, India

F. Thakkar et al. (eds.), *Proceedings of the International e-Conference on Intelligent Systems and Signal Processing*, Advances in Intelligent Systems and Computing 1370,
https://doi.org/10.1007/978-981-16-2123-9_3

Routing is a technique for the transmission of data from the source node to the sink which if done vigorously increases the network lifetime by reducing its energy consumption in radio communication. In order to make the routing protocol reliable, some issues are taken such as network structure, type of network (homogenous or heterogenous), connectivity, data aggregation and node deployment. Generally, WSN routing protocols are categorized as flat routing, cluster-based routing and location-based routing [1]. In flat routing, direct transmission of data from sensor node to sink is done; in cluster-based routing, there is no direct transmission of data from the sensor node to the sink, but data is aggregated to the cluster-head (CH) (a node with more energy as compared to sensor node) and cluster-head sends the data to the sink, and location-based routing nodes use restricted data from the chosen location and decide the transmission route accordingly. In this paper, we are only taking cluster-based routing protocols using the game theoretical approach.

During the last few years, game theory has been applied to wireless sensor networks specially for resolving issues in routing techniques like selection of cluster-head termed as clustering game [2], determining the optimal route [3], deciding whether to forward the receive packet or not known as forwarding game, energy balancing by maintaining the allocation problem of member nodes and load balancing of cluster-heads. In this article, we discuss how the game theoretical approach can be set up to sort out several issues in routing protocols in WSN and emphasized which type of game is appropriate for particular routing protocol for getting its best utility function. Rest of the paper is organized as follows: fundamentals of game theory and its application in the cluster-based routing protocol with basic equations are given in Sects. 2 and 3, respectively. Section 4 gives the assessment of different cluster-based routing protocol using the game theoretical approach. And it ends with the conclusion in Sect. 5.

2 Basic Concept of Game Theory

Here, we elaborate the fundamentals of game theory. It is one of the optimization techniques, a very powerful logical and mathematical tool for analyzing conflicts or cooperation between the entities and predicting the behavior of selfish entities using equilibrium analysis. The game is formulated by a set of players, a set of action or strategy for each player, and the utility function or payoff of the players [4]. It is modeled as

$$G = (P, Si, Ui) \tag{1}$$

where $P = \{1, 2, \ldots n\}$ is a set of players.

S1, S2, …, Sn are sets called strategy for player 1, 2, … n, respectively.

Ui: $S1 \times S2 \times \ldots \times Sn \rightarrow R$ for $I = 1, 2, \ldots, n$ are mapping called utility functions.

Games are classified as Non-cooperative games, Cooperative games and Cooperation enforcement games. For the taxonomy of game, the reader is prompted to [5]. Some very basic types of games used in wireless sensor networks and terms generally are

(i) Non-cooperative games: In non-cooperative games,, the strategy of an individual player does not coordinate with the action of the other players, that is they play for their own interest.
(ii) Cooperative games: In cooperative games, players cooperate with each other to get the maximum benefit in terms of utility function.
(iii) Cooperation enforcement games: In this game, selfish nodes are enforced to play in cooperation to achieve maximum outcome [5].
(iv) Nash equilibrium: It is a state in a game where the player has achieved its maximum and cannot get better than this [4].

3 Game Theory in Cluster-Based Routing Protocol for WSN

In this section first, we introduce the fundamentals of game theory used in wireless sensor network and then discuss all the game theoretical approaches that has been used for improving the network lifetime using clustering techniques for the routing protocol in the past 10 years.

3.1 Fundamentals of Game Theory for WSN

A game used in the selection of cluster-head is known as a Clustering game [2] and is modeled as

$$\mathrm{CG} = (\mathrm{P}, \mathrm{S}, \mathrm{U}) \tag{2}$$

where P is set of sensor nodes.

S is the set of actions taken by the sensor nodes.
U is the utility function of the sensor nodes.

In a clustering game, players are the sensor nodes, strategy is the set of actions taken by the sensor node and utility function is the difference between the award and penalty achieved by the sensor nodes. Now, what is award and penalty in WSN; let us consider any arbitrary node *i;* if node *i* has low energy and selects itself as a cluster-head, then it will be its penalty; similarly, when node *i* has high energy and selects itself as a cluster-head, then it will be its award and thus the utility function [6] of this node *i* is given as

$$\mathrm{U}i = \mathrm{a}i - \mathrm{p}i \tag{3}$$

where U*i* is the utility function of node *i,*

a*i* is the award of node *i and*

p*i* is the penalty of node *i.*

The basic concept of game theory implies that the payoff matrix for a two-player clustering game will be as shown below.

$$\begin{array}{lll} & \text{Yes for CH} & \text{No for CH} \\ \text{Yes for CH} & (g-c,\ g-c) & (g-c,\ g) \\ \text{No for CH} & (g,\ g-c) & (0,0) \end{array} \tag{4}$$

The strategy set consists of two elements; a sensor decides to present itself as CH or not. If the sensor says Yes to be the cluster-head (CH), then let it be represented as {Y} and if it does not then let it be represented as {N}. Thus, the strategy space is defined as {Y, N}. g is the gain for delivery and c is the energy consumed in becoming a cluster-head. If the game is played for M players, let the strategies followed by them be given as s = {s1, s2, ... sm}. If none of the players say yes to become a CH, then their payoff will be zero as shown in the above matrix. This means no data can be sent to the Base station. Thus, the utility function of any node *i* is in the following form:

$$\mathrm{U}i(\mathrm{s}) = \begin{cases} 0, & \text{if sj} = \text{N for j belongs to M} \\ g-c, & \text{if si} = \text{Y} \\ g, & \text{if si} = \text{N for j belongs to M and sj} = \text{Y} \end{cases} \tag{5}$$

This clustering game is indeed helpful in the analysis of Nash Equilibrium.

3.2 *Various Game Theory Approaches Presented for Cluster-Based Routing Protocol in WSN*

Koltsidas and Pavlidou [2] gave a remarkable change in the existing cluster-based routing protocols by formulating clustering mechanism called Clustered Routing of Selfish Sensors [CROSS]. In this, the author has supposed that each sensor node acts selfishly to conserve its energy for increasing the network lifetime. They have listed some propositions which state that

(i) For the symmetrical clustering game, the set of strategies if all the nodes Declared and if all the nodes did Not Declare themselves as cluster-heads can never form the Nash equilibrium. The existence of the Nash equilibrium is only possible if a single player declares to be the cluster-head and all other player do not declare.
(ii) Nash equilibria only exist if the player plays mixed strategies, that is, no Nash equilibria exist for pure strategies in symmetrical clustering game.

In CROSS the author have worked on few points like selection of cluster-head, transmission of data from sensor node to cluster-head, then after that from the cluster-head to the base station and most importantly expected pay off and price of anarchy for the WSN using non-cooperation game theory in order to save energy of the nodes and increase network lifetime. The only drawback of using CROSS is it is not sure that after a specific number of rounds all nodes have served as cluster-heads for exactly one time.

In the same year, after a few months a survey on Game Theory application in wireless networks was done by Charilas and Panagopoulos [3] stating forwarding games and how to determine optimal route for the source node to send the data to the destination node. Here they also discussed Min–Max Game and to calculate the utility function of all games as discussed in Eq. 5.

In [7], the author has proposed a new protocol termed as Hybrid Homogenous LEACH Protocol (HHO LEACH) formulating with learning automata and Bayesian coalition games. They are adaptive learning techniques in which previous action is described by conditional probability and performs its task from the constant feedback from the environment. If the network is small in size and the base station is at the center of the network, then the transmission of data from the sensor network to the base station is done through the Direct Diffusion method. For the direct diffusion method, the reader can read [8]. And if the network is large and the base station is at the corner of the network, then the cluster-based routing protocol is used. According to the distance-based threshold, the network is divided into different regions. This distance-based threshold is used by the players to take an action for coalition game formation.

Efficient Energy Consumption routing Protocol (EECP) achieves good performance in terms of network lifetime using game theory [9]. EECP has three stages of working forming the cluster and choosing CH from the sensor node, gathering data from sensor node to a cluster-head termed as intra-communication and then transmission of data from CH to the base station termed as inter-communication. A node is selected as CH if it has the highest weight in cluster [10]. Cooperative game theory is being applied. Here a pre-defined transmitting distance for each node is calculated; if the calculated value is greater than the value which is already set, then it signifies successful transmission and gives the strategy as one or else, if the calculated value is less the defined value then the strategy is zero. Next the transmission only takes place when the strategy is one or remains ideal.

Hot spot problem, optimal size of cluster and selection of cluster-head are the main problems discussed in [6]. There are many applications in which data is unevenly distributed, and so the nodes over there will get exhausted fast; this is known as the hot spot problem. This will surely affect the network lifetime. The remedy for the hot spot problem and the selection of the cluster-head is done through Evolutionary game theory and optimal size of the cluster is obtained through the OCS algorithm. OCS (Optimal Cluster Size) Algorithm. This algorithm states that the size of the cluster should be small when it is near the Base station or the size of the cluster is calculated according to its distance from the Base station. Energy-Efficient clustering Routing Protocol based on Evolutionary Game theory [EEREG] works on clustering games for the selection of cluster-head.

In [11], the cluster is formed then cluster-head is selected, every sensor node in its cluster is given a time slot to transmit data to the cluster-head creating a setup state and then cluster-head transmits data to the base station. Fine number of CHs are determined if the distribution of CHs is uneven, and if the number of CHs are less than the pre-defined value then new CH will automatically be generated. If two or more clusters intersect then the common area produced, it is called cross coverage. Here the authors have applied game theory to determine the allocation of sensor nodes and then properly distributed the CH for balancing load which finally balances the energy in the network. They compared the result with the most reliable localized game theoretical clustering algorithm [12] and hybrid game theory-based distributed clustering algorithm [13] and proved that their protocol gives better performance.

The ECAGT algorithm [14] also focuses on CH selection and cluster formation but in a different style; here they use a localized method to ensure that CH are evenly distributed. In the setup phase, they created a communication radius of Rc for every sensor node and each sensor node only communicates to its neighbor within the radius defined resulting in proper cluster formation. Then after applying clustering game and through mixed strategy, Nash equilibrium analysis selection of CH is done.

Lin and Wang [15] proposed a new energy-efficient clustering algorithm combined with game theory and dual cluster-head mechanism (ECGD). They worked to reduce the energy overhead that resulted in changing of CH in its cluster through the dual cluster-head mechanism is based on the concept of the ratio of the node's remaining energy to its distance to CH expressed as $R_{e\text{-}d}$, even distribution of CH and a new concept of backup cluster-head (BCH). If the value of ratio $R_{e\text{-}d}$ is large for a sensor node, then it is selected as the CH and the very next value to it is selected as BCH which leads to save energy during broadcasting. By applying non-cooperation game model, inter-cluster communication is done and intra-cluster transmission is achieved by time slot based on TDMA. Nash equilibrium point is achieved through which balanced energy consumption for the prevention of hot spot problem is obtained in the network. Simulation results show that ECGD gives better performance in terms of energy efficiencies when compared to conventional routing protocols like LEACH [10], PEGASIS [16] and TEEN [17]. In [18], the authors adopt the data replication process of game theory to reduce possible network disconnection, first order energy model was taken, CH selection was done by clustering game method, if there is a disconnection between CH and the sink all the data of that cluster would be lost. Now to solve this issue, another sensor node called candidate CH was adopted, having all the replicas of the data of CH in it. Here the author offers bid to the player to get proper candidate having larger residual energy, less cost for data replication and less communication cost with base station which they have achieved by using second price sealed auction. For the knowledge of auction reader may refer [4].

4 Assessment of Different Cluster-Based Routing Protocols Built on Game Theory Approach

In this section, we summarize all the cluster-based routing protocols which are being developed using game theory. As aforementioned in Sect. 2, games are of different types and have different routing metrics for performance analysis. The main focus of each protocol is to provide an energy-efficient routing path by proper formation of the cluster and then proper selection of cluster-head preventing uneven distribution of energy consumption for prolonged network lifetime. Table 1 shows the mapping of different cluster-based routing protocols with games and their characteristics.

Table 1 Mapping of different cluster-based routing protocols with games, features and their limitations

S. No.	Name of protocol	Type of game	Features	Limitations
1	CROSS [2] Cluster routing of selfish sensor	Non-cooperation	• Prove Nash equilibrium • For equal energy consumption zero rule probability (ZRP) is proposed	Not suitable for practical applications
2	EECP [7] Efficient energy communication routing protocol	Cooperation	• Homogenous network architecture • Multihop transmission	For each node to know the exact time, it will transmit data to avoid collision during transmission that can increase energy consumption
3	CCNG [8] Clustering competition node game	Non-cooperation	• Transmission capacity is restricted to its radius • Concept of cross coverage helps in proper allocation of CH	Not suitable for multihop transmission
4	ECAGT [9] Energy-efficient clustering algorithm based on game theory	Non-cooperation	• For proper distribution of CH, localized method is adopted • Communication radius is set for every sensor node, and they communicate only with their neighbors	Uniform transmission radius for all sensor node
5	EERGT [5] Energy-efficient clustering routing protocol based on evolutionary game theory	Evolutionary	• Aims to ease hot spot problem through optimal cluster size • Mathematical model is presented for traffic load distribution	Complex routing process
6	ECGD [10] Energy-efficient clustering algorithm combined game theory and dual cluster-head	Non-cooperation	• Dual CH mechanism is used • Reduces energy overhead during the process of CH rotation	Congestion because of multiple nodes not discussed

(continued)

Table 1 (continued)

S. No.	Name of protocol	Type of game	Features	Limitations
7	HHO LEACH [6] Hybrid Homogeneous low-energy adaptive clustering hierarchy	Coalition game	• Protocol is formulated with learning automate and BCG	Not suitable for heterogenous network

5 Conclusion

In this paper, the cluster-based routing protocol using game theory has been review and elaborated extensively. It has been explained how to form cluster, selection of CH through clustering game and emphasizing how to make utility function for different game models. It has been investigated that cooperative games are used for node deployment and residual energy analysis whereas non-cooperative games have been extensively used for designing of routing protocol for energy balancing, high energy efficiencies and extended network lifetime.

Evolutionary games have been applied where sensors with imperfect rationality needs to cooperate with other nodes. It is observed that the next generation technology that is 6G technology will be using non-terrestrial network component for network topology and artificial intelligences which will be moreover sensor-based networks, and use of game theory in the cluster-based routing protocol for WSN will give a breakthrough performance.

References

1. S.P. Singh, S.C. Sharma, A survey on cluster-based routing protocols in wireless sensor network, in *International Conference on Advanced Computing Technologies and Applications (ICACTA)* (2015)
2. G. Koltsidas, F.-N. Pavlidou, A game theoretical approach to clustering of ad-hoc and sensor networks. Telecommun. Syst. **47**, 81–93 (2011)
3. D.E. Charilas, A.D. Panagopoulos, A survey on game theory applications in wireless networks. Comput. Netw. **54**(18), 3421–3430 (2010)
4. Y. Narahari, *Game Theory and Mechanism Design* (IISc Press, World Scientific, 2014)
5. T. AlSkaif, M.G. Zapata, B. Bellata, Game theory energy efficiency in wireless sensor networks: latest trends. J. Netw. Comput. Appl.
6. D. Lin, Q. Wang, D. Lin, Y. Deng, An energy efficient clustering routing protocol based on evolutionary game theory in wireless sensor networks. Int. J. Distrib. Sens. Netw., 12 pages. Article ID 409503 (2015)
7. S. Tyagi, S. Tanwar, S.K. Gupta, N. Kumar, S. Misra, J.P.C. Rodrigues, S. Ullah, Bayesian coalition game based optimized clustering in WSN, in *IEEE ICC 2015—Mobile and Wireless Network Symposium*
8. C. Intanagoonwiwat, R. Govindan, D. Estrin et al., Directed diffusion for wireless sensor networking. IEEE ACM T Netw. **11**(1), 2–16 (2003)

9. P. Raja, P. Dananjan, Game theory-based efficient energy consumption routing protocol to enhance the lifetime of WSN. Int. J. Inf. Commun. Technol. **8**(4) (2016)
10. W.R. Heinzelman, A. Chandrakasan, H. Balakrishnan, Energy efficient communication protocol for wireless micro SNs, in *Proceedings of the IEEE Hawaii International Conference on System Sciences*, Maui, HI, USA, Jan 2000, pp.1–10
11. X. Wu, X. Zeng, B. Fang, L. Yang, W. Zhang, An energy-balance and game theory-based cluster formation method for wireless sensor networks. Int. J. Distrib. Sens. Netw. **13**(8) (2017)
12. D. Xie, Q. Sun, Q. Zhou et al., An efficient clustering protocol for wireless sensor networks based on localized game theoretical approach. Int. J. Distrib. Sens. Netw. **9**(8), 476313 (2013)
13. L. Yang, Y. Lu, Y. Zhong et al., A hybrid, game theory based and distributed clustering protocol for wireless sensor networks. Wirel. Netw. **22**(3), 1007–1021 (2016)
14. Q. Liu, M. Liu, Energy-efficient clustering algorithm based on game theory for wireless sensor networks. Int. J. Distrib. Sens. Netw. **13**(8) (2017)
15. D. Lin, Q. Wang, An energy-efficient clustering algorithm combined game theory and dual-cluster-head mechanism for WSNs. IEEE Access (2019)
16. S. Lindsey, C.S. Raghavendra, PEGASIS: power efficient gathering in sensor information system, in *Proceedings, IEEE Aerospace Conference*, Mar 2002, pp. 1125–1130
17. A. Manjeshwar, D.P. Agrawal, TEEN: a routing protocol for enhanced efficiency in wireless sensor network, in *International Symposium on Parallel and Distributed Processing*, Apr 2001, pp. 2009–2015
18. Z. Xu, Y. Yin, X. Chen, J. Wang, A game theory-based clustering approach for wireless sensor networks, NGCIT. ASTL **27**, 58–66 (2013)

Advanced Digital Signal Processing for Interference Mitigation in Very High Throughput Satellite

Neeraj Mishra, Deepak Mishra, Nagendra Gajjar, and Kiran Parmar

Abstract The telecom market is eyeing the reasserted interest towards satellite communication to deliver the future ultra-high rate demand. These requirements lead to the excess reuse and sharing of limited frequency resources. In the field of satellite communication, ultra/very high throughput satellites (V/HTS) with multiple and aliquot spot-beams are the key solutions for distributing the future high data rate services. The performance of V/HTS is limited due to inter-system and intra-system generated interference rather than classical system generated thermal noise. The paper presents the literature survey of the mitigation techniques for interference generated due to excessive/full frequency reuse in V/HTS. The implementation aspects and progress of digital signal processing for MIMO based approaches (Precoding, Non-Orthogonal-based Multiple Access, and Predistortion) in V/HTS are also reported.

Keywords CM · FSS · MIMO · NOMA · Precoding · Predistortion · SATCOM · SIC · SUD · V/HTS

1 Introduction

During the last decade, the cellular communication systems have evolved rapidly, thanks to inclusion of new digital radio access standards and ease of availability of software-defined radios. Today, the third generation partnership project (3GPP) with Long-Term Evolution (LTE)-enabled 4G cellular standard communication is fully functional in the world and is promoting progressively upgradable services to users [1]. In the new technology development front, next generation radio access 5G

N. Mishra (✉) · D. Mishra
Space Applications Centre, ISRO, Ahmedabad 380058, India

N. Gajjar
Institute of Technology, Nirma University, Ahmedabad 382481, India

K. Parmar
Adani Institute of Infrastructure Engineering, Gandhinagar 382421, India

F. Thakkar et al. (eds.), *Proceedings of the International e-Conference on Intelligent Systems and Signal Processing*, Advances in Intelligent Systems and Computing 1370,
https://doi.org/10.1007/978-981-16-2123-9_4

mobile system will further enhance the broadband experience with the high peak data rate, increased area capacity, and better spectral efficiency [2]. On the other side, since birth, the satellite communication industry has witnessed the uninterrupted progress due to their unique pervasive style of content delivery to end-users. However, free space propagation loss, rain attenuation, and long duration of development cycle associated with the satellite communication (SATCOM) system makes the SATCOM architecture a redundant option compared to the terrestrial communication system. Fixed Satellite Service (FSS) are currently facing tremendous pressure from the terrestrial competitors for low-cost high-quality offers to support high data rate demands. Apart from this, one of the peculiar cases is in the areas where terrestrial networks offer low-quality services or do not exist at all. In these scenarios, satellite operators are facing deep challenges to offer high data rate solutions to their consumers in order to be complementary and/or competitive to the terrestrial networks. The homologue to the 5G development, the continuing accelerated high data demand, has opened new possibilities for innovative solutions with ultra-high or very high throughput satellites (V/HTS) [3–8]. On the further extent, to cater to the real-time excessive data rate demand, satellite communication has incorporated the use of unconventional technologies that were originally developed for terrestrial communications, e.g. spot beam configuration, multiple-input multiple-output (MIMO) processing, aggressive resource utilization, and interference mitigation. As the aggressive frequency reuse and huge beam configuration improves the system throughput for HTS, however, it leads to subsystem-generated interference. This interference severely degrades the link quality and needs to be mitigated in the chain. The various papers are available in the literature that describe the use of processing techniques for V/HTS for interference mitigation. However, most of the papers target one of the mitigation techniques and lack in providing ready reference for the designers. With these frameworks, an attempt is made to compile the two most attractive advances digital signal-processing techniques of MIMO for next generation V/HTS. It focuses on the two prime technologies with their overview, design, and implementation aspects in V/HTS case. The rest of this article is structured as follows. Secttion 2 describes the motivation for V/HTS in terms of the advancement of MIMO processing algorithms in satellite. Section 3 provides an overview of precoding techniques. Section 4 discusses the research done in the area of the Non-Orthogonal Multiple Access scheme. Section 5 discusses signal processing related to the predistortion algorithm in V/HTS. Section 6 concludes the article.

2 Motivation of the Signal Processing in V/HTS

To support the eruptive data demand, various innovative approaches are emerged in recent years [4–9]. One such example is the incorporation of DVB-S2X standard in the V/HTS network. The standard operates under highly efficient adaptive coding and modulation schemes, thus making significant contribution in improving the spectral efficiencies of digital satellite broadcast systems [10]. Despite all the achievements of the current SATCOM design, existing systems are far from the future goals of terabyte capacity. One of the main obstacles in current generation satellites are namely the system generated interferences (i.e. intra-beam or inter-beam interferences). These interferences are inherently associated due to the inclusion of the huge number of spot beams and massive frequency reuse in these beams. The next generations V/HTS are opting even more aggressive frequency reuse, e.g. full frequency reuse (all spot beams with same frequency band operation) or frequency reuse factor less than 3. This approach greatly enhances the system throughput, however, in this case the V/HTS is an interference-limited system rather than a thermal noise limited system. In the full frequency reuse scenario, the median of the Signal-to-Noise and Interference Ratio (SINR) is around 0 dB and in the range of − 5–15 dB. Similarly, for frequency reuse 2, the median of SINR is 8 dB and in the range of 0–25 dB. While for frequency reuse 3, the median of SINR is 25 dB and in ranges from 10 to 30 dB. Apart from these, as per conventionally used frequency reuse factor 4, the median of SINR is 30 dB and lies in the range of 10–40 dB [11]. This interference combines with the other interference generated by different means. The conventional approach to diminish interference is inadequate and insufficient. Hence, novel signal processing alternatives that exploit the structure of the co-channel interference are needed to be explored and adopted in the SATCOM systems. One such solution is the use of MIMO technology in V/HTS. The success of MIMO technology in terrestrial applications in realizing the current generation gigabit wireless is well known. This technology is now standardized for IEEE 802.11n, 802.16 m, 802.16e, 802.22, 802.20, 3GPP releases 7, 8 (LTE), etc. The salient feature of this MIMO processing is that it provides additional degrees of freedom through transmit diversity, multiuser diversity gains, spatial multiplexing gain, and interference reduction with no additional constraint on transmitted power and bandwidth. However, the application of terrestrial MIMO is not directly absorbable in the V/HTS scenario. This is due to Line-Of-Sight (LOS) communication, overloaded nature of the SATCOM system, pilot contamination, and co-channel interference, which does not decrease with increase in the number of beams [12]. For MIMO processing, the Channel matrix (CM) for MIMO must have both LOS and Non-Line-Of-Sight (NLOS) components. In a single satellite single gateway system due to strong LOS path and no nearby scattering, this CM component is rank deficient. In this scenario, transmit diversity due to MIMO cannot be achieved. To achieve the transmit diversity in the satellite system; the multiple transmitter with large geo-separation transmitting and receiving at the same frequency system is required. In this scenario, diversity gain is achievable as described by various authors [12, 13]. Further, Christian Hofmann and et al. demonstrated that MIMO transmit diversity gain in strong LOS channel is possible with particular antenna geometries. This idea has been applied to SATCOM systems in [14, 15], and a criterion has been developed for the optimal positioning of the MIMO antenna elements on Earth and in Geostationary Earth Orbit (GEO). It was also shown that key design parameters are inter-antenna spacing for onboard system and separation of ground terminals. This leads to mainly three concepts: (a) multiple antennas onboard in single-satellite, or (b) multiple satellites at different orbital positions, or (c) multiple gateways multi-beam satellite configuration operating at the same frequency. The considered V/HTS scenario in the paper is case 'c' for multibeam satellite with multigateway network transmitting the same frequency with different information.

3 Precoding

Continuing from the previous discussion on MU-MIMO processing, a direct analogy exists between the forward link of a multibeam satellite channel and the Broadcast Channel (BC) of a multiple-user multiple-input multiple-output (MU-MIMO) system. The Dirty Paper Coding (DPC) governs the throughput region for such systems [15, 16]. The DPC works to achieve the same sum rate for a BC in MU-MIMO as the co-channel interference is not present. Unfortunately, it is a nonlinear complex technique and practically not feasible. The restriction of implementation of DPC leads to the suboptimum approach known as precoding [17, 18]. In V/HTS scenarios, the precoding is advanced signal technique applied at gateways in forward link, exploiting the spatial degree of freedom offered by multiple antenna arrays and spot beam architecture of the onboard payload [19]. The main aim of precoding is to counteract the multibeam interference so that noncomplex user terminals can maintain a high SNR; even the same carrier frequency is present in the adjacent beam. A precondition for successful precoding is that the user receivers have to provide high-quality reports of their channel conditions (amplitude and phase) back to the gateways. Consider the forward link operation with full frequency reuse described in Fig. 1. In normal operation, the gateway (GW) will transmit the composite signal containing signals for all the users. The receiver terminal will experience the severely degraded reception of the signal due to the presence of the same frequency in the adjacent beam. In this situation, user receiver will not able to detect the intended signal. The precoding can establish the reliable link in this scenario. Figure 2 shows the system architecture for precoding in V/HTS. This figure shows the full frequency

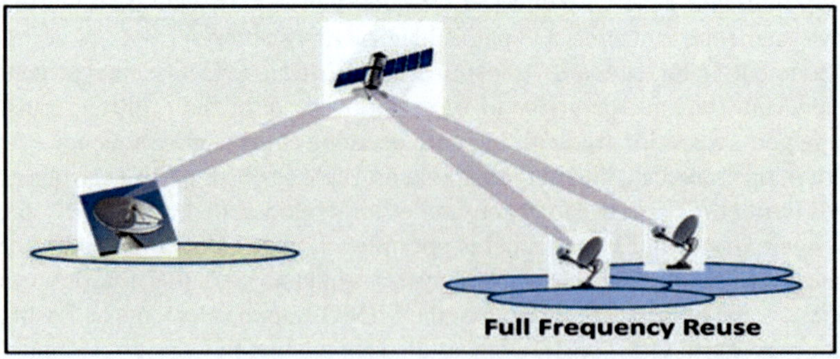

Fig. 1 Full frequency reuse architecture

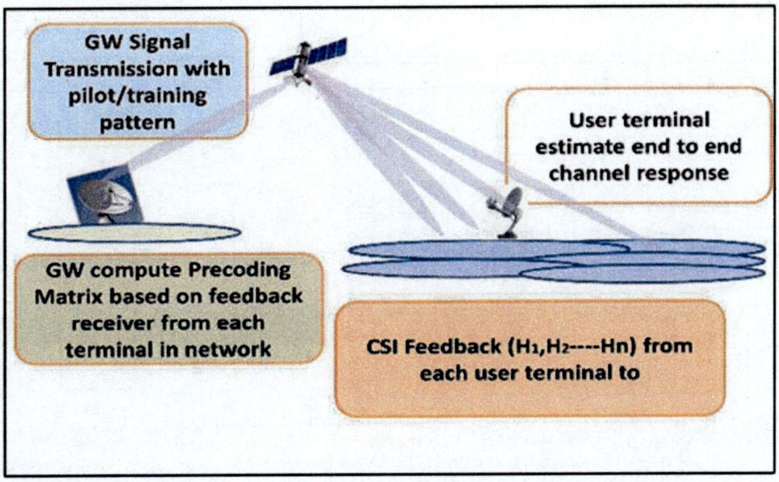

Fig. 2 System implementation of precoding in satellite network

reuse case in satellite network. Similar to previous situation, user receiver will experience the severe interference due to full frequency reuse. To mitigate this interference-limited link, each user terminal will estimate the precise Channel State Information (CSI) condition of the link. The components of CSI are received amplitude and phase values. The measured CSI is feedback to the gateway. The gateway will generate the precoding matrix and precodes the signal before the transmission to satellite. The user terminal will now receive the precoded signal. The precoded signal has inverse channel response. The user receiver can easily decode this signal without any additional processing. Figure 3 shows the positioning of the precoding function in the signal processing chain at the gateway. The user raw data streams are processed separately for baseband processing, source coding, channel coding, and modulation mapping. The individual streams are weighted as per precoding matrix. The gateway calculates the precoding matrix based on the received CSI and target optimization goal for the link.

After the precoding, precoded streams are then band-limited and combined to form single stream. This composite stream is up-converted and transmitted to satellite. In V/HTS multibeam configuration, a gateway is associated with a set of few beams. This gateway will have CSI of the users of these beams only. Therefore, there should be communication amongst the gateways for proper exchange of users' CSI from all beams in the network for perfect interference cancellation. The general input output equation at any receiver kth in the network is given in Eq. 1.

$$y = h_k' x + \eta \tag{1}$$

In Eq. 1, h_k' donate the conjugate transpose of the matrix. The h_k is the channel vector coefficient (composed of antenna gains, propagation losses, and phase shifts)

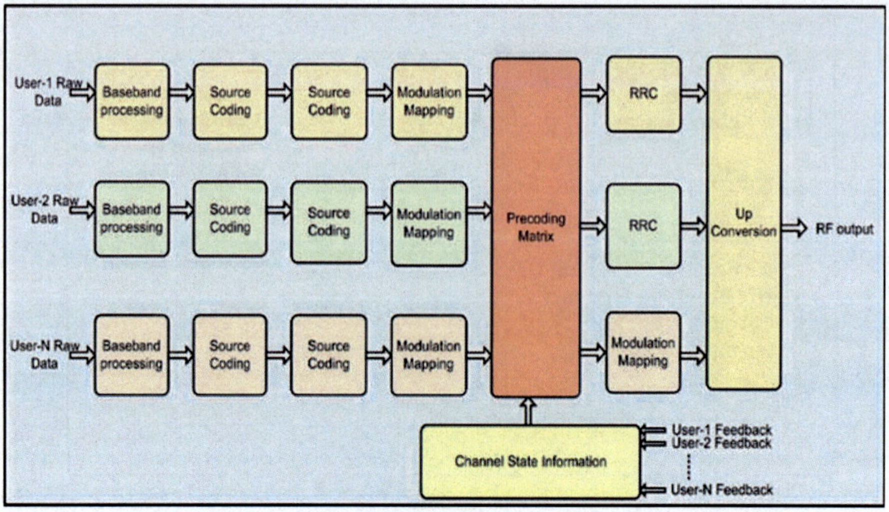

Fig. 3 Gateway processing in precoding

between the receivers and gateway. The h_k is a $1 \times N$ column matrix. The satellite is a transparent payload performing only the frequency conversion and amplification operation. N is the number of transmitting feed in the satellite. The x is the row vector with $N \times 1$ dimension. The η is the zero mean additive white Gaussian noise (AWGN) measured at the kth receive antenna. The Eq. 2 describes the baseband block-fading model of the signal at receiver end.

$$\mathbf{Y} = \mathbf{Hx} + n \tag{2}$$

Equation 3 describes the dimension of H matrix. The K is the total number of users in the network.

$$H = [h_1, h_2, h_3, h_4 \ldots h_K]' \tag{3}$$

The coefficient calculation of H is complex procedure. Table 1 shows the few optimization strategies for the computation of H. Apart from these, various authors [20, 21] have considered the optimization of the H matrix via an alternating projection technique with a Semi-Definite Relaxation (SDR) procedure. SDR is adequate for small to medium coverage areas (i.e. small to medium number of beams). However, if the larger number of beams and/or users are targeted, nonconvex optimization alternatives like SDR fail because of the immense computational complexity. This opens potential avenues for future research in efficient H matrix computation. Consider the load for computation of H matrix for Viasat-3 satellite. Viasat-3 is 1000-beams per colour satellite and considers average 1000 users per beam. The size of H matrix will be of 10^6 components. Further, these values must be calculated and periodically

Table 1 Precoding algorithms

Precoding algorithm	Function	Remarks
Linear precoding	Zero forcing (ZF)	The simplest suboptimal techniques that pre-filters the transmit signal vector
Linear precoding	Regularized zero forcing (R-ZF)	R-ZF extends ZF by taking into account the noise variance in order to improve performance in the low SNR regime
Linear precoding	Opportunistic beamforming (OB)	In OB, each user selects amongst predefined random beam formers from a codebook based on both instantaneous channel state information (CSI) and long-term channel statistics
Nonlinear	ZF-THP-Fi	The energy of the precoded signal is kept almost at the same level as the original one but each user experiences different level of SNR
Nonlinear	ZF-THP-Sa	The energy of the precoded signal is kept almost at the same level as the original one and each user is associated with the same level of SNR
Nonlinear	ZF-THP-Jo	The energy of the precoded signal is kept almost at the same level as the original one with optimization of different or equal level of SNR

updated for all the users. One of the solutions to reduce the complexity of H is to implement the partial precoding at gateway and partial in satellite. This is known as hybrid precoding. Figure 4 shows the full precoding implementation at gateway. Figure 5 shows the hybrid precoding implementation in the satellite. Equation 4 represents the factorization of channel matrix H

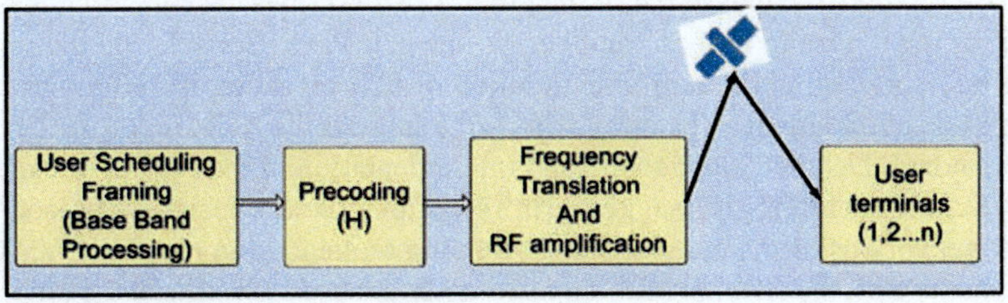

Fig. 4 Ground-based precoding

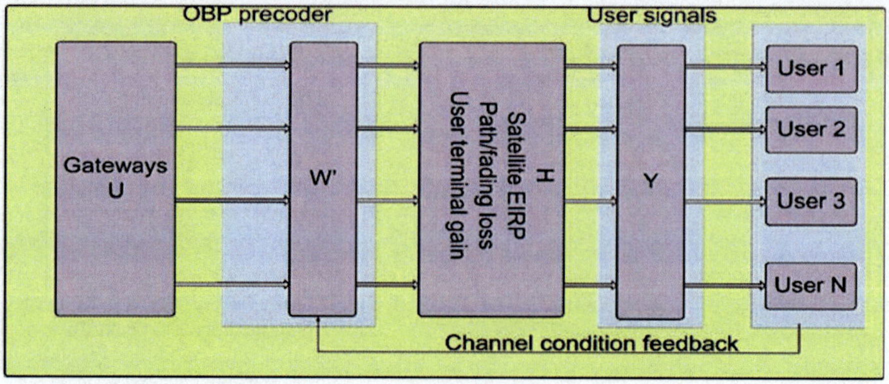

Fig. 5 Hybrid precoding

$$H = WU \tag{4}$$

where U is the $N \times K$ matrix for gateway precoding and W is the $K \times K$ matrix for onboard precoding. Here, W operation seizes to beamforming processing. Thus, the gateway has to calculate and transmit U precoded data instead of H precoded data. This reduces the feeder link requirement by H/U factor. However, the interference mitigation process acts on the K user beams directly instead of all the N radiating elements. This reduces the available degrees of freedom for the case where the number of radiating elements are higher than the number of user beams. The penalty paid in this approach is reduction in achieved throughput. This reduction is up to 20% as per [11].

The success of precoding is based on precise estimation and prompt reporting of CSI from the user terminal to gateways [21]. The DVB-S2X provides a framework for the procedure. The standard provides the super frame structure for the estimation of CSI while reporting of the CSI is through DVB-RCS or terrestrial link. The super frame structure uses Walsh-Hadamard sequence embedded into super frame format specification 2 and 3. The receive terminal on reception of super frame calculates the complex numbers representing the channel impairments in terms of amplitude and phase vectors. The clause E.4 of the DVB-S2X describes the signalling message format for the reporting of these numbers to the gateways. Considering the reliable precoding operation, maximum time duration of CSI reporting is 500 ms from the transmission of the signal by gateway. This is the same as the ACM reporting duration of the standard. Detailed architecture and implementation of the system is described in [10]. Schwarz et al. [15] has suggested different architecture design for precoding in satellite network using the transmission of training sequence compared to DVB-S2X. To obtain the CSI, the channel sounding strategies are used to estimate the amplitude and phase of the channel vectors. The gateways will transmit orthogonal training sequences like the Constant Amplitude Zero Auto-Correlation (CAZAC) sequence in the forward link [13, 22]. Using this method, the phase and amplitude information

Nu in a frame	Throughput ideal	Precoding Gain	Estimation error	Precoding Gain
2	77,56 Gbps	139 %	50,60 Gbps	57 % (-82 %)
5	64,94 Gbps	104 %	45,43 Gbps	42 % (-62 %)
10	51,46 Gbps	67 %	37,24 Gbps	21 % (-46 %)

Table C.9: CSI Errors, 500 MBaud

Nu in a frame	Throughput ideal	Precoding Gain	Estimation error	Precoding Gain
2	566,03 Gbps	174 %	328,88 Gbps	60 % (-114 %)
5	459,81 Gbps	127 %	295,99 Gbps	45 % (-82 %)
10	353,69 Gbps	79 %	239,11 Gbps	22 % (-57 %)

Fig. 6 Precoding gain with outdated CSI [10]

of the channel coefficients can be estimated via a cyclic cross-correlation of the received signal at the receiver as proposed in [14]. This information is forwarded to the gateways via the return link through satellite or via terrestrial link for enabling the pre-processing of the transmit signals. A small part of the available resources in the system must be allocated for the transmission of a training sequence and the estimation of the channel apart from additional complexity for precoding at the gateway. However, the impact on the overall system is very less from an operator perspective [10]. The Round Trip Delay (RTD) in V/HTS is also an important factor while implementing the precoding. With the RTD of 500 ms for GEO V/HTS, the channel conditions might affect the CSI that is reported to the gateway. A detailed analysis of the imperfection due to the outdated CSI, reported in [10] and [23], shows 21–174% precoding gain compared to the conventional system.

Figure 6 is referred from [10]. The results are encouraging to use the precoding in V/HTS apart from having the imperfect condition.

The standardization of precoding in next generation mobile communication has opened the probability of the use of precoding in V/HTS. In this direction, the European Space Agency (ESA) has started the Satellite Network of Experts III (SatNEx III) program. The SatNEx III is working on the SATCOM architecture for practical use of aggressive frequency reuse in HTS network. They have proposed a hybrid space-ground processing architecture (DIGISAT) payload. The payload is targeted for hybrid space-ground beamforming, MIMO and MIMO-MUD, Precoding, as well as Digital Feeder link [24]. The DIGISAT is aimed to strike the balance between the onboard complexity (due to the use of single frequency in payload) and on ground processing at gateways. The simulation measurements considering significant practical impairments show that precoding provides 40–140% more gain compared to the conventional 4-colour frequency reuse network [25].

The section provided the overview of the system aspects and related signal processing for precoding in V/HTS. The excellent research works have been reported in the field of precoding in V/HTS [17–19, 26, 27, 30, 31]. Signal processing at gateway, CSI estimation procedure, and CSI feedback mechanism are the main components of the precoding-based SATCOM link. Apart from the delayed CSI due to RTD, reporting mechanism and CSI exchange process between the gateways, the onboard complexity is a major concern using precoding in the full frequency scenario. This requires the separate receive/transmit chain per beam compared to shared receive/transmit chain in conventional 4-colour frequency reuse. This increases the payload mass, weight, and complexity. The issue is of prime emphasis while incorporating the precoding in V/HTS.

4 Non-orthogonal Multiple Access

In the previous section, we have presented the technique in which all the processing was implemented at gateways, thus making the simpler receiver to work in aggressive frequency reuse environment. In this section, we will show that the incorporation of little complexity in the receiver can further increase the system throughput.

Traditionally, HTSs are employing the Orthogonal Multiple Access (OMA) to utilize the resources amongst the beams and users. For fixed assignment services, frequency division multiple access or time division multiple access schemes are preferred OMA options while for dynamic assignment, demand assigned resource allocation schemes OMA are used in the architectures. In conventional scenario, these schemes provide good resource management with minimal interference. However, in these OMA techniques, the system allocates the resources to a particular user at the accessing instant. This leads to the bottleneck in achieving the maximum sum rate (maximizing the throughput of the network) of the system. Therefore, the OMA schemes cannot provide an enhanced performance with high efficiency to meet the explosive growth of traffic demand for future V/HTS satellite networks.

Recent studies and development in resource access schemes show that the Non-Orthogonal Multiple Access Scheme (NOMA) can be a potential candidate for increasing the system throughput while simultaneously serving massive user demand [31, 32]. One of the variants of NOMA, power domain NOMA, works on the principle of large imbalance in the user link quality and in such conditions outperforms the FDMA- or TDMA-based OMA scheme. In the paper, NOMA refers to the power domain NOMA. It has shown that at large difference in the received power level, NOMA clearly outperforms the OMA techniques. This is due to the inherited capability of dynamically adopting the system parameter as per traffic demand and receiver's CSI. Due to the capability of working in power imbalance environment, NOMA has attracted a lot of attention as a promising technology for the 5th generation new radio standards as well as in satellite access techniques. NOMA arranges the user in a natural way from the strongest to the weakest transmitted signal power using superposition coding so that user terminals can simultaneously use the same time and frequency slots. The abilities of NOMA to operate in a highly interfering situation has increased the idea of using this access scheme in V/HTS employing full or half frequency reuse amongst the spot beams. However, the direct implementation of terrestrial NOMA is also not applicable and needs to be envisaged for V/HTS MIMO BC. The first attempt towards this was made by Joroughi et al. [32]. In V/HTS, due to LOS communication and low scattering nature of the satellite channel, the

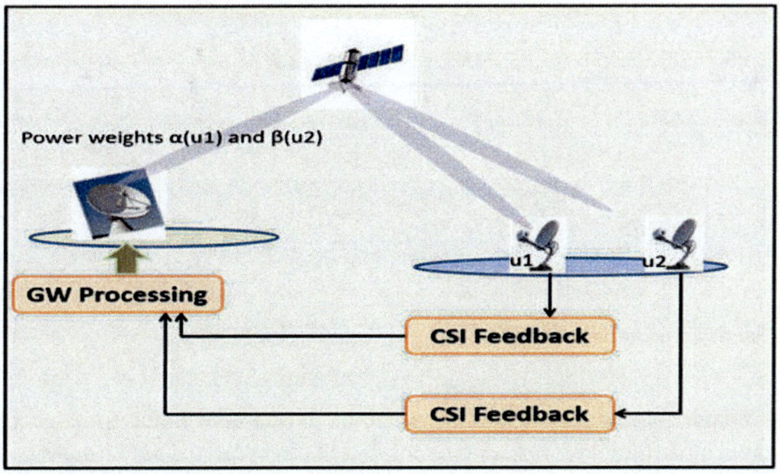

Fig. 7 System architecture for NOMA in satellite in intra-beam communication

received signal does not show high level of variation in signal power. However, due to the use of heterogeneous user receiver, the variation in SNR is seen and it may be exploited for use of NOMA in V/HTS. The situation arises due to variation in received signal gain within beam, different receiver antenna sizes associated gain, and target services requirement (fixed vs. mobile or different QoS services). These scenarios make NOMA applicability more pronounced in satellite network.

In the paper, two working scenarios with intra-beam and inter-beam operations are presented to explain the applicability of NOMA in V/HTS. Figure 7 shows the typical architecture of one of the spot beams in the conventional satellite network with intra-beam operation in NOMA. Consider an intrabeam scenario with two users u1 and u2. The user u1 is at beam centre and user u2 is near to the beam edge. The user u1 is accessing the satellite with low data rate requirement while user u2 is handling high data rate requirement. Since in the normal operation, the satellite will broadcast the same power to both the users. Thus, user u2 may experience less link margin while user u1 might be working with additional margin due to beam footprint. This is not an appreciable situation. NOMA in this scenario can provide additional gain in the achievable throughput of the system. In forward link, the gateway will arrange the user u1 and user u2 signals using supposition coding. In this architecture, both the users u1 and u2 will report their CSI to the respective gateway. The CSI data will be sent back to GW either through DVB-RCS or terrestrial link. The assumed reporting mechanism of operation will be quite similar to ACM followed in DVB-S2/DVB-S2x. After receiving the complete CSI, the gateway will arrange the user intended signal as per the availability of resource and received link condition. Figure 8 shows the signal processing at gateway for user 1 raw data (signal intended for u1) and user 2 raw data (signal intended for u2). User 1 raw data and user 2 raw data is independently processed through baseband processing (data formatting and source

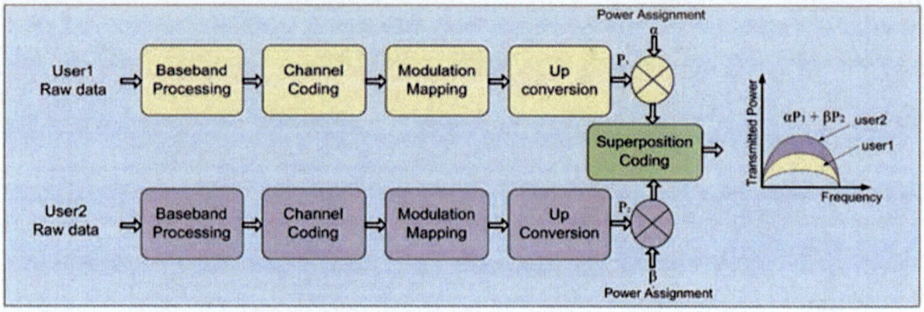

Fig. 8 Gateway processing for NOMA

coding), channel coding, modulation mapping, and up-conversion to respective IF frequency. The different power assignments based on reported CSI and superposition coding combine the generated signals P_1 and P_2. The combined signal is up-converted to RF frequency before being transmitted to the satellite. It is clear from Fig. 7 that the user u1 and user u2 will use the same frequency band. Equation 5 shows the total transmitted power P in terms of P_1 and P_2,

$$P = \alpha P_1 + \beta P_2 \tag{5}$$

Figure 9 shows the signal processing at the user end. After the transmission of NOMA composite signal, user u2 will receive more power compared to previous situation. The user u2 will apply Signal User Detection (SUD) technique to recover its intended message. On the user u1 side, it will receive not only its intended signal but also the u2 signal appearing as interference signal to it. User u1 will apply the Successive Interference Cancellation (SIC) technique to recover the intended signal. User u1 recovers the user u2 RF spectrum from the composite received signal. It will subtract the reconstructed u2 signal form received composite signal to recover the intended message. Thus, the receiver has to incorporate significant signal processing at their end. This extra processing increases the complexity in the receivers. Further

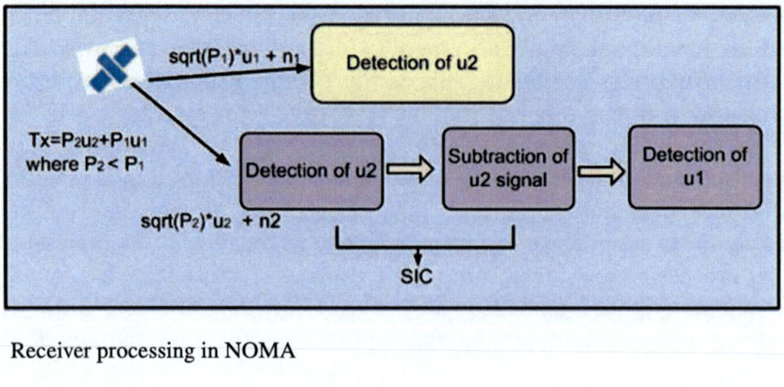

Fig. 9 Receiver processing in NOMA

proper mechanism of reporting the CSI back to the gateway is also to be incorporated in the design architecture. More intended work are required on the above issues before the practical inclusion of NOMA in V/HTS.

Now the paper will present the second working scenario for NOMA related to inter-beam NOMA. Figure 10 shows the inter-beam communication (between two spot beams with two user) with full frequency reuse. Suppose user p1 is in beam A and user p2 in beam B. Due to the same frequency/aggressive frequency reuse in operation, both users will experience the strong signal interference. This two-beam and two-user scenario with no beam cooperation draws an analogy to two-user interference channel (IC). The Han Kobayashi (HK) capacity inner bound characterizes this IC. The solution strategy for recovering the message is based on the splitting each user information into two sections. Figure 11 shows the signal processing at gateway. The intended messages form user p1 and p2 will be divided into two sections. The nomenclature for the sections are public and private.

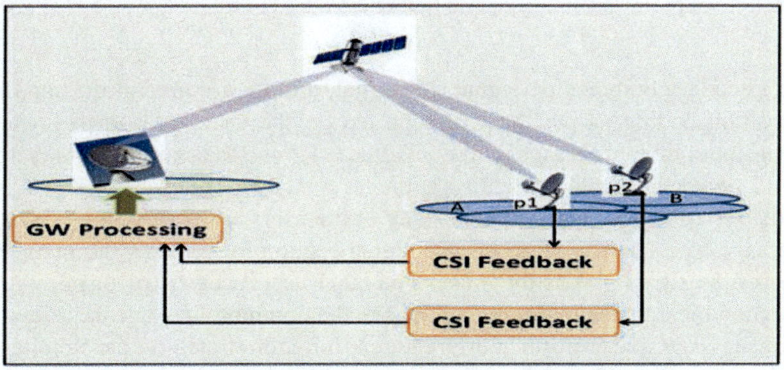

Fig. 10 Gateway processing in Inter-beam communication

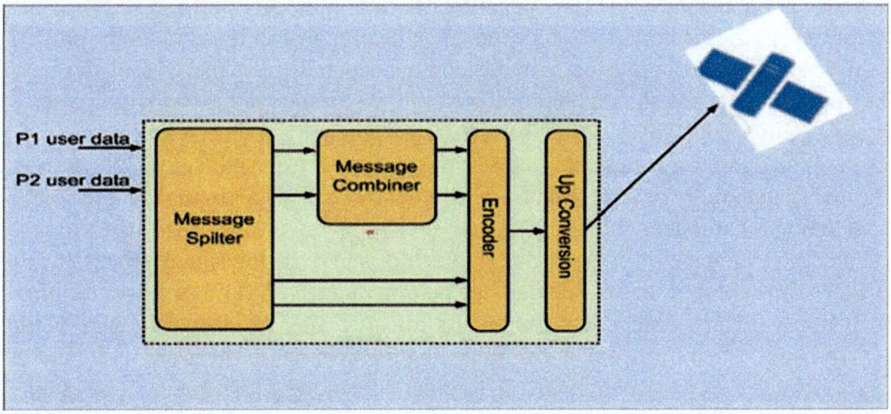

Fig. 11 Gateway processing for multibeam NOMA

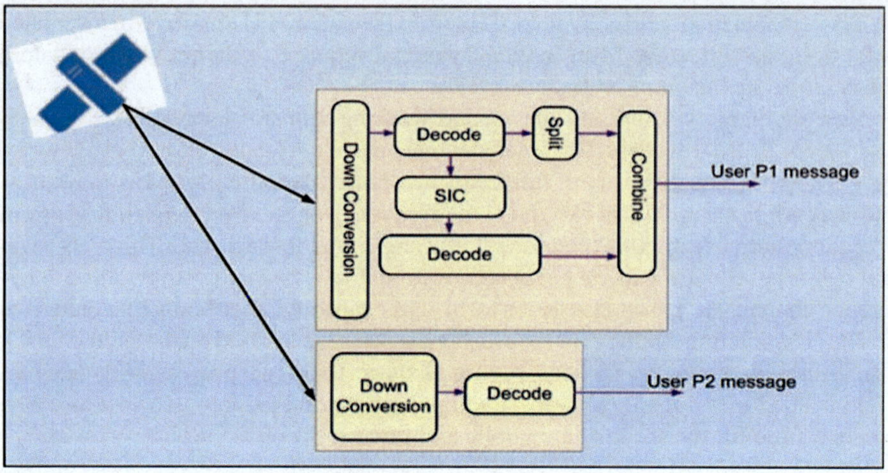

Fig. 12 Receiver processing for multibeam NOMA

The Encoder combines p1 signal, p2 signal, and their common message using superposition coding as per the CSI reported by the users p1 and p2. After the encoding, the combine message is up-converted to RF and transmitted via the satellite. Figure 12 shows the signal processing at the receiver end. The received signal is first down-converted at both receiver ends. One of the users either p1 or p2 will recover the message by SUD using the private signal section message as the main signal and rest of the signal part as the noise. The other user receiver (p1 or p2) will take the common message as interference, decode the common message, regenerate the common message, and perform interference cancellation strategy as per simultaneous non-unique decoding. This process is known as the rate splitting approach described in detail in [32, 33].

In centralized processing, where the different gateway has full access via CSI exchange in the network, the HK capacity bound is no longer valid and BC becomes the reference. In this bounded condition, each beam pair can independently process the user signals and apply the suitable coding depending on the CSI of the user. The author in [34] has shown the implementation approach in the beam cooperation scenario with partial CSI. The CSI parameter is signal power received at user end. In this case, interference is not completely cancelled at the transmit side and thus, user terminals must jointly decode two non-orthogonal signals by resorting to interference cancellation techniques.

Precoding along with NOMA is another domain area for signal processing in V/HTS to be explored. In this scheme, the gateway has the CSI of each user terminal of all beams. Thus, precoding counteracts towards inter-beam interference while NOMA deals with intra-beam interference. In this scenario, the network has to incorporate advanced signal processing such as precoding along with superposition coding at the gateway while using successive interference cancellation, multiuser detection and simultaneous non-unique user decoding algorithms at receiver terminal [33, 34].

The scenario was studied under the Satnex IV Network of Excellence, from the European Space Agency. The system level evaluation of a multibeam satellite system was done with 245 feed elements serving 245 beams with the maximum power per feed of 55 W and a power amplifier operating with 5 dB output back-off. Two benchmark cases have been reported: the four-frequency reuse scheme and the precoding with single layer transmission. The numerical results show that even without any type of scheduling, precoding with a multilayer transmission presents a substantial sum rate gain compared to the benchmark case. In this case as shown in Fig. 13, a 109% and 384% gain is obtained with respect to the single layer precoding case and the four-frequency scheme.

In this section, the paper has presented the potential use of NOMA in V/HTS network. However, various aspects are still yet to be defined and need more work. A few of them are the impact of contribution due to channel estimation error at receiver end, the effect of timing offset amongst the transmitted signal in different beams, and massive increase in complexity of the MUD scheme when more than two users participate in NOMA. The success of NOMA relies on the feedback mechanism and delay in the CSI reporting from the user. These areas need more attention before incorporating NOMA in V/HTS. Incorporation must include upper layer aspect such as

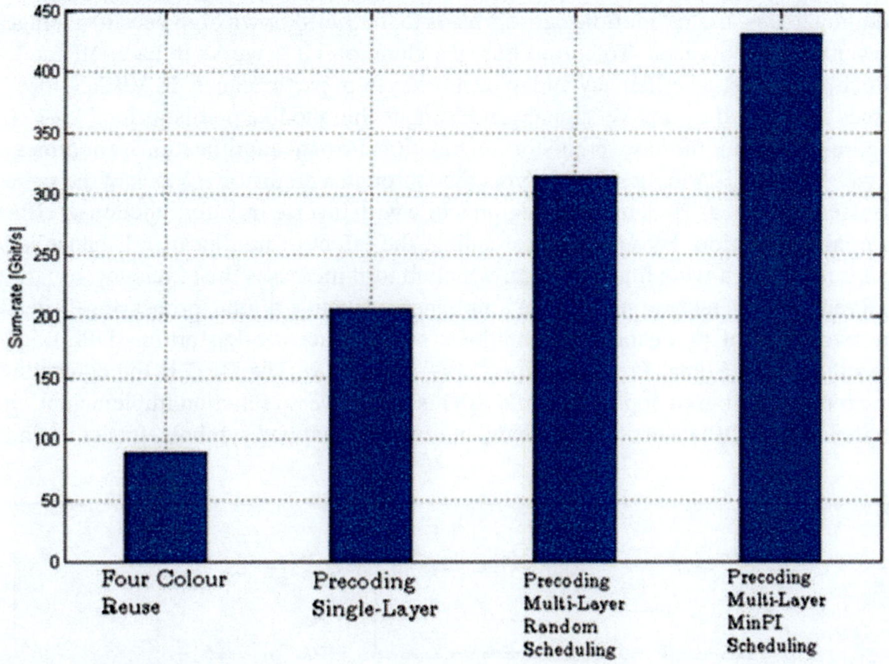

Fig. 13 Sum rate evaluation for multibeam precoding transmission jointly with NOMA and different scheduling strategies [32]

non-uniform traffic amongst the beam, fading of satellite channel, automatic coding, modulation, and multiuser multi-group framing encapsulation of V/HTS.

5 Predistortion

To enhance the spectral efficiency in current generation satellites, video broadcasting standard DVB-S2X presents a framework for growing use of the higher order modulation (up to 256 APSK) and more sturdy forward error correction schemes (use of long frame LDPC) through different MODCOD options [10]. This has made immediate improvement of around 51% compared to previous standard DVB-S2. The goal is to consolidate higher order modulation, coding, and signal handling. This is to accomplish the most noteworthy conceivable bits/Hz/W for combating the accelerating data demand. Nonetheless, there are two essential issues while utilizing higher order modulation in the satellite: one is a power-limited satellite system and the other is the highly nonlinear satellite channel. Further, it is much required to maximize the onboard power efficiency of the High Power Amplifier (HPA).

The HPA must function closer to the saturation for higher efficiency, however, the distortion outcomes are more prominent in this region with multicarrier operation. In addition to this, use of multiple carriers leads to the well-known high peak to average power ratio phenomena. To curtail this, the situation HPA works in back-off mode which, degrades the efficiency and in turn the system performance. In V/HTS, these issues are tackled by a novel signal processing technique like predistortion [35–37]. Figure 14 explains the basic predistortion function. Prior to amplification by nonlinear device, typical HPA input signal is processed through a predistorter block of the same transfer function as HPA in linear region while with inverse transfer function of HPA in nonlinear region. By virtue of cascading, the effect of nonlinearity is cancelled and it results in a wide linear system, which in turn increases the efficiency.

Predistortion techniques can be implemented using digital processing. There are two types of predistortion techniques; one is Data Pre-Distortion (DPD) and the other is the Signal Pre-Distortion (SPD) technique. The DPD is the constellation remapping-based approach while SPD is the converse function implementation approach. The DPD works on baseband modulation mapped symbols. It tries to find

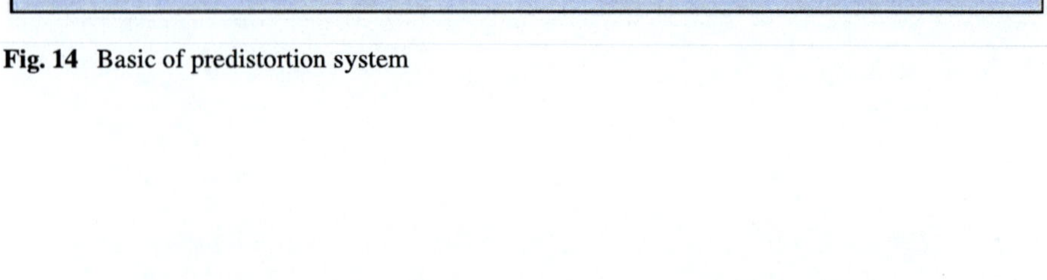

Fig. 14 Basic of predistortion system

out the error between the input signal constellation and amplified signal constellation points. The processed errors are used to distort the input signal constellation in such a way that after the amplification, the output signal achieves the same constellation as it has processed through ideal amplifier. Depending on the implementation strategy, the DPD distortion functions are preloaded in lookup table LUT or dynamically changed on line. The dynamic DPD has to generate the error function by processing the current symbol, previous symbols as well as estimated futuristic upcoming symbols. This approach requires huge onboard memory. For M constellation size and k length symbol processing, dynamic processing requires $M * k$ memory depth. The complexity increases by manifold in multicarrier scenarios where the error function for all the carriers after the modulations has to be processed.

The SPD approach employs converse function of I/O characteristic of the system (typical HPA). The converse function is mapped in the form of a transfer function of the channel. This is the equation-based approach, which requires the bounded input signal to avoid unstable condition in the HPA. The SPD processing is based on waveform processing, hence it is more advantageous with DVB-S2 or DVB-S2X format. However, due to continuous nature of the transfer function, bandwidth of the SPD processed signal will not be the same and may violate the stringent requirement of spectrum mask required before uplinking. An additional filtering should be applied before uplinking the signal, which is a high power and high frequency signal [10, 25]. On the implementation side, the predistortion can be placed on ground or in satellite depending on the implementation complexity of the SATCOM chain. Figures 15 and 16 show the implementation approach for the ground-based technique and onboard technique. The ground-based predistortion suffers from the high spectral regrowth compared to the onboard-based approach. This spectral growth is to be controlled to meet stringent requirement of the emission mask of the transmitted signal. However, the penalty in the onboard design is the requirement of additional mass and power. A detailed analysis considering the pros and cons of both the implementation techniques is evaluated in [36]. Figure 17 shows the empirical digital processing diagram for onboard digital non-regenerative payload. The signal output at IMUX is $s(t)$ and $r(n)$ is the digital sample generated via ADC defined as in Eq. 6.

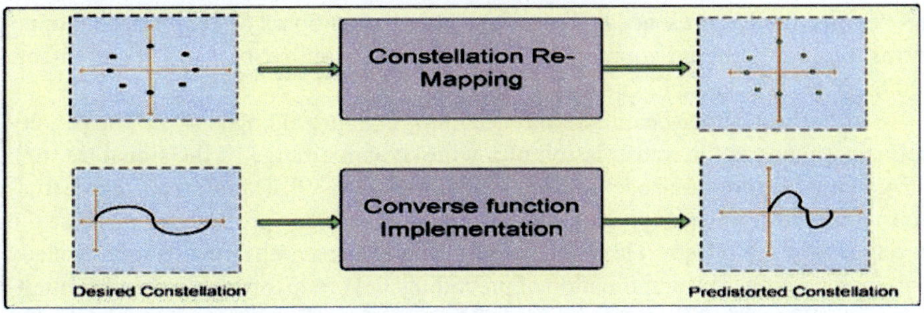

Fig. 15 SPD versus DPD predistortion functionality

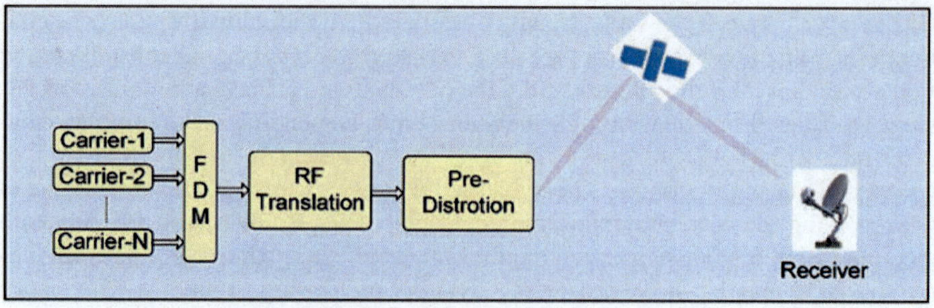

Fig. 16 Ground-based predistortion System

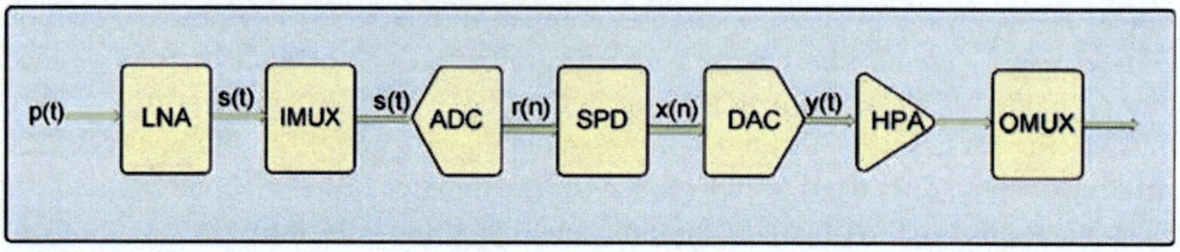

Fig. 17 Onboard predistortion system

$$r(n) = s(nT_s) \tag{6}$$

where T_s is the sampling instant and $x(n)$ is the output of SPD processing which is defined as

$$x(n) = f(r(n)) \tag{7}$$

The DAC converts $x(n)$ to pre-processed analog signal $x(t)$ which will be further amplified by TWTA to produce $y(t)$. The amplified signal is downlinked after passing through OMUX in the chain. The SPD function ($f(r(n))$) models the inverse characteristics of TWTA in digital domain to compensate the nonlinear effect of the TWTA. It has been verified in presence or absence of jitter that onboard SPD approach outperforms the ground-based counterpart both in terms of output back-off requirements and SINR performance (2 or 4 dB) [36].

To mitigate the inter-beam interference in V/HTS, precoding is used. The penalty in this approach is the transmission of a wide dynamic range of the signal through HPA, which requires high back-off mode operation [38, 39]. Therefore, in these situations, joint precoding and predistortion techniques are being proposed to increase the throughput of the system. This is one of the attractive research domains for advanced signal processing. The combination of precoding and predistortion needs more attention for full/fractional frequency reuse and fractional predistortion systems in V/HTS scenario.

6 Conclusion

In this literature survey, a comprehensive summary of three important signal processing techniques related to interference mitigation for future V/HTS are presented. The study was done for considering the V/HTS scenario with aggressive frequency reuse. The work done and the open issues related to the MIMO algorithm for interference mitigation were presented in brief, and it has shown that proper implementation of these signal processing techniques can provide significant improvement in system throughput.

The paper has tried to accom,modate the work done in the field of design, implementation, and system aspects of precoding in V/HTS. The signal processing at gateway, CSI estimation process, and feedback mechanism to gateway are reported in detail. Further, various practical issues, e.g. computation and exchange of precoding matrix for large number of beams, CSI reporting delay due to RTD, and payload complexity are explained with the research done in this field.

Being a MUD technique, the incorporation of NOMA in V/HTS is not as straightforward. The research done by various authors and the ESA initiative towards the inclusion of NOMA in V/HTS are referred to for explanatory cases. The two-beam two-user case scenario is discussed with the example for inter-beam and intra-beam SATCOM. The design, implementation, and system aspect of NOMA are explained with associated signal processing.

Apart from these, the paper brings out the importance of onboard predistortion in V/HTS. The various authors have concluded that predistortion can significantly reduce the back-off requirement of onboard HPAs in the V/HTS. The development in these areas is quite new, however; the milestone works done in this field are summarized in the article.

References

1. https://www.etsi.org/technologies/mobile/4g
2. A. Gohil, H. Modi, S.K. Patel, 5G technology of mobile communication: a survey, in *2013 International Conference on Intelligent Systems and Signal Processing (ISSP)*, Gujarat, 2013, pp. 288–292. https://doi.org/10.1109/ISSP.2013.6526920
3. H. Fenech, A. Tomatis, S. Amos, V. Soumpholphakdy, D. Serrano-Velarde, Future high throughput satellite systems, in *2012 IEEE First AESS European Conference on Satellite Telecommunications (ESTEL)*, Rome, 2012, pp. 1–7. https://doi.org/10.1109/ESTEL.2012.6400117
4. Y. Vasavada, R. Gopal, C. Ravishankar, G. Zakaria, N. Ben Ammar, Architectures for next generation high throughput satellite systems. Int. J. Satellite Commun. Netw. **34**(4), 523–546 (2016)
5. M. Hasan, C. Bianchi, Ka band enabling technologies for high throughput satellite (HTS) communications. Int. J. Satell. Commun. Netw. **34**(4), 483–501 (2016)
6. L.R. Shet, G.P. Reddy, I. Ahmad, P. Killedar, P.K. Gupta, T.K. Anuradha, Challenges and configuration of ISRO's future Q/V band satellite, in *2016 International Conference on Wireless Communications, Signal Processing and Networking (WiSPNET)*, Chennai, 2016, pp. 674–679

7. D. Minoli, *Innovations in Satellite Communications Technology* (John Wiley & Sons, Inc., 2015)
8. T. Delamotte, K. Storek, A. Knopp, MIMO processing for satellites in the 5G Era, in *2019 IEEE 2nd 5G World Forum (5GATEWAYF)*, Dresden, Germany, 2019, pp. 629–635
9. P. Arapoglou, K. Liolis, M. Bertinelli, A. Panagopoulos, P. Cottis, R. De Gaudenzi, MIMO over satellite: a review. IEEE Commun. Surv. Tutor. **13**(1), 27–51 (Ist Quarter 2011)
10. Second generation framing structure, channel coding and modulation systems for broadcasting, interactive services, news gathering and other broadband satellite applications; part II: S2-extensions (S2X), DVB Blue Book A171–2, March 2014
11. A.I. Pérez-Neira, M.A. Vazquez, S. Maleki, M.R.B. Shankar, S. Chatzinotas, Signal processing for high throughput satellite systems: challenges in new interference-limited scenarios (2018)
12. A. Pérez-Neria, M.A. Lagunas, M.A. Vázquez, High throughput satellites in 5G and MIMO interference limited communications, in *MATEC Web of Conferences 2017*
13. R. Schwarz, A. Knopp, D. Ogermann, C. Hofmann, B. Lank, Optimum-capacity MIMO satellite link for fixed and mobile services, in *Smart Antennas, 2008.* International ITG Workshop on WSA 2008, pp. 209–216 (2008)
14. C. Hofmann, K. Storek, R.T. Schwarz, A. Knopp, Spatial MIMO over satellite: a proof of concept, in *2016 IEEE International Conference on Communications (ICC)*, Kuala Lumpur, 2016, pp. 1–6. https://doi.org/10.1109/ICC.2016.7510945
15. R.T. Schwarz, T. Delamotte, K. Storek, A. Knopp, MIMO applications for multibeam satellites. IEEE Trans. Broadcast. **65**(4), 664–681 (2019)
16. M. Costa, Writing on dirty paper (Corresp.). IEEE Trans. Inf. Theory **29**(3), 439–441 (1983)
17. J. Lee, N. Jindal, Dirty paper coding vs. linear precoding for MIMO broadcast channels, in *2006 Fortieth Asilomar Conference on Signals, Systems and Computers, CA*, 2006, pp. 779–783
18. M.Á. Vázquez, A. Pérez-Neira, D. Christopoulos, S. Chatzinotas, B. Ottersten, P.-D. Arapoglou et al., Precoding in multibeam satellite communications: present and future challenges. IEEE Wireless Commun. **23**(6), 88–95 (2016)
19. V. Joroughi, M.Á. Vázquez, A.I. Pérez-Neira, Precoding in multigateway multibeam satellite systems. IEEE Trans. Wireless Commun. **15**(7), 4944–4956 (2016)
20. G. Zheng et al., Generic optimization of linear precoding in multibeam satellite systems. IEEE Trans. Wireless Commun. **11**(6), 2308–2320 (2012)
21. V. Joroughi, M.Á. Vázquez, A.I. Pérez-Neira, Generalized multicast multibeam precoding for satellite communications. IEEE Trans. Wireless Commun. **16**(2), 952–966 (2017)
22. D. Christopoulos, S. Chatzinotas, B. Ottersten, Multicast multigroup precoding and user scheduling for frame-based satellite communications. IEEE Trans. Wirel. Commun. **14**(9), 4695–4707 (2015)
23. M. Ángel Vázquez, M.R. Bhavani Shankar, C.I. Kourogiorgas, P.-D. Arapoglou, V. Icolari, S. Chatzinotas, A.D. Panagopoulos, A.I. Pérez-Neira, Precoding, scheduling, and link adaptation in mobile interactive multibeam satellite systems. IEEE J. Select. Areas Commun. **36**(5) (2018)
24. J. Arnau-Yanez, M. Bergmann, E.A. Candreva, G.E. Corazza, R. de Gaudenzi, B. Devillers, W. Gappmair, F. Lombardo, C. Mosquera, A. Pérez-Neira, I. Thibault, A. Vanelli-Coralli, Hybrid space-ground processing for high-capacity multi-beam satellite systems, in *IEEE Globecom 2011 Proceedings*
25. P.-D. Arapoglou, A. Ginesi, S. Cioni, S. Erl, F. Clazzer, S. Andrenacci, A. Vanelli Coralli, DVB-S2X-enabled precoding for high throughput satellite systems. Int. J. Satell. Commun. Netw. **34**(3), 439–455 (2016)
26. V. Joroughi, M.R. Bhavani Shankar, S. Maleki, S. Chatzinotas, J. Grotz, B. Ottersten, Precoder design for multibeam mobile satellite systems, in *2018 52nd Asilomar Conference on Signals, Systems, and Computers*, CA, USA, 2018, pp. 456–460
27. D. Christopoulos, S. Chatzinotas, G. Zheng et al., Linear and nonlinear techniques for multibeam joint processing in satellite communications. Wireless Com. Netw. **2012**, 162 (2012)
28. G. Scutari, F. Facchinei, L. Lampariello, Parallel and distributed methods for constrained nonconvex optimization—part I: theory. IEEE Trans. Signal Process. **65**(8), 1929–1944 (2017)

29. X. Jiang, Z. Zhong, X. Liu, H.C. So, Robust matrix completion via alternating projection. IEEE Signal Process. Lett. **24**(5), 579–583 (2017)
30. V. Joroughi, M.R.B. Shankar, S. Maleki, S. Chatzinotas, J. Grotz, B. Ottersten, On-board precoding in a multiple gateway multibeam satellite system, in *2018 IEEE 88th Vehicular Technology Conference (VTC-Fall)*, Chicago, IL, USA, 2018, pp. 1–5
31. Y. Liu, Z. Qin, M. Elkashlan, Z. Ding, A. Nallanathan, L. Hanzo, Nonorthogonal multiple access for 5G and beyond. Proc. IEEE **105**(12), 2347–2381 (2017)
32. A. Perez-Neira, M. Caus, A. Vazquez, N. Alagha, NOMA schemes for multibeam satellite communications (2018). https://arxiv.org/ftp/arxiv/papers/1810/1810.08440.pdf
33. M. Caus, M.Á. Vázquez, A. Pérez-Neira, NOMA and interference limited satellite scenarios, in *50th Asilomar Conference on Signals, Systems and Computers, CA*, 2016, pp. 497–501
34. X. Yan et al., The application of power-domain non-orthogonal multiple access in satellite communication networks. IEEE Access **7**, 63531–63539 (2019). https://doi.org/10.1109/ACCESS.2019.2917060
35. L. Guan, A. Zhu, Green communications: digital predistortion for wideband RF power amplifiers. IEEE Micro. Mag. **15**(7), 84–99 (2014)
36. N. Mazzali, M.R. Bhavani Shankar, B. Ottersten, On-board signal predistortion for digital transparent satellites, in *2015 IEEE 16th International Workshop on Signal Processing Advances in Wireless Communications (SPAWC)*, Stockholm, 2015, pp. 535–539
37. Kelly, M. Allegue-Martnez, P.-D. Arapoglou et al., Bandwidth-constrained digital pre-compensation technique for multi-carrier satellite communications. Int. J. Satell. Commun. Network. (2015)
38. Spano, M. Alodeh, S. Chatzinotas, J. Krause, B. Ottersten, Spatial PAPR reduction in symbol-level precoding for the multi-beam satellite downlink, in *2017 IEEE 18th International Workshop on Signal Processing Advances in Wireless Communications* (2017)
39. Alvarez-Diaz, C. Mosquera, M. Neri, G.E. Corazza, Joint precoding and predistortion techniques for satellite telecommunication systems, in *2005 2nd International Symposium on Wireless Communication Systems, Siena*, 2005, pp. 688–692

Low-Power Endoscopic Image Compression Algorithms Using Modified Golomb Codes

Kinde A. Fante and Basabi Bhaumik

Abstract The design of an image compressor for wireless capsule endoscopy is challenging due to the stringent constraints on the size and lifetime of the battery. In this paper, we present a single-pass endoscopic image compression algorithm which combines differential pulse-coded modulation (DPCM), color space transformation, sub-sampling, uniform quantization, and Golomb codes. Two methods are employed in order to improve the performance of existing DPCM-based endoscopic image compression algorithms. These are applying two-dimensional sub-sampling scheme for chrominance components of the image and modifying Golomb codes. Computationally, efficient hardware architecture is proposed which uses computationally simple operations: binary shift operations, logical operations, and addition and subtraction operations. It doesn't need buffer memory. The hardware implementation of the architecture consumes 4.89 μW power for transmitting two full-color images of 256×256 resolution. Experimental results show that the proposed image compression method achieves a good balance among compression ratio (11.02), reconstructed image quality (40.55 dB of peak signal-to-noise ratio), and computational complexity. The obtained performance improvement is significant as compared to the existing image compression methods for wireless capsule endoscopy application.

Keywords Wireless capsule endoscopy · Differential pulse-coded modulation · Sub-sampling · Image compression · Golomb code

---

K. A. Fante (✉)
Faculty of Electrical and Computer Engineering, Jimma Institute of Technology, Jimma University, P. O Box 378 Jimma, Ethiopia
e-mail: kinde.anlay@ju.edu.et

B. Bhaumik
Department of Electrical Engineering, Indian Institute of Technology Delhi, Hauz Khas, New Delhi 110016, Delhi, India
e-mail: bhaumk@ee.iitd.ac.in

F. Thakkar et al. (eds.), *Proceedings of the International e-Conference on Intelligent Systems and Signal Processing*, Advances in Intelligent Systems and Computing 1370,
https://doi.org/10.1007/978-981-16-2123-9_5

1 Introduction

The invention of wireless capsule endoscopy (WCE) [11] has been a breakthrough in non-invasive diagnosis of gastrointestinal abnormalities [18, 22]. It can also be used for the assessment of small bowel transit time [9]. As shown in Fig. 1, the swallowable capsule consists of image sensor, battery, lighting source (LEDs), image compressor, and RF transmitter. After being swallowed by the patient, the capsule moves through the gastrointestinal tract with the help of peristalsis. While moving in an uncontrollable way, it captures color images of the inner wall of the gastrointestinal tract and sends it via a wireless link to a data recorder which is worn by the patient around the waist. The doctors can download and view the recorded image on a computer with image processing software.

Even after 15 years of its appearance, WCE has several unsolved challenges (locomotive system [17], computer-aided pathology detection [10], image compression [15], etc.). More than 90% of the battery power of the capsule is consumed by the RF transmitter [26]. The inclusion of suitable image compressor inside WCE is a viable option to reduce the power consumption of the RF transmitter. Researchers tried to optimize the general-purpose image compression algorithms by exploiting the unique features of the images of the human digestive system. Among the techniques, DCT-based [16] and DPCM-based methods [3, 13, 14] have been extensively explored. The DCT-based algorithms are computational, more complex, and require high memory on the encoder side when it works with an image sensor which sends the pixel values in a raster-scan fashion. Even though it gives high compression rate, the overhead is quite high. On the other hand, the DPCM-based image compression algorithms look more attractive for this domain of images because of their low computational complexity and low memory requirement. However, their compression rate is inferior to the DCT-based methods.

In the literature, the existing DPCM-based endoscopic image compression methods employ one-dimensional (1-D) prediction function and sub-sampling of chrominance components of an image [3, 13, 14]. Using two-dimensional (2-D) prediction function in the DPCM method gives higher compression rate than the 1-D prediction function but it requires to store at least one row of the previously encoded pixel values. Hence, the choice of using only 1-D prediction function in DPCM methods is to

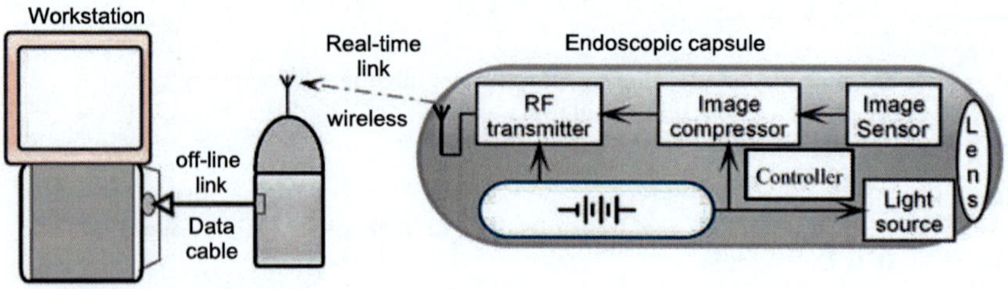

Fig. 1 The block diagram of capsule endoscopy system

reduce the computational complexity and memory requirement overheads [14] when the image compressor is integrated with an image sensor that sends pixel values in a raster-scan fashion [21]. One of the methods that has been widely employed to improve the performance of DPCM-based endoscopic image compression methods is sub-sampling chrominance components of the image. This method is efficient as well as computationally simple which makes it ideally suitable for WCE applications [3, 4, 14]. The choice of 1-D sub-sampling of the chrominance components of an image is not proved experimentally in the literature, at least relevant to endoscopic images, so far.

In this work, we show that the 2-D image chroma component sub-sampling scheme improves both image quality and compression rate as compared to its 1-D counterpart when used in the DPCM-based endoscopic image compression method that employs the 1-D prediction function. In addition to this, the length-limited Golomb-Rice code has been extensively used in endoscopic image compression algorithms [3, 13, 14]. The maximum length of the code is limited to 32. In hardware implementation, all the codes are 32-bit codes. Converting the 32-bit code which is a parallel data into a serial bit requires a clock which runs 32 times faster than the clock used in an image compressor in order to interface the encoder with the RF transmitter. Even though, the parallel-to-serial (P2S) converter circuit is a simple circuit and is needed merely for interfacing the two modules (image compressor and RF Tx), it consumes a significant amount of power. For example, the P2S converter circuit consumes more than 17 times the power consumed by an image compressor in [14]. This is due to the fact that the power consumption of integrated circuits is a non-linear function of the frequency of operation [23]. In this work, we address this issue using optimized exponential Golomb codes which can reduce the clock frequency of the parallel-to-serial converter by one-half and hence the power consumption.

The remaining part of the paper is organized as follows. In Sect. 2, the proposed image compression algorithm is presented. The experimental results are discussed in Sect. 3. The power consumption analysis of the proposed image compressor is detailed in Sect. 4. The conclusion is given in Sect. 5.

2 The Proposed Image Compression Algorithm

The block diagram of the proposed image compression algorithm, which is an optimal combination of computationally simple methods, is shown in Fig. 2. The image compressor employs scalar quantization, color space transformation, sub-sampling, differential pulse-coded modulation, and modified Golomb encoder. The rationale for choosing these methods and the optimal parameters of each method is illustrated in the following sections. The sub-sampling operation and the entropy encoder are given more emphasis in this work.

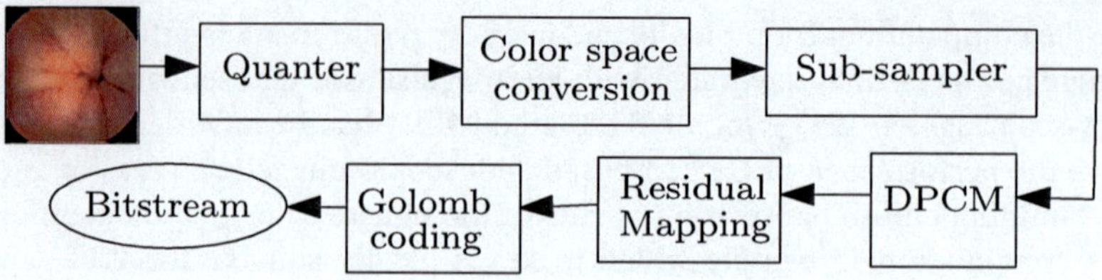

Fig. 2 Block diagram of the proposed image compression algorithm

2.1 Uniform Quantization

We uniformly quantize the RGB components of a color endoscopic image in order to improve the compression rate as detailed in [3]. We use a mid-rise quantizer with the quantization step of four. This is the same as removing the two least significant bits form every pixel. It introduces uniformly distributed error in $[-2, 2]$ but improves the compression efficiency significantly. The expected value of the quantization error is zero and the variance is 1.33. This method is computationally very simple and efficient [3].

2.2 Color Space Transformation

We use the reversible color transform recommended in the JPEG-2000 image compression standard [1] which is given by (1). Its chroma components are U_j and V_j, and the luminance component is Y_j. The chrominance components have low frequency contents that makes them suitable to apply sub-sampling and differential pulse-coded modulation techniques.

$$\begin{aligned} Y_j &= \lfloor \frac{R + 2G + B}{4} \rfloor \\ U_j &= R - G \\ V_j &= G - B. \end{aligned} \tag{1}$$

The inverse transformation of (1) is given by (2).

$$\begin{aligned} G &= Y_j - \lfloor \frac{U_j - V_j}{4} \rfloor \\ R &= U_j + G \\ B &= G - V_j. \end{aligned} \tag{2}$$

Since it is a reversible color space transformation equation, this color space doesn't introduce a rounding-off error. The uniform appearance of the human digestive system and high correlation between the three color channels of endoscopic image make this color transformation suitable for endoscopic image compression. The detailed study of the suitability of this color space transformation for the compression of endoscopic images can be found in [3].

2.3 The Sub-sampling Schemes

Once the chrominance and luminance components of the color image are separated, it's a common practice to down-sample the chrominance components in order to improve the efficiency of the image compression algorithms. This method is very efficient in terms of computational simplicity and efficiency for endoscopic image compression [3, 13, 14]. This is due to the fact that the chrominance components of the endoscopic image are smooth signals and the sub-sampling operation introduces a little artifact.

The existing DPCM-based endoscopic image compression algorithms [3, 13, 14] employ a 1-D sub-sampling scheme. It reduces only the horizontal resolution of the image. There is no obvious reason for using this scheme in endoscopic applications as the decompression process is done offline on the receiver side. Especially, the prediction function used in the DPCM scheme which is preferable for endoscopic image compression is the horizontal prediction function as detailed Sect. 2.4. The efficiency of this method highly depends on the correlation of pixel values in the horizontal direction. The down-sampling operation which reduces the horizontal resolution of the image reduces the correlation of pixels in that direction. Even though the sub-sampling operation is used to increase the compression efficiency, it reduces the compression efficiency that can be obtained using the DPCM method. This is the drawback of employing thesub-sampling operation and prediction function of the DPCM method in the same direction.

The drawback of performing sub-sampling and prediction of DPCM in a horizontal direction can be alleviated using two different approaches. The first approach is to perform the sub-sampling in the vertical direction and the prediction of the DPCM in the horizontal direction. The second one is to perform the sub-sampling in both horizontal and vertical directions, and restrict the prediction of the DPCM method to the horizontal direction. Note that using the vertical, diagonal, or 2-D prediction functions of DPCM increases the computational complexity and buffer requirement of the image compressor significantly as detailed in [14]. However, using a 2-D sub-sampling operation increases neither the computational complexity nor the buffer requirements of the image compressor as discussed in Sect. 4.1. In the following section, we will investigate the impact of different sub-sampling schemes on the reconstructed image quality and the correlation of pixel values.

(a) Impact of Sub-sampling on the Image Quality:
Sub-sampling introduces distortion to the reconstructed image. For a high down-sampling factor, the 1-D down-sampling operation introduces high distortion. We

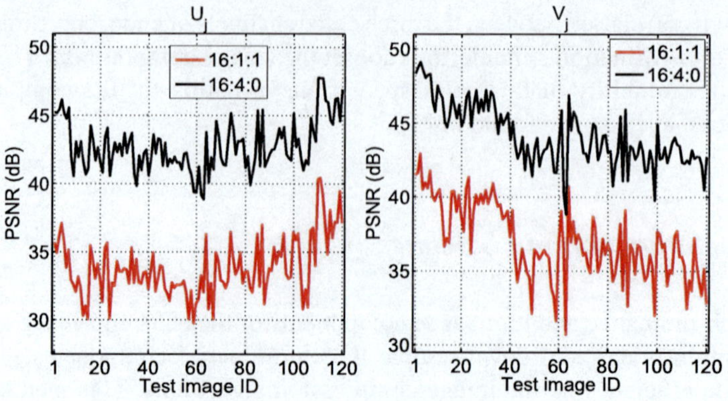

Fig. 3 The comparison of the reconstructed image quality of the chrominance components of 120 endoscopic images from 16:1:1 and 16:4:0 sub-sampling schemes

compared the peak signal-to-noise ratio (PSNR) [5] of two 1-D sub-sampling schemes with two 2-D sub-sampling schemes in Table 1. The 1-D sub-sampling schemes are 4:1:1 and 16:1:1 that reduce the resolution of the chrominance components by a factor of 4 and 16, respectively, in the horizontal direction. The 2-D sub-sampling schemes are 4:2:0 and 16:4:0 that reduce the resolution of the chrominance components of the image by a factor of two and four, respectively, in both horizontal and vertical directions. The 4:1:1 and 4:2:0 sub-sampling schemes reduce the resolution of the image by the same factor (four) but in different directions. Similarly, both the 16:1:1 and 16:4:0 sub-sampling schemes down-sample the chrominance component of the image by a factor of 16 but in different directions. The dataset used for this experiment consists of 120 images obtained from Gastrolab [6]. The experiment is done, first, by down-sampling the chrominance components of the image. Then the image is reconstructed by up-sampling the down-sampled component using the cubic interpolation method. The PSNR value is used to compare the quality of the original and reconstructed image components.

The comparison of 16:1:1 and 16:4:0 sub-sampling schemes for U_j and V_j is shown in Fig. 3. The average PSNR value is shown in Table 1. The difference in reconstructed image quality, which can be achieved using 1-D and 2-D sub-sampling schemes, increases when the down-sampling factor increases.

(b) Impact of Sub-sampling on the Correlation of Pixel Values:
Sub-sampling operation reduces the spatial correlation of pixels in its neighborhood. As shown in Table 1, there is no significant difference between the statistical properties of 1-D and 2-D sub-sampling schemes. This is due to the fact that average entropy and average magnitude are global properties of the image. The impact of the sub-sampling can be more clearly demonstrated using the 1-D DPCM method which is a widely employed method for WCE image compression. The statistical properties that are obtained from the differential pixel values of the 1-D DPCM scheme incorporate the local properties of the image and they help us to see the difference

Table 1 The average statistical properties of the down-sampled chrominance components of 120 endoscopic images

Sub-sampling scheme	Avg. PSNR (dB)		Avg. Entropy (bits/pixel)		Average magnitude	
	U_j	V_j	U_j	V_j	U_j	V_j
4:1:1 (1-D)	45.46	46.87	5.96	5.25	24.57	13.40
4:2:0 (2-D)	48.43	49.16	5.98	5.25	24.54	13.39
16:1:1 (1-D)	33.83	37.10	5.94	5.26	24.67	13.51
16:4:0 (2-D)	42.75	44.17	5.99	5.25	24.64	13.44

Table 2 The average statistical properties of the down-sampled differential pixel values of chrominance components of 120 endoscopic images

Sub-sampling scheme	Entropy (bits/pixel)		Average magnitude	
	dU_j	dV_j	dU_j	dV_j
4:1:1 (1-D)	3.30	3.14	2.17	1.75
4:2:0 (2-D)	2.56	2.50	1.23	1.06
16:1:1 (1-D)	4.80	4.39	6.14	4.34
16:4:0 (2-D)	3.29	3.13	2.15	1.74

that can be obtained when using the 1-D and 2-D sub-sampling schemes as shown in Table 2. The impact of the sub-sampling on the correlation of pixel values is further illustrated in Sect. 2.4.

2.4 Differential Pulse-Coded Modulation (DPCM)

The differential pulse-coded modulation is one of the most widely used methods for the compression of endoscopic images. Its simplicity makes it an attractive option for this application. The basic idea is to encode the difference between the current pixel and a predicted pixel. The prediction error, denoted by dX, is given by

$$dX_{i,j} = X_{i,j} - X_p. \quad (3)$$

The predicted pixel value X_p is obtained from the previously encoded neighborhood pixels. As shown in [3], the suitable prediction is given by $X_p = X_{i,j-1}$, where i and j are the row and column indexes of the image array, respectively. This prediction function doesn't need buffer memory and gives a good compression rate. The compression efficiency is dependent on the correlation of the pixel values in the horizontal direction (across columns).

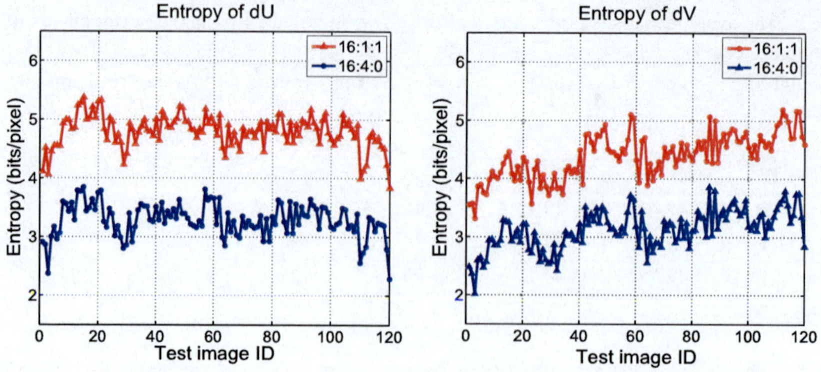

Fig. 4 The comparison of the entropy of differential pixel values of down-sampled chrominance components of 120 endoscopic images using 16:1:1 and 16:4:0 sub-sampling schemes

When we use a one-dimensional sub-sampling scheme in the horizontal direction of the chrominance components of an image, the correlation between neighborhood pixels reduces in that particular direction. This reduces the compression efficiency of the encoder. Therefore, performing sub-sampling and prediction in the same direction has a negative effect on the compression efficiency as shown Fig. 4. The entropy of the differential pixel value of the original image component is the lowest and that of the component down-sampled using the 1-D sub-sampling scheme is the highest. From the experimental results, we can conclude that the 2-D sub-sampling scheme improves both the compression efficiency as well as the image quality of the DPCM-based image compression algorithms as shown in Tables 1 and 2.

2.5 *Golomb Coding*

Golomb code is optimal code for encoding non-negative integers with exponentially decreasing probability distributions [7]. To encode differential pixel values, which can be either positive or negative, the following mapping function is used [24].

$$mdx = \begin{cases} 2dx & \text{if}\, d \geq 0, \\ 2|dx| - 1 & \text{if}\, dx < 0. \end{cases} \tag{4}$$

Golomb code is a parametric code. Given mdX and Golomb code with parameter m, where $m > 0$, the Golomb code of mdX with respect to the m is a concatenation of the unary code of quotient $q = \lfloor mdX/m \rfloor$ and the binary representation of the remainder $r = mdX \ \% \ m$. There are many Golomb codes, and the key step in their effective application is the choice of the parameter m. Among the Golomb codes, the Golomb-Rice codes are very popular, where the code parameter m is given by $m = 2^k, k \geq 0$ [19]. Given mdX, its Golomb-Rice code length is given by

$$code\_length(mdX) = \lfloor \frac{mdX}{2^k} \rfloor + k + 1. \quad (5)$$

Golomb-Rice code has a very simple hardware realization in which the division operation can be performed using binary shift operation. The code parameter can be chosen in different ways. For instance, in JPEG-LS [20] k is adaptively determined from the statistics of the previously encoded pixel values using 365 contexts. To reduce the high computational complexity and memory requirement, a fixed value of k is adopted in [14].

Due to its simple hardware realization, Golomb-Rice code has been used for encoding endoscopic images [3, 4, 14]. So far, two variants of Golomb-Rice codes have been used in endoscopic image compression algorithms. These are non-adaptive Golomb-Rice code [14] and the locally adaptive Golomb-Rice code [3, 4]. In addition to the previously proposed codes, we present two additional variants of Golomb codes which are suitable for wireless capsule endoscopy application in this paper. These are exponential Golomb (exp-Golomb) code [25] and fast-adaptive Golomb-Rice code [8]. The non-adaptive, locally adaptive, and fast-adaptive Golomb-Rice codes differ only in the way the value of the code parameter k is calculated. The properties of the exponential Golomb code slightly differ from the Golomb-Rice codes as detailed in Sect. 2.5. The comparative analysis of these four variants of Golomb codes in terms of compression ratio, computational complexity, and power consumption is presented in the following.

(a)Non-Adaptive Golomb-Rice Code:
The parameter of the code, k, is predetermined and fixed throughout the coding process. In [14], the k value is chosen to be 2 for coding the luminance component and 0 for chrominance components of the endoscopic image. Despite its low computational complexity, the optimal parameter k that yields the shortest possible average code length is dependent on the distribution of the differential pixel values. The distribution of the differential pixel values of endoscopic images depends on the parts of the GI tract and its condition such as the presence of abnormalities, faeces, and bubbles.

(b) Locally Adaptive Golomb-Rice Code:
The parameter k is adaptively determined from the local properties of the image [3]. It is calculated from the statistics of the 8 previously encoded values for each component of the endoscopic image. The context C is maintained in two registers, A_c and N_c, per color image channel so that the image can be encoded in sample interleaved manner. A_c stores the accumulated magnitude of the prediction error and N_c counts how many contexts are encountered so far. These context variables are used to determine k using (6).

$$k = min\{k : 2^k.N_c \geq A_c\}. \quad (6)$$

Every time a given prediction error value (dX) is encoded, its magnitude is added to A_c and N_c is incremented by one. Both A_c and N_c are halved whenever N_c exceeds 8.

The computational complexity and memory requirement of this Golomb-Rice code are higher than the non-adaptive code. However, its adaptation property makes this code superior in terms of compression ratio [3].

(c) Fast-adaptive Golomb-Rice Code:
This code determines the code parameter, k, depending on the previously encoded pixel value and its previous value [8]. This code is adaptive, computationally simple, and needs to store only a single pixel value for the determination of its parameter k. It is a high-order code which determines the sub-optimal code parameter from the statistical properties of the image. Let k_n and k_{n-1} be the parameters of the previous and current coding steps, respectively, and mdX_{n-1} be the mapped differential pixel value of the previous coding step; then, the adaptive Golomb-Rice code parameter is given as follows.

$$k_n = \begin{cases} k_{n-1} - 1, & \text{if } 0.5 \times 2^{k_{n-1}} < mdX_{n-1}, \\ k_{n-1} + 1, & \text{if } 2.5 \times 2^{k_{n-1}} \geq mdX_{n-1}. \\ k_{n-1}, & otherwise \end{cases} \tag{7}$$

The initial value of k is 0. This code has both the required properties, adaptation, and low computational complexity. It was proposed for the compression of images in other applications. It can be further optimized for coding color endoscopic images.

(d) Exponential Golomb Code:
Golomb-Rice code [7] divides the source alphabet into groups of equal size whereas, in exponential Golomb (exp-Golomb) code [25] the group size increases exponentially. Exp-Golomb codes can be defined for different values of k. In this work, we use a k value of 0. Hereafter when we say exp-Golomb code, we mean an exponential Golomb code with a k value of 0 unless explicitly stated. The exponential increment of the group size in exp-Golomb code helps us to encode large numbers using shorter codes as compared to its Golomb-Rice counterpart.

The exponential Golomb codes also consist of two parts, unary code followed by fixed-length code. The part which is represented in unary code is called the group number of the value. The exponential Golomb code group number (g) of a non-negative integer (n) is calculated using (8).

$$g = \lfloor log_2(n + 1) \rfloor. \tag{8}$$

The group number is represented in unary code. The number of elements (N) in a group g is obtained using (9).

$$N = 2^g. \tag{9}$$

Then the elements in a given group are represented using g number of bits in binary form. The final code is obtained by concatenating the unary code of the group number and the binary code representation of the group elements. For instance, if we take a value of 3, then we get a group number of 2. Its unary code is given by 110. The elements in group 2 are 3, 4, 5, and 6. These elements are represented by 00, 01, 10, and 11, respectively, in binary form within the group (index in the group). Hence, the exp-Golomb code of 3 becomes 11000. The length of exp-Golomb code for a given non-negative integer n is given by (10).

$$code\_length(n) = 2 \times \lfloor log_2(n+1) \rfloor + 1. \quad (10)$$

From (10), we can see that the length of the code of a given non-negative integer is logarithmically related with its magnitude.

Now let us use the existing endoscopic image compression algorithms [3] and [14] to compare the properties of these four Golomb codes. The Golomb-Rice code used in [14] is non-adaptive and the one that was presented in [3] is locally adaptive. The exp-Golomb code, which is proposed in this paper, is also a non-adaptive code. In these image compression algorithms, the input pixel values are represented in 8 bits. The prediction error range is [-255, 255] in [14] and [-64, 64] in [3]. After mapping, the range of the prediction error is [0, 510] in [14] and [0, 128] in [3]. The Golomb-Rice codes limit the maximum length of the code to 32 in order to avoid code expansion. The exponential Golomb code divides this set of values into 9 groups as shown in Table 3. The group which contains the largest value can be represented using 17 bits. The other appealing property of the exponential Golomb code is that the smallest value in the source alphabet (0) is represented using a single bit. This is possible only when we use unary code. But the length of the unary code increases linearly with the magnitude of the source. For instance, 510 is encoded using 511 bits using unary code if we use it without limiting the length of the code and exp-Golomb code uses only 17 bits. The comparison of the code length of non-adaptive Golomb codes is shown in Fig. 5. We can see that the property of the exp-Golomb code resembles the property of unary code while encoding small values. And it resembles the property of Golomb-Rice code with a large value of k while encoding large values.

The comparison of the properties of the adaptive Golomb-Rice codes is depicted in Fig. 6. The adaptation of the code parameter k for linearly increasing non-negative integer value is shown in the figure. The two codes have almost identical adaptation properties while encoding smaller values. And the locally adaptive Golomb-Rice code adapts its parameter faster than the fast-adaptive Golomb-Rice code when the values to be encoded are large.

Table 3 The groups of the endoscopic image processed pixel values to be encoded using Golomb code

mdX values	Group No. (g)	mdX values	Group No. (g)
0	0	31–62	5
1–2	1	63–126	6
3–6	2	127–254	7
7–14	3	255–510	8
15–30	4		

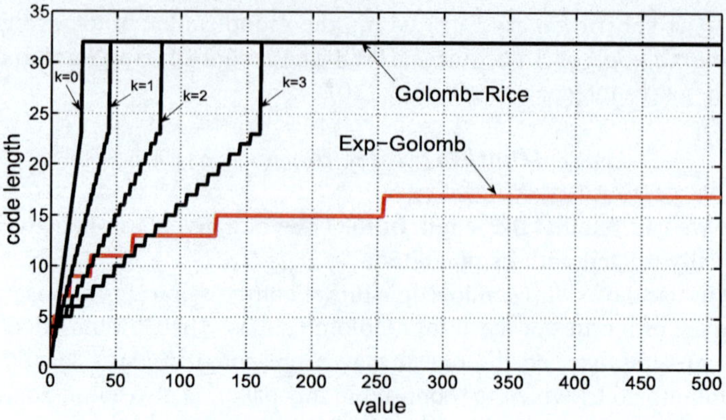

Fig. 5 The comparison of the code length of length-limited Golomb-Rice and Exp-Golomb codes

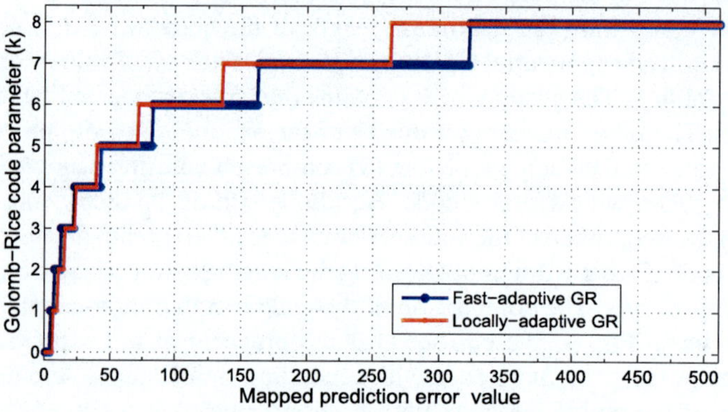

Fig. 6 The adaptation of the code parameter *k* of locally adaptive Golomb-Rice and fast-adaptive Golomb-Rice code for linearly increasing symbol

(e) Modified Exponential Golomb Code:

The number of bits required to represent the last group shown in Table 3 is 17. This number is not a hardware friendly number when we consider that the two clocks (the clock used in the image compressor and the one used in the P2S converter as shown in Fig. 13) are generated from the same clock source. We need a number which has a power-of-two form. In endoscopic images, the occurrence of large prediction error is rare. By utilizing this property, we modify the group shown in Table 3 so that the maximum number in the group can be represented using 16 bits. This can be done by merging group number 6, 7, and 8 when the maximum mapped prediction error is 510 [14]. The group number of the newly formed group becomes the group number of group 6 shown in Table 3 and it is represented by 1111110. The fixed-length code for this newly formed group is represented using 9 bits. The 9 bits can represent values in the range from 0 to 511. Hence, the last group in the modified exp-

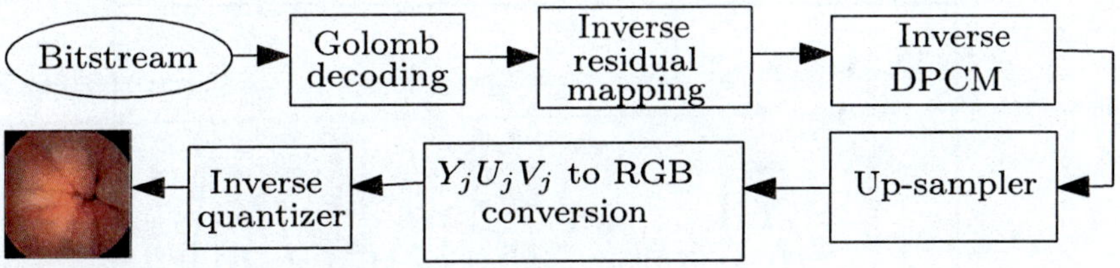

Fig. 7 The functional block diagram of the decoder

Golomb code consists of values ranging from 63 to 510. Of course, it can represent up to a maximum value of 573. The total number of bits required to represent the newly formed group is 16, which is our target. The total number of groups in the source becomes seven now. The modified exp-Golomb code can help us to reduce the maximum clock frequency of the parallel-to-serial converter by half.

Similarly, for the image compression algorithm [3] with mapped prediction error [0, 128], the maximum mapped prediction error is in group 7. So, it can be represented using 15 bits. It is also possible to approximate the extreme prediction errors to reduce the code length. Here, the prediction errors 64 and −64 can be approximated by 63 and −63, respectively, which results in mapped prediction error [0, 126]. The maximum value lies in group 6 and it can be represented using 13 bits.

2.6 Decoding

The functional block diagram of the decoder is depicted in Fig. 7, which is the inverse of the encoding process. The up-sampling operation is performed using the cubic interpolation method and the inverse color space transformation is given by (2). The dequantization is done by multiplying the quantized value by four. The error from the reconstructed result of the quantization operation is uniformly distributed in [−2, 2]. The error from the sub-sampling operation is random and depends on the frequency content of the image.

3 Experimental Results and Discussions

The dataset used in this experiment consists of 120 endoscopic color images obtained from Gastrolab [6]. It consists of images taken from the esophagus to the end of the colon. We believe that this dataset is a good representative of the whole human GI system.

Let us first see the impact of sub-sampling on image quality in detail. The reconstructed (up-sampled) image quality depends on the employed up-sampling algorithm. We compared the quality of the reconstructed images for three interpolation

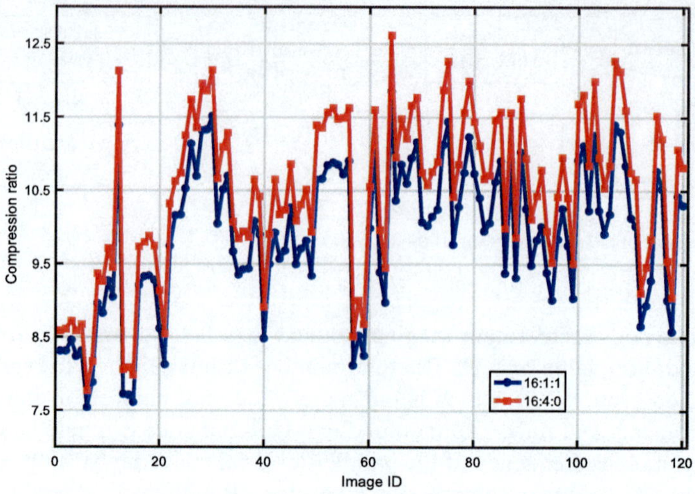

Fig. 8 The comparison of the compression ratio of the proposed image compressor which uses the modified exp-Golomb code and sub-sampling scheme of 16:4:0 or 16:1:1

methods: nearest neighborhood, bilinear, and cubic interpolation methods. The cubic interpolation method gives the highest PSNR value. In all these three up-sampling methods, the 2-D sub-sampling scheme outperforms the 1-D sub-sampling scheme for the same down-sampling factor. The image quality improvement obtained by employing the 2-D sub-sampling scheme can help us to increase the down-sampling factor in order to increase the compression rate. For instance, in [3] the 4:1:1 sub-sampling scheme with the quantizer value of four is used to achieve the PSNR value of 40.66 dB. However, the 4:2:0 sub-sampling scheme, which reduces the overall image resolution by the same factor as that of the 4:1:1 sub-sampling scheme, achieves a PSNR value of 42.19 dB.

The compression ratio of the proposed image compression algorithm on the test dataset is shown in Fig. 8. For all the test images, the 16:4:0 sub-sampling mode gives a higher compression ratio than the 16:1:1 sub-sampling mode. On average, the proposed image compressor gives a compression ratio of 10.025 and 9.40, when 16:4:0 and 16:1:1 sub-sampling modes are employed, respectively. This shows that the 2-D sub-sampling scheme improves the compression ratio by 0.63, on average, as compared to the 1-D sub-sampling scheme. Average PSNR value of 40.55 dB is obtained when we use 16:4:0 with the quantizer value of four as shown in Fig. 10. When we use the 16:1:1 sub-sampling scheme with a quantizer value of four, we obtain an average PSNR value of 36.29 dB, in which some of the images will have PSNR value which is less than the minimum value required for the accurate diagnosis of abnormalities using medical images [2, 12].

The PSNR value versus compression ratio of the proposed image compression algorithm, which uses exp-Golomb code, is shown in Figs. 9 and 10 for three quantizer (Q) values. The figure is drawn for 1-D five sub-sampling schemes: 2:1:1, 4:1:1,

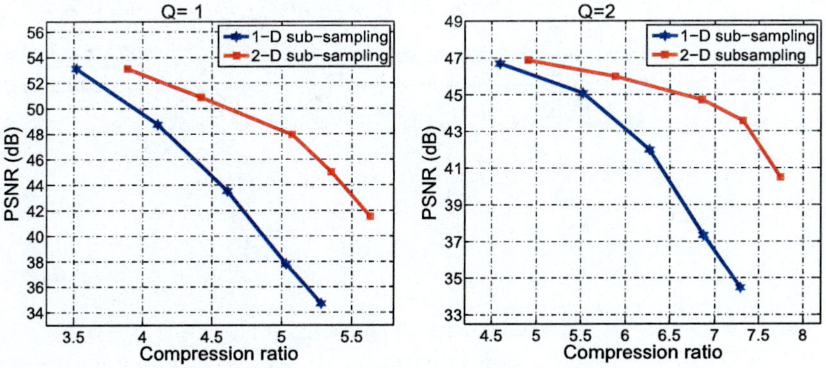

Fig. 9 The comparison of compression ratio of the proposed image compressor with different quantizer values when 16:1:1 (1-D) or 16:4:0 (2-D) sub-sampling schemes is used

Fig. 10 The comparison of compression ratio of the proposed image compressor with quantizer values of four when 16:1:1 (1-D) or 16:4:0 (2-D) sub-sampling schemes is used

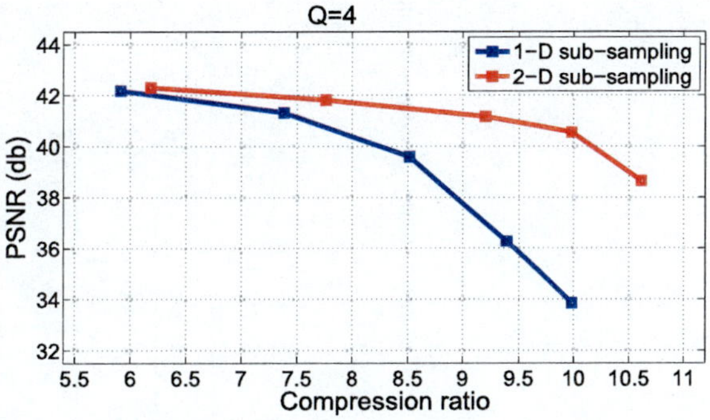

8:1:1, 16:1:1, and 32:1:1 and for the 2-D sub-sampling schemes: 2:1:0, 4:2:0, 8:4:2, 16:4:0, and 32:8:4. As we go from the lowest to the highest sub-sampling factor, the performance of the 2-D sub-sampling scheme improves more as compared to its 1-D counterpart in both PSNR and compression ratio. Especially, the performance improvement we achieve in terms of image quality is more significant. The 16:4:0 when used with quantizer value (Q) of four (4) gives the optimal combination in terms of both compression rate and PSNR value.

We also compared the performance of the proposed image compressor with two existing algorithms. As shown in Table 4, the performance of the three image compression algorithms is evaluated using both the original algorithm and by replacing the entropy coder in each algorithm with one of the four Golomb codes described in the above. In all the three algorithms, the non-adaptive Golomb-Rice code gives the poorest performance in terms of compression ratio. In [14], the highest compression is obtained using the fast-adaptive Golomb-Rice. In [3], both locally adaptive and fast-adaptive Golomb-Rice codes achieve comparable compression ratio. In the proposed image compression algorithm, both locally adaptive and fast-adaptive Golomb-Rice codes achieve comparable compression ratios. For all the studied algo-

Table 4 Comparison of the compression ratio and standard deviation of endoscopic image compression methods

Algorithms	Sub-sampling scheme	Compression ratio (standard deviation) using								Overall PSNR (dB)
		Non-adaptive Golomb-Rice code		Locally adaptive Golomb-Rice code		Fast-adaptive Golomb-Rice code		Exp-Golomb code		
		32 bit	16 bit	32 bit	16 bit	32 bit	16 bit	17 bit	16 bit	
Khan et al. [14]	8:1:2	4.8984 (0.3710)	4.8455 (0.3891)	5.3107 (1.5861)	5.0884 (1.6219)	5.469 (0.476)	4.9564 (0.4652)	4.9564 (0.5098)	4.9505 (0.5113)	48.2
Fante et al. [3]	4:1:1	6.5674 (0.3126)	6.5312 (0.3303)	8.1353 (1.2091)	7.9700 (1.309)	8.0957 (0.6883)	7.8354 (0.7176)	7.4830 (0.6721)	7.4830 (0.6721)	41.27
Proposed	16:4:0	8.8569 (0.5415)	8.7883 (0.5694)	10.9995 (1.1469)	10.7871 (1.2419)	11.0198 (1.1357)	10.6907 (1.1839)	10.0253 (1.1208)	10.0253 (1.1208)	40.55

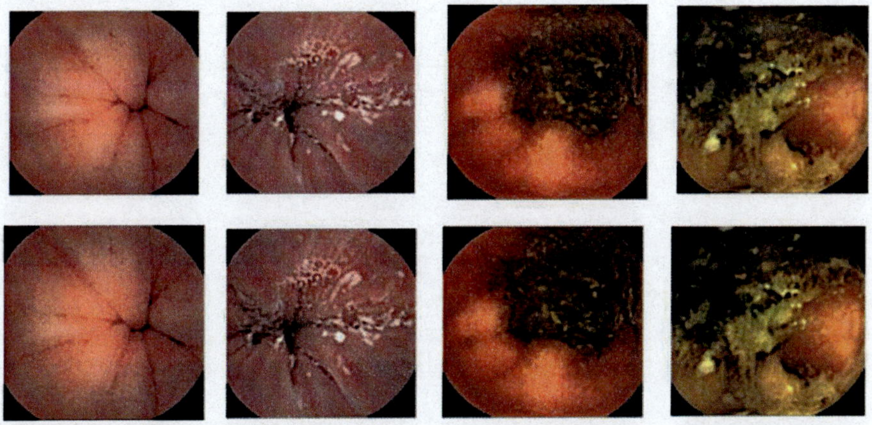

(a) CR = 11.69, PSNR=40.13 dB　(b) CR = 9.24, PSNR=41.04 dB　(c) CR=10.29, PSNR=40.25　(d) CR=8.58, PSNR=40.81

Fig. 11 The compression ratio (CR) and PSNR value of four reconstructed endoscopic images among the 120 test images. The top images are the original images and the bottom ones are the reconstructed images

rithms, using the exp-Golomb code achieve compression ratio which is close to that can be achieved using adaptive Golomb-Rice codes. From this simulation results, we can conclude that fast-adaptive Golomb-Rice code and exp-Golomb code are good candidates for wireless capsule endoscopy application. The reconstructed images do not show visually noticeable artifacts as shown in Fig. 11.

4 Analysis of Computational Complexity and Power Consumption

In this section, we present the computational complexity and power consumption analysis of chroma component sub-sampling and variants of Golomb codes. The hardware architecture of the proposed image compressor with serializer is shown in Fig. 12.

4.1 Sub-sampling

The hardware realization of the sub-sampling operations is a trivial task. The implementation of 16:1:1 and 16:4:0 sub-sampling schemes using digital logic gates are shown in Fig. 13. It uses multiplexers and logic gates. The 16:4:0 sub-sampling scheme realization uses one more NAND gate as compared to the 16:1:1 sub-

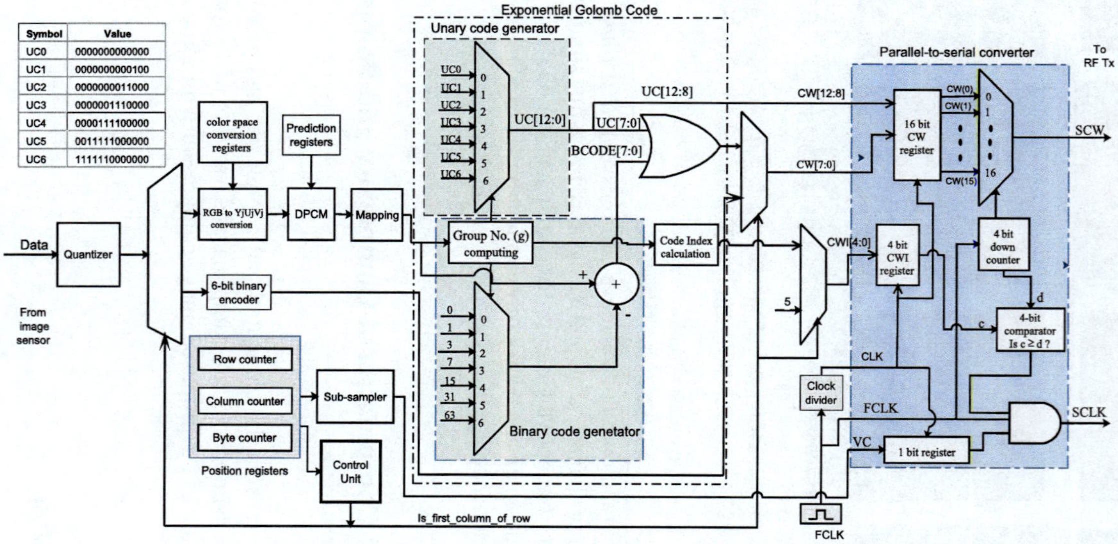

Fig. 12 The hardware architecture of the proposed image compressor which uses exp-Golomb code with parallel-to-serial converter

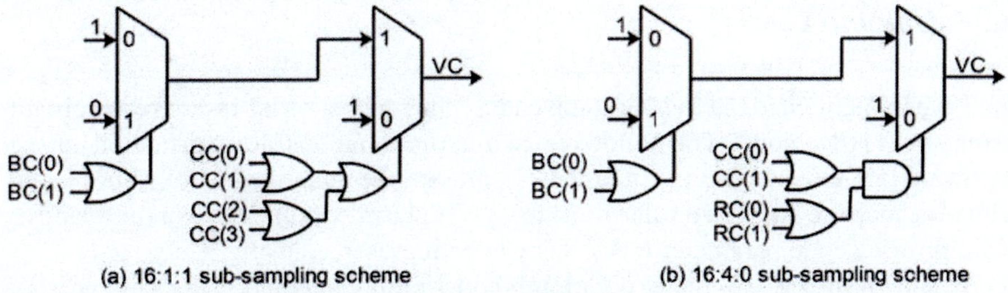

Fig. 13 The comparison of hardware architecture of 16:1:1 and 16:4:0 sub-sampling schemes

sampling scheme. This shows that the two realizations require almost equivalent resources. The signal denoted by BC is a byte counter, CC is a column counter, and RC is a row counter of the image array. A single pixel value of a color image is represented using 3 bytes. The first byte represents the luminance component and it is encoded without down-sampling, whereas the remaining two bytes represent the two chrominance components which are down-sampled before entropy encoding.

As shown in Fig. 13, in the 16:1:1 sub-sampling scheme only the pixels in 0th, 16th, 32nd, ... columns of all the rows are encoded. The columns which are multiples of 16 are encoded and the remaining pixel values are dropped. It is required to check whether the column counter is a multiple of 16 or not. The VC value is set to zero for dropped pixel values. This is the same as checking whether the four LSBs of a column counter are all zeros or not.

In the 16:4:0 sub-sampling scheme, only the pixels in 0th, 4th, 8th, ..., M rows and 0th, 4th, 8th, ..., N columns of the chrominance components of the image are entropy encoded. The other pixels are dropped. Hence, it is required to check whether the row and column counters are multiples of 4 or not. This is the same as checking whether the two LSB of the column and row counters are all zero or not. The hardware architecture of endoscopic image compression methods that use Golomb-Rice was presented in [3, 14]. In this paper, we present the hardware architecture of an endoscopic image compression scheme that uses the exp-Golomb code. The hardware architecture of the proposed image compressor is shown in Fig. 12. The overall computational complexity of the image compressor is comparable to those in [3, 14]. The difference between the proposed algorithm and the already existing methods is the computational complexity and memory requirement of the Golomb codes. The method used to determine the code parameter k determines the computational complexity, memory requirement, and the power consumption of the code. The maximum length of the code generated by the Golomb codes determines the memory requirement and power consumption of the parallel-to-serial converter circuit.

4.2 Golomb Codes

We have implemented the four Golomb codes along with a serial-to-converter circuit using the VHDL code. The following two approximations are considered in the hardware implementation: (1) In order to represent the quantized pixel values using 6 bits, we map the quantized value of 64 to 63, (2) The maximum value of the adaptive Golomb-Rice code parameter k is 7 that can be represented using 3 bits.

As shown in Fig. 12, the exp-Golomb code is implemented using comparison, subtraction, and bitwise logical operations. Let S = $\{0, 1, 3, 7, 15, 31, 63\}$ be a set consisting of the first elements of the seven groups as shown in Table 3. The binary code (BCODE) of an input non-negative integer n is given by $n - S(g)$, where g is the group number of n. The final code of the exp-Golomb encoder is obtained using the bitwise logical OR operation of the BCODE and the unary code of the group number of n. The first columns of the sampled rows are encoded using a 6-bit binary code since it is a boundary pixel. The code word index (CWI) is given by $2 \times g$ and it can be implemented using the logical shift operation. The CWI holds the index of the last non-zero bit in the CW from right to left. The implementation of the group number (g) is depicted in Fig. 14. The logarithm operation given by (8) is implemented using multiplexers and 6 comparators.

The VHDL codes are then synthesized and mapped to 130 nm standard CMOS process using Synopsys Design Compiler. The power consumption of the circuits is evaluated for encoding two color images, with a resolution of 256×256, per second. The comparison of the hardware realization of the Golomb codes with a serializer is shown in Table 5. As shown in Fig. 12, the implementation of the exp-Golomb code doesn't need any logarithm or division operations. Among the Golomb codes, the exp-Golomb and non-adaptive Golomb codes need no registers and the locally adaptive Golomb-Rice code needs the highest number of registers. The serializer circuit used with exp-Golomb code needs the least number of registers as compared to the existing implementation of Golomb-Rice codes [3, 14].

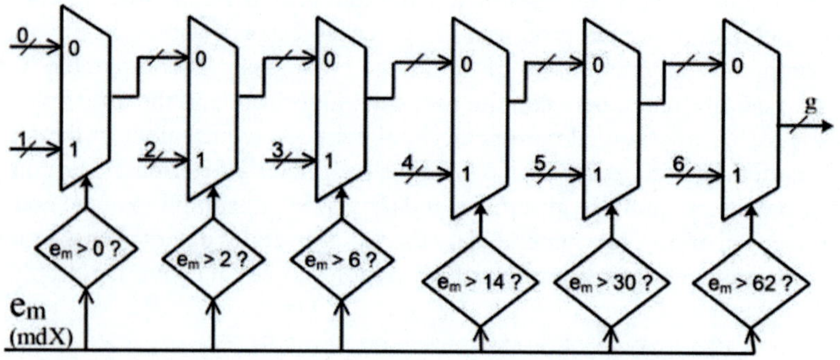

Fig. 14 The implementation details of group number computation (8)

Table 5 Comparison of hardware realization details of four variants of Golomb codes with serializer in the proposed endoscopic image compressor

Types of Golomb Code	Module	No. of 1-bit registers	No. of cells	Clock frequency	Power consumption (μW)
Non-adaptive GR code [14]	Encoder	0	193	CLK	0.61
	Serializer	43	89	$32 \times CLK$	3.76
	Encoder + Serializer	43	287	CLK, $32 \times CLK$	4.37
Locally adaptive GR code [3]	Encoder	50	555	CLK	2.36
	Serializer	43	89	$32 \times CLK$	3.76
	Encoder + Serializer	93	655	CLK, $32 \times CLK$	6.12
Fast-adaptive GR code [8] (proposed)	Encoder	9	374	CLK	1.13
	Serializer	43	89	$32 \times CLK$	3.76
	Encoder + Serializer	52	464	CLK, $32 \times CLK$	4.89
Modified Exp-Golomb code [25] (proposed)	Encoder	0	143	CLK	0.44
	Serializer	25	56	$16 \times CLK$	2.50
	Encoder + Serializer	25	202	CLK, $16 \times CLK$	4.20

The locally adaptive Golomb-Rice code has the highest power consumption because it uses a relatively large number of memory elements. The power consumption of the non-adaptive and fast-adaptive Golomb-Rice codes is comparable. The exp-Golomb code has the least power consumption. As shown in Table 5, by using exp-Golomb code instead of Golomb-Rice we can reduce the frequency of operation of the parallel-to-serial (P2S) converter by half. Since the power consumption of CMOS circuits depends on the frequency of operation, using exp-Golomb code reduces the power consumption of the P2S converter by 33.51% as compared to the Golomb-Rice codes.

The sub-sampling scheme used in this work neither increases nor decreases the power consumption of the circuits presented in the existing endoscopic image compression algorithms [3, 14]. It increases either the reconstructed image quality or/and compression ratio. The two modified Golomb codes proposed in this work can improve the compression ratio of the algorithm in [14] without increasing the power consumption. These codes reduce the power consumption of [3] without significant reduction on the compression ratio as shown in Tables 4 and 5.

The power consumption of the proposed image compressor, which uses fast-adaptive Golomb-Rice code or exp-Golomb code, has almost the same power consumption as that of [14]. The proposed image compressor has lower power consumption than [3]. The power consumption of the serializer circuit used along with non-adaptive, locally adaptive, and fast-adaptive Golomb-Rice codes can be reduced by limiting the maximum code length to 16 bits as shown in Table 4. Reducing the maximum code length has an insignificant impact on the compression ratio. The power consumption of the serializer circuit used with the image compressor that uses exp-Golomb code is less than those used in [3, 14]. Hence, the proposed endoscopic image compression algorithm outperforms the existing algorithms.

5 Conclusion

This paper has presented improved the DPCM-based image compression algorithm for wireless capsule endoscopy. The algorithm uses a combination of computationally simple methods consisting of uniform quantization, reversible color space transformation, 2-D sub-sampling, DPCM, and modified Golomb codes. It is a single-pass method and has low computational complexity. The performance of the proposed endoscopic image compression algorithm has been thoroughly evaluated in terms of compression ratio, image quality, and power consumption.

The implementation of the proposed hardware architecture of the algorithm doesn't require buffer memory. It has low power consumption since the architecture does not involve any divisions, multiplications, or logarithm operations. It gives an average compression ratio of more than 10 with 40.55 dB of PSNR. The performance of the proposed image compression algorithm is significantly better than the existing methods in terms of both compression ratio and power consumption. This makes it a good candidate for wireless capsule endoscopy applications.

The future work includes the implementation of the entire wireless capsule endoscopy system, carrying out in vivo and ex vivo trials for validating the system. The impact of distortion introduced due to compression on the diagnosis accuracy of different GI abnormalities will be investigated in the future.

Acknowledgements The authors would like to thank Sanjeev Jain for his help in VHDL code synthesis using Synopsys Design Compiler at VLSI laboratory of IID Delhi.

References

1. T. Acharya, P.S. Tsai, *JPEG2000 Standard for Image Compression: Concepts, Algorithms and VLSI Architectures* (Wiley, 2005)
2. P. Cosman, R. Gray, R. Olshen, Evaluating quality of compressed medical images: Snr, subjective rating, and diagnostic accuracy. Proc. IEEE **82**(6), 919–932 (1994). https://doi.org/10.1109/5.286196 (Jun)
3. K. Fante, B. Bhaumik, S. Chatterjee, Design and implementation of computationally efficient image compressor for wireless capsule endoscopy. Circuits Syst. Signal Process. **35**(5), 1677–1708 (2015). https://doi.org/10.1007/s00034-015-0136-z
4. K.A. Fante, B. Bhaumik, S. Chatterjee, A low-power color mosaic image compressor based on optimal combination of 1-d discrete wavelet packet transform and dpcm for wireless capsule endoscopy, in *BIODEVICES* (2015), pp. 190–197. https://doi.org/10.5220/0005284701900197
5. K.A. Fante, F. Abdurahman, M.T. Gemeda, An ingenious application-specific quality assessment methods for compressed wireless capsule endoscopy images. Trans. Environ. Electr. Eng. **4**(1), 18–24 (2020)
6. Gastrolab, http://www.gastrolab.net. Accessed 2016
7. S. Golomb, Run-length encodings (corresp.). IEEE Trans. Inf. Theory **12**(3), 399–401 (1966). https://doi.org/10.1109/TIT.1966.1053907
8. K. He, Fast-adaptive near-lossless image compression. J. Electron. Imaging **25**(3), 033022 (2016)
9. R.A. Hejazi, M. Bashashati, M. Saadi, Z.D. Mulla, I. Sarosik, R.W. McCullam, M.J. Zuckerman, Video capsule endoscopy: a tool for assessment of small bowel transit time. Front. Med. **3**(6), 1–5 (2016)
10. D.K. Iakovidis, A. Koulaouzidis, Software for enhanced video capsule endoscopy: challenges for essential progress. Nat. Rev. Gastroenterol. Hepatol. **12**(3), 172–186 (2015)
11. G. Iddan, G. Meron, A. Glukhovsky, P. Swain, Wireless capsule endoscopy. Nature **405**, 417 (2000)
12. R.S.H. Istepanian, N. Philip, M. Martini, N. Amso, P. Shorvon, Subjective and objective quality assessment in wireless teleultrasonography imaging, in: 30th Annual International Conference of the IEEE Engineering in Medicine and Biology Society, 2008, EMBS 2008 (2008), pp. 5346–5349. https://doi.org/10.1109/IEMBS.2008.4650422
13. T.H. Khan, K. Wahid, Subsample-based image compression for capsule endoscopy. J. Real-Time Image Process. **8**(1), 5–19 (2011). https://doi.org/10.1155/2011/343787
14. T. Khan, K. Wahid, Low power and low complexity compressor for video capsule endoscopy. IEEE Trans. Circuits Syst. Video Technol. **21**(10), 1534–1546 (2011). https://doi.org/10.1109/TCSVT.2011.2163985
15. A. Koulaouzidis, D.K. Iakovidis, A. Karargyris, E. Rondonotti, Wireless endoscopy in 2020: will it still be a capsule? World J. Gastroenterol. **21**(17), 5119–5130 (2015)
16. M. Lin, L. Dung, A subsample-based low-power image compressor for capsule gastrointestinal endoscopy. EURASIP J. Adv. Signal Process. **2011** (2011). https://doi.org/10.1155/2011/257095
17. L. Liu, S. Towfighian, A. Hila, A review of locomotion systems for capsule endoscopy. IEEE Rev. Biomed. Eng. **8**, 138–151 (2015)
18. B.F. Mustafa, M. Samaan, L. Langmead, M. Khasraw, Small bowel video capsule endoscopy: an overview. Expert Rev. Gastroenterol. Hepatol. **7**(4), 323–329 (2013)
19. R.F. Rice, *Some Practical Universal Noiseless Coding Techniques—Part III*. Tech. Rep. JPL-91-3 (Jet Propulsion Laboratory, Pasadena, CA, 1991)
20. A. Savakis, M. Piorun, Benchmarking and hardware implementation of jpeg-ls, in *2002 International Conference on Image Processing*, vol. 2 (IEEE, 2002), pp. II–949
21. O.C. Tube, Omnivision, http://www.ovt.com/products/category.php?id=21. Accessed 2016
22. C. Van de Bruaene, D. De Looze, P. Hindryckx, Small bowel capsule endoscopy: where are we after almost 15 years of use? World J. Gastrointest. Endosc. **7**(1), 13–36 (2015)

23. A. Wang, C. Sodini, On the energy efficiency of wireless transceivers, in *IEEE International Conference on Communications, 2006, ICC '06*, vol. 8 (2006), pp. 3783–3788. https://doi.org/10.1109/ICC.2006.255661
24. M.J. Weinberger, G. Seroussi, G. Sapiro, Loco-i: a low complexity, context-based, lossless image compression algorithm, in *Data Compression Conference, 1996, DCC'96. Proceedings* (IEEE, 1996), pp. 140–149
25. J. Wen, J. Villasenor, Structured prefix codes for quantized low-shape-parameter generalized Gaussian sources. IEEE Trans. Inf. Theory **45**(4), 1307–1314 (1999). https://doi.org/10.1109/18.761289 (May)
26. X. Xie, G. Li, X. Chen, X. Li, Z. Wang, A low-power digital ic design inside the wireless endoscopic capsule. IEEE J. Solid-State Circuits **41**(11), 2390–2400 (2006). https://doi.org/10.1109/JSSC.2006.882884

Image Steganography Using Ridgelet Transform and SVD

Mansi Subhedar

Abstract Image steganography involves embedding the secret information into a cover media without creating noticeable changes in it and keeping its presence hidden. In this paper, transform domain image steganography is proposed that benefits from the ridgelet transform and SVD matrix decomposition technique. The principal component obtained by singular value decomposition of ridgelet coefficients acts as an embedding location. Stego image is obtained by embedding the secret image in it. Experimental results in terms of imperceptibility and robustness are verified using standard image quality metrics. Simulation results validate superior visual quality with a large payload. Proposed approach is also robust to several image processing attacks like noise addition, geometric transformations, histogram equalisation, JPEG compression, etc.

Keywords Information hiding · Image steganography · Ridgelet transform · Singular Value Decomposition

1 Introduction

Privacy and security of the information transferred over the Internet is an important aspect in today's digital era. Image Steganography is a covert communication that hides the secret information in a cover image such that its presence is not noticed. The performance measures of image steganography are imperceptibility, robustness, and undetectability. Image watermarking, another branch of information hiding, fails when the watermark is removed or replaced, however, image steganography fails when the presence of hidden information is detected. In watermarking, the focus is on the robustness but in steganography, undetectability and payload capacity are the significant parameters that influence the efficacy of the steganography scheme. These critical differences separate the field of image watermarking and image steganogra-

M. Subhedar (✉)
Department of E & TC Engineering, Pillai HOC College of Engineering and Technology, Rasayani, Raigad, Maharashtra, India

F. Thakkar et al. (eds.), *Proceedings of the International e-Conference on Intelligent Systems and Signal Processing*, Advances in Intelligent Systems and Computing 1370,
https://doi.org/10.1007/978-981-16-2123-9_6

phy, both being disciplines of information hiding [1]. Imperceptibility confirms no visual evidence of hidden information. Though there seems no visual evidence of tamper in the original media, steganalyser can apply some statistical checks to reveal the hidden secret information.

In cryptography, the secret message is altered so as to make its meaning obscure; hence, breaking cryptography is being able to decrypt the message. Creating a suspicious file itself will destroy the purpose of steganography as it attracts the eavesdropper's attention. Steganalysis deals with the identification of hidden information from the stego image. A steganography scheme must ensure poor detection accuracy against steganalyser. A variety of steganalysis techniques to detect both spatial and transform domain image steganography techniques can be found in the literature. These schemes apply statistical checks on the stego images to identify the contents hidden inside stego as embedding distortion will affect the image properties. Thus, hiding locations are of significant importance to achieve imperceptibility as well as undetectability [2].

Many transform domain image steganography methods are presented in the literature. A false and multi-secret steganography scheme was discussed by Ogiela and Koptyra [3]. This scheme employs multiple secret messages. Hence, it will be difficult for the intruder to detect the true secret. Unless a true secret is revealed, the scheme could not be broken and there will be no information loss. Mohammed et al. has presented a Discrete Wavelet Transform (DWT)-based image steganography scheme in [4]. A cover image of size 512×512 was embedded with 12288 bits of information. Imperceptibility results were found to be acceptable with a stego of superior quality. Integer Wavelet Transform (IWT)-based steganography scheme was presented in [5]. The Pixel Value Differencing (PVD) technique was employed with two wavelet coefficients in order to embed the secret information. Obtained results for Lena, Baboon, and Peppers images show that PSNR value ranges from 39 to 40 dB for the capacity of 81,326, 82,933, and 81,337 bytes, respectively. Another Discrete Cosine Transform (DCT)-based scheme was proposed in [6]. PSNR value of 32.2 dB was obtained. Pramanik et al. proposed the DWT-based technique that benefits from Genetic algorithm and is a hybid model [7]. The DCT domain edge-based data hiding scheme is proposed in [8] and embedding is done in edge pixels of the cover image.

A contourlet transform-based steganographic approach was discussed by Subhedar and Mankar [9]. The 8×8 block of contourlet coefficients that possess energy higher than mean energy are selected for embedding secret bits. This scheme has the benefits of advance capacity estimation. PSNR values are found to be in the range of 39–45 dB for a payload of 24,927 bits. A similar approach is presented by Sajedi and Jamzad [10] that employs the contourlet transform. Embedding is performed in non-smooth regions belonging to edge areas. The average PSNR is in the range of 38–40.6 dB for a capacity of 12,000–3000 bits. Steganalysis results are obtained by employing wavelet-based and feature-based steganalyser. Hajduk et al. proposed a steganography approach using QR code and DWT. The Least Significant Bit (LSB)-based embedding is performed [11] and PSNR of 50.24 dB is obtained for embedding 141,744 Kbytes of data in the cover. Achieving better imperceptibility while remaining undetectable is the key for secure steganography scheme design. The remaining part of this article is arranged as follows. Section 2 summarises the fundamentals of ridgelet transform. Section 3 reviews features of SVD. Section 4 covers the proposed steganography algorithm details. Experimental results are discussed in Sect. 5. Concluding remarks are covered in Sect. 6.

2 Ridgelet Transform

Images can be effectively modelled and analysed. Moreover, redundancy in digital images further helps to opt for digital images as a medium to carry private information over an insecure communication channel. The performance of the transform domain image steganography scheme depends on the selection of a suitable type of transform as transform coefficients play the role of embedding locations. In this paper, ridgelet transform is employed to implement the embedding scheme. Ridgelets offer efficient representation of images with edges [12, 13]. This transform also performs better for smoothness along contours and solves the problem of sub-band mixing which occurs in wavelets. It can be computed by wavelet analysis in the Radon domain as indicated in Fig. 1. The continuous Ridgelet transform of $s \in L^2(\Re^2)$ is defined as

$$r(a, b, \theta) = \int_{\Re^2} \psi_{a,b,\theta}(x) s(x) dx \quad (1)$$

where $x = (x_1, x_2) \in \Re_2$, $\psi_{a,b,\theta}(x)$ is the Ridgelet 2-D function. The ridgelet coefficients r_s of s are represented by 1-D wavelet transform to the projections of Radon transform where the direction θ is constant and t is varying.

$$r_s(\theta, a, b) = \int_{R} \psi_{a,b}(t) R_s(\theta, t) dt \quad (2)$$

3 Singular Value Decomposition

Singular value decomposition technique is widely used for the analysis of matrices and can be defined with the help of submatrices, U, S, and V^T where U and V are orthogonal square matrices and S is rectangular diagonal matrix. Any real matrix A can be represented as

Fig. 1 Relationship between transforms

$$A = USV^T \tag{3}$$

$$A = \begin{bmatrix} u_{1,1} & .. & u_{1,N} \\ ... & .. & ... \\ u_{N,1} & ... & u_{N,N} \end{bmatrix} \begin{bmatrix} \sigma_1 & .. & 0 \\ ... & .. & ... \\ 0 & .. & \sigma_N \end{bmatrix} \begin{bmatrix} v_{1,1} & .. & v_{1,N} \\ ... & .. & ... \\ v_{N,1} & ... & v_{N,N} \end{bmatrix}$$

Singular values possess intrinsic algebraic properties and have very good stability. They have a wide variety of applications in fields like image compression, steganography, and watermarking.

Many SVD-based steganography schemes can be found in the literature. These schemes can be classified into two categories; pure SVD and hybrid schemes. Pure SVD schemes employ singular values of the cover image for data hiding whereas hybrid schemes employ a combination of image transform and SVD. Here, singular values of transform coefficients act as secret data carriers. SVD-based schemes offer better imperceptibility, however, the major flaw in these methods is the false positive issue that creates the ambiguity situation while extraction of the secret message at the receiver. Due to false positive issue, exact retrieval of the secret message cannot be guaranteed. This problem is discussed in detail in [14, 15]. To address this issue, the principal component is preferred for embedding rather than only singular values. It helps to avoid ambiguity while achieving better imperceptibility.

4 Proposed Steganography Approach

Proposed scheme benefits from the advantages of ridgelets over wavelets and singular value decomposition. Figure 2a exhibits the framework for proposed steganography approach and Fig. 2b explains the secret image retrieval. Let 'C' denote the original cover image 'S' denote the secret image to be embedded in the cover.

4.1 Embedding Algorithm

A stepwise approach to embedding algorithm can be explained as

1. Apply level 1 ridgelet transform on cover.

$$r \Leftarrow \Im\{C\}, \tag{4}$$

   where $\Im\{.\}$ denotes the ridgelet decomposition and r denote ridgelet coefficients.

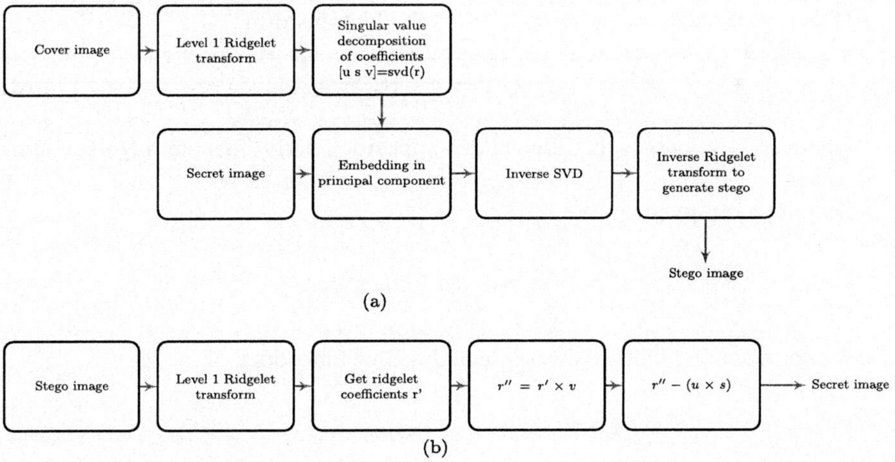

Fig. 2 Framework of proposed steganography system

2. Further, decompose ridgelet coefficients with SVD.

$$r \Rightarrow U \Sigma V^T \tag{5}$$

3. Perform embedding in principal component with suitable scale factor and obtain modified ridgelet coefficients.

$$r' \Leftarrow C_{U\Sigma} + \alpha \times S \tag{6}$$

   Scaling factor also known as embedding strength is decided experimentally to achieve better imperceptibility. It may vary for different images.
4. Apply inverse SVD and level 1 inverse ridgelet transformation. It results in stego image.

$$C' \Leftarrow \Im^{-1}\{r'\} \tag{7}$$

   where $\Im^{\{-1\}}$ is inverse ridgelet transformation.

4.2 Extraction Algorithm

To check exact retrieval of secret embedded in cover image, extraction algorithm is designed and is shown in Fig. 4b. Stepwise approach to extraction procedure is explained as follows.

1. Decompose stego image using level 1 ridgelet transform.

$$r'' \Leftarrow \Im^{-1}\{C'\}, \tag{8}$$

where $\Im^{-1}\{.\}$ denotes the ridgelet decomposition and r'' denote ridgelet coefficients.

2. Obtain secret image as

$$S' \Leftarrow \frac{1}{\alpha}\{C'_{U\Sigma} - C_{U\Sigma}\} \tag{9}$$

Use the same scaling factor as was selected while embedding.

5 Experimental Results

In this section, performance of proposed scheme is verified using measures like perceptual assessment and robustness to image processing operations.

5.1 *Perceptual Measurement*

Simulation results are obtained using test images from USC-SIPI Image Database [16] and some of them are depicted in Fig. 3. It includes Barbara, Lena, Elaine, Airplane, Lake, Baboon, Peppers, and Splash. Cover images are grayscale images of size 512 × 512. Here, the cameraman image is used as a secret image. It is of size 512 × 512. Figure 4 shows the stego images obtained by hiding a secret image in the cover image. Steps followed for embedding and extraction are mentioned in the proposed approach section. Subjective measurement of the stego images shown in Fig. 4 suggests there are no visible distortions in obtained stego images and it does not indicate any traces of tamper. Table 1 shows imperceptibility results using image quality measures. It can be seen that all image quality measures possess values in the desired range. It is known that structural similarity, image fidelity, correlation coefficient, and universal quality indicator measures must have a value very close to 1 to validate the similarity between the cover and the stego image. The same is achieved and can be confirmed from Table 1. These simulation results confirm better imperceptibility. Figure 4 also shows the extracted secret image from the stego image.

Fig. 3 **a**–**g** Cover images used in proposed scheme. **h** Secret image

(a) PSNR=55.5998 dB (b) PSNR=57.3496 dB

(c) PSNR=58.5593 dB

Fig. 4 Stego images **a–c**. **a** Lake. **b** Tiffany. **c** Elaine. **d** Extracted secret

Table 1 Image quality assessment for stego images

Image↓/Metric⟶	PSNR	RMSE	MSSIM	IF	FSIM	NC	UQI
Splash	55.87	0.40	0.99	0.99	0.99	0.99	0.99
Tiffany	57.34	0.34	0.99	0.99	0.99	0.99	0.99
Baboon	57.90	0.32	0.99	0.99	0.99	0.99	0.99
Lena	57.18	0.35	0.99	0.99	0.99	0.99	0.99
Airplane	58.45	0.30	0.99	0.99	0.99	0.99	0.99
Lake	55.59	0.42	0.99	0.99	0.99	0.99	0.99
Peppers	58.61	0.29	0.99	0.99	0.99	0.99	0.99
Elaine	58.55	0.30	0.99	0.99	0.99	0.99	0.99
Barbara	56.13	0.39	0.99	0.99	0.99	0.99	0.99
Living room	56.60	0.37	0.99	0.99	0.99	0.99	0.99

Table 2 PSNR, NC, and BER values for attacked stego

Attack ↓/Image ⟶	Elaine			Splash			Tiffany			Baboon		
	PSNR	NC	BER	PSNR	NC	BER	PSNR	NC	BER	PSNR	NC	BER
Rotate (80°)	26.88	0.48	0.14	27.32	0.69	0.14	27.31	0.75	0.14	27.31	0.60	0.14
Crop	29.84	0.57	0	25.30	0.61	0	25.86	0.41	0	25.86	0.53	0.0006
Resize (1024, 1024)	45.84	0.99	0	50.49	0.99	0	46.61	0.99	0	46.61	0.99	0
Gaussian noise (0, 0.01)	28.71	0.90	0.0016	28.86	0.88	0.02	28.72	0.87	0.0005	28.72	0.89	0.02
Salt and Pepper noise (0.01)	46.50	0.99	0.0049	47.03	0.99	0.0051	46.29	0.99	0.0051	46.29	0.99	0.0041
Speckle noise (0.02)	29.02	0.92	0	30.23	0.91	0	28.07	0.82	0.0003	28.07	0.92	0
Poison noise	31.36	0.96	0	32.23	0.94	0	30.31	0.94	0.0005	30.31	0.96	0
Sharpen	30.39	0.95	0.0025	34.43	0.96	0.0067	32.37	0.96	0.0014	32.37	0.92	0.01
Reduce colour	55.65	0.99	0	65.37	0.99	0	54.37	0.99	0.0003	54.37	0.99	0
Histogram equalisation	27.23	0.99	0.01	31.25	0.99	0.01	24.45	0.99	0.01	24.45	0.99	0.01
Median filter [5 5]	37.06	0.99	0	41.19	0.99	0	38.10	0.98	0.0003	38.10	0.91	0
Wiener filter [5 5]	37.40	0.90	0	41.96	0.99	0	38.09	0.99	0	38.09	0.95	0
Soft thresholding	27.32	0.90	0	31.61	0.94	0	19.24	0.35	0.0003	19.24	0.8804	0
Hard thresholding	54.32	0.99	0	52.86	1	0	53.76	0.99	0.0003	53.76	0.99	0

5.2 *Robustness to Attacks*

Robustness of stego image while it is being transferred over the network must be examined. Unintended users may try to intercept stego image to extract or modify the hidden contents. To verify the robustness to image manipulation, stego is passed through variety of image processing operations and its impact is observed. Several image processing operations like image rotation by 80°, sharpening an image, addition of different types of noise like Gaussian, salt and pepper, speckle and poison noise, cropping an image, histogram equalisation, and JPEG compression are tested. Image quality metrics like PSNR, NC, and BER are preferred to verify the robustness. Simulation results for robustness are presented in Table 2. Ideally, PSNR should be as high as possible, NC should be close to 1, and BER should be 0. Obtained values are very close to these standard values suggesting strong robustness of stego images.

6 Conclusion

This work presents transform domain image steganography algorithm using ridgelet transform and SVD. The secret image is embedded in principal component of ridgelet coefficients of cover. Choice of embedding location does not cause false positive issue which is usually observed while using SVD methods. Percetual measurement of stego images confirms better imperceptibility obtained using proposed scheme. Robustness to image processing operations was verified using image quality measures. Strong robustness as shown by simulation results in Table 2 further validates the proposed scheme.

References

1. M.S. Subhedar, V.H. Mankar, Current status and key issues in image steganography: a survey. Comput. Sci. Rev. **13–14**, 95–113 (2014)
2. B. Li, J. He, J. Huang, Y.Q. Shi, A survey on image steganography and steganalysis. Int. J. Inf. Hid. Multimed. Sig. Process. **2**(2), 142–172 (2011)
3. M.R. Ogiela, K. Koptyra, False and multi-secret steganography in digital images. Soft Comput. **19**(11), 3331–3339 (2015)
4. M.H. Shirafkan, E. Akhtarkavan, J. Vahidi, A image steganography scheme based on discrete wavelet transform using lattice vector quantization and reed Solomon encoding, in *2nd International Conference on Knowledge Based Engineering and Innovation* (2015)
5. A.K. Gulve, M.S. Joshi, An image steganography method hiding secret data into coefficients of integer wavelet transform using pixel value differencing approach, in *Hindawi Mathematical Problems in Engineering* (2015). Article ID 684824
6. T. Rabie, I. Kamel, Multimed Tools Appl. (2016). https://doi.org/10.1007/s11042-016-3301-x
7. S. Pramanik, R.P. Singh, R. Ghosh, Application of bi-orthogonal wavelet transform and genetic algorithm in image steganography. Multimed. Tools Appl. **79**, 17463–17482 (2020). https://doi.org/10.1007/s11042-020-08676-1

8. N. Ayub, A. Selwal, An improved image steganography technique using edge based data hiding in DCT domain. J. Interdis. Math. **23**(2), 357–366 (2020). https://doi.org/10.1080/09720502.2020.1731949
9. M.S. Subhedar, V.H. Mankar, Performance evaluation of image steganography based on cover selection and Contourlet transform, in *International Conference on Cloud & Ubiquitous Computing & Emerging Technologies (CUBE)* (2013). https://doi.org/10.1109/CUBE.2013.39
10. H. Sajedi, M. Jamzad, Adaptive steganography method based on Contourlet transform, in *9th International Conference on Signal Processing* (2008). https://doi.org/10.1109/ICOSP.2008.4697237
11. V. Hajduk, M. Broda, O. Kova, D. Levicky, Image steganography with using QR code and cryptography, in *26th Conference on Radio elektronika* (2016)
12. P. Carrlle, E. Andres, Discrete analytical Ridgelet transform. Sig. Process., Elsevier **84**(11), 2165–2173 (2004)
13. M.N. Do, M. Vetterli, The finite Ridgelet transform for image representation. IEEE Trans. Image Process. **12**(1), 16–28 (2003). https://doi.org/10.1109/TIP.2002.806252
14. J.M. Guo, H. Prasetyo, Security analyses of the watermarking scheme based on redundant discrete wavelet transform and singular value decomposition. AEU Int. J. Electron. Commun. **68**(9), 816–834 (2014)
15. M.S. Subhedar, V.H. Mankar, Image steganography using redundant discrete wavelet transform and QR factorization. Comput. Electri. Eng. https://doi.org/10.1016/j.compeleceng.2016.04.017
16. USC-SIPI (1997). http://sipi.usc.edu/database

HW-CMA and HW-LS-CMA Blind Learning Method for Intelligent Antenna System

Ashwini Pandhare and Khyati Zalawadia

Abstract The trend in the conception of future satellite system and modern communication systems are undoubtedly in the direction of the use of digital beamformer. These systems will include many digital links. These beamformer needs to control their weights adaptively for build intelligent systems. For controlling these weights, various blind and nonblind algorithm apply as per applications. Blind learning process is based on constant modulus algorithm (CMA) and least square CMA (LS-CMA) analyses. Also, to improve converge area and side lob level reduction, modified Hamming window-based algorithm to constant modulus algorithm and least square simulate using MATLAB. These algorithms are validated for linear, circular and planar antenna array geometry.

Keywords Intelligent antenna · Adaptive beamforming · CMA · LS-CMA · HW-CMA · HW-LS-CMA

1 Smart Antenna

Smart antenna is subdivided in two types: first one is switch beamforming. Switch beamforming contains a set of sensors which are arranged in the different array geometry. By combining that array elements form a switch beam pattern that follows the desired user. Smart antenna is an antenna which can adjust its own beam pattern in particular direction with minimizing interference from undesired user. Smart antenna has two objective DOA (direction of arrival estimation) and beamforming [1]. The process of adding the sensors signal and then update the radiation pattern in a desired called as digital beamforming [2–4]. Most advanced type of smart antenna is "adaptive beamforming". Adaptive Beamforming having a strong signal processing capability to automatically change the radiation pattern in accordance with the changing signal environment. Forming beam in desired direction and nulling the pattern in undesired direction, is obtain by constantly multiplying the incoming signal with

A. Pandhare · K. Zalawadia (✉)
Babaria Institute of Technology, Vadodara, India

F. Thakkar et al. (eds.), *Proceedings of the International e-Conference on Intelligent Systems and Signal Processing*, Advances in Intelligent Systems and Computing 1370,
https://doi.org/10.1007/978-981-16-2123-9_7

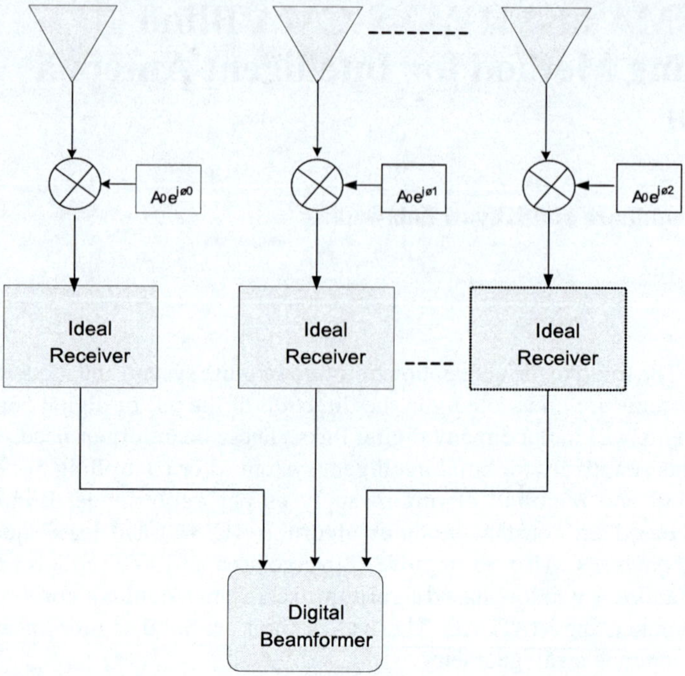

Fig. 1 Digital beam former

complex weights and then combine them to obtain the desired pattern. Weights are processed by using various adaptive beamforming algorithms [5, 6]. So for adaptive beamforming, adaptive algorithms are very important (Fig. 1).

2 Adaptive Beamforming

Beamforming which is digitally controlled, provide connection between array antenna system and advance digital systems. Main benefit of digital beamforming is that once the Radio Frequency information is captured in digital sequence, digital signal processing techniques and algorithms are prepared for the spatial domain. Digital beamforming first convert RF signal of each sensor into I and Q channels. These digital baseband signals are shown as the amplitudes and phases of the received signal of each sensor. For computing weights and I and Q data, adaptive beam formers are more popular.

An adaptive beam former able to self optimize the array sample by way of adjusting and control weights till prescribed goal characteristic is glad. The means by which the optimization is executed is targeted through a set of rules designed for that reason. [7, 8].

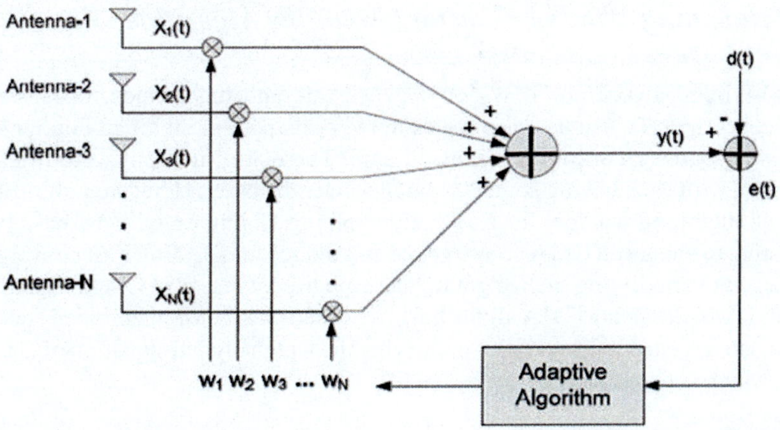

Fig. 2 Adaptive beamforming

Historically, adaptive beamforming has been employed more often in sonar and radar systems. The invention was started of the intermediate frequency side lobe canceller in 1959, by Howells [2, 9]. The idea of an entirely adaptive array was developed in 1965, by Applebaum. This algorithm is primarily based on the general issue of maximization of SNR on the array output [10].

In this figure, uniform linear array antenna receives the signal $x_n(t)$. Then $x_n(t)$ multiplied with weight vector W_N, which has to be update I and Q of incoming signal x(t).This modified weighted signals are combine and produce output y(t). Then adaptive beamforming algorithm is used to observe the error between y(t) and d(t) as shown in Fig. 2. The e(t) is, minimizing by updating weight using adaptive algorithm [11–13].

3 Blind Adaptive Beamforming Algorithms

3.1 Constant Modulus Algorithm

Some adaptive beamforming algorithms are working on to reduce and minimize the error between a reference signal and the desired output. For reference signal-based algorithm, prior knowledge of the nature of the arriving signals is needed. In such case where a reference signal is not available, the beamforming is called as blind beamforming. In many wireless communication, radar signals are frequency-or Phase-modulated signals. These types of modulator design on assumption of constant amplitude. Hence, the signal is consider as a constant magnitude or modulus.

$$w(m+1) = w(m) + \mu \tag{1}$$

3.2 Hamming Window Constant Modulus Algorithm

As CMA blind algorithm, slow convergence rate and higher side lobe level are major concerned. To handle these issues in radiation pattern modified Fourier series windowing-based algorithm is discussed here. There are variety of windows-based algorithm, with each having some advantages over the other. Here, we will study the effect of Hamming window for CMA. By applying Hamming window-based logic on existing to enhanced CMA convergence and reduce the SLL [6]. Fast convergence is achieved by modifying the weight update equation of classical CMA. Equation to determine weights using CMA algorithm is modified with below mentioned equation.

Fast convergence is achieved by modifying the weight update equation of classical CMA, which is expressed as

$$w(m+1) = w(m) + h(N)\mu\left(1 - \frac{1}{|y(m)|}\right)y^{H}(m)x(m) \tag{2}$$

$$Where, h(N) = hanningprameter$$

Here w (m) is the Hamming weight, which can be expressed as

$$w(m) = 0.54 - 0.46, cos\left(2\pi\frac{m}{N}\right)0 \leq m \leq N \tag{3}$$

3.3 Least Square- Constant Modulus Algorithm

One severe disadvantage of the Godard CMA algorithm is the high convergence time. This type of algorithm is not capable of handlling high speed and advanced communication systems. To overcome this limitation, high speed algorithm was explored by using nonlinear least-squares method. The least-squares method is also known as the Gauss method. This method is referred to as the LS-CMA algorithm and is also known as an auto regressive estimation method.

$$w(m+1) = {R_{xx}}^{-1}(m)r_{xd}(m) \tag{4}$$

3.4 *Hamming Window Least Square Constant Modulus Algorithm*

LS-CMA method does not require improvement in convergence speed. In LS-CMA major improvement is required to improve the side lobe level. Hence, by modify the LS-CMA using Hamming window-based algorithm with following equations.

Array correlation matrix and the correlation vector is defined as

$$R_{xx}(m) = \frac{X(m)X^H(m)}{K}$$

$$r_{xd}(m) = \frac{X(m)d^H(m)}{K} \tag{5}$$

$$w(m+1) = {R_{xx}}^{-1}(m)r_{xd}(m)$$

$$w(m+1) = h(N).{R_{xx}}^{-1}(m)r_{xd}(m) \tag{6}$$

Hamming weight is expressed as

$$w(m) = 0.54 - 0.46,\, cos\left(2\pi \frac{m}{N}\right) 0 \le m \le N. \tag{7}$$

4 Simulation Result

Simulation result produced by using Matlab software and planer antenna geometry has been used to analyze various blind algorithms.

4.1 *Constant Modulus Algorithm*

For uniform linear array, Rectangular plots show that CMA can successfully create the radiation pattern as per given specification (Fig. 3).

4.2 *Hamming Window-CMA*

See (Fig. 4).

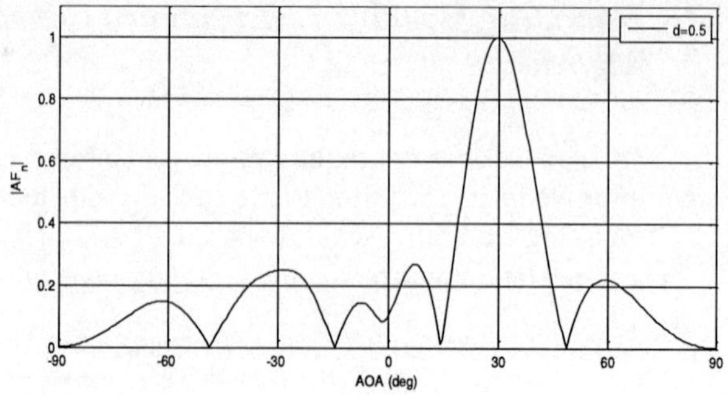

Fig. 3 AOA versus normalized AF_n desired at 30° and null at 10° using CMA algorithm for d = 0.5 & N = 8

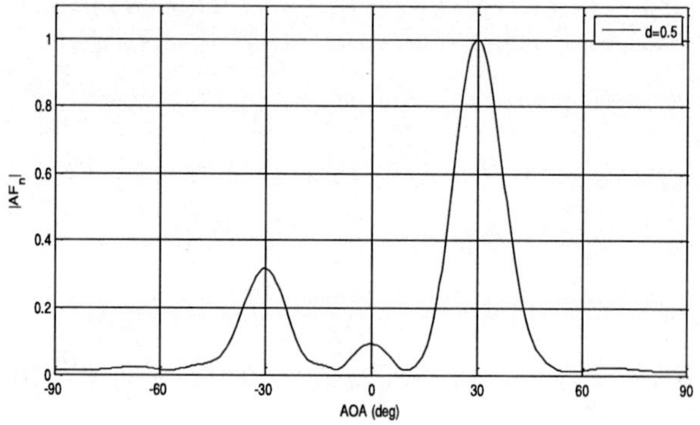

Fig. 4 AOA versus normalized AF_n desired at 30° and null at 10° using HW-CMA algorithm for d = 0.5 & N = 8

4.3 Least Square CMA

See (Fig. 5).

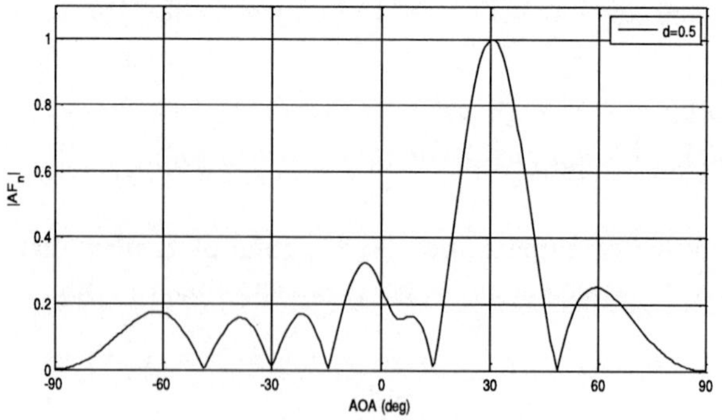

Fig. 5 AOA versus normalized AF_n desired at 30° and null at 10° using LS-CMA algorithm for d = 0.5 & N = 8

4.4 Hamming Window–LS-CMA

See (Figs. 6 and 7, Table 1).

Above figures shows result of HW-CMA algorithm for uniform linear array with N = 8 array element and spacing between array elements 0.5. In this, Angle of arrival vs. array factor graph shows narrow beam width towards desired angle, i.e., 30° and null at -30° and 0°. In HW-CMA, when increase in spacing between elements beam width becomes narrower, but wrong beam formed increases the interference. So

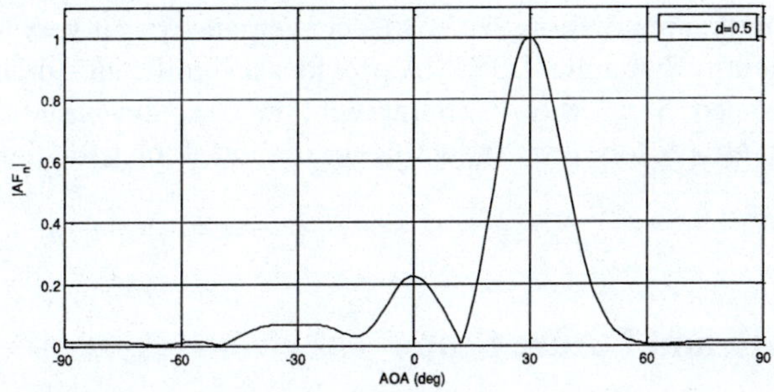

Fig. 6 AOA versus normalized AF_n desired at 30° and null at 10° using HW-LS-CMA algorithm for d = 0.5 & N = 8

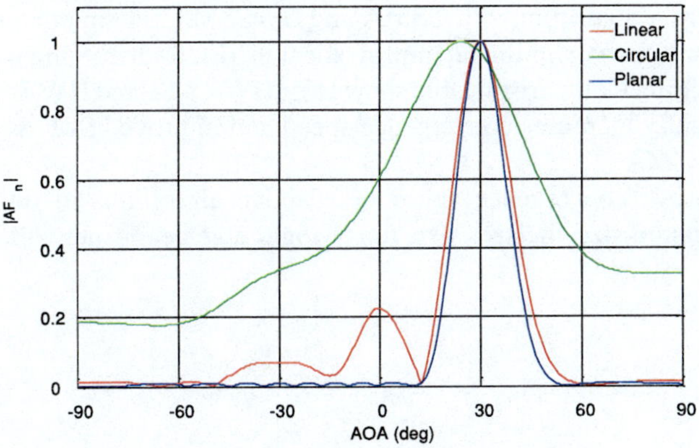

Fig. 7 AOA versus normalized AF_n Beam desired at 30° and null at $I_1 = -30^0$, $I_2 = 0^0$ using HW-LS-CMA algorithm comparison with different array geometry

Table 1 Result summary of various algorithms

	CMA	LS-CMA	HW-CMA	HW-LS-CMA
Iteration	90	8	6	2
Side lobe level	0.270	0.324	0.09	0.224

consider the most appropriate distance d = 0.5 λ. When increase number of elements beam width becomes narrower and also side lobes are decreases. HW-CMA using different array geometries show that the beam width of planar array is narrower than linear array and circular array. And also side lobes decrease than other geometry. The simulation results show that the convergence rate of HW-CMA is significantly improved as compared to classical CMA. CMA algorithm requires 90 iterations, but HW-CMA requires only 6 iterations. Previous CMA algorithm shows SLL 0.270. Hence, HW-CMA gives an improvement of about 0.09.

All four blind algorithm CMA, LS-CMA, HW-CMA, and HW-LS-CMA were analyzed for design beamformer as shown in figures. From the above table, CMA requires more number of iterations, i.e., CMA convergence rate is very low, especially when environment is dymamic. LS-CMA provides a significant side lobe reduction with moderate speed. Same way, hamming window selection-based algorithms are capable of reducing side lobe level in significant amount along with high convergence rate.

5 Conclusion and Future Scope

As per the above simulation results, CMA and LS-CMA produce higher side lobe level for high speed, especially for OFDM modulation using these low level SLL needs. This paper compares the performance of CMA (HW-CMA) and Hamming window LS-CMA algorithm with CSMA and LS-CSMA for smart antenna system. Simulation result shows significant improvement in SLL for hamming window-based adaptive algorithm. Comparison table shows that HW-CMA and HW-LS-CMA algorithm significantly increases convergence speed and improved SLL as compared to CMA and LS-CMA.

So, to improve convergence speed of adaptive algorithms in future, one can approach new step-size incisive like fuzzy logic and neural network-based algorithms.

References

1. D.N. Patel, Comparative analysis of adaptive beamforming algorithm LMS, SMI and RLS for ULA smart antenna, in *International Conference on Communication and Signal Processing* (2016), pp. 1029–1033
2. J. Litva, T.K.Y. Lo, *Digital Beamforming in Wireless Communication* (Artech House Boston, London, 1996)
3. K.R. Zalawadia, T.V. Doshi, U.D. Dalal, Adaptive beam former design using RLS algorithm for smart antenna system, in *International Conference on Computational Intelligence and Communication Networks* (Gwalior, 2011), pp. 102–106
4. A.A. Awan, Performance comparisons of fixed and adaptive beamforming techniques for 4G smart antennas, in *International Conference on Communication, Computing and Digital Systems* (2017), pp. 17–20

5. T.S. Rappaport, *Wireless Communications: Principles & Practice*, 2nd edn. Prentice Hall of India (2007)
6. B. Frank, *Gross* (Mc Graw Hill Publication, Smart Antennas for Wireless Communications, 2005)
7. J.W. Yoo, An improved NLMS algorithm in sparse systems against noisy input signals. IEEE Trans. Circuits Syst.—II: Express Briefs **62**(3), 271–275 (2015)
8. K.R. Zalawadia, T.V. Doshi, U.D. Dalal, Study and design narrow band phase shift beam former, in *2013 International Conference on Intelligent Systems and Signal Processing (ISSP)* (2013), pp. 238–241
9. M. Jiang, Adaptive beamforming for vector-sensor arrays based on a reweighted zero-attracting quaternion-valued LMS algorithm. IEEE Trans. Circuits Syst.—II: Express Briefs **63**(3), 274–278 (2016)
10. J.A. Srar, Adaptive array beamforming using a combined LMS-LMS algorithm. IEEE Trans. Antennas Propag. **58**(11), 3545–3557 (2010)
11. R.M. Shubair, Adaptive beamforming using variable step-size LMS algorithm with novel ULA array configuration, in *IEEE Conference* (2013), pp. 650–654
12. R.M. Shubair, A convergence study of Adaptive beamforming algorithms used in smart antenna system, in *IEEE Conference* (2005)
13. R.M. Shubair, W. Jessmi, Performance analysis of SMI Adaptive beamforming arrays for smart antenna systems, in *IEEE Conference* (2005), pp. 311–313

Performance Evaluation of Prediction Algorithm-Based Tracking Methods in a Recovery of a Lost Target Using Wireless Sensor Network

Alpesh Sankaliya and Maulin Joshi

Abstract The prediction algorithm can help select the foremost nodes to supervise the target position by preserving network energy. The unreliable prediction mechanism may hamper tracking by causing temporary loss of target, which is a major issue in critical applications. However, very few methods are available in the literature to unfold target loss issues during tracking. This paper evaluates the effect of various prediction algorithms, including regressive neural networks on sensor network energy consumption. The Kalman filter, Interacting Multiple Model, and Time Delay Neural Network-based tracking methods are compared for different target motion scenarios. The simulation results show the numerical analysis of all prediction algorithms in terms of tracking accuracy, tracking energy, recovery energy, total network energy used, and instance of a target lost.

Keywords Wireless Sensor Network (WSN) · Target tracking · Prediction algorithm · Lost target recovery · Dynamic neural network · Time Delay Neural Network (TDNN)

1 Introduction

The Wireless Sensor Network (WSN) has been successfully applied in target tracking applications. The target may be animal, enemy vehicle, or human being. This application aims to keep continuous track of target locations and provide its information whenever being asked. An interesting part of WSN is that it does not require any human intervention. As sensor nodes have limited resources like memory, processing capability, communication and sensing range, bandwidth, power supply, etc., which makes it difficult to achieve perfect tracking for a longer time of period [3].

The tracking algorithm must have fewer computations to save processing power and should be energy-efficient to save the energy of sensor node. The major amount of energy is saved by switching sensor nodes between active and sleep mode. The

A. Sankaliya (✉) · M. Joshi
Gujarat Technological University, Gujarat, India
e-mail: maulin.joshi@scet.ac.in

F. Thakkar et al. (eds.), *Proceedings of the International e-Conference on Intelligent Systems and Signal Processing*, Advances in Intelligent Systems and Computing 1370,
https://doi.org/10.1007/978-981-16-2123-9_8

prediction algorithm helps to decide which nodes should be activated and which should be in sleep mode. Still, critical tracking applications cannot rely on prediction algorithms, except it is accurate enough and can work with linear and nonlinear target motions [14].

Moreover, the tracking algorithm is vulnerable to loss of target due to inaccurate prediction, coverage hole, node failure, communication failure, atmospheric effects, noise, etc., and a target remains unobserved for the amount of time. This paper compares the performance of autoregressive neural network—Time Delay Neural Network-based tracking algorithm with existing algorithms in case of target loss scenarios. The target lost instances are compared with state of the art methods in terms of number of target lost instances, energy used in tracking a target, energy used in recovery of a lost target, and number of nodes awakened.

The remaining paper content is structured as follows: Sect. 2 describes the existing tracking and recovery methods in the literature. Section 3 briefs the tracking and recovery algorithm. Section 4 deals with performance evaluation and results. The conclusion is presented in Sect. 5.

2 Related Work

The recovery mechanism is an important part of the target tracking application. If the target tracking algorithm fails, then only a robust recovery mechanism can relocate a target on time. The target might get lost because of several reasons as listed below [2, 11].

(a) Network Hole: The network hole is a dark part of a network that has no network coverage. The network holes are created in situations where the sensor nodes are deployed randomly. In many applications, nodes are distributed via airplane in an area of interest. So, the target remains untracked for a particular time while passing through the network hole.

(b) Sensor Node Failure: The sensor nodes are deployed in a remote area where direct access is difficult. The nodes are left unattended once they are installed and even it is not possible to recharge their battery. The depletion of battery is a major reason for node failure. Besides, node failure also occurs by cause of physical damages and atmospheric effects. For example, nodes get damaged in a fire, when deployed for forest fire detection. Node failure may result in a networking hole.

(c) Localization error: The sensor nodes collaboratively try to find the position of a target as it gets detected. In many applications, nodes are not equipped with any kind of a positioning system. Inaccurate node position leads to inaccurate target localization. The wrong estimation of a target location leads to loss of target.

(d) An abrupt change in velocity: If a target is moving with an abrupt change in its velocity, it might get undetected by sensor nodes and hence it will be missed. If the speed of the target is more than the sensing period of a sensor node, the target just flees the sensing area by the time nodes sense it.

(e) Prediction error: In hybrid tracking algorithms, the prediction algorithm helps to activate nodes nearest to the target location. But, an inaccurate prediction will activate the wrong group of sensor nodes or clusters for further tracking process. So, an actual target position cannot be obtained, and it results in target loss.
(f) Communication failure: Proper time synchronization is necessary for successful communication in a network. If intermediate nodes create a delay in the dissemination of target information, it may also result in target loss.

In literature, significant work has been carried out to develop an accurate tracking mechanism [4]. The tracking framework contains target detection, sensing, localization, prediction, and communication process. Most of the tracking methods assume clustering-based network architecture as it saves energy by reducing communication distance between nodes [1]. The communication energy depends on the distance directly. As a target enters into the surveillance area, nodes lying on the boundary detect and senses it. They alert the cluster head about intervention immediately. The cluster head localizes a target and activates a required number of sensor nodes to keep track of a target. The cluster head can take the help of a reliable prediction mechanism to awaken prominent nodes in advance in this complete process.

In literature, the Kalman filter is a widely accepted solution for target tracking applications. It predicts the next state based on prior information, calculates error based on current observation, and applies a correction to predicted state. It involves a simple computation to predict the next state. The Kalman filter works well in for linear target motion, but fails in case of nonlinear target movement. Hence, the probability of target loss is high in Kalman based tracking approaches [7, 13]. For nonlinear target movement, researchers developed extended Kalman filter (EKF). The particle filter is proposed [16], which can work in the case of a nonlinear and non-gaussian system. Some papers also use constant gain Kalman filter, which calculates optimized Kalman filter gain based on trajectory.

Neural network-based approaches are also proposed in the literature to increase the accuracy of a prediction algorithm. The neural network is trained with RSSI and location database. Once it is trained, it is used to find location of a target in real-time operation [8]. In [12], SVM is used to classify the type of a target. If the target motion is linear, linear filter is used, if the target motion is non-liear, second nonlinear.

The recovery framework involves a target relocation and communication process. In [17], a clustering-based distributed tracking protocol is proposed, which uses a RADAR beam for recovery. The sensor nodes which were tracking a target switches from normal beam to high beam in case of a target loss. If it fails, second-level recovery is initiated. In the second level of recovery, the search region is increased. The all the nodes in the circle of radius (p) distance are activated. Otherwise, the nth stage of recovery is initiated in which the new circle radius is changed to $(2n - 3)$. It is difficult to implement RADAR beam in tiny sensor nodes and it will consume more energy in addition to recovery energy.

The recovery method in [15] predicts the next target location using target kinematics. It awakens the senor nodes in a hierarchical manner to recover a lost target. This method activates a large number of nodes and hence hamper network energy consumption.

The clustering-based recovery mechanism is proposed in the paper [9]. The cluster head tries to relocate a target in current cluster first. If target is not found, then the current CH wakes up all its single hop clusters. If it still missing, then two hop clusters are activated, and this process is repeated till the target is recovered. This method also activates a large number of sensor nodes.

The novel method proposed in paper [5] predicts the next possible location of a target in case of target loss. To relocate a target, only single cluster nearest to the predicted position (PL) is activated instead of all the single hop clusters. If the target is not detected here, the two hop clusters nearer to the predicted location are activated. If the target is not found here also, then three hop clusters are activated in third level of recovery. In this way, less numbers are activated to relocate the target. These method also awakens lots of sensor nodes.

In this paper, we defined a new parameter called $Probability\ of\ Recovery\ P_r$. It is assumed that recovery process will be executed in a sequential manner and level by level. The level word indicates here search range. The detailed recovery algorithm is explained in proposed algorithm.

3 Target Tracking in Wireless Sensor Network

The conventional target tracking process has main three steps: (1) Target detection, (2) Localization, and (3) Communication. As sensor network detects the target, a minimum of three sensor nodes will localize the target and its information will be sent to a central entity. This process is repeated until target leaves the sensing area.

3.1 Network Model

It is assumed that all sensor nodes are deployed randomly in a tracking area. The central entity called base station will collect all the information from sensor nodes. The sensor nodes and base station are immobile. All sensor nodes have equal energy level and they can change their mode to sleep mode to save energy. All nodes are well synchronized and can measure distance information to target. The nature of target is non-cooperative and its movement is unknown in advance.

3.2 The Target Tracking Process

The hybrid tracking methods combine prediction algorithm into conventional target tracking framework to save network energy.The main idea of prediction algorithm is predict future location in advance and to activate an subarea of sensor network that target is likely to visit. The detailed tracking framework is explained below.

3.2.1 Target Detection

The tracking is initiated with the current location estimation. The nodes lying on a boundary detect the target's intervention. If the minimum three nodes lying on a boundary can detect target simultaneously, then they send distance to target information to cluster head that a target likely to visit. The cluster head performs the trilateration to localize a target. Once a target is located, the cluster head of a cluster predicts the next location of a target.

3.2.2 Target Location Prediction

The current cluster head will predict the next location of a target based on available target movement history. Here, in this paper, time-delayed neural network is to a prediction target location. The TDNN-based tracking framework [10] is compared with Kalman based approach [9] and IMM-based framework [13].

The current cluster head performs the following steps: (a) predicting the next location of a target, (b) finding cluster and it's head that target is likely to visit, (c) waking up the nearest nodes to a target's predicted location.

3.2.3 Recovery Phase

It is executed in three steps: declaration of a missing target, relocation, and hibernation or switching to the sleep mode.

1. Declaration of missing target: If the activated group of nodes are not able to track a target location, a target loss will be declared by the head of current cluster. This is an outcome of inaccurate prediction performed an algorithm. Target may also get lost if the predicted location is not surrounded by enough number of sensors or nodes lacks enough resources.
2. Relocation: Once a target is lost, it must be recovered immediately in critical applications. It is assumed that relocation process will be executed in a sequential manner. First, level 1 nodes are awakened to relocate the target. If it fails, then level 2 nodes are activated and it is repeated until target is found. The recovery of a target mainly depends on speed of a target. If target is moving slowly, then it can be relocated easily with less number of nodes. If the target is moving faster, search span should be increased, and hence more number of nodes are required to catch the target. In the case of non-cooperative target, it's very difficult to comment on future target trajectory and its speed. In addition to speed of a target, parameters like sensing time, communication delay, etc., may also effect the recovery time. It is difficult to simulate this kind of recovery method with simultaneous problems. The following are the key points about new recovery framework:

   - It is assumed that target recovery is done in levels.
   - In each level, search region is increased gadually to relocate the target.

- The parameter called "probability of recovery" is defined that indicate the chance of target recovery in each level.
- In first level, nodes in initial $searchrange$ is activated. If target is not found here, then second level is initiated. In second level, $searchrange$ is increased as $(level \cdot searchrange)$. The number of levels and search region is kept increasing until the target is found.
- If the probability of target recovery is high, it indicates that target can be recovered easily with less number of nodes.
- If the probability of target recovery is less, it indicates that target remains untracked for a longer time and more number of nodes will be activated to relocate the target.

The target relocation method used by WSN in form of algorithm is explained in Algorithm 1 shown below.

Algorithm 1: Target Relocation Algorithm

```
if target get lost then
    activate all nodes in current cluster
    if target not found then
        Initiate target recovery;
        Set search range = communication range;
        Set probability of recovery;
        while target not found do
            Draw a cicle of radius search range with center as current cluster head;
            Wake up all clusters available in circle;
            n = random(0,1);
            if n ≤ probability of recovery then
                Declare target found;
                Send target found message to all awakened clusters;
                Handover tracking to nearest cluster
            else
                search range = search range + communication range;
            end
        end
    else
        Activate best nodes for tracking
    end
else
    Keep Tracking
end
```

3. Hibernation: All the active nodes involved in recovery go to sleep mode as soon as target found message is received. The new cluster will continue further tracking process.

These three steps will be continuously repeated until target exits sensing area.

Table 1 Sensor network parameters

Parameter	Value
Sensor area	100 x 100 m^2
Sensor nodes	930
Initial energy of nodes	2 J
Speed of target vehicle	0–10 m^2
Sensing aange	10 m
Communication aange	20 m

Table 2 Energy consumption parameters

Parameter	Value
Sensing energy (E_{sense})	9.6×10^{-3} J/Second
Processing energy (E_{pro})	4.0×10^{-9} J/Instruction
Receiving energy (E_{recv})	0.88×10^{-3} J/Byte
Transmission energy (E_{trans})	5.76×10^{-3} J/Byte
Sleep energy (E_{sleep})	0.33×10^{-3} J/Second
Message size D	64 Byte
Instruction size IS	1000

4 Performance Evaluation and Results

The proposed algorithm has been simulated in MATLAB. Various target trajectories are been used for simulation. The experimental set up includes 600 nodes randomly deployed in the field of area 100m $\times$ 100m.The communication range of all sensor nodes is 20 meters. As soon as the target enters into the field, sensor nodes detect the presence, target tracking process starts. Table 1 shows the network parameters used in simulations. Table 2 shows the energy model and energy used for each process of a sensor node [6].

The Fig. 1 shows the wireless sensor network with clusters. It contains total 24 number of clusters. It shows an area of each cluster containing member nodes. The cluster head of each cluster is also highlighted by "CH" with its ID.

4.1 Performance Parameters

1. Average Tracking Error (ATE): It is calculated from a reference trajectory and trajectory tracked by the filter. If t is the reference trajectory and t' is the tracked trajectory using a prediction filter with m time stamps, then average tracking error be calculated as follows: (Eqs. 1–3).

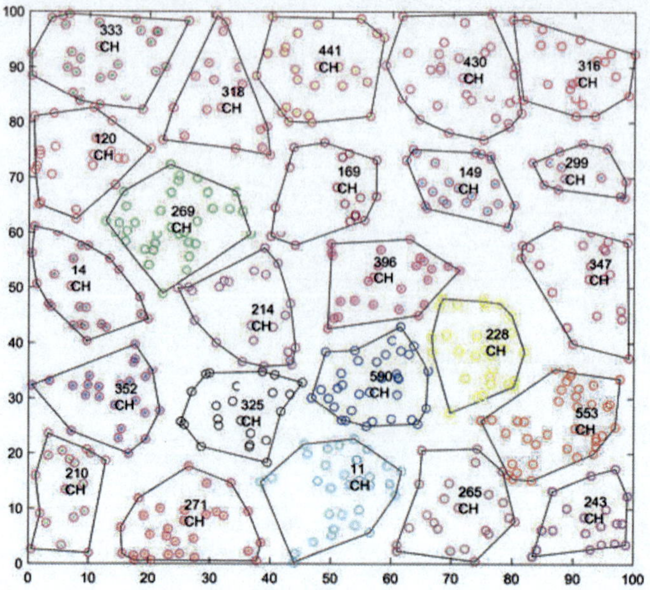

Fig. 1 Clustered wireless sensor network

$$t = \lceil (tx_1, ty_1), (tx_2, ty_2), \ldots\ldots (tx_m, ty_m) \rceil \tag{1}$$

$$t' = \left[\left(t'x_1, t'y_1\right), \left(t'x_2, t'y_2\right), \ldots\ldots \left(t'x_m, t'y_m\right) \right] \tag{2}$$

$$ATE = \frac{1}{m} \sum_{j=0}^{m} \sqrt[2]{\left(tx_j - t'x_j\right)^2 + \left(ty_j - t'y_j\right)^2} \tag{3}$$

2. Total Network Energy Consumed (TNEC): t is a summation of energy used by each node for tracking target in the sensor network. It includes sensing energy, data transmission and reception energy, and processing energy and sleep energy.
3. A number of times target lost (NTL): This number indicates the count of target lost instances on a given trajectory due to the inaccurate prediction. Zero count indicates successful tracking without loss of a target.

4.2 Energy Calculation

Each sensor node spends its energy to track the moving target continuously. The active sensor node performs operations like sensing, processing, transmission, and reception of data. The other sensor nodes remain in sleep mode, and they consume significantly less energy compared to active nodes. The per bit energy values are defined in Table 2.

The energy used by the node in transmission (E_t) and receiving (E_r) is given as Eqs. 4 and 5.

$$E_t = E_{trans} \cdot D \tag{4}$$

$$E_r = E_{recv} \cdot D \tag{5}$$

The energy used by a node in sensing target (E_s) and processing (E_p) is given as Eqs. 6 and 7.

$$E_s = E_{sense} \cdot time\_of\_sensing. \tag{6}$$

$$E_p = E_{pro} \cdot IS \tag{7}$$

The energy used by nodes who remains in sleep mode (E_{sl}) is given as Eq. 8.

$$E_{sl} = E_{sleep} \cdot sleep\_time. \tag{8}$$

The total tracking energy ($E_{tracking}$) is calculated as Eq. 9, assuming that out of total N nodes, N_1 is the number of nodes that have participated in target tracking process.

$$E_{tracking} = N_1 \cdot \left(E_t + E_r + E_s + E_p\right) + (N - N_1) \cdot E_{sl} \tag{9}$$

The recovery energy ($E_{recovery}$) of a target, assuming that N_2 is the number of nodes that have participated in the relocation of a target is given as Eq. 10.

$$E_{recovery} = N_2 \cdot \left(E_t + E_r + E_s + E_p\right) \tag{10}$$

The total energy consumed (E_{total}) by the network is given as Eq. 11.

$$E_{total} = E_{recovery} + E_{tracking} \tag{11}$$

4.3 Simulation Results

This section shows the importance of an accurate prediction algorithm in target tracking application. The target loss occurs due to high prediction error. The prediction algorithm does prediction of next target location and activates set of nodes to handover tracking process. If prediction algorithm fails to predict accurate location, wrong set of nodes will be activate which will not be able to track a target.

Figure 2 shows the target lost instances in case of Kalman filter-based tracking algorithm. In this case, the target is assumed moving with a low velocity range from standstill (0 m/sec) to low (5 m/sec) and can follow any linear or nonlinear trajectory traversing through the sensor network. The probability of recovery (Pr) is initialize

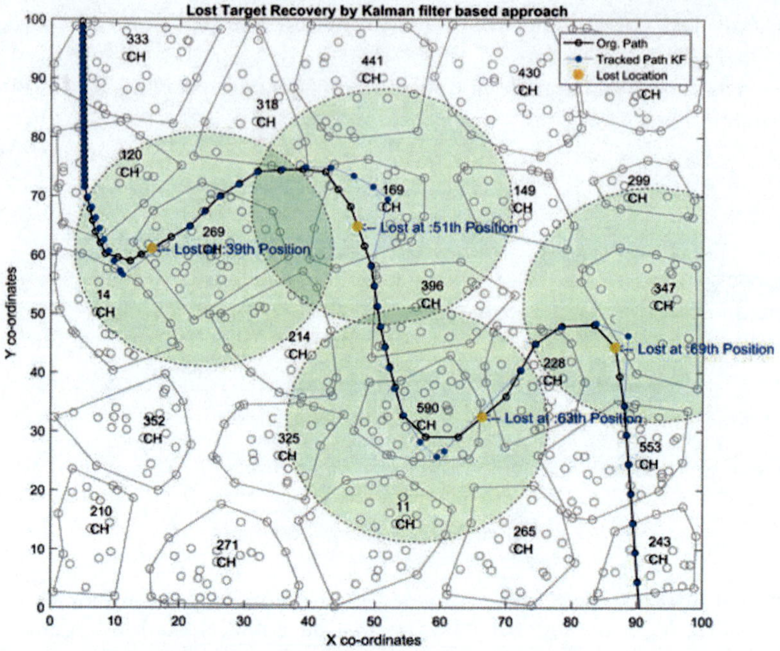

Fig. 2 Trajectory-1: lost target instances in case of Kalman filter, $P_r = 0.95$

as 0.95. The Kalman filter does 4 times wrong prediction and hence target lost occurs 4 times. The Kalman filter uses a constant velocity model to do prediction of target location. In our experimentation, non-cooperative target is assumed, which keep changing it's velocity while transversing through WSN. Once the target changes its velocity or direction, the Kalman filter prediction goes wrong, and error starts increasing that results intto target loss. The light green circle in Fig. 2 shows the search area in which all nodes were activated to relocate the target. The Kalman filter is not able to track a nonlinear trajectory as it uses constant velocity model by default. Hence, it fails to mimic nonlinear motion of target trajectory. It can be seen that target instances 39, 51, 63, and 69 are the point where target takes turn, and at the same location KF filter unable to track a target.

Figure 3 shows the target lost instances in case of IMM model-based tracking algorithm. The IMM model uses two kalman filters in parallel: one for constant velocity target and second for constant acceleration target. Still, its unable to track a target when a target switches its velocity. The limitations of Kalman filter are inherited in to IMM model. Though, the IMM model uses two Kalman filter, it is bounded to user defined switching probability that is used to combine output of two Kalman filter. If it is defined incorrectly, the IMM model fails to do correct prediction. The target instances 64, 66, and 68 shows that target has increased its velocity for few times. For the same instances, IMM model fails to track target and target loss occurs. Here, switching between two models creates ambiguity and generates error. The IMM model also lost target 4 times.

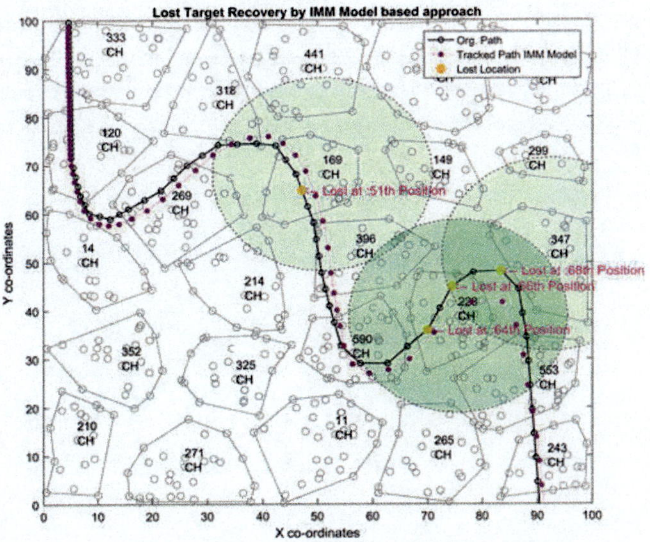

Fig. 3 Trajectory-1: lost target instances in case of IMM model, $P_r = 0.95$

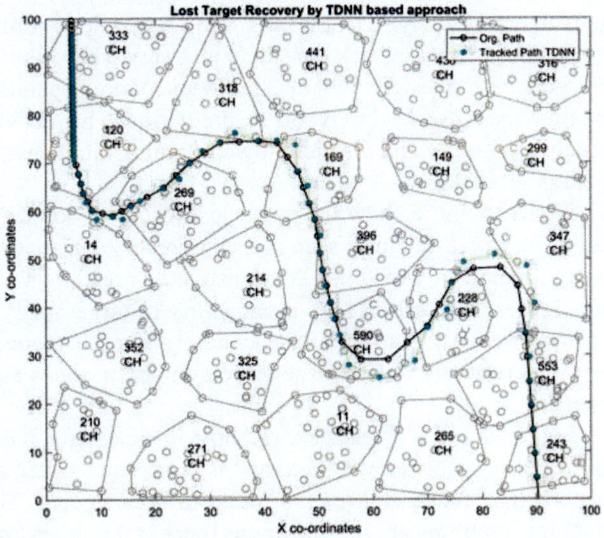

Fig. 4 Trajectory-1: lost target instances in case of TDNN model, $P_r = 0.95$

Figure 4 shows the target lost instances for TDNN-based tracking algorithm. It shows that in case of TDNN, the target is not lost for a single time. The TDNN is autoregressive neural network that does prediction based on previous and current samples. In this simulation, TDNN is trained to predict using last two samples and current sample. The TDNN gives so accurate prediction and hence prove its usefulness in target tracking application.

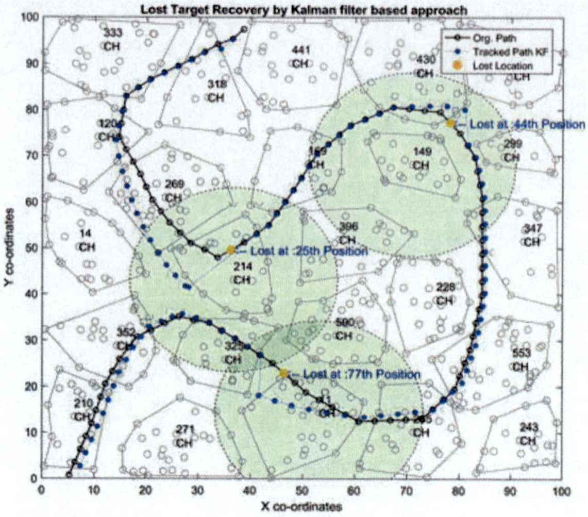

Fig. 5 Trajectory-2: lost target instances in case of KF model, $P_r = 0.95$

For non-cooperative target, the neural network should balance netween stability and plasticity. The stable neural network can work well as a predictor for linear movement of a target. But it fails when target do nonlinear motion with maneuver. Ihe stable neural network considers just previously learned information. The neural network with plasticity will also not work. More dependency on the current states of a target may also lead to incorrect prediction. The TDNN neural network is able to balance statbility and plasticity by combining tap delays (memory elements) with a neural network.

Figures 5 and 6 show the target lost instances for KF and IMM-based approach for trajectory-2. It can be seen that in trajectory-2 also, target is lost many times. The search region in which nodes are activated are shown as light green circle in both the figures. Because of high proabability of target recovery, the target is recovered in first level only. The default seach region is of 20 meters here. The less probability of target recovery will activate more number of nodes and consumes more network energy.

Figures 7 and 8 show the target lost instances for KF and IMM-based approach for trajectory-3. It shows that target is not being tracked by both the algorithms, and hence target lost instances are found in both figures. The graph for TDNN-based tracking agoithms are not shown, as there is no target loss in TDNN approach. The Table 3 shows the numerical analysis of different prediction algorithms. The Kalman-based apporach lost the target almost 4 times due to prediction error and due to that it consumes significant energy in recovery of a lost target. The Kalman-based approach gives tracking error of 4.89 m for trajectory-1, 4.16 m for trajectory-2, and 2.43 m for trajectory-3. The KF consumes 364.73 J of network energy for recovery, which is almost double as comapred to energy used in tracking process. Similary, for trajectory-2, the KF consumes (389.87/135.84 = 2.87) times and IMM consumes (493.20/247.61 = 1.99) times more energy as compared to tracking process energy.

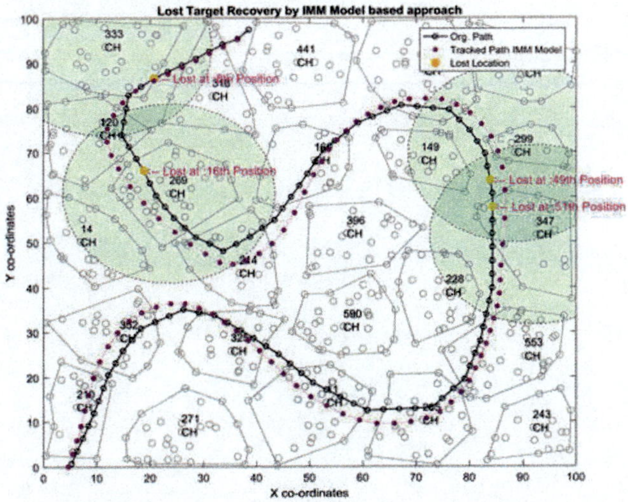

Fig. 6 Trajectory-2: lost target instances in case of IMM model, $P_r = 0.95$

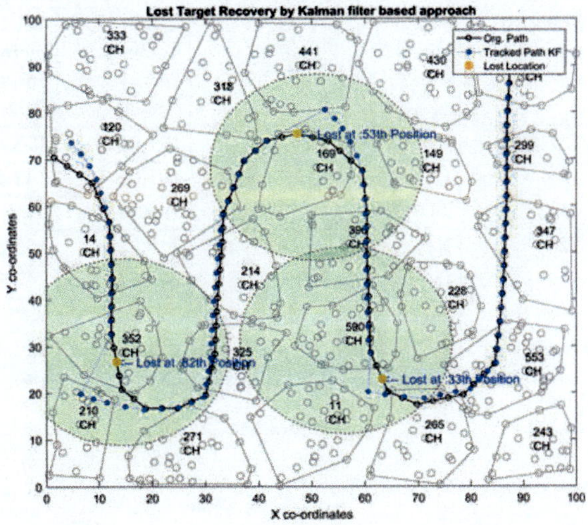

Fig. 7 Trajectory-3: lost target instances in case of KF model, $P_r = 0.95$

It can be seen that TDNN-based approach does not use any recovery energy as target is not being lost single time. The prediction error of TDNN is very high as comapred to KF and IMM algorithms.

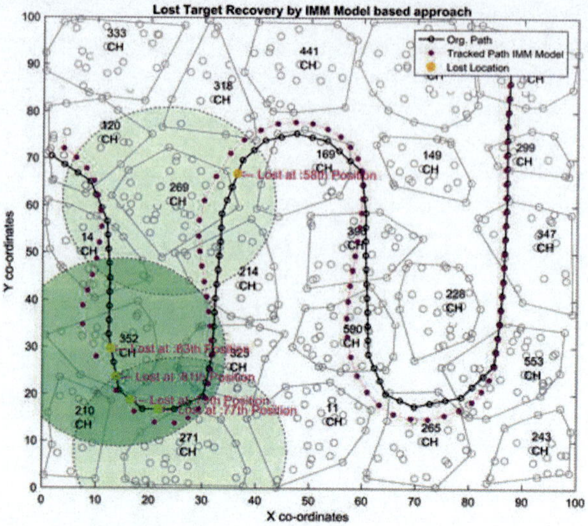

Fig. 8 Trajectory-3: lost target instances in case of IMM model, $P_r = 0.95$

Table 3 Performance comparison of TDNN with other filters in case of target loss instances

Track	Filter	NTL	Total tracking energy (Joule)	Total recovery energy (Joule)	Total energy (Joule)	Average RMSE (m)
T1	KF	4	191.06	173.67	364.73	4.89
	IMM	4	191.05	143.57	334.63	6.27
	TDNN	0	195.50	0	195.49	0.59
T2	KF	3	248.027	135.84	389.87	4.16
	IMM	4	245.59	247.61	493.20	5.97
	TDNN	0	250.16	0	250.16	0.39
T3	KF	3	245.46	150.46	395.92	2.43
	IMM	5	240.64	159.92	400.55	4.24
	TDNN	0	247.53	0	247.53	0.36

5 Conclusion

This paper evaluates the effect of the prediction algorithm on the target tracking application of WSN. If the prediction-based tracking algorithm fails, the target gets lost, and an efficient recovery mechanism is required that relocate the target using minimum network resources. This paper defines the correlation between the network energy used by different prediction algorithms and the probability of target recovery. The TDNN neural network with memory elements in terms of time delays is perfectly able to balance between stability and plasticity which makes it completely suitable for non-cooperative target tracking application. The TDNN algorithm makes an accurate prediction as compared to other algorithms, and hence minimum energy is utilized by TDNN-based approach for target tracking. The target lost instances is almost zero in the TDNN-based approach, which is found frequent in KF and IMM-based approaches. The TDNN approach saves network energy that may be wasted in recovering the target with less prediction error.

References

1. T. Ahmad, A.M. Abbas, EEAC: an energy efficient adaptive cluster based target tracking in wireless sensor networks. J. Interdiscip. Math. **23**(2), 379–392 (2020). ISSN 09720502, https://doi.org/10.1080/09720502.2020.1731951
2. E.Z. Asmaa, R. Said, K. Lahoucine, Review of recovery techniques to recapture lost targets in wireless sensor networks, in *Proceedings of 2016 International Conference on Electrical and Information Technologies, ICEIT 2016* (2016), pp. 208–213. https://doi.org/10.1109/EITech.2016.7519591
3. D.P. Bhagat, H. Soni, Target tracking in a wireless sensor network using a multi-step KF-PSO model. Int. J. Comput. Appl. 1–12 (2019). ISSN 19257074. https://doi.org/10.1080/1206212X.2018.1564172
4. O. Demigha, W.K. Hidouci, T. Ahmed, On Energy efficiency in collaborative target tracking in wireless sensor network: a review. IEEE Commun. Surv. Tutor. **15**(3), 1210–1222 (2013). ISSN 1553877X. https://doi.org/10.1109/SURV.2012.042512.00030. http://www.scopus.com/inward/record.url?eid=2-s2.0-84881316895&partnerID=tZOtx3y1
5. A. Gupta, S. Patil, M. Zaveri, Lost target recovery in wireless sensor network using tracking, in *Proceedings—International Conference on Communication Systems and Network Technologies, CSNT* (2012), pp. 352–356. https://doi.org/10.1109/CSNT.2012.83
6. B. Jiang, K. Han, B. Ravindran, H. Cho, Energy efficient sleep scheduling based on moving directions in target tracking sensor network, in *IPDPS Miami 2008—Proceedings of the 22nd IEEE International Parallel and Distributed Processing Symposium, Program and CD-ROM* (2008). https://doi.org/10.1109/IPDPS.2008.4536330
7. S.R. Jondhale, R.S. Deshpande, Kalman filtering framework-based real time target tracking in wireless sensor networks using generalized regression neural networks. IEEE Sens. J. **19**(1), 224–233 (2019). ISSN 1530437X. https://doi.org/10.1109/JSEN.2018.2873357
8. S.R. Jondhale, R. Shubair, R.P. Labade, J. Lloret, P.R. Gunjal, *Application of Supervised Learning Approach for Target Localization in Wireless Sensor Network*, vol. 1132 (Springer International Publishing, 2020). ISBN 9783030403058. https://doi.org/10.1007/978-3-030-40305-8_24
9. A. Khare, K.M. Sivalingam, On recovery of lost targets in a cluster-based wireless sensor network, in *2011 IEEE International Conference on Pervasive Computing and Communications Workshops, PERCOM Workshops 2011* (2011), pp. 208–213. https://doi.org/10.1109/PERCOMW.2011.5766870
10. J.H. Munjani, M. Joshi, Target tracking in WSN using time delay neural network, in *IEEE Region 10 Annual International Conference, Proceedings/TENCON* (2017), pp. 3839–3845. ISBN 9781509025961. https://doi.org/10.1109/TENCON.2016.7848783
11. S. Patil, A. Gupta, M. Zaveri, Recovery of lost target using target tracking in event driven clustered wireless sensor network. J. Comput. Netw. Commun. (2014) ISSN 2090715X. https://doi.org/10.1155/2014/923479
12. X. Wang, X. Liu, Z. Wang, R. Li, Y. Wu, SVM + KF Target tracking strategy using the signal, 1–25 (2020)

13. Y. Wang, X. Feng, Maneuvering target tracking in wireless sensor network with range only measurement. J. Phys.: Conf. Ser. **1325**(1) (2019a). ISSN 17426596. https://doi.org/10.1088/1742-6596/1325/1/012171
14. Y. Wang, X. Feng, Maneuvering target tracking in wireless sensor network with range only measurement. J. Phys.: Conf. Ser. **1325**(1) (2019b.) ISSN 17426596. https://doi.org/10.1088/1742-6596/1325/1/012171
15. Y. Xu, J. Winter, W.C. Lee, Prediction-based strategies for energy saving in object tracking sensor networks, in *Proceedings—2004 IEEE International Conference on Mobile Data Management* (2004), pp. 346–357. https://doi.org/10.1109/mdm.2004.1263084
16. Y. Xu, K. Xu, J. Wan, Z. Xiong, Y. Li, Research on particle filter tracking method based on Kalman filter, in *Proceedings of 2018 2nd IEEE Advanced Information Management, Communicates, Electronic and Automation Control Conference, IMCEC 2018* (2018), (Imcec): 1564–1568. https://doi.org/10.1109/IMCEC.2018.8469578
17. H. Yang, B. Sikdar, A protocol for tracking mobile targets using sensor networks, in *Proceedings of the 1st IEEE International Workshop on Sensor Network Protocols and Applications, SNPA 2003* (2003), pp. 71–81. https://doi.org/10.1109/SNPA.2003.1203358

An Efficient Convolutional Neural Network for Acute Pain Recognition Using HRV Features

Saranya Devi Subramaniam and Brindha Dass

Abstract The pain of any individual can be felt only through experience. Therefore, clinicians are using self-report measures to assess acute pain after surgery. Thus, there is a need to predict and monitor acute pain effectively without self-report measures. In this study, Heart Rate Variability (HRV) from the ECG signal is used to objectively assess the acute pain. The BioVid Heat Pain Database is utilized to evaluate the acute pain recognition task. HRV features such as time domain, frequency domain, and nonlinear features are extracted and are utilized for this study. These features might be impacted with outliers, therefore, applied normal distribution and removed the outlines using standard deviation. After the data cleaning process, the best pain associated features are selected using One-way ANOVA and Post Hoc test. Finally, the pain level classification is performed using the modified Convolutional Neural Network (CNN) and compared its performance with other baseline systems such as ELM and SVM. It reveals that the proposed system outperforms the other baseline systems with 99.2% classification accuracy.

Keywords Biomedical signal processing · Learning systems · Feedforward neural networks · Statistical analysis

1 Introduction

Pain is a critical problem, where the healing rate after surgery is directly impacted. Acute pain is subjective, short lived, and complex in nature. Therefore, measuring acute pain has been always a challenging task. In spite of its influence, acute pain is often underestimated and untreated [6, 24]. From the literature, this can be proved that the pain is related to autonomic dysfunction [22, 23]. The Autonomic Nervous System (ANS) helps to sense the pain. Its action is directly controlled by the human brainstem. It consists of two sub-systems; they are Parasympathetic Nervous System (PNS) and Sympathetic Nervous System (SNS). The PNS is stimulated during rest

S. D. Subramaniam (✉) · B. Dass
Department of Biomedical Engineering, PSG College of Technology, Coimbatore 641004, India

F. Thakkar et al. (eds.), *Proceedings of the International e-Conference on Intelligent Systems and Signal Processing*, Advances in Intelligent Systems and Computing 1370,
https://doi.org/10.1007/978-981-16-2123-9_9

and the SNS is stimulated during stress. In hospitals, caretakers are using self-report measures such as Visual Analogue Scale (VAS), Faces Pain Scale-Revised (FPS-R), Numeric Rating Scale (NRS) to determine pain. But these measures are not useful as they are subjective and not stable. Therefore, an accurate and stable automated computer-based system is required to monitor patient's pain.

Thus, to develop a computer-based pain recognition system, many researchers have concentrated their research on physiological signals to determine and monitor acute pain [14]. In this study, HRV is extracted from ECG signal and chosen to develop an automated computer-based system to compute an acute pain index. HRV is the measure of time difference between two consecutive heart beats. It can measure the ANS activity indirectly and helps to maintain the balances between PNS and SNS activity [2, 4, 13].

1.1 Related Works

There are many research works related to classification of pain using physiological signals, especially using ECG signals and few research articles related to it are explained in the following passages. Bourguignon et al. [23] had proposed a study to understand the relationship between HRV and pain generated by rheumatoid arthritis. It shows that the HRV decreases when pain increases and the changes in HRV signal was mainly influenced by age. Chang et al. [14] had studied a relation between postoperative pain and HRV. In this study, the standard self-report measures such as the Visual Analogue Scale (VAS) and the short-form McGill pain questionnaire (SF-MPQ) are used to measure the human acute pain. The SF-MPQ could effectively measure postoperative pain when compared with the VAS and the LF/HF ratio feature of HRV correlates well with SF-MPQ.

Cho et al. [4] had developed a stress level measuring system using physiological signals such as peak characteristics of Electro Dermal Activity (EDA), Photoplethesmogram (PPG), and changes in the Skin Temperature (SKT). In this work, different levels of stress (baseline, mild stress, moderate stress, severe stress, and recovery) are generated using virtual reality device. For each stress level, physiological signals (PPG, EDA, and SKT) are captured and their respective pain related features are extracted. The most statistically significant features are selected using an ANOVA test. Finally, these extracted features are fed to the ELM classifier to classify the different levels of stress. This method has achieved 95% of classification accuracy.

De Jonckheere et al. [12] had used HRV analysis for pain measurement during a surgical procedure under general anesthesia. HRV analysis helps in optimized drug delivery and also acts as an indicator of stress. Sesay et al. [17] had focused on the relation between the acute Pain, HRV, and NRS. HRV is chosen for the study as it is the indirect measure of ANS activity. In this work, HRV power spectral features are extracted to predict the onset of pain and found optimal threshold values for LF feature as 298 ms2 and for LF/HF ratio feature as 3.12 with no appreciable change in HF values.

The time, frequency, and nonlinear domain features are the main features of HRV which describes the ANS activity directly [3, 16]. To extract these features, the 'time taken between two consecutive heart beats' is called as Normal-Normal (NN) Interval, which needs to be determined [8]. According to theories, pain is directly or indirectly impacting ANS activity.

Hence, the above mentioned features are extracted and are utilized for pain analysis. To select the best associated features of pain, feature selection method is applied.

In Schumann et al. [1] proposal, linear discriminant feature selection method was applied to classify different cardiovascular diseases. Time, frequency, and nonlinear domain parameters of HRV are utilized for the study. Out of these parameters, time domain features have given good result in discriminating cardiac diseases.

Wan-Hui et al. [25] had proposed an emotion recognition system to classify the Joy and Sad emotions using ECG signal. To extract the emotion related information from the ECG signal, its main features such as R peak, R-R interval, P-P interval, T-T interval, and QRS segment are to be extracted. All these extracted features might not contain emotion information. Hence, applied feature selection process uses Tabu search to select the best combination of ECG features for classification. Finally, the fisher projection classification algorithm is utilized to classify the data into different emotion clusters and achieved 82% of accuracy.

Even though good feature selection methods are used to attain the best results, until and unless the proper classifiers are taken into account; the best classification results cannot be achieved. The following passages discuss about the research articles which describe in detail about the classification algorithms that can be used for the pain classification task. Kim et al. [13] had developed an emotion recognition system using biosignals such as ECG, Skin Temperature (SKT), and EDA. The BIOPAC MP100 system was used to measure all the required signals. From the ECG signal, parameters like time, frequency, and nonlinear features are extracted. From EDA, SCR amplitude was chosen, and finally mean and maximum values of SKT are used for the study. The SVM pattern classifier was used to classify three emotion levels and achieved the classification accuracy of 78.4% for user dependent system and 61.8% for user independent system.

Zheng et al. [2] presented a work to explore the effect of negative emotions using ECG signal. In this study, 20 university students with a mean age of 24 years were involved. The ECG signal is measured for these subjects, while playing the video clips of negative emotions. Self-Assessment Manikin (SAM) is the assessment tool to analyze the response of the emotional response of individuals during data acquisition. To analyze the emotional response of individuals, HRV features (Approximate Entropy (ApEn), Mean R amplitude (MRA), Mean R-R interval (MRRI), SDNN, and RMSSD) are extracted from ECG signals. The authors have divided the work into two stages, one is to classify the collected data into positive emotions, negative emotions, and neutral and then further classifying the negative emotions into emotional stress and non-emotional stress. The Fuzzy C Means and Euclidean distance measure is used to visualize the distance between each class. It is found that HR parameter is the best feature in the classification of emotional stress.

Chu et al. [26] had developed an objective method of assessing pain. In this study, seven different electrical stimuli are used to induce pain on subjects and acquired physiological signals such as BVP, ECG, and SC during stimulation. The feature reduction was performed using a hybrid approach, i.e., combining fisher projection and PCA methods to select the best features of pain. Finally, linear discriminant analysis was used to quantify the intensity of pain. This algorithm has achieved 98% of classification accuracy.

1.2 Workflow

From the literature survey, it is clearly understood that all the features (time, frequency and nonlinear) of HRV are related to ANS activity and cannot be neglected. Therefore, all the features of HRV are extracted and are used for the classification. In this study, extracted features might be impacted with outliers. Therefore, the normal distribution is applied and removed the outliers using standard deviation. This process can be used to clean data for further process. To select the best features of pain, one of the filter-based method called one-way ANOVA and post hoc test is used. In this analysis, if the calculated F-value is higher, then there is a difference between the groups. Thus, the selected features are proving to be the statistically significant features ($P < 0.05$). Finally, classification is performed by using the proposed algorithm.

The Convolutional Neural Network is one of the popular deep learning architectures used to extract higher level features [7]. But in this study, the original CNN is modified by using some feature extraction techniques to achieve good classification accuracy with a smaller number of epochs [16]. Figure 1 represents the overall proposed work.

The BioVid Heat Pain Database is publicly available pain dataset. Hence, this database is used for the study. In this work, ECG signal is used to develop pain

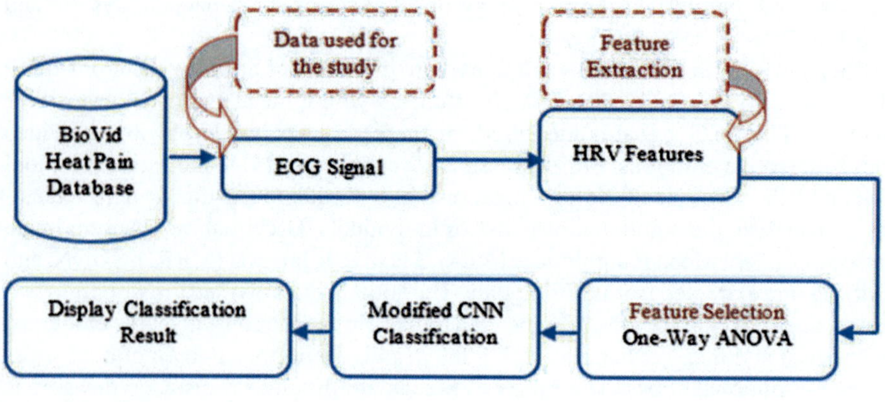

Fig. 1 Block diagram of the proposed work

recognition system. HRV features of ECG signal is used as the input for the pain recognition task. One-way ANOVA is used as the feature selection method to choose required pain information task. Here, modified shallow CNN algorithm is used to perform pain recognition task.

2 A Pain Recognition System

Pain can be measured using physiological signals like ECG, EMG, EEG, SC, PPG, etc. In this work, HRV from ECG signal is considered to develop a human acute pain recognition system [10].

2.1 Dataset Used

The BioVid Heat Pain Database (BVDB) [21] is a freely available multimodal acute pain dataset, collected in collaboration with the Medical Psychology group of the University of Ulm and the Neuro-Information Technology group of the University of Magdeburg. This database consists of both bio-potential signals (Galvanic Skin Response (GSR), ECG, and EMG at Trapezius muscle) and frontal video signals. It has been developed by subjecting 4 different levels of heat stimulus (BL1 (Base Line—No pain, PA1 (Pain Level 1—Mild Pain, PA2 (Pain Level 2)—Moderate pain, PA3 (Pain Level 3)—Severe pain, PA4 (Pain Level 4)—Very severe pain) to 90 participants (with different age groups) under a controlled environment. Each level of heat stimuli is administered to subjects 20 times in a random manner for maximum of 4 s and given resting time of 8–12 s between each stimulus. In this paper, fourteen subjects from BVDB (Part-A) is chosen for the study. Every subject has five different level of pain with 100 recordings of both video and biosignals with proper VAS labeling.

2.2 Feature Extraction

As mentioned in the outline of the work, time domain, frequency domain, and nonlinear features of HRV are extracted for the study [7, 10, 15, 20, 27,]. All these features can be calculated with the help of NN interval. NN interval is measured from the ECG signal using Pan Tompkins peak detection algorithm [9]. With the help of this parameter, time domain features such as meanNN (ms)—mean of NN, minNN (ms)—minimum of NN, maxNN (ms)—maximum of NN, RMSSD (ms)—Root Mean Square of Successive Differences, SDNN (ms)—Standard Deviation of NN Interval, pNN50 (%)—Number of pairs of adjacent NN intervals differing by more than 50 ms divided by the total number of all NN intervals are extracted.

These features quantify the changes in the NN interval and detect the complete change of the heartbeat in the time domain. Frequency domain features such as TP—Total Power, LF (ms^2)—Power in the low frequency range (LFR) (0.04–0.15 Hz), HF (ms^2)—Power in high frequency range (HFR) (0.15–0.4 Hz), and LF/HF—Ratio between LF and HF range powers are extracted from the power spectrum density prepared by using Lomb spectrum modeling [18]. In general, the sympathetic and parasympathetic branches of the ANS have changing time constants. But the frequency components like the HF—the measure of parasympathetic activity, LF—reflects both sympathetic and parasympathetic activity, LF/HF ratio—when it is high, it reflects the dominance of sympathetic activity; when it is low, it reflects the dominance of parasympathetic activity are directly reflecting the activity of the ANS. Lastly, nonlinear features such as SD1 (ms)—Standard deviation of the Poincare plot perpendicular to the line-of-identity, SD2 (ms)—Standard deviation of the Poincare plot along the line-of-identity are extracted. This Poincaré analysis is performed to measure the self-similarity and evaluates the dynamics of a system [19].

2.3 *Feature Reduction*

All the extracted HRV features might contain outliers, so the given data are applied to the normal distribution and removed the outliers using standard deviation (i.e., by applying the concept of removing the data outside mean ±3SD). After this cleaning process, the resulted data is utilized as the input for feature selection and classification processes directly. The next step is to select the best pain associated features and this can be achieved using one-way ANOVA test and post hoc test [4, 11, 21]. One-way ANOVA test is capable of analyzing whether the mean value difference between the two or more independent groups is statistically significant ($P < 0.05$) or not. This can be confirmed by analyzing the F-value. The higher the F-value indicates that the groups can be well distinguished. The use of one-way ANOVA test will be difficult to select the accurate hypothesis due to a multiple comparison problem. Therefore, post hoc test step is followed as a prerequisite for performing one-way ANOVA test. Finally, after applying these tests, four features were selected as statistically significant features they are average HR, minNN, SD2, and mean NN. Table 1 provides the F values for all the statistically significant features.

Table 1 One-way ANOVA test results

S. No	Parameter	F-value
1	Average HR	7.23
2	minNN	6.99
3	SD2	5.55
4	meanNN	5.23

2.4 *A Pain Recognition System Using Modified CNN Approach*

The Convolutional neural network (CNN) is one type of artificial neural network, mainly designed to minimize the data in the preprocessing stage. It recognizes patterns effectively even with diverse variations, geometric distortions, and with the transformations. The CNN architecture follows three concepts such as a local receptive field, shared weights, and spatial sub sampling to easily recognize the inputs with different degree of shift, scale, and distortion [1].

The proposed method is slightly modified by adding features of HRV in the input layer. This modified One-Dimensional CNN (1-D CNN) algorithm consists of convolutional layer, flatten layer, and an output layer. The input matrix to the CNN layer consists of 1280 records with 4 HRV features. With this input matrix, convolutional layer performs feature extraction process and sends the complex vector to the flatten layer. This layer converts the output of the convolutional layer to a 1-D feature vector. Finally, the output layer is connected to the last convolutional layer, to perform classification. In this layer, the huge feature vector will be reduced to a vector of five as there are five classes. This is achieved by using a softmax function in the output layer. Figure 2 represents the proposed 1-D CNN network. Compared to the conventional 1-D CNN method, the modified 1-D CNN approach learns the features faster and performs classification with better accuracy.

3 Results and Discussion

In this study, cleaned data is fed to one-way ANOVA test and found average HR, minNN, SD2, and meanNN are statistically significant features for the study ($P < 0.05$). All these features were expressed in terms of mean ± SD. From this analysis, it is found that, the average HR was increased from 72.10 ± 8.13 to 77.03 ± 8.29 as pain level increases, the minNN was reduced from 0.791642 ± 0.101613 to 0.731018 ± 0.083969 as pain level increases, the SD2 was increased from 0.039143 ± 0.024886 to 0.052209 ± 0.027664 as pain level increases, the meanNN was reduced from 0.838229 ± 0.095671 to 0.791135 ± 0.084744 as pain level increases. Before providing the data to the classifier, it is better to remove the skewness if found from the feature data. To achieve this, log transformation is applied and finally used this data for classification.

Fig. 2 The proposed convolutional neural network

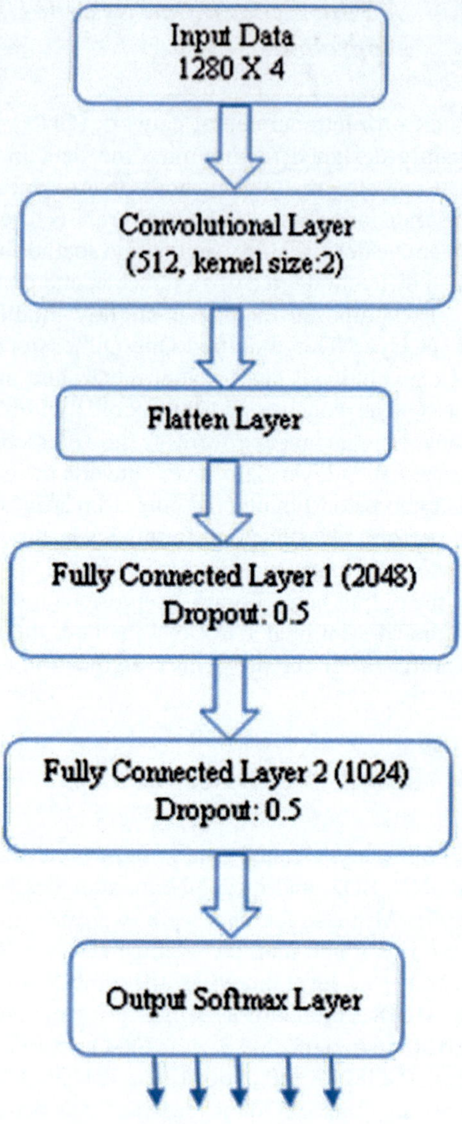

3.1 Computational Speed

In this study, windows server 2012R2 is used to develop python-based machine learning algorithm to assess pain. CNN algorithm has taken 31.8642 s to complete the classification task.

3.2 Comparison with Baseline Algorithms

The methods that are presented in [4] and [5] utilize ELM and SVM as classifiers to perform stress and emotion level classification. Therefore, the same algorithms are used as baseline algorithms to test the performance of the modified CNN approach.

Compared to the baseline algorithms, modified CNN has achieved better classification accuracy of 99.2%.

Tables 2 and 3 shows the classification metrics for the proposed method in comparison with the other existing methods. The performance of the classification algorithms, tests on 285 samples.

The performance of the SVM, ELM, and modified CNN algorithms are evaluated using confusion matrix (Figs. 3, 4, and 5—confusion matrix results of all algorithms,

Table 2 Classification metrics for test dataset

S. No	Classification algorithm	Classification error	MSE	MAE
1	SVM	0.0245	0.0245	0.0245
2	ELM	0.0350	0.0350	0.0350
3	Modified CNN	0.0070	0.0070	0.0070

Table 3 Classification accuracy for test dataset

S. No	Classification algorithm	Training samples	Testing samples	Accuracy (%)
1	SVM	995	285	97.54
2	ELM	995	285	96.49
3	Modified CNN	995	285	99.20

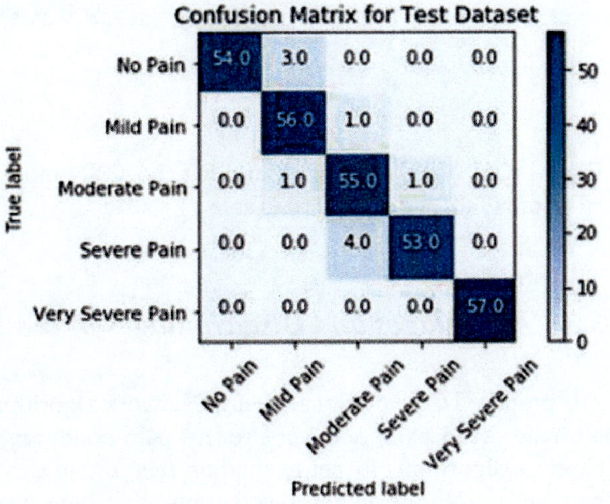

Fig. 3 Confusion matrix—SVM algorithm

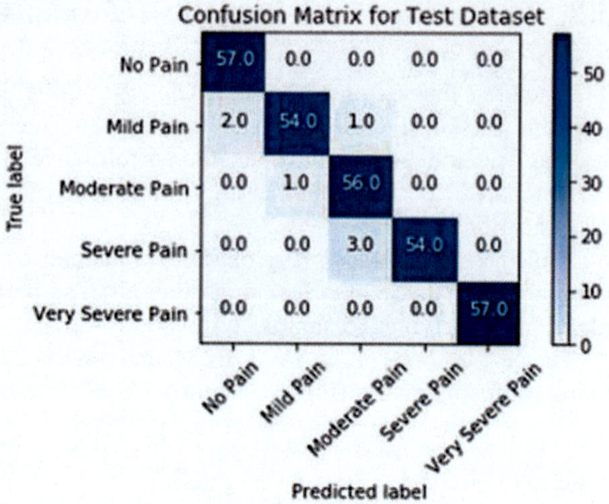

Fig. 4 Confusion matrix—ELM algorithm

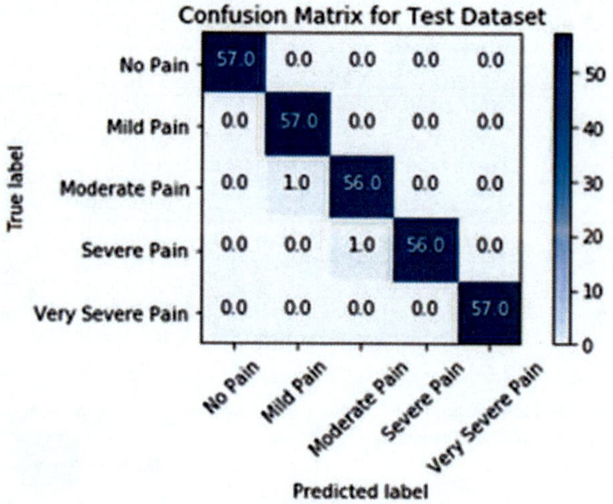

Fig. 5 Confusion matrix—modified CNN algorithm

Fig. 6—CNN model accuracy and Fig. 7—CNN model loss), precision, Recall, and F1-score (Tables 4, 5, and 6) (Table 7).

3.3 Advantages and Disadvantages of the Proposed System

The proposed Convolutional Neural Network algorithm can effectively differentiate No pain, Mild pain, and Very severe pain conditions when compared with other baseline algorithms as per evaluation metrics. In this study, the selection of CNN parameters (No. of hidden layers, optimizers and epochs) is important to achieve

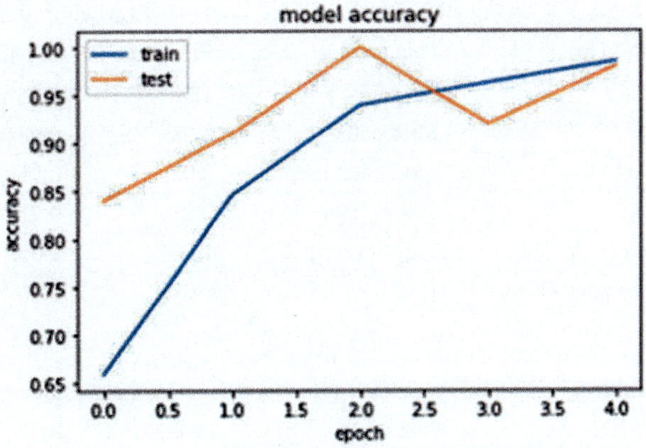

Fig. 6 Model accuracy—proposed CNN algorithm

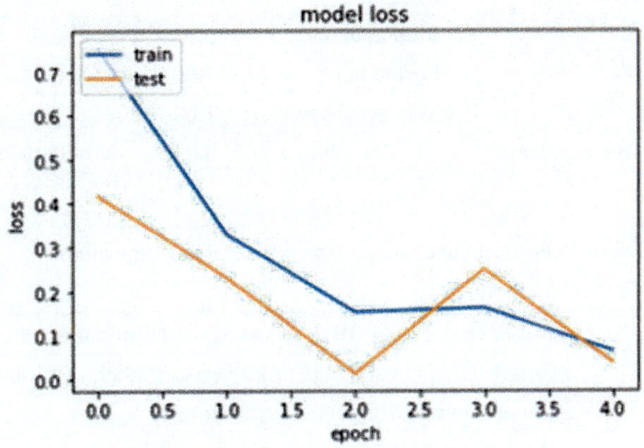

Fig. 7 Model loss—proposed CNN algorithm

Table 4 Classification metrics for test dataset—SVM algorithm

S. No	Pain level	Precision	Recall	F1-score	Test data
1	No pain	0.97	1.00	0.98	57
2	Mild pain	0.98	0.95	0.96	57
3	Moderate pain	0.93	0.98	0.96	57
4	Severe pain	1.00	0.95	0.97	57
5	Very severe pain	1.00	1.00	1.00	57
Average/Total		0.98	0.98	0.98	285

good results. But selection of these parameters is based on trial and error, as there is no definite solution. Therefore, CNN parameters and sample size are to be properly identified to achieve the best classification result. The performance of the proposed algorithm should be tested with the real-time data. Since, deep learning network is

Table 5 Classification metrics for test dataset—ELM algorithm

S. No	Pain level	Precision	Recall	F1-score	Test data
1	No pain	1.00	0.95	0.97	57
2	Mild pain	0.93	0.98	0.96	57
3	Moderate pain	0.92	0.96	0.94	57
4	Severe pain	0.98	0.93	0.95	57
5	Very severe pain	1.00	1.00	1.00	57
Average/Total		0.97	0.96	0.96	285

Table 6 Classification metrics for test data—CNN algorithm

S. No	Pain level	Precision	Recall	F1-score	Test data
1	No pain	1.00	1.00	1.00	57
2	Mild pain	0.98	1.00	0.99	57
3	Moderate pain	0.98	0.98	0.98	57
4	Severe pain	1.00	0.98	0.99	57
5	Very severe pain	1.00	1.00	1.00	57
Average/Total		0.99	0.99	0.99	285

Table 7 Performance comparison with existing algorithms and the proposed algorithm

Sl. No	Method	ECG (%)
1	Walter et al. (2014) BL1 versus others (Without feature extraction)	43.05
2	Patrick Thiam et al. (2019) BL1 versus (Others) (Without Feature Extraction)	36.54
3	Our approach (with feature extraction)	99.20
4	Our approach SVM (with feature extraction)	97.54
5	Our approach ELM (with feature extraction)	96.49

used, its complexity is unknown. Therefore, it has to be remodeled using eigenvalues and eigenvectors.

4 Conclusion

This work proposes an HRV based objective method to classify five different acute pain levels. Here, HRV features such as time, frequency, and nonlinear features are extracted from ECG signal and used for classification. In this study, the original CNN algorithm is modified by using feature extraction techniques to achieve the best classification accuracy of 99.20%. Hence, it is clear that the proposed CNN classification algorithm can be used for the measurement of pain index from HRV. It is known that the HRV is influenced by a great variety of factors such as age, sex, or medication intake. Therefore, the future scope of the study may be implementing the same study by considering these factors to improve both the onset and monitoring of postoperative pain.

Acknowledgements We are very grateful to the Department of Science and Technology (DST), Govt. Of India, for funding our research work and we thank them for the same. We would like to thank the Department of Biomedical Engineering, PSG college of Technology, Coimbatore – 04 for their guidance and support for the completion of this attempt. We would also like to express our sincere thankfulness to PSGIMSR, for their guidance and help for the ethical clearance approval and collecting database from PSG Hospital, Coimbatore-04.

Conflict of Interest There are no conflicts of interest.

References

1. A. Schumann, N. Wessel, A. Schirdewan, K.J. Osterziel, Potential of feature selection methods in heart rate variability analysis for the classification of different cardiovascular diseases. Stat. Med. **21**, 2225–2242 (2002) [Online]. https://doi.org/10.1002/sim.979
2. B.S. Zheng, M. Murugappan, S. Yaacob, FCM clustering of emotional stress using ECG features, in *2013 International Conference on Communication and Signal Processing*, (2013), pp. 305–309 [Online]. https://doi.org/10.1109/iccsp.2013.6577064
3. C. Broucqsault-Dédrie, J. De Jonckheere, M. Jeanne, S. Nseir, Measurement of heart rate variability to assess pain in sedated critically ill patients: a prospective observational study. PLoS One **11**(1), 1–11 (2016) [Online]. https://doi.org/10.1371/journal.pone.0147720
4. D. Cho, J. Ham, J. Oh, J. Park, S. Kim, N.-K. Lee, B. Lee, Detection of stress levels from biosignals measured in virtual reality environments using a kernel-based extreme learning machine. Sensors **17**(10), 1–18 (2017) [Online]. https://doi.org/10.3390/s17102435
5. E.-H. Jang, B.-J. Park, S.-H. Kim, Y. Eum, J. Sohn, Identification of the optimal emotion recognition algorithm using physiological signals, in *Proceedings of the 2011 2nd International Conference on Engineering and Industries (ICEI)* (2011), pp. 1–6
6. H.E. Steffen Walter, S. Gruss, The biovid heat pain database data for the advancement and systematic validation of an automated pain recognition system, in *IEEE International Conference on Cybernetics (CYBCONF)* (2013), pp. 128–131 [Online]. https://doi.org/10.1109/CYBConf.2013.6617456
7. I. Bornoiu, O. Grigore, Kohonen neural network stress detection using only electrodermal activity features. Adv. Electr. Comput. Eng. **14**(3), 71–78 (2014) [Online]. https://doi.org/10.4316/AECE.2014.03009
8. J. Koenig, M.N. Jarczok, R.J. Ellis, T.K. Hillecke, J.F. Thayer, Heart rate variability and experimentally induced pain in healthy adults: a systematic review. Eur. J. Pain. **18**(3), 301–314 (2014) [Online]. https://doi.org/10.1002/j.1532-2149.2013.00379.x
9. J. Pan, J. Willis, A real-time QRS detection algorithm. IEEE Trans. Biomed. Eng. **32**(3), 230–236 (1985) [Online]. https://doi.org/10.1109/TBME.1985.325532
10. J. Ye, K. Lee, J. Lin, C. Chuang, Observing continuous change in heart rate variability and photoplethysmography-derived parameters during the process of pain production/relief with thermal stimuli. J. Pain Res. **10**, 527–533 (2017) [Online]. https://doi.org/10.2147/JPR.S129287
11. J. Zhang, D. Dean, D. Nosco, D. Strathopulos, M. Floros, Effect of chiropractic care on heart rate variability and pain in a multisite clinical study. Manip. Physiol. Ther. **29**(4), 267–274 (2006) [Online]. https://doi.org/10.1016/j.jmpt.2006.03.010

12. J. De Jonckheere, D. Rommel, J.L. Nandrino, M. Jeanne, R. Logier, Heart rate variability analysis as an index of emotion regulation processes: interest of the Analgesia Nociception Index (ANI), in *Proceedings of the 34th Annual International Conference of the IEEE Engineering in Medicine and Biology Society*, pp. 3432–3435 (2012) [Online]. https://doi.org/10.1109/EMBC.2012.6346703
13. K.H. Kim, S.W. Bang, S.R. Kim, Emotion recognition system using short-term monitoring of physiological signals. Med. Biol. Eng. Comput. **42**(3), 419–427 (2004) [Online]. https://doi.org/10.1007/BF02344719
14. L. Chang, T. Ma, S. Tsay, G. Jong, Relationships between pain intensity and heart rate variability in patients after abdominal surgery: a pilot study. Chin. Med. J. (Engl). **125**(11), 1964–1969 (2012)
15. M. Kächele, P. Thiam, M. Amirian, F. Schwenker, G. Palm, Methods for person-centered continuous pain intensity assessment from bio-physiological channels. IEEE J. Sel. Top. Signal Process. **10**(5), 854–864 (2016) [Online]. https://doi.org/10.1109/JSTSP.2016.2535962
16. M. Mega, D. Indra, A. Murni, Cattle race classification using gray level co-occurrence matrix convolutional neural networks. Procedia Comput. Sci. **59**, 493–502 (2015) [Online]. https://doi.org/10.1016/j.procs.2015.07.525
17. M. Sesay, et al., Responses of heart rate variability to acute pain after minor spinal surgery: optimal thresholds and correlation with the numeric rating scale. J. Neurosurg. Anesthesiol. **27**(2), 148–154 (2015) [Online]. https://doi.org/10.1097/ANA.0000000000000102
18. N.R. Lomb, Least-squares frequency analysis of unequally spaced data. Astrophys. Space Sci. **39**(2), 447–462 (1976) [Online]. https://doi.org/10.1007/BF00648343
19. P. Melillo, M. Bracale, L. Pecchia, Nonlinear heart rate variability features for real-life stress detection. Case study: students under stress due to university examination. Biomed. Eng. Online **10**(96), 1–13 (2011) [Online]. https://doi.org/10.1186/1475-925X-10-96
20. S. Zulj, R. Magjarevic, D. Miklavcic, T. Jarm, Matlab-based tool for ECG and HRV analysis. Biomed. Signal Process. Control **10**, 108–116 (2014) [Online]. https://doi.org/10.1016/j.bspc.2014.01.011
21. S.D. Subramaniam, B. Dass, Automated nociceptive pain assessment using physiological signals and a hybrid deep learning network. IEEE Sens. J. (2020) [Online]. https://doi.org/10.1109/JSEN.2020.3023656
22. S.D. Subramaniam, B. Doss, L.D. Chanderasekar, A. Madhavan, A.M. Rosary, Scope of physiological and behavioural pain assessment techniques in children—a review. Healthc. Technol. Lett. **5**, 124–129 (2018)
23. V.B.C. Bourguignon, D. Taibi, A. Taylor, D. Lake, Pain, heart rate variability, and inflammatory indicators in women with rheumatoid arthritis. J. Pain **6**(3), 24 (2005) [Online]. https://doi.org/10.1016/j.jpain.2005.01.094
24. W. Aziz, F.S. Schlindwein, M. Wailoo, T. Biala, Heart rate variability analysis of normal and growth restricted children. Clin. Auto. Res. **22**, 91–97 (2012) [Online]. https://doi.org/10.1007/s10286-011-0149-z
25. W. Wan-hui, Q. Yu-Hui, L. Guang-Yuan, Electrocardiography recording, feature extraction and classification for emotion recognition, in *Proceedings of the 2009 WRI World Congress on Computer Science and Information Engineering* (2009), pp. 168–172 [Online]. https://doi.org/10.1109/CSIE.2009.130
26. Y. Chu, X. Zhao, J. Han, Y. Su, Physiological signal-based method for measurement of pain intensity. Front. Neurosci. **11**, 1–13 (2017) [Online]. https://doi.org/10.3389/fnins.2017.00279
27. J. Zhai, A. Barreto, Stress detection in computer users through non-invasive monitoring of physiological signals. Biomed. Sci. Instrum. **42**, 495–500 (2006)

Design and Development of LSTM-RNN Model for the Prediction of RR Intervals in ECG Signals

B. Dhananjay, N. Prasanna Venkatesh, Arya Bhardwaj, and J. Sivaraman

Abstract Traditional machine learning models are common to classify Electrocardiogram (ECG) signals. In comparison, ECG prediction with such models is rare. The proposed work aims to predict the RR Interval (RRI) for a 10 sec. Normal Sinus Rhythm (NSR) ECG signals by designing and developing a Long Short Term Memory (LSTM)-Recurrent Neural Network (RNN) based model. The ECG signals from the Chapman University and Shaoxing People's hospital have 10 sec. of sample data of each patient used in this study. Signal processing techniques to detect the R-peaks and RRI were calculated further and provided as the model's input. LSTM-RNN model based on univariate time series predicted the RRI for the subsequent 20 ECG cycles. The model underwent 500 epochs of training before the accurate ECG prediction. The model's error value started at 0.8 and ended at 0.000001, which provided a good fit for the predicted data. The LSTM-RNN model predicted the RRI from the 10th to the 30th ECG cycle based on 9 RRI. The model's error rate decreased exponentially as the epoch's number increased, stating the model's accuracy, helping to determine the intra-individual Heart Rate Variability (HRV).

Keywords Heart rate variability · Long short term memory · Recurrent neural network · RR interval · Univariate time series

1 Introduction

Electrocardiogram (ECG) records the heart's electrical activity generated by cardiac myocytes to diagnose and detect cardiac abnormalities. Early detection and ECG signal classification are essential for patients' clinical care to decrease heart disease mortality rates. Type and prediction of cardiac myopotential signals using machine learning techniques can substantially influence physicians in clinical diagnosis [1]. Cardiac anomalies lead to fatal diseases by affecting blood vessel functionality and

B. Dhananjay · N. P. Venkatesh · A. Bhardwaj · J. Sivaraman (✉)
Bio-signals and Medical Instrumentation Laboratory, Department of Biotechnology and Medical Engineering, National Institute of Technology, Rourkela, Odisha, India
e-mail: jsiva@nitrkl.ac.in

F. Thakkar et al. (eds.), *Proceedings of the International e-Conference on Intelligent Systems and Signal Processing*, Advances in Intelligent Systems and Computing 1370,
https://doi.org/10.1007/978-981-16-2123-9_10

cause Coronary Artery Disease (CAD) in patients, especially in adults and older adults, and are considered a prominent cause of mortality worldwide. World Health Organization statistics, in 2016, published that heart diseases led to about 18 million deaths. Therefore, early heart disease prediction effectively prevents cardiac patients before a heart attack or stroke [2]. Heart Rate Variability (HRV) acts as a measure of non-invasive and parasympathetic activity, indicating a possibility of sudden cardiac death [3]. Various machine learning techniques help to find and diagnose different heart abnormalities across human beings and are classified based on multiple methods like K-Nearest Neighbor algorithm (KNN), Decision Tree (DT), Genetic Algorithm (GA), and Naive Bayes Classifier (NB) [4]. Few studies [5–12] have also predicted the ECG signal by developing pure mathematical models. The Convolutional Neural Network (CNN) helps diagnose CAD using two and five seconds durations of ECG signal segments [13].

Yildirim et al. [14] took 1000 patients' ECG data, and they fragmented the ECG signals of 1 min each to 10 sec to classify different cardiac arrhythmias by developing a 1D CNN machine learning model. Mathews et al. [15] developed a deep machine learning model with a combination of Restricted Boltzmann Machine (RBM) and Deep Belief Networks (DBN) to classify ventricular and supraventricular heartbeats. The accuracy of ventricular and supraventricular ectopic beats classification was 93.63% and 95.57%, respectively [15]. Hannun et al. [16] developed a Deep Neural Network (DNN) to classify 12 rhythms observed in ECG from a clinical dataset of 91,232 single-lead ECG recordings and a total of 53,877 patients. Rahhal et al. [17] proposed an unsupervised DNN model to classify ECG signals based on Stacked Denoising Auto Encoders (SDAE). Sannino et al. [18] developed DNN to detect heartbeat type, and the model is composed of 7 hidden layers. To classify the ECG signal based on heartbeat, Acharya et al. [19] designed and developed a CNN model of 9 layers with 94% accuracy. Andersen et al. [20] proposed a deep learning model with a combination of CNN and Recurrent Neural Network (RNN) to classify and detect atrial fibrillation in real-time with the help of publicly available databases.

The current research trend focuses less on observational study and more on computational analysis. Hence, signal processing and machine learning techniques help in the automated interpretation of the ECG signal and cardiovascular disease classification. This work proposes a LSTM-RNN based Univariate time series analysis model to predict the RR Interval (RRI) of the impending heart rate based on the preceding available RRI (ms). The prediction helps to analyze the performance of intra-individual variations in the HRV.

2 Methodology

2.1 *ECG Signal and Database*

The ECG signal considered in this work is Chapman University and Shaoxing People's hospital database having standard 12-lead ECG signals. The database has 10,646 ECGs of different patients, and all the signals sampled at 500 Hz. The length of ECG recording available for each patient in the database is 10 sec. The main focus was on the Lead-II ECG signal for the proposed work. Figure 1 depicts the input of a Normal Sinus Rhythm (NSR) volunteers ECG signal downloaded from the database.

2.2 *Signal Processing*

The downloaded ECG signal underwent a signal processing technique to remove the noise by developing a 10 point window. After signal averaging, R-peak detection was performed on the ECG signal to calculate the RRI. The representation of database ECG signals shown in amplitude and time domain in microvolts (μV) and milliseconds (ms). Figure 2a illustrates the raw NSR ECG signal, and Fig. 2b depicts the NSR ECG signal after filtering and detecting R-peaks.

2.3 *Implementation*

The principle behind the design and development of a LSTM-RNN model is based on univariate time series forecasting. As the downloaded ECG signal sample size is 10 sec, the observed and recorded, RRI's were nine samples. The input time series to the LSTM-RNN model was practical and recorded RRI, and Eq. 1 represents the model's input. The designed algorithm sets the number of time steps to be considered to 5 for the organized and developed model to predict the current output, as shown in Eq. 2. As from the downloaded ECG signal, the observed and recorded RRI's were

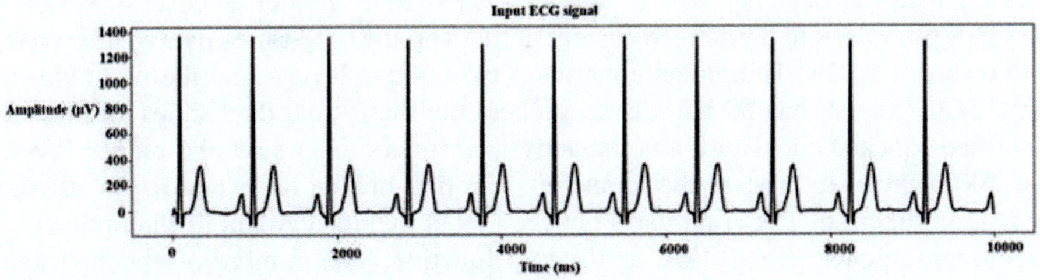

Fig. 1 Input ECG signal of a NSR volunteer

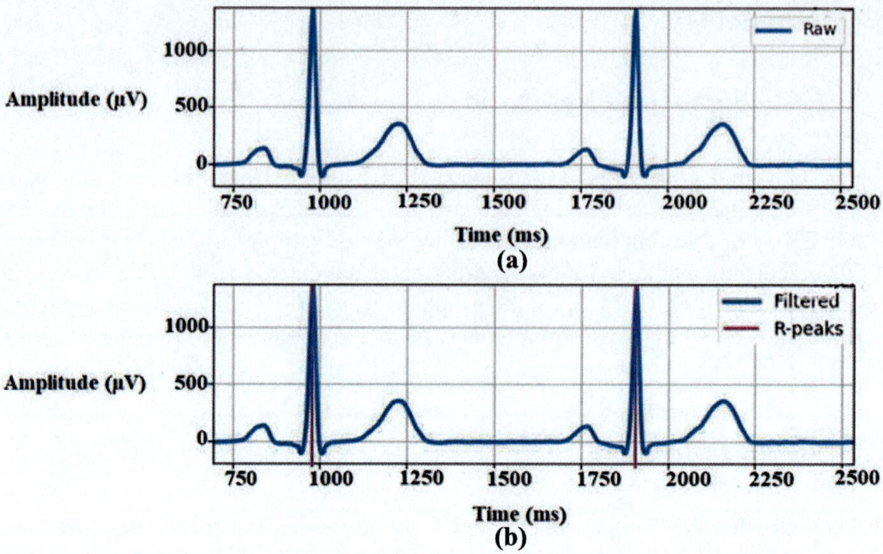

Fig. 2 **a** Raw ECG signal of a NSR volunteer. **b** Filtered and R-peaks detected ECG signal of a NSR volunteer

9. In this proposed study, the model predicted the next 20 RRI's, i.e., from the 10th ECG cycle to the 30th ECG cycle, by the LSTM-RNN model.

$$Input = (RRI_{x-9}, RRI_{x-8}, RRI_{x-7}, RRI_{x-6}, RRI_{x-5}, RRI_{x-4}, RRI_{x-3}, RRI_{x-2}, RRI_{x-1}) \quad (1)$$

$$Output_x = (RRI_{x-5}, RRI_{x-4}, RRI_{x-3}, RRI_{x-2}, RRI_{x-1})LSTM - RNNmodel \quad (2)$$

2.4 LSTM-RNN Model Building

The input to the designed and developed LSTM-RNN model consists of several previous time steps to predict the current output and the time series data of observed and recorded RRI's. The model consists of two hidden layers, and the first hidden layer of the model has 50 neurons to process the input time data series and has a Rectified Linear Unit (ReLU) as the activation function. The second hidden layer has 100 neurons to process the input from the first hidden layer and ReLU as the activation function. The compilation of the model included Adam as the optimizer and Mean Square Error (MSE) as the loss function. The number of epochs used to validate the model was 500. For every period, the MSE calculated the loss rate

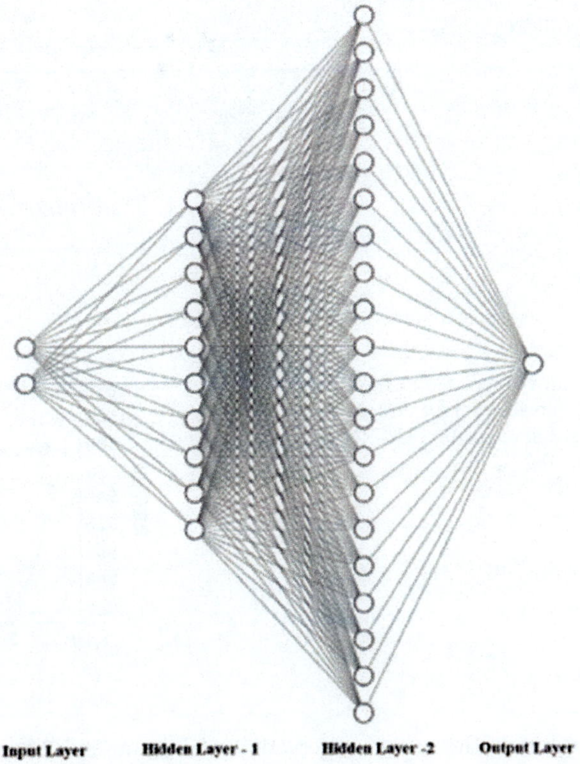

Fig. 3 The designed and developed LSTM-RNN model

of the model. Equation 3 represents the working of the ReLU activation function. Equation 4 showcases Adam optimizer's principle of updating weights in the LSTM-RNN model [21]. Figure 3 shows the LSTM-RNN model for predicting the RRI from the 10th ECG cycle to the 30th ECG cycle.

$$\varnothing(z) = \max(0, z) \tag{3}$$

$$Predicted\ weight = Previous\ weight - Step\ size * Estimator \tag{4}$$

3 Results

Figure 4 shows the heart rate plot of the NSR ECG signal considered in the study. The plot's x-axis represents the time in seconds (s), and the y-axis represents the heart rate in beats per minute (bpm). The figure displays the heart rate observed in a 10 sec NSR ECG signal along with RRI (ms) calculated for each time interval. The

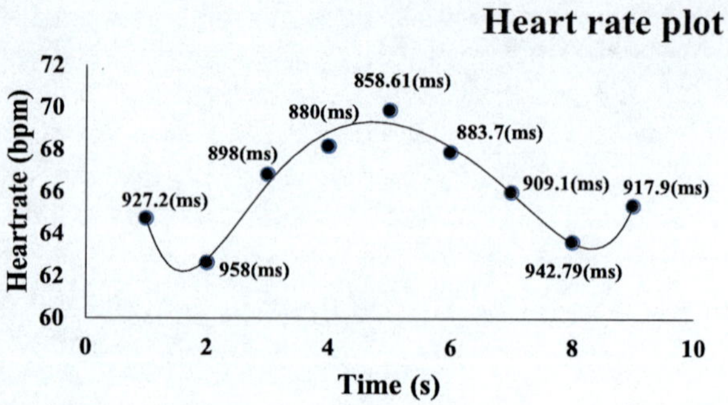

Fig. 4 Heart rate plot of the NSR ECG signal

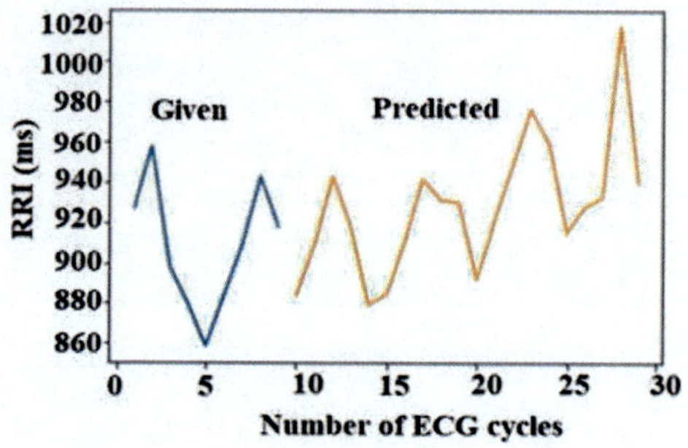

Fig. 5 Predicted ECG RR intervals from the designed and developed LSTM-RNN model

minimum heart rate observed was 62 bpm, and the maximum heart rate observed was 70 bpm. Heart rate calculation was based on the observed and calculated from the NSR ECG signal.

Figure 5 represented the given input RRI's (ms) and predicted output RRI's (ms) of the model. The graph's x-axis denotes the number of ECG cycles, and the y-axis of the diagram depicts the RRI's. The LSTM-RNN model input was RRI of the first 9 ECG cycles, and the model predicted the RRI for the next 20 ECG cycles. Figure 6 showcases the error rate given by the model. The x-axis displays the number of epochs and the y-axis represents the Error values. Figure 6a shows the model's error rate for 0 to 250 and Fig. 6b depicts the error rate of the model for the epochs 251 to 500. The error rate is continually decreasing in both figures.

4 Discussion

Chauhan et al. [22] developed a LSTM model to identify the ECG signal's anomaly. The developed model [22] is a two-stage pipeline model where step 1 represents the LSTM network, and step 2 represents LR, MLP, and SVM. In the proposed work, the input to the designed and developed LSTM-RNN model consists of several previous time steps to predict the current output and the time series data of observed

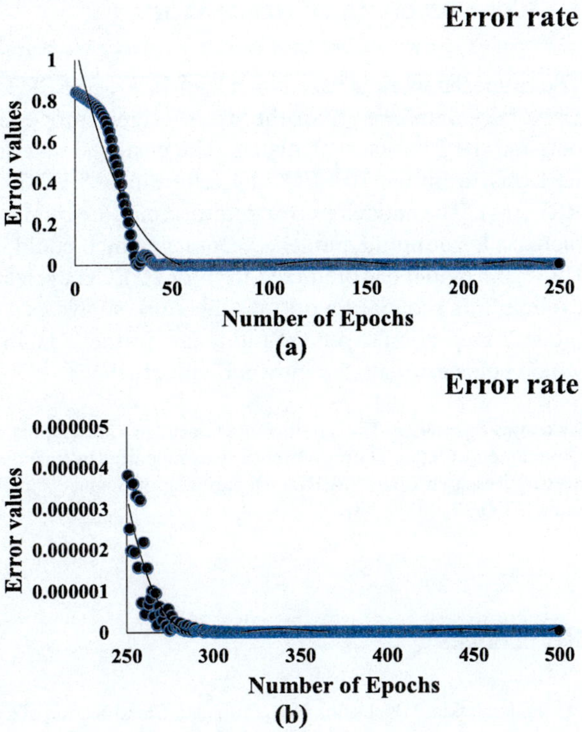

Fig. 6 **a** The error rate of the model from 0 to 250 Epochs. **b** The error rate of the model from 250 to 500 Epochs

and recorded RRI's. The model consists of two hidden layers, and the first hidden layer of the model has 50 neurons to process the input time data series and has a ReLU as the activation function. The second hidden layer has 100 neurons to process the input from the first hidden layer and ReLU as the activation function. Raghunath et al. [23] proposed a DNN to predict a patient's mortality from 12-lead ECG recordings, whereas in this work, a deep learning based LSTM-RNN model is proposed and to predict a NSR volunteer's intra-individual HRV from Lead–II ECG recordings. Bhavya et al. [24], in their work, concluded that LSTM/ Bi- LSTM is the type of model excessively used to predict the time series data in the healthcare sector with an average accuracy of 0.85. In the proposed work, the accuracy of the developed LSTM-RNN model is 0.99. The developed model's error rate decreased exponentially as the epoch's number increased, stating the model's accuracy, which could help determine the intra-individual HRV. Gao et al. [25] proposed a new deep learning model consisting of CNN to predict the multivariate time series model in their work, considering ECG signals, galvanic skin response, and respiration rate. In the proposed work, a deep learning-based model LSTM-RNN model predicts univariate time series RRI. Su et al. [26] developed the LSTM-RNN model to predict arterial blood pressure where root mean square error was used to calculate the error rate, and the model's accuracy is 0.95. The proposed work calculated the error rate based on MSE and the developed model's accuracy is 0.99.

5 Conclusion and Future Work

The proposed work is based on a lead-II 10 sec NSR ECG signal from the Chapman University database. With the use of signal processing, noise removal is carried out, and the RRI was calculated. The designed and developed model predicted the RRI (ms) from the 10th ECG cycle to the 30th ECG cycle based on the previous 9 RRI (ms). The model's error rate decreased exponentially as the epoch's number increased, stating the model's accuracy, which could determine the intra-individual HRV. The model has predicted RRI for 20 ECG cycles from the 10th to the 30th ECG cycle in this proposed work. As the work showcased is a preliminary analysis, the model's prediction to predict future and further RRI in ECG cycles can be extended, which helps estimate the intra-individual HRV.

Acknowledgements The authors acknowledge the support from the Ministry of Education, Government of India. The present study was supported by financial grants from the Science and Engineering Research Board (SERB), Department of Science and Technology (DST), Government of India (EEQ/2019/000148).

References

1. H. Jambukia, V.K. Dabhi, H.B. Prajapati, Classification of ECG signals using machine learning techniques: a survey. ICACEA **2015**, 714–721 (2015). https://doi.org/10.1109/ICACEA.2015.7164783
2. H. Ahmed, E.M.G. Younis, A. Hendawi, A.A. Ali, Heart disease identification from patients social posts, machine learning solution on spark. Futur. Gener. Comput. Syst. **111**, 714–722 (2020). https://doi.org/10.1016/j.future.2019.09.056
3. F. Azuaje, Predicting coronary disease risk based on short-term RR interval measurements: a neural network approach. Artif. Intell. Med. **15**(3), 275–297 (1999). https://doi.org/10.1016/S0933-3657(98)00058-X
4. N.M. Lutimath, C. Chethan, B.S. Pol, Prediction of heart disease using machine learning. Int. J. Recent. Technol. Eng. **8**(2), 474–477 (2019). https://doi.org/10.35940/ijrte.B1081.0982S1019
5. J. Sivaraman, G. Uma, S. Venkatesan, M. Umapathy, K.N. Keshav, A study on atrial Ta wave morphology in healthy subjects: an approach using P wave signal-averaging method. J. Med. Imaging Health Inform. **4**(5), 675–680 (2014). https://doi.org/10.1166/jmihi.2014.1306
6. J. Sivaraman, S. Venkatesan, R. Periyasamy, J. Joseph, Modified limb lead ECG system effects on electrocardiographic wave amplitudes and frontal plane axis in sinus rhythm subjects. Anatol. J. Cardiol. **17**(1), 46–54 (2017). https://doi.org/10.14744/AnatolJCardiol.2016.6843
7. J. Sivaraman, R. John, Effects of sinus rhythm on atrial ECG components using a modified limb lead system. ISPCC **2017**, 527–530 (2017). https://doi.org/10.1109/ISPCC.2017.8269735
8. S. Karimulla, J. Sivaraman, The role and significance of atrial ECG components in standard and modified lead systems, in ed. by P.K. Mallick, P. Meher, A. Majumder, S.K. Das. ESIC 2020, vol. 686 (2020) https://doi.org/10.1007/978-981-15-7031-5_33
9. A. Jyothsana, J. Sivaraman, A study on stability analysis of QT interval dynamics of ECG using ARMAX model, in ed. by P.K. Mallick. P. Meher, A. Majumder, S.K. Das. ESIC 2020, vol. 686 (2020). https://doi.org/10.1007/978-981-15-7031-5_29
10. A. Jyothsana, B. Arya, J. Sivaraman, Stability analysis on the effects of heart rate variability and premature activation of atrial ECG dynamics using ARMAX model. Phys. Eng. Sci. Med. **43**, 1361–1370 (2020). https://doi.org/10.1007/s13246-020-00940-w

11. J. Sivaraman, G. Uma, P. Langley, M. Umapathy, S. Venkatesan, G. Palanikumar, A study on stability analysis of atrial repolarization variability using ARX model in sinus rhythm and atrial tachycardia ECGs. Comput. Meth. Prog. Bio. **137**, 341–351 (2016). https://doi.org/10.1016/j.cmpb.2016.10.005
12. B. Dhananjay, J. Sivaraman, The role of heart rate variability in atrial ECG components of normal sinus rhythm and sinus tachycardia subjects, in ed. by S. Satapathy, V. Bhateja, B. Janakiramaiah, Y.W. Chen. Advanced Intelligent, vol. 1171 (2021). https://doi.org/10.1007/978-981-15-5400-1_61
13. U.R. Acharya, H. Fujita, O.S. Lih, M. Adam, J.H. Tan, C.K. Chua, Automated detection of coronary artery disease using different durations of ECG segments with a convolutional neural network. Knowl. Based Syst. **132**, 62–71 (2017). https://doi.org/10.1016/j.knosys.2017.06.003
14. O. Yildirim, P. Plawaik, R.-S. Tan, U.R. Acharya, Arrhythmia detection using a deep convolutional neural network with long-duration ECG signals. Comput. Biol. Med. **102**, 411–420 (2018). https://doi.org/10.1016/j.compbiomed.2018.09.009
15. S.K. Mathews, C. Kambhamettu, K.E. Barner, A novel application of deep learning for single-lead ECG classification. Comput. Biol. Med. **99**, 53–62 (2018). https://doi.org/10.1016/j.compbiomed.2018.05.013
16. A.Y. Hannun, P. Rajpurkar, M. Haghpanahi, G.H. Tison, C. Bourn, M.P. Turakhia, A.Y. Ng, Cardiologist-level arrhythmia detection and classification in ambulatory electrocardiograms using a deep neural network. Nat. Med. **25**(1), 65–69 (2019). https://doi.org/10.1038/s41591-018-0268-3
17. M.M. Rahhal, Y. Bazi, H. Al Hichri, Deep learning approach for active classification of electrocardiogram signals. Inf. Sci. **345**, 340–354 (2016). https://doi.org/10.1016/j.ins.2016.01.082
18. G. Sannino, G. De. Pietro, A deep learning approach for ECG-based heartbeat classification for arrhythmia detection. Future Gener. Comp. Sys. **86**, 446–455 (2018). https://doi.org/10.1016/j.future.2018.03.057
19. U.R. Acharya, L.O. Shu, Y. Hagiwara, A deep convolutional neural network model to classify heartbeats. Comput. Biol. Med. **89**, 389–396 (2017). https://doi.org/10.1016/j.compbiomed.2017.08.022
20. R.S. Andersen, A. Peimanker, S. Puthusserypady, A deep learning approach for real-time detection of atrial fibrillation. Expert Syst. Appl. **115**, 465–473 (2019). https://doi.org/10.1016/j.eswa.2018.08.011
21. A. Minchole, B. Rodriguez, Artificial intelligence for the electrocardiogram. Nat. Med. **25**(1), 22–23 (2019). https://doi.org/10.1038/s41591-018-0306-1
22. S. Chauhan, L. Vig, S. Ahmad, ECG anomaly class identification using LSTM error profile modeling. Comput. Biol. Med. **109**, 14–21 (2019). https://doi.org/10.1016/j.compbiomed.2019.04.009
23. S. Raghunath, A.E. UlloaCerna, L. Jing, Prediction of mortality from 12-lead electrocardiogram voltage data using a deep neural network. Nat. Med. **26**, 886–891 (2020). https://doi.org/10.1038/s41591-020-0870-z
24. S. Bhavya, A.S. Pillai, Prediction models in healthcare using deep learning, in ed. by A. Abraham, M. Jabbar, S. Tiwari, I. Jesus. SoCPaR 2019. AISC 2019, vol. 1182 (2019) https://doi.org/10.1007/978-3-030-49345-5_21
25. J. Gao, Y.L. Murphey, H. Zhu, Multivariate time series prediction of lane changing behavior using deep neural network. Appl. Intell. **48**, 3523–3537 (2018). https://doi.org/10.1007/s10489-018-1163-9
26. P. Su, X. Ding, Y. Zhang, J. Liu, F. Miao, Long-term blood pressure prediction with deep recurrent neural networks. IEEE EMBS **2018**, 323–328 (2018). https://doi.org/10.1109/BHI.2018.8333434

FHSS Signals Classification by Linear Discriminant in a Multi-signal Environment

Muhammad Turyalai Khan, Ahmad Zuri Sha'ameri, Muhammad Mun'im Ahmad Zabidi, and Chun Choon Chia

Abstract Frequency-hopping spread spectrum (FHSS) spreads the signal over a large bandwidth where the carrier frequencies change quickly according to a pseudorandom number making signal classification difficult. Furthermore, classification becomes more complex with the presence of additive white Gaussian noise (AWGN) and interference due to background signals. In this paper, a linear discriminant (LD) method based on the Euclidean distance is proposed for the classification of FHSS signals in the presence of AWGN and background signal. Probability of correct classification (PCC) of the FHSS signals is performed by the LD method for the signal-to-noise ratio (SNR) range of −6 to 15 dB. Results show that the proposed method has achieved 90% detection rate at the SNR range of −1.6 to 3.5 dB in the presence of AWGN only, while its performance is degraded to 0.9 to 12 dB when the background signal is present.

Keywords Frequency-hopping spread spectrum · Linear discriminant · Probability of correct classification

1 Introduction

Multi-signal environment comprises various kinds of wireless technologies which share a common frequency band [1]. For example, the sharing of the 2.4 GHz frequency between the Bluetooth and Wi-Fi [2]. Depending on the wireless technology, the signals may have either a fixed or variable carrier frequency. Most important there should not be any overlap of carrier frequency between the various wireless technologies which could cause interference between the various users. Thus, a spectrum monitoring system can be utilized to manage the use of carrier frequency between the various wireless technologies [3].

Frequency-hopping spread spectrum (FHSS) is a spread spectrum technique where the frequency changes randomly according to a pattern known to both the

M. T. Khan (✉) · A. Z. Sha'ameri · M. M. A. Zabidi · C. C. Chia
Division of Electronic and Computer Engineering, School of Electrical Engineering, Faculty of Engineering, Universiti Teknologi Malaysia, 81310 Johor Bahru, Johor, Malaysia

F. Thakkar et al. (eds.), *Proceedings of the International e-Conference on Intelligent Systems and Signal Processing*, Advances in Intelligent Systems and Computing 1370,
https://doi.org/10.1007/978-981-16-2123-9_11

communicating parties for the synchronized communication [1]. The pseudo-noise (PN) sequence is used to generate the random number to change the carrier frequency [4]. FHSS is used in both military and civilian applications, for example, drones [5], military radios like SINCGARS [6], and low-power applications such as Bluetooth and LoRa [7]. Drones use FHSS to avoid interference from other drones by using a hopping sequence while the military uses it to avoid eavesdropping, jamming, and interception by an adversary [8]. The Federal Communications Commission (FCC) allowed the use of FHSS systems in unlicensed frequency bands which can be used as a rogue device [9] such as drones [5]. Recently occurred events related to drone abuses results in the 50% cut in oil production at Abqaiq, Saudi Arabia, on Sep. 14, 2019 [10], and drugs smuggling at Kranji, Singapore, on Jun. 17, 2020 [11].

A suitable method is required to analyze time-varying signals such as FHSS signals for which time-frequency (TF) analysis is appropriate as it presents the signal jointly in the time and frequency domain [12]. There are two general classes of time-frequency distributions (TFDs): linear TFD and quadratic TFD (QTFD). The linear TFD produces an amplitude representation in time and frequency, whereas the QTFD produces a representation in energy and power. The linear TFD such as the short-time Fourier transform (STFT) [13] is used to obtain the time-frequency representation (TFR) of FHSS signals and estimate their parameters such as hop frequency and hop duration. The problem that relates to the TFR obtained from the STFT or spectrogram is the uncertainty principle which is the trade-off between time and frequency resolution. The wavelet transform is another class of the linear TFD, it leads to the concept of multi-resolution analysis [14] which performs like different filters for a signal. It provides good frequency resolution at the low frequency range of a signal, while low frequency resolution at the high frequency range [15]. Due to this property, the wavelet transform is not suitable for the FHSS signals because the frequency content is distributed at the very high frequency (VHF) band. The QTFD, in general, provides high resolution in time and frequency, provided the cross-terms are attenuated unlike the STFT, but at the cost of high computational complexity (CC) [12, 16]. Conversely, the STFT has four times less CC than the QTFD [17]. Parameters such as hop frequency and hop duration can be estimated using the QTFD. Two methods were used in [18] to estimate the FHSS parameters which are the TF moments method and the instantaneous frequency estimation method. The results show that the instantaneous frequency method can accurately estimate the FHSS parameters and meet the Cramer–Rao lower bound (CRLB) at signal-to-noise ratio (SNR) of −3 dB for hop frequency and 0 dB for hop duration, while the performance of TF moments degraded for both hop frequency and hop duration at SNR above 1 dB and 4 dB, respectively. Due to low CC, the spectrogram method is used in this paper because it is important for the FHSS applications to implement algorithms in real-time and on embedded systems.

Once the signal parameters are estimated, the next step is to classify the signal of interest. Statistical modeling is a formalization of the correlation between variables in the data in the shape of mathematical expressions, where outcomes are predicted through a statistical model by finding the relationships between variables [19]. Examples of the statistical classifiers are rule-based classifier [17], linear discriminant

(LD) analysis based on distance classifiers [20–22], and maximum likelihood classifier [23]. Classification of speech signals is performed in [21] by using the wavelet transform and Euclidean distance classifier, where 91.67% classification accuracy is achieved. Similarly, four classes of speech are classified by extracting appropriate features and using different statistical classifiers [22]. Automatic modulation classification based on likelihood-ratio methods is discussed in [23], where the maximum likelihood classifier achieved 90% accuracy at 10 dB of SNR for M-ary phase-shift keying (MPSK) and quadrature amplitude modulation (QAM) signals.

The objective of this paper is to develop an LD method based on the Euclidean distance for the classification of FHSS signals in a multi-signal environment. Thus, the remainder of this paper is organized as follows: Sect. 2 defines the signal model and problem statement. Section 3 discusses the TFD, parameter estimation, and describe the methodology of classifying FHSS signals in the multi-signal environment. The results are presented in Sect. 4 and the findings are concluded in Sect. 5.

2 Signal Model

In this section, the model of FHSS signals and the orthogonal frequency-division multiplexing (OFDM) signal used in this paper are described. The wireless technologies based on FHSS such as used by drones and Bluetooth often utilize 2.4 GHz Industrial, Scientific, and Medical (ISM) bands which occupy a wide bandwidth of 100 MHz [5, 24]. Thus, a radio-frequency (RF) signal with the 2.4 GHz frequency band is chosen, but the work in this paper can be employed to other frequency bands such as 5.8 GHz or any other frequency bands. The OFDM signal represents the Wi-Fi signal that will be utilized as the background signal. In this paper, the modulation methods used for the FHSS and OFDM are binary phase-shift keying (BPSK) and quadrature phase-shift keying (QPSK), respectively. The FHSS signal is expressed as [25]

$$s(t) = a(t)e^{j(2\pi f_c(t)t)}, 0 \leq t \leq T \tag{1}$$

where $a(t)$ is the information bearing signal, T is the signal duration, and $f_c(t)$ is the time-varying channel frequency.

The OFDM signal typically used in Wi-Fi is expressed as [26]

$$v(t) = A\sum_{k=1}^{64} a(t)e^{j2\pi(f_c+f_k)t}, 0 \leq t \leq T \tag{2}$$

where $a(t)$ is the information bearing signal, f_k is the subcarrier frequency, k is the frequency index in range of $1 \leq k \leq 64$, f_c is the constant channel frequency, and A is the signal gain. It is to be noted that the subcarrier spacing for Wi-Fi is

$\Delta f_k = 312.5$ KHz and number of subcarriers is 64 according to IEEE802.11 b/g/n standard.

In practice, the received signal is corrupted by noise and interfered with background signals that can be expressed as

$$y(t) = s(t) + v(t) + n(t) \tag{3}$$

where $s(t)$ is the FHSS signal, $v(t)$ is the background signal, and $n(t)$ is the additive white Gaussian noise (AWGN).

The three cases of interference for simulation purpose are considered as follows:

(i) Case 1: FHSS + AWGN.
(ii) Case 2: FHSS + AWGN + $1/\sqrt{2}$ OFDM.
(iii) Case 3: FHSS + AWGN + OFDM.

The OFDM signal in Case 2 and 3 is included at the frequency of 27 MHz. In Case 2, multiplication by $1/\sqrt{2}$ makes the power of the OFDM signal at half with the power of the FHSS signal.

2.1 *FHSS Signals Parameters*

The parameters of the five FHSS signals are shown in Table 1, where every FHSS signal is different from each other in terms of hop frequencies and hop duration. Among the five FHSS signals, the FHSS_1 occupy the shortest hop duration which will capture 16 hop frequencies within the signal length of 32 μs. In comparison, the FHSS_2 signal will capture less hop frequencies (9 hop frequencies) due to the long hop duration. Thus, the number of hop frequencies depends on the hop duration and the signal length.

OFDM Signal Parameters. In this paper, the OFDM signal is used as the background signal which can be utilized by 4G cellular technology, Wi-Fi, and digital audio/video broadcasting. It represents a Wi-Fi signal for the subcarrier spacing of 312.5 kHz, 64 number of subcarriers, and bandwidth of 20 MHz for each channel. Table 2 shows the parameters of the OFDM signal that are used in this work.

Table 1 Parameters of the FHSS signals: hop duration (T_{hop}) and hop frequency (f_c)

Signal	Hop duration (μs)	No. of hop frequencies	Hop frequency (MHz)
FHSS_1	2	16	$f_c = k \times 6$ MHz; where $k = 1, 2, \ldots, 16$
FHSS_2	3.5	9	$f_c = k \times 7$ MHz; where $k = 1, 2, \ldots, 9$
FHSS_3	5	6	$f_c = k \times 15$ MHz; where $k = 1, 2, \ldots, 6$
FHSS_4	6.5	4	$f_c = k \times 7$ MHz; where $k = 1, 2, 3, 4$
FHSS_5	8	4	$f_c = k \times 15$ MHz; where $k = 1, 2, 3, 4$

Table 2 Parameters of the OFDM signal

Signal	Signal length (μs)	No. of subcarriers	Bandwidth (MHz)	Carrier frequency (MHz)
OFDM	32	64	20	27

Problem Statement. The carrier frequency of an FHSS signal quickly changes with time unlike a fixed-frequency signal where frequency remains unchanged. In conjunction with noise in the RF spectrum, the variation in carrier frequencies over a wide frequency range makes the classification of FHSS signals challenging [25, 27]. Moreover, interference occurs between the FHSS-based wireless technology (such as drones and Bluetooth) and fixed-frequency wireless technology (such as Wi-Fi and Zigbee) due to the sharing of a common frequency band such as 2.4 GHz. Collectively with noise modeled as AWGN inherently exist in wireless communication, correct classification of FHSS signal together with the need to scan a large frequency range within a short time becomes a complex problem [5, 28].

3 Analysis and Classification of FHSS Signals

This section describes the methodology of classifying the FHSS signals in the presence of AWGN and the background signal. The analysis of the FHSS signals is performed by using TFD, which is followed by the parameter estimation and the LD method. The implementation of the algorithm is based on MATLAB simulation.

3.1 Time-Frequency Distribution

The frequency band of interest is within a frequency band of 2.4 GHz and bandwidth of 100 MHz. The RF signal is down converted to the intermediate frequency (IF) at 50 MHz by using spectrogram TF analysis. Figure 1 shows the change in the frequency from RF to IF.

The spectrogram is used to obtain the TFR from the IF which can be expressed as follows [12]:

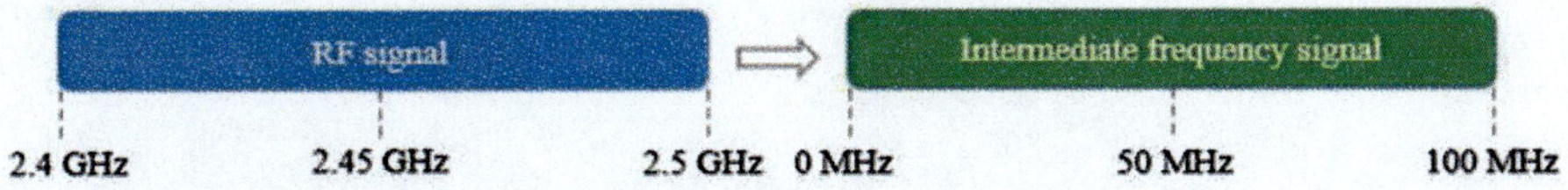

Fig. 1 Spectrogram analysis bandwidth

$$X(n,k) = \sum_{m=0}^{N-1} x(m)w(m-n)e^{-j2\pi km/N_w} \tag{4}$$

$$S_x(n,k) = |X(n,k)|^2 \tag{5}$$

where $x(n)$ is the signal, N is the signal length, $w(n)$ is the window function with length N_w, and $S_x(n,k)$ is the TFR.

The FHSS signals need to be discriminated from the background signal for which a threshold value is required. The threshold is acquired from a baseline signal, which is without the FHSS signals existence, but could comprise any composition of the background signals or at least AWGN. The baseline signal will be acquired from the beginning, while the derivation of the baseline signal's power spectrum will be performed by frequency marginal of TFR [12, 18]:

$$P_x(k) = \sum_{n=0}^{\infty} S_x(n,k) \tag{6}$$

As AWGN exist over the frequency range, a default threshold P_{T_def} is acquired by taking the peak value of the AWGN's power spectrum. It occurs at a certain frequency range for background signal such as the OFDM that has more peak power than AWGN. Thus, a higher threshold is needed at the frequency where the OFDM signal exists which is expressed as

$$P_T(k) = \max(P_{T\_def}, P_B(k)) \tag{7}$$

where $P_T(k)$ is the threshold at k frequency while $P_B(k)$ is the maximum power level of background signal other than AWGN.

The threshold setting $P_T(k)$ across the spectrum is shown in Fig. 2 where the threshold setting at frequency band hold by the OFDM is adaptively adjusted to its

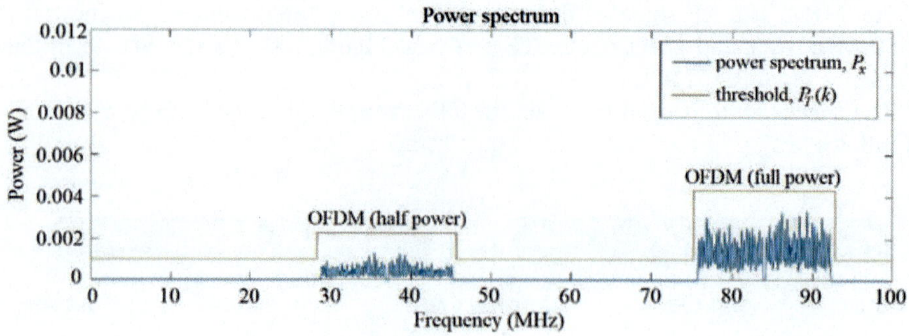

Fig. 2 Threshold setting from baseline signal

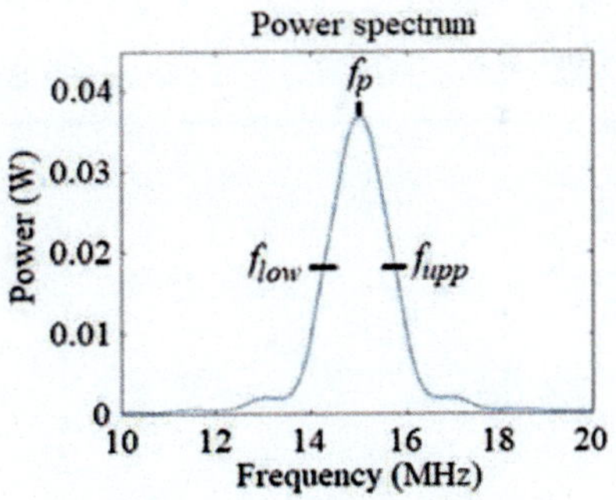

Fig. 3 Power spectrum from TFR

peak value. In this example, the threshold values chosen for certain frequency bands are:

(a) For 0–28 MHz, 45–75 MHz, and 93–100 MHz, the threshold is 0.001 W.
(b) For 28–45 MHz, the threshold is 0.002 W.
(c) For 75–93 MHz, the threshold is 0.004 W.

Parameter Estimation. The parameters of the FHSS signals such as hop frequency and hop duration are estimated according to the methods proposed in [29]. The estimated threshold is implemented on the TFR to separate out the background signals before estimating the hop frequency. Thereafter, frequency marginal is used to acquire the power spectrum from the TFR and then the hop frequencies are estimated from its peaks as shown in Fig. 3. The estimated hop frequency and signal bandwidth is expressed as

$$\text{Estimated hop frequency: } \hat{f}_c = f_p \tag{8}$$

$$\text{Estimated signal bandwidth: } \hat{f}_{BW} = f_{upp} - f_{low} \tag{9}$$

where f_p is the peak frequency, f_{low} and f_{upp} are the lower and upper frequencies measured from 50% power of the peak, which is equivalent to the −3 dB.

The TFR is analyzed within a frequency range to estimate the hop duration at a specific hop frequency and the instantaneous power can be expressed as

$$P_t(n) = \sum_{k=f_{low}}^{f_{upp}} S_x(n,k) \tag{10}$$

where f_{low} and f_{upp} are the lower and upper frequency limit respectively within the hop frequency estimated in Eq. 8.

The instantaneous power at every estimated hop frequency is determined and used to estimate the hop duration by measuring the pulse width at reference level of 50% from the waveform peak value as shown in Fig. 4. The cut-off point of 50% power is used to obtain the transition between the pass band and stop band.

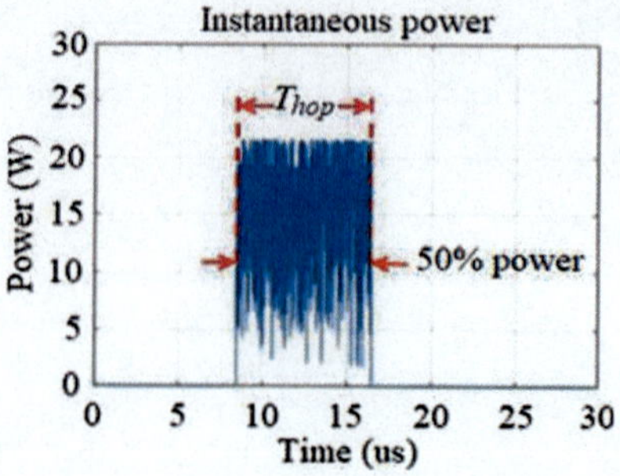

Fig. 4 Instantaneous power from TFR

Classification by LD. Classification of FHSS signals in the presence of AWGN and the background signal is performed by the LD method. *LD is a function of a linear combination of the components of feature vector which is expressed as* [30]

$$g(\mathbf{x}) = \mathbf{w}^{\mathrm{T}}\mathbf{x} + w_0 \tag{11}$$

where **w** is weight vector, **x** is a feature vector and w_0 the bias.

The Euclidean distance can be used to measure the distance between patterns. Minimum Euclidean distance among feature space representations is used as basis for recognition decisions [20, 30]. The Euclidean distance is expressed as

$$d(\mathbf{x}, \mathbf{y}) = \sqrt{\sum_{i=1}^{l} |x_i - y_i|^2} \tag{12}$$

where **x** and **y** are the two vectors, and l is the number of elements in each vector.

The minimum Euclidean distance also known as the minimum distance classifier is obtained by using Eq. 12, which can be expressed as [30]

$$m = \min\left[d(\mathbf{x}, \mathbf{y})\right] \tag{13}$$

where m is the minimum Euclidean distance between two vectors **x** and **y**.

Following is the pseudo code of the LD method to classify the five FHSS signals.

```
Input: ideal FHSS signals xi, where i = 1, 2, …, 5; esti-
mated FHSS signal y;
Output: classified signal z;
Compute the Euclidean distances between the estimated and
ideal FHSS signals: d = [norm (y - xi)];
Find a minimum distance among the computed Euclidean dis-
tances: m = min (d);
if (m = first element of d); z = FHSS_1;
elseif (m = second element of d); z = FHSS_2;
elseif (m = third element of d); z = FHSS_3;
elseif (m = fourth element of d); z = FHSS_4;
elseif (m = fifth element of d); z = FHSS_5;
else z = unknown;
end
```

4 Results

In this section, the TFR of the FHSS signals in the multi-signal environment is discussed and shown. Afterward, probability of correct classification (PCC) of the FHSS signals is performed for the SNR range of −6 to 15 dB for all the three cases mentioned in Sect. 2. The results obtained will be further discussed based on the SNR range at 90% PCC.

4.1 Time-Frequency Representation of FHSS Signals

Figure 5 shows the TFRs of the FHSS signals in the presence of AWGN and the background signal. The background signal is the OFDM (Wi-Fi) signal which is included at the frequency of 27 MHz. The FHSS_1 has the shortest hop duration (2 μs), while the FHSS_5 has the longest hop duration (8 μs) which results in capturing more hops (16 frequencies) and fewer hops (4 frequencies), respectively, within the signal length of 32 μs. The signals defined in Tables 1 and 2 are described by the TFR.

Probability of Correct Classification of FHSS Signals. Figure 6 shows the plots of PCC versus SNR for all three cases. In Case 1, interference is modeled as AWGN where the FHSS_1 signal with the maximum number of hop frequencies (16 frequencies) performed least, while the FHSS_4 signal with the minimum number of hop frequencies (4 frequencies) performed best. In Case 2, besides AWGN, the OFDM signal is used as the background signal where its power is half with the power of the FHSS signal. Figure 6(b) shows that the performance of the LD method is degraded for all the five FHSS signals, especially for the FHSS_2 because its 3 frequencies are overlapping with the background signal as shown in Fig. 5(b), while the number of overlapping frequencies of other FHSS signals is only 1 or 2. Similarly,

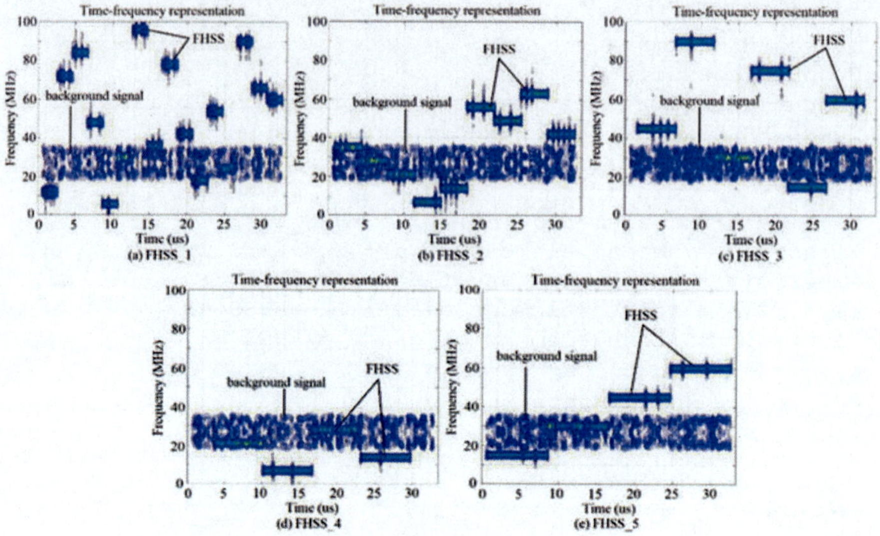

Fig. 5 TFRs of the FHSS signals in the multi-signal environment

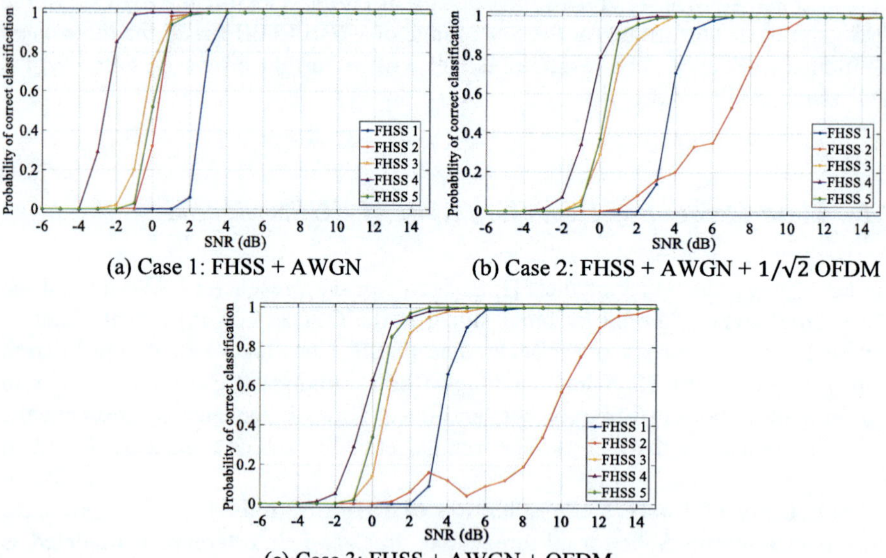

Fig. 6 Probability of correct classification of FHSS signals for all the three cases

Table 3 SNR (dB) range at 90% PCC for all three cases

Case 1: FHSS + AWGN	Case 2: FHSS + AWGN + $1/\sqrt{2}$ OFDM	Case 3: FHSS + AWGN + OFDM
$-1.6 \leq \text{SNR} \leq 3.5$	$0.6 \leq \text{SNR} \leq 8.9$	$0.9 \leq \text{SNR} \leq 12$

in Case 3, the power of the OFDM signal is equal to the power of the FHSS signal. More power of the background signal affects the performance of the LD method more.

Table 3 presents the SNR range at 90% PCC for the Case 1, 2, and 3 as follows: −1.6 to 3.5 dB, 0.6 to 8.9 dB, and 0.9 to 12 dB, respectively. Hence, the best performance of the LD method is for Case 1, while the worst performance is for Case 3. Thus, it can be concluded that the performance of the LD method in the multi-signal environment is inconsistent and degraded significantly in Case 3.

5 Conclusion

Classification of FHSS signals in the multi-signal environment has been achieved by using an LD method based on the Euclidean distance. In this method, the received signal was down converted to the IF signal with the frequency range of 0 to 100 MHz. Thereafter, the spectrogram is produced by the squared magnitude of the STFT to obtain the TFR of the FHSS signals. Parameters of the FHSS signal such as hop frequency and hop duration were estimated from the TFR.

The PCC of the FHSS signals for the SNR range of −6 to 15 dB was performed. Based on the SNR range at 90% PCC, the best performance of the LD method was in Case 1, which was −1.6 to 3.5 dB, where AWGN existed only while its performance was degraded in both Case 2 and 3 to 0.6 to 8.9 dB and 0.9 to 12 dB, respectively, where the background signal also existed besides AWGN. Future work will focus on the implementation of machine learning methods to improve the PCC of the FHSS signals in a multi-signal environment.

References

1. J.G. Proakis, M. Salehi, *Digital Communications*, 5th edn. (McGraw-Hill, New York, NY, USA, 2001)
2. E. Fernández de Gorostiza, J. Berzosa, J. Mabe, R. Cortiñas, A method for dynamically selecting the best frequency hopping technique in industrial wireless sensor network applications. Sensors **18**(2) (2018)
3. A. Boyacı, A.R. Ekti, S. Yarkan, M.A. Aydın, Monitoring, surveillance, and management of the electromagnetic spectrum: current issues in electromagnetic spectrum monitoring. Electrica **18**(1), 100–108 (2018)

4. T.J. Rouphael, *RF and Digital Signal Processing for Software-Defined Radio: A Multi-Standard Multi-Mode Approach* (Elsevier, Amsterdam, The Netherlands, 2009)
5. H. Shin, K. Choi, Y. Park, J. Choi, Y. Kim, Security analysis of FHSS-type drone controller, in *Information Security Applications WISA 2015, LNCS*, vol. 9503, ed. by H. Kim, D. Choi (Springer, Cham, 2015), pp. 240–253
6. J.H. Gass, P.J. Curry, C.J. Langford, An application of turbo trellis-coded modulation to tactical communications, in *MILCOM 1999. IEEE Military Communications. Conference Proceedings (Cat No. 99CH36341)*, pp. 530–533 (IEEE, Atlantic City, NJ, USA, 1999)
7. S. Kahveci, T. Atasoy, From Wireless Personal Area Network (WPAN) to Long Range (LoRa) Technology, in *2019 11th International Conference on Electrical and Electronics Engineering (ELECO)*, pp. 679–681 (IEEE, Bursa, Turkey, 2019)
8. Rohde and Schwarz GmbH and Co KG, Solution for Aviation, https://scdn.rohde-schwarz.com/ur/pws/dl_downloads/dl_common_library/dl_brochures_and_datasheets/pdf_1/Aviation_bro_en_3609-3536-42_v0100_120dpi.pdf. Accessed 7 Oct 2020
9. D.D. Coleman, D.A. Westcott, *CWNA: Certified Wireless Network Administrator Official Study Guide: CWNA-106*, 4th edn. (Wiley, Indianapolis, Indiana, USA, 2014)
10. Y. Li, Saudi oil production cut by 50% after drones attack crude facilities, https://www.cnbc.com/2019/09/14/saudi-arabia-is-shutting-down-half-of-its-oil-production-after-drone-attack-wsj-says.html. Accessed 7 Oct 2020
11. A. Wong, Singaporeans arrested for smuggling drugs with a drone from Malaysia, https://www.soyacincau.com/2020/06/21/singaporean-smuggle-drugs-dji-mavic-air-2-drone-jb-kranji/. Accessed 7 Oct 2020
12. B. Boashash, *Time-Frequency Signal Analysis and Processing: A Comprehensive Reference*, 2nd edn. (Elsevier, Amsterdam, The Netherlands, 2015)
13. W. Fu, Y. Hei, X. Li, UBSS and blind parameters estimation algorithms for synchronous orthogonal FH signals. J. Syst. Eng. Electron. **25**(6), 911–920 (2014)
14. S.G. Mallat, A theory for multiresolution signal decomposition: the wavelet representation. IEEE Trans. Pattern Anal. Mach. Intell. **11**(7), 674–693 (1989)
15. A.Z. Sha'ameri, Analysis and parameter estimation of time-varying signals: theory and methods. J. Phys.: Conf. Series, 012091. IOP Publishing, Purwokerto, Indonesia (2019)
16. F. Hlawatsch, G.F. Boudreaux-Bartels, Linear and quadratic time-frequency signal representations. IEEE Signal Process. Mag. **9**(2), 21–67 (1992)
17. A.A. Ahmad, A.Z. Sha'ameri, Classification of airborne radar signals based on pulse feature estimation using time-frequency analysis. Defence S&T Technical Bulletin **8**(2), 103–120 (2015)
18. A. Kanaa, A.Z. Sha'ameri, A robust parameter estimation of FHSS signals using time–frequency analysis in a non-cooperative environment. Phys. Commun. **26**, 9–20 (2018)
19. B. Ratner, *Statistical and Machine-Learning Data Mining: Techniques for Better Predictive Modeling and Analysis of Big Data*, 3rd edn. (CRC Press. Boca Raton, FL, USA, 2017)
20. N.H. Barnouti, S.S.M. Al-Dabbagh, W.E. Matti, M.A.S. Naser, Face detection and recognition using Viola-Jones with PCA-LDA and square Euclidean distance. Int. J. Adv. Comput. Sci. Appl. (IJACSA) **7**(5), 371–377 (2016)
21. T. Lambrou, P. Kudumakis, R. Speller, M. Sandler, A. Linney, Classification of audio signals using statistical features on time and wavelet transform domains, in *Proceedings of the 1998 IEEE International Conference on Acoustics, Speech and Signal Processing, ICASSP'98* (Cat. No. 98CH36181), pp. 3621–3624 (IEEE, Seattle, WA, USA, 1998)
22. M. Büchler, S. Allegro, S. Launer, N. Dillier, Sound classification in hearing aids inspired by auditory scene analysis. EURASIP J. Adv. Signal Process. **2005**(18), 2991–3002 (2005)
23. J.L. Xu, W. Su, M. Zhou, Likelihood-ratio approaches to automatic modulation classification. IEEE Trans. Syst. Man Cybern. Part C (Applications and Reviews) **41**(4), 455–469 (2011)
24. Rohde and Schwarz GmbH and Co KG.: Protecting the Sky: Signal Monitoring of Radio Controlled Civilian Unmanned Aerial Vehicles and Possible Countermeasures, http://www.rohde-schwarz-usa.com/rs/324-UVH-477/images/Drone Monitoring Whitepaper.pdf. Accessed 14 Oct 2020

25. F. Liu, M.W. Marcellin, N.A. Goodman, A. Bilgin, Compressive sampling for detection of frequency-hopping spread spectrum signals. IEEE Trans. Signal Process. **64**(21), 5513–5524 (2016)
26. H. Schulze, C. Luders, *Theory and Applications of OFDM and CDMA: Wideband Wireless Communications* (Wiley, Chichester, England, 2005)
27. F. Liu, M.W. Marcellin, N.A. Goodman, A. Bilgin, Compressive detection of frequency-hopping spread spectrum signals, in *Compressive Sensing II*, pp. 87170P–87170P–6 (SPIE, Baltimore, Maryland, USA, 2013)
28. B. Kaplan, I. Kahraman, A. Görçin, H.A. Çırpan, A.R. Ekti, Measurement based FHSS-type drone controller detection at 2.4 GHz: an STFT approach, in *2020 IEEE 91st Vehicular Technology Conference (VTC2020-Spring)*, pp. 1–6 (IEEE, Antwerp, Belgium, 2020)
29. C.C. Chia, A.Z. Sha'ameri, Adaptive window size and stepped frequency scan spectrogram analysis for drone signal detection in multi-signal environment. Defence S&T Tech. Bull. **13**(1), 41–60 (2020)
30. R.O. Duda, P.E. Hart, D.G. Stork, *Pattern Classification*, 2nd edn. (Wiley, New York, NY, USA, 2012)

Non-invasive Thyroid Detection Using Thermal Imaging Technique

A. R. Reshma Ruth Pauline, T. Rajalakshmi, Sai P. Vijay, S. Rajalakshmi, R. Jai Reethikha, and U. Snekhalatha

Abstract Thyroid disorder is due to variation in TSH hormones which helps the human body by controlling metabolism. Thyroid hormones are generally generated in response to a different hormone which is released by pituitary gland. There are four main types of thyroid disorders: Hyperthyroidism (too much thyroid hormone), Hypothyroidism (too little thyroid hormone), Benign (non-cancerous). A thyroid disorders is an abnormal growth of cells within the thyroid gland or inside the throat, which can be cancerous (malign). Thermal distribution in human body is a natural indicator of abnormalities. Thermal imaging is a non – invasive screening method for monitoring the distribution of body temperature. In the view of study we propose classification of Thyroid abnormalities using thermal image. The proposed technique is based on the following computational methods; the median filter for pre-processing, Active Contour segmentation is used to segment the selected Region of Interest and then the features are extracted. Feature extraction was done from the taken images and correlation was estimated between the normal subjects and the thyroid subjects. Various classifiers have been tested for their accuracy and some classifiers such as Multilayer Perceptron, Bayes Net and KNN showed high accuracy. The accuracy estimated by these classifiers was tested in Weka tool and their ROC curves with AUC scores have been derived. These classifiers can be used for feature classification and be efficiently used for the performance of non-invasive thyroid diagnosis.

Keywords Thermal imaging · Feature extraction · Feature selection · Segmentation · Classification

A. R. Reshma Ruth Pauline (✉) · S. P. Vijay · S. Rajalakshmi · R. Jai Reethikha · U. Snekhalatha
Department of Biomedical Engineering, SRM IST, Chennai, India

T. Rajalakshmi (✉)
Department of Electronics and Communication Engineering, SRM IST, Kattankulathur, India
e-mail: rajalakt@srmist.edu.in

F. Thakkar et al. (eds.), *Proceedings of the International e-Conference on Intelligent Systems and Signal Processing*, Advances in Intelligent Systems and Computing 1370,
https://doi.org/10.1007/978-981-16-2123-9_12

1 Introduction

The thyroid hormone secreting gland is a gland in the shape of a butterfly that lies low on the front side of the windpipe of a person. A person's thyroid lies under their Adam's apple, along the front side of the neck. The thyroid has two side portions which are attached together by isthmus (a bridge like structure in the middle). The gland, Reddish-Brown in colour, is rich with blood circulating vessels. The thyroid gland secretes numerous hormones, which are commonly called as the thyroid hormones. The main hormone secreted is Thyroxine that is also called the T4 hormone. These thyroid hormones take action through the full body, inducing metabolism, evolution and advance, and body heat. Thyroid disorders are among the most common endocrine related disorders worldwide. India too, is no exception. In India, there has been a substantial burden of thyroid diseases. According to a prediction from different studies done on thyroid disease, it has been estimated that about 42 million people in India are suffering from these thyroid diseases. Thyroid diseases are diverse from other various diseases in the terms of their effortlessness of diagnosis, the accessibility of a medical treatment for it, and the comparative visibility that even a small swelling of the thyroid offers to the treating physician. Early diagnosis and treatments still remains the cornerstone of a management. Gopinath et al. [1] researched that Thermography can be used for detecting and classifying thyroid abnormalities by simple mapping with the body temperature. A new method has been proposed, where first median filter is applied and histogram equalization is done for noise removal and Otsu segmentation is done to extract the region of interest. Then Gabor filter is applied for feature extraction from the thermal images and results are discussed. Sivanandam et al. [2] researched that ANN classifier has more sensitivity than SVM classifier and that this method of detecting thyroid using thermography can be used similar with using mammography. Thermal camera could capture the heat emanating from the human body since it's non-traumatic and time saving method. Valanarasi Antony Santiagu Vaz et al. [3] researched that the thyroid images classified using MLPNN efficiently produces sensitivity of 93% for three-folds and 89% in the case of tenfolds. This study was planned to slice the thyroid images using a logic which was fuzzy and to categorize the images as thyroid and nonthyroid.

Yende et al. [4] researched that Ultrasound images are more dominant and a reasonable tool for the clinical analysis of thyroid. But this method is time over-riding in segmentation of the region and the approximation volume of the thyroid regions in the CT images is costly. They have also said that the influence of noise in the images sources the segmentation outcome of the thyroid glands regions of the ultrasound images inefficient and inaccurate. Thus they used the process of image enlargement and regions growing were useful to the reclamation of the precise shape of the thyroids gland in practice. Pallavi Mahajan et al. [5] researched that thermography might be the best suitable technique to detect thyroid disease. It can be used because it is non-contactive, non-invasive, non-traumatic and an easy method to map the skin temperature. Aweda et al. [6] researched that the mean skin temperature of normal person and the skin temperature of a person having thyroid showed temperature variations. Comparison and differences obtained from the thermograms of benign, as well as malignant cases, were also demonstrated. The values obtained have been recorded. Lekshmy Ashok et al. [7] researched that as high sensitivity is achieved, this method of infrared thermography can be used to diagnose thyroid subjects among normal. The features extracted could be classified by using other various networks for higher and better efficiency. Apart from the various concentrated region, the features could be extracted from palm regions and it could still be classified using neural network for efficiency. Barry Ivans Samuels et al. [8] researched about a preliminary study of the various clinical applications of thermography to diseases related to the thyroid gland. In the past, cold nodules signifying the benign disease used to be differentiated from those representing malignancy only by a surgical removal and a subsequent histological examination. Thermography represents a major additional aid in the pre-surgical diagnosis of such instances. Pallavi Mahajan et al. [9] researched that the use of Bayesian Classifier gives 81.81% accuracy for classifying the different cases of Hypothyroid, Hyperthyroid and the Normal subjects. Thus, thermography could be said the best and easy technique for detecting thyroid gland disease as it's non-contact, non-invasive, non-traumatic and a meek method of plotting the body skin temperatures. Ahdy Helmy et al. [10] researched that thermography can be used as an alternative to more expensive, intrusive medical imaging methods for analyzing thyroid anomalies. It can thus be established that the thermograms which are obtained can be used to roughly define the structure of thyroid gland.

Ms. Priti Dhaygude et al. [11] researched about the various thyroid hormones and their diseases and diagnosis. Research has also been done on various techniques used to detect the disorder which will reduce the false diagnosis of the diseases. Various Classifiers can enhance the results obtained such as cancerous/non-cancerous. Farshad Bahramian et al. [12] researched on a three-dimensional model human neck and the internal organs of the thyroid gland. A transient analysis was done to get the temperature differences on the skin in neck and concluded that thyroid gland generated more heat because of the higher metabolic rate and higher blood perfusion compared to other organs surrounding it. The thermal image can give estimation from the shape of the gland. Thus, thermal imaging can be used for detecting thyroid disorders. Jianning Chi et al. [13] researched on a fine-tuned thyroid classification method as an already present deep convolutional neural network GoogLeNet. Their research showed that deep learning methods can improve the performance of assessing the thyroid nodules as a computer aided diagnostic system. Hailiang Li et al. [14] researched various strategies for improving the ability of faster R-CNN for detecting cancer spots in the thyroid images. They validated methods like layer concantenation and Spatial constrained layer. Combination of all the strategies gave the best result. Gonzalez et al. [15] researched that there are different aspects that are there in thermograms of the neck region that are caused by the involuntary movement of patients during the time of an infrared thyroid testing. The methods to reduce this have been discussed in the paper.

The proposed study focuses on the method of detecting thyroid using thermography, as it is non-invasive with more accuracy and is time saving. It is aimed at the early detection of thyroid in people of all ages as there is no use of any radiations. Digital Infrared Thermal Imaging (DITI) was aimed as a non-invasive system that could detect and display the comparative temperature deviations in patients having thyroid nodules, especially the non-palpables. DITI is one of the valuable supports for ultrasound, specifically in the case of thyroid hypo echoic nodule and it is used in amalgamation with other imaging methods, it could play a spare role in the non-invasive diagnosis of the thyroid nodules. The paper is ordered as follows. Section 2 describes the methodology used, features used, software algorithm, and feature classification. Section 3 describes the results that were derived from the study and the discussion. Section 4 deals with the conclusion of the study.

2 Methodology

In the proposed study, Thermal Imaging technique was used to capture the images of normal and thyroid individuals. Thermal camera is used to capture the images for both normal and thyroid individuals. This camera detects the heat energy emitted by the subject. The heat energy is converted into an electronic signal which is then processed to create thermal images. This also enables temperature calculation for the images. In the proposed study, we have used thermal camera to capture the image because it is non-invasive and time saving. On an average, 20 subjects were taken for thyroid and normal subjects each. The FLIR (Forward Looking InfraRed) tool was used to process the images which were captured using the thermal camera and feature extraction is done to get the results with proper accuracy. After the images were taken, they were processed using MATLAB. In Matlab, the images were segmented using Active Contour Segmentation and features were extracted. After the images were segmented, the images were processed in the Weka software, which is a collection of various machine learning set of rules which can either be applied directly to a dataset or called from your own Java code. In this study, we used the Weka software for comparing the different classifiers used and the values and results have been discussed in the coming section. Figure 1 shows the proposed flow diagram that is incorporated in the following study. In this study, the capturing of images using a Thermal camera was done in a completely dark room. The images taken were processed in MATLAB, where the images were read using the Read command. Then the image was converted to grey image and resized. These grey resized images were processed by a Median Filter. A Median Filter is a non-linear digital filtering technique, which is often used to remove noise from an image or a signal. Such noise reduction method is a typical pre-processing step to improve the results and accuracy of later processing (for example, performing edge detection on an image). Median filter is considered good compared to the other filters because, unlike averaging filter, which ruins the edges of an image by only blurring it to remove the noise, median

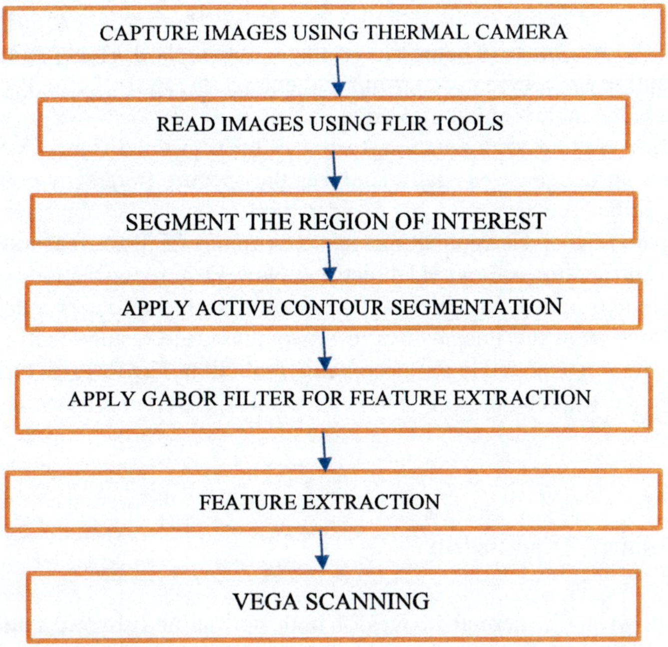

Fig. 1 Proposed flow diagram

filter also removes the noise without disturbing the edges. Thus, median filter is the best and only filter to remove noise.

After the noise was filtered, the Histogram and Histogram equalization of the images were generated. A histogram of an image represents the type of bar plot for the numeric data that group the data into bins. After a Histogram object is created, there can be modifications made in the aspects of the histogram by changing its property values. This is mainly useful for quickly modifying the properties of the bins and thus changing the display of the image. A histogram also offers a way to display the frequency of incidences of data along an interval. When the data presented in the histogram is used, it is easy to regulate statistical information.Histogram Equalization is a computer image processing technique which is used often to improve the contrasts in images. It also accomplishes this by efficiently spreading out the most frequent intensity values, i.e. broadening out the intensity range of the image. Histogram Equalization is performed because it assigns the intensity values of pixels in the input image such that the output image contains an even scattering of intensities. It improves the contrast and helps to obtain a uniform histogram. This technique can also be used on a whole image or just on a part of an image.

After Histogram and its Equalization have been performed, the image is segmented using Active Contour Segmentation method. The active contours technique, which is called Snakes, is an iterative region-growing image segmentation algorithm. Using this active contour algorithm, it is easy to state the initial curves on an image and then the use of the active contour function evolves the curves towards the object boundaries. The advantage of using this segmentation method is that it is autonomous and self-adapting in their search for minimal energy. It can track moving objects in temporal and also in the spatial direction.

After active contour segmentation, Gabor Filter is applied for extracting useful features from an image. It basically analyzes the specific frequency content in the image in specific direction in a localized region or region of analysis. It is also applied for pattern analysis application and texture analysis. It has real and imaginary component representing orthogonal directions. Now, Feature Extraction is performed for the segmented image, in which, the algorithm is used to detect and isolate the desired features from the image. Due to this process, the results can be improved accordingly. It is also related to dimensionality reduction. The features that extracted are total variance, total standard deviation, correlation, covariance, skewness, kurtosis and entropy.

3 Results and Discussion

In this study, we took thermal images of both normal and thyroid patients using a thermal camera and processed them using Matlab. The statistical features were extracted from the segmented images and tabulated. The values obtained from the feature extraction were compared and tabulated. The differences in the values have been discussed. Figures 2, 3, 4, 5, 6 and 7 depicts the input image, Noised image, median filtered image, histogram equivalized image, histogram of median filtered image and segmented image for normal and thyroid individuals respectively.

The following statistical features were extracted from the obtained segmented images of both normal and thyroid subjects and the values extracted are depicted in Table 1.

Here, the Mean value is the sum of the values of the data set divided by the number of images taken. The Standard Deviation is the measure of the amount of variation in a set of values. Correlation is the process by which a filter mask is moved that

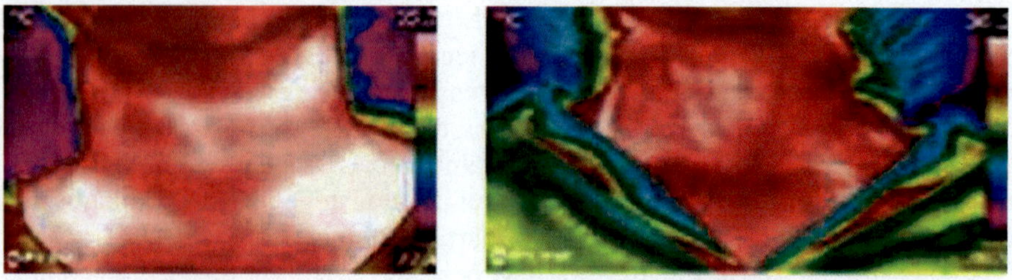

(a) (b)

Fig. 2 **a** Input image—Normal subject. **b** Input image—Thyroid subject

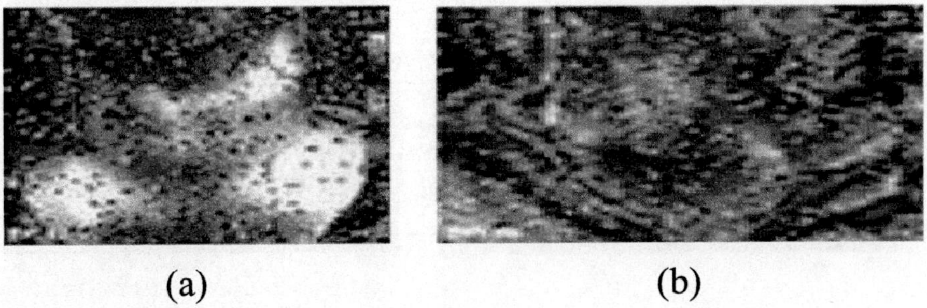

Fig. 3 **a** Noised Image—Normal subject. **b** Noised Image—Thyroid subject

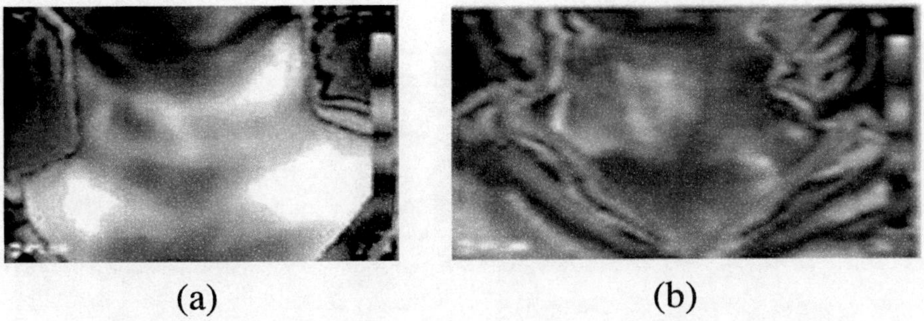

Fig. 4 **a** Median Filtered Image—Normal subject. **b** Median Filtered Image—Thyroid subject

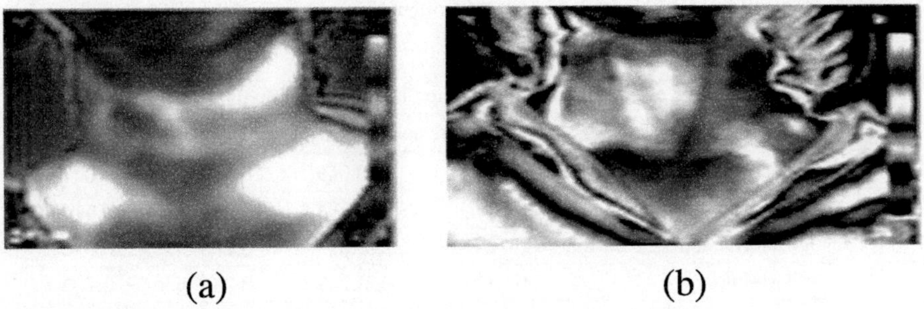

Fig. 5 **a** Histogram Equalized Image—Normal subject. **b** Histogram Equalized Image—Thyroid subject

is often referred to as kernel above the image and then computing the sum of the products at each location. This can also be said as, the starting value of the correlation corresponds to the zero shift of the filter, and the second value corresponds to one unit of shift and so on. Covariance can be defined as the measure of finding by how much the changes done in the one variable are allied with the changes in the second variable. Skewness can be defined as the measure of lack of symmetry. A distribution,

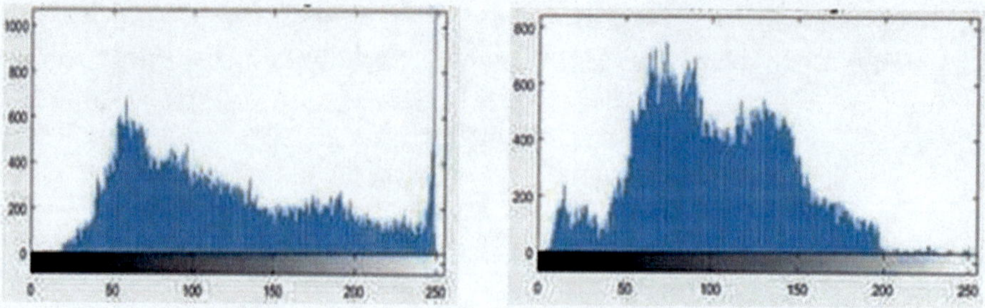

(a) (b)

Fig. 6 **a** Histogram of median filtered image of normal subject. **b** Histogram of median filtered image of thyroid subject

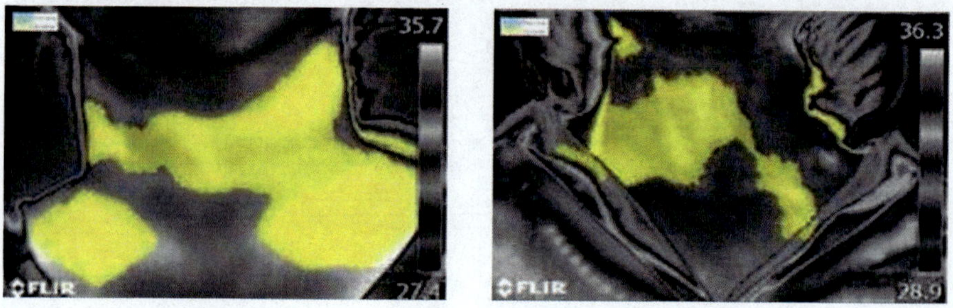

(a) (b)

Fig. 7 **a** Segmented Image—Normal subject. **b** Segmented Image—Thyroid subject

Table 1 Statistical features extracted from normal and thyroid individuals

S. No.	Features extracted	Normal (Mean ± S.D)		Thyroid (Mean ± S.D)	
1.	Mean Value	0.632	± 0.069	0.562	± 0.0443
2.	Total Variance	0.057	± 0.034	0.0827	± 0.0145
3.	Total Standard Deviation	0.227	± 0.074	0.286	± 0.0267
4.	Correlation	0.281	± 0.241	0.236	± 0.103
5.	Covariance	891.26 ± 758.928		928.83	± 586.978
6.	Skewness	178.491	± 398.448	186.001	± 415.259
7.	Kurtosis	178.491	± 356.383	186.001	± 371.419
8.	Entropy	7.199	± 0.286	7.6483	± 0.1938

or better called a dataset, is said to have the property of symmetric if it looks the same from the right and left of the centre point. Kurtosis is defined as the measuring tool to detect whether the data are heavy tailed or light tailed relative to a level of normal distribution. Entropy is defined as the statistical amount of randomness in an image that can be used to describe its texture.

In addition to this, the features that were obtained for normal dataset and thyroid datasets were used for performing classification using Weka software tool.

The classifier analysis was done based on tenfold cross validation method. The results of the classifiers that were tested are shown in Table 2. The table provides the output parameters such as Precision, Recall, F-Measure, Accuracy, Specificity, Sensitivity, Negative Predictive Value (NPV) and Positive Predictive Value (PPV) values in terms of percentage. Hence, it is observed from the table that KNN algorithm shows an accuracy of 93.3% whose specificity and sensitivity values are found to be higher in comparison with other classifiers like Bayes net and multilayer perceptron classifier.

From the Tables 3, 4 and 5, it is observed that the accuracy of Bayes Net Classifier is 90.74%, which is efficient. The Accuracy of Multilayer Perceptron is 92.59%, which is more efficient. The highly efficient one out of the classifiers tested is using the KNN classifier, which has 96.3%. The Negative Predictive Values and the Positive Predictive Values of the classifiers have been noted down.

An ROC curve is defined as a graphical plot which illustrates the diagnostic ability of a binary classifier system. The ROC curves have been acquired for the thyroid class values or based on thyroid threshold values. In addition to the previous results, the ROC curves for all the classifiers have also been acquired. An ROC curve is defined as a graphical plot which illustrates the diagnostic ability of a binary classifier system.

The ROC curves have been acquired for the thyroid class values or based on thyroid threshold values. It is usually generated by plotting the true positive rates against the false positive rates. ROC curves commonly represent the tradeoff that exists between the sensitivity and the specificity. ROC curves are studied based on its AUC score. AUC score refers to the area Under Curve". The score of AUC is defined based on the usefulness of the curve. If the AUC score is more, then it is more useful as a classifier. Similarly, if the AUC score is less, then it is comparatively less useful. Figure 8, 9 and 10 depicts the ROC curve obtained for Bayes net, KNN net and Multilayer perceptron classifier, respectively.

From Table 6, it is observed that the AUC scores are above 0.9 (90%), which shows a good accuracy value. All these observations state that though there are many techniques and methods to detect thyroid disorder, this method of thermal imaging is more efficient, time saving and has a good accuracy level.

Table 2 Classifier –Output

Classifier	Precision (%)	Recall (%)	F-Measure (%)	Accuracy (%)	Specificity (%)	Sensitivity (%)	NPV (%)	PPV (%)
Bayes Net	91.4	90.7	90.8	90.7	85.1	96.3	95.8	86.6
Multilayer Perceptron	93.7	92.6	92.6	92.5	85.7	1	1	86.6
KNN	96.6	96.3	96.3	96.3	92.3	1	1	93.3

Table 3 Confusion Matrix of Bayes Net Classifier

Bayes Net	Thyroid	Normal	Accuracy 0.9074
Thyroid	26	4	Sensitivity 0.963
Normal	1	23	Specificity 0.8519
	PPV 0.8667	NPV 0.9583	

Table 4 Confusion Matrix of KNN Classifier

KNN	Thyroid	Normal	Accuracy 0.963
Thyroid	28	2	Sensitivity 1
Normal	0	24	Specificity 0.9231
	PPV 0.9333	NPV 1	

Table 5 Confusion Matrix of Multilayer Perceptron

Multilayer Perceptron	Thyroid	Normal	Accuracy 0.9259
Thyroid	26	4	Sensitivity 1
Normal	0	24	Specificity 0.8571
	PPV 0.8667	NPV 1	

4 Conclusion

This study aims at developing a non-invasive technique to detect thyroid. For the proposed study, the work was carried out using Thermal Imaging technique and the hotspot region was segmented using Active Contour Segmentation technique. Depending on the hotspots obtained from each image, the images have been classified as Normal and Thyroid subjects. Images with Even distribution of Hotspot were classified as Normal subjects and Images with Uneven distribution of Hotspots were classified as Thyroid subjects. After the Active Contour Segmentation is performed, the features were extracted from the hotspots and the values were compared and observed. The ROC-AUC scores were compared. Using this, we will be able to classify the subjects as Normal or Thyroid. Accuracy of Bayes Net Classifier is 90.74%, which is efficient. The Accuracy of Multilayer Perceptron is 92.59%, which is more efficient. The highly efficient one out of the classifiers tested is the KNN

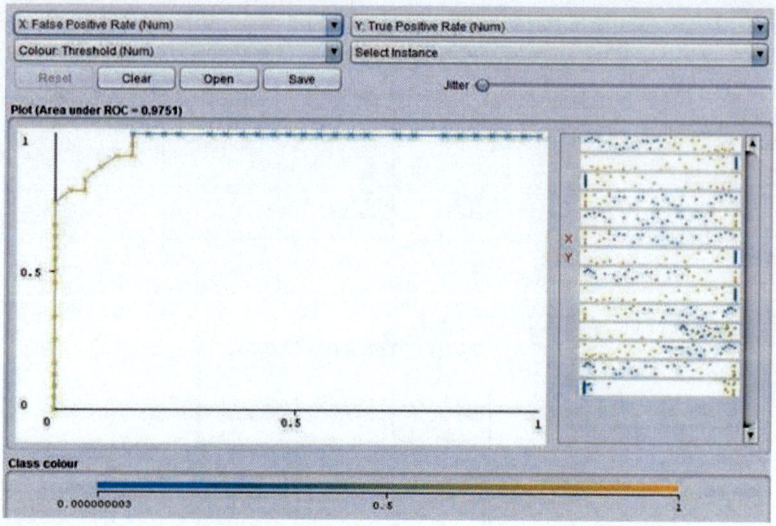

Fig. 8 ROC Curve of Bayes Net Classifier

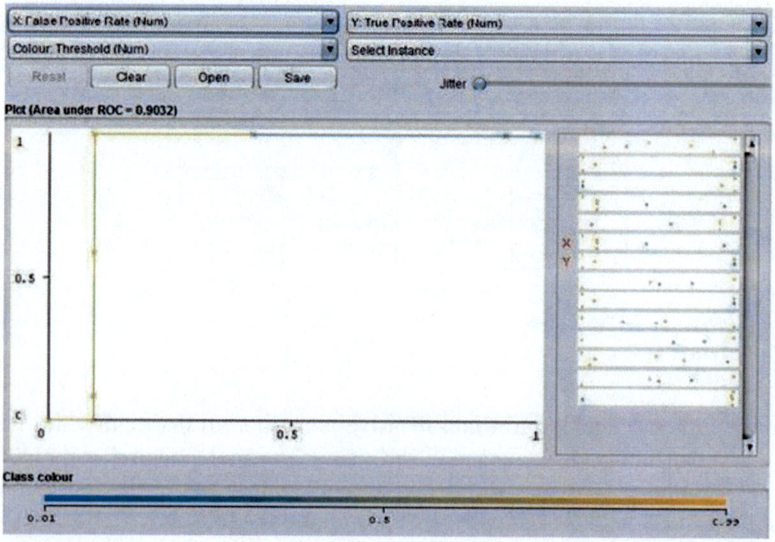

Fig. 9 ROC Curve of KNN Classifier

classifier, which has 96.3%. In the proposed study, we have completed the feature extraction and images have been classified.

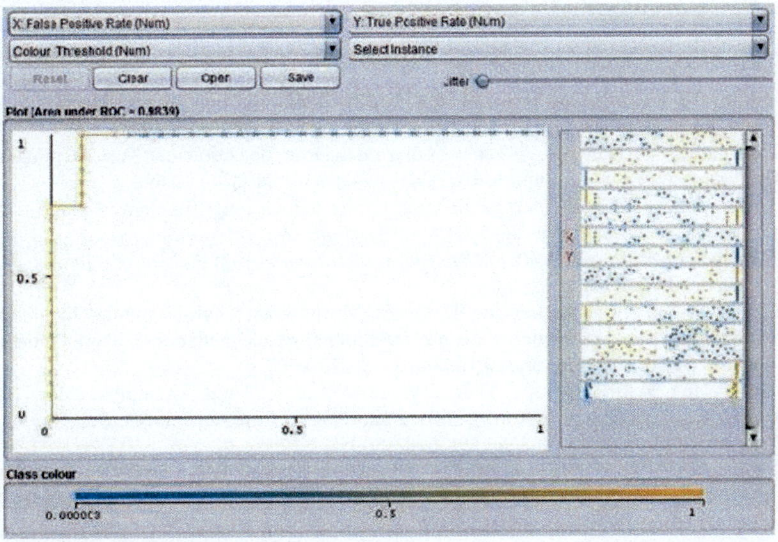

Fig. 10 ROC Curve of Multilayer Perceptron Classifier

Table 6 ROC-AUC Scores

Classifier Tested	ROC—AUC SCORE
BAYES NET	0.975
KNN	0.921
MULTILAYER PERCEPTRON	0.974

References

1. M.P. Gopinathm, S. Prabu, Classification of Thyroid Abnormalities on Thermal Image: A Study and Approach", Special Issue (SCMDSA). IIOABJ **7**(5), 41–57
2. L. Ashok, S.Sivanandam, Diagnosis of hypothyroidism using infrared thermography. Int. J. Pure Appl. Math. **119**(7), 1085–1092 (2018). ISSN: 1311-8080 (printed version); ISSN: 1314-3395 (on-line version)
3. V. Antony S. Vaz, S. Ammal, Diagnosis of hypo and hyperthyroid using MLPN network. Int. J. Innov. Res. Sci. Eng. Technol. (An ISO 3297: 2007 Certified Organization) **3**(7) (2014)
4. G.S.Yende, K.M. Punwatkar, Pusad, Review: thyroid segmentation and volume estimation using image processing. Int. Res. J. Eng. Technol. (IRJET) **3** (2016). e-ISSN: 2395-0056 www.irjet.net, p-ISSN: 2395-0072
5. P. Mahajan, S. Madhe, Morphological feature extraction of thermal images for thyroid detection. Int. J. Electron. Commun. Comput. Eng. **5**(4). ISSN2249–071X
6. M.A. Aweda, A.O. Adeyomoye, G.A. Abe, Thermographic analysis of thyroid diseases at the Lagos university teaching hospital, Nigeria. ISSN: 0976-8610 CODEN (USA): AASRFC 2027
7. L. Ashok, S. Sivanandam, Diagnosis of Thyroid Disorder using Infrared Thermography, in *International Conference on Electronics, Communication and Aerospace Technology ICECA*, 2017
8. M.D. Barry Ivan Samuels, Thermography: a valuable tool in the detection of thyroid disease. Radiology **102**, 53-62 (1972)

9. P. Mahajan, S. Madhe, Hypo and Hyperthyroid disorder detection from thermal images using Bayesian Classifier, in *2014 International Conference on Advances in Communication and Computing Technologies*
10. A. Helmy, M. Holdman, M. Rizkalla, Application of thermography for noninvasive diagnosis of thyroid gland disease. IEEE Trans. Biomed. Eng. **55**(3) (2008)
11. P. Dhaygude, S.M. Handore, A review of thyroid disorder detection using medical images. Int. J. Recent Innov. Trends Commun. **2**(12), 4194–4197. ISSN: 2321-8169
12. F. Bahramian, A. Mojra, Thermal imaging of the human neck for thyroid gland detection based on CT scan images, in *2017 24th national and 2nd International Iranian Conference on Biomedical Engineering (ICBME), Amirkabir University of Technology, Tehran, Iran*, 30 November–1 December 2017
13. J. Chi, E. Walia, P. Babyn, J. Wang, G. Groot, M. Eramian, Thyroid nodule classification in ultrasound images by fine-tuning deep convolutional neural network. J. Digit. Imaging **30**, 477–486 (2017). https://doi.org/10.1007/s10278-017-9997-y
14. H. Li, J. Weng, Y. Shi, W. Gu, Y. Mao, Y. Wang, W. Liu, J. Zhang, An improved deep learning approach for detection of thyroid papillary cancer in ultrasound images. Sci. Rep. **8**, 6600 (2018). https://doi.org/10.1038/s41598-018-25005-7
15. J.R. González, Y. Pupo, M. Hernandez, A. Conci, T. Machenry, W. Fiirst, On image registration for study of thyroid disorders by infrared exams, in *International Conference on IP, Composite Vision, and Pattern Recognition, IPCV'18*

Non Orthogonal Multiple Access Techniques for Next Generation Wireless Networks: A Review

Najuk Parekh and Rutvij Joshi

Abstract Future needs of wireless and cellular communication in current, as well as the next decade, require to support many services like system average throughput, mobile traffic explosion, live HD streaming, enhanced location-based services (LBS), M2M communication, cloud computing, etc. These bring many challenges in capacity, coverage, data rate, and spectral efficiency. Traditional Orthogonal Multiple Access techniques (OMA) can't fulfill these basic requirements of next generation networks. However, by allocating the same radio resources to many users, the system capacity and user throughput can be improved. Non-Orthogonal Multiple Access technique has recently been recommended for the fifth generation (5G) and beyond fifth generation networks (5G+) which can address such issues. By non-orthogonally allocating many users, the capacity and spectral efficiency can be improved. NOMA scheme can be categorized into two main classes: Power domain and Code Domain. This survey paper centers on the Power domain NOMA technique with superposition coding plus Successive interference cancellation at transmitter and receiver, respectively, as well as sum-rate analysis of NOMA in uplink and downlink. The blend of NOMA with MIMO and massive MIMO in addition to the performance of NOMA in the cellular network has been discussed and analyzed.

Keywords NOMA · Sum-rate analysis · MIMO-NOMA

1 Introduction

Massive connectivity, high spectral efficiency, plus low latency has become a stringent requirement for 5G and beyond 5G networks. On which network should support four major applications such as enhanced mobile broadband (eMBB), massive

N. Parekh (✉)
The Maharaja Sayajirao University of Baroda, Vadodara, Gujarat, India
e-mail: parekh.najuk-eed@msubaroda.ac.in

R. Joshi
A. D. Patel Institute of Technology, New Vallabh Vidyanagar, Anand, Gujarat, India

F. Thakkar et al. (eds.), *Proceedings of the International e-Conference on Intelligent Systems and Signal Processing*, Advances in Intelligent Systems and Computing 1370,
https://doi.org/10.1007/978-981-16-2123-9_13

machine-type communication (mMTC), ultra-reliable, and low latency type communication (URLLC), as well as an enhanced vehicle to everything (eV2X). According to the IMT report for beyond 4G networks [1], there will be a requirement for a new technique that addresses a great number of users at the same time and enhance system throughput performance by allocating the same radio resources to multiple users. This technique should also address issues of massive connectivity, reliability, and low latency.

With an exponentially increasing plea of users such as in cellular networks and to achieve the objective of "anybody who at anywhere can communicate with anyone at any time by any network." [1], there is a need to find new multiple access techniques. For this, Researchers have switched from orthogonality to non-orthogonality to accommodate more number of users and to improve system capacity with high throughput. In this survey, new multiple access techniques, which are known as Non-Orthogonal Multiple Access techniques have been discussed. This technique can support more number of users by non-orthogonally allocating resources.

It has been discussed in IMT Report [2] that "several Non-Orthogonal multiple Access schemes exist for 5G networks such as power domain NOMA (PD-NOMA) proposed by NTT-DoCoMo., Low-density spreading-CDMA, Interleaving-based multiple access scheme (IDMA) proposed by NOKIA, Multiple User shared access (MUSA) by ZTE and Pattern Division Multiple Access (PDMA) by Huawei."

Mainly NOMA can be categorized into two types: PD-NOMA (Power domain NOMA) and CD-NOMA (Code domain NOMA). The PD-NOMA technique uses power domain multiplexing to multiplex multiple users by allocating different power levels. Formerly in LTE networks, it was known as multiuser superposition transmission (MUST). PD-NOMA scheme employs superposition coding at the transmitter side and SIC at the receiver side. At the receiver side, the user can opt general multi-user detection technique (MUD) or interference cancellation technique (IC) such as parallel interference cancelation (PIC) or successive interference cancelation (SIC). Depending on, if all received signals are similar (in terms of power level), PIC is used and if received signals are at different power levels, SIC is used. Among which SIC technique is best for the PD-NOMA scheme as there are different power levels assigned to the users due to channel conditions. The following Table 1 summarizes the PD-NOMA scheme with the technical contribution and future work of the given reference.

In CD-NOMA, there exists different ways in which it (code domain-NOMA) is classified. According to [11] CD-NOMA technique can be distinguished based on a single carrier and multicarrier. Interleave Division Multiple Access (IDMA), Low-dense spread CDMA (LDS-CDMA), lattice –partition-based multiple access (LPMA), Multi-user superposition transmission (MUST) which falls under single carrier category while low-Density Spread OFDM (LDS-OFDM), Sparse Code Multiple Access (SCMA), Pattern Division Multiple Access (PDMA), and multiuser shared access (MUSA), which falls under multi-carrier technique. In [12], M. Le et al. classified CD-NOMA in the main two categories as dense NOMA and low-dense NOMA. In dense noma further, it is categorized as DS-CDMA and IDMA, while

Table 1 Summary of technical contribution in PD-NOMA technique

References	Technical contribution	Future work
[3]	Fast and simple user pairing algorithm which forms user pair with highest PF metric and use of a pre-configured ratio of power allocation	The proposed algorithm can be applied to other antenna technologies and channel models for future work
[4]	SC-FDMA for PD-NOMA UFMC system for PAPR reduction	
[5]	BER analysis for PD-NOMA in Rician and Rayleigh fading channel for many modulation schemes	AMC based MIMO system
[6]	Two major ordering schemes are analyzed and associated with both direct and relay path	
[7]	The mathematical formulation of NOMA resource allocation problems	
[8]	PD-NOMA scheme in a single cell and multi-cell scenario with the design of the System model	Resource allocation and SIC implementation
[9]	Studied capacity analysis, power allocation schemes, user fairness, and user pairing schemes in NOMA	
[10]	Outline of the application of PD-NOMA to numerous satellite architecture for the main advantage of coverage efficiency requirements and accessibility	

in low-dense NOMA, there are LDS-CDMA, SAMA, MUSA, and PDMA. The following Table 2 summarizes the work done by authors in CD-NOMA.

Compared to PD-NOMA, CD-NOMA techniques have several advantages like it has low block error rate, improvement incapacity, and improvement in spectral efficiency, etc. For which it requires a high transmission bandwidth which is not easily

Table 2 Summary of technical contribution in CD-NOMA technique

References	Technical contribution
[13]	Categorized and summarize transceiver block diagram of NOMA schemes
[14]	The resource allocation problem of each scheme is formulated with a solution
[12]	Reviewed CD-NOMA technique from an information theory perspective and derived unified framework
[15]	A novel technique compared to LDS–CDMA, SCMA is proposed
[16]	Unique LDS structure is proposed for transmission and detection of symbol synchronous communication over memoryless Gaussian channel
[17]	Propose a new NOMA scheme based on code pattern, called Pattern Division Multiple Access

applicable to the current system [18]. This leads to some disadvantages like interference problem, design, and optimization of pattern/code. However, in PD-NOMA there require no changes in the current network. PD-NOMA can easily integrate with other technologies like MIMO, Massive MIMO, mm-wave communication, visible light communication, etc. This review paper is focused on the PD-NOMA scenario.

It has been reviewed that NOMA can flexibly combine with other technology to further increase the capacity and enhance the system performance, such as NOMA with MIMO, mm-wave NOMA, Co-Operative NOMA, Cognitive communication, Mobile edge computing [19, 20], etc.

In [13], Z. Wu et al. differentiate the NOMA scheme based on the signal transmission which is Scrambling-based NOMA, Interleaving-based NOMA, Spreading-based NOMA, and Coding-based NOMA, where the author has done simulation based on the above technique with MMSE-SIC, MMSE-PIC or MPA algorithm at the receiver side. The extensive survey of NOMA in the future generation of Wireless networks, in 5G and beyond, has been discussed by [21], where the author has taken a "scenario of single-carrier and multi-carrier NOMA, MIMO-NOMA, cooperative NOMA, and mm-wave NOMA.".

The new waveform of 5G and alternative Multiple Access are discussed, compared and analyzed by [22], in [23], the author has shown, by using NOMA principle uplink spectral efficiency improves by 20% and in downlink throughput improve by 30% than conventional multiple access. Based on the instantaneous CSI, the allocation of two different technologies OMA and NOMA to the user in the downlink multi-user and the multi-carrier system has been done by [6].

In [24], Cai et al. focus mainly on "different Orthogonal and Non-Orthogonal Multiple Access techniques with their performance in terms of spectral efficiency, Out of Band leakage (OOB), and Bit Error Rate." In [25], the Author proposes MIMO/NOMA using SIC and IRC, where the author mentions that "basic signal waveform can be based on OFDM or DFT-Spread OFDM." In [26] Author has done an extensive survey on theoretical NOMA principles with multiple antennae aided NOMA design with other technologies. Investigation on PD-NOMA and CD-NOMA with a new introduction of an adaptive based solution of SoDeMa (Software Defined Radio SDR for multiple access) has been proposed depending on the requirement of user and network is done in [21].

Motivated by these aforementioned challenges in wireless technology and the NOMA system, this survey explicitly focuses on the basic concept of NOMA, major issues, challenges, and integration with other technology like MIMO and performance of NOMA in the cellular network.

The remainder of this paper is organized as follows Sect. 2 describes basic concepts of NOMA with superposition coding and successive interference cancellation and sum rate analysis of NOMA in uplink as well as downlink. Section 3 deals with the integration of NOMA with MIMO and the performance of NOMA in cellular communication with sum-rate analysis. Every section has a discussion and outlook with future aspects. Finally, Sect. 4 provides the conclusive remarks of the paper.

2 Basic Concept

In this section, the basic concept of NOMA technology with SIC at the receiver and superposition coding (SC) at the transmitter with sum rate analysis of NOMA in uplink and downlink scenario has been discussed. The NOMA concept in power domain multiplexing has been used, where multiple users are multiplexed in the power domain with time/frequency. It has been analyzed NOMA as of two types Fixed NOMA and Ordered NOMA, in the former one, the decoding order of users is fixed with statistical CSI, while in the latter one, the decoding of users is based on instantaneous CSI where a weaker user is decoded first.

Superposition Coding In [27], the author firstly introduces the concept of "simultaneous communication of information from one source to many receivers" which is known as the broadcasting of information. The author has taken binary symmetric channels and Gaussian channels. Superimposing of the signals in which he compares the technique of superimposing high rate information on a low rate with conventional Time sharing or Time multiplexing technique. Another coding strategy for the broadcast channel is designed and evaluate practically by [28], the author has proved SC can provide a gain in spectral efficiency than conventional orthogonal time-division multiplexing. Vanka et al. have designed system architecture and implement it on GNU Radio/USRP for packet error constraints.

Successive Interference Cancellation SIC which is abbreviated as successive interference cancellation is also known as successive decoding, perform at receiver after SC at the transmitter, this is firstly proposed by [27], the main theme of decoding users are performing non-orthogonally that is not all users are decoded at once, but one by one by canceling other users signal. So they are successively decoded, however, channel ordering is required according to the signal strength. Here interference cancelation follows the procedure as "it first enables the user with a stronger link to decode the user with a weaker link than it decodes the weaker user signal at the stronger user side and subtracts the interference. Finally, stronger user decodes the signal without suffering interference from the weaker user [26]." In this scheme, one main practical drawback is error propagation; if an error occurs in decoding a particular user's signal, it can cascade down to remaining users, which affects further signal decoding.

We know that the NOMA protocol uses the SIC technique at the receiver side. A lot of work has been produced which describes the SIC principle with NOMA technology. There exists key component technology with NOMA in which SIC is performed whether it is uplink or downlink at the receiver side. In the downlink, SIC performs at the user end, while in the uplink it performs at the B.S receiver [29]. Generally, in SIC decoding: the ordering of users is fixed irrespective of their instantaneous channel condition, [30] proposes a Dynamic Ordering scheme in SIC which performs decoding of the users according to instantaneous channel CSI. The author analyzed the outage probability of three user systems in a closed-form expression.

An additional degree of freedom in signal processing to increase the throughput of wireless communication network at the transmitter can be achieved by improper

Gaussian signaling which is addressed by [31], where the author has done "sum-rate maximization for downlink NOMA system in the presence of imperfect SIC". In [25] author investigates the performance and scheme of SIC for DL NOMA mutual with 2 by 2 open-loop single-user MIMO. Where Saito et al. deals with inter-user plus inter-stream interference. New SIC techniques, code word level interference cancellation (CWIC), and symbol level SIC (SLIC) have been anticipated and equated with ideal SIC.

In a traditional decoding scheme, "the later decoding of the previous user can only be processed after the successful decoding of the previous user, an advanced SIC decoding strategy is proposed to achieve outage balancing for the later decoded users in [32]". With improper Gaussian Signaling (IGS) under OMA or NOMA, the "design of transmit beamforming to maximize the users' minimum throughput subject to transmit power constraints are addressed by [33]".

In [34], author proposed a "Multiple Decision aided Successive interference cancellation (MD-SIC) strategy with relatively low computational complexity to improve the performance of SIC receiver in the NOMA system". As a replacement for choosing one codeword for each user in conventional SIC detection, the proposed algorithm introduces multiple code-words as the candidate to conflict error propagation. In [35], author investigates the joint optimization of SIC ordering in NB-IoT devices as well as computation resource allocation. He compares the proposed work with existing multiple access techniques. In [36], the author has analyzed outage capacities of SIC and JD (Joint Detection) in the event of single–block transmission over a two-user Gaussian multiple-access channel with partial channel state information at the transmitter. In the JD algorithm, the signals which are transmitted from multiple users were concurrently decoded using sparse coding plus iterative multi-user MPA.

2.1 *Sum-Rate Analysis of NOMA in Downlink [18]*

In this rate analysis [18], two-user scenarios have been taken, UE1 which is at a far distance from the base station, and UE2 which is nearer to the base station. BS creates a superposed signal which contains data for both users UE1 and UE2, with power allocated differently as α_1 and α_2. Where $\alpha_1 + \alpha_2 = 1$. Channel gain $|h_1|^2$ is for UE1 and $|h_2|^2$ is for UE2. Here, $|h_2|^2 \geq |h_1|^2$ means UE2 is a stronger user than UE1 which allocates more power to the UE1 than UE2 i.e. $\alpha_1 \geq \alpha_2$ The superposed signals are

$$s_s = \sqrt{\alpha_1 \rho_s s_1} + \sqrt{\alpha_2 \rho_s s_2} \tag{1}$$

Here ρ_s denotes total transmit SNR. In the downlink, SIC is accomplished at the user end. At UE1, the received signal is

$$y_1 = h_1\left(\sqrt{\alpha_1 \rho_s} s_1 + \sqrt{\alpha_2 \rho_s} s_2\right) + n_1 \tag{2}$$

$$= h_1\sqrt{\alpha_1 \rho_s} s_1 + h_1\sqrt{\alpha_2 \rho_s} s_2 + n_1 \tag{3}$$

where n_1 is the zero-mean complex additive Gaussian noise with the variance of σ^2; at UE1.

The SINR for decoding data s_1 at UE1 is

$$SINR_{u_1}^{s_1} = \frac{\alpha_1 \rho_s |h_1|^2}{\alpha_2 \rho_s |h_1|^2 + n_1} \tag{4}$$

The achievable rate of UE1 is

$$\log_2\left(1 + SINR_{u_1}^{s_1}\right) = \log_2\left(1 + \frac{\alpha_1 \rho_s |h_1|^2}{\alpha_2 \rho_s |h_1|^2 + n_1}\right) \tag{5}$$

Now at user 2, the received signal will be

$$y_2 = h_2\left(\sqrt{\alpha_1 \rho_s} s_1 + \sqrt{\alpha_2 \rho_s} s_2\right) + n_2 \tag{6}$$

$$= h_2\sqrt{\alpha_1 \rho_s} s_1 + h_2\sqrt{\alpha_2 \rho_s} s_2 + n_2 \tag{7}$$

Here, SIC will perform on user 2 where user 1 will be decoded first than and interference of that will be successively removed from UE2 which results in decoding of UE2 data.

So, SINR for decoding s_1 at UE2

$$SINR_{u_2}^{s_1} = \frac{\alpha_1 \rho_s |h_2|^2}{\alpha_2 \rho_s |h_2|^2 + n_2} \tag{8}$$

After decoding UE1 and canceling interference successfully the received signal at UE2 will be

$$y_2 = h_2\sqrt{\alpha_2 \rho_s} s_2 + n_2 \tag{9}$$

SNR for decoding s_2 at UE2 is

$$SNR_{u_2}^{s_2} = \alpha_2 \rho_s |h_2|^2 \tag{10}$$

The sum-rate of NOMA $|h_2|^2 \geq |h_1|^2$ can be calculated as

$$R_{u_1}^{s_1} + R_{u_2}^{s_2} = \log_2\left(1 + \frac{\alpha_1 \rho_s |h_1|^2}{\alpha_2 \rho_s |h_1|^2 + n_1}\right) + \log_2(1 + \alpha_2 \rho_s |h_2|^2) \tag{11}$$

At high SNR that is $\rho_s \rightarrow \infty$, the term $\frac{\alpha_1\rho_s|h_1|^2}{\alpha_2\rho_s|h_1|^2+n_1} \rightarrow \frac{\alpha_1}{\alpha_2}$ and by further simplification

$$NOMA_{SUM-RATE} = \log_2\left(\frac{1}{\alpha_2} + \rho_s|h_2|^2\right) \geq \log_2\left(\rho_s|h_2|^2\right) \tag{12}$$

$$NOMA_{SUM-RATE} \geq \log_2\left(\rho_s|h_2|^2\right) \tag{13}$$

In the OMA system, the resources are allocated equally with equal power distribution among users, the received signals for the OMA downlink system for user 1 and user 2 are

$$\begin{aligned} y_1 &= h_1\sqrt{\rho_s}s_1 + n_1 \\ y_2 &= h_2\sqrt{\rho_s}s_2 + n_2 \end{aligned} \tag{14}$$

The SNR for decoding signals s_1 and s_2 UE1 and UE2 are

$$\begin{aligned} SN\tilde{R}_{u_1}^{s_1} &= \rho_s|h_1|^2 \\ SN\tilde{R}_{u_2}^{s_2} &= \rho_s|h_2|^2 \end{aligned} \tag{15}$$

The sum rate of OMA downlink for comparison purpose is given as

$$\tilde{R}_{u_1}^{s_1} + \tilde{R}_{u_2}^{s_2} = \frac{1}{2}\left(1 + \rho_s|h_1|^2\right) + \frac{1}{2}\left(1 + \rho_s|h_2|^2\right) \tag{16}$$

Which is approximately equal to

$$OMA_{SUM-RATE} \approx \frac{1}{2}\log_2\left(\rho_s|h_2|^2\right) \tag{17}$$

It is clearly seen from Eqs. (13) and (14) that

$$\begin{aligned} \log_2\left(\rho_s|h_2|^2\right) &>> \frac{1}{2}\log_2\left(\rho_s|h_2|^2\right) \\ NOMA_{SUM-RATE} &>> OMA_{SUM-RATE} \end{aligned} \tag{18}$$

Thus, the sum data rate of the NOMA downlink is higher than the OMA downlink.

2.2 *Sum-Rate Analysis of NOMA in Uplink [18]*

In the uplink, the base station receives superposed signals of both users that is UE1 and UE2. Here, UE1 is considered as the strongest user, and UE2 is considered as a weak user. At BS, stronger user will be decoded first and then SIC is performed on the weaker user. It will first decode the UE1 signal while considering UE2's signal as noise, then SIC performs on UE2's signal. The received signal at BS for both the users is

$$y_{B.S} = h_1\sqrt{\alpha_1\rho_s}s_1 + h_2\sqrt{\alpha_2\rho_s}s_2 + n \tag{19}$$

The B.S decodes s_1 first, followed by s_2 using SIC as

$$SINR_{B.S}^{s_1} = \frac{\alpha_1\rho_s|h_1|^2}{1+\alpha_2\rho_s|h_2|^2} \tag{20}$$

The resulting signal after decoding and removing signal s_1 is

$$y_2 = h_2\sqrt{\alpha_2\rho_s}s_2 + n \tag{21}$$

SNR for UE2 is calculated as

$$SNR_{B.S}^{s_2} = \alpha_2\rho_s|h_2|^2 \tag{22}$$

The sum-rate of NOMA for uplink $|h_1|^2 \geq |h_2|^2$ can be calculated as

$$R_{B.S}^{s_1} + R_{B.S}^{s_2} = \log_2\left(1 + \frac{\alpha_1\rho_s|h_1|^2}{1+\alpha_2\rho_s|h_2|^2}\right) + \log_2\left(1+\alpha_2\rho_s|h_2|^2\right) \tag{23}$$

At high SNR that is $\rho_s \to \infty$,

$$NOMA_{SUM-RATE} \approx \log_2\left(\rho_s|h_1|^2 + \rho_s|h_2|^2\right) \tag{24}$$

While in the OMA system, the sum rate is given as

$$OMA_{SUM-RATE} \approx \frac{1}{2}\log_2\left(\rho_s|h_1|^2\right) + \frac{1}{2}\log_2\left(\rho_s|h_2|^2\right) \tag{25}$$

$$\leq \log_2\left(\rho_s|h_2|^2\right).$$

It is clearly seen that $NOMA_{SUM-RATE} \geq OMA_{SUM-RATE}$ for the uplink network.

Discussion and Outlook. It has been realized that NOMA outperformance OMA in sum rate analysis in uplink as well as downlink, Most of the work of SIC falls under a downlink scenario with three users, where they mostly focus on user decoding and SIC ordering, [30] propose dynamic ordering of users. Improper Gaussian signaling has been considered for OMA as well as NOMA to maximize the throughput of the user. In the future, work can be done on user scheduling and error propagation. Work can also be done on the MIMO-NOMA scheme where there is a more number of users on the receiver side where the issue of latency and error propagation can be addressed.

3 NOMA with MIMO

This section deals with the integration of NOMA with other technologies. As per the IMT report [1], the fifth generation and beyond the fifth generation network must support a large number of connections. For this, several new technologies have been developed in the past decade and some are in research. Examples of these are multiple-input multiple-output (MIMO), Massive MIMO, Cognitive Radio, mm-wave, Co-Operative Communication, etc. In this section, we have reviewed MIMO-NOMA and NOMA in Cellular Communication.

3.1 MIMO-NOMA

MIMO is the key technology which can improve the spectral efficiency. Generally, MIMO techniques can be characterized into SU-MIMO and MU-MIMO [37] which can exploit the spatial domain, and combination with NOMA can further exploit the power domain. To enhance system capacity and spectral efficiency, these two techniques can be integrated. There have been several research work that is carried out in the context of MIMO-NOMA. This paper mainly focuses on MIMO-NOMA and NOMA for the cellular network in the uplink and downlink scenario.

It has been seen in literature that NOMA overtakes OMA in terms of sum channel capacity also larger sum Ergodic capacity [38, 39]. While the combination of MIMO with NOMA gives a spatial degree of freedom to meet the requirements of 5G standards [40]. MIMO-NOMA uses superposition coding (SC) at the transmitter plus successive interference cancellation (SIC) at the receiver with multiple antennas, which can be further categorized and analyzed with uplink and downlink scenarios separately.

Real-world experimental trials of open-loop and closed-loop MIMO with NOMA has been performed by [37], where A. Benjebbour et al. have used advanced NOMA with CWIC and IC receiver technique to enhance the spectrum efficiency, they have considered two transmit CRS (Common Reference Signal) and 8 transmit DMRS(Demodulation Reference Signal)-based transmit scheme. Power allocation policies are the main feature of the NOMA scheme which has been addressed by [40] for small packet communication in the IoT, as well as in [41] author design a new precoding and detection matrices for MIMO-NOMA with a static set of power allocation.

In [26], Liu et al. focus on a different perspective of MIMO-NOMA beamforming techniques such as centralized and coordinated beamforming. He discussed the inter-cluster interference-free scheme and inter-cluster interference tolerant scheme. Rate regions of the MIMO-NOMA network and resource allocation problem for single-carrier and multi-carrier has also been investigated. Capacity and Ergodic Capacity assessment between MIMO-NOMA as well as MIMO-OMA has been done in [38] and [39], respectively. Later, the author has considered Rayleigh faded channel with knowledge of CSIT. In [42], Cui et al. have considered the system utility maximization delinquent of downlink MIMO-NOMA system for two types of imperfect CSI: CDI and channel estimation uncertainty.

User pairing in the MIMO-NOMA DL transmission system using different linear beamforming techniques has been discussed in [43], where Park et al. has formulated sum spectral efficiency maximization. "Low complexity and capacity approaching solution for practical coded MIMO-NOMA system has been discussed by [44]", where the receiver contains a multiuser linear MMSE detector besides bank of single-user message passing decoder.

The sum-rate optimization problem for the MIMO-NOMA system is studied through a total transmit power constraint and the minutest rate constraint of a weak user in [45]. A novel joint user clustering and power allocation algorithm has been investigated and compared with the MIMO-OMA system by [46]. Whether NOMA or Massive MIMO using ZF has been related in terms of the highest average sum-rate by [47].

NOMA in massive MIMO. By using massive numbers of antennas at the transmitter side, using rich scattering space domain, massive MIMO has gain attention in 5G technology to address massive connectivity, reduced latency, and improvement in system gain [48]. However, PD-NOMA which uses a non-orthogonal concept connect more number of users [8]. Integration of massive MIMO and NOMA can further improve system capacity, spectral efficiency, and outage probability. The receive antenna selection (RAS) for massive MIMO-NOMA uplink channel has been derived in sum rate analysis and capacity region in [49], where the author has considered Rayleigh fading channel. One bit feedback and user orders are done in [50], where the author has simplified massive MIMO network into multiple SISO-NOMA network, two scenarios with one-bit feedback and perfect user ordering have been considered for analysis.

The disadvantage of error propagation in SIC can deteriorate the performance of NOMA system and while we consider massive MIMO, this can be a major problem, in [48], author has derived closed-form expression for outage probability and Ergodic capacity for multi-user multi-cluster massive MIMO-NOMA system. It has been proved that if residual error propagation is high, OMA is a better option than NOMA. In [51], the author has analyzed sum-rate gains for multi-cell massive MIMO NOMA under the issue of user clustering and pilot assignment schemes. Two cases, under loading and overloading, have been considered. NOMA requires CSI knowledge at the receiver which can be removed by applying blind belief propagation (BP) detection for Non-Coherent NOMA with massive MIMO [52], author has used LDS for overloaded users, the scheme doesn't require knowledge of CSI. NOMA/OMA mode switching with the allocation of pilots to each cluster with orthogonality has been proposed in [53] where the author has analyzed DL of NOMA based cell-free massive MIMO system.

Practical channel measurement of massive MIMO-NOMA system in an indoor and outdoor environment has been done in [54], the author has evaluated system performance in terms of sum-rate capacity for ZF and MRT scheme. In [55] user pairing and pair scheduling algorithms are proposed to achieve sum-rate and outage performance. The author has design ICC matrix to mitigate inter pair interference for receiving SINRs of each user in each pair. [47] discuss, what the role of NOMA in massive MIMO is? as both serve the same purpose in terms of non-orthogonality. The author has provided performance analysis and comparison between NOMA and ZF based massive MIMO beamforming schemes with NLOS channels. It has been seen that massive MIMO outperforms NOMA in sum rate analysis. In [56], the author has investigated intracluster pilot contamination, inter-cluster interference, imperfect SIC, and imperfect CSI. The achievable DL sum rate of massive MIMO-NOMA with underlay spectrum sharing has been derived.

NOMA in Cellular Communication. NOMA has been considered as a favorable enabling technique for 5G and beyond 5G cellular networks [19, 21, 22]. Till now Orthogonal Multiple Access (OMA) is deliberated as a state of the art wireless communication technique. These techniques, for example, FDMA, TDMA, CDMA, and OFDMA is used in 1G, 2G, 3G, and 4G, respectively. They utilize orthogonal resources (time, frequency, or code) to address multiple users. In OMA, there is no intra cell interference and receiver complexity is likewise very less. Nonetheless, the number of served users is less which is restricted to the number of subcarriers. This can be solved by utilizing the same spectrum resources non-orthogonally with the cost of increasing intra cell interference.

The key distinction and operational principle of NOMA in the cellular network have been exploited by [57, 58] where H. Tabassum et al. have "derived the NOMA dominant condition for each user in a two-user NOMA cluster" which has been compared with a random user selection scheme. The author has also focused on open research issues such as SIC error propagation, multi-cell NOMA transmission, User grouping/ scheduling, power allocation, and cooperation in NOMA. BER analysis in various fading channel conditions such as AWGN, Rayleigh, and Rician has been done in [5], where Sadia et al. concluded that the performance of this system develops significantly with appropriate channel coding. To combat interference, a real interference cancellation scheme, i.e. zero forcing is use. In the future, this study can be a base for the improvement of adaptive modulation and coding scheme.

In [59] Yang et al. propose a novel NOMA scheme which is dynamic NOMA (D-NOMA), which works on both uplinks as well as downlink case of power allocation strategy. The analytical expression has been derived for outage probability and the average rate in uplink plus downlink NOMA. In the future D-NOMA, power allocation algorithm can be combined with user pairing/clustering algorithms, even the scheme can be extended to the MIMO scenario. Various beamforming techniques for the downlink MISO NOMA system with perfect CSI has been discussed and analyzed in [60].

In [61], Ali et al. demonstrate a comparison between proposed CoMP-NOMA models with traditional CoMP-OMA systems this scenario can be extended to Multiple BS antennas. Ali et al. [62] Discussed user grouping as well as the power allocation problem in uplink and downlink NOMA for multi-user NOMA system. Ali et al. have also mentioned issues related to SIC error processing plus inter-cell interference, power allocation solution should be extended for dense cellular networks. Random Beamforming and intra beam SIC in downlink cellular MIMO have been discussed by [63].

NOMA scheme works better in near far user when the amount of base station antennas are the same or more than the number of users, Ali et al. formulate cluster for more number of a user than the number of transmit antenna. Novel inter-cluster ZF-BF technique for downlink NOMA can eliminate inter-cluster interference which maximizes the overall throughput of the system. Power allocation for the downlink SISO-NOMA system has been addressed by [64]. It has obtained optimal power allocation in closed form. User grouping is an essential requirement to carry out the NOMA scheme which has been addressed by [65] to maximize the sum rate. In [66] author investigates system-level output performance of NOMA with MMSE-SIC-based linear filtering in the cellular uplink system.

3.2 *Sum-Rate Analysis of the MIMO-NOMA [38, 39, 41]*

In this system, it uses the NOMA scheme by SC at transmitter plus SIC at receiver for signal detection is shown in the figure. One B.S uses two users in a one-time slot, each user has N receive antennas and B.S has M transmit antennas. B.S transmits symbol vectors $S_1 \in C^N$ and $S_2 \in C^N$ where $\left(E\left(S_1 S_1^H\right) = E\left(S_2 S_2^H\right) = I_N\right)$ to UE1 and UE2 correspondingly at the same time and frequency slot via power domain separation. The signal at the B.S transmitter can be written as

$$S = V_1 S_1 + V_2 S_2 \tag{26}$$

where $V_1 \in C^{M \times N}$ $V_2 \in C^{M \times N}$ and are the precoding matrix for UE1 and UE2, respectively. We have considered spatially uncorrelated frequency flat Rayleigh fading channel,$H_1 H_2$ and are i.i.d with $CN\left(0, \sigma_{H_1}^2\right)$ and $CN\left(0, \sigma_{H_2}^2\right)$ respectively. n_1, n_2 are normalized AWGN noise at UE1 and UE2, respectively $n_1, n_2 \sim CN(0, I)$. According to the NOMA scheme at UE2, which is the weakest user, SIC will perform on it. Here UE1 will decode its own symbol s_1 by treating s_2 as interfering noise, then the achievable rate of UE1 is constrained by

$$R_{S_1}^{u_1} \leq \log_2 \det\left(I + \left(I + H_1 Q_2 H_1^H\right)^{-1} H_1 Q_1 H_1^H\right) \tag{27}$$

where $Q_1 = V_1 V_1^H, Q_2 = V_2 V_2^H$ are transmit covariance matrix for UE1 and UE2, respectively. At UE2, it will first decipher the symbol vector s_1 via treating interference initiated by s_2 as noise and then cancel the part of received signal caused by s_1 from receive signal and decode s_2 from the remaining part of the received signal.

The achievable rate of UE2 can be written as

$$R_{S_2}^{u_2} \leq \log_2 \det\left(I + H_2 Q_2 H_2^H\right) \tag{28}$$

And at the same time, the rate of UE1 is upper bound by

$$R_{S_1}^{u_2} \leq \log_2 \det\left(I + \left(I + H_2 Q_2 H_2^H\right)^{-1} H_2 Q_1 H_2^H\right) \tag{29}$$

To ensure that UE2 can successfully decode s_1. The achievable rate region for UE1 and UE2 are given as

$$R_1^{MIMO-NOMA} \leq \min\left\{\begin{array}{l} \log_2 \det\left(I + \left(I + H_1 Q_2 H_1^H\right)^{-1} H_1 Q_1 H_1^H\right), \\ \log_2 \det\left(I + \left(I + H_2 Q_2 H_2^H\right)^{-1} H_2 Q_1 H_2^H\right) \end{array}\right\}$$
$$R_2^{MIMO-NOMA} \leq \log_2 \det\left(I + H_2 Q_2 H_2^H\right) \tag{30}$$

Discussion and Outlook. Existing research of MIMO-NOMA mainly focuses on open-loop MIMO transmission where authors have to assume M = N that is the sum of transmit antennas is equals to the sum of receive antennas or there should be more transmit antenna than the receiver. Closed-loop MIMO transmission with limited channel feedback is an open issue for the more number of users on the receiver side. Also, user grouping and scheduling is the prime requirement for successful MIMO-NOMA implementation. In [26], the author suggests SIC ordering as an open issue to be addressed for the MIMO-NOMA network. There is also work to be done for optimal NOMA transmission for the multi-cell network. For a large network, when several users are more than the transmitter, SIC ordering and user clustering can be considered as an issue to reduce receiver computational complexity.

Existing work on NOMA cellular network is mostly done in the downlink NOMA system, there are some papers [57, 58, 62, 64], in which they focus on equally Uplink and Downlink NOMA systems. User grouping, Clustering, and power allocation strategies are the main problems with which future work has to be done. The multi-cell dense cellular network is an open issue that is yet to be resolved. The main problem of implementing the NOMA system is to mitigate Inter/ Intra-cluster interference and Intra/Intercell interference. The design of the uplink reference signal for channel estimation may also be considered as future work.

4 Conclusion

NOMA is an important enabling multiple access technique in the 5G network and 5G+. A lot of work is in progress to fulfill system requirements such as high system throughput, low latency, besides, massive connectivity. In this review report, different aspects of NOMA, which include its basic concept, types of NOMA, and integration of NOMA with additional technologies such as NOMA with MIMO, NOMA with massive MIMO, and NOMA in Cellular communication has been considered. All aspects are extensively reviewed and compared. NOMA works better when there is a large distinction of channel gain among users, which can be one of the limitations of the NOMA technique. To optimize the NOMA system, user grouping and scheduling is necessary as well as SIC ordering is a must to further reduce interference. Large work has been done and in progress to optimize the system in terms of power allocation strategies, SIC ordering, user grouping, and clustering, which have been mentioned in the future work at the end of each section with discussion and outlook.

References

1. ITU-R, IMT vision—framework and overall objectives of the future development of IMT for 2020 and beyond. Itu-R M.2083-0 (2015), https://www.itu.int/dms_pubrec/itu-r/rec/m/R-REC-M
2. International Telecommunications Union, Future technology trends of terrestrial IMT systems. Int. Telecommun. Union (2014)
3. J. He, Z. Tang, Z. Che, Fast and efficient user pairing and power allocation algorithm for non-orthogonal multiple access in cellular networks. Electron. Lett. **52**(25), 2065–2067 (2016)
4. A. Singh, K.K. Naik, C.R.S. Kumar, Impact of SC-FDMA and pilots on PAPR and performance of power domain NOMA-UFMC system, in *International Conference on Ubiquitous and Future Networks, ICUFN*, vol. 2018 (2018), pp. 507–511
5. H. Sadia, M. Zeeshan, S.A. Sheikh, Performance analysis of downlink power domain NOMA under fading channels, in *12th International Conference on ELEKTRO 2018* (2018), pp. 1–6
6. J. Zhang, J. Ge, Q. Ni, M. Wen, Y. Zhang, Performance analysis of user ordering schemes in cooperative power-domain non-orthogonal multiple access network. IEEE Access **6**(c), 47319–47331 (2018)
7. L. Lei, D. Yuan, C.K. Ho, S. Sun, Power and channel allocation for non-orthogonal multiple access in 5G systems: tractability and computation. IEEE Trans. Wirel. Commun. **15**(12), 8580–8594 (2016)
8. A. Kassir, R.A. Dziyauddin, H.M. Kaidi, M.A. Mohd Izhar, Power domain non-orthogonal multiple access: a review, in *2018 2nd International Conference on Telematics and Future Generation Networks, TAFGEN 2018* (2018), pp. 66–71
9. S.M.R. Islam, N. Avazov, O.A. Dobre, K.S. Kwak, Power-domain non-orthogonal multiple access (NOMA) in 5G systems: potentials and challenges. IEEE Commun. Surv. Tutor. **19**(2), 721–742 (2017)
10. X. Yan et al., The application of power-domain non-orthogonal multiple access in satellite communication networks. IEEE Access **7**, 63531–63539 (2019)
11. L. Dai, B. Wang, Z. Ding, Z. Wang, S. Chen, L. Hanzo, A survey of non-orthogonal multiple access for 5G. IEEE Commun. Surv. Tutor. **20**(3), 2294–2323 (2018)
12. M.T.P. Le, G.C. Ferrante, G. Caso, L. De Nardis, M.G. Di Benedetto, On information-theoretic limits of code domain NOMA for 5G. IET Commun. **12**(15), 1864–1871 (2018)
13. Z. Wu, K. Lu, C. Jiang, X. Shao, Comprehensive study and comparison on 5G NOMA schemes. IEEE Access **6**, 18511–18519 (2018)
14. M. Moltafet, N.M. Yamchi, M.R. Javan, P. Azmi, Comparison study between PD-NOMA and SCMA. IEEE Trans. Veh. Technol. **67**(2), 1830–1834 (2018)

15. H. Nikopour, H. Baligh, Sparse code multiple access, in *IEEE International Symposium on Personal, Indoor and Mobile Radio Communications, PIMRC* (2013), pp. 332–336
16. R. Hoshyar, F.P. Wathan, R. Tafazolli, Novel low-density signature for synchronous CDMA systems over AWGN channel. IEEE Trans. Signal Process. **56**(4), 1616–1626 (2008)
17. S. Chen, B. Ren, Q. Gao, S. Kang, S. Sun, K. Niu, Pattern division multiple access-a novel nonorthogonal multiple access for fifth-generation radio networks. IEEE Trans. Veh. Technol. **66**(4), 3185–3196 (2017)
18. M. Aldababsa, M. Toka, S. Gökçeli, G.K. Kurt, O. Kucur, A tutorial on nonorthogonal multiple access for 5G and beyond. Wirel. Commun. Mob. Comput. **2018** (2018)
19. Z. Ding, X. Lei, G.K. Karagiannidis, R. Schober, J. Yuan, V.K. Bhargava, A survey on non-orthogonal multiple access for 5G networks: research challenges and future trends. IEEE J. Sel. Areas Commun. **35**(10), 2181–2195 (2017)
20. M. Vaezi, G. Amarasuriya, Y. Liu, A. Arafa, F. Fang, Z. Ding, Interplay between NOMA and other emerging technologies: a survey. **7731**(c), 1–20 (2019)
21. L. Dai, B. Wang, Y. Yuan, S. Han, I. Chih-Lin, Z. Wang, Non-orthogonal multiple access for 5G: solutions, challenges, opportunities, and future research trends. IEEE Commun. Mag. **53**(9), 74–81 (2015)
22. Y. Tao, L. Liu, S. Liu, Z. Zhang, A survey: several technologies of non-orthogonal transmission for 5G. China Commun. **12**(10), 1–15 (2015)
23. M. Baghani, S. Parsaeefard, M. Derakhshani, W. Saad, Dynamic non-orthogonal multiple access and orthogonal multiple access in 5G wireless networks. IEEE Trans. Commun. **67**(9), 6360–6373 (2019)
24. Y. Cai, Z. Qin, F. Cui, G.Y. Li, J.A. McCann, Modulation and multiple access for 5G networks. IEEE Commun. Surv. Tutor. **20**(1), 629–646 (2018)
25. K. Saito, A. Benjebbour, Y. Kishiyama, Y. Okumura, T. Nakamura, Performance and design of SIC receiver for downlink NOMA with open-loop SU-MIMO, in *2015 IEEE International Conference on Communications Workshops, ICCW 2015* (2015), pp. 1161–1165
26. Y. Liu, H. Xing, C. Pan, A. Nallanathan, M. Elkashlan, L. Hanzo, Multiple-antenna-assisted non-orthogonal multiple access. IEEE Wirel. Commun. **25**(2), 17–23 (2018)
27. T.M. Cover, Broadcast channels. IEEE Trans. Inf. Theory **18**(1), 2–14 (1972)
28. S. Vanka, S. Srinivasa, Z. Gong, P. Vizi, K. Stamatiou, M. Haenggi, Superposition coding strategies: design and experimental evaluation. IEEE Trans. Wirel. Commun. **11**(7), 2628–2639 (2012)
29. K. Higuchi, A. Benjebbour, S. Member, Non-orthogonal multiple access (NOMA) with successive. IEICE Trans. Commun. **E98-B**(3), 403–414 (2015)
30. Y. Gao, B. Xia, K. Xiao, Z. Chen, X. Li, S. Zhang, Theoretical analysis of the dynamic decode ordering SIC receiver for uplink NOMA systems. IEEE Commun. Lett. **21**(10), 2246–2249 (2017)
31. I.A. Mahady, E. Bedeer, S. Ikki, H. Yanikomeroglu, Sum-rate maximization of NOMA systems under imperfect successive interference cancellation. IEEE Commun. Lett. **23**(3), 474–477 (2019)
32. B. Xia, J. Wang, K. Xiao, Y. Gao, Y. Yao, S. Ma, Outage performance analysis for the advanced SIC receiver in wireless NOMA systems. IEEE Trans. Veh. Technol. **67**(7), 6711–6715 (2018)
33. H.D. Tuan, A.A. Nasir, H.H. Nguyen, T.Q. Duong, H.V. Poor, Non-orthogonal multiple access with improper gaussian signaling. IEEE J. Sel. Top. Signal Process. **13**(3), 496–507 (2019)
34. B. Ling, C. Dong, J. Dai, J. Lin, Multiple decision aided successive interference cancellation receiver for NOMA systems. IEEE Wirel. Commun. Lett. **6**(4), 498–501 (2017)
35. L.P. Qian, A. Feng, Y. Huang, Y. Wu, B. Ji, Z. Shi, Optimal SIC ordering and computation resource allocation in MEC-aware NOMA NB-IoT networks. IEEE Internet Things J. **6**(2), 2806–2816 (2019)
36. S. Chen, K. Peng, H. Jin, J. Song, Analysis of outage capacity of NOMA: SIC vs. JD. Tsinghua Sci. Technol. **21**(5), 538–543 (2016)
37. A. Benjebbour, Y. Kishiyama, Combination of NOMA and MIMO: concept and experimental trials, in *2018 25th International Conference on Telecommunications (ICT 2018)* (2018), pp. 433–438

38. Y. Liu, G. Pan, H. Zhang, M. Song, On the capacity comparison between MIMO-NOMA and MIMO-OMA. IEEE Access **4**, 2123–2129 (2016)
39. Q. Sun, S. Han, I. Chin-Lin, Z. Pan, On the ergodic capacity of MIMO NOMA systems. IEEE Wirel. Commun. Lett. **4**(4), 405–408 (2015)
40. Z. Ding, L. Dai, H.V. Poor, MIMO-NOMA design for small packet transmission in the internet of things. IEEE Access **4**, 1393–1405 (2016)
41. Z. Ding, F. Adachi, H.V. Poor, The application of MIMO to non-orthogonal multiple access. IEEE Trans. Wirel. Commun. **15**(1), 537–552 (2016)
42. J. Cui, Z. Ding, P. Fan, Outage probability constrained MIMO-NOMA designs under imperfect CSI. IEEE Trans. Wirel. Commun. **17**(12), 8239–8255 (2018)
43. S. Park, A.Q. Truong, T.H. Nguyen, Power control for sum spectral efficiency optimization in MIMO-NOMA systems with linear beamforming. IEEE Access **7**, 10593–10605 (2019)
44. Y. Chi, L. Liu, G. Song, C. Yuen, Y.L. Guan, Y. Li, Practical MIMO-NOMA: low complexity and capacity-approaching solution. IEEE Trans. Wirel. Commun. **17**(9), 6251–6264 (2018)
45. M. Tian, Q. Zhang, S. Zhao, Q. Li, J. Qin, Secrecy sum rate optimization for downlink MIMO nonorthogonal multiple access systems. IEEE Signal Process. Lett. **24**(8), 1113–1117 (2017)
46. S. Chinnadurai et al., User clustering and robust beamforming design in multicell MIMO-NOMA system for 5G communications. AEU Int. J. Electron. Commun. **78**, 181–191 (2017)
47. K. Senel, H.V. Cheng, E. Björnson, E.G. Larsson, What role can NOMA play in massive MIMO? IEEE J. Sel. Top. Signal Process. **13**(3), 597–611 (2019)
48. A.S. De Sena et al., Massive MIMO-NOMA networks with imperfect SIC: design and fairness enhancement. IEEE Trans. Wirel. Commun. **19**(9), 6100–6115 (2020)
49. W.A. Al-Hussaibi, F.H. Ali, Efficient user clustering, receive antenna selection, and power allocation algorithms for massive MIMO-NOMA systems. IEEE Access **7**, 31865–31882 (2019)
50. Z. Ding, H.V. Poor, Design of massive-MIMO-NOMA with limited feedback. IEEE Signal Process. Lett. **23**(5), 629–633 (2016)
51. D. Kudathanthirige, G.A.A. Baduge, NOMA-aided multicell downlink massive MIMO. IEEE J. Sel. Top. Signal Process. **13**(3), 612–627 (2019)
52. T. Wang, L. Shi, K. Cai, L. Tian, S. Zhang, Non-coherent NOMA with massive MIMO. IEEE Wirel. Commun. Lett. **9**(2), 134–138 (2020)
53. M. Bashar, K. Cumanan, A.G. Burr, H.Q. Ngo, L. Hanzo, P. Xiao, On the performance of cell-free massive MIMO relying on adaptive NOMA/OMA mode-switching. IEEE Trans. Commun. **68**(2), 792–810 (2020)
54. E.P. Simon, J. Farah, P. Laly, Performance evaluation of massive MIMO with beamforming and nonorthogonal multiple access based on practical channel measurements. IEEE Antennas Wirel. Propag. Lett. **18**(6), 1263–1267 (2019)
55. X. Chen, F.K. Gong, G. Li, H. Zhang, P. Song, User pairing and pair scheduling in massive MIMO-NOMA systems. IEEE Commun. Lett. **22**(4), 788–791 (2018)
56. Y. Li, G.A.A. Baduge, Underlay spectrum-sharing massive MIMO NOMA. IEEE Commun. Lett. **23**(1), 116–119 (2019)
57. H. Tabassum, M.S. Ali, E. Hossain, M.J. Hossain, D.I. Kim, Non-orthogonal multiple access (NOMA) in cellular uplink and downlink: challenges and enabling techniques (2016), pp. 1–8
58. H. Tabassum, M.S. Ali, E. Hossain, M.J. Hossain, D.I. Kim, Uplink Vs. downlink NOMA in cellular networks: challenges and research directions, in *IEEE Vehicular Technology Conference*, vol. 2017, June (2017)
59. Z. Yang, W. Xu, C. Pan, Y. Pan, M. Chen, On the optimality of power allocation for NOMA downlinks with individual QoS constraints. IEEE Commun. Lett. **21**(7), 1649–1652 (2017)
60. F. Alavi, K. Cumanan, Z. Ding, A.G. Burr, Beamforming techniques for nonorthogonal multiple access in 5G cellular networks. IEEE Trans. Veh. Technol. **67**(10), 9474–9487 (2018)
61. M.S. Ali, E. Hossain, A. Al-Dweik, D.I. Kim, Downlink power allocation for CoMP-NOMA in multi-cell networks. IEEE Trans. Commun. **66**(9), 3982–3998 (2018)
62. M.S. Ali, E. Hossain, D.I. Kim, Non-orthogonal multiple access (NOMA) for downlink multiuser MIMO systems: user clustering, beamforming, and power allocation. IEEE Access **5**, 565–577 (2017)

63. Y. Hayashi, Y. Kishiyama, K. Higuchi, Investigations on power allocation among beams in nonorthogonal access with random beamforming and intra-beam SIC for cellular MIMO downlink, in *IEEE Vehicular Technology Conference* (2013)
64. Z. Yang, Z. Ding, P. Fan, N. Al-Dhahir, A general power allocation scheme to guarantee quality of service in downlink and uplink NOMA systems. IEEE Trans. Wirel. Commun. **15**(11), 7244–7257 (2016)
65. J.M. Kang, I.M. Kim, Optimal user grouping for downlink NOMA. IEEE Wirel. Commun. Lett. **7**(5), 724–727 (2018)
66. Y. Endo, Y. Kishiyama, K. Higuchi, Uplink non-orthogonal access with MMSE-SIC in the presence of inter-cell interference, in *Proceedings International Symposium on Wireless Communication Systems* (2012, pp. 261–265

Triple Band Circular Patch Antenna Using Complimentary Split Ring Resonators

Falguni Raval

Abstract This paper presents a triple band microstrip circular antenna which uses two different sized complementary split ring resonators in the ground plane. The proposed antenna works on 2.4, 3.58, and 5.58 GHz frequencies. The complementary split ring resonators stimulate 2.4 and 3.58 GHz in antenna. The complementary split ring resonators have minor effect on 5.5 GHz frequency. The prototype antenna is fabricated and tested. There is good matching observed between simulated and measured results. The proposed antenna is simple to construct.

Keywords Circular patch antenna · Complementary split ring resonators (CSRRs) · Split ring resonator (SRR)

1 Introduction

Microstrip patch has recently become desirable for antenna designs because of its small scale, basic construction, and simple production and incorporation with the circuit. A lot of microstrip patch antenna designs have been reported in wireless communications. Furthermore, it also offers an opportunity of multi-band operations. The main benefit of it is that number of frequencies can be incorporated in single antenna.

Different architectures can be used to provide mutli-band frequency operation. A broad monopole with two narrow slots of various lengths is used in [1–3] to have a triple band. In [4], the U-shaped slot stimulate two extra frequencies in the partial ground plane and the radiating element with split ring slot. In [5], S-shaped resonators are used for achieving three band. In [6], three resonating elements are shaped out above the ground plane to have the triple band notch characteristics for a compact monopole antenna. In [7], rectangular microstrip patch antennas are partially

F. Raval (✉)
V. T. Patel Department of Electronics & Communication Engineering, Chandubhai S Patel Institute of Technology, Charortar University of Science and Technology(CHARUSAT), Changa, India
e-mail: falguniraval.ec@charusat.ac.in

F. Thakkar et al. (eds.), *Proceedings of the International e-Conference on Intelligent Systems and Signal Processing*, Advances in Intelligent Systems and Computing 1370,
https://doi.org/10.1007/978-981-16-2123-9_14

loaded with ENG metamaterial. In [8], the ground plane is changed by loading E shaped and U-shaped strip and two inverted L-shaped protruding strips are used in radiating element for triple band achievement. In [9], in order to acquire triple band, complementary split ring resonator (CSRR) is loaded into radiating element. The properties of dual bands [10–12] and triple bands [4, 13] can also be accomplished using structures inspired by metamaterials. In [14], triple band notching is achieved by CSRR loading in radiation patch, defected microstrip line and defected ground plane. Two distinct geometrical radiating structures are used and the truncated ground plane is used for the characteristics of the triple band [15]. A crinkle-fractal-structure monopole antenna for DCS/WiMAX and WLAN applications is recorded in [16]. In [17], modified rectangular SRR and quarter wavelength rectangular slot are introduced in a radiating patch to obtain triple band.

In the presented work, a triple band patch antenna is designed using two different sized CSRRs in the ground plane of a conventional patch antenna. This antenna is simple in structure.

2 Split Ring Resonator

Figure 1 shows the split ring resonator which is made up of two conducting rings which has common center and it has split on opposite sides [18]. It has distributed L and C and it acts as LC resonator [19]. Based on the concept of duality and Babinet's principle, CSRR is the complementary structure of SRR. The process of resonance is the similar for SRR and CSRR [19]. The CSRR has an electromagnetic behavior that is dual of the SRR [19]. The resonance frequency of it can be changed by changing its geometric parameters (g, s, r1, and r2) [19]. In Table 1, the sizes of the split ring resonators used are specified.

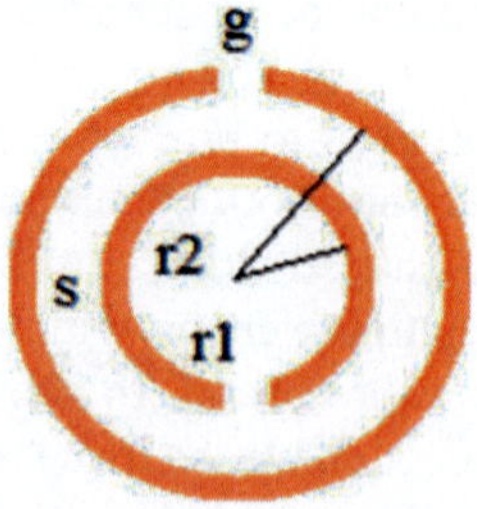

Fig. 1 Split ring resonator

Table 1 Dimensions of split ring resonators

	Split ring resonator 1 (mm)	Split ring resonator 2 (mm)
Radius r1 (mm)	3.052	2.17
Radius r2 (mm)	4.578	3.255
Spacing between ring S (mm)	0.654	0.465
Split gap g (mm)	1.744	1.4
Thickness of rings (mm)	0.872	0.62

3 Design

The shape of patch antenna is circular. The dielectric material used is RT Duroid-5880 whose dielectric constant is 2.2. The thickness of the material is 0.83 mm. The transmission line feeding technique is used in this design. The width of feed line is varied for impedance matching. The design frequency is 5.5 GHz. The radius of the circular patch antenna is calculated using design Eqs. (1) and (2) [20] mentioned below. The patch antenna has a radius of 10.24 mm. The feed line length and width are 12.76 mm and 1 mm, respectively. The dielectric substrate size is 46 × 46 mm.

Radius

$$a = \frac{F}{\sqrt{1 + \frac{2h}{\varepsilon_r F \pi}(ln\frac{\pi F}{2h} + 1.7726}} \quad (1)$$

and

$$F = \frac{8.791 \times 10^9}{f_r\sqrt{\varepsilon_r}} \quad (2)$$

where h is height of substrate.

ε_r dielectric constant of the substrate.

fr is resonance frequency.

4 Arrangement of CSRRs

The arrangement of CSRRs is shown in Fig. 2a. As shown in Fig. 2a, the large CSRR1 is located 10.4 mm away from line AB. Similarly, smaller CSRR2 is located 0.71 mm away from line AB, as shown in Fig. 2a. In order to have desired frequency; the placement of CSRR is changed. The patch antenna with CSRRs is fabricated. Figure 2b shows top view of fabricated patch antenna with CSRRs and Fig. 2c shows bottom layer of fabricated patch antenna with CSRRs.

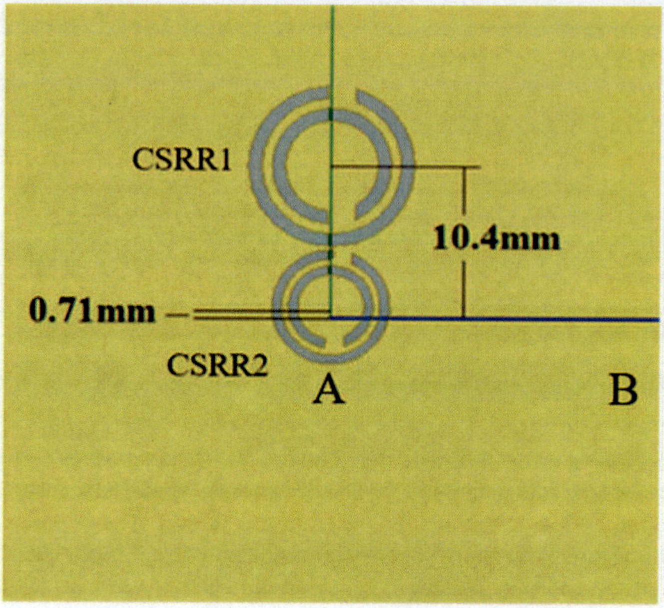

(a) Configuration of CSRRs in ground plane

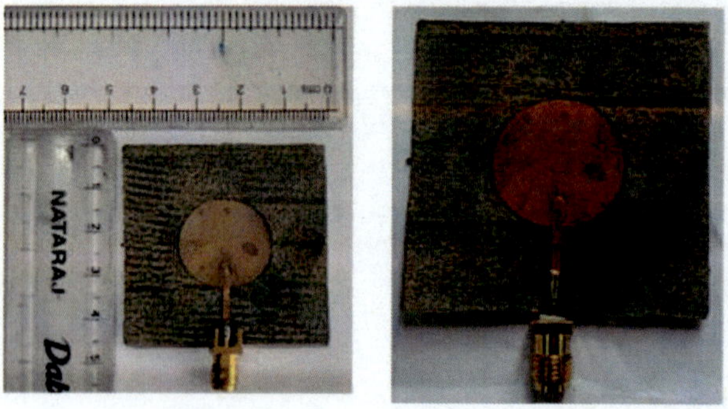

(b) Top view of Proposed antenna

(c) bottom view of Proposed antenna

Fig. 2 Fabricated prototype antenna

5 Simulation, Measurement, and Results

The simulation is carried out using high frequency structure simulator. Figure 3 shows placement of individual CSRR and combination of CSRRs in ground plane of antennas and the simulated reflection coefficient is shown in Fig. 4. It is observed that the antenna with larger CSRR1 introduces new resonating frequency 2.41 GHz in antenna as shown in Fig. 4. Same way, antenna with smaller CSRR2 also introduces

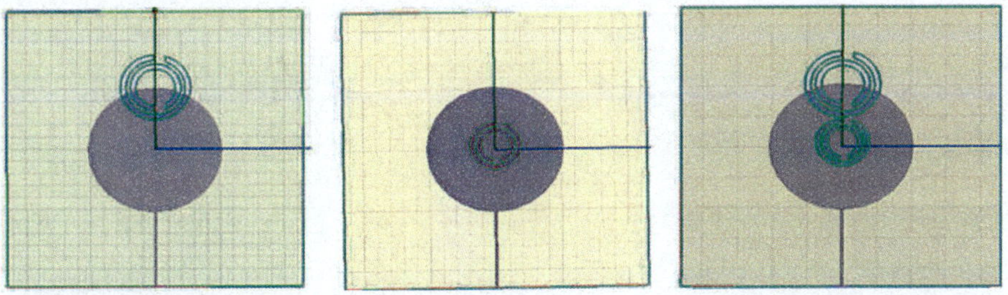

Fig. 3 Structures of antennas with different numbers of CSRRs

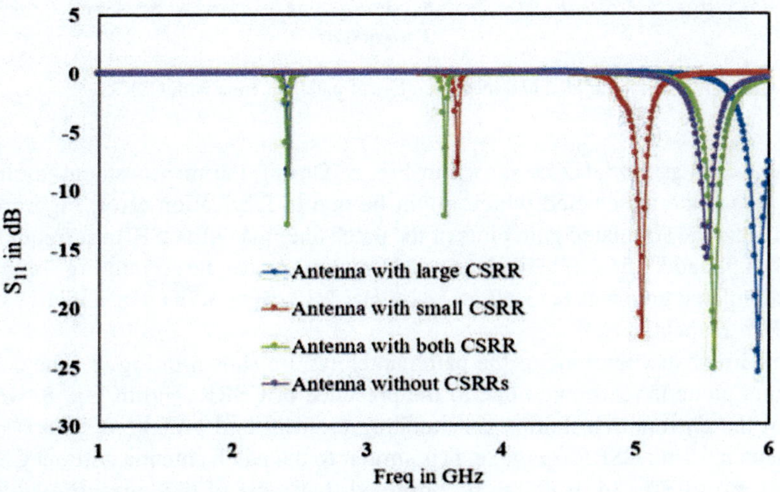

Fig. 4 Simulated S_{11} for different CSRRs

new operating frequency at 3.58 GHz as shown in Fig. 4. The antenna acts as a triple band antenna, integrating all CSRRs into the antenna, as seen in Fig. 4. The placement of the CSRRs is varied. The CSRRs are failed to stimulate new frequencies if they are placed near the feed line.

The conventional antenna is working on 5.5 GHz as shown in Fig. 4. The simulated reflection coefficient is −13.78 dB at 5.5 GHz. E-plane and H-plane gain plots are shown in Fig. 6a. Figure 7a shows that the peak gain of this antenna is 8.22 dB.

The antenna with CSRRs is fabricated and tested using network analyzer. Figure 5 shows comparison graph of simulated and measured reflection coefficient of antenna with CSRRs. The simulated reflection coefficients are −12.91 dB at 2.41 GHz, −12.24 dB at 3.58 GHz, and −25.25 dB at 5.58 GHz as shown in Fig. 5. The measured reflection coefficients are −17.21 dB at 2.43 GHz, −12.26 dB at 3.63 GHz,

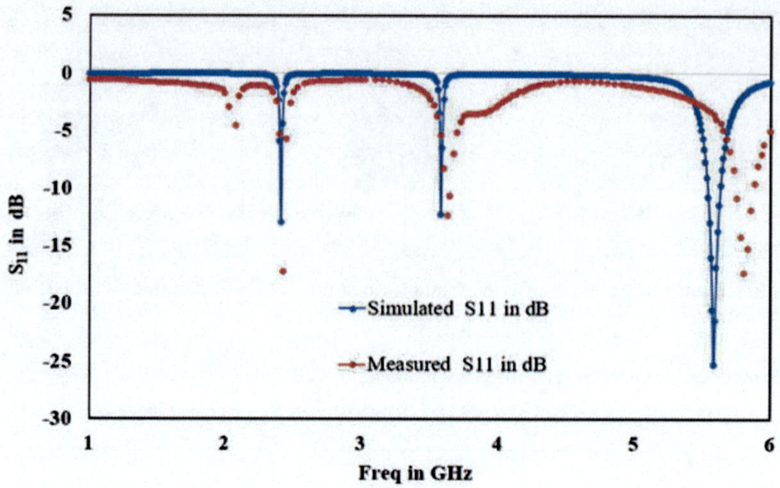

Fig. 5 Comparison of simulated and measured S_{11} of patch antenna with CSRRs

and −17.8 dB at 5.8 GHz as shown in Fig. 5. During the measurement, a change in the resonance is detected which might be due to fabrication error. Figure 6b–d demonstrate the simulated gain plots of the patch antenna with CSRRs at frequencies of 2.41, 5.58, and 3.58 GHz (E-plane and H-plane). As can be seen in Fig. 7c, b, and d, the simulated gain was 8.13 dB at 2.41 GHz, 7.17 dB at 5.58 GHz, and 4.77 dB at 3.58 GHz, respectively.

The current distributions of the patch antennas are shown in Fig. 8. The current in ground plane is rearranged due to the presence of CSRRs. From Fig. 8, we can see that the current distribution on radiating element and ground at 5.58 GHz of patch antenna with CSRRs (Fig. 8e, f) is similar to the patch antenna without CSRRs (Fig. 8a, b). In Fig. 8d, h, it can be observed that most of the current is gathered near CSRRs in ground plane, which explains new frequencies are stimulated due to CSRRs. Based on the above analysis, it can be concluded that the two frequencies 2.41 and 3.58 GHz are stimulated because of CSRRs in the ground. There is a small frequency shift observed because of the coupling between the CSRRs and the patch. Thus, resonant frequencies can be altered by modifying the geometric parameters of the CSRRs, which makes antenna design simple.

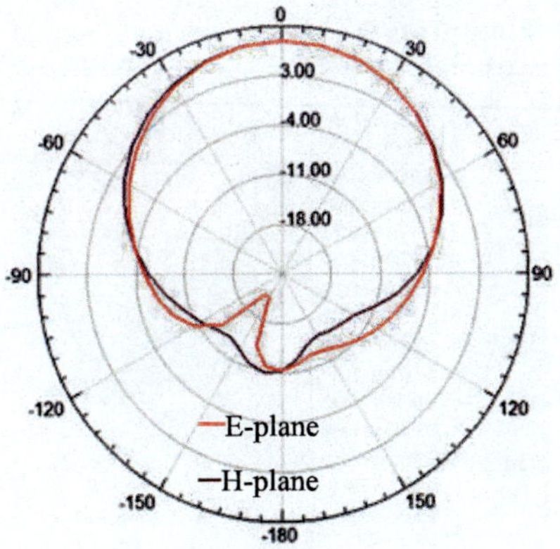

(a) patch antenna without CSRRs

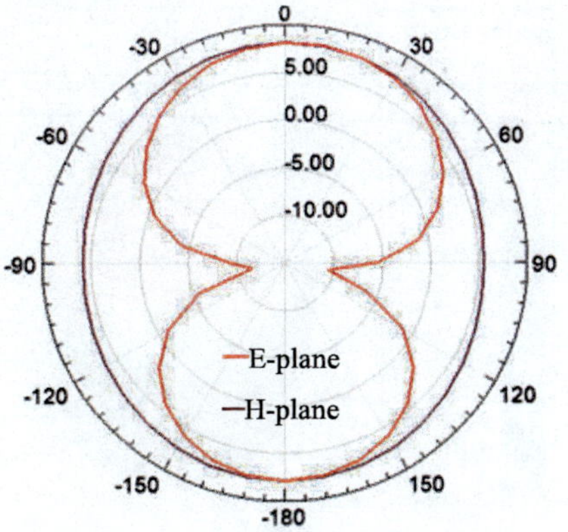

(b) patch antenna with CSRRs at 2.41GHz

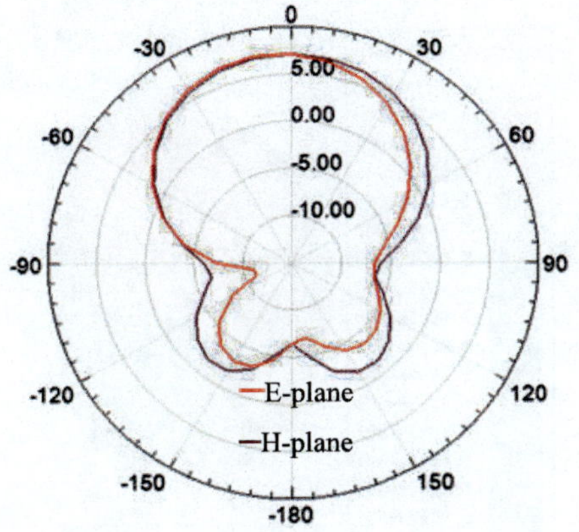

(c) patch antenna with CSRRs at 5.58GHz

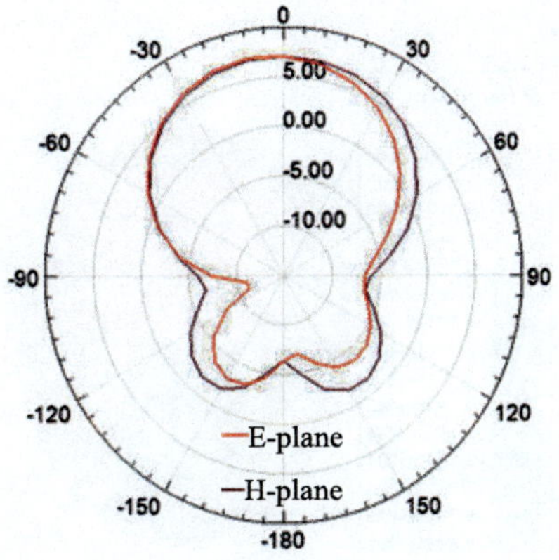

(d) patch antenna with CSRRs at 3.58 GHz

Fig. 6 Radiation pattern (—E-plane, —H-plane)

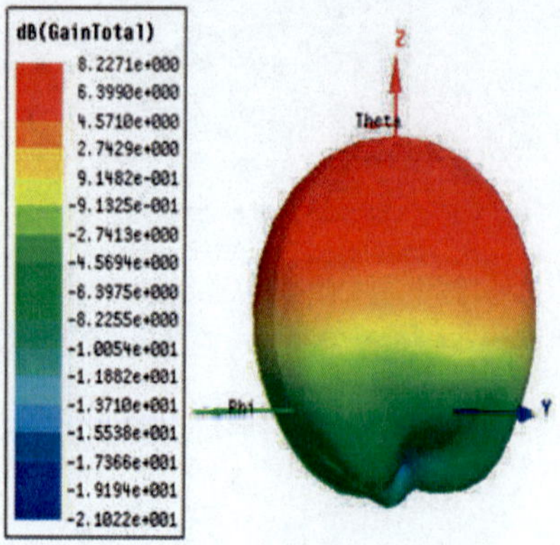

(a) patch antenna without CSRRs

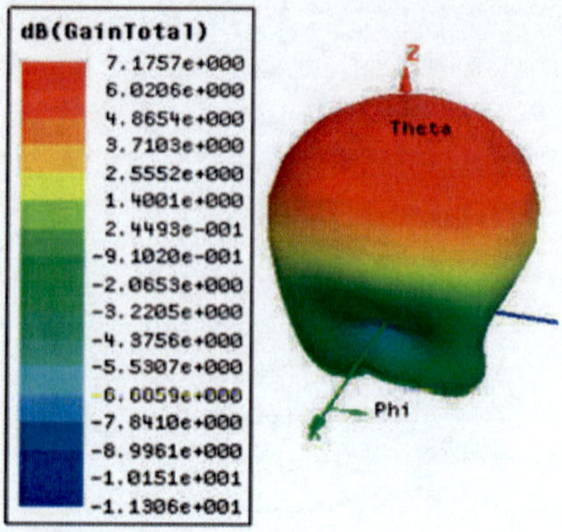

(b) patch antenna with CSRRs at 5.58 GHz

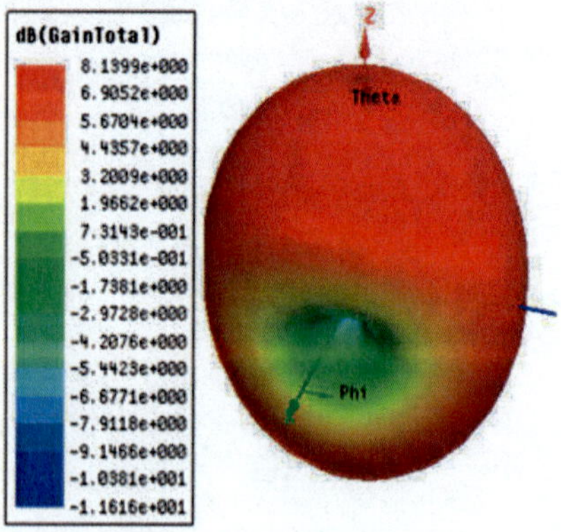

(c) patch antenna with CSRRs at 2.41GHz

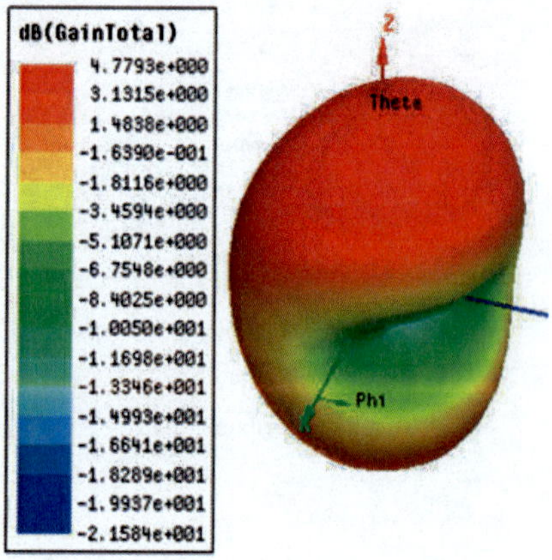

(d) patch antenna with CSRRs at 3.58

Fig. 7 3D Gain plot

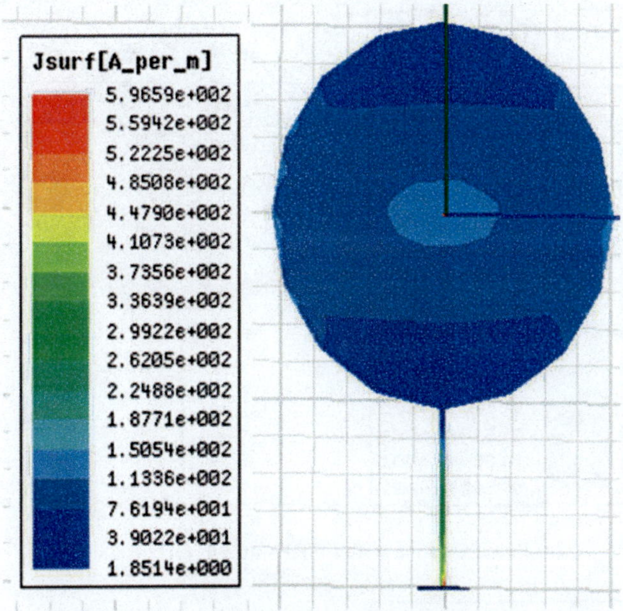

(a) Patch antenna without CSRRs

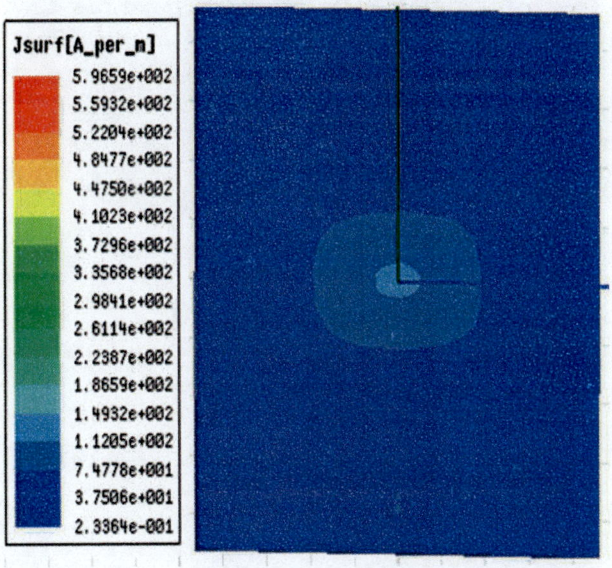

(b) Ground plane of patch antenna without CSRRs

Fig. 8 Current distribution of patch antennas

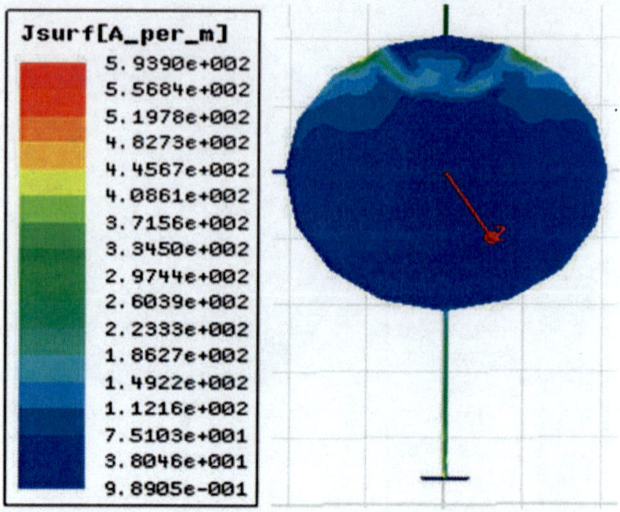

(c) Patch antenna with CSRRs at 2.41GHz (patch)

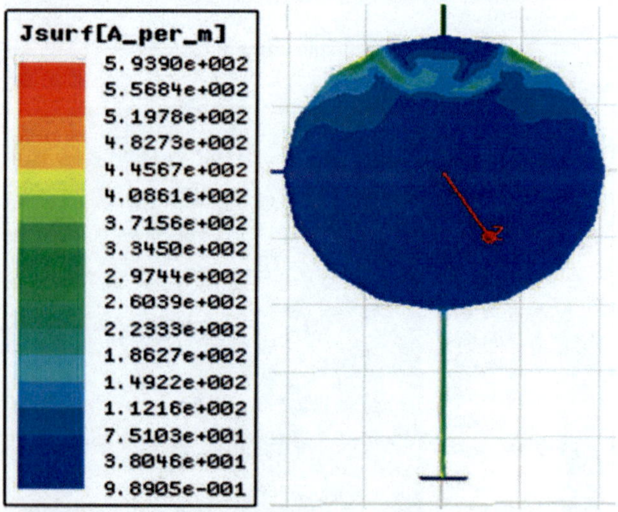

(d) Ground plane of patch antenna with CSRRs at 2.41GHz

Fig. 8 (continued)

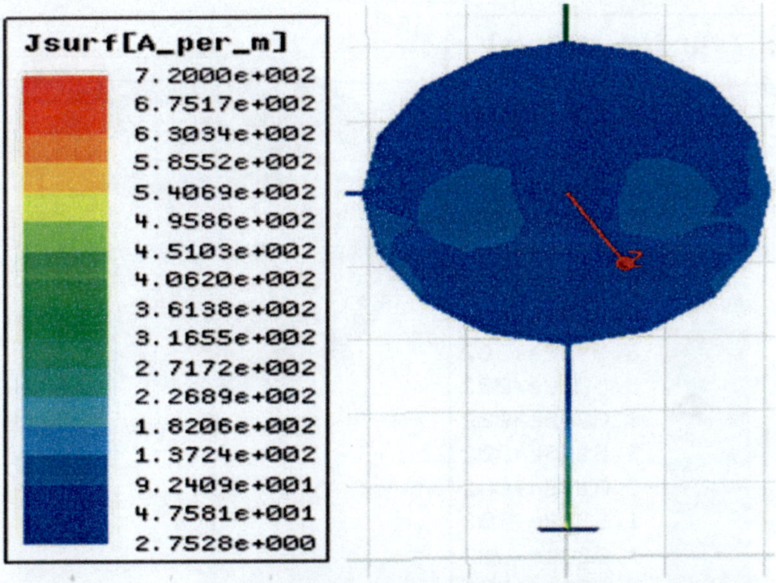

(e) Patch antenna with CSRRs at 5.58 GHz

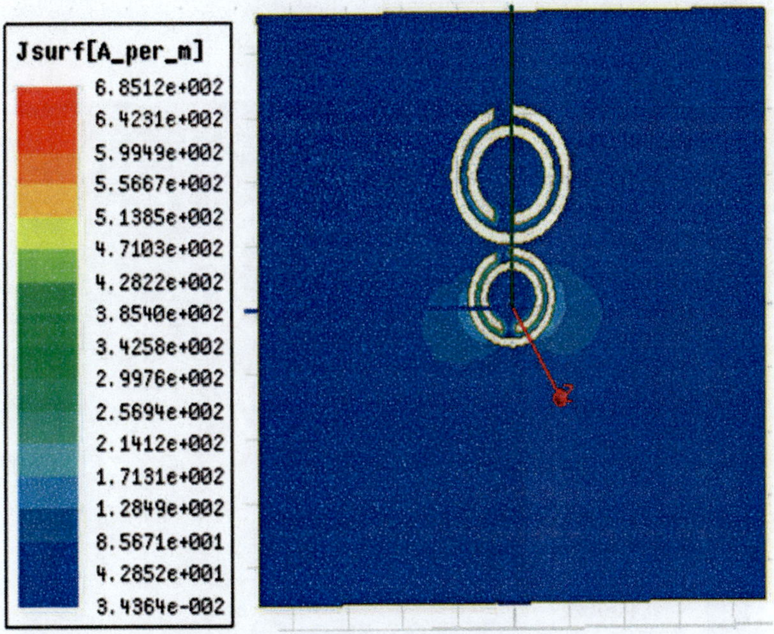

(f) Ground plane of patch antenna with CSRRs at 5.58GHz

Fig. 8 (continued)

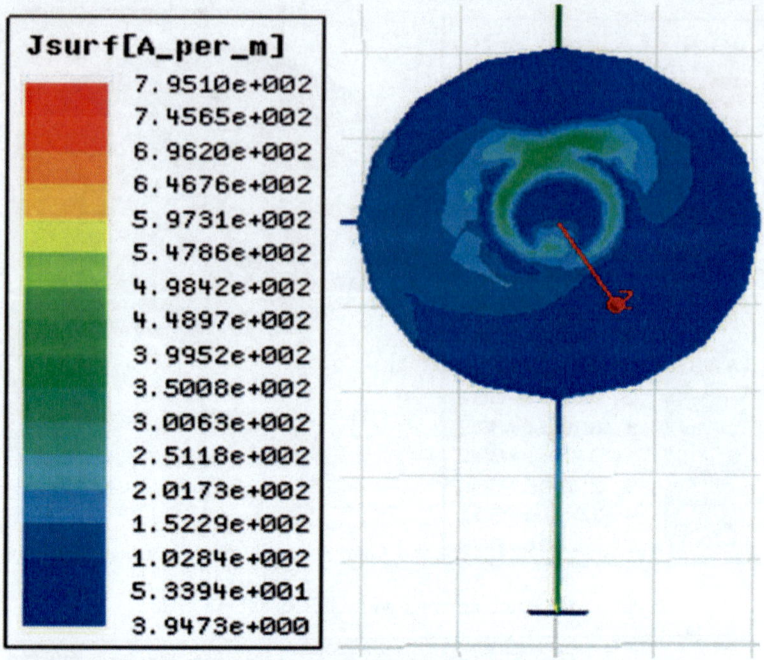

(g) Patch antenna with CSRRs at 3.58GHz

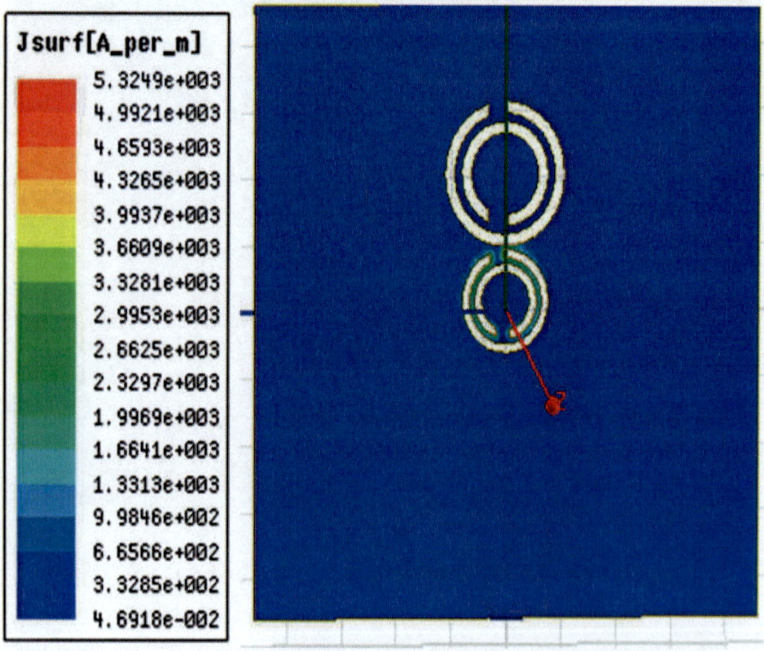

(h) Ground plane of patch antenna with CSRRs at 3.58GHz

Fig. 8 (continued)

6 Conclusion

In this article, by using two separate sized CSRRs in the ground plane, triple band characteristics are obtained. The proposed antenna operates at 2.41, 3.58, and 5.58 GHz because of the presence of two CSRRs in the ground plane. Measured results are good in agreement with simulated results. There is a difference in the resonance frequencies found due to errors related to fabrication.

References

1. L.Y. Cai, G. Zeng, H.C. Yang, Compact triple band antenna for blue tooth/WiMAX/WLAN applications, in *Proceedings of International Symposium on Signals, Systems and Electronics*, Nanjing, China (2010)
2. D.D. Krishna, M. Gopikrishna, C.K. Aanandan, A CPW-fed triple band monopole antenna for WiMAX/WLAN applications, in *Proceedings of the 38th European Microwave Conference*, Amsterdam, Netherlands (2008)
3. H.-W. Liu, C.-H. Ku, C.-F. Yang, Novel CPW-fed planar monopole antenna forWiMAX/WLAN applications. IEEE Antennas Wirel. Propag. Lett. **9**, 240–243 (2010)
4. H.M. El Misilmani, M. Al-Husseini, K.Y. Kabalan, A. El-Hajj, A simple miniaturized triple band antenna for WLAN/WiMAX applications, in *PIERS Proceedings*, Moscow, Russia (2012)
5. G. Du, X. Tang, F. Xiao, Tri-band metamaterial-inspired monopole antenna with modified s-shaped resonator. Prog. Electromagn. Res. Lett. **23**, 39–48 (2011)
6. M.T. Islam, R. Azim, A.T. Mobashsher, Triple band-notched planar UWB antenna using parasitic strips. Prog. Electromagn. Res. **129**, 161–179 (2012)
7. M.R.C. Mahdy, M.R.A. Zuboraj, A.A.N. Ovi, M. A. Matin, Novel design of triple band rectangular patch antenna loaded with metamaterial. Prog. Electromagn. Res. Lett. **21**, 99–107 (2011)
8. T. Wang, Y.-Z. Yin, J. Yang, Y.-L. Zhang, J.-J. Xie, Compact triple band antenna using defected ground structure for WLAN/WIMAX applications. Prog. Electromagn. Res. Lett. **35**, 155–164 (2012)
9. Y. Li, Q. Feng, A compact tri-band monopole antenna with metamaterial loaded for WLAN/WiMAX applications. J. Electromagn. Waves Appl. **27**(6), 772–782 (2013)
10. Y. Xie, L. Li, C. Zhu, C.H. Liang, A novel dual-band patch antenna with complementary split ring resonators embedded in the ground plane. Prog. Electromagn. Res. Lett. **25**, 117–126 (2011)
11. F. Raval, Y.P. Kosta, H. Joshi, Dual-band patch antenna with complementary split ring resonators loaded modified ground plane. Int. J. Appl. Electromagn. Mech. **43**(3), 227–235 (2013)
12. F. Raval, Y.P. Kosta, Dual-band wearable antenna using split ring resonator. Int. J. Waves Complex Random Media **26**(2), 235–242 (2016)
13. F. Raval, Y.P. Kosta, H. Joshi, triple band circular patch antenna using rising sun shaped metamaterial structure. Int. J. Appl. Electromagn. Mech. **46**(3), 501–509 (2014)
14. A. Yadav, S. Agrawal, R.P. Yadav, SRR and S-shape slot loaded triple band notched UWB antenna. AEU Int. J. Electron. Commun. **79**, 192–198 (2017)
15. S. Ullah, S. Hayat, A. Umar, U. Ali, F.A. Tahir, J.A. Flint, Design, fabrication and measurement of triple band frequency reconfigurable antennas for portable wireless communications. Int. J. Electron. Commun. (2017). https://doi.org/10.1016/j.aeue.2017.07.028
16. P. Beigi, P. Mohammadi, A novel small triple band monopole antenna with crinkle fractal-structure. AEU Int. J. Electron. Commun. **70**, 1382–1387 (2016)

17. V. Rajeshkumar, S. Raghavan, A compact metamaterial inspired triple band antenna for reconfigurable WLAN/WiMAX applications. AEU Int. J. Electron. Commun. **69**, 274–280 (2015)
18. D.R. Smith, W.J. Padilla, D.C. Vier, S.C. Nemat-Nasser, S. Schultz, Composite medium with simultaneously negative permeability and permittivity. Phys. Rev. Lett. **84**, 18 (2000)
19. J.D. Baena, J. Bonache, F. Martín, R.M. Sillero, F. Falcone, T. Lopetegi, M.A. Laso, J. García-García, I. Gil, M.F. Portillo, M. Sorolla, Equivalent-circuit models for split-ring resonators and complementary split-ring resonators coupled to planar transmission lines. IEEE Trans. Microw. Theory Techn. **53**(4) (2005)
20. C.A. Balanis, *Antenna Theory and Design* (Wiley, Hoboken, 1997)

Features Analysis of Electroencephalography (EEG) for Mindfulness Meditation Effect on Cancer Patients Toward Stress Level

En Yu Khor, Chee Chin Lim, Yen Fook Chong, and Poh Foong Lee

Abstract Cancer diagnosis can cause patients hard to accept the fact which easily lead to the emotional upheaval and affecting the mental healthiness, moreover brings negative impact on their quality of life. Consequently, strategies aimed at reducing psychological symptoms and encouraging relaxation are required. Mindfulness meditation has become one of the ways to treat physical and psychological symptoms together with conventional allopathic care. In this study, the meditation intervention effect toward the psychological perceived stress is investigated by using Electroencephalogram (EEG). Thus, a portable and lightweight Muse Headband was used as the EEG data acquisition device to collect 13 cancer patients. Besides, KNN, SVM and Bagged Tree classifiers was used to conduct classification to analyze EEG of the cancer patients and compared the accuracy between three classifiers for mindfulness meditation effect and the perceived stress level. The outcome shows that the Bagged Tree is the most suitable classifier for the classification, with the highest accuracy of classification is 66.67% compared to KNN (62.65%) and SVM (55.45%). The meditation practises were found be significantly affecting beta and gamma band of cancer patients, with the highest accuracy of the classification of meditation effect at both bands are 66.21% and 66.67%, respectively. Lastly, perceived stress level was found correlated with beta and gamma band as well, with the highest accuracy

E. Y. Khor · C. C. Lim (✉) · Y. F. Chong
Faculty of Electronic Engineering Technology, Universiti Malaysia Perlis, 02600 Arau, Perlis, Malaysia
e-mail: cclim@unimap.edu.my

C. C. Lim · Y. F. Chong
Sports Engineering Research Centre (SERC), Universiti Malaysia Perlis, 02600 Arau, Perlis, Malaysia

P. F. Lee
Lee Kong Chien Faculty of Engineering & Science, Universiti Tunku Abdul Rahman (UTAR), Jalan Sungai Long, Bandar Sungai Long, Cheras, 43000 Kajang, Selangor, Malaysia

F. Thakkar et al. (eds.), *Proceedings of the International e-Conference on Intelligent Systems and Signal Processing*, Advances in Intelligent Systems and Computing 1370,
https://doi.org/10.1007/978-981-16-2123-9_15

of the classification of perceived stress level at both bands are 59.29% and 66%, respectively. In short, we can conclude that perceived stress level is interrelated with meditation practice.

Keywords EEG · Mindfulness · Meditation · Cancer and stress

1 Introduction

Cancer diagnosis is an intense stressful experience. Many people are afraid of cancer diagnosis due to its life-threatening nature and the severe treatment's side effects. Consequently, strategies aimed at reducing psychological symptoms and encouraging relaxation are required. Meditation is a form of mental practise aimed at enhancing the core psychological abilities of an individual, such as mindful and emotional self-regulation. Meditation comprises of various diverse exercises, which include meditation or mindfulness meditation, Mantra meditation, Yoga, Tai Chi, Qigong [1]. In the current study, the effect of mindfulness meditation on the biological and psychological parameters of cancer patients will be investigated. Electroencephalogram (EEG) is chosen to assess the perceived stress and the meditation intervention effect.

2 Literature Review

Boyle et al. [2] proved that depressive symptoms and depression are strengthened by reducing rumination and strengthening the self-kindness in younger cancer survivors after 6 months Mindful Awareness Practices (MAP). Additionally, S. Rahmani, et al. [3] investigated mindfulness meditation effects on 24 breast cancer patients in terms of their mental fatigue severity and life quality using Fatigue Severity Scale, Global Life Quality of Cancer Patient and Specific Life Quality of Cancer Patient questionnaires.

Saeed et al. [4] conducted an experiment to investigate the perceived stress recognition by using PSS 10 labels and expert evaluation. EMOTIV Insight headset was used to collect 33 healthy subjects' EEG data for 3 min. EMOTIV Insight headset consists of 5 EEG sensors at AF3, AF4, T7, T8 and Pz locations of the brain. Millstine et al. [5] used the Muse Headband to provide guided meditation practices to breast cancer patients from the time of diagnosis to three months after surgical therapy. Thirty women (aged 20–75 years) with newly diagnosed breast cancer were enrolled in this experiment.

Based on Nagar and Sethia [6], thresholding was used to remove data above 100 μV. Data above a value of 100 μV may be regarded as noise caused by the signal of EOG (eye-blinking) or signal produced by muscle contraction called EMG [7]. Fiscon et al. [8] applied Fast Fourier Transform (FFT) and the Wavelet Transform (WT) for estimating EEG spectrum. Mehmood and Lee [9] evaluated the selected feature extraction method through SVM and KNN. The highest accuracy was obtained with KNN with k = 3, which was 61%, whereas SVM only obtained the accuracy with 38.9% in this research.

3 Methodology

3.1 Data Collection

Thirteen subjects (1 male and 12 females) with age group of 42–77 years (mean age, 55 years) participated in this study. Cancer patients were enrolled in a programme "Turn Cancer into Strength Meditation Retreat" for this study. This programme was held for a week to develop a healthier mind through meditation practise to face various challenges of life. These 3 cancer patients included 10 breast cancer female patients, 2 ovarian cancer female patients and 1 nasopharyngeal carcinoma male patients.

Each participant was escorted to a small area with less disturbance and noise, where they were seated and introduced to the experimental procedure and asked to sign the consent form. Subjects were then directed to fill in the PSS 10 questionnaire, which was designed by expert psychologists to measure the level of perceived stress. Once the PSS 10 questionnaire was answered, the pre-intervention EEG data of subjects were recorded for a duration of 5 min sitting in a relaxed position with eyes closed as shown in Fig. 1. These data acquisition steps were done on the 1st day of programme pre-intervention and 5th day programme post-intervention.

During pre-intervention and post-intervention, data were collected using Muse Headband at a sampling frequency of 256 Hz as shown in Fig. 2. There are five channels for Muse Headband located at Fpz, AF7, AF8, TP9 and TP10 as shown in Fig. 3. Muse Headband was connected to an iPad through Bluetooth connection

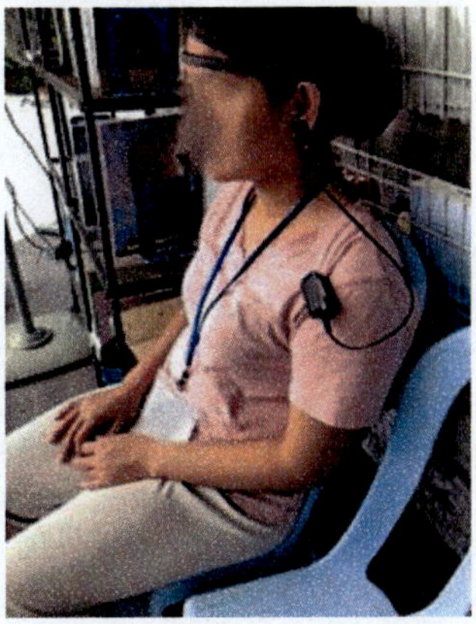

Fig. 1 EEG acquisition using Muse Headband

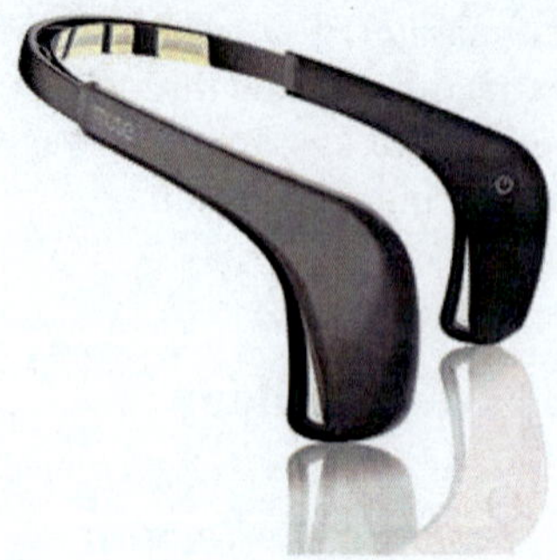

Fig. 2 Interaxon Muse 2 Headband

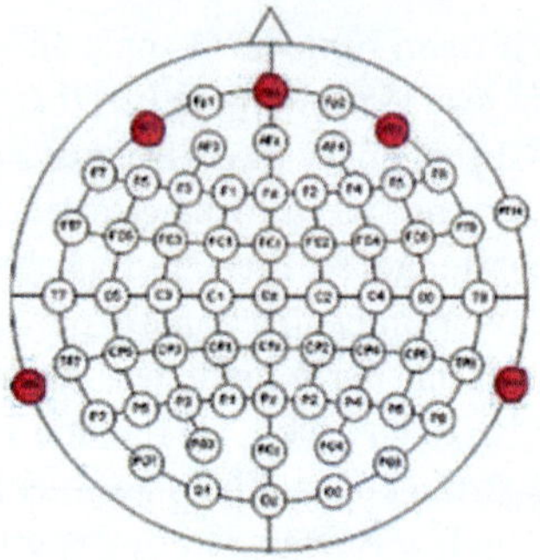

Fig. 3 Position of EEG channels selected for electrodes of Muse Headband

for signal recording and transmission. A mobile application, Muse Direct, was used to record EEG data on iPad, which was transferred to a Muse Direct Cloud for further signal processing. The 5 minutes collected data were segmented from 1 minutes to 3 minutes. Then, The segmented data were filtered by using 50 Hz notch filter and 6th order Butterworth filter. FFT and DWT algorithms were applied for feature extraction. Feature selection and classification of EEG frequency bands for meditation effect and perceived stress level were condutred by KNN, SVM and Bagged Tress classifiers.

3.2 *Data Segmentation and Pre-processing*

EEG signal is considered as non-stationary signals. The data collected need to be segmented to allow smoother signal processing algorithms [10]. Therefore, the data is segmented into three segments which have three times intervals as shown in Fig. 4. After segmentation, pre-processing is applied to the three segmented raw EEG data to remove the noises and artefacts. In this section, notch filter 50 Hz and 6th order Butterworth filter is implemented. The 50 Hz notch filter is used to ensure that strong AC lines signal which would interrupt the EEG data signal is rejected perfectly. The filtered EEG signal by notch filter was further pre-processing by using

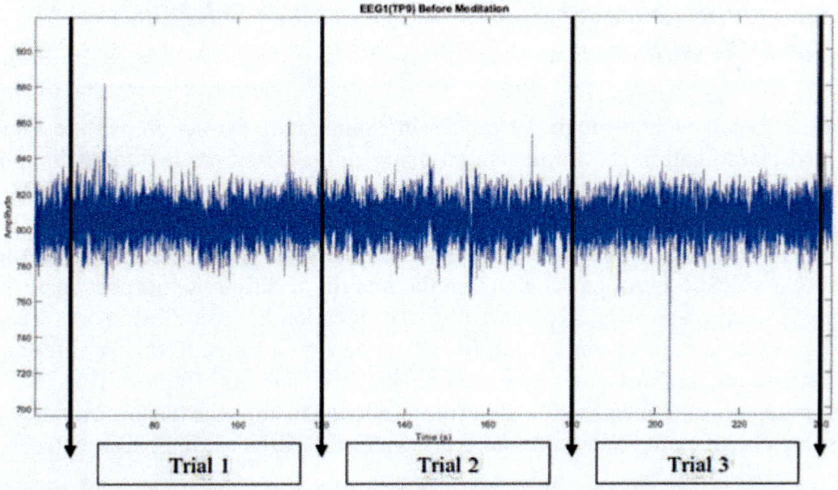

Trials	1	2	3
Data points	15361	30721	46081
Time(s)	60-120	120-180	180-240

Fig. 4 EEG signal segmentation

6th Butterworth filter to obtain five different frequency bands (Theta, Delta, Alpha, Beta and Gamma).

3.3 *Feature Extraction*

Fast Fourier Transform (FFT) is used to extract the simple statistical features such as mean, median, variance, standard deviation, power energy, minimum and maximum value, which can provide useful information about the changes in EEG signals [11].

DWT is defined on the basis of a multiscale representation of features. Every DWT scale is a specific brain-signal representation [12]. Signals on the wavelets are divided into approximation and detail. The number of decomposition levels is chosen according to the dominant frequency component of the EEG signal. In the present study, the EEG signals will be decomposed into five levels corresponding to 5 different EEG frequency bands using db2.

3.4 Machine Learning Algorithm for Feature Selection and Classification

After the signals went through pre-processing and feature extraction, feature selection and classification are implemented using three classifiers including Support Vector Machine (SVM), K-Nearest Neighbours (KNN) and Bagged Tree. SVM is designed explicitly for the supervised solution of a binary classification issue. The primary purpose of SVM is to establish an optimal separate hyperplane (OSH) for binary classification problems, such that the margin of differentiation between two groups is maximized. KNN is a learning classification based on instances, where training is stored in its original form. Based on the stored cases. KNN classifies new cases based on similarity measure such as distance function. Bagged Tree (Bootstrap aggregating) is an ensemble algorithm designed to enhance the performance of statistical classification and regression. It reduces the variance of a decision tree.

3.5 Experimental Frameworks

The collected EEG signal will be analyzed through three main frameworks as the following:

Framework 1: Feature extraction and feature selection for mindfulness meditation effect and perceived stress level. There were 8 features extracted by FFT and DWT from each EEG frequency bands, which are, minimum, maximum, mean, standard deviation, variance, power, energy and entropy. After all the features were collected, feature selection was proceeded using classifiers SVM, KNN and Bagged Tree. This step was applied to EEG signal obtained before and after mindfulness meditation. Besides, all the subjects were labelled and grouped into stress and non-stress group using PSS 10 score that was obtained before and after the experiment. Table 2 demonstrates the formula of threshold value used to label the subjects, where μ represents mean and σ represents standard deviation of the PSS 10 score collected before and after meditation.

Framework 2: Analysis of EEG frequency bands of the cancer patients. The features that produced high accuracy for classification of EEG bands signal before and after meditation in Framework 1 were selected. These selected features were separated into five EEG frequency bands (Delta, Theta, Alpha, beta and gamma) and fed into the classifiers SVM, KNN and Bagged tree for classification. In this framework, the effect of pre- and post-intervention of meditation on each EEG frequency bands can be investigated.

Framework 3: Classification of perceived stress level of the EEG cancer patient. The features that produced high accuracy for classification perceived stress level of EEG signal Framework 1 were selected. These selected features were separated into five EEG frequency bands (Delta, Theta, Alpha, beta and gamma) and fed

Table 1 Accuracy for classification for FFT and DWT extracted features toward meditation intervention

Feature	Accuracy of cassification for FFT				Accuracy of cassification for DWT			
	KNN	SVM	Bagged Tree	Average accuracy	KNN	SVM	Bagged Tree	Average accuracy
MIN	46.90	55.80	59.94	54.2	48.39	49.04	49.34	48.92
MAX	67.74	51.60	65.06	61.5	59.38	58.65	67.63	61.87
Mean	59.43	56.73	64.74	60.3	54.84	55.45	65.1	58.46
STD	58.06	54.17	69.23	60.5	54.84	52.56	69.87	59.09
VAR	46.67	49.68	68.91	55.1	48.39	53.21	69.23	56.94
Energy	53.33	51.28	71.79	58.8	50.00	47.44	67.95	55.13
Power	50.00	51.30	61.79	54.4	56.25	47.44	67.95	57.21
Entropy	59.38	55.45	68.59	61.1	53.13	53.53	65.80	57.49
STD + Entropy	64.52	57.69	73.08	65.1	58.06	55.13	72.12	61.77
STD + Entropy + MAX	61.29	57.69	72.21	63.73	68.9	62.82	74.36	68.69
STD + Entropy + Mean	64.52	59.29	71.47	65.09	62.5	60.89	68.59	63.99

into the classifiers SVM, KNN and Bagged tree for classification. In this framework, the effect of perceived stress level on each EEG frequency bands can be investigated.

4 Result and Discussion

4.1 *Feature Selection for Classification of Meditation Effect and Perceived Stress*

After the FFT and DWT feature extraction was done, the project proceeded to the step of feature selection. In order to find the features of signal that are significantly affected by meditation intervention, every feature extracted by FFT and DWT algorithms were tested using classifiers SVM, KNN and Bagged Tree as shown in Table 1.

After the FFT and DWT feature extraction was done, the project proceeded to the step of feature selection. In order to find the features of signal that are significantly affected by meditation intervention, every feature extracted by FFT and DWT algorithms were tested using classifiers SVM, KNN and Bagged Tree as shown in Table 1.

Table 2 The range of scores for the subjects labelling and grouping. (μ = 12.08, σ = 5.17)

Group	Threshold	Range (Score)
Less sress	0 to μ-$\frac{\sigma}{2}$	0–10
Stress	[μ-$\frac{\sigma}{2}$] + 1 to 40	11–40

Based on Table 1, the highest average accuracy of feature was achieved by maximum (61.5%), followed by standard deviation (61.1%) and entropy (60.5%) using FFT method. Besides that, the average accuracy for the features is lower than 62%. Hence, the hybrid features is proposed to achieve better accuracy as tabulated in Table 1. Therefore, the hybrid feature [13] among the maximum, standard deviation and entropy is combined which giving the promising average accuracy above 60%. The selected features from FFT and DWT were the same for both methods. Based on DWT and FFT features extraction algorithm, Bagged Tree is the best classifier for features classification for meditation effect since it achieves the highest accuracy among three classifiers. Bagged Tree is capable to enhance the performance of classification of statistical and regression for the statistical features due to its stability and ability to resist outlier suits well in this project [14, 15].

The perceived stress level is validated by using PSS 10 score, the features extracted by FFT and DWT algorithms are labelled and grouped into stress group and less stress group as shown in Table 2. The mean and standard deviation of PSS 10 score is calculated from the 13 subjects, which are μ = 12.08 and σ = 5.17.

Three classifiers were employed on the extracted features to classify the perceived stress level, which classify into stress group and less stress group. The highest average accuracy was achieved by maximum (64.1%), followed by entropy (62.6%) and standard deviation (60.4%) using FFT as in Table 3. Whereas, highest average accuracy using DWT was achieved by entropy (65.52%), followed by mean (64.54%) and standard deviation (63.97%). Therefore, feature selected among the eight extracted features are maximum, mean, standard deviation and entropy. However, the hybrid feature selected are STD + Entropy, MAX + STD + Entropy and Mean + STD + Entropy.

In the case of FFT and DWT features extraction algorithm, Bagged Tree is the best classifiers for features classification for perceived stress level since it achieves the highest accuracy among three classifiers. The accuracy of classification for all features using Bagged Tree exceeds 65% for both FFT and DWT. In FFT-based features, the highest accuracy of classification achieved by using Bagged Tree is 73.08%. However, KNN and SVM able to achieve the highest accuracy of 66.67% and 66.67% respectively. Next, the highest accuracy of classification achieved for the DWT-based features by using Bagged Tree, KNN and SVM is 75.32%, 65.63% and 66.35% respectively. Hence, the Bagged Tree is the outperform classifier for perceived stress compared to KNN and SVM classifiers. Bagged Tree is the best

Table 3 Classification accuracy for perceived stress level using FFT and DWT extracted features

Feature	Accuracy of classification using FFT				Accuracy of classification using DWT			
	KNN	SVM	Bagged Tree	Average accuracy	KNN	SVM	Bagged Tree	Average accuracy
MIN	45.16	49.68	49.36	48.1	51.61	54.49	49.37	51.82
MAX	66.67	59.3	66.35	64.1	61.29	61.86	66.35	63.17
Mean	58.06	54.8	64.74	59.2	58.06	61.54	74.04	64.54
STD	59.32	54.48	67.35	60.4	65.63	55.77	70.51	63.97
VAR	54.84	53.85	68.27	59.0	53.3	54.81	71.79	59.97
Energy	53.13	53.53	69.55	58.7	54.84	56.41	74.04	61.76
Power	48.39	53.85	66.03	56.1	53.33	55.77	74.03	61.04
Entropy	60	59.3	68.59	62.6	59.38	65.71	71.47	65.52
STD + Entropy	64.52	57.37	73.76	65.22	59.38	66.35	75.32	78.32
STD + Entropy + MAX	58.06	66.67	71.47	65.4	64.52	62.82	71.03	66.12
STD + Entropy + Mean	58.06	66.35	73.08	65.83	61.29	65.38	68.71	65.13

classifier for feature classification for perceived stress since it achieves the highest accuracy among three classifiers.

4.2 *Analysis of EEG Frequency Band Signal for Mindfulness Meditation Effect to the Cancer Patients Before and After Mindfulness Meditation*

The selected features and features combination were separated into five frequency bands. The selected features and features combination of specific brainwave were fed into classifiers KNN, SVM and Bagged Tree to classify brainwaves for meditation effect. In Figs. 5 and 6, there are three types of comparison that are discussed. The first comparison is to compare between five brainwaves and meditation effect. By the first comparison, the brainwaves that is significantly affected by mindfulness meditation are decided. Secondly, Figs. 5 and 6 compares the selected features from FFT and meditation effect. In this comparison, the extracted feature from FFT method that can optimize the performance of EEG bands classification for meditation effect is selected. Finally, the comparison between classifiers performance to figure out which classifier works the best for the EEG bands classification for meditation effect.

According to Fig. 5, it clearly shows that gamma band has the highest accuracy (62.5%) in MAX+ STD + entropy features of classification for meditation effect

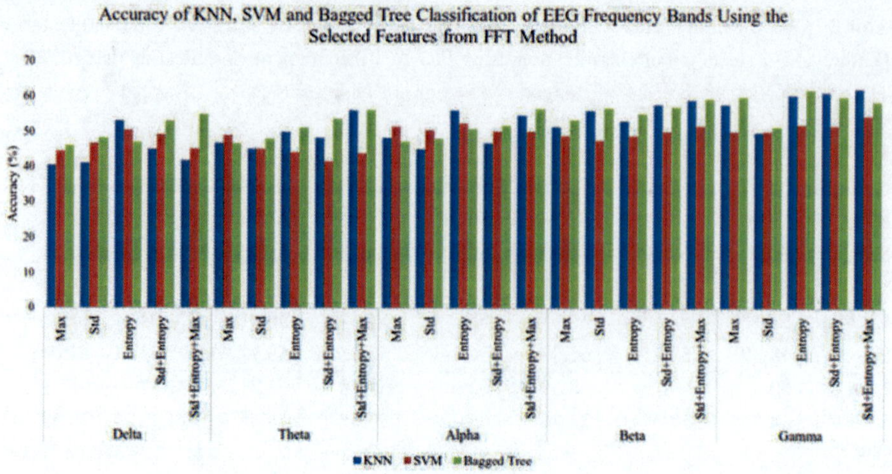

Fig. 5 Accuracy of KNN, SVM and Bagged Tree classification of EEG frequency bands for meditation effect using the selected features from DWT method

compared to other brainwaves. Besides, beta band achieved the second highest accuracy (59.62%) in MAX + STD + entropy classification of EEG bands for meditation effect. Therefore, Figure 5 proved that, by the classification using features extracted by FFT method, beta and gamma bands are the brainwaves that is most affected by meditation effect. Studies have found that the meditation has a correlation with human cognition and gamma-related perception function in meditation practises [16–18]. Beta band provides more information on the subject transition emotionally over other brainwaves, which is associated with meditation effect that promotes the emotional balance and well-being of people [19, 20].

By comparison of the selected features, which are extracted using FFT method, and meditation effect, the features combination of MAX + STD + entropy is the best feature for all EEG bands classification of meditation effect. For Delta band, it obtained 55.13%; Theta band obtained 56.4%; Alpha band obtained 56.73%; beta band obtained 59.62% and gamma band obtained 62.5%. In summary, features combination of MAX + STD + entropy contain the most useful information for the classification of meditation effect.

According to Fig. 6, it clearly shows that the gamma band has the highest accuracy (66.67%) in MAX + STD + entropy features of classification for meditation effect compared to other brainwaves. Besides, beta band achieved the second highest accuracy (66.21%) in MAX + STD + entropy classification of EEG bands for meditation effect. Therefore, Fig. 6 has proved that, by the classification using features extracted by DWT method, beta and gamma bands are the brainwaves that is most affected by meditation effect. Studies proved that beta and gamma bands are associated with meditation effect, where gamma band is correlated with human cognition while beta band is correlated with human emotion [16–20].

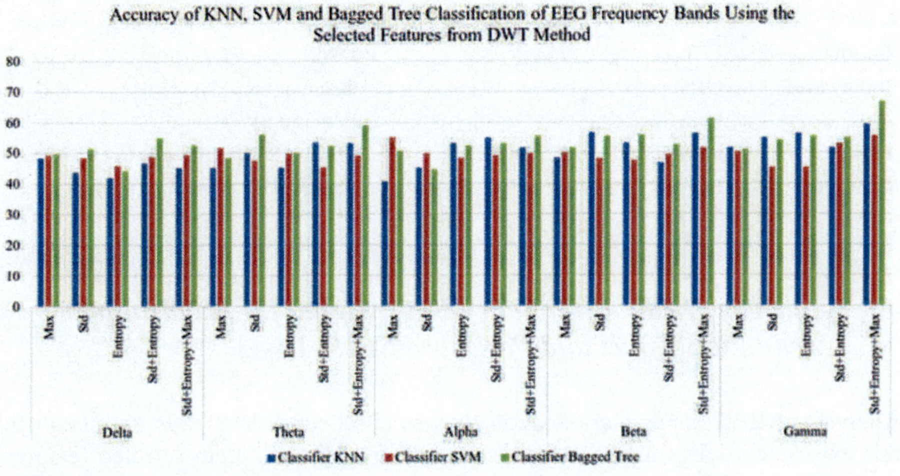

Fig. 6 Accuracy of KNN, SVM and Bagged Tree classification of EEG frequency bands for meditation effect using the selected features from DWT method

By comparison of the selected features, which are extracted using DWT method and meditation effect, the features combination of STD + entropy is the best feature for all EEG bands classification of meditation effect except Delta and Theta band. Alpha band obtained 58.33%; beta band obtained 59.29%; gamma band obtained 66%. Delta achieves the highest accuracy of classification, which is 58% for features MAX + STD + entropy, while Theta band achieves the highest accuracy of classification, which is 58% for features standard deviation. Although Delta and Theta bands obtain the highest accuracy for the features that are not the same with other bands, but the difference of the accuracy is very small and can be neglected. In summary, features combination of STD + entropy contain the most useful information for the classification of perceived stress level.

By using both FFT and DWT types of features, the result of analysis of EEG bands classification is the same: gamma and beta bands are strongly associated with meditation effect. The features combination MAX + STD + entropy gives the highest accuracy of the classification for meditation effect. Besides, Bagged Tree classifiers performs the best for the classification among three classifiers. Table 4 shows the highest accuracy of classification of EEG bands classification for meditation effect in term of brainwaves, features and classifier.

Table 4 The highest accuracy of classification of EEG bands classification for meditation effect

Result	FFT	DWT
Brainwaves	Gamma (62.5%)	Gamma (66.67%)
	Beta (59.62%)	Beta (66.21%)
Features	MAX + STD + Entropy (62.5%)	MAX + STD + Entropy (61.9%)
Classifier	Bagged Tree Classifier (62.18%)	Bagged Tree Classifier (61.5%)

4.3 Analysis of EEG Frequency Band Signal for Mindfulness Meditation Effect and Perceived Stress Level

Analysis of EEG bands classification for perceives stress level was discussed. In this experiment, classification was also conducted by using both selected features from FFT and DWT. By using PSS 10 score, the features extracted by FFT and DWT algorithms are labelled and grouped into stress group and less stress group. The mean and standard deviation of PSS 10 score is calculated from the 13 subjects, which are $\mu = 12.08$ and $\sigma = 5.17$ as in Table 2.

According to Fig. 7, it clearly shows that the gamma band has the highest accuracy (66%) in STD + entropy features classification for perceived stress level compared to other brainwaves. Besides, beta band achieved the second highest accuracy (59.29%) in STD + entropy features classification of EEG bands for perceived stress level. Therefore, Fig. 7 proved that, by the classification using features extracted by FFT method, beta and gamma bands are the brainwaves that is most effected by perceived stress level. Luijcks et al. has investigated how perceived stress affect EEG frequency bands activity [21]. The gamma activity was associated with experimentally induced stress in anxious people [22, 23]. Jena has proved that beta wave was associated with the increment of stress level [24].

According to Fig. 8, DWT method clearly shows that the Delta band has the highest accuracy (61.5%) in entropy features classification for perceived stress level compared to other brainwaves. Besides, beta band achieved the second highest accuracy (61.9%) in MAX + STD + entropy feature classification of EEG bands for perceived stress level. Therefore, Fig. 7 proved that, by the classification using features extracted by DWT method, Delta and beta bands are the brainwaves that is most effected by perceived stress level. Studies implicated that heightened delta activity reflects the stress component of human pain responsivity [25, 26], while beta wave was associated with the increment of stress level [24, 27, 28]. These studies proven that delta and beta bands are the most affected because cancer patients are normally suffering from the pain and stress events simultaneously.

By comparison of the selected features from DWT and perceived stress level, the features combination of MAX + STD + entropy is the best feature for Theta and beta bands classification of perceived stress level (54.8% and 61.9%, respectively). For Delta band, it obtained 61.5% for entropy; Alpha band obtained 59.3% for standard deviation and gamma band obtained 57.1% for STD + entropy. In summary, features

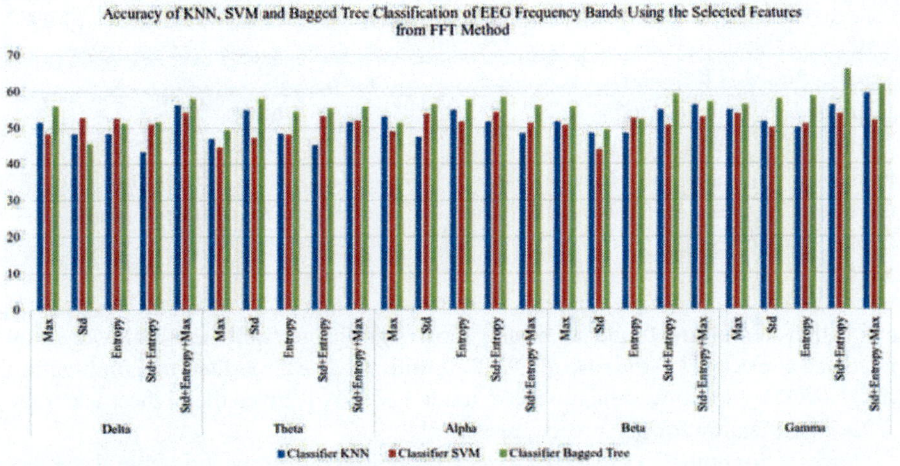

Fig. 7 Accuracy of KNN, SVM and Bagged Tree classification of EEG frequency bands for perceives stress level using the selected features from FFT method

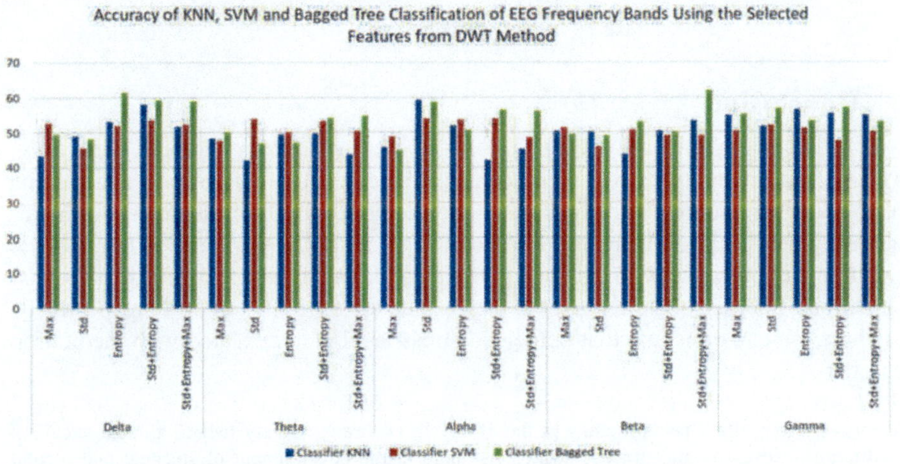

Fig. 8 Accuracy of KNN, SVM and Bagged Tree classification of EEG frequency bands for perceives stress level using the selected features from DWT method

combination of MAX + STD + entropy contain the most useful information for the classification of perceived stress level since there are two brainwaves that achieve the highest accuracy of EEG bands classification for perceived stress level using this features combination.

In the case of features from FFT, it found that perceived stress level significantly affects beta and gamma band. The two-features combination STD + entropy gives the highest accuracy of the classification for perceived stress level. In contrast, Delta

Table 5 The highest accuracy of classification of EEG bands classification for perceived stress level

Result	FFT	DWT
Brainwaves	Gamma (66%)	Delta (61.5%)
	Beta (59.29%)	Beta (61.9%)
Features	STD + Entropy (59.29%)	MAX + STD + Entropy (61.9%)
Classifier	Bagged Tree classifier (66%)	Bagged Tree classifier (61.5%)

and gamma band are found to be the most significantly affected brainwaves by perceived stress level in the case of DWT algorithms. The three-features combination MAX + STD + entropy, which were extracted by DWT, gives the highest accuracy of the classification for perceived stress level.

However, for both FFT and DWT cases, Bagged Tree is the most suitable classifiers for the classification foe perceived stress level. Table 5 shows the highest accuracy of classification of EEG bands classification for perceived stress level in terms of brainwaves, features and classifier.

5 Conclusion

The meditation practises were significantly affected the brainwaves beta and gamma band of cancer patients. The highest accuracy achieved for the classification of meditation effect at both bands are 66.21% and 66.67%, respectively. Perceived stress level was found to correlate with beta and gamma band, the highest accuracy for the classification of perceived stress level at both bands are 59.29% and 66%, respectively. In short, we can conclude that perceived stress level is interrelated with meditation practise.

Acknowledgements Special thanks to the Malaysia Dhamma Society (MDS) for the excellent meditation instruction and practise, which has help in the development of spiritual and mental growth related to scientific research in this project.

References

1. M.B. Ospina et al., *Meditation Practices for Health: State of the Research* (Alberta, Canada, 2007)
2. C.C. Boyle, A.L. Stanton, P.A. Ganz, C.M. Crespi, J.E. Bower, Improvements in emotion regulation following mindfulness meditation: effects on depressive symptoms and perceived stress in younger breast cancer survivors. J. Consult. Clin. Psychol. **85**(4), 397–402 (2017)

3. S. Rahmani, S. Talepasand, The effect of group mindfulness-based stress reduction program and conscious yoga on the fatigue severity and global and specific life quality in women with breast cancer. Med. J. Islam. Repub. Iran **29** (2015)
4. S.M.U. Saeed, S.M. Anwar, H. Khalid, M. Majid, U. Bagci, *Electroencephalography based classification of long-term stress using psychological labeling* J. Biomed. Heal, Informat. (2019)
5. D.M. Millstine et al., Use of a wearable EEG headband as a meditation device for women with newly diagnosed breast cancer: a randomized controlled trial. Integr. Cancer Ther. **18**, 1–10 (2019)
6. P. Nagar, D. Sethia, *Brain Mapping Based Stress Identification Using Portable EEG Based Device* (India, 2019)
7. M. Teplan, M. Teplan, Fundamental of EEG measurement. Meas. Sci. Rev. **2**(2) (2002)
8. G. Fiscon et al., Combining EEG signal processing with supervised methods for Alzheimer's patients classification. BMC Med. Inform. Decis. Mak. **18**(1) (2018)
9. R.M. Mehmood, H.J. Lee, Emotion classification of EEG brain signal using SVM and KNN, in *International Conference on Multimedia and Expo Workshops* (2015)
10. L. Wong, W. Abdulla, Time-frequency evaluation of segmentation methods for neonatal EEG signals
11. A.S. Al-Fahoum, A.A. Al-Fraihat, Methods of EEG signal features extraction using linear analysis in frequency and time-frequency domains. ISRN Neurosci. **2014**, 1–7 (2014)
12. N. Hazarika, J. Zhu, C. Ah, C. Tsoi, A. Sergejew, *Classification of EEG Signals Using the Wavelet Transform*
13. T.K.G.J.R. Panda, P.S. Khobragade, P.D. Jambhule, S.N. Jengthe, P.R. Pal, Classification of EEG signal using wavelet transform and support vector machine for epileptic seizure diction, in *Systems in Medicine and Biology* (2010), pp. 405–408
14. S.-K.T. Chung-Yen Liao, R.-C. Chen, Emotion stress detection using EEG signal and deep learning technologies, in *Proceedings of IEEE International Conference on Applied System Innovation* (2018), pp. 90–93
15. T.H.T. Vo, T. Luu-Nha Dang, N.V.T. Nguyen, T. Van Huynh, Classification electroencephalography using machine learning, in *Proceedings—2019 19th International Symposium on Communications and Information Technologies. ISCIT* (2019), pp. 14–19
16. F. Ferrarelli et al., Experienced mindfulness meditators exhibit higher parietal-occipital EEG gamma activity during NREM sleep. PLoS One **8**(8) (2013)
17. A. Berkovich-Ohana, J. Glicksohn, A. Goldstein, Mindfulness-induced changes in gamma band activity—implications for the default mode network, self-reference and attention. Clin. Neurophysiol. **123**(4), 700–710 (2012).
18. C. Braboszcz, B. Rael Cahn, J. Levy, M. Fernandez, A. Delorme, Increased gamma brainwave amplitude compared to control in three different meditation traditions. PLoS One **12**(1) (2017)
19. S.M.M Murugappan, *Human Emotion Recognition Through Short Time Electroencephalogram (EEG) Signals Using Fast Fourier Transform (FFT)* (2013), pp. 289–294
20. N. Jadhav, R. Manthalkar, Y. Joshi, Effect of meditation on emotional response: an EEG-based study. Biomed. Signal Process. Control **34**, 101–113 (2017)
21. R. Luijcks, C.J. Vossen, H.J. Hermens, J. Van Os, R. Lousberg, The influence of perceived stress on cortical reactivity: a proof-of-principle study. PLoS One **10**(6) (2015)
22. P. Cipresso et al. (2012) EEG alpha asymmetry in virtual environments for the assessment of stress-related disorders, in *Studies in Health Technology and Informatics*, vol. 173, pp. 102–104 (2012)
23. V. Miskovic, A.R. Ashbaugh, D.L. Santesso, R.E. McCabe, M.M. Antony, L.A. Schmidt (2010) Frontal brain oscillations and social anxiety: a cross-frequency spectral analysis during baseline and speech anticipation. Biol. Psychol. **83**(2), 125–132 (2010)
24. S. Jena, Stress and its effect on EEG. Int. J. Med. Sci. Public Heal. **4**(11), 1493 (2015)
25. A.C.N. Chen, S.F. Dworkin, J. Haug, J. Gehrig, *Clinical Section Topographic Brain Measures of Human Pain and Pain Responsivity* (1989)

26. N.W. Bin, S.A. Awang, C.Y. Fook, L.C. Chin, O.Z. Ying, A study of informative EEG channel and brain region for typing activity. J. Phys. Conf. Ser. **1372**(1), 012008 (2019)
27. L.C. Chin, A.M. Zazid, C.Y. Fook, V. Vijean, S.A. Awang, M. Affandi, L.S. Chee, Differentiate characteristic EEG tobacco smoking and non-smoking. J. Phys. Conf. Ser. **1372**(1), 012055 (2019)
28. F.N.M. Yusof, L.C. Chin, C.Y. Fook, A.S.A. Nasir, V. Vijean, Objective analysis of heart rate variability during meditation and rest stage, in *6th International Conference on Green Design and Manufacture* (2020)

Effects of Menstrual Cycle on Atrial ECG Components

Ankita Padhan, Arya Bhardwaj, and J. Sivaraman

Abstract Cyclic variation of sex hormones occurs throughout the menstrual cycle in reproductive women. This variation causes physiological alteration of cardiovascular activities and ultimately affects Electrocardiogram (ECG) parameters. The aim of this study is to analyze the changes in the atrial ECG in reproductive women during menstruation. Thirty-five healthy women volunteers aged between 20–25 with regular menstrual cycles were enlisted for this study. ECG was recorded during the first phase of menstrual cycle (study group) and other phases of menstrual cycle from 14th to 26th day (control group) using the EDAN SE-1010 PC ECG system. Among all atrial ECG components, the P wave duration (PWD) decreased in the study group. Menstruation caused alteration in the P wave axis, and P wave amplitude in ECG Leads II and aVF. PR Interval, PR segment, and PP Interval did not show significant variation between the groups. Hence, cyclic alteration of sex hormones and electrolytes concentration may cause decreased PWD at the time of menstruation. Therefore, reproductive women have a lower occurrence of Atrial Fibrillation (AF).

Keywords Atrial fibrillation · Electrocardiogram · Hormonal flux · Menstrual cycle · P wave

1 Introduction

Systematic alteration in the release of female steroid hormones like progesterone and estrogen affects periodic ovarian variation. This periodic variation is termed as menstrual cycle, and it is restricted to primates. Over 200 psychological, physiological, and behavioral alterations occur in the body due to the menstrual cycle [1]. Typically, this cycle repeats every 21–35 days (approximately 28 days). Based on the discharge of the ovum, menstrual cycle can be divided into four different phases, i.e.

A. Padhan · A. Bhardwaj · J. Sivaraman (✉)
Bio-signals and Medical Instrumentation Laboratory, Department of Biotechnology and Medical Engineering, National Institute of Technology, Rourkela, Odisha 769008, India
e-mail: jsiva@nitrkl.ac.in

F. Thakkar et al. (eds.), *Proceedings of the International e-Conference on Intelligent Systems and Signal Processing*, Advances in Intelligent Systems and Computing 1370,
https://doi.org/10.1007/978-981-16-2123-9_16

(i) menstrual phase, (ii) follicular phase, (iii) ovulatory phase, and (iv) luteal phase [2]. Variability in the discharge of female steroid hormones released from ovaries like progesterone, estrogen, and hormones released from the anterior pituitary like LH, FSH occurs throughout the menstrual cycle. The secretion of progesterone and estrogen is lowest in the menstruation phase; however, estrogen secretion gradually increases towards the end of the follicular phase [3], to release the ovum in ovulatory phase. Discharge of progesterone hormone increases in the luteal phase [3].

The level of female steroid hormones in blood plasma is lowest during the menstrual phase of the cycle compared to all other phases. A study has shown that electrical impulse conductivity, rhythmicity of heart, and blood clotting are influenced by the female steroid hormones [4]. The inter-relationship between the uterus and hypothalamohypophyseal ovarian axis is responsible for physiological variation throughout the menstrual cycle, and this process is also controlled by hormonal flux [4]. Hormonal variations affect the electrophysiology of the cardiovascular system in humans. Hormonal fluctuations during different phases of the menstrual cycle could stimulate the ion channels, affecting the electrophysiological characteristics of the heart [3]. On the other hand, the molecular mechanism behind this alteration is a change in the mRNA level responsible for coding different ion channels. Menarche, pregnancy, and menopause causes menstrual irregularity due to hormonal variation. Moreover, stress, anxiety, and excessive weight loss also cause menstrual irregularity by interrupting the normal hypothalamo-pituitary-endocrine axis, which ultimately leads to hormonal fluctuation [5].

Menstrual cycle influences the normal physiological, psychological, and behavioral outcome of the body. Steroid hormone levels in postmenopausal women differs from reproductively active women, which causes an increase in the incidence of many cardiovascular diseases like sudden cardiac death [6] and Atrial Fibrillation (AF) [7]. Studies show that walls of blood vessels in the heart, including capillaries, contain several nuclear and cytoplasmic receptors for estrogen and progesterone. This indicates that sex hormones could directly or indirectly affect the cardiovascular system [6]. The occurrence of several cardiac diseases varies throughout the menstrual cycle due to the variation in the hormonal flux. However, the occurrence of ventricular tachycardia, Torsade de Pointes (TdP) increase in the reproductively active women [8]. Incidences of drug-induced cardiac arrhythmia increases during the menstrual phase and ovulation [9], ventricular arrhythmia and supra-ventricular tachycardia during the luteal phase [10], and increased prevalence of AF during the post-menstrual phase [11]. Similarly, the incidences of TdP, sudden cardiac death, ventricular tachycardia, and many other disorders vary throughout the cycle [11].

Several studies show that fluctuation in ion channels during the menstrual cycle leads to variation in atrial and ventricular ECG components along with the Heart Rate Variability (HRV). Earlier studies have highlighted the importance of atrial ECG components in clinical diagnosis [12–15]. No significant variation was noted for P wave amplitude [1], PR dispersion, PR interval throughout the cycle [11]; however, few studies show that P wave Dispersion (PD), an essential indicator of AF, is affected by the menstrual cycle [16]. Many investigations have been done concerning the change in ventricular components of ECG and HRV at different phases of the menstrual cycle. Thus, in this observational study, changes in the atrial ECG components during the lowest hormonal fluctuation, i.e., menstrual phase, and highest hormonal fluctuation, i.e., luteal and follicular phase were investigated.

2 Methodology

2.1 *Study Population*

Thirty-five healthy, eumenorrheic, non-obese women volunteers in the age range of 20 to 25 years were selected for this study. The study was divided into two groups, i.e., ECG was recorded on the 2nd or 3rd day of menstruation (study group) in 21 volunteers, and in 31 volunteers, reading was taken at any time between 14th and 26th day of menstrual cycle (control group). To prevent diurnal fluctuation, ECG was recorded at a particular time of the day, i.e., between 8 to10 pm. The study details were explained and each volunteer filled an informed consent form before participation in the study.

2.2 *Data Acquisition*

All ECGs were recorded using four Ag–AgCl clamp electrodes with a digital EDAN SE-1010 PC ECG system. The sampling rate of the machine was 1000 samples per second, and a frequency range of 0.05–150 Hz. The data was recorded at 10 mm/mV gain and 25 mm/s paper speed. However, the paper speed can be adjusted from 2.5 mm/mV to 100 mm/mV and 5 mm/s to 200 mm/s, for better offline annotations. Wave amplitude, duration, frontal plane axis were analyzed.

2.3 *Statistical Analysis*

The mean and standard deviation values were calculated from the ECG data and mentioned as mean $\pm$ SD. Unpaired Student t-test was used to compare the data between two groups of menstrual cycle. To calculate the normal distribution of individual data, the Kolmogorov Smirnov test was performed. $P < 0.05$ was considered as statistically significant. Statistical evaluation of data was performed using Sigma plot.

3 Results

ECGs were recorded in 21 female volunteers during the 2nd or 3rd day of menstruation and in 31 female volunteers between 14th and 26th day of menstrual cycle, as shown in Fig. 1a and b.

The median of PWD was comparatively more in the control group (Median = 88 ms) than the study group (Median = 83 ms), as shown in Fig. 2a. However, the median value of the PR segment (control group = 47 ms, study group = 46 ms) and PR interval (control group = 133 ms, study group = 129 ms) were similar in both groups as shown in Fig. 2b, c. Though the box-whisker plot of PD shows a slightly higher median value in the study group (Median = 19 ms) than the control group (Median = 16 ms); however, it was not significantly different as shown in Fig. 2d. The interquartile of PR segment and PD is higher in the control group than the study group.

No statistically significant difference was found for the P wave axis and P wave amplitudes in both groups, as shown in Tables 1 and 2, respectively. However, lower

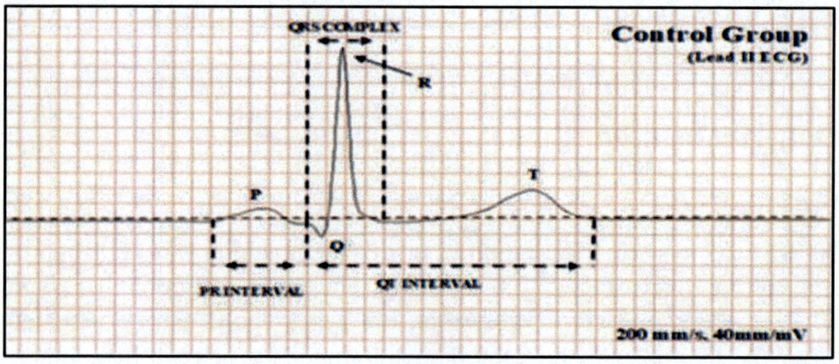

(a)

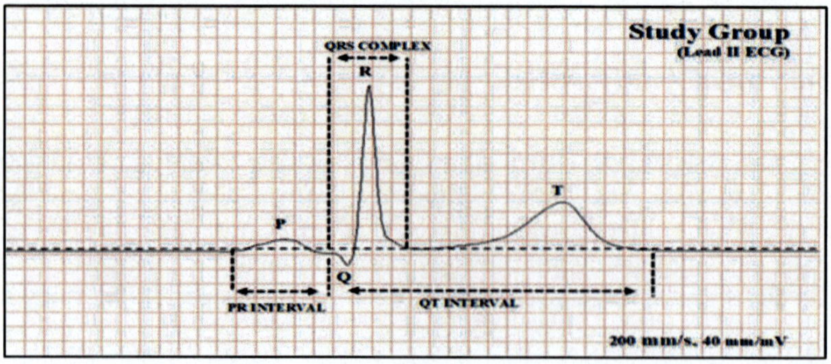

(b)

Fig. 1 Single beat ECG at a paper speed of 200 mm/s, 40 mm/mV in **a** Control group **b** Study group

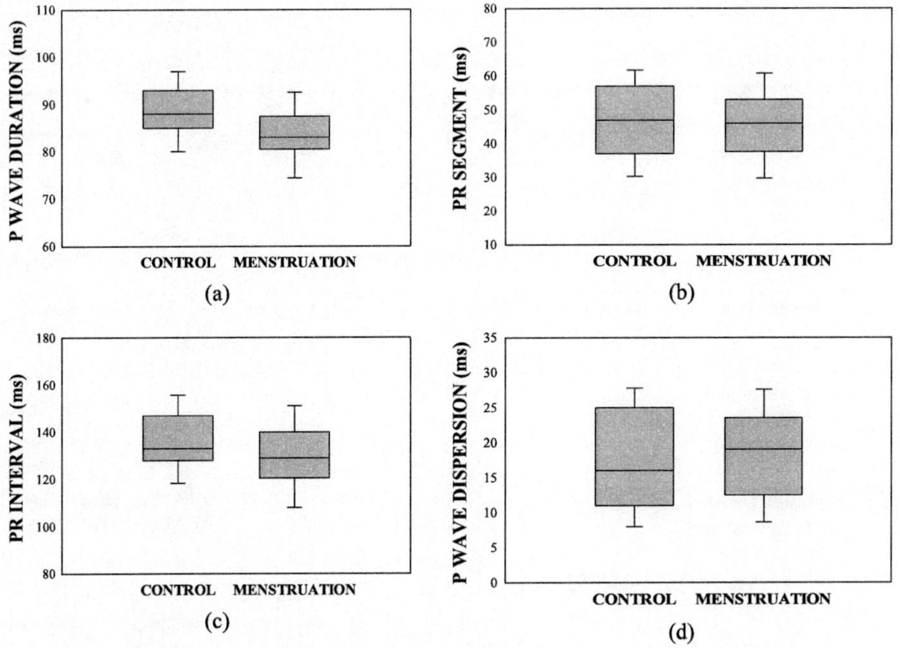

Fig. 2 Box Whisker plot of **a** P wave duration, **b** PR segment, **c** PR interval, and **d** P wave dispersion

Table 1 Change in the P wave axis (in degree) between the control and study group (Positive and negative values indicate an increase or decrease in angles)

P-wave axis changes						
Parameters	Control		Study		Difference in the mean value	P-value*
	Mean ± SD	Range	Mean ± SD	Range		
P wave	50 ± 19	−3 to 78	44 ± 16	14–76	−6	>0.05

All values are in degree; Difference = Study—Control group; *Unpaired sample t-test

Table 2 Change in P wave amplitude (μV) between the control and study group (Positive and negative values represent an increase and decrease in amplitudes)

Parameters	Leads	Group				% change in the mean value		P-value*
		Control		Study				
P wave amplitude		Mean	SD	Mean	SD	Mean	%	
	I	60	13.89	63	11.2	3	5	>0.05
	II	91	21.71	89	21.78	−2	−2	>0.05
	III	51	16.66	52	13.85	1	2	>0.05
	aVR	−77	16.1	−75	12.47	2	−3	>0.05
	aVL	32	8.87	35	7.22	3	9	>0.05
	aVF	71	21.08	68	16.56	−3	−4	>0.05

All values are in microvolt (μV); *unpaired sample t-test

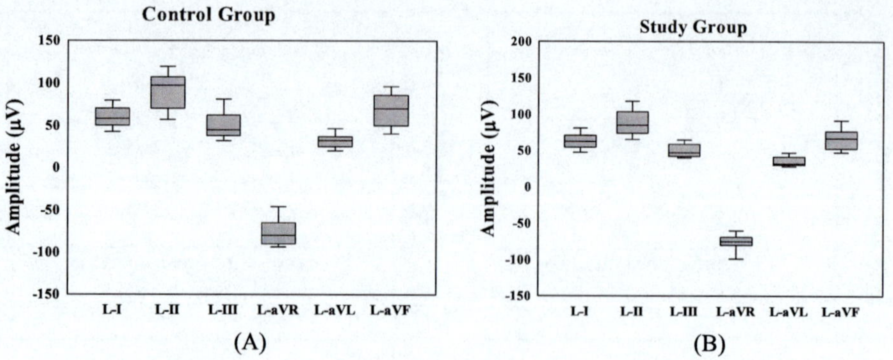

Fig. 3 The box-whisker plot of P wave amplitude for each lead in both control and study groups

Fig. 4 The bar-chart of P wave amplitude for each lead between the control and study groups

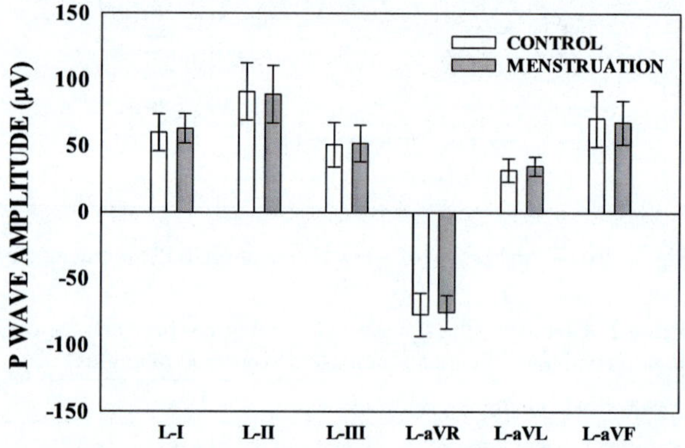

P wave amplitude in the inferior leads (II, aVF) and a higher value in the lateral leads (I, aVL) was noted in women volunteers undergoing menstruation, as shown in Table 2.

A higher median and interquartile range were noted for inferior leads (II, III, and aVF) in the control group compared to the study group, as shown in Fig. 3. However, no significant difference was found in the lateral leads (I and aVL). Similarly, no difference was noted for the P wave amplitude between control and study groups, as shown in Fig. 4.

4 Discussion

The menstrual cycle is not an independent physiological process; instead, it is a part of periodic histochemical alterations in all reproductively active females. Different phases of the menstrual cycle occur due to changes in the level of female sex hormones. These alterations in the level of sex hormones also affect the cardiovascular activity of reproductively active females and ultimately result in morphological variations in the ECG pattern.

In the present study, both groups of women did not show any significant variation in the PR interval and PR segment, though there was a significant decrease in the PWD, as shown in Fig. 2. It is known that prolong PWD is an essential predictor of AF [16]. A study noted that men have longer PWD than women and have nearly two-fold higher chance of developing AF at young age [17]. Nevertheless, after menopause, women have an equivalent potential to develop AF as men. The increased hormonal flux of FSH and LH in blood serum may be the reason for increasing incidence of AF after menopause among women. In the current study, it is found that reproductive women have varying PWD throughout the menstrual cycle, with the decreased value at the time of menstruation. It is also known that hormonal flux is lowest during menstruation, and hence this could be the reason for decreased PWD. Hence, decreased PWD may reduce the chances of AF occurrence. A study proved PD as a non-invasive indicator of AF [12]. They also noted the increase in PD value at the time of luteal phase compared to the other two phases of the menstrual cycle, indicates higher incidence of AF at the time of luteal phase in young women [16]. However, in the current study, no variations were found in PD value throughout the menstrual cycle.

The electrophysiological axis has a direct relationship with the wave amplitude. In the current study, there is a decrease in P wave axis during menstruation, without any significant difference. Hence, no statistically significant variation was found in the P wave amplitude of the frontal plane ECG leads (Table 2). From the findings of this study, during menstrual phase, an increase in the P wave amplitude were noted in the lateral leads I and aVL, which indicates that menstruation may affect the lateral side of the heart. Similarly, leads II and aVF are also affected by the menstrual cycle, and they represent the inferior section of the heart. But the effect on the inferior part of the heart is opposite to that of the lateral part. Variation in the value of atrial amplitude has a direct impact on the concentration of K^+ ion known as hyperkalemia [18]. Electrophysiological variations in the ECG reading is influenced by several factors, including hormonal flux and ionic imbalance in the blood serum. Many studies proved that the receptors for female sex hormones like estrogen and progesterone are also found in the walls of blood vessels, including capillaries [6]. Hence, the occurrence of hormonal flux throughout the menstrual cycle may also affect the electrophysiology of the heart. Similarly, electrolytes concentration in blood serum also varies throughout the menstrual cycle, which directly affects P wave amplitude.

Although many studies have been carried out showing relationship between various ECG parameters and different phases of menstrual cycle, however, few studies exist on the effects of menstural cycle on atrial ECG components. Previous studies show, PD increases significantly at the time of luteal phase than follicular phase of the menstrual cycle as P_{min} decreases significantly at the time of luteal phase [16] but not much variation in PD was found in the present study. On the other hand, few studies show that PR interval [1, 18] and PR dispersion [11] do not show any significant variation throughout the menstrual cycle, which is similar to the findings of this study. In the present study, the value of P wave amplitude in both groups did not change significantly, which is similar to the results of the previous study [1]. No previous studies have been carried out showing the relationship between the menstrual cycle and PWD, however, the current study showed a significant decrease in the PWD at the time of menstruation which could be a preliminary insight for atrial ECG variations in women.

5 Conclusion

The cyclic fluctuation of hormones and electrolytes concentration during the menstrual cycle alters several electrophysiological parameters, hence affecting the cardiovascular system. Increased PWD is an indicator of AF, however, PWD decreases during menstruation indicates that reproductive women are less prone to AF compared to menopausal women. Moreover, variation in electrolyte concentration during the menstrual cycle affects P wave amplitude. Further studies on a larger population are needed to figure out the normal limits of atrial ECG parameters at different phases of the menstrual cycle in reproductive women.

Acknowledgements The authors acknowledge the support from the Ministry of Education, Government of India. The present study was supported by financial grants from the Science and Engineering Research Board (SERB), Department of Science and Technology (DST), Government of India (EEQ/2019/000148).

References

1. S. Khan, J. Prakash J, M.R.U. Beg, M. Kumar, G. Hussain, R. Dixit, S. Sinhi, To study the effect of different phases of menstrual cycle on ECG & blood pressure in healthy young adult females. J. Med. Sci. Clin. Res. (2016). https://doi.org/10.18535/jmscr/v4i5.07
2. G.J. Tortora, B.H. Derrickson, *Principles of Anatomy & Physiology* (2014), pp. 1069–1073
3. B.M.P. Gowd, P.D. Thompson, Effect of female sex on cardiac arrhythmias. Cardiol. Rev. **20**, 297–303 (2012). https://doi.org/10.1097/CRD.0b013e318259294b
4. B. Gupta, C. Batham, Effect of various phases of menstrual cycle on electrocardiogram in 18–22 years healthy females. Glob. J. Res. Anal. **8**, 15–17 (2019). https://doi.org/10.36106/gjra
5. M. Pullaganti, S. Kumar, J.P. Gurja, K.K. Chintala, Association of stress with heart rate variability in different phases of the menstrual cycle. Natl. J. Physiol. Pharm. Pharmacol. **9**, 256–260 (2019). https://doi.org/10.5455/njppp.2019.9.0101922012019
6. C. Kim, C.E. Fahrenbruch, L.A. Cobb, M.S. Eisenberg, Out-of-hospital cardiac arrest in men and women. Circulation **104**, 2699–2703 (2001). https://doi.org/10.1161/hc4701.099784
7. K.E. Barrett, S.M. Barman, S. Boitano, H.L. Brooks, Cardiovascular physiology, in *Ganong's Review of Medical Physiology* (2016), pp. 332–415
8. R.R. Makkar, B.S. Fromm, R.T. Steinman, M.D. Meissner, M.H. Lehmann, Female gender as a risk factor for torsades de pointes associated with cardiovascular drugs. JAMA J. Am. Med. Assoc. **270**, 2590–2597 (1993). https://doi.org/10.1001/jama.1993.03510210076031

9. R. Lokeshwaraiah, A. Gorkal, C. Ganashree, A. Chandran, Comparative study of heart rate, QT and QTc intervals during the different phases of the menstrual cycle. Int. J. Heal. Allied. Sci. **3**, 154–158 (2014). https://doi.org/10.4103/2278-344x.138589
10. M.M. Schreuder, M. Sunamura, J.E. Roeters van Lennep, Supraventricular tachycardia and the menstrual cycle. Case Rep. Women's Heal. **24**, e00153 (2019). https://doi.org/10.1016/j.crwh.2019.e00153
11. M. Dogan, Electrophysiological changes during menstrual cycle, in *Sex and Cardiac Electrophysiology* (Elsevier, 2020), pp 223–229
12. S. Jayaraman, V. Sangareddi, R. Periyasamy, J. Joseph, R.M. Shanmugam, Modified limb lead ECG system effects on electrocardiographic wave amplitudes and frontal plane axis in sinus rhythm subjects. Anatol. J. Cardiol. **17**, 46–54 (2017). https://doi.org/10.14744/AnatolJCardiol.2016.6843
13. R. John, J. Sivaraman, Effects of sinus rhythm on atrial ECG components using modified limb lead system, in *4th IEEE International Conference on Signal Processing, Computing and Control, ISPCC 2017*, pp. 527–530. https://doi.org/10.1109/ISPCC.2017.8269735
14. J. Sivaraman, G. Uma, P. Langley, M. Umapathy, S. Venkatesan, G. Palanikumar, A study on stability analysis of atrial repolarization variability using ARX model in sinus rhythm and atrial tachycardia ECGs. Comput. Methods Programs Biomed. **137**, 341–351 (2016). https://doi.org/10.1016/j.cmpb.2016.10.005
15. J. Sivaraman, G. Uma, M. Umapathy, A modified chest leads for minimization of ventricular activity in electrocardiograms, in *2012 International Conference on Biomedical Engineering* (2012), pp. 79–82. https://doi.org/10.1109/ICoBE.2012.6178959
16. T. Karabag, V. Hanci, M. Aydin, S.M. Dogan, I.O. Turan, N. Yildirim, N.E. Gudul, Influence of menstrual cycle on P wave dispersion. Int. Heart J. **52**, 23–26 (2011). https://doi.org/10.1536/ihj.52.23
17. D. Ko, F. Rahman, R.B. Schnabel, X. Yin, E.J. Benjamin, I.E. Christophersen, Atrial fibrillation in women: epidemiology, pathophysiology, presentation, and prognosis. Nat. Rev. Cardiol. **13**, 321–332 (2016). https://doi.org/10.1038/nrcardio.2016.45
18. M. Longmore, I. Wilkinson, A. Baldwin, E. Wallin, *Oxford Handbook of Clinical Medicine*. (Oxford University Press, 2014).https://doi.org/10.1093/med/9780199609628.001.0001

Performance Assessment of Waveform Modulation Methods for Fifth-Generation Wireless Systems

Getachew H. Geleta, Kinde A. Fante, and Dereje M. Molla

Abstract The 5G cellular network systems feature several innovative strategies such as cognitive radio, large scale antenna arrays installed on macro base stations, extensive adoption of small cells, use of mm-wave communications for short-range links, cloud-based radio access network, etc. In addition to this, 5G cellular networks have more stringent requirements than LTE in terms of data rates, energy efficiency, and latency. However, all these strategies are impacted by the modulation format used at the physical layer. In spite of their success in LTE, the existing modulation scheme such as orthogonal frequency division multiplexing (OFDM) cannot fulfill all the requirements of 5G wireless networks. Hence, the choice of efficient waveform modulation scheme is crucial to meet the requirements of the new 5G cellular network systems. In this paper, we analyze and compare the performance of OFDM, Universal filtered multicarrier (UFMC), and Filter bank multicarrier (FBMC) in terms of peak-to-average power ratio (PAPR) and bit error rate (BER). A detail performance analysis is done by varying the parameters that affect the performance of these the modulation techniques such as number of subcarriers, number of FFT, guard bands, and mapping schemes. The simulation results show that FBMC outperforms both OFDM and UFMC in terms of BER, which makes it more suitable in noisy wireless channels. In terms of PAPR, UFMC gives best performance among the three modulation schemes. Hence, it is a preferable modulation scheme in applications where the power consumption of transmitter power amplifier is critical. The contribution of this work is significant in supporting the selection of efficient modulation technique for the upcoming 5G wireless system standards.

Keywords 5G · BER · FBMC · PAPR · UFMC

G. H. Geleta (✉)
Institute of Technology, Woldia University, 400, Woldia, Ethiopia

K. A. Fante
Institute of Technology, Jimma University, 378 Jimma, Ethiopia
e-mail: kinde.anlay@ju.edu.et

D. M. Molla
Institute of Technology, Hawassa University, 05 Hawassa, Ethiopia

F. Thakkar et al. (eds.), *Proceedings of the International e-Conference on Intelligent Systems and Signal Processing*, Advances in Intelligent Systems and Computing 1370,
https://doi.org/10.1007/978-981-16-2123-9_17

1 Introduction

1.1 Background

The impact of wireless communication systems in our daily life is increasing from time to time. From the introduction of First-Generation mobile networks (1G) up till now, we have witnessed the exponential growth in number of users and the supported applications. In order to fulfill the increasing users' demand, Fifth-Generation wireless systems (5G) is proposed for next telecommunications standards beyond the current Fourth Generation (4G/IMT-Advanced Standard). The 5G aim shows that it will incorporate several innovative ideas which allow a higher density of mobile broadband users, support device-to-device communication, more reliable, and massive machine communications [1, 2].

The important characteristics that are envisioned in 5G wireless systems are higher mobile data volume per geographical area, more connected devices, higher data rate, lower energy consumption, and lower latency. In addition to this, it is expected to support new use cases, such as the IoT, as well as broadcast-like services and lifeline communication in times of natural disaster [2–11].

The Fourth-Generation (4G) Mobile networks standard uses Orthogonal Frequency Division Multiplexing (OFDM) in the core of its physical layer. However, OFDM has several limitations when the scenarios foreseen for future 5G wireless networks are considered. Among the shortcoming of OFDM, low spectral efficiency, high latency, and high out of band emission are critical. Every symbol in OFDM requires insertion of Cyclic Prefix (CP) which reduces the spectral efficiency and increases latency. Furthermore, the sensitivity of OFDM on time and frequency synchronization errors is very high which increases the complexity of receiver design. Due to the rectangular pulse shaping techniques, which is employed in OFDM, it has high Out-of-Band (OOB) emission. Thus, when the scenarios and requirements of 5G wireless systems are considered, OFDM has limitations [3–11].

In order to overcome the limitations of OFDM, many waveform formats have been proposed recently by the research community. These waveforms can be grouped into two main classes: Cyclically-Prefixed OFDM and non-cyclically-prefixed OFDM. The waveform formats in the first group, such as filtered OFDM (f-OFDM) and windowed OFDM (W-OFDM) [2], keep the orthogonality property and attempt to alleviate the limitations of OFDM. Those in the second group initially remove the orthogonality property to obtain improved spectral and temporal characteristics. This is a major paradigm shift in the context of efficient waveform design, which may cause compatibility issues. So far, the new modulation schemes which overcome the limitations of OFDM include f-OFDM, Filter Bank Multicarrier (FBMC), and Universal Filtered Multicarrier (UFMC). These waveform formats are candidate waveforms for 5G wireless systems [3–14].

From literature, the previous works on this topic include the comparative performance analysis of OFDM, UFMC, and FBMC in terms of spectral efficiency (SE) and power spectral density (PSD) [4–10]. These performance metrics do not give the complete picture of the performance of these modulation techniques. In addition to this, the performance variation, in terms of the PSD and BER that could be obtained by tuning the parameters of these modulation techniques was not done so far. In order to fill these research gaps, in this paper, we study the performance of OFDM, FBMC, and UFMC using peak-to-average power ratio (PAPR) and bit error rate (BER) as performance metrics. Moreover, we evaluate the performance of these modulation techniques by tuning their corresponding parameters in a reasonable range of values. This paper is the extension of our previous work [7].

The contributions of this study are two. These are (1) the evaluation of performance metrics of OFDM, UFMC, and FBMC by tuning the corresponding parameters. Accordingly, the parameters that strongly influence the performance, in terms of PAPR and BER, of these modulation methods will be identified. (2) We show that there is performance trade-off when a particular modulation method is employed for any application.

This paper has four sections. Introduction is given in Sect. 1. The System models of the candidate modulation methods for 5G wireless systems are detailed in Sect. 2. In Sect. 3, the performance analysis of the OFDM, FBMC, and UFMC is provided. Finally, in Sect. 4, the conclusion and recommendations are given.

2 Candidate Modulation Methods for 5G Wireless Systems

In this section, the details of system models for the performance analysis of OFDM, FBMC, and UFMC, are provided. The transceiver structures are given and the parameters of the modulation methods are also defined.

Orthogonal Frequency Division Multiplexing (OFDM)

The simplified transmitter and receiver of OFDM system block diagram is shown in Fig. 1. By applying different modulation techniques, the signal is mapped into a suitable constellation. Then, this serial data is then converted into parallel data stream using serial to parallel converter (S/P) module, to which OFDM is performed. It consists of N sub carriers which carries the symbols. An OFDM transmitter involves applying an IFFT block. In order to reduce ISI, a cyclic prefix is added to the output. Then it processed to a serial output using parallel to serial (P/S) converter, which is then passed through the corresponding channel. Then at the receiver, the data

Fig. 1 OFDM Transceiver block diagram

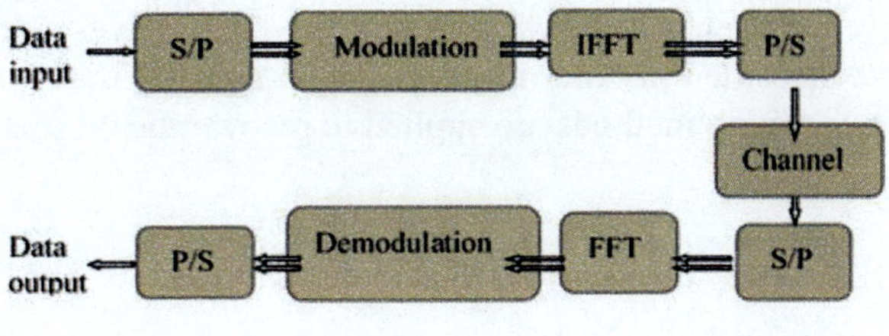

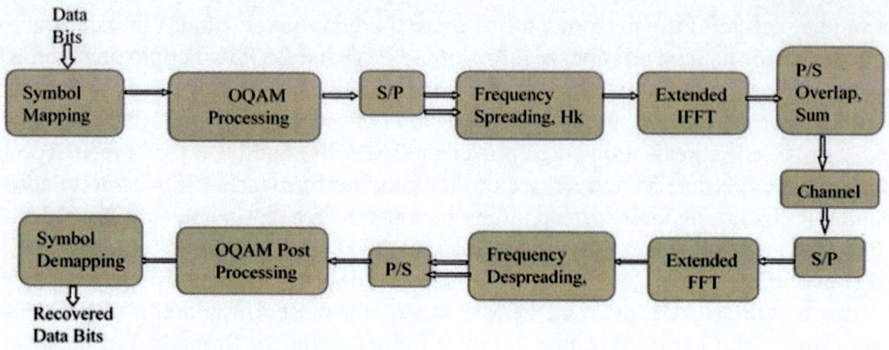

Fig. 2 FBMC Transceiver block diagram

is converted into parallel input and the cyclic prefix is removed. So, this is then subjected to FFT [7–9].

Filter Bank Multicarrier (FBMC)

FBMC is similar to OFDM in format, whereas in FBMC the subcarriers are filtered to suppress side lobe and limit the band. FBMC is preceded either with Quadrature Amplitude Modulation (QAM) or with that of Offset QAM (OQAM) modulation formats. As shown in FBMC Transceiver block diagram Fig. 2, the input signal is first converted from serial to parallel (S/P) form and then filtered using synthesis filter bank. It is then converted back to serial form. On the receiver side, Fig. 2, the signal that passed through channel is converted to parallel form by serial to parallel converter (S/P) and it is filtered using analysis filter bank. The final output is obtained in serial form [5–9].

Universal Filter Multicarrier (UFMC)

The simplified block diagram of UFMC transceiver is depicted in Fig. 3. UFMC modulation technique divides the full band subcarriers into sub bands. The number of subcarriers in each sub band is fixed. For a given transmission, not all the sub bands are required. In this method, each sub band is processed separately. For each sub band, N-point Inverse Fast Fourier Transform (IFFT) is computed by inserting zeros for the carriers which are unallocated [5–9].

The receiver of UFMC computes 2 N-point FFT of the signal, which is received from the channel. In between the successive symbols of IFFT, a guard interval of zeros is inserted in order to avoid the inter symbol interference (ISI) which is cause by filter delay in the transmitter. On the receiver side, the even subcarrier points are discarded in order to get N length signal. Finally, the Equalization and symbol de-mapping methods are applied to recover the original data bits [5–7].

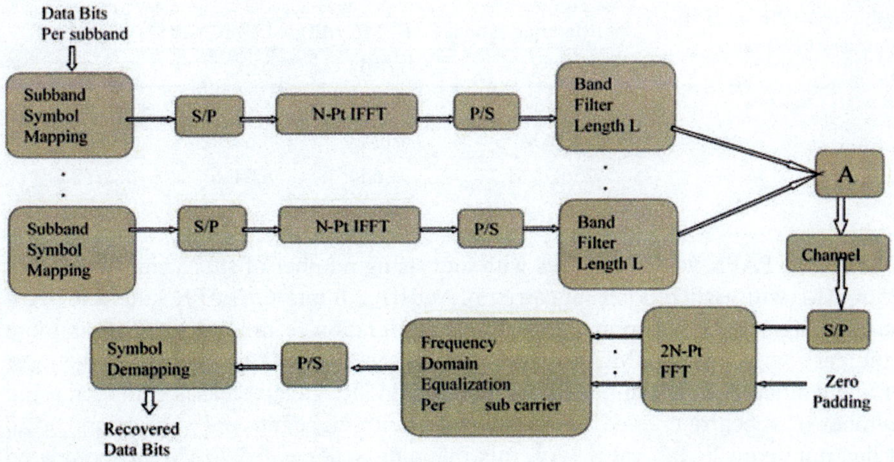

Fig. 3 UFMC Transceiver block diagram

3 Performance Analysis

3.1 PAPR Comparison of OFDM, UFMC and FBMC

PAPR is the maximum amplitude squared divided by the RMS value squared power. It plays a vital role in signal processing applications. The system is more energy efficient if the PAPR is less. However, the modulation techniques with multiple inputs have higher PAPR, whereas those with single input have lower PAPR [9, 10].

A small value of PAPR has prodigiously significance to the battery powered mobile terminals. Because it permits reducing the back-off power amplifier, for certain transmission rate thus lengthens the mobile terminal operating time as compared with the large PAPR modulation schemes. Moreover, it allows raising the coverage at cell edges. A high PAPR makes the power amplifier work with large IBOs, resulting in inefficient use of the amplifier.

The following comparative analysis of UFMC, FBMC, and OFDM is done for 200 subcarriers for different or various types of mapping schemes, such as 4QAM, 16QAM, 64QAM, and 256QAM. Therefore, the MATLAB simulation results for 200 subcarriers of FBMC with 11 dB SNR value and also for UFMC and OFDM with 8 dB SNR value is listed out in the following tables.

As shown in Table 1, for mapping schemes having 512 number of FFT; PAPR value of FBMC reduces by increasing number of bits per subcarrier except for the number of subcarriers equal to eight at which point it increases again although it is smaller than the PAPR value with two bits per subcarrier. In FBMC, 64QAM has minimum PAPR value with number of FFT equal to 512 as compared with UFMC and OFDM of 64QAM mapping scheme. In the case of UFMC, the PAPR values haven't shown consistent behavior with increasing number of bits per subcarriers. In

Table 1 PAPR values for number of FFT 512

Mapping scheme	FBMC (dB)	UFMC (dB)	OFDM (dB)
4QAM	9.6503	9.04	8.4377
16QAM	8.7711	8.2379	8.8843
64QAM	8.1743	8.6229	9.9269
256QAM	9.3654	8.0416	7.2553

OFDM, the PAPR value increases with increasing number of subcarriers except for 256QAM (with eight bits per subcarriers). And it has minimum PAPR value 256QAM at mapping scheme as compared with all modulation techniques. UFMC is more efficient compared to FBMC except for 64QAM. With 64QAM, FBMC is the most efficient modulation technique. In OFDM, the PAPR value increases with increasing number of subcarriers except for 256QAM (with eight bits per subcarriers). And it has minimum PAPR value with this mapping scheme (256QAM) as compared with all modulation techniques. UFMC is efficient as compared with FBMC except for 64QAM. With 64QAM, FBMC is efficient as compared with both UFMC and OFDM.

As shown in Table 2, FBMC the PAPR values haven't shown consistent pattern of change with increasing number of bits per subcarriers. In other words, the PAPR values decreases from 16 to 4QAM and it increases from 64 to 16QAM. The 16QAM mapping scheme achieves the minimum PAPR value. In UFMC and OFDM, the PAPR value is increasing with increasing number of subcarriers except for 256QAM (with eight bits per subcarriers). The minimum value of PAPR is achieved at 256QAM mapping scheme. This implies with 256QAM and 1024 number of FFT, OFDM achieves the least PPR value and FBMC achieves the highest PPR value.

As shown in Table 3, for mapping schemes with 2048 number of FFT, the FBMC PAPR values do not show consistent pattern of change with increasing number of bits per subcarriers. In this case, 4QAM has minimum PAPR value. Thus, FBMC is more

Table 2 PAPR values for number of FFT 1024

Mapping scheme	FBMC (dB)	UFMC (dB)	OFDM (dB)
4QAM	9.3883	7.7732	8.4377
16QAM	8.8375	8.2967	8.8843
64QAM	8.9976	9.6202	9.9269
256QAM	8.9561	7.5373	7.4133

Table 3 PAPR values for number of FFT 2048

Mapping scheme	FBMC (dB)	UFMC (dB)	OFDM (dB)
4QAM	9.501	8.1008	8.5165
16QAM	10.3033	8.0195	8.8843
64QAM	9.754	9.2209	9.9269
256QAM	10.3408	7.3563	7.4833

efficient at low mapping schemes in terms of PPR for the given setting. Similarly, in UFMC, the PPR value does not show consistent pattern when we change the mapping schemes. At 256QAM and 2048 number of FFTs, the UFMC achieves the least value of PPR. In OFDM, the PAPR value is increasing with increasing number of bits per subcarrier except for the number of subcarriers equal to eight. The minimum value of PPR is achieved at 256QAM mapping scheme.

Generally, the system can be efficient by using either 64QAM for 512 number of FFT equal or 16QAM for 1024 number of FFT, or 4QAM for 2048 number of FFT for FBMC. The least value of PPR is obtained from OFDM with 256 QAM and 512 number of FFT.

3.2 BER Comparison of OFDM, UFMC and FBMC

The BER of each modulation techniques were simulated by using parameters that yield the best performance of SE, PSD [7], and PAPR. Thus, to simulate the BER of FBMC the following system parameters of Table 4 have been considered initially. To simulate the BER of UFMC and OFDM for the initial condition, the following system parameters have been considered as shown in Table 5. The BER versus SNR value of FBMC, UFMC, and OFDM with those parameters are shown as in Figs. 4 and 5, respectively.

Table 4 BER simulation parameter values of FBMC

Parameters of simulation	Values used
K, symbols of overlapping	4
NFFT, number of FFT	512
Guard bands	212
Symbols Number	100
4QAM, bits per subcarrier	2
SNR	15 dB

Table 5 BER simulation parameter values of UFMC and OFDM

Simulation parameters	Value
Number of subcarriers	288
Number of FFT points (N_{FFT}) for OFDM	1024
Number of FFT points (N_{FFT}) for UFMC	512
Filter length for UFMC	43
Cyclic prefix length for OFDM	43
Symbols number	200
Bits per subcarrier	4 (16QAM)
SNR	15 dB

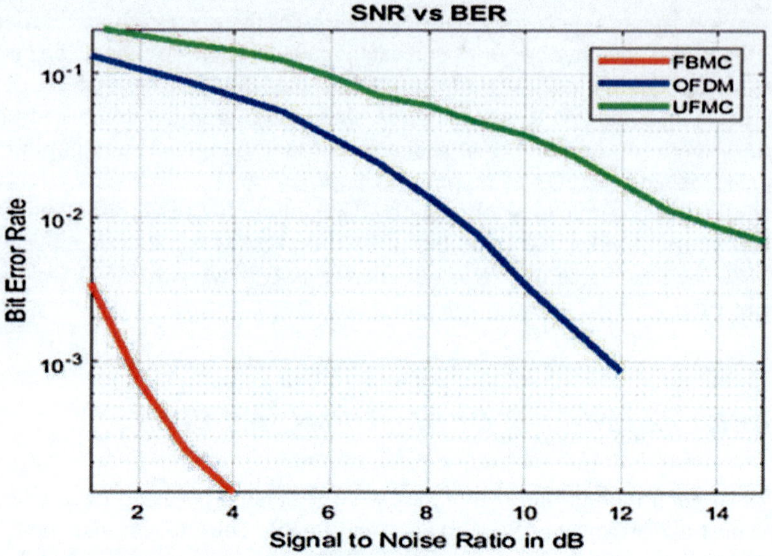

Fig. 4 BER versus SNR of FBMC, OFDM, and UFMC

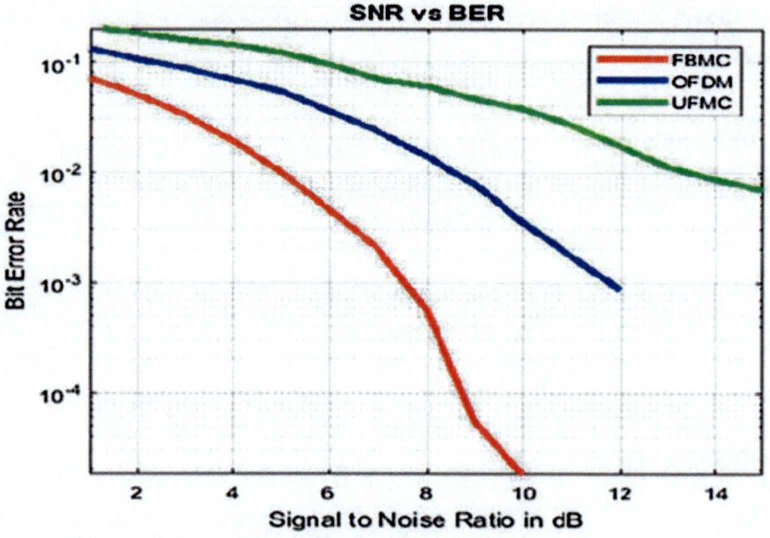

Fig. 5 BER with increasing number of FFT of FBMC to 1024

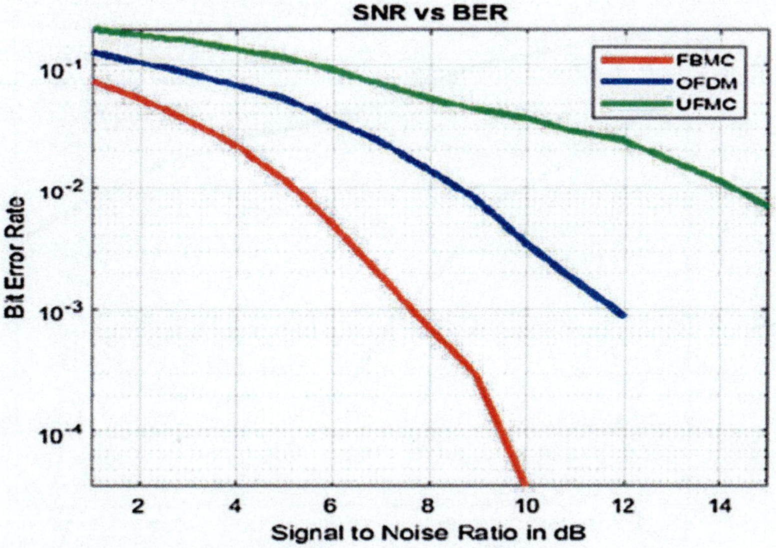

Fig. 6 BER with decreasing number of guard bands of FBMC to 102

As shown in Fig. 4, the BER versus SNR graphs of FBMC, OFDM, and UFMC, the BER value of each modulation technique decreases with increasing SNR. The BER value of FBMC is closer to zero for SNR value of 4 dB. In this case, FBMC shows best performance among all the modulation techniques. OFDM shows better BER performance than UFMC.

Figure 5 shows the BER versus SNR graphs of FBMC, OFDM, and UFMC. Similar to the previous case, FBMC shows the best BER performance and UFMC achieves the least performance.

As shown in Fig. 6, the BER versus SNR graphs of FBMC, OFDM, and UFMC, the BER value of FBMC is closer to zero around 10 dB and it is smaller than that of OFDM and UFMC. Whereas, the BER value of OFDM is smaller than that of UFMC.

As shown in Fig. 7, the BER versus SNR graphs of FBMC, OFDM, and UFMC, the BER value of FBMC is closer to zero around 4 dB and it is smallest as compared with OFDM and UFMC. FBMC shows as best performance and UFMC achieves the worst performance in terms of BER. Moreover, as compared with Fig. 4, the BER value of OFDM is decreasing with increasing number of FFT. In other words, OFDM BER performance improves with increasing number of FFT.

As shown in Fig. 8, the BER value of FBMC is closer to zero around 4 dB. Similar to the previous cases, FBMC shows best performance and UFMC achieves the worst performance in terms of BER. Moreover, as compared to Figs. 4 and 8, the BER value of OFDM increases by increasing number of subcarriers. Meaning, the performance of OFDM deteriorates with increasing number of subcarriers.

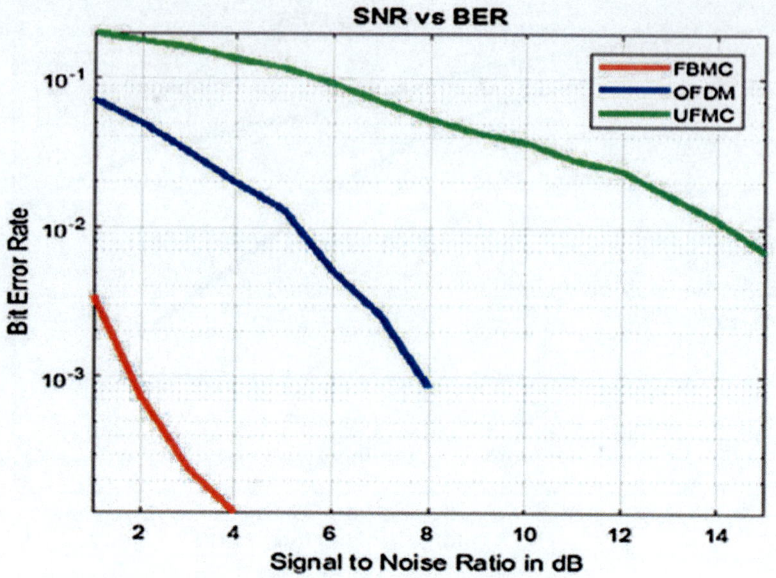

Fig. 7 BER with number of FFT of OFDM to 2048

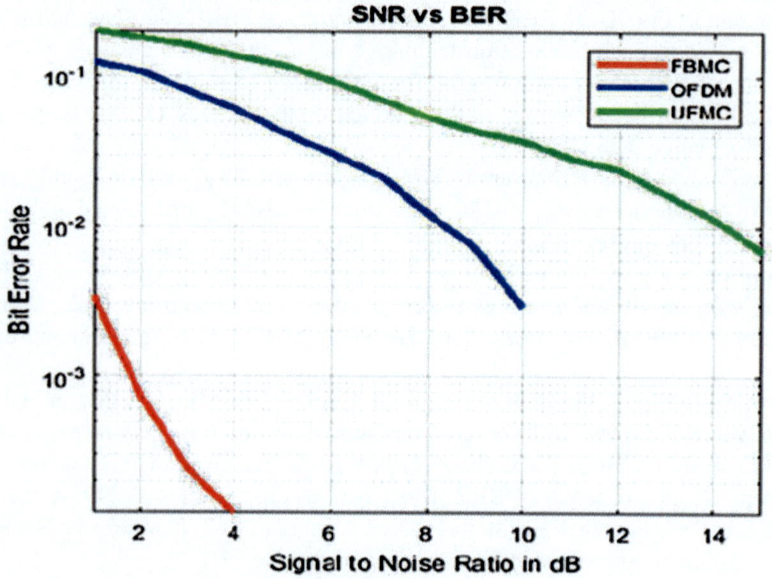

Fig. 8 BER with number of subcarriers to 512

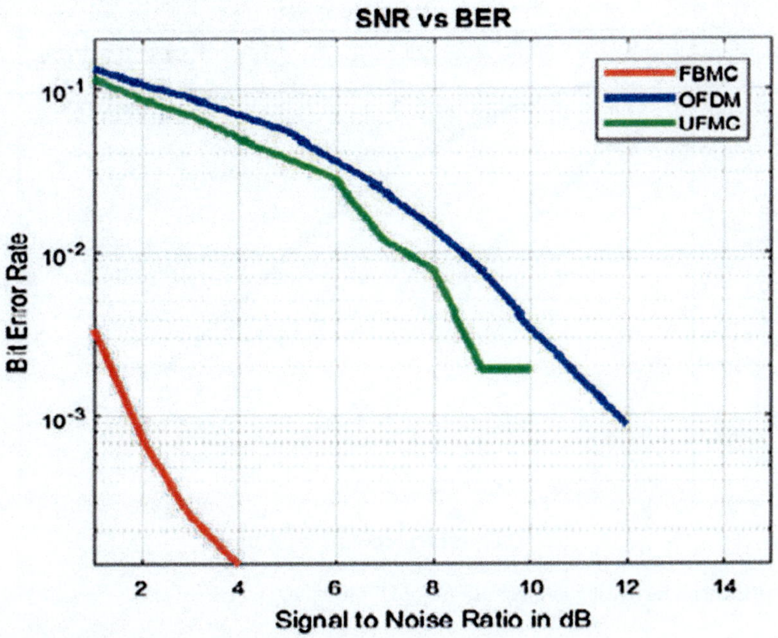

Fig. 9 BER with decreasing number of subcarriers of UFMC to 128

As shown Fig. 9 BER, FBMC shows best performance and OFDM achieves the worst performance in terms of BER. Moreover, as compared to Fig. 4, the BER value of UFMC reduces with decreasing number of subcarriers. In other words, UFMC performance improves with decreasing number of subcarriers.

As shown in Fig. 10, the FBMC achieves the best performance and OFDM achieves the worst performance. Moreover, as compared to Fig. 4, the BER value of UFMC reduces with increasing number of FFT. In other words, performance of UFMC reduces with increasing number of FFT.

Generally, the BER value is decreasing with increasing SNR value. In addition to this, the BER value of FBMC is closer to zero before OFDM and UFMC; and it has the smaller BER value than both OFDM and UFMC. Thus, FBMC has the best performance. FBMC BER performance can be improved by reducing number of FFT as well as through increasing guard bands. And the BER value of OFDM can be enhanced by increasing number of FFT and by decreasing number of subcarriers. Moreover, the BER value of UFMC can also be improved by decreasing number of subcarriers and increasing number of FFT.

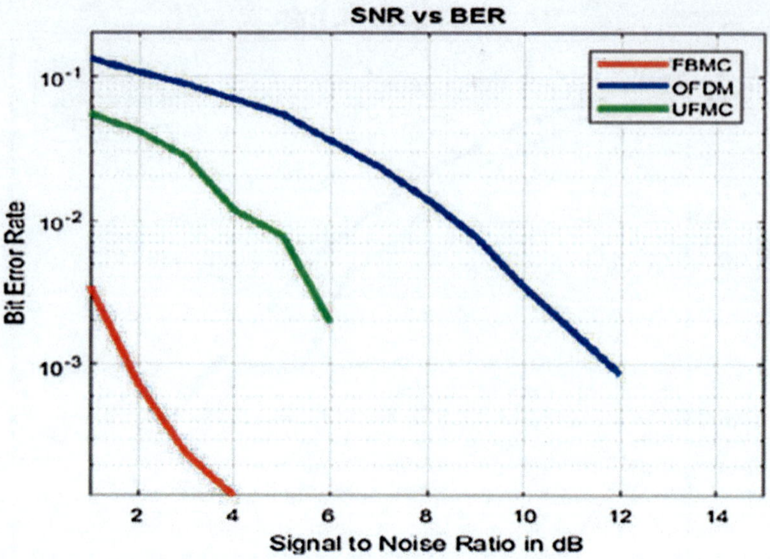

Fig. 10 BER with increasing number of FFT of UFMC to 1024

4 Conclusions

In this paper, the comparative performance analysis of three new modulation techniques, namely OFDM, UFMC, and FBMC, are performed in terms of PAPR and BER. The simulation results show that both FBMC and UFMC overcome the limitations of OFDM. The impact of varying the parameters of these the modulation techniques, such as number of subcarriers, number of FFT, guard bands, and mapping schemes, on the performance metrics has been analyzed in detail. The performance of UFMC is superior than that of OFDM and FBMC in terms of PAPR which makes it preferable for energy-constrained applications. However, FBMC is the best among the three modulation techniques in terms of BER, which makes it suitable for communications systems in noisy wireless channels.

In the future, we will further explore the performance analysis of additional 5G modulation techniques by considering additional parameters such as latency and data rates.

References

1. GPP, Study on scenarios and requirements for next generation access technologies. Technical report 38.913, February (2016)
2. G. Wunder et al., 5G NOW: Non-orthogonal, asynchronous waveforms for future mobile applications. IEEE Commun. **52**, 97–105 (2014)

3. Y. Cai, Z. Qin, et al., *Modulation and Multiple Access for 5G Networks*, China (2017)
4. E. Öztürk, E. Basar, H.A. Çirpan, *Generalized Frequency Division Multiplexing With Flexible Index Modulation* (Istanbul, Turkey, 2017)
5. P. Naga Rani, C. Santhi Rani, *UFMC: The 5G Modulation Technique* (Andhra Pradesh, India, 2016)
6. P. Kansal, A.K. Shankhwar, *FBMC versus OFDM Waveform Contenders for 5G Wireless Communication Systems, India* (2017)
7. G.H. Geleta, D.M. Molla, *Kinde, International Conference on Advances of Science and Technology ICAST 2018: Advances of Science and Technology* (Springer Science and Business Media LLC, 2019), pp. 503–518
8. P. Banelli, S. Buzzi et al., *Modulation Formats and Waveforms for the Physical Layer of 5G Wireless Networks: Who Will BE the Heir of OFDM?* (2014)
9. Y. Liu, X. Chen et al., *Waveform Candidates for 5G Networks: Analysis and Comparison* (Beijing, China, 2016)
10. R. Gerzaguet, N. Bartzoudis et al., *The 5G Candidate Waveform Race: A Comparison of Complexity and Performance* (CEA-Leti, Minatec Campus, Grenoble, France, 2017)
11. A.R. Reyhani, A. Farhang, B. Farhang-Boroujeny, Circularly Pulse-Shaped Waveforms for 5G: Options and Comparisons. ECE Department, University of Utah, USA (2015)
12. M. Van Eeckhaute, A. Bourdoux et al., *Performance of Emerging Multi-Carrier Waveforms for 5G Asynchronous Communications* (Belgium, Brussels, 2017)
13. J. Abdoli, et al., Filtered OFDM: a new waveform for future wireless systems, in *Signal Processing Advances Wireless Communication* (Stockholm, Sweden, 2015)
14. Y. El Gholb, N. El Imrani, H. Ghenniouiet, 5G: an idea whose time has come. Int. J. Sci. Eng. Res. **8**(8) (2017)

Design and Fabrication of CPW Fed Jeans Antenna Working at 2.4 GHz ISM Band

Poonam Thanki and Falguni Raval

Abstract The wearable antenna working at 2.4 GHz for Industrial, Scientific and Medical (ISM) radio bands has been presented in this paper. The proposed antenna is having a coplanar waveguide (CPW) feeding on the jeans as the substrate material. This antenna is a hexagonal shaped monopole having heptagonal cut inside. The antenna has a size of dimensions of 80 × 97.48 mm^2. This compact size antenna is suitable for wearable applications. The antenna parametric analysis is carried out to observe the performance of the antenna. The simulated and measured results are compared and they match with each other.

Keywords CPW · ISM band · Wearable antenna

1 Introduction

One of the main research topics in antennas for body-warn communications is wearable, fabric-based antennas. In wireless communication systems, compact wireless devices are required. Wearable antenna incorporated in textile material will help in monitoring the particular body parameter. It is very important to monitor the health of the people involved in security operation, mining, military and some other hazardous environment. In such cases, health data of such people can be communicated to doctors and other medical staff using wearable antennas. So development has happened in the wearable systems by utilizing the textile materials [1]. The dielectric constant of textile material is generally less which reduces the surface wave losses and improves the bandwidth of the antenna [2]. While designing wearable antenna, care must be taken in choosing the textile material as its thickness and permittivity (εr) are the deciding factor to design high-efficiency antennas [3, 4]. These antennas are

P. Thanki (✉) · F. Raval
V.T. Patel Department of Electronics & Communication Engineering, Chandubhai S. Patel Institute of Technology, Charotar University of Science and Technology, CHARUSAT, Changa, India
e-mail: poonamthanki.ec@charusat.ac.in

F. Thakkar et al. (eds.), *Proceedings of the International e-Conference on Intelligent Systems and Signal Processing*, Advances in Intelligent Systems and Computing 1370,
https://doi.org/10.1007/978-981-16-2123-9_18

having wide range of applications in the area of tracking, navigation, telemedicine, public safety, defense, portable communication and health monitoring [5, 6].

In [7], CPW fed antenna is designed at 2.4 GHz frequency on jeans substrate. In [8], Dual band wearable rectangular patch is designed and simulated on jeans substrate around operating frequency of 0.9 and 2.4 GHz. In [9], Comparison of 2.4 GHz Wearable Antennas on Three Textile Substrates is carried out in order to check its performance characteristics. In [10], T shaped dual band microstrip antenna fed by coaxial feed for wearable applications in the ISM band (2.4–2.5, 5.7–5.8 GHz) is proposed. In [11], S shape structured metamaterial antenna for 2.4 GHz is designed which increases the probable application in the field of wireless communication. In [12], wearable antenna using jeans as substrate is proposed at 2.45 GHz frequency. In [13], rectangular patch having partial ground wearable antenna working on frequency of 2.4 GHz application is proposed.

This work intends to present antenna fabricated on jeans substrate having dielectric constant 1.7. This antenna is having coplanar waveguide feed which gives wide bandwidth. The proposed wearable antenna working at 2.4 GHz ISM band is suitable for health care monitoring.

2 Antenna Design

The proposed antenna is CPW fed antenna in which jeans is used as substrate. The hexagonal shaped slot is created on ground plane. Inside this hexagonal slot, monopole of hexagonal shaped is placed. This hexagonal shaped monopole is having heptagonal cut inside as shown in Fig. 1. Jeans is used as dielectric material which is having dielectric constant of 1.7 [5] and loss tangent of 0.025 [5]. The geometry of the wearable antenna is shown in Fig. 1.

The dimensions of antenna are: Substrate width W = 80 mm, Substrate length L = 97.48 mm, Hexagon monopole width (b) = 23 mm, Heptagon width (a) = 14.5 mm, Outer Hexagon slot width (k) = 30 mm, Hexagon monopole feed line width (W_f) = 3.045 mm, gap width = 1 mm each side, substrate height = 1.6 mm.

3 Parametric Study

The design and optimization of proposed antenna is performed by using the simulation tool Ansoft High Frequency Structure Simulator. The Hexagonal monopole and heptagon cut are the deciding factors in optimization process. The width and position of Hexagon and Heptagon are changed in order to see the effect on S_{11}.

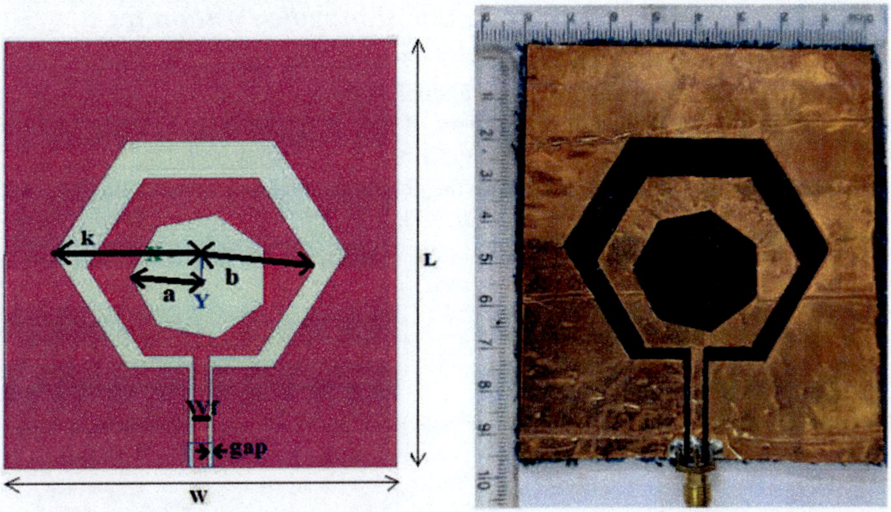

Fig. 1 Simulated and fabricated antenna

3.1 *Effect of Change in Monopole Width (b)*

Change in monopole width causes change in operating frequency as well as S_{11} changes significantly.

Figure 2 shows the change in S_{11} with the change in hexagonal monopole width (b) value. The other parameters [substrate width: 80 mm, length: 97.48 mm, hexagonal monopole centre (c): 2.8 mm, heptagon width (a):14.5 mm, heptagon centre position (yp):5 mm, heptagon position (cy):5 mm] are kept constant. It is observed from the graph that S_{11} is −23.19 dB at 2.36 GHz for b = 23 mm, S_{11} is −19.16 dB at 2.22 GHz for b = 24.5 mm and S_{11} is −20.30 dB at 2.16 GHz for b = 25 mm.

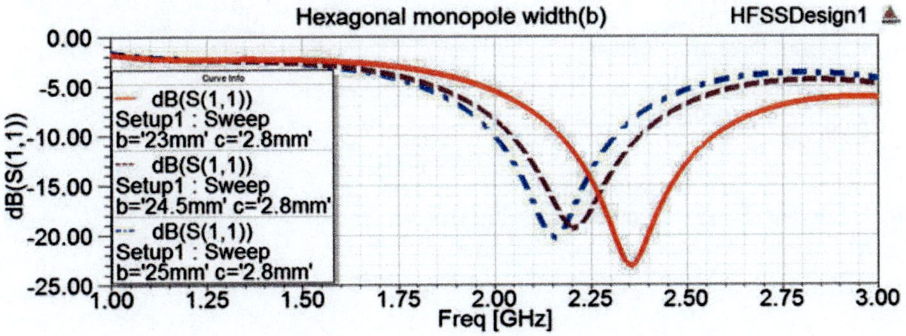

Fig. 2 Effect of change in Hexagonal monopole width (b) on S_{11}

3.2 *Effect of Change in Hexagonal Monopole Position (c)*

Figure 3 show the effect of change in hexagonal monopole position on S_{11}. Hexagonal monopole position (c) refers to movement of monopole along Y axis from centre position. Variation in monopole position affects the frequency and S_{11}.

Figure 3 shows the change in S_{11} with the change in hexagonal monopole position (c) value. The other parameters [Substrate width: 80 mm, length: 97.48 mm, hexagonal monopole width (b):23 mm, heptagon width (a):14.5 mm, heptagon centre position (yp):5 mm, heptagon position (cy):5 mm] are kept constant. It is observed from the graph that S_{11} is -19.25 dB at 2.44 GHz c = 1.4 mm, S_{11} is −23.19 dB at 2.36 GHz for c = 2.8 mm and S_{11} is −25.23 dB at 2.32 GHz for c = 3.8 mm.

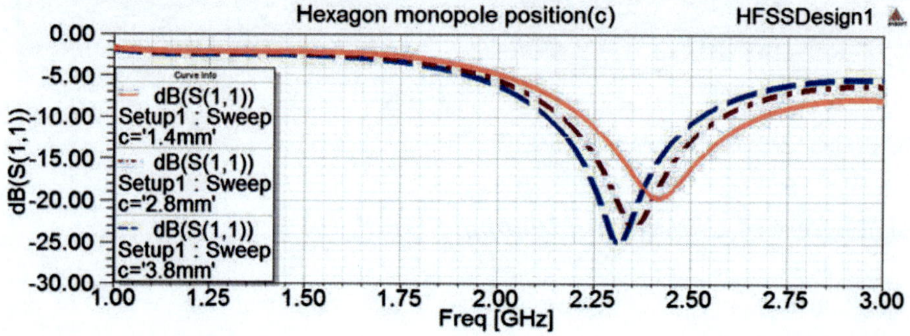

Fig. 3 Effect of change in Hexagonal monopole position (c) on S_{11}

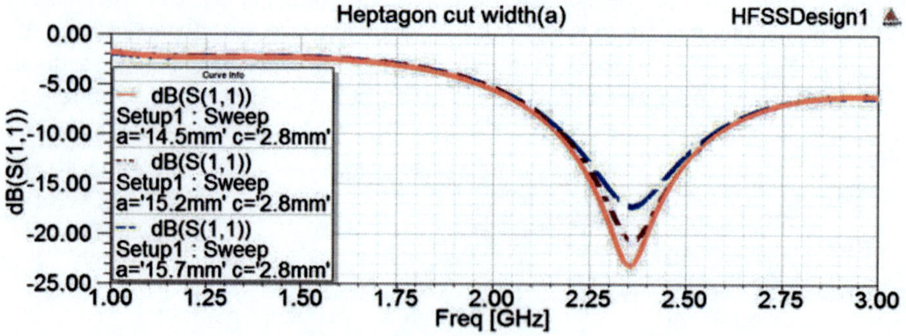

Fig. 4 Effect of change in Heptagon cut width (a) on S_{11}

3.3 *Effect of Change in Heptagon Width (a)*

Figure 4 shows the effect of change in heptagon width. Due to change in width of heptagon, S_{11} changes considerably.

Figure 4 shows the change in S_{11} with the change in heptagon cut width (a) value. The other parameters [Substrate width: 80 mm, length: 97.48 mm, hexagonal monopole width (b):23 mm, hexagonal monopole centre position (c):2.8 mm, heptagon centre position (yp):5 mm, heptagon position (cy):5 mm] are kept constant. It is observed from the graph that S_{11} is −23.19 dB at 2.36 GHz for a = 14.5 mm, S_{11} is −20.68 dB at 2.36 GHz for a = 15.2 mm and $S_{11}$is −17.20 dB at 2.36 GHz for a = 15.7 mm.

3.4 *Effect of Rotation in Heptagon (cy)*

Figure 5 shows the effect of rotation in heptagon position on S_{11}. Heptagon position (cy) refers to rotation of heptagon in XY plane at different angle. If the heptagon is rotated around XY axis at different angle, it gives change in S_{11}. Value of Heptagon centre position is yp = 5 mm. For this centre position value of cy = 5 mm refers to 0°, cy = 3 mm refers to 43.5° and cy = 9 mm refers to 68° angle position.

Figure 5 shows the change in $S_{11}$with the change in heptagon position (cy) value. The other parameters [Substrate width: 80 mm, length: 97.48 mm, hexagonal monopole width (b):23 mm, hexagonal monopole centre position (c):2.8 mm, heptagon centre position (yp):5 mm, heptagon width (a):14.5 mm] are kept constant. It is observed from the graph that S_{11} is −23.19 dB at 2.36 GHz for cy = 5 mm, S_{11} is −22.72 dB at 2.36 GHz for cy = 3 mm and S_{11} is −16.13 dB at 2.36 GHz for cy = 9 mm.

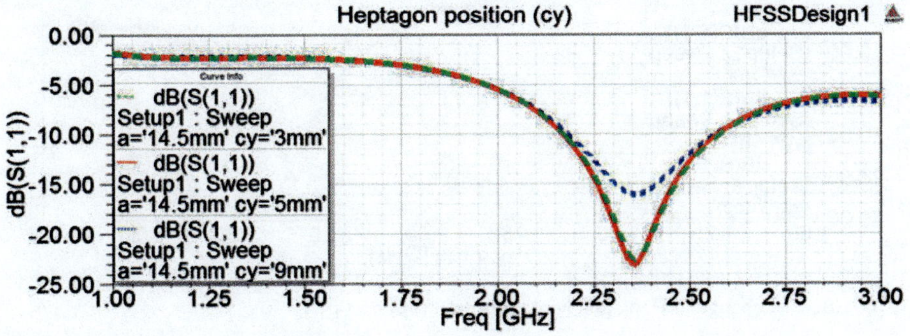

Fig. 5 Effect of change in Heptagon position (cy) on S_{11}

3.5 Effect of Change in Heptagon Centre Position (yp)

Figure 6 shows the effect of change in heptagon centre position on S_{11}. Heptagon centre position (yp) refers to movement of heptagon along Y axis. If the heptagon is moved along Y axis, it changes $S_{11.}$ Heptagon centre position is deciding factor in optimization process. By properly choosing its value, the resonating frequency can be obtained.

Figure 6 shows the change in $S_{11}$with the change in heptagon centre position (yp) value. The other parameters [Substrate width: 80 mm, length: 97.48 mm, hexagonal monopole width (b): 23 mm, hexagonal monopole centre position (c):2.8 mm, heptagon position (cy):5 mm, heptagon width (a):14.5 mm] are kept constant. It is observed from the graph that S_{11} is −23.19 dB at 2.36 GHz for yp = 5 mm, S_{11} is −23.64 dB at 2.34 GHz for yp = 4 mm and S_{11} is −25.05 dB at 2.34 GHz for yp = 3 mm. For yp = 5 mm, the resonance achieved is nearest to 2.4 GHz which is desired frequency. So yp = 5 mm is selected.

4 Fabrication

After successful simulation, the antenna is fabricated as shown in Fig. 1. Jeans is used as substrate for the antenna. The thickness of single layer of jeans is 0.54 mm. So, three layers of jeans are sewed at the edges to achieve required thickness of 1.6 mm. The conductive copper sheet is used for hexagonal monopole and ground plane. The soldering of SMA connector is done by using normal soldering technique. To avoid the chances of burning, extra care is taken during stitching and soldering of jeans.

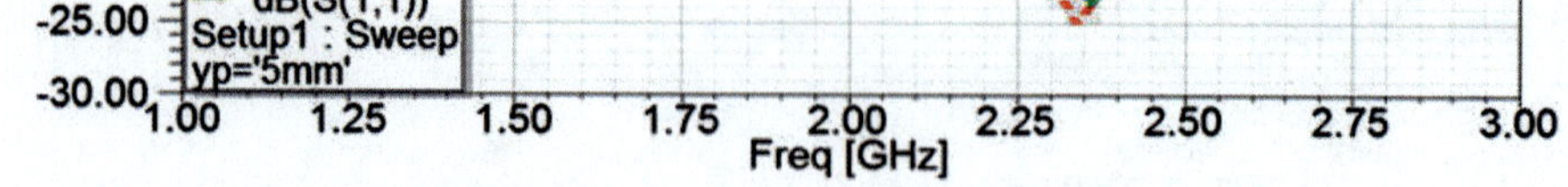

Fig. 6 Effect of change in Heptagon centre position (yp) on S_{11}

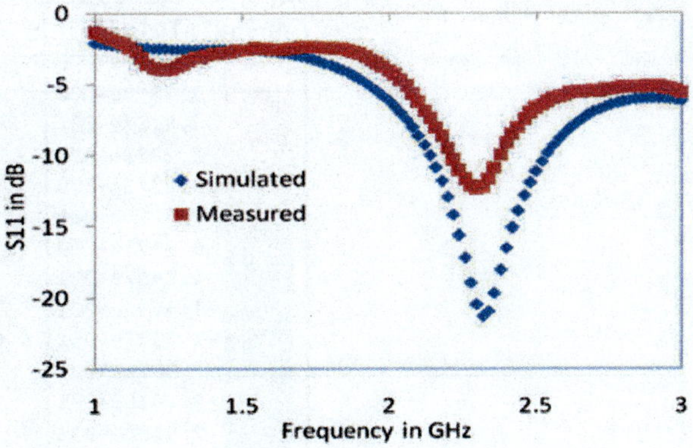

Fig. 7 S_{11} of the simulation and measurement

5 Measurement and Results

5.1 *Reflection Coefficient*

The prototype antenna is tested using network analyzer (Agilent N9912A). Figure 7 shows the comparison of simulated and measured S11 of proposed antenna. It can be seen that simulated S11 at 2.4 GHz is −23.1970 dB. The measured S11 at the same frequency is −12.27 dB. It shows a good matching between the results of simulation and measurement.

5.2 *Gain*

The prototype antenna is having 5.3210 dB gain at working frequency 2.4 GHz as shown in Fig. 8.

5.3 *Radiation Pattern*

The E plane radiation pattern and H plane radiation pattern are measured by placing proposed antenna in anechoic chamber. There is a good matching between the results of simulation and measurement (Figs. 9 and 10).

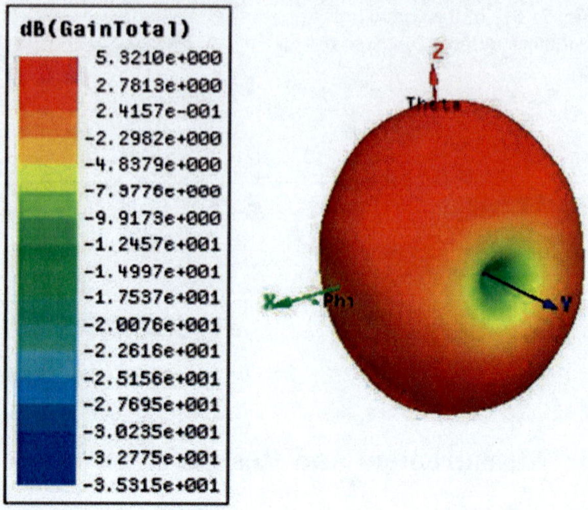

Fig. 8 Simulated 3D gain

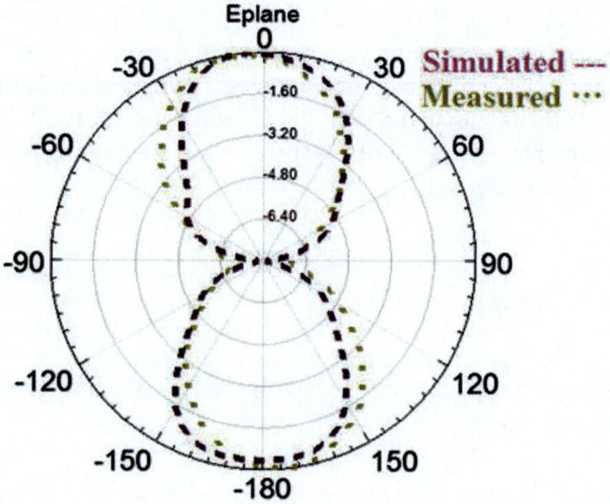

Fig. 9 E plane of simulation and measurement

6 Conclusion

In this paper, CPW fed Hexagonal monopole with heptagonal cut antenna on jeans substrate operating at 2.4 GHz ISM band of compact size is fabricated and measured. Parametric study of various parameters of antenna shows the effect of parameter change on return loss. The simulated S_{11} of antenna is −23.19 dB and measured S_{11} is −12.27 dB. The fractional bandwidth is 17.5%. The maximum gain of proposed antenna is 5.32 dB.

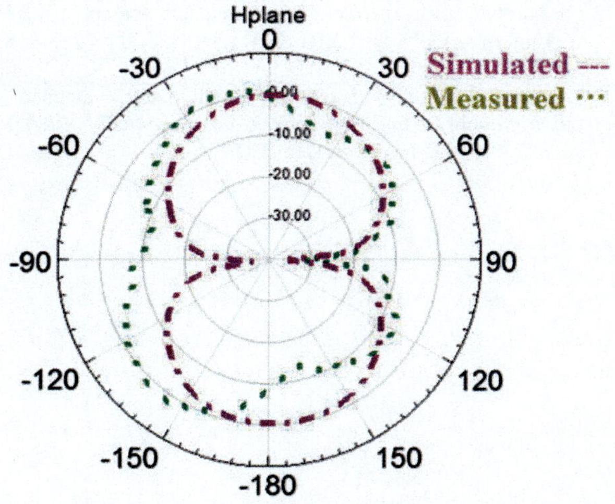

Fig. 10 H plane of simulation and measurement

References

1. M.A.R. Osman, M.K.A. Rahim, M. Azfar, N.A. Samsuri, F. Zubir, K. Kamardin, Design, implementation and performance of ultra wideband textile antenna, Prog. Electromagn. Res. B **27**, 307–325 (2011)
2. M.A.R. Osman, M.K.A. Rahim, M.A. Azfar, K. Kamardin, F. Zubir, N.A. Samsuri, Design and analysis UWB wearable textile antenna, in *Proceedings of the 5th European Conference on Antennas and Propagation (EUCAP) IEEE Conference Publication* (2011), pp. 530–533
3. S. Sankaralingam, B. Gupta, Development of textile antennas for body wearable applications and investigations on their performance under bent conditions. Prog. Electromagn. Res. B **22**, 53–71 (2010)
4. C.A. Winterhalter, J. Teverovsky, W. Horowitz, V. Sharma, K. Lee, *Wearable Electro-Textiles For Battlefield Awareness, Mechanical Testing of Electro-Textile Cables and Connectors* (Materials Research Society, Shur, Wilson, Urban, Ed., Warrendale PA, 2003)
5. N. Singh, A.K. Singh, V.K. Singh, Design and performance of wearable ultra wide band textile antenna for medical application. Research article De gruyter Open Eng. **5**, 117–123 (2015). https://doi.org/10.1515/eng-2015-0012
6. M.A.R. Osman, M.K.A. Rahim, N.A. Samsuri, M.K. Elbasheer, M.E. Ali, Textile UWB antenna bending and wet performances. Int. J. Antennas Propag. **2012**(251682) (2012). https://doi.org/10.1155/2012/251682
7. Z. Muhammad, S.M. Shah, Z.Z. Abidin, A.Y.I. Asyhap, S.M. Mustam, Y. Ma: CPW-fed wearable antenna at 2.4 GHz ISM band, in *American Institute of Physic*, 020003-1, ISBN: 978-0-7354-1563-8
8. P. Muthu Kannan, V. Palanisamy, Dual band rectangular patch wearable antenna on jeans material. Int. J. Eng. Technol. **3**(6), 442–446 (2011–2012)
9. M.I. Ahmed, M.F. Ahmed, A.A. Shaalan, Investigation and comparison of 2.4 GHz wearable antennas on three textile substrates and its performance characteristics. Open J. Antennas Propag. **5**(3), 110–120
10. S. Singh, R.K. Gangwar, S. Agarwal, A dualband T-shaped microstrip antenna for wearable application. Int. J. Electron. Electr. Eng. (2014)
11. V. Dwivedi, Y.P. Costa, R. Jyoti, An investigation and application issues of miniaturized compact MSTPA for RF communication system using metamaterial: a study, in *IEEE International RF and Microwave Conference Proceedings* (Malasiya, 2008). 2–4 Dec 2008

12. S. Purohit, F. Raval, Wearable-textile patch antenna using jeans as substrate at 2.45 GHz. Int. J. Eng. Res. Technol. (IJERT) **3**(5) (2014). ISSN: 2278-0181
13. W.N.N.W. Marzudi, Z.Z. Abidin, S.H. Dahlan, K.N. Ramli, H.A. Majid, M.R. Kamarudin, Rectangular patch with partial ground wearable antenna for 2.4 ghz applications, in *IEEE International RF and Microwave Conference (RFM 2015)* (Riverside Majestic Hotel, Kuching, 2015). 14–16 Sarawak on Dec 2015

Effect of Artifacts on the Interpretation of EEG-Based Functional Connectivity Estimation Using Partial Directed Coherence

Taha Mahmoud Al-Naimi, Ahmad Zuri Sha'ameri, and Norlaili Mat Safri

Abstract The proposed work investigates the design of an algorithm to enhance the functional connectivity for human brain using electroencephalogram (EEG) signals. The partial directed coherence (PDC) approach was used to estimate the right analysis of the functional connectivity. However, the presence of artifacts due to eye movement/blinking and muscles movement affects the correct estimation of functional connectivity. Therefore, the time-average differencing (TAD) method was developed to remove the artifacts. Two methods based on discrete wavelet transform (DWT) denoising and moving average filter are used, and comparison is made to access their effectiveness. Between the two methods, the TAD MA method removes all artifacts and ensures correct estimation of functional connectivity using PDC.

Keywords Artifacts removal · Functional connectivity · Partial directed coherence (PDC) · Time-average differencing (TAD)

1 Introduction

EEG is widely used for research purposes due to the simplicity of using it and offers a millisecond range of resolution that is unachievable with computerized tomography (CT) or magnetic resonance imaging (MRI) [1]. It has been used to detect the electrical activity of the human brain using electrodes attached to the scalp based on the international 10–20 system [2]. Furthermore, the recordings can determine the active parts, the flow of information, and diagnosing disorders of the brain.

The connectivity between the brain areas estimated using PDC [3] depends on the tasks performed by the subjects. The low magnitude of EEG signals within 40–100 μV [4] makes it sensitive to contamination by non-cerebral signals such as ocular artifacts, muscular artifacts, heart rate variation artifacts, and impedance fluctuation artifacts. Even though many trustworthy solutions exist to deal with most

T. M. Al-Naimi (✉) · A. Z. Sha'ameri · N. M. Safri
Division of Electronic and Computer Engineering, School of Electrical Engineering, Faculty of Engineering, Universiti Teknologi Malaysia, Johor Bahru 81310, Johor, Malaysia
e-mail: mataha3@graduate.utm.my

F. Thakkar et al. (eds.), *Proceedings of the International e-Conference on Intelligent Systems and Signal Processing*, Advances in Intelligent Systems and Computing 1370,
https://doi.org/10.1007/978-981-16-2123-9_19

EEG artifacts [5], it is challenging for humans to avoid ocular artifacts. Moreover, eye blink occurs as larger peaks with amplitude ten times higher than the EEG and is the most dominant over other artifacts. Generally, a human involuntarily blinks at least once each five seconds, and the activity of blinking lasts about 100–400 μs [5]. Furthermore, the levator scapulae muscle causes muscular artifacts near the occipital as a result of moving the neck or by looking around [6].

There are many methods developed to remove the affection of artifacts from the EEG signals, such as using extra reference signal [7] or by using software such as Brain Electrical Source Analysis (BESA) [8]. However, the most used methods vary on the unsupervised learning algorithm that knew as Blind Source Separation (BSS) techniques to detect and remove muscular artifacts [9]. There are commonly known methods that adopted the BSS algorithm to remove artifacts viz: Principal Component Analysis (PCA), Independent Component Analysis (ICA), and Canonical correlation analysis (CCA). Yet, the possibility of removing valuable information with the artifacts is high [10]. In this paper, the TAD algorithm was developed to remove the effect of the artifact from EEG recordings. This method was developed based on the characteristics of wavelet denoising and moving average filtering technique. Moreover, it will be shown that the TAD algorithm will remove the artifacts and improve the estimation of function connectivity for a given task using PDC.

The organization of this paper is as follows: In Sect. 2, the signal mode is described, and it includes EEG signal characteristics, artifacts characteristics, EEG signal measurement procedure, and the problem statements. Section 3 describes the estimation of functional connectivity from EEG signals using PDC. Section 4 presents TAD algorithm that is used to remove the artifacts. Section 5 presents the simulation results, including the affection of artifacts on the estimation of functional connectivity using PDC, and the consequences of using the proposed method, while Sect. 6 describes the conclusion.

2 Signal Model

In this section, EEG signal characteristics and measurement procedure are described. Also, the EEG signals are modeled as MVAR process followed by the cause of artifacts and the characteristics of each type are explained. Finally, the problem caused by artifacts and MVAR are presented in the last sub-section.

2.1 EEG Signals

The recording of EEG signals can be done invasively that requires surgery known as intracranial electroencephalogram (iEEG), or by non-invasive (EEG) [11]. Both methods are used to measure the electrical activities produced by the firing of neurons in the human brain. Also, recordings can be used to determine the activity of the subjects' brain and the direction of information during tasks [4]. The range of amplitude and frequency for the EEG signal is from 40 to 100 μv and from 0 to 100 Hz, respectively [5]. Besides, the frequency bands covers are as follows: Delta (0–4 Hz), Theta (4–8 Hz), Alpha (8–13 Hz), Beta (13–31 Hz), Gamma (31–51 Hz), and high Gamma (51–above 100 Hz) [12].

2.2 Multivariate Autoregressive Modeling

EEG signals have no well-known characteristics to be generated; therefore, signals that had been used for simulation in some articles depend on clean EEG segments and generally selected visually with the help of neurophysiologist [13]. However, selecting elements from different EEG signals affects the connectivity between the channels. For this purpose, Multivariate Autoregressive (MVAR) process was used to model and simulate EEG signals for multiple channels. The MVAR process is similar to the AR process, except the observed signals could be as the result of a linear combination of signals from other channels.

For multiple channels prediction, Granger causality (GC) states that a time series $x_1(n)$ could be caused by another time series $x_2(n)$ containing information, which helps to predict $x_1(n)$ [14]. The MVAR process can be expressed, as shown in Eq. 1 [15]:

$$\begin{bmatrix} x_1(n) \\ x_2(n) \\ \vdots \\ x_N(n) \end{bmatrix} = \sum_{r=1}^{N} A_r \begin{bmatrix} x_1(n-r) \\ x_2(n-r) \\ \vdots \\ x_N(n-r) \end{bmatrix} + \begin{bmatrix} w_1(n) \\ w_2(n) \\ \vdots \\ w_N(n) \end{bmatrix} \tag{1}$$

where $w_i(n)$ is white uncorrelated innovations noises, $x_1(n)$ to $x_N(n)$ are the stationary processes for each channel, and A_r is the coefficients of the MVAR as shown in Eq. 2:

$$A_r = \begin{bmatrix} a_{11}(r) & a_{12}(r) & \cdots & \cdots & a_{1N}(r) \\ \vdots & \vdots & \vdots & \vdots & \vdots \\ \vdots & \vdots & \vdots & a_{ij}(r) & \vdots \\ \vdots & \vdots & \vdots & \vdots & \vdots \\ a_{N1}(r) & \cdots & \cdots & \cdots & a_{NN}(r) \end{bmatrix} \tag{2}$$

where the coefficients $a_{ij}(r)$ are the interaction effect due to $x_j(n-r)$ on $x_i(n)$.

2.3 Artifacts

The 10–20 system of electrode placement is used to collect EEG signals from subjects' scalp, as shown in Fig. 1. The placements depend on measuring scalp from the nasion to inion through vertex once, and through the pre-auricular point the other time [16]. After that, each measured length was divided into 10 and 20% to place the electrodes on particular points.

The EEG recordings are always corrupted by unwanted signals known as artifacts. Artifacts are unwanted signals captured during EEG recordings that are caused by external sources such as unsatisfactory technology, connections loss, the power line interference at frequency 50/60 Hz, or external electronic devices that can create an electromagnetic field. Also, it can be caused by internal sources that arise from the body activities such as muscles movement "muscular artifacts," eye blinking/movement "ocular artifacts," and cardiac artifacts [4]. Blinking amplitude is relatively higher than the amplitude of EEG signals as with peaks of more than 800 μv [17]. The artifacts occur in durations of 200–400 ms and a period of 15–20 per minute and eye-ball movement artifacts occur at frequency of below 5 Hz.

On the other hand, the frequency of muscular movements such as head shaking nodding, and swallowing range between 20 and 50 Hz and overlapped with EEG signals [18]. There are various methods for detecting and removing the artifacts automatically, such as using extra reference signal [7], thresholding [5], and using deep learning [17]. Furthermore, some articles depend on filtering the recorded signal [19] and BSS methods to detect and remove muscular artifacts [9].

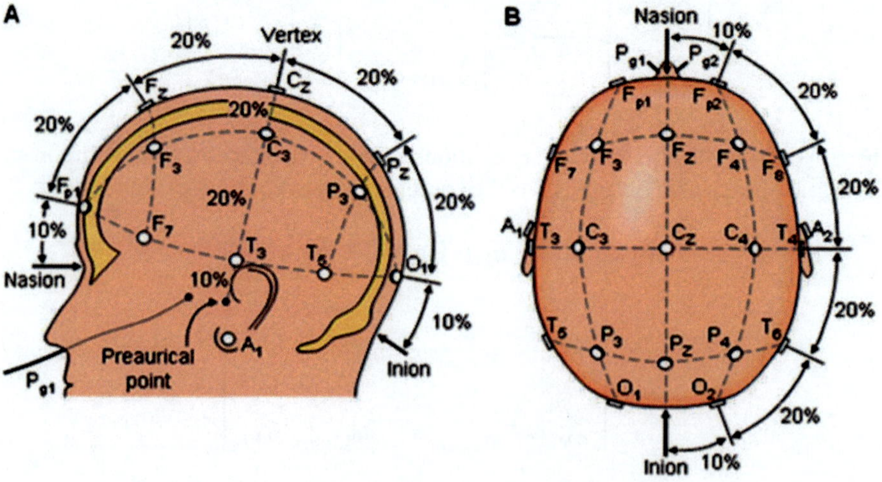

Fig. 1 Electrode points based on the 10–20 electrode placement system [2]

2.4 Problem Statements

The MVAR process models the connectivity between the channels [15]. As mentioned in the section before, EEG recordings contain additional components, "artifacts" due to muscular movements and eyes blinking/movement that will conflict with the signal and creates new interaction between the channels. Thus, artifacts need to be removed before estimating the connectivity between the channels. Moreover, using additional electrodes for eyes and muscles as reference signals is not suggested because it will cause discomfort for the subjects [17]. For this paper, an MVAR model for five channels [15] that includes artifacts is selected and described as follows:

$$\begin{aligned}
x_1(n) &= 0.95\sqrt{2}x_1(n-1) + 0.9025x_1(n-2) + b(n) + m(n) + w_1(n) \\
x_2(n) &= 0.5x_1(n-2) + w_2(n) \\
x_3(n) &= -0.4x_1(n-3) + w_3(n) \\
x_4(n) &= -0.5x_1(n-2) + 0.25\sqrt{2}x_4(n-1) + 0.25\sqrt{2}x_5(n-1) + w_4(n) \\
x_5(n) &= -0.25\sqrt{2}x_4(n-1) + 0.25\sqrt{2}x_5(n-1) + b(n) + m(n) + w_5(n)
\end{aligned} \tag{3}$$

where $b(n)$ represents the blinking artifacts and $m(n)$ represents the muscular artifacts.

The blinking effect is assumed a convolution between a linear time-invariant system $h(n)$ and pulse train $x(n)$ expressed as

$$b(n) = \sum_{m=-\infty}^{\infty} h(m)x(n-m) \tag{4}$$

where $h(n)$ is a Hamming function of length 200 samples and $x(n)$ represent the train of pulses defined as

$$x(n) = \begin{bmatrix} 1 & \underbrace{0 \ \ 0 \ \ \cdots \ \ 0}_{16 \text{ segments}} & 1 & \cdots \end{bmatrix} \tag{5}$$

where the mask and space for each period of the pulse train are 126 and 1875 samples, respectively. Therefore, the duration of each pulse in the blinking is 302 samples, and the period is 2130 samples. Based on Eqs. 4 and 5, the simulated eye blinking signal is shown in Fig. 2.

Muscular artifacts were simulated by using the principles of electromyography (EMG) signals. It was based on the action potential (AP) in the muscle fiber that is caused by depolarization and repolarization of the muscle [20]. Moreover, the essential mechanism influencing the magnitude and density of it are the requirements of motor unit action potential (MUAP) and its firing frequencies as shown in Fig. 3. The simulated EMG signal can be expressed as

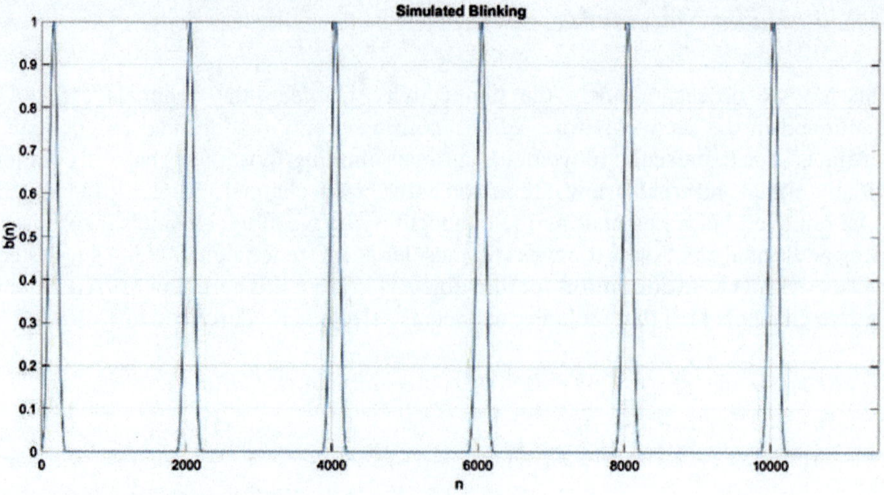

Fig. 2 Simulated eye blinking signal using MATLAB

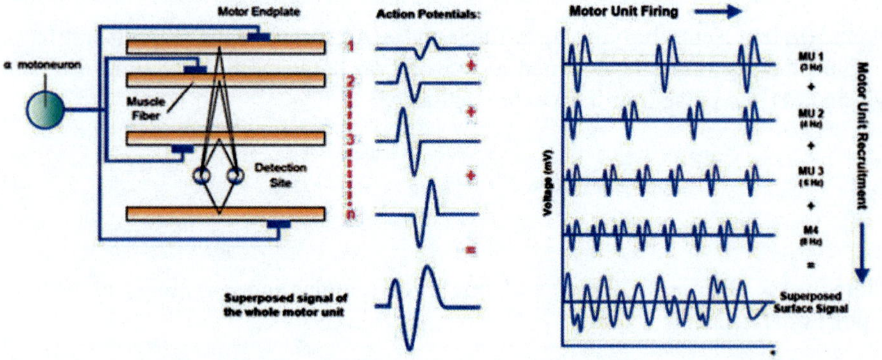

a) Generation of MUAP b) Requirement and firing frequency

Fig. 3 Requirements and firing frequencies of MUAP [21]

$$m(n) = \sum_{r=0}^{N-1} h(r)w(n-r) \tag{6}$$

$$h(n) = \sum_{l=0}^{M} A(l)\sin(2\pi f_l n) \tag{7}$$

where $m(n)$ represents the simulated EMG signal, $h(n)$ for the (MUAP), $w(n)$ for point processed (firing impulse), and N is the number of motor unit firing. While M

is the number of muscle fiber action potentials, $A(l)$ represents the l-th amplitude for each action potential, and f_l is the firing frequency.

3 PDC Analysis

Complex neural network nodes connectivities can be caused either by direct or indirect connections depending on the interactions between the nodes [22]. Cross-correlation or cross coherence could be used to verify the connectivity of two signal sources by describing how two signals are similar to each other in the time or frequency domain. However, both methods could not assess the presence of a third or more connectivity, where the dependence came from [23]. For this purpose, methods were developed, such as DTF and PDC derived from the GC concepts [24].

Both PDC and DTF methods depend on the prediction coefficients of the MVAR model for multiple channels, while GC depends on the error prediction of the MVAR [22]. Although DTF and PDC play a vital role in the analysis of the EEG signal, the DTF method cannot detect the right directional connectivity when the technique is applied for more than two channels [25]. Nevertheless, PDC provides the causal relation between the channels in the frequency domain and focuses on direct and indirect interrelations [26]. Furthermore, PDC computes the coherence between two of k channels while avoiding volume conduction. It can also ensure measuring the directional flow of information and can be described as follows:

$$\pi_{ij}(f) = \frac{a_{ij}(f)}{\sqrt{a_j(f)a_j(f)}} \tag{8}$$

where $a_{ij}(f)$ represents the difference between the identity matrix of n-elements and the Fourier transform of the coefficient series. Moreover, $\pi_{ij}(f)$ represents the ratio of information flow from channel $i \leftarrow j$ and information flow from j to all the channels at frequency f.

PDC values will be in the interval 0,1 because of the normalization in Eq. 8. The results of the PDC analysis based on the MVAR model defined in Eq. 3 are shown in Fig. 4. It shows the interactions between signals from different channels and their coherence level in frequency. The vertical axis represents magnitude, and horizontal axis represents frequency in samples.

It is observed that the diagonal plots have a higher magnitude compare to the others because it is related to the same channel. For the off-diagonal channels, the direction of information is from channel $(1 \rightarrow 2)$, $(1 \rightarrow 3)$, $(1 \rightarrow 4)$, and $(4 \leftrightarrow 5)$. The connectivities caused by artifacts can be seen between channels $(1 \leftarrow 4)$, $(1 \leftarrow 5)$, and $(5 \leftarrow 1)$. After that, the connectivity caused by artifacts makes the detection of the sources and destinations hard.

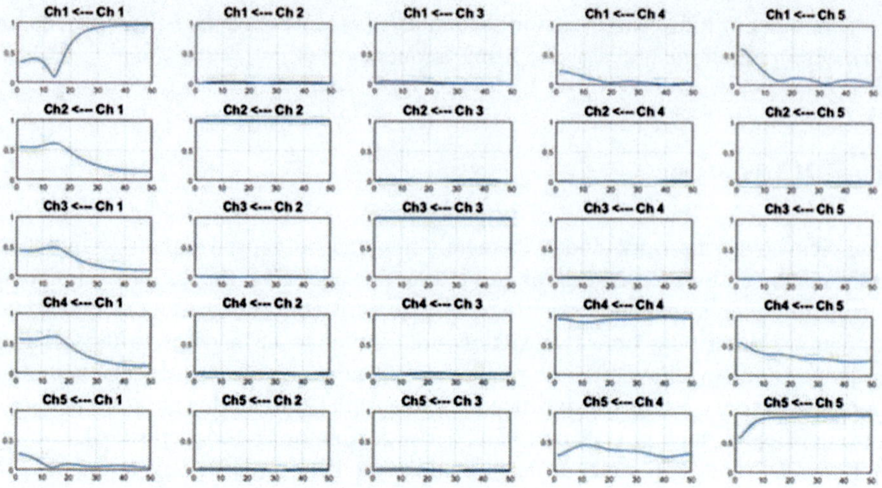

Fig. 4 PDC analysis of EEG signal generated from Eq. 3

4 Time-Average Differencing (TAD)

TAD is developed to remove major artifacts from EEG signals such as ocular artifacts (eye blinking/movement) and muscular artifacts and preserve the EEG signals. Figure 5 shows the basic structure for TAD. The function represented by $h(n)$ is a linear time-invariant that can be implemented by any filtering or denoising algorithms. For this paper, $h(n)$ is implemented using moving average (MA) filtering and discrete wavelet transform (DWT) denoising.

In the above figure, $x(n)$ is the input signal—in this case, it is the raw EEG recording received from one of the channels—while $h(n)$ is the linear time-invariant system, and the output is $x_{sm}(n)$. The clean EEG recording $y(n)$ is the difference between system output $x_{sm}(n)$ and the input signal $x(n)$.

Fig. 5 General diagram of TAD algorithm

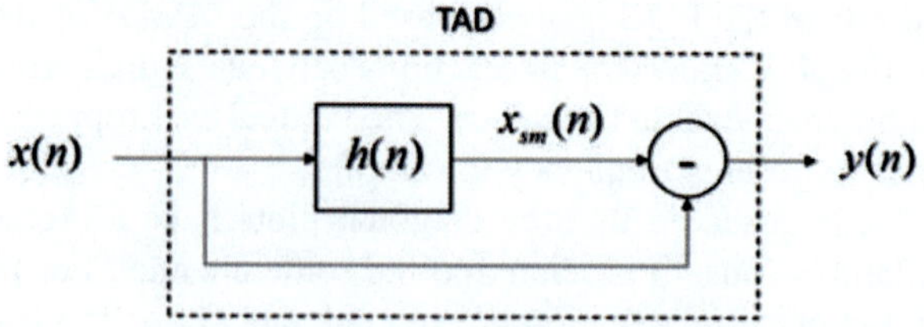

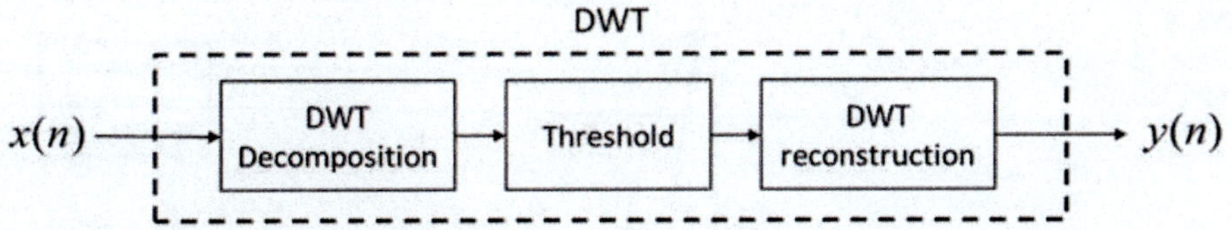

Fig. 6 DWT denoising process [7]

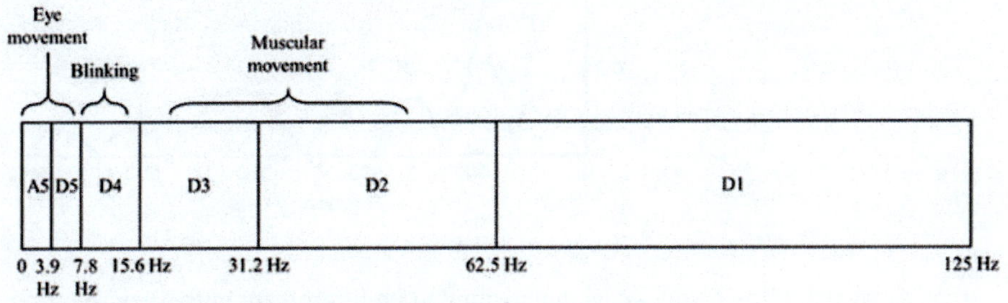

Fig. 7 DWT decomposition levels and the presence of artifacts

4.1 DWT Denoising

DWT decomposes the signal by using basis functions known as wavelets [27]. After that, the signal passes through a high-pass filter and low-pass filter followed by down-sampling algorithm to obtain the detail coefficients and approximate coefficient, respectively. Then, the denoising is performed by thresholding at the appropriate decomposition level to ensure the removal of artifacts without loss of information from the signal. The DWT denoising procedure is described in Fig. 6.

For EEG signal, the band of frequency is 0.5–120 Hz, and the artifacts frequencies are in ranges of (0–7 Hz) for eye movement, (8–13 Hz) for blinking, and (20–50 Hz) for muscular activity [18]. Therefore, DWT decomposition should be five and the artifacts caused by eye movement, blinking and muscular should appear at decomposition levels 5, 4, and 3, respectively. After that, the statistical thresholding is used for denoising each decomposition level of each EEG channel recorded with artifacts. The decomposition levels and the presence of artifacts are described as shown in Fig. 7.

4.2 MA Filtering

The MA filter is the simplest filter that uses a rectangular window function to smooth the signals from noise depending on the filter length [28]. The filter length is inversely proportional to the frequency, as shown in Fig. 8 below [29].

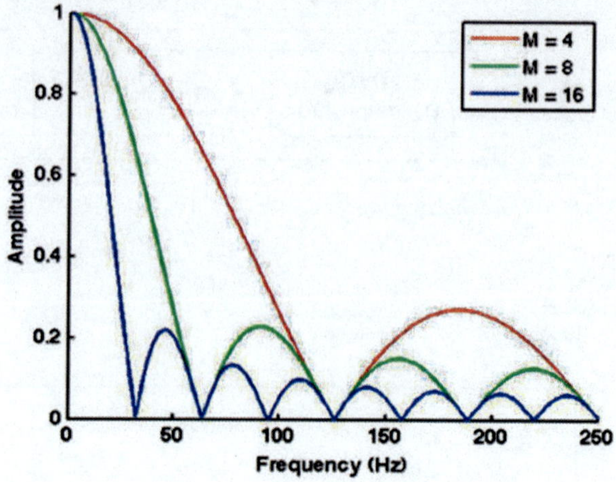

Fig. 8 Frequency response of the MA filter for various filter lengths

In other words, if the frequencies that needed to be filtered are high, then the filter length should be made small. For that reason, the selected filter length will decide how much the artifacts will pass through the MA filter. The presence of artifacts in power spectral is described as shown in Fig. 9.

The MA filter window size can be estimated from the highest artifacts' frequency that allows the signal to pass through the filter and calculates the mean values of the

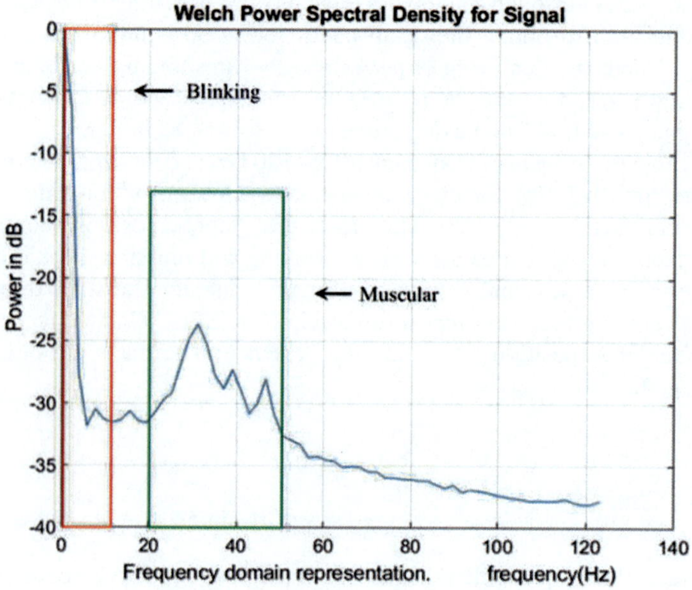

Fig. 9 Power spectral and the presence of artifacts

artifacts such as ocular and muscular artifacts. Moreover, the smoothed signal can be extracted from the recorded EEG data without removing any valuable information from it by knowing the size of the filter from Eq. 8 [29] below:

$$N = \frac{F_s}{F} \tag{9}$$

where N is the filter size, while F_s is the sampling frequency and the highest artifacts' frequency is F.

From the formula above, the estimated size for the MA filter is (5) that will allow the mean values of the artifacts to pass through. After that, the signal will be subtracted from the recorded EEG signal to remove the affection of artifacts from it.

5 Results

This section compares the estimation of functional connectivity before and after the removal of artifacts. The performance measures used are based on cumulative density function and direct comparison of the estimated function connectivity.

5.1 Performance Measure

From the characteristic of MVAR, it can be seen that the clean EEG signal should be Gaussian distributed that is symmetric about the mean value. For that case, the PDF of each channel is compared before and after removing the artifacts. Figure 10 shows the PDF before and after artifact removal using TAD DWT denoising and TAD MA filter algorithm. As shown in Fig. 10a, the artifacts caused by muscular and blinking produce a non-symmetric distribution and appear in the range (−20-−5) and (5–30). However, for Fig. 10b, c the consequences of artifacts are removed.

The comparison between both PDF results and Gaussian distribution can be represented using CDF for both standard normal Gaussian and the simulated signals as shown in Fig. 11. Similar to the PDF, the results show that the CDF after artifact removal is closer to the standard normal CDF. The result for the TAD MA filter is closer to standard normal CDF compared to TAD DWT denoising.

5.2 Effects of Artifacts on PDC

The effect of the artifacts can be seen in Fig. 12 below for various thresholds where the x-axis represents sources, and the y-axis represents the destinations.

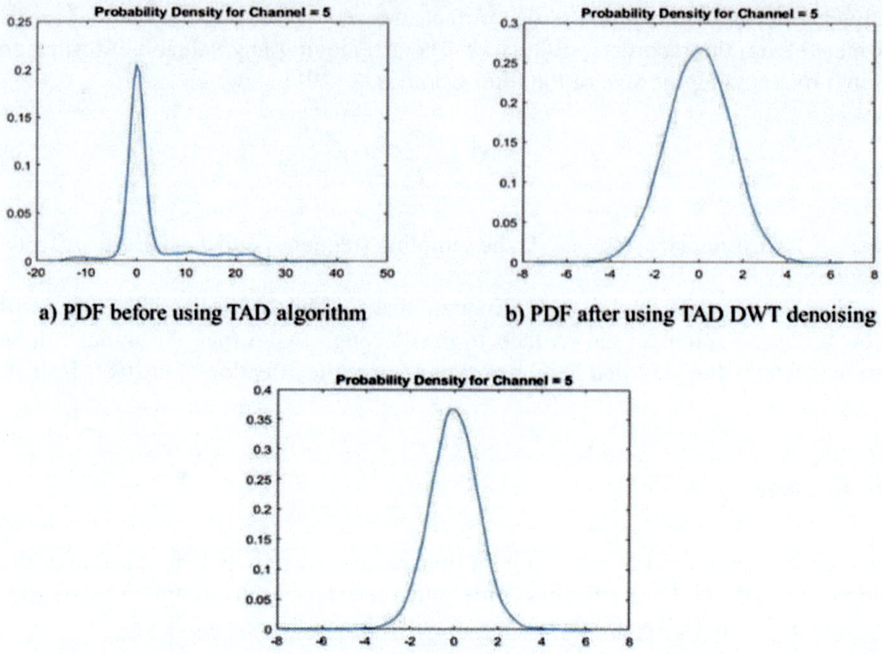

a) PDF before using TAD algorithm

b) PDF after using TAD DWT denoising

c) PDF after using TAD MA filter

Fig. 10 TAD MA filter and TAD DWT denoising

The interactions between the channels in Fig. 12 are regular connections, those are channel 1 to channels 2, 3, and 4, and channels 4 and 5 from/to each other. As well, the effect of artifacts is ($x_1 \leftarrow x_5$) for threshold (0.2) and for threshold (0.05) connectivities are from ($x_1 \leftarrow x_4$) and ($x_5 \leftarrow x_1$).

The removal of artifacts by the TAD algorithm enables the correct estimation of the functional connectivity from PDC analysis. The functional connectivity analysis after using TAD DWT denoising can be seen in Fig. 13.

The results in Fig. 13 show that connectivities caused by artifacts have been removed those are ($x_1 \leftarrow x_4$), ($x_1 \leftarrow x_5$), and ($x_5 \leftarrow x_1$) for thresholds (0.2) and (0.1). However, additional connectivities occurred due to down-sampling and up-sampling in the DWT during decomposition and reconstruction, respectively, that distorts the signal resolution [30].

Figure 14 shows the estimated functional connectivity after the TAD MA algorithm is applied. The results show the difference in the functional connectivity for each threshold after removing the artifacts from the simulated signals for thresholds (0.2) and (0.1) as compared with Fig. 11. Also, the results for threshold (0.05) show that the method can reduce the effect of artifacts to observe the good connectivities caused by connections between the channels only.

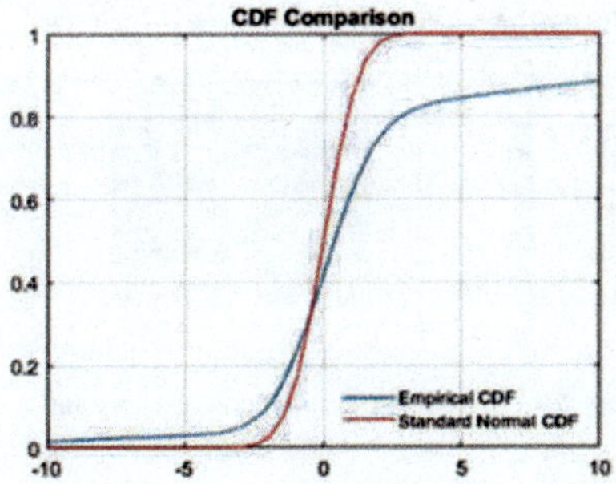

a) CDF before using TAD algorithm

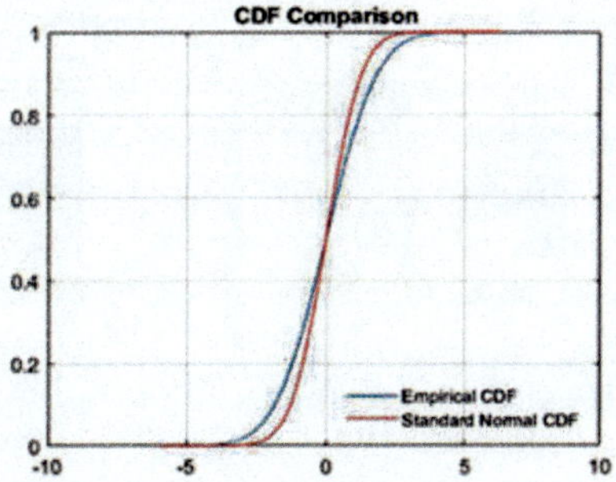

b) CDF after using TAD DWT denoising

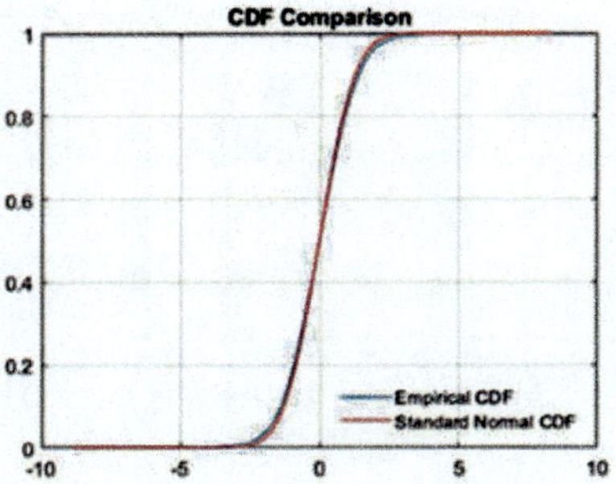

c) CDF after using TAD MA filter

Fig. 11 CDF of simulated EEG signal before and after using TAD algorithm

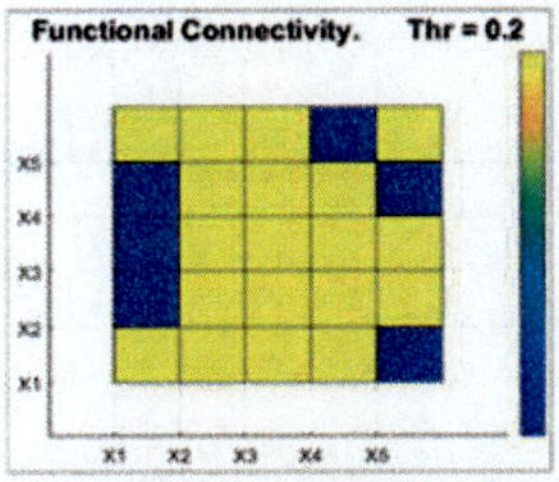

a) Functional connectivity for Threshold = 0.2

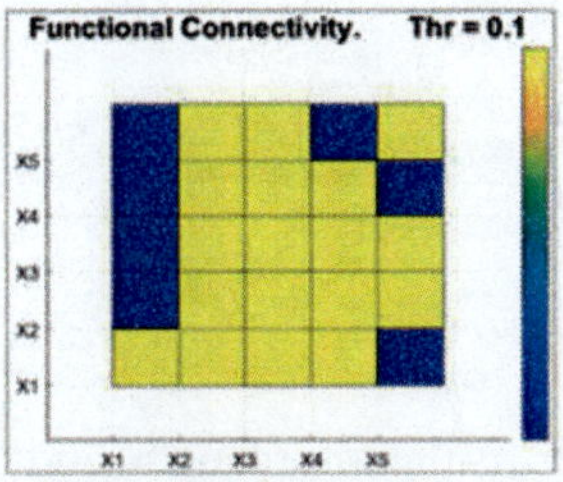

b) Functional connectivity for Threshold = 0.1

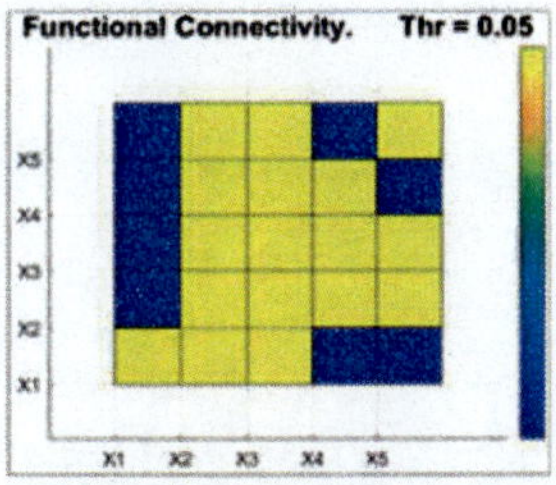

c) Functional connectivity for Threshold = 0.05

Fig. 12 Functional connectivity for Eq. 3 for various thresholds

5.3 Computational Complexity

The difference between TAD DWT denoising and TAD MA filter methods is the computation complexity whereby MA is more straightforward as compared to the DWT method. For the TAD DWT denoising method, assuming the decomposition

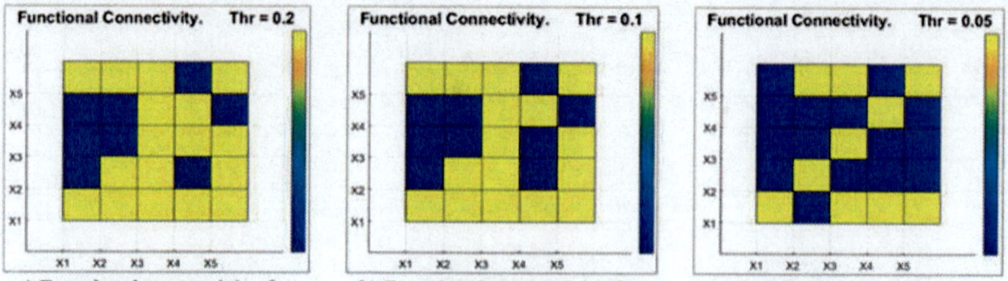

a) Functional connectivity for Threshold = 0.2

b) Functional connectivity for Threshold = 0.1

c) Functional connectivity for Threshold = 0.05

Fig. 13 Functional connectivity estimated using PDC after applying TAD DWT denoising

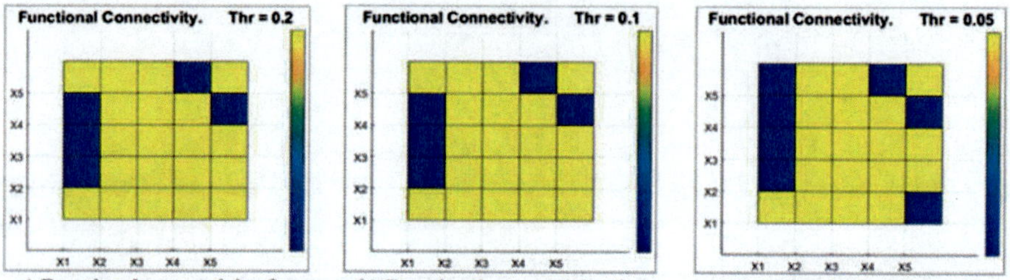

a) Functional connectivity for Threshold = 0.2

b) Functional connectivity for Threshold = 0.1

c) Functional connectivity for Threshold = 0.05

Fig. 14 Functional connectivity estimated using PDC after applying TAD MA filter

level, wavelet basis function, and thresholding method known, whereas only the size of the filter is required for the TAD MA filter. The computation complexity for both methods can be computed as follows [31]:

$$C_{\mathrm{DWT}} = 4N_h N\left[1 - 0.5^L\right] \tag{10}$$

$$C_{\mathrm{MA}} = N_h N \tag{11}$$

where N is the signal length, N_h is the filter length, and L is the decomposition level.

Assuming that: $L = 5$, $N = 30{,}000$, and $N_h = 5$, the results become $C_{\mathrm{DWT}} = 581{,}250$ and $C_{\mathrm{MA}} = 150{,}000$.

The results show that the difference in complexity for DWT denoising is double times MA filtering complexity. Since the TAD MA filter is capable to remove the artifacts from the MVAR signals with lower complexity, thus the TAD MA is more suitable for artifact removal from EEG recordings.

6 Conclusion

It is very crucial to know the flow of information in the human brain in order for us to understand the functions, the active parts, and the connections of each region during tasks. For the aforementioned reasons, EEG signals were used because EEG continues to be a valuable tool for research and diagnosis purposes. Moreover, PDC approach was employed to analyze the functional connectivity because PDC is focused on direct and indirect interaction. Furthermore, the PDC method is able to detect the right directional connectivity for more than two channels. However, the artifacts are caused by external or internal sources producing additional interactions between the channels. Therefore, the TAD algorithm was developed to remove the affection of the artifacts from the EEG signals. Moreover, the TAD algorithm enhanced the analysis of functional connectivity, which contributes to a better understanding of the flow of information between each region of the human brain.

References

1. K. Wang, W. Li, L. Dong, L. Zou, C. Wang, Clustering-constrained ICA for ballistocardiogram artifacts removal in simultaneous EEG-fMRI. Front. Neurosci. **12**, 1–13 (2018)
2. U. Herwig, P. Satrapi, C. Schönfeldt-Lecuona, Using the international 10–20 EEG system for positioning of transcranial magnetic stimulation. Brain Topogr. **16**(2), 95–99 (2003)
3. J.A. Gaxiola-Tirado, R. Salazar-Varas, D. Gutierrez, Using the partial directed coherence to assess functional connectivity in electroencephalography data for brain-computer interfaces. IEEE Trans. Cogn. Dev. Syst. **10**(3), 776–783 (2018)
4. A. Savelainen, *An Introduction to EEG Artifacts. Independent Research Projects in Applied Mathematics* (2010)
5. S.R. Sreeja, R.R. Sahay, D. Samanta, P. Mitra, Removal of eye blink artifacts from EEG signals using sparsity. IEEE J. Biomed. Health Inform. **22**(5), 1362–1372 (2018)
6. J. Tortora, *Principles of Anatomy and Physiology*, 2nd edn. (John Wiley & Sons, Australia, 2018)
7. S. Khatun, S. Member, R. Mahajan, S. Member, Comparative study of wavelet-based unsupervised ocular artifact removal techniques for single-channel EEG data. IEEE J. Transl. Eng. Health Med. **4**, 1–8 (2016)
8. M. Mazher, A.A. Abd, A.S. Malik, A.H. Ullah, An EEG-based cognitive load assessment in multimedia learning using feature extraction and partial directed coherence. IEEE Access **5**, 14819–14829 (2017)
9. E. Oja, Independent component analysis: algorithms and applications. Neural Netw. **13**(4–5), 411–430 (2000)
10. X. Jiang, G.B. Bian, Z. Tian, Removal of artifacts from EEG signals: a review. Sensors **19**(5), 987 (2019)
11. R.W. Homan, J. Herman, P. Purdy, Cerebral location of international 10–20 system electrode placement. Electroencephalogr. Clin. Neurophysiol. **66**(4), 376–382 (1987)
12. N.M. Safri, A.Z. Sha'ameri, N.A. Samah, S. Daliman, Resolving gender difference in problem solving based on the analysis of electroencephalogram (EEG) signals. Int. J. Integr. Eng. **10**(7), 90–101 (2018)

13. X. Chen, A. Liu, J. Chiang, Z.J. Wang, M.J. McKeown, R.K. Ward, Removing muscle artifacts from EEG data: multichannel or single-channel techniques? IEEE Sens. J. **16**(7), 1986–1997 (2015)
14. L.A. Baccala, D.Y. Takahashi, K. Sameshima, Directed transfer function: unified asymptotic theory and some of its implications. IEEE Trans. Biomed. Eng. **63**(12), 2450–2460 (2016)
15. K. Sameshima, L.A. Baccala, Partial directed coherence: a new concept in neural structure determination. Biol. Cybern. **84**(6), 463–474 (2001)
16. J.S. Kumar, P. Bhuvaneswari, Analysis of electroencephalography (EEG) signals and its categorization—a study. Procedia Eng. **38**, 2525–2536 (2012)
17. B. Yang, K. Duan, C. Fan, C. Hu, J. Wang, Automatic ocular artifacts removal in EEG using deep learning. Biomed. Signal Process. Control **43**, 148–158 (2018)
18. Y. Zou, S. Member, V. Nathan, S. Member, Automatic identification of artifact-related independent components for artifact removal in EEG recordings. IEEE J. Biomed. Health Inform. **20**(1), 73–81 (2016)
19. L. Frølich, I. Dowding, Removal of muscular artifacts in EEG signals: a comparison of linear decomposition methods. Brain Inf. **5**(1), 13–22 (2018)
20. M.B.I. Reaz, M.S. Hussain, F. Mohd-Yasin, Techniques of EMG signal analysis: detection, processing, classification and applications. Biol. Proced. Online **8**(1), 11–35 (2006)
21. P. Konrad, *The ABC of EMG—Noraxon. A Pracicat Introduction to Kinesiological Electromyography* (2005)
22. M.H. Wu, R.E. Frye, G. Zouridakis, A comparison of multivariate causality based measures of effective connectivity. Comput. Biol. Med. **41**(12), 1132–1141 (2011)
23. K.J. Blinowska, Review of the methods of determination of directed connectivity from multichannel data. Med. Biol. Eng. Comput. **49**(5), 521–529 (2011)
24. Z. Djordjević, A. Jovanović, A. Perović, Brain connectivity measure—the direct transfer function—advantages and weak points, in *2012 IEEE 10th Jubil Int Symp Intell Syst Informatics SISY 2012* (2012), pp. 93–97
25. B. Schelter, M. Winterhalder, M. Eichler, M. Peifer, B. Hellwig, B. Guschlbauer, C.H. Lücking, R. Dahlhaus, J. Timmer, Testing for directed influences among neural signals using partial directed coherence. J. Neurosci. Methods **152**(1–2), 210–219 (2006)
26. L. Barnett, A.K. Seth, The MVGC multivariate Granger causality toolbox: a new approach to Granger-causal inference. J. Neurosci. Methods **223**, 50–68 (2014)
27. S. Khatun, R. Mahajan, B.I. Morshed, Comparative study of wavelet-based unsupervised ocular artifact removal techniques for single-channel EEG data. IEEE J. Transl. Eng. Health Med. **4**, 1–8 (2016)
28. H. Azami, K. Mohammadi, B. Bozorgtabar, An improved signal segmentation using moving average and Savitzky-Golay filter. J. Signal Inf. Process. **3**, 39–44 (2012)
29. S.W. Smith, *The Scientist and Engineer's Guide to Digital Signal Processing* (2003), pp 277–284.
30. X. Zhou, C. Zhou, B.G. Stewart, Comparisons of discrete wavelet transform, wavelet packet transform and stationary wavelet transform in denoising PD measurement data, in *Conf Rec 2006 IEEE Int Symp Electr Insul* (2006), pp 237–240
31. S. Hosur, A.H. Tewfik, Wavelet transform domain adaptive FIR filtering. IEEE Trans. Signal Process. **45**(3), 617–630 (1997)

Development of Optimal Corrected PTa Interval Formula for Different Heart Rates

Shaik Karimulla, Arya Bhardwaj, J. Sivaraman, and B. Dhananjay

Abstract The standard 12-lead Electrocardiogram (ECG) system is the most commonly used technique for obtaining the electrocardiographic signal to evaluate the heart's electrical activity. In the surface ECG, atrial repolarization (Ta wave) is not observed. However, during the exercise stress test, ST-segment depression validates the existence of Ta wave.To record the Ta wave and enhance atrial depolarization (P wave) in Sinus Rhythm (SR) ECG, Modified Limb Lead (MLL) system is used. PTa Interval (PTaI) represents the duration from the beginning of P wave to the end of Ta wave. This study aims to develop a corrected PTaI (PTa_c) formula to correct the PTaI from MLL ECGs for different heart rates. ECGs were recorded from 35 volunteers in SR and Sinus Tachycardia (ST) condition of mean age 24 ± 5 years using EDAN PC ECG system. Regression analysis was implemented on the recorded data to derive the slope for PP Interval (PPI) vs. PTaI plots in both groups. In SR group, PPI correlated well with PTaI ($r = 0.48$) compared to the ST group ($r = 0.17$). A new corrected PTaI (PTa_cN) formula was developed from the slope values of SR and ST and compared with the previously developed PTa_c formula. The developed PTa_c formula of this study showed accurate results with the least slope in different heart rates. Implementation of new PTa_c formula in automatic algorithms further improves the clinical diagnosis related to atrial ECG components.

Keywords Atrial repolarization · Electrocardiogram · PTa interval · Sinus rhythm · Sinus tachycardia

1 Introduction

Standard 12-lead Electrocardiogram (ECG) is the most commonly used lead system to record the heart's electrical activity. The placement of electrodes provides the view of impulse propagation primarily of ventricles. Also, atrial depolarization (P

S. Karimulla · A. Bhardwaj · J. Sivaraman (✉) · B. Dhananjay
Bio-signals and Medical Instrumentation Laboratory, Department of Biotechnology and Medical Engineering, National Institute of Technology, Rourkela, Odisha, India
e-mail: jsiva@nitrkl.ac.in

F. Thakkar et al. (eds.), *Proceedings of the International e-Conference on Intelligent Systems and Signal Processing*, Advances in Intelligent Systems and Computing 1370,
https://doi.org/10.1007/978-981-16-2123-9_20

wave) is observed; however, atrial repolarization (Ta wave) is not witnessed, as it is largely hidden by following QRS complex; moreover, both waves have very low amplitude during normal Sinus Rhythm (SR) [1]. Hence, the 12-lead ECG system focuses more on ventricular disorders; whereas, studies on atrial ECG components are rare. However, the study of atrial components is significant for determination of atrial-related disorders.

Sprague and White described the presence of Ta wave in Atrio-Ventricular Block (AVB) patients where the QRS complex and Ta wave are uncoupled [2]. They noted that the complete Ta wave was observed in AVB with the duration of 0.32 to 0.42 s and was opposite in direction to the P wave. Further studies have also investigated the presence of Ta wave in AVB conditions and noted that Ta was 2-3 times longer than P wave and was in opposite polarity [3–5]. Langley et al. [6] identified the presence of Ta wave in SR and atrial fibrillation ECGs using signal-averaged analysis techniques. They also noted that the characteristics of Ta wave did not differ significantly between the groups. Ihara et al. observed the Ta wave in PR interval employing 64 lead system in SR [7]. Wang et al. explained the imaging patterns of Ta wave in a healthy subject using 256-carbon electrode mapping approach, and documented that both the P and Ta wave follow the same sequence of propagation [8]. Furthermore, a new lead system referred to as Modified Limb Lead (MLL) recorded the Ta wave in surface ECG of SR by placing the limb electrodes on human torso [9, 10]. The MLL system provides the scope for determining the atrial disorders as it enhances the amplitude of atrial components and also witnesses the Ta wave [11, 12].

PTa Interval (PTaI) represents the duration from the P wave beginning till the Ta wave end, which is homologous to the ventricular counterpart, QT Interval (QTI). Many studies have verified the importance of clinical diagnosis of QTI and its impact on life expectancy. Similarly, atrial sinus cycle length affects the duration of P, Ta, and PTaI [13–16]. As the heart rate increases, the PP Interval (PPI) and PTaI shortens; while, as heart rate decreases the PPI and PTaI gets prolongs; therefore, PPI and PTaI shows a positive correlation. Hence, it is crucial to correct the PTaI at 60 bpm to estimate the corrected PTaI (PTa_c) as the heart rate changes. During diagnostic purposes, (PTa_c) is derived from recorded PTaI which is independent of PPI. Therefore, this study aims to develop optimum PTaI correction formulas for varying heart rate; hence, it prevents the overestimation of PTa_c for different heart rhythm.

2 Methods

2.1 Study Population

ECGs were recorded from 35 volunteers with the mean age of 24 ± 5 years in SR and Sinus Tachycardia (ST) conditions using the MLL system. No volunteers were having any medical issues, and written consent was taken before recording.

2.2 Data Collection

ECGs were recorded with EDAN SE-1010 PC ECG instrument using Ag-AgCl electrodes placed on body surface in SR and ST conditions. ECGs recorded for 60 s with a frequency range of 0.05–150 Hz. Noise filtering of recorded ECGs is done by processing the signals through EDAN software at 25 Hz. ST data recorded after the physical exercise was performed by the SR volunteers.

2.3 Statistical Analysis

The recorded ECG data is stated as mean ± standard deviation. Pearson correlation was used to determine the relation between PPI and PTaI. $P < 0.05$ is considered as statistically significant. All the statistical analyses were carried out in sigma plot software. The linear regression technique is used to obtain the equation in a scatter plot. The most frequently used form of regression is linear regression, which fits data according to mathematical principles. The regression plot shows slope and intercept values and can be expressed as:

$$\mathrm{Y} = a + b * x \tag{1}$$

Y represents the dependent variable, x represents independent variables, a represents intercept, and b represents slope [17].

3 Results

The PPI and PTaI vary in linear relation, as the heart rate changes. The average heart rate of SR and ST was 74 ± 5 bpm, 101 ± 2 bpm, respectively, and the values of PPI and PTaI of both groups are shown in Table 1.

Table 1 Measurements of PPI and PTaI of the SR and ST groups

Leads	Sinus Rhythm Group		Sinus Tachycardia Group	
	PP Interval (ms)	PTa Interval (ms)	PP Interval (ms)	PTa Interval (ms)
I	814 ± 48	182 ± 15	614 ± 22	118 ± 20
II	818 ± 55	184 ± 11	618 ± 17	124 ± 17
aVR	813 ± 66	179 ± 10	619 ± 32	124 ± 30
aVL	820 ± 77	177 ± 11	616 ± 26	126 ± 23
aVF	819 ± 67	174 ± 14	614 ± 31	125 ± 18

*values represented as Mean ± SD

Correction of PTaI is required for the approximation of PTaI, particularly in abnormal conditions to minimize the misdiagnosis. Debbas et al. [18] developed a PTa_c formula which required PPI, PTaI, and heart rate values from AVB patients. The previously developed PTa correction formula by Debbas et al. is given as:

$$PTa_c = PTa - 0.24(PP - 1000) \tag{2}$$

The best PTa_c value is more dependent on the recorded PTaI and should be more independent of the PPI. The correction formula showing the least slope between PPI and PTa_c is a more efficient formula for PTa correction.

3.1 Development of New PTa Correction Formula

For developing optimum PTa_c formula, ECGs were recorded in SR and ST conditions. The main aim is to develop a PTa_c formula, suitable for different heart rates like SR and ST. Linear regression plot is derived from PPI and PTaI values, and the slope value is analyzed for each data. The formula can be derived by taking the mean value of slope from SR and ST groups. The slope values in SR and ST are found to be 0.097 and −0.051 respectively, as shown in Fig. 1. From the above analysis, individual formulas derived for SR and ST from the slope value between PPI and PTaI are shown in Eqs. 3 and 4, respectively.

$$PTa_cSR = PTa - 0.097(PP - 1000) \tag{3}$$

$$PTa_cST = PTa - 0.051(PP - 1000) \tag{4}$$

The mean value of the slope is -0.023. From the above data, a new corrected PTaI (PTa_cN) formula is developed as expressed in Eq. 5.

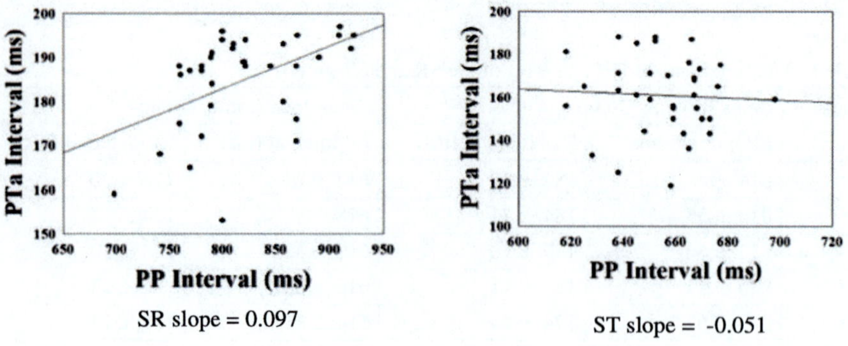

Fig. 1 Scatter plot between PPI and PTaI for SR and ST

Table 2 Slope values with the previous and new individual formula for SR and ST

Groups	Slope in %	
SR	PTa_c	PTa_cSR
	14.1	0.26
ST	PTa_c	PTa_cST
	28.8	10.4

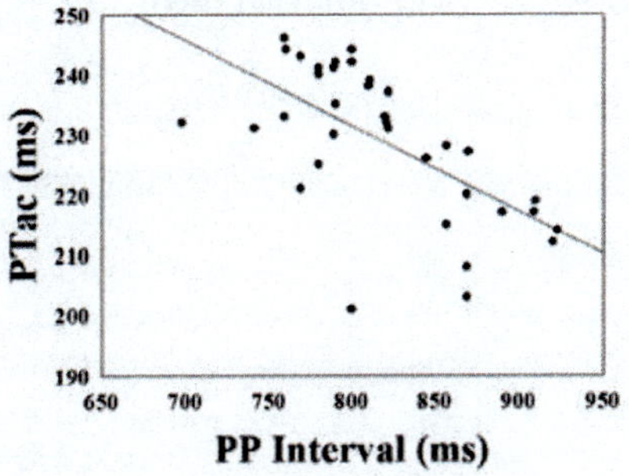

Slope = -0.141 Slope = 0.075

Fig. 2 Relationship between the previous formula and the new corrected PTa_cN formula in SR data

$$PTa_cN = PTa - 0.023(PP - 1000) \tag{5}$$

The PTa_c analysis for SR and ST (Eqs. 3 and 4, respectively) was compared to the formula developed by Debbas et al. (Eq. 2). From the results of Table 2, it is noted that the individual formulas developed in this study showed better results and are more optimum for different heart conditions. The relationship of PPI vs. PTa_c plots of SR and ST are shown in Figs. 2 and 3, respectively.

It is evident from Figs. 2 and 3 and Table 3 that the newly developed standard formula for SR and ST volunteers is showing the least slope value than the previously developed formula, which signifies that PTa_c calculated from the new formula is more independent of PPI.

4 Discussion

The presence of the Ta wave in the PRI is denoted as PTaI. Changes in the PR segment morphology are found in myocardial infarction of the right atrium. More often, a change in the PRI is found in pre-excitation or AV conduction blockage. Therefore, correct evaluation of the PR segment and PTaI has high diagnostic importance. In

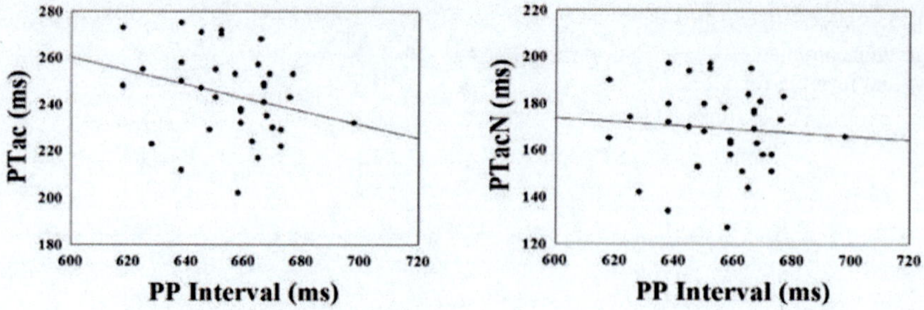

Slope = -0.288 Slope = -0.0786

Fig. 3 Relationship between the previous formula and the new corrected PTa_cN formula in ST data

Table 3 Slope values of previous and new standard formula in SR and ST

Study Group	Slope in %	
	PTa_c	PTa_cN
Sinus Rhythm	14.1	7.5
Sinus Tachycardia	28.8	7.8

the retrograde conduction, where the transmission of cardiac electrical impulse propagation is backward from the AV node into the atria, the amplitude of Ta wave is larger than the sinus P wave. ECG studies on PTaI are rare, and not many studies have investigated the relation of PPI with Ta and PTaI. The study on the effects of heart rate and its correlation with the P wave, Ta wave, and PTaI is rare. Although the atrial ECG studies are scarce, one earlier study by Debbas et al. [18] documented the relationship between PPI with P wave, Ta wave, and PTaI in SR drug-induced AVB patients.

Intensive studies have been done on QT correction formula and many authors proposed different formulas for QT correction at 60 bpm and also at different heart rates. Few correction formulas are given as Bazett, Frederica, Framingham, Hodges, Rautaharju, and many more. These formulas are implemented for QT correction in automated ECG algorithms. A new PTa correction formula is required to correct PTaIs at different heart rates; as previously, PTa_c was developed only in AVB condition. In this study, new PTa_c formula was developed and validated in SR and ST conditions.

Moreover, in study [18], a new PTa_c formula was derived to correct the PTaI at varying heart rate. The derivation of PTa_c includes extraction of duration values of P, Ta, PPI, and PTaI from ECGs and then using regression analysis, slope values were derived by analysis between PPI vs. P, PPI vs. Ta, and PPI vs. PTaI. It was concluded that the PPI vs. P showed more scattering of data, and no clear relation was observed. An excellent linear relation was observed in plots of PPI vs. Ta, and PPI vs. PTaI with a slope of 0.24. The derived PTa_c formula in [18] was PTa_c = PTaI - 0.24 (PP

- 1000). From the findings of this study in SR and ST volunteers, a PTa_cN formula was developed, which found to be better than the previously developed formula.

Sivaraman et al. studied the effects of PTaI in age and different heart rates [13] and found that the PPI was affecting the PTaI, similar to RR interval relation with QTI. A further study documented the importance of PTaI with different heart rates [15]. The stability of the PTaI was documented using system identification models in SR volunteers and atrial tachycardia patients, and found the reason for the unstable segments in the PTaI [16]. It was noted from that study, the heart rate variability was one of the critical factors causing PTaI instability in SR volunteers and suggested that more complex models are needed to shed light on PTaI segments further. This study noted that PTaI varies between the evaluated groups, and it was found to be consistent with the previous study on PTaI. The purpose of the corrected PTaI formula is primarily to adjust for different heart rates, which is crucial in understanding the mechanism of the PTaI in a more extensive clinical study.

5 Conclusion

In the present study, individual PTa_c formula was developed for SR and ST conditions which are independent of heart rates. Further, a common PTa_cN formula was derived from the slope values of the study population for better optimization and validation. The newly developed standard PTa_c formula in SR and ST conditions would help to minimize the misdiagnosis of atrial disorders and in other atrial arrhythmia conditions. The developed formula could reduce the overestimation of PTaI, which may lead to improper clinical impacts. Implementation of common PTa_cN formula in automated ECG algorithms may improve PTaI estimation accuracy to provide specific clinical interpretation. It is also warranted that studies on the sinus bradycardia condition would further improve the optimization and validation of the developed formula.

Acknowledgements The authors acknowledge the support from the Ministry of Education, Government of India. The present study was supported by financial grants from the Science and Engineering Research Board (SERB), Department of Science and Technology (DST), Government of India (EEQ/2019/000148).

References

1. K.L. Briggs, A digital approach to cardiac cycle. IEEE Eng. Med. Biol. **13**, 454–456 (1994). https://doi.org/10.1109/51.310982
2. H.B. Sprague, P.D. White, Clinical observations on the T wave of the auricle appearing in the human electrocardiogram. J. Clin. Invest. **1**, 389–402 (1925). https://doi.org/10.1172/jci100020

3. D.I. Abramson, N.M. Fenichel, C. Shookhoff, A study of electrical activity in the auricles. Am. Heart J. **15**, 471–481 (1938). https://doi.org/10.1016/s0002-8703(38)90301-1
4. T. Joao, A. Victor, O. De Martin, Atrial repolarization-its importance in clinical electrocardiography. Circulation **22**, 635–644 (1960). https://doi.org/10.1161/01.cir.22.4.635
5. H. Hayashi, M. Okajima, K. Yamada, Atrial T (Ta) wave and atrial gradient in patients with A-V block. Am. Heart J. **91**, 689–698 (1976). https://doi.org/10.1016/s0002-8703(76)80533-9
6. P. Langley, A. Murray, Analysis of the atrial repolarisation phase of the ECG in health and in atrial fibrillation. Comput. Cardiol. **34**, 785–788 (2007). https://doi.org/10.1109/cic.2007.4745603
7. Z. Ihara, A. van Oosterom, R. Hoekema, Atrial repolarization as observable during the PQ interval. J. Electrocardiol. **39**, 290–297 (2006). https://doi.org/10.1016/j.jelectrocard.2005.12.001
8. Y. Wang, Y. Rudy, Electrocardiographic imaging (ECGI) of normal human atrial repolarization. Heart Rhythm. **6**, 582–583 (2009). https://doi.org/10.1016/j.hrthm.2008.07.024
9. S. Jayaraman, V. Sangareddi, R. Periyasamy, J. Joseph, R.M. Shanmugam, Modified limb lead ECG system effects on electrocardiographic wave amplitudes and frontal plane axis in sinus rhythm subjects. Anatol. J. Cardiol. **17**, 46–54. https://doi.org/10.14744/anatoljcardiol.2016.6843 (2017)
10. J. Sivaraman, G. Uma, M. Umapathy, A modified chest leads for minimization of ventricular activity in electrocardiograms, in *IEEE International Conference on Biomedical Engineering*, pp. 79–82. https://doi.org/10.1109/icobe.2012.6178959 (2012)
11. J. Sivaraman, G. Uma, S. Venkatesan, M. Umapathy, M.S. Ravi, Unmasking of atrial repolarization waves using asimple modified limb lead system. Anatol. J. Cardiol. **15**, 605–610 (2015). https://doi.org/10.5152/akd.2014.5695
12. J. Sivaraman, G. Uma, S. Venkatesan, M. Umapathy, V.E. Dhandapani, Normal limits of ECG measurements related to atrial activity using a modified limb lead system. Anatol. J. Cardiol. **15**, 2–6 (2015). https://doi.org/10.5152/akd.2014.5155
13. J. Sivaraman, G. Uma, S. Venkatesan, M. Umapathy, N. Keshav kumar, A study on atrial Ta wave morphology inhealthy group: An approach using P wave signal-averaging method. J. Med. Imaging Health Inform. **4**, 675–680 (2014). https://doi.org/10.1166/jmihi.2014.1306
14. J. Sivaraman, R. John, Effects of sinus rhythm on atrial ECG components using modified limb lead system, in *4th International Conference on Signal Processing Computing and Control*, pp. 527–530. https://doi.org/10.1109/ispcc.2017.8269735 (2017)
15. J. Sivaraman, G. Uma, P. Langley, M. Umapathy, A study on stability analysis of atrial repolarization variability using ARX model in sinus rhythm and atrial tachycardia ECGs. Comput. Methods Programs Biomed. **137**, 341–351 (2016). https://doi.org/10.1016/j.cmpb.2016.10.005
16. A. Jyothsana, B. Arya, J. Sivaraman, Stability analysis on the effects of heart rate variability and premature activation of atrial ECG dynamics using ARMAX model. Phys. Eng. Sci. Med. **43**, 1361–1370 (2020). https://doi.org/10.1007/s13246-020-00940-w
17. A. Schneider, G. Hommel, M. Blettner, Linear regression analysis. Dtsch Arztebl Int. **107**, 776–782 (2010). https://doi.org/10.3238/arztebl.2010.0776
18. N.M.G. Debbas, S.H.D. Jackson, D. De Jonghe, A. Robert, A.J. Camm, Human atrial repolarization: Effects ofsinus rate, pacing and drugs on the surface electrocardiogram. J. Am. Coll. Cardiol. **33**, 358–365 (1999). https://doi.org/10.1016/s0735-1097(98)00580-4

Medical Image Fusion Using Lifting Wavelet and Fractional Bird Swarm Optimization

Jayant Bhardwaj and Abhijit Nayak

Abstract This paper concentrates on an eminent technology of medical imaging data science called image fusion. An unique pixel-level image fusion method is presented here. The various multimodal MRI brain images are taken from BRATS database and an effectively fused and a comparatively more informative image called fused image is obtained. The two multimodal images are first decomposed by Haar Wavelet to obtain high- and low-frequency coefficients. Intermediately Bayesian fusion is performed on these extracted information-rich coefficients. Later on proposed fusion rule is adopted which is basically an optimization technique called fractional bird swarm optimization (Fractional-BSA). It is observed that the proposed scheme called fractional BSA-Bayesian Fusion outperformed the other contemporary existing methods of image fusions like wavelet and HW fusion, SWT and NSCT and NSCT. The better results of assessment parameters like Mutual information, peak signal to noise ratio (PSNR) and root mean square error (RMSE) are proving the merit of proposed method.

Keyword Image fusion · Bayesian · Wavelet transform

1 Introduction

Every medical modality has some specific feature to be concerned. No single modality is sufficient to deliver maximum of information and a potential decision about an ailment [1]. Therefore According to the representation point of view, different image fusion methods are categorized in pixel level, feature level, and decision level [2]. An important pixel-level image fusion scheme is being proposed here for brain magnetic resonance imaging (MRI) images of different modalities. The

J. Bhardwaj (✉)
USICT & BPIT, Delhi, India
e-mail: jayantbhardwaj89@gmail.com

A. Nayak
Bhagwan Parshuram Institute of Technology (BPIT), Delhi, India

F. Thakkar et al. (eds.), *Proceedings of the International e-Conference on Intelligent Systems and Signal Processing*, Advances in Intelligent Systems and Computing 1370,
https://doi.org/10.1007/978-981-16-2123-9_21

important modalities of magnetic resonance imaging (MRI) used here are Flair, T1, T1C, and T2 [3]. Every modality has some specific direction of feature extraction with unique property of diagnosis. For example, T1-MRI modality expresses anatomical structure-based details of various tissues while T 2-MRI modality is providing details of pathological tissues only [4]. From last two decades, several methods for medical image fusion (IF) have been proposed. These methods are generally classified into transform domain and spatial domain [5]. Due to ease in implementation and detail information preservation multiscale, transform techniques are quite popular in recent years. Gradient pyramid, Laplacian pyramid, and various wavelet transform methods are few examples of this. Few more transform techniques like shearlet transform and non-sub-sampled contourlet transform (NSCT) are becoming more popular these days [6, 7]. Advanced wavelet transforms like lifting (LWT) and static wavelet transform (SWT) with the cascaded form are also enhancing the technique of fusion greatly [2]. Various anomalies around singularities like pseudo-Gibbs noise should be removed by a fusion method, and this problem is removed at an extent by some high generation wavelet like lifting wavelet transform (LWT). In this paper, we are proposing a pixel-level fusion method of optimization employing very popular Bayesian scheme of signal processing [8]. The Bayesian parameters are optimally controlled by an improve optimization scheme called fractional bird swarm optimization (BSA).

2 Lifting Wavelet Transform

Lifting waveletis a second-generation wavelet used here for image decomposition [10, 13]. It is basically a discrete wavelet transform (DWT) which can be viewed as a predictor-error decomposition. The "predictors" are the scaling coefficients at a given scale (*j*) that are used for the analysis of data at the next higher resolution or scale (*j-1*). The wavelet coefficients thus generated are prediction errors [4]. LWT is factoring the wavelet transforms into some steps called Lifting [9]. The LWT process an image in the iteration of the three steps called lazy, predict and update. In Lazy wavelet transform step, the image x[n] is explained in Eq. (1)

$$\mathrm{x[n]R,\ (nZ)} \tag{1}$$

divided into its even and odd polyphase components

$$\mathrm{x_e[n]\ and\ x_o[n]} \tag{2}$$

respectively, where

$$\mathrm{x_e[n] = x[2n]} \tag{3}$$

$$x_o[n] = x[2n + 1] \tag{4}$$

In the Predict step of LWT, the neighboring even coefficients as in Eq. (2) are used to predict odd polyphase coefficient, and the detail wavelet coefficients (high pass) are generated as an error generated in the process of predicting the odd samples from the even samples as in Eqs. 3 to 5.

$$d = x_o - P(x_e) \tag{5}$$

Now the odd components can be calculated as

$$x_o = P(x_e) + d \tag{6}$$

In the last update step of LWT, the even set is updated here using the wavelet coefficients to compute the scaling function coefficients (low pass) as in Eq. (6). An operator U called update operator is applied to detail coefficients obtained in the predicting step as in given Eq. (7).

$$S = x_e + U(d) \tag{7}$$

It reproduces x_e as in Eq. (8).

$$x_e = s - U(d) \tag{8}$$

LWT is a fast implementation of Wavelet Transform because of its almost similar nature of operation on low pass and high pass filtering [10]. Perfect reconstruction of original image is it's another attracting property [17, 18]. It also saves lots of auxiliary memory and have an advantage of simple inverse transform property [18]. It also reduces computational complexity by a factor of two. A separable wavelet transform is implemented on images by first applying 1-D wavelet transform along the columns and then along the rows of an image. This provides 1-level wavelet decomposition that consists of four components labeled as LL, LH, HL, and HH, respectively [6, 7, 9, 10].

3 Framework of Proposed Method

The Bayesian fusion approach uses the sub-images corresponding to the source images. $I_1(x, y)$ and $I_2(x, y)$ are employed for acquiring the fused image Q , and the fusion is performed at the pixel levels such that the original details of the source images are used for fusion (Fig. 1).

The fused image provides the significant information for clinical diagnosis. The individual pixels in the sub-images of $I_1(x, y)$ and $I_2(x, y)$ are fused for which

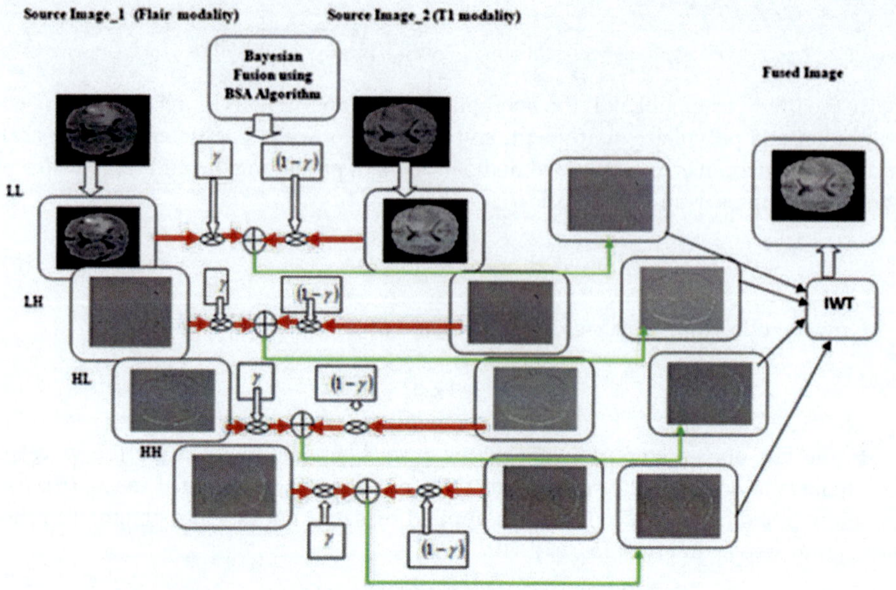

Fig. 1 Framework of Fractional-BSA-Bayesian Fusion

the Bayesian parameter. γ is employed which is determined optimally using the proposed algorithm. The fusion of the low sub-band corresponding to $I_1(x, y)$ with low sub-band of $I_2(x, y)$ is done using the Bayesian parameter γ and at the same time, the high-level bands belonging to $I_1(x, y)$ and $I_2(x, y)$ are fused as follows. The sub-bands for generating the fused image are represented as Eqs. (9) to (12).

$$Q^{LL}(x, y) = \gamma\, I_1^{LL}(x, y) + (1 - \gamma)\, I_2^{LL}(x, y) \tag{9}$$

$$Q^{LH}(x, y) = \gamma\, I_1^{LH}(x, y) + (1 - \gamma)\, I_2^{LH}(x, y) \tag{10}$$

$$Q^{HL}(x, y) = \gamma\, I_1^{HL}(x, y) + (1 - \gamma)\, I_2^{HL}(x, y) \tag{11}$$

$$Q^{HH}(x, y) = \gamma\, I_1^{HH}(x, y) + (1 - \gamma)\, I_2^{HH}(x, y) \tag{12}$$

where γ represents the Bayesian factor. $Q^{LL}(x, y)$, $Q^{LH}(x, y)$, $Q^{HL}(x, y)$, and $Q^{HH}(x, y)$ are the LL, LH, HL, and HH band of the fused medical image [3]. The Bayesian factor is tuned optimally using the proposed Fractional-BSA in such a way that the value of the factor renders the effective fusion in order to acquire good quality [9].

3.1 *Optimizing the Bayesian Approach Using the Fractional-BSA*

The ultimate aim of the proposed Fractional-BSA-based Bayesian Fusion approach is to determine the optimal value of the Bayesian parameter, which is computed, based on the characteristics of the input images. Initially, the Bayesian model is developed using the pixels in the image and using the model, the class value is set, which is selected optimally using the optimization algorithm [1–8]. Let us have a deep insight over the Bayesian approach. Once the sub-images are fused using the Bayesian approach, the Bayesian model is developed for all the four sub-images for which all the pixels in the sub-images are interpreted. The dimension of the training model is denoted as, $[K \times (c \times 8)]$. The solution model is represented as

$$M = \{M_1, M_2, ..., M_k, ..., M_v\} \quad (13)$$

where v indicates the total solution models and M stands for the Bayesian model consisting of v number of models and v possess the dimension $[(B \times B)].[(B \times B)]$ refers to the total number of pixels in the image, which is given as $K = [B \times B]$. Let M_k be the kth solution model, which corresponds to a pixel with respect to a class. The data is established pixel-wise for all the four sub-images and the training model is developed by finding the mean and variance for individual pixel of the sub-images with respect to the given c number of classes [11]. Thus, the dimension of a solution vector or the dimension of the training model for the individual pixel is given as $[1 \times (c \times 8)]$. Therefore, it is clear that the role of the optimization is to derive the optimal solution in such a way to enable the optimal fusion of the medical image. The Bayesian factor is determined using fractional-BSA, which is referred to as the class label in the Bayesian model. The proposed fractional-BSA is the modification of the standard BSA with the fractional concept, and the advantage of the proposed method is that the computational overhead associated with the standard BSA is overcome through the integration of the fractional concept in such a way that solutions of previous iterations are interpreted for updating the position in the new iteration. The accuracy and stability of the proposed algorithm is better compared with the existing algorithms and in addition, the convergence time and accuracy is balanced among each other with a higher performance. Moreover, the proposed algorithm is capable of adjusting with the different search methodologies and there is a perfect balance in the exploration and exploitation phases. Additionally, the diversity of the method is better and avoids the prematurity of the solutions.

3.1.1 Solution Encoding

In this step, a better representation of the solution will be determined using the proposed scheme that is Fractional-BSA. The solution vector with the dimension $[1 \times (c \times 8)]$ is having mean and variance of the individual pixel in the image with respect to the class labels. The class label for pixels will be derived using Bayesian approach.

3.1.2 Bayesian Objective Function

The fitness of the solution is computed for deriving the optimal solution and the fitness is evaluated based on the mutual information, which should be maximal for the best solution. The objective function based on mutual information is determined using the following formula as in Eq. (14),

$$f = h\,(I_1) + h\,(I_2) - h\,(I_1,\ I_2) \tag{14}$$

where $h\,(I_1)$ and $h\,(I_2)$ are the individual entropies of the image I_1 and I_2, respectively, and $h\,(I_1,\ I_2)$ specify the joint entropy of the image. The joint entropy is computed as in Eq. (15),

$$h\,(I_1,\ I_2) = -\sum\nolimits_{x,\,y} \rho_{I_1,\,I_2}(x,\ y)\ \log\ \rho_{I_1,\,I_2}(x,\ y) \tag{15}$$

It is interesting to note that the joint entropy between the images reduce due to the one-to-one mapping pixels in I_1 and the corresponding counterparts in I_2. However, joint entropy decreases when the statistical relationship between the images minimizes. Therefore, the objective function aims at the maximal mutual information for the selection of the optimal solution.

3.1.3 Proposed Fractional-BSA Algorithm

The proposed fractional-BSA is based on the social interactions and behaviors of birds, which includes the foraging, vigilance, and flight behaviors. The social behaviors of the birds are deliberated below: The first condition is regarding the stochastic decision for switching between the vigilance and the foraging behavior [8]. Yet another interesting phenomenon is regarding the foraging behavior such that during the foraging mechanism, the bird keeps in record the previous best experience acquired along with the best experience of the swarm in order to search their food in the present [14]. Likewise, the information is carried to the entire swarm and speaking about the vigilance, the bird moves to the center of the swarm, which could be affected by the interference occurring during the competition in the swarm [7,

8]. The most significant mechanism is birds shift between the places during production and scrounging and the birds with highest reserve, lowest reserve, and others are referred to as producer, scrounger, and the rest of the birds choose randomly between producing and scrounging [15]. The producers play a major role in searching food, whereas the scroungers follow the producers for their food. The foraging behavior of the birds is modeled as follows. Foraging behavior: As per the social behavior of bird, the best experience in the past for the individual bird and the swarm is recalled for update in the foraging behavior [16]. Thus, the foraging behavior is given as in Eq. (16),

$$A_{b,d}^{\tau+1} = A_{b,d}^{\tau} + \left(P_{b,d}^{\tau} - A_{b,d}^{\tau}\right) \times p_1 \times rand\,(0,\ 1) + \left(G_d - A_{b,d}^{\tau}\right) \times p_2 \times rand\,(0,\ 1) \quad (16)$$

where b refers to the total birds in the population and it is given as $b \in 1, \ldots, m$ and m stands for the total number of birds in the population. The dimensional space is denoted as d, which is given as $d \in 1, \ldots, S$ and τ specifies the time-step. The previous best position of the bth bird in d^{th} dimension is notated as $P_{b,d}^{\tau}$ and the previous best position of the bird swarm is denoted as G_d. The position of the bth bird in d^{th} dimension is notated as $A_{b,d}^{\tau}$, $r\,and\,(0,\ 1)$ specifies the independent members distributed in (0, 1) with p_1 and p_2 being the positive numbers, presenting the cognitive as well as social accelerated parameters. Whenever the uniform random number (0, 1) lies below the constant $const(0,\ 1)$, the bird continues foraging for discovering the food or else the birds continue with the vigilance behavior in the swarm [7]. The vigilance behavior of the birds is modeled as Eq. (17),

$$A_{b,d}^{\tau+1} = A_{b,d}^{\tau} + J_1\left(\mu_d - A_{b,d}^{\tau}\right) \times rand\,(0,\ 1) + J_2\left(P_{id} - A_{b,d}^{\tau}\right) \times rand\,(-1,\ 1) \quad (17)$$

The J_1 and J_2 are further described as per Eqs. (18) and (19)

$$J_1 = r_1 \times \exp\left(-\frac{F_b^p}{\sum F + B} \times m\right); i \neq b; 1 \leq i \leq m \quad (18)$$

$$J_2 = r_2 \times \exp\left[\left(\frac{F_b^p - F_i^p}{\left|F_i^p - F_{ib}^p + \beta\right|}\right)\frac{n \times F_i^p}{\sum F + \beta}\right] \quad (19)$$

where F_b^p refers to the value of best fitness of ith bird, $\sum F$ refers to the sum of the swarm's best fitness measure, β signifies the smallest constant, and μ_d is the mean of dth element of average position of whole swarm. Let r_1 and r_2 are the positive constants varying between 0 and 2 and m specify the total birds in the population. During the vigilance, the bird moves to the center in such a way that J_1 and $r\,and\,(0,\ 1)$ is not more than one and at the same time, the direct impact of the interference is measured with J_2. Whenever the fitness of the random i^{th} bird is better than the bth bird, the J_2 is said to be greater than r_2, which implies that the bird suffers from interference and more specifically, b^{th} bird faces more interference compared with b^{th} bird. The above Eq. (19) describes the standard equation of the BSA in the vigilance behavior, which is modified using the fractional theory in order to recall the behavior of the birds in the previous iterations such that the optimization becomes effective in deciding the Bayesian parameter. The vigilance mechanism of the birds in Fractional-BSA is modeled as Eqs. (20) and (21),

$$A_{b,d}^{\tau+1} - A_{b,d}^{\tau} = +J_1\left(\mu_d - A_{b,d}^{\tau}\right) \times r\,and\,(0,\ 1) + J_2\left(P_{id} - A_{b,d}^{\tau}\right) \times r\,and\,(-1,\ 1) \quad (20)$$

$$\partial^{\alpha}\left[A_{b,d}^{\tau+1}\right] = J_1\left(\mu_d - A_{b,d}^{\tau}\right) \times r\,and\,(0,\ 1) + J_2\left(P_{id} - A_{b,d}^{\tau}\right) \times r\,and\,(-1,\ 1) \quad (21)$$

where $\partial^{\alpha}\left[A_{b,d}^{\tau+1}\right]$ is the fractional term and is defined based on the fractional theory in [13]. The inclusion of the fractional concept adds the historical solutions in the update equation so that the position update occurs effectively with better convergence time. This section displays the results of the proposed Fractional-BSA-based Bayesian Fusion approach and the comparative analysis reveals the effectiveness of the proposed image fusion as shown in Figs. 2 and 3 with Tables 1 and 2.

The experimentation is performed in MATLAB with the BRATS database that is BRATS 2015 [10–12]. The multimodal scans provide four modalities, T1, T1C,

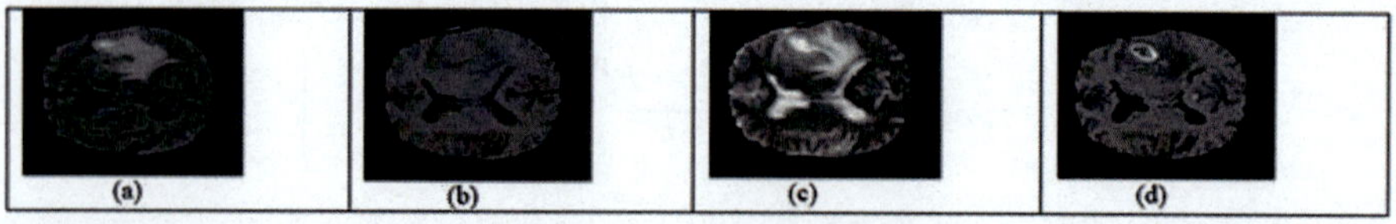

Fig. 2 Data set 1 multimodal images **a** Flair **b** T1 **c** T2 **d** T1C

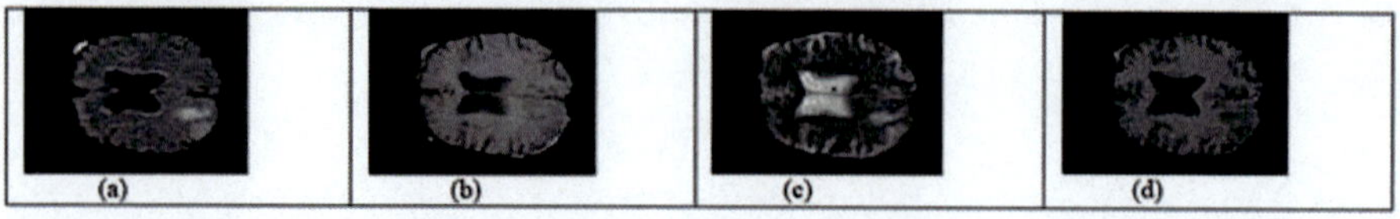

Fig. 3 Data set 2 multimodal images **a** Flair **b** T1 **c** T2 **d** T1C

Table 1 Comparative analysis using dataset-1

Metrics	Wavelet + HW Fusion	SWT + NSCT	NSCT	Proposed fractional BSA-bayesian fusion
Mutual information	1.4673	1 4669	1.4299	1.5765
PSNR (in ffi)	37.8289	37.8848	36.8904	44.0957
RMSE	6.5101	6.5281	6.5409	5.4940

Table 2 Comparative analysis using dataset-2

Metrics	Wavelet + HW Fusion	SWT + NSCT	NSCT	Proposed fractional-BSA-bayesian fusion
Mutual information	1.4612	1.4550	1.4508	1.4960
PSNR (in dB)	32.991	33.470	32.498	35.541
RMSE	10.0053	10.0510	9.9689	9.6404

Flair, and T2 and these are determined using various clinical protocols and scanners. BRATS-2015 [10–12] is the dataset with about 300 low- and high-grade glioma tumors with cystic and Gd-enhanced core tumor regions [11].

Now consider the following Fig. 4a, b, c for data set 1 image fusion. Fig. 5a, b, c describes the performance for BRATS Data set-2 images.

Now we can consider the following Fig. 6a, b, c as a comparison of proposed methodology performance with other methodologies like wavelet plus Helo-whale fusion, SWT plus NSCT, and with only NSCT for Data set 1 images. Similarly, Fig. 7a, b, c are for data set-2 images (Fig. 7).

Fig. 4 Performance analysis using the dataset-1, **a** Mutual Information, **b** PSNR, **c** RMSE

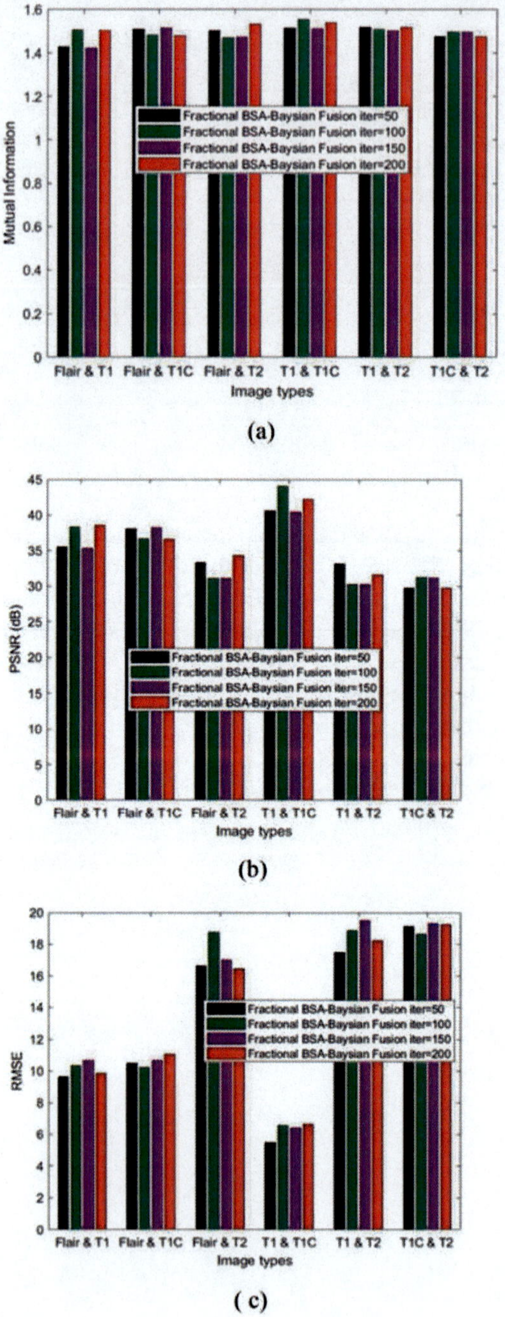

Fig. 5 Performance analysis using the dataset-2, **a** Mutual Information, **b** PSNR, **c** RMSE

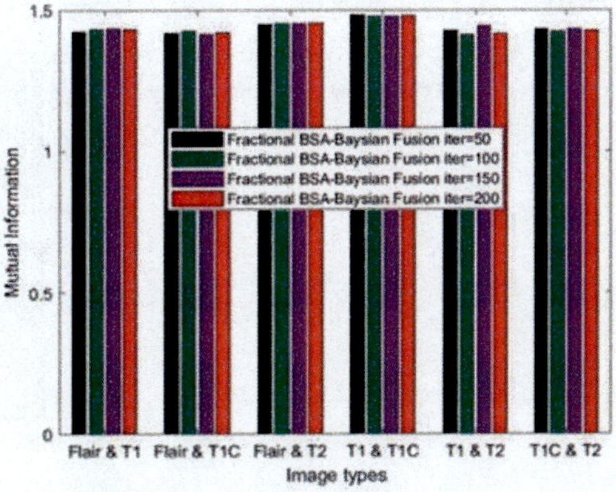

(a)

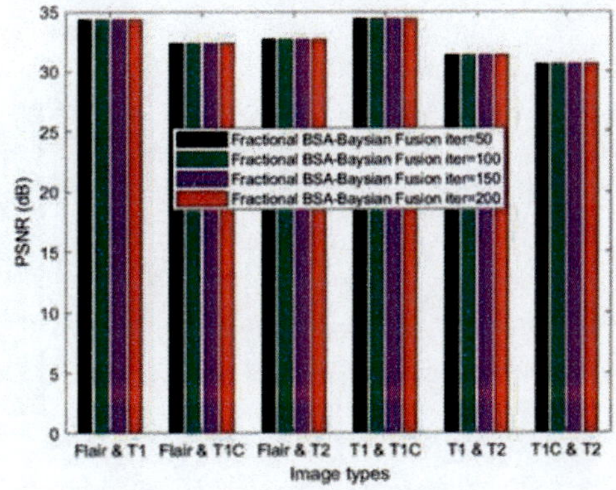

(b)

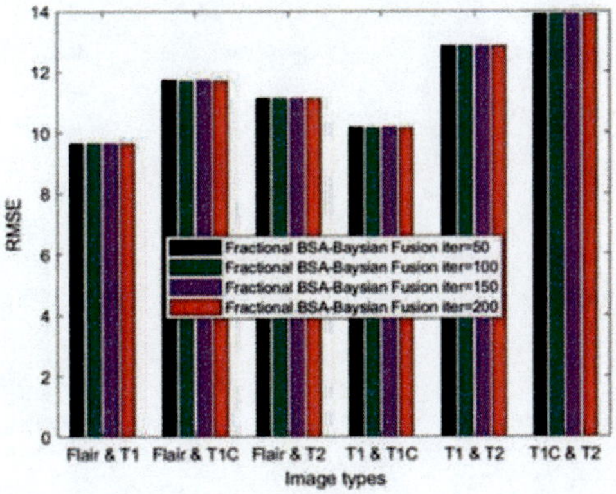

(c)

Fig. 6 Comparative analysis using the dataset-1, **a** MI **b** PSNR **c** RMSE

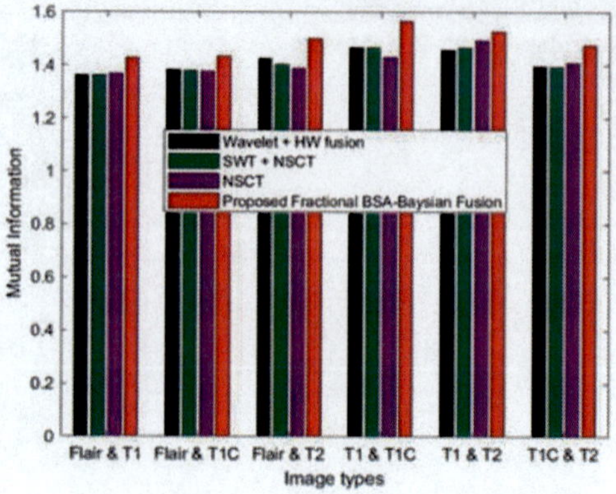

(a)

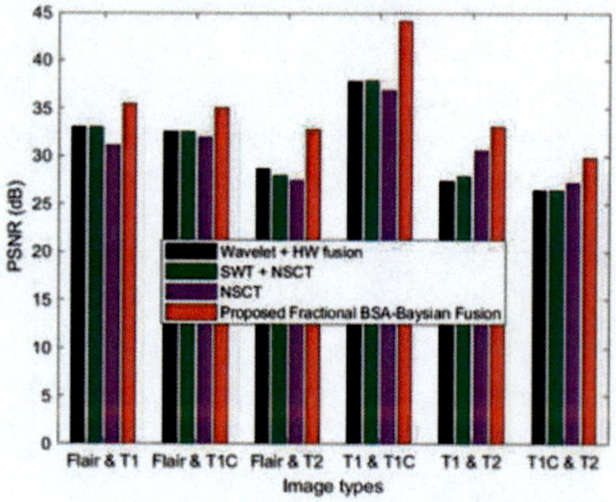

(b)

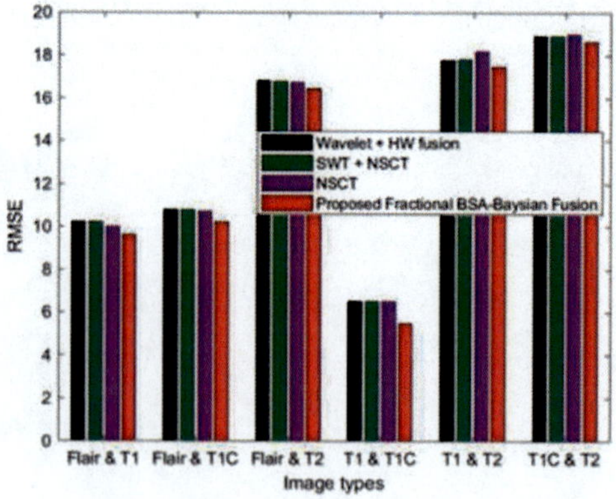

(c)

Fig. 7 Comparative analysis using the dataset-2, **a** Mutual Information, **b** PSNR, **c** RMSE

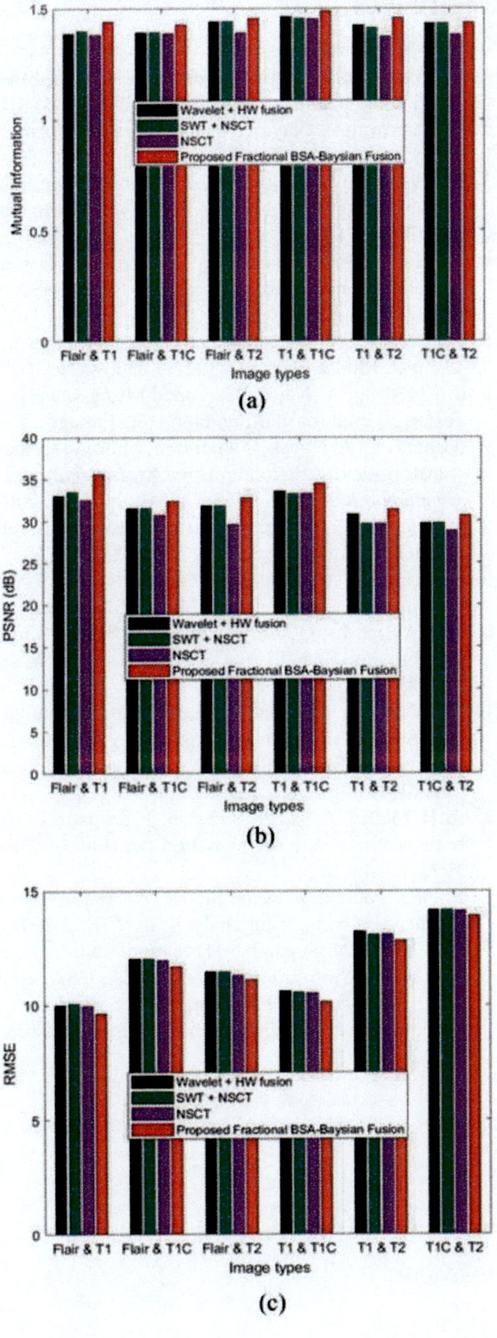

References

1. J. Bhardwaj, A. Nayak, A discrete wavelet transform and bird swarm optimized Bayesian multimodal medical image fusion. Helix **10**(1), 07–12 (2020)
2. J. Bhardwaj, A. Nayak, Haar wavelet transform–based optimal Bayesian method formedical image fusion. Med. Biol. Eng. Comput. **58**, 2397–2411 (2020)
3. D. Ebenezer, J. Anithaa, K.K. Kamaleshwaranb, I. Rani, Optimum spectrum mask based medical image fusion using Gray Wolf Optimization. Biomed. Signal Process. Control **34**, 36–43 (2017)
4. S. Majumdar, J. Bharadwaj, Feature level fusion of multimodal images using Haar lifting wavelet transform. World Acad. Sci. Eng. Technol. Int. J. Comput. Inform. Eng. **8**(6), 1023–1027 (2014)
5. S.P. Yadav, S. Yadav, Fusion of medical images in wavelet domain: a hybrid implementation. Comput. Model. Eng. Sci. **122**(1), :303–321 (2020)
6. J. Bhardwaj, A. Nayak, Cascaded lifting wavelet and contourlet framework based dual stage fusion scheme for multimodal medical images. J. Electr. Electron. Syst. (2018)
7. J Bhardwaj, A. Nayak, D. Gambhir, Multimodal medical image fusion based on discrete wavelet transform and genetic algorithm, in *International Conference on Innovative Computing and Communications*, vol. 1165. (AISC, Springer, 2021), pp. 1047–1057
8. P. Chai, X.L.Z. Zhang, Image fusion using quaternion wavelet transform and multiple features. IEEE Trans. Image Process. **17**(4), 500–511 (2017)
9. J. Bhardwaj, A. Nayak, Lifting wavelet transform based ultrasound image fusion scheme, in *Imaging and Applied Optics 2018*, OSA Technical Digest (2018)
10. Jayant Bhardwaj, Abhijit Nayak, Kulvinder Singh: Feature Level Fusion of Gray Dentistry Imagesusing Haar Lifting Wavelet Transform, *Imaging and Applied Optics 2017*, Optical Society of America (2017).
11. X.B. Mengab, X.Z. Gaoc, L. Lu, Y. Liu, H. Zhanga, Anewbio-inspired optimisation algorithm: Bird Swarm Algorithm. J. Exp. Theor. Artif. Intell. (2016)
12. Multimodal Brain Images. https://www.med.upenn.edu/sbia/brats2018/data.html. Accessed Sept (2019)
13. .B.H. Menze, A. Jakab, S. Bauer, J. Kalpathy-Cramer, K. Farahani, J. Kirby, The multimodal brain tumor image segmentation benchmark (BRATS). IEEE Trans. Med. Imaging **34**(10), 1993–2024 (2015)
14. F. Nian, W. Li, X. Sun, M. Li, An improved particle swarm optimization application to independent component analysis, in *ICIECS 2009* (2009) pp. 1–4
15. J.J. Liang, P.N. Suganthan, Dynamic multi-swarm particle swarm optimizer, in *IEEE Swarm Intelligence Symposium* (Pasadena, CA, USA, 2005), pp. 124–129
16. J.J. Liang, A.K. Qin, P.N. Suganthan, S. Baskar, Comprehensive learning particle swarm optimizer for global optimization of multimodal functions. IEEE Trans. Evol. Comput. **10**(3), 281–295 (2006)

Review on Beamforming Techniques for Millimeter Wave Massive MIMO

Gayatri D. Londhe and Vaibhav S. Hendre

Abstract To meet the demands of the increasing wireless data traffic, there is a need to explore the new spectrum region to meet the demands. This leads to the exploration of the millimeter wave band which will be able to handle the traffic. The principle target of this paper is to give survey of the diverse beamforming procedures for the massive MIMO framework. Classification and comparison of different optimal beamforming techniques are reviewed for energy as well as for spectral efficiency, throughput, inter-cell, and intra-cell interference to decide which technique gives the better performance for the millimeter Wave massive MIMO system. By analyzing different beamforming algorithms, an optimal beamforming algorithm is suggested that will enhance massive MIMO system performances and also satisfy the demands of future generation of wireless communication system.

Keywords Massive MIMO · mmWave · Beamforming · Adaptive beamforming · Hybrid beamforming

1 Introduction

Huge increase in the users increases the telecommunication traffic, and hence there is a need to design the cellular network. There is upgradation in the wireless communication system in every decade which has lead to the different technology generations. Now we are heading toward 5G in order to handle the huge traffic and increased data requirements. The 5G networks should support 1000x higher mobile traffic as compared to 4G system, 100x increased typical user data, 10x reduced energy consumption of network, 100x more connected devices, and 5x lowered end-to-end latency [1, 2]. To achieve the 5G design goals, it is required to design the ultra-dense

G. D. Londhe (✉) · V. S. Hendre
Department of Electronics and Telecommunication Engineering, G.H. Raisoni College of Engineering and Management, Wagholi, Pune, India

G. D. Londhe
Dr. D.Y. Patil Institute of Technology, Pimpri, Pune, India

F. Thakkar et al. (eds.), *Proceedings of the International e-Conference on Intelligent Systems and Signal Processing*, Advances in Intelligent Systems and Computing 1370,
https://doi.org/10.1007/978-981-16-2123-9_22

networks, i.e., small cell technology which helps to increase the network capacity. In order to satisfy 5G requirements, improvement in energy along with spectral efficiency is necessary with existing 4G network. The frequencies ranging from 600 MHz to 3 GHz are used by 4G network is being overcrowded, as these bands of frequencies are widely used for various wireless communication applications. To meet the 5G requirement, it is required to increase the bandwidth. For that purpose, it is required to shift from the existing band to the unused millimeter wave frequency bands ranging from 30 to 300 GHz which provides ample amount of bandwidth. To increase the spectrum efficiency, more number of antennas can be used, i.e., massive MIMO should be deployed [2].

As millimeter waves are having short wavelength, the spacing between antennas gets reduced. Hence, in the given area, multiple antennas may get accommodated which provides abundant spatial processing gain. Massive MIMO is used in combination with mmWave bandwidth to achieve expected gain. Mmwave frequencies can be used for indoor wireless applications or for outdoor wireless applications. Although the available bandwidth is large, the millimeter frequency bands are prone to various losses like diffraction and blockage, path loss, rain fading, foliage loss, and atmospheric absorption [1]. The architecture of 5G network in association with mmWave massive MIMO is presented in Fig. 1. Due to deployment of small cell, mmWave can be effectively used for communication.

The architecture uses bands of frequencies belonging to mmWave and microwave range. The frequency bands are dependent on factors like path loss characteristics of different frequency bands, applications, and channels. The increase in carrier frequency leads to increase in the path loss. Hence, microwave band frequencies are utilized for long-range outdoor communication. Also frequencies at 60 and 180 GHz

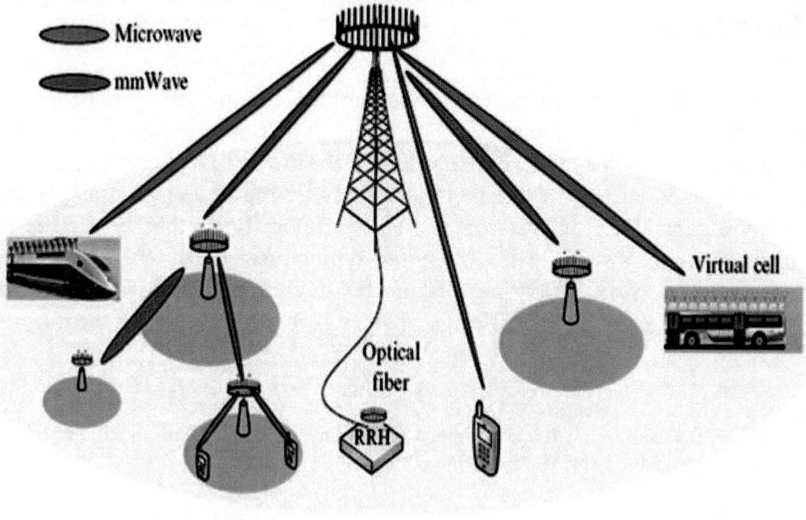

Fig. 1 Architecture of 5G network [2]

are attenuated drastically due to molecular and atmospheric absorption which are best suited in case of indoor communication [2]. Precoders and detectors employed in massive MIMO system operating on mmWave band contain antennas in large quantity. Because of which massive MIMO system can utilize the resources available for improvement in energy and spectral efficiency by using beamforming technique, as it directs energy toward desired terminals and suppresses interference. Smart antenna is used for this purpose. Smart antenna system is very important in wireless communication as it minimizes interference because of co-channel and enhances system capacity. The smart antenna system is used at mobile station terminal and base station terminal to enhance wireless mobile connection and system performance by managing the signal appropriately. Smart antenna system is the combination of arrays of antennas along with signal processing techniques. The smart antenna helps to direct the core beam in the preferred direction where it is being aimed and discard the signals interfering at the matching frequency from other directions by keeping the antenna at the same position. This can be done by varying the amplitudes and phases of the signal and then transmit the signal by the transmitter. The MIMO channels will carry the different variations of the signal transmitted such that the signals at the user intended are added constructively while at other users they are added destructively [1, 3].

The massive MIMO has helped in the improvement of energy efficiency and system capacity because of the use of antenna in large quantity. In massive MIMO system, interference problem arises due to use of antennas in large quantity. This problem can be overcomed by deployment of beamforming antennas. Beamforming is defined as signal processing technique which is being applied to the several antennas arrays at precoder and detector end to transmit or receive different signals at a time at several desired point. The organizations like LTE advanced and long-term evolution (LTE) operators are already using the beamforming technique. The 5G wireless communication system focused mainly on beamforming solutions for massive MIMO system [3]. For various network performance challenges for MIMO relay networks, the review of beamforming techniques is given for interference mitigation, complexity and power consumption, channel modeling, diversity issues, channel estimation and feedback, degree of freedom and spectral efficiency. Different topologies are considered for the challenges in implementation of beamforming in MIMO relay networks and the parameters like MSE minimization and SNR maximization are taken into consideration [4].

In [5], summary of different signal processing techniques by using mmWave for communication systems is given and MIMO architectures have been identified to the different approaches like analog beamforming, one-bit architecture, and hybrid precoding and combining. Also beam training protocol and channel estimation algorithm are reviewed in detail. In [6], it gives the review on architecture of hybrid beamforming, HetNet, and signal processing. The paper takes into consideration analog as well as digital processing of signal for different UE/BS designs and HetNet, fully as well as partially connected hybrid beamforming architecture, resource management. The paper entirely presented the crucial features of hybrid beamforming, such as energy efficiency, and hardware disabilities. This paper presents a brief overview of various beamforming techniques that are applied in massive MIMO system. The paper focuses on comparative study of various beamforming techniques applied in massive MIMO system and determines optimal beamforming that may be useful in massive MIMO system.

In Sect. 2, beamforming technique importance is highlighted. Section 3 gives the brief idea of various types of beamforming techniques and algorithms. Basic switched beamforming and adaptive beamforming algorithms are compared to figure out the best one for massive MIMO. Also different adaptive algorithms like blind and nonblind algorithms are presented. Analogue, digital, and hybrid beamforming techniques are discussed in detailed with their pros and cons for massive MIMO implementation. Section 4 provides the conclusions.

2 Beamforming Benefits

The gains that MIMO topologies offered due to transmitter and receiver diversity leads to their use in wireless standards like LTE, LTE advanced, IEEE 802.11n, 802.16e, 3GPP. For the improvement of network capacity, transformation of the network from multiuser MIMO to massive MIMO is necessary. Massive MIMO contains number of antennas at BS as well as at UE. In massive MIMO system, each antenna is loaded with the company of different signals for increasing the multiplexing gain or diversity gain by loading identical signal on various antennas. Adaptive arrays with mmWave technologies are widely developed in urban environments. Massive MIMO helps to enhance the network efficiency. But the use of antennas in abundant quantity gives rise to the problem of signal interference. To resolve this problem smart antennas are used. Smart antenna facilitates beamforming. By using beamforming a well-organized data path is recognized to reach the intended user also nearby interferences are minimized. By making arrangement of elements in an array precisely, beams pointed in the particular direction are added whereas others are ignored. Beam management helps in rejection and avoidance of interferences, SNR improvement, and network efficiency which leads to the enhancement in the performance of 5G network. Beamforming is an optional name for spatial filtering where an antennas array is provided with analog or digital signal processing, to obstruct the reception of signal from specified directions. The beamforming techniques provide improved spectral efficiency, better energy efficiency, and system security.

2.1 *Increased Spectral Efficiency*

Spectral efficiency is expressed in bits/s/Hz. MmWave massive MIMO consists of antennas in large amount which enhances the capacity of the data to be transmitted. By using beamforming technique, the signal can be directed toward the intended user and minimizing the interference. The main aim of beamforming is to improve the SNR by increasing the received power of signal as well as reducing the power of interfering signal which helps to improve the capacity. At receiver end, ricean factor gain is improved because of multipath dispersion with suitable antenna elements arrangement to minimize RMS delay propagation leads to the improved SNR [7]. The active cell sectoring efficiency is upgraded by beamforming techniques instead of using antennas that radiate in all directions at base stations. As compared to omni-directional antenna, sectored antenna is preferred for beamforming [3].

2.2 *Good Energy Efficiency*

The unit of energy efficiency is bits/Joule. Energy efficiency can be expressed as proportion of spectral efficiency and transmit power. Massive MIMO system operating on mmWave band has drastically reduced uplink and downlink transmit power as large antenna arrays are using hybrid beamforming by coherent combining with increased antenna aperture. In uplink, EE is increased by minimizing the UE transmit power as it is inversely proportional to antennas quantity present at base station [6]. For a very large MIMO system, there is increase in spectral efficiency in range of one or two which simultaneously improves the energy efficiency in order of three by using linear processing like MRC and ZF by obtaining channel estimates by using uplink pilots [8]. For wireless communication systems, the importance of energy efficiency was explored by [6, 8, 9]. For the improvement of energy efficiency, the consumption of power at BS must be reduced by using power control beamforming techniques. Energy and spectral efficiency are having tradeoff among them. With perfect channel state information for uplink, the increment in spectral efficiency leads to the decrement in energy efficiency. Energy and spectral efficiency increases together provided with imperfect CSI, in the small transmit power regime, while efficiency of energy decreases as spectral efficiency increases with high transmit power. The downlink tradeoff in comparison to MF precoder, ZF provides good results in situation with low energy and high spectral efficiency, whereas the opposite hold onto for situation with high energy and low spectral efficiency [8].

2.3 *System Security*

By using beamforming only the required signal will be selected by the receiver and other signals will be neglected from the given signal because beamforming helps to transmit signal by directing it toward the desired user. Therefore, the chances of eavesdropping of the transmitted signal will be lower as compared to the signal transmission using conventional method [3]. In mmWave massive MIMO technology, mmWave bands frequency is used for signal reception and transmission. These waves having higher frequency are applicable for short distances as they are heavily attenuated due to their propagation characteristics. Massive MIMO helps to deploy large number of antennas. As frequency increases antenna size decreases and spacing between antennas also reduces which leads to the development of compact antennas, which can be easily deployed at UE terminal also. Smart antennas are used that lead to processing of the signal and formation of beams toward intended direction to minimize the interference.

The balance between SE and EE is achieved by optimizing multiuser beamforming for a massive MIMO broadcast channel. Many of the works are focusing on EE maximization; a new term called resource efficiency (RE) is coined, which is the weighted sum of SE and EE. Considering MAC-BC duality, a resource-efficient multiuser beamforming design is proposed in [9].

3 Beamforming Techniques

The beamforming techniques are applied to smart antenna system for the transmission and reception of the signal in targeted direction. The smart antennas are provided with the signal processing algorithm that is used to determine the signal's directions of arrival (DOA) which are used for judging beamforming vectors. By using the given vectors, one can follow the required signal transmitted by mobile station. UMTS, WCDMA, LTE, LTE Advanced use smart antennas system. The incoming signals direction of arrival is determined by using techniques like estimation of signal parameters via rotational invariance techniques (ESPRIT), multiple signal classification (MUSIC) technique and the matrix pencil method and its derivatives [3].

Beamforming techniques are mainly divided into narrowband beamforming and broadband beamforming based on the signal bandwidth. The major beamforming techniques are categorized into switched beamforming, adaptive beamforming, analog beamforming, digital beamforming, and hybrid Beamforming as shown in Fig. 2.

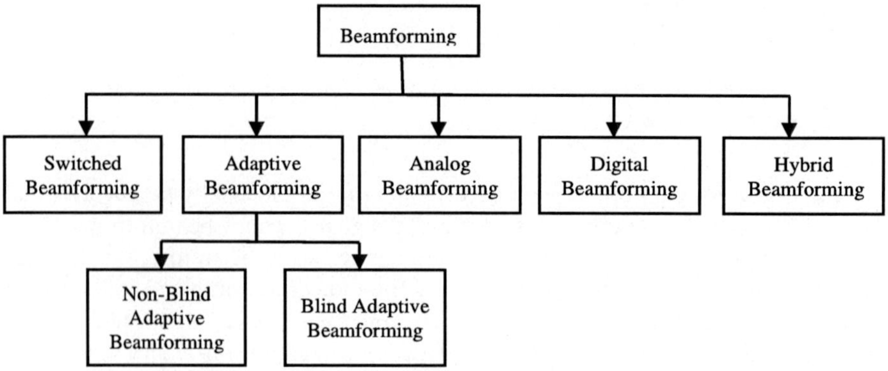

Fig. 2 Categorization of beamforming techniques

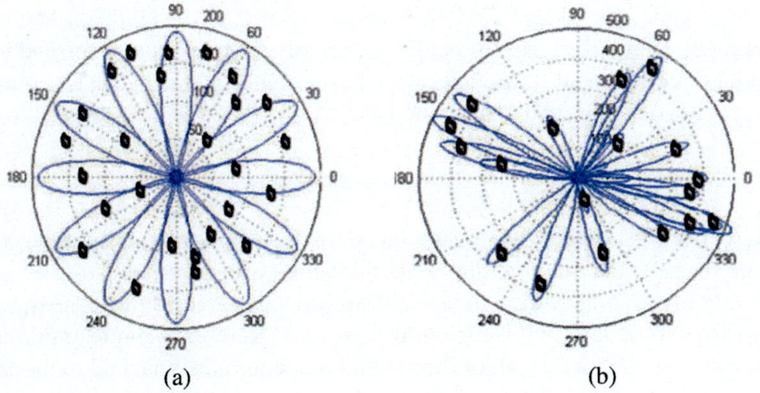

Fig. 3 **a** Switched beamforming. **b** Adaptive beamforming [3]

3.1 Switched Beamforming

In switched beamforming, predefined beams are formed using fixed beamforming network where most of the beams are not pointed toward desired direction. Switched beamforming is based on Butler matrix that uses fixed beamforming. Crossovers, phase shifters, and hybrid couplers are the components of butler matrix [3]. Also the beam covers more number of mobile stations as shown in Fig. 3a.

In [10], beamforming was achieved in different directions by using 4 × 4 Butler matrix switched beam array at three different operating frequencies at 2, 6.4, and 8.5 GHz. By using these frequencies, radiation pattern is analyzed for different port excitation. Switched beamforming is easy to implement and inexpensive and gives better coverage capacity but it suffers from the problem of interference and it is difficult to differentiate between desired and interfering signal.

3.2 Adaptive Beamforming

Fixed beamforming is applicable for specified angle of arrival sources, but if the sources angle of arrival is changing then there is requirement to modify the array weights constantly according to the change in environment which is termed as adaptive beamforming. An adaptive processor is used in adaptive array system. As the source moves the adaptive processor modifies and compensates the weights of array.

At the BS, the signals detected at the arrays of antennas are multiplied with the complex weights. The changes in the weights are instantaneously added up. In an adaptive beamforming depending on the DOA, the required beam can be directed in the direction of targeted signal, whereas the unwanted signals are nullified. Adaptive beamforming gives good coverage and is immune to interference as compared to switched beamforming that also gives better coverage. It is complex and expensive. Figure 3b shows the adaptive beamforming in which beams are formed toward the desired directions and null toward the interfering sequences [3]. The output of adaptive array $y(t)$ in general is described as

$$y(t) = w^H x(t) \tag{1}$$

In Eq. (1), w^H is complex conjugate transpose for weight vector. Iterative method is adopted for calculation of weight to get the output of array $y(t)$.

Nonblind adaptive algorithms as well as blind adaptive algorithm are the two major classification of the adaptive beamforming algorithm. Reference signal or pilot signal are used in nonblind adaptive algorithm to find the required signal and to modify the complex weights. At the end of every loop, the response of the weights is compared to the reference signal or pilot signal to produce the error signal. According to the error signal, the weights are modified to reduce the error. Blind adaptive algorithm has no reference or pilot signal to modify the complex weights. Instead it uses some of the properties of the required signal that are known.

The comparison of different nonblind algorithms like conjugate gradient method (CGM), recursive least squares (RLS), sample matrix inversion (SMI), and least mean square (LMS) is done in [11]. The comparison is done on the base of beamforming, null steering capacity, and speed of convergence. Comparison states that out of all the algorithms mentioned the algorithm having fastest convergence and high null depths to give good performance is CGM algorithm, whereas RLS has best beamforming in terms of convergence speed but it cannot manage the side lobe level. Therefore, RLS algorithm is improved by mixing it with Dolph-Chebyshev method which helps to obtain the desired focused beam by controlling the side lobe level (SLL) and suppressing the unwanted side lobes is proposed in [12]. The performance analysis of nonblind algorithm like, least mean square (LMS), data reuse least mean square (DRLMS), and decision feedback equalizer-based least mean square (DFELMS) beamforming algorithms is done in [13]. The DRLMS algorithm helps to direct the beam in the direction of required signal with increased gain and reduced interference and side lobe levels in comparison with DFELMS and LMS algorithms. The data accessible is reutilized again and again in every loop for updation of weight. DRLMS gives better convergence and reduces the error among required and output signal and enhance system throughput and performance.

Blind adaptive algorithm is not requiring the reference signal hence it is not required to adapt the array weights. Constant modulus algorithm (CMA) as well as least square constant modulus algorithm (LS-CMA) are the well-known blind adaptive algorithm. In [14], the evaluation of downlink transmission performance done at user terminal by using simple control is proposed by the massive MIMO transmission system. By using CMA at user terminals, interfering signals can be minimized. The use of millimeter wave (mmwave) communications for adaptive beamforming is explored in [15]. Antenna with uniform linear array (ULA) is provided with signal processing technique for beamforming in indoor surrounding for different users. Alphabet-matched constant modulus algorithm (CMA) is used to provide proper channel state information which helps to direct beam of data toward the intended user within uplink communication. Pre-coder used the same channel state information in the downlink for beamforming for multiuser system. Nonblind adaptive algorithm leads to the low transmission efficiency. To enhance the transmission efficiency, blind adaptive algorithm is used which does not require the training sequences for beamforming. In [16], on the basis of CM algorithm a semi-blind adaptive beamforming is proposed. Semi-blind adaptive algorithm gives good performance as compared to nonblind adaptive algorithm, provided with training sequence in the first loop.

Based on null steering linearly constrained minimum variance (LCMV) beamforming as well as minimum variance distortion less response (MVDR) beamforming are the well-known adaptive algorithm. The SINR is enhanced by using null steering antennas as it places null in the interfering direction and also main lobe is pointed toward the desired direction. Firefly algorithm (FA) is used for SINR improvement in LCMV beamforming. In this method, SINR is improved by optimizing the power to give better performance to the user. This method effectively nullifies the interferences and stabilizes the power [17].

In MVDR beamforming algorithm, the variance of beam response is reduced by choosing the antenna weights while maintaining the constant gain level so that beam directs to the desired direction. Interfering signals coming from unwanted directions are terminated. To improve the strength of MVDR beamformer, various techniques are available but diagonal loading and its extensions are broadly used. The latest dynamic mutated artificial immune system (DM-AIS) is proposed in [18] which controls the null steering of interfering signals and increases the SINR for required signal which helps to improve performance of MVDR beamforming. Some of the properties state that for massive MIMO system MVDR is best suited as compared to LCMV. The inter-cell interference is reduced, and there is enhancement in the massive MIMO system performance as DOA for the given algorithm is evaluated with respect to azimuth as well as elevation angle (2D-DOA) that enhances the correctness of the results [3].

3.3 Analog, Digital, Hybrid Analog/Digital Beamforming

In analog beamforming, phases of the signal transmitted are controlled by using phase shifters. Only phases are varied amplitude remain unchanged. Provided with the exact information of the CSI at both base station and user terminal, in an analog beamformer N antennas are provided at BS with single RF chain to transmit only one data stream at user terminal which consists of M antennas with single RF chain. Since only signal phases are varied in analog beamforming, hence it goes through various performance losses. Also it is applicable to single user and single stream of data [1]. Figure 4a shows the analog beamforming structure it consists of single RF chain and multiple phase shifters. In digital beamforming, signals amplitudes as well as phases are controlled and it can be applied to single-user as well as multiuser systems. After crossover RF chain, the frequency of the carrier signal processed is

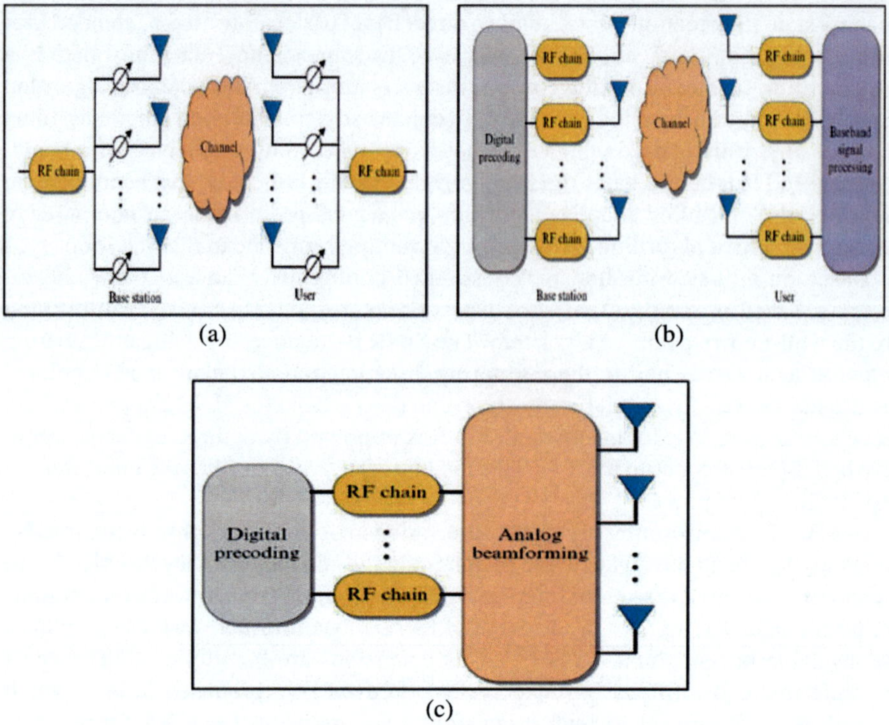

Fig. 4 **a** Analog beamforming. **b** Digital beamforming. **c** Hybrid beamforming [19]

raised by using digital-to-analog converters, power amplifiers along with mixers as it is required to control signal phase and amplitude.

After that the antenna elements are combined with RF chain responses. For this, a committed RF chain is required to be provided to each element of antenna array. To realize such a system is expensive as massive MIMO system consists of antennas in larger quantity. Figure 4b shows the structure of digital beamforming implementation.

It is expensive to deploy digital beamforming as number of RF chains is required in larger quantity. Analog beamforming is cost effective as only phase shifter is required but performance is poor as compared to digital beamforming because amplitude of the phase shifters cannot be varied. Hence, a optimal solution is required which gives better performance as well as cost effective. To achieve better mixing of analog and digital, beamforming is done which is called hybrid beamforming.

Hybrid beamforming is a good technique for massive MIMO system operating in mmWave band. Because it required less number of RF chain which minimizes the energy consumption and cost and gives better performance when compared to digital beamforming. Digital pre-coder consists of RF chains in of small number which is used to cancel interference followed by analog beamformer consisting of phase shifters to enhance the gain of antenna array [1, 19]. The implementation of hybrid beamforming is shown in Fig. 4c.

Hybrid beamforming architectures can be classified into fully connected and partially or sub-connected beamforming architecture. In fully connected design, every RF chain is linked to antennas of every base station through phase shifters, whereas in sub-connected architecture every RF chain is connected to only subset of base station antennas. The diagrammatic representation of fully and partially connected architecture is shown in Fig. 5a, b, respectively.

Consider N_t transmit antenna and N_{RF} RF chains then signal processing path in case of fully connected architecture is $N_t \times N^2_{RF}$ and for sub-connected architecture it is $N_t \times N_{RF}$. In case of fully connected architecture, beamforming gain is N_{RF} times larger than partially or sub-connected architecture. The main aim of hybrid beamforming architectures is to minimize the complexity of hardware along with signal processing and to get output same as that of digital beamforming [6].

In [20], a hybrid beamforming architecture with less complexity to maximize the energy efficiency of the system for downlink multiuser condition is proposed. Fully connected architecture consisting of phase shifters of high resolution consumes more circuitry power as compared to fully digital (FD) counterpart if users increase beyond a certain threshold level. For such case, the fully digital architecture is having high energy efficiency. As compared to both fully connected and FD structure, partially connected architecture consumes less circuitry power. FD architecture gives high spectral and energy efficiency as compared to hybrid architecture.

The comparative analysis of various beamforming techniques is given in Table 1.

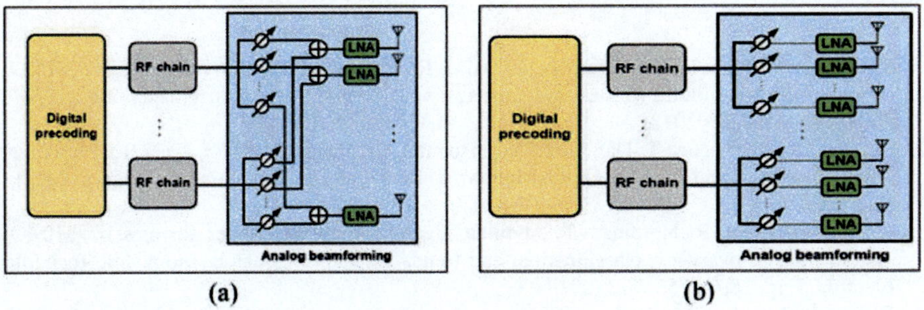

Fig. 5 **a** Fully connected architecture. **b** Sub-connected architecture [1]

Table 1 Comparative analysis of various beamforming techniques

Features	Analog beamforming	Digital beamforming	Hybrid beamforming
Cost	Low	High	Medium
Number of users	Single user	Multi user	Multi user
Number of streams	Single stream	Multi stream	Multi stream
Signal control capability	Phase control	Phase and amplitude control	Phase and amplitude control
Energy consumption	Low	High	Medium
Performance	Low	Optimal	Near optimal
Hardware requirement	low	High	Medium
Implementation	Unsuitable	Impracticable	Practicable and realistic

4 Conclusion

This paper gives the brief survey of different beamforming techniques. It gives the review of recent research on different beamforming algorithm applicable to massive MIMO system. The comparison between switched beamforming and adaptive beamforming is done. Adaptive beamforming is more immune to interferences as compared to switched beamforming. Research on different nonblind and blind adaptive algorithm is done. Algorithms are compared in terms of convergence rate, complexity, side lobe level, and null steering performance. Hybrid beamforming which is formed by combining analog and digital beamforming gives good performance for mmWave band. Optimal beamforming can be achieved by combining hybrid beamforming with any one adaptive algorithm such as MVDR, or by combining any two algorithms which give the exact DOA to form the beam exactly in desired directions.

References

1. S.A. Busari, K.M.S. Huq, S. Mumtaz, L. Dai, J. Rodriguez, Millimeter-wave massive MIMO communication for future wireless systems: a survey. IEEE Commun. Surv. Tutor. **20**(2), 836–869 (Second quarter 2018)
2. S. Mumtaz, J. Rodriguez, L. Dai, Introduction to mmWave massive MIMO, in *mmWave massive MIMO: A Paradigm for 5G*, ed. by S. Mumtaz, J. Rodriguez, L. Dai. (Academic Press, 2017), pp. 1–18
3. E. Ali, M. Ismail, R. Nordin. N.F. Abdulah, Beamforming techniques for massive MIMO systems in 5G: overview, classification, and trends for future research. Front. Inf. Technol. Electron. Eng. **18**(6), 753–772 (2017)
4. D. Vouyioukas, A survey on beamforming techniques for wireless MIMO relay networks. Int. J. Ant. Propagat. 1–21 (2013)

5. R.W. Heath, N. González-Prelcic, S. Rangan, W. Roh, A.M. Sayeed, An overview of signal processing techniques for millimeter wave MIMO systems. IEEE J. Select. Topics Signal Process. **10**(3), 436–453 (2016)
6. I. Ahmed et al., A survey on hybrid beamforming techniques in 5G: architecture and system model perspectives. IEEE Commun. Surv. Tutor. **20**(4), 3060–3097 (Fourthquarter 2018)
7. H. Manai, L. Ben Hadj Slama, R. Bouallegue, Interference management by adaptive beamforming algorithm in massive MIMO networks, in *2019 15th International Wireless Communications & Mobile Computing Conference (IWCMC)*, Tangier, Morocco, pp. 49–54 (2019)
8. H.Q. Ngo, E.G. Larsson, T.L. Marzetta, Energy and spectral efficiency of very large multiuser MIMO systems. IEEE Trans. Commun. **61**(4), 1436–1449 (2013)
9. Y. Huang, S. He, J. Wang, J. Zhu, Spectral and energy efficiency tradeoff for massive MIMO. IEEE Trans. Veh. Technol. **67**(8), 6991–7002 (2018)
10. S. Sivasundarapandian, Performance analysis of multi-band multiple beamforming butler matrix for smart antenna systems, in *2015 International Conference on Robotics, Automation, Control and Embedded Systems (RACE)*, Chennai, pp. 1–5 (2015)
11. P. Saxena, A.G. Kothari, Performance analysis of adaptive beamforming algorithms for smart antennas. IERI Procedia **10**, 131–137 (2014)
12. H.S. Vu, K.T. Truong, L.T. Bang, V.Y. Vu, M.T. Le, An investigation of adaptive digital beamforming antenna for gNodeB 5G, in *2019 International Conference on Advanced Technologies for Communications (ATC)*, Hanoi, Vietnam, pp. 221–224 (2019)
13. A. Rathore, D.K. Panda, DRLMS, DFELMS, LMS adaptive beamforming algorithms for smart antenna system, in *2017 International Conference on Information, Communication, Instrumentation and Control (ICICIC)*, Indore, pp. 1–5 (2017)
14. S. Ogawa, K. Nishimori, R. Taniguchi, T. Hiraguri, Blind beamforming method by terminal station for downlink multi-beam massive MIMO, in *2017 International Symposium on Antennas and Propagation (ISAP)*, Phuket, pp. 1–2 (2017)
15. P.O. Taiwo, A. Cole-Rhodes, Adaptive beamforming for multiple-access millimeter wave communications: invited presentation, in *2019 53rd Annual Conference on Information Sciences and Systems (CISS)*, Baltimore, MD, USA, pp. 1–6 (2019)
16. Y. Zhao, B. Ai, Y. Liu, D. Fei, D. Li, Semi-blind adaptive beamforming based on constant modulus algorithm for smart antennas, in *2017 4th International Conference on Information, Cybernetics and Computational Social Systems (ICCSS)*, Dalian, pp. 286–289 (2017)
17. C. Doroody, S.K. Tiong, S. Darzi, SINR improvement using Firefly Algorithm (FA) for Linear Constrained Minimum Variance (LCMV) beamforming technique, in *2015 International Conference on Computer, Communications, and Control Technology (I4CT)*, Kuching, pp. 441–445 (2015)
18. T.S. Kiong, S.B. Salem, J.K.S. Paw, K.P. Sankar, S. Darzi, Minimum variance distortionless response beamformer with enhanced nulling level control via dynamic mutated artificial immune system. Scient. World J. 1–9 (2014)
19. X. Gao, L. Dai, Z. Gao, T. Xie, Z. Wang, Precoding for mmWave massive MIMO, in *mmWave massive MIMO: A Paradigm for 5G*, ed. By S. Mumtaz, J. Rodriguez, L. Dai. (Academic Press, 2017), pp. 79–111
20. K. Ardah, G. Fodor, Y.C.B. Silva, W.C. Freitas, A.L.F. de Almeida, Hybrid analog-digital beamforming design for SE and EE maximization in massive MIMO networks. IEEE Trans. Veh. Technol. **69**(1), 377–389 (2020)

False Arrhythmia Alarm Detection Through ECG Signals

Vikneswaran Vijean, Gunashareene R. Pavinthiran, Yen Fook Chong, Chee Chin Lim, Saidatul Ardeenawati Awang, and Hariharan Muthusamy

Abstract Patients with life-threatening illness and injuries are often admitted to intensive care units (ICU) as they require constant support from life support machines and medical care. Their vital signs need to be continuously monitored, and electrocardiogram (ECG) signals are often used for this purpose. Any irregular ECG signals or arrhythmia will be closely monitored in intensive care units (ICU) and timely medical intervention is provided to critical patients. However, occurrence of false arrhythmia alarm hinders the monitoring process. Therefore, this research focuses on predicting the false alarms associated with arrhythmic ECG signals. Discrete wavelet Transform (DWT)-based non-linear entropies and morphological features are utilized for the study. Due to the imbalanced nature of the dataset, adaptive synthetic sampling (ADASYN) technique was used to create synthetic data. The principal component analysis (PCA)-based dimensionally reduced features were trained with supervised classifier such as fuzzy K-nearest neighbour (FKNN), Random Forest, and AdaBoost classifiers to quantify predictive rate of the false alarm. The proposed method was able to provide accuracies above 80% for all the testing instances.

Keywords Arrhythmia · Electrocardiogram · Wavelets

1 Introduction

Although modern Intensive Care Units (ICU) are equipped with alarms for arrhythmia detection, these monitors are not 100% foolproof and are still subjected to

V. Vijean (✉) · Y. F. Chong · C. C. Lim · S. A. Awang
BioSIM Research Group, Sports Engineering Research Centre, University Malaysia Perlis, Kangar, Malaysia
e-mail: vikneswaran@unimap.edu.my

G. R. Pavinthiran
School of Mechatronics, University Malaysia Perlis, Kangar, Malaysia

H. Muthusamy
National Institute of Technology Uttarakhand, Srinagar, India

F. Thakkar et al. (eds.), *Proceedings of the International e-Conference on Intelligent Systems and Signal Processing*, Advances in Intelligent Systems and Computing 1370,
https://doi.org/10.1007/978-981-16-2123-9_23

false alarms. False alarms are needless alarms that are given by error or on an occasion on which danger is perceived but fails to materialize such as misinterpreted patient's movement like muscle tremor, baseline shift, or wandering [1]. In the medicine field, false alarm is known as the alarm without therapeutic or clinical consequence [2]. False alarms commonly occur in ICUs and are generally accepted as the logic of 'better safe than sorry' is practiced in hospital monitoring systems. However, high false alarm rates may bring the consequence of 'Crying Wolf Phenomenon' in which, the real critical alarms are ignored. Since the patients in ICUs are in critical state and require round the clock supervision, any missed critical alarms may lead to patient injury or even death. Therefore, the need for machine learning-based alarm systems for monitoring the ECG signals are required to reduce the instances of false alarms.

2 Literature Review

From the literature, researches have proposed various methods of processing the ECG signals for developing machine learning-based systems that can distinguish between the real and false alarms. In order to remove common noise such as baseline wander, Kalidas et al. [3] suggested removing the components of signals less than 1 Hz from the rest of the signals by using the low-pass Butterworth filter. Pacemaker spiking is also common noise that disrupts the effectiveness of the algorithm. Based on Ansari et al. [4], the pacemaker spiking can be removed using the statistical approach and application of median filter of wide span of 2.4 ms without compromising QRS width of 60–100 ms, whereas for feature extraction method, Fallet et al. [5] had used Mathematical Morphology (MM) approach using Adaptive Structuring Element (AMM) for their QRS complex extraction algorithm. They had also used spectral purity index (SPI), to access the quality of the extracted features to classify the presence of true arrhythmia events. Rodrigues and Couto [6] had, however, proposed a beat polarity identifier that is based on QRS complexes for false alarm detection. Relying on the four different types of polarity models, they were able to provide predictive accuracy of 79.02% for real time and 78.28% for retrospective detection. For classification of the signals, commonly used classification methods are Random Forest classifier [7, 8], Linear Discriminant Analysis [9, 10], Support Vector Machine [9], Binary Classification tree [9], and Neural Networks [11]. Each of the methods proposed has its own advantages, although direct comparison is not possible due to variation between the testing instances and dataset used. Therefore, this research investigates the use of DWT-based entropies and morphological features from the ECG samples as predictive measure for false alarm detection. The DWT was chosen as it is an effective decomposition method that allows the features of the ECG signal to be represented in multiple resolutions with less computational complexity. A publicly available Arrhythmia database from PhysioNet database is utilized for this purpose.

3 Data Pre-processing

3.1 Dataset

The data that was used in this project was acquired from Physionet 'Computing in Cardiology Challenge Database' that can be accessed from https://archive.physionet.org/challenge/2015/. The data consists of 750 samples from the training set. It encompasses 456 false alarmed signals and 294 of true alarm signals. The provided data was multimodal, consisting of photo-plethysmogram (PPG), Arterial Blood Pressure (ABP), and ECG signals. Only the first two leads of ECG recordings were used for this research. There were two types of records provided called real time (5 min) and retrospective record (5 min and 30 s). Each of these records were also annotated by a group of experts on the true and false alarm and the type of arrhythmias that the waveform indicates for reference purposes.

Data segment used for the analysis of the ECG waveform in the algorithm was only 10 s of data prior to the onset of alarm, from the 4th minute 50th second to 5th minute. The 30 s post-alarm period (retrospective category) was not taken into analysis as to standardize the duration of the data analysis.

3.2 Pre-processing

The baseline wander are caused by respiration and perspiration of patient causing the iso-electric line of the signal change position from reference line. They commonly exist in the range of 0.15–0.3 Hz [2, 7, 8]. To eliminate this noise, initially the QRS complex and the P waves have been removed from the signal using a median filter with 200 ms width [8]. Then, the second median filter of 600 ms width was applied on the PQRS-free signal to eliminate the T-waves. The output of the second median filter was just the baseline wander noise which was subtracted from the original raw ECG signal to get baseline noise free signal as demonstrated on Fig. 1 where blue is the raw signal and red indicates the signal after filtering. The effect of the filter is apparent where the blue original line is effectively placed on the zero reference line represented by the red line.

Pacemaker spiking exists among patients who wore pacemakers for their heart beat initiation where the existence of this spikes interrupts the QRS peak detection where the spiking would also be considered as peaks. The method of Rodrigues and Couto [6] was adopted with a slight modification on this algorithm in order to remove the noise. The algorithm let x_n as pacemaker spike sample, if the ECG signal is $S(x_n)$ as in (1):

$$\left|(S(x_n)) - \frac{s(x_{n-1}) - s(x_{n+1})}{2}\right| > 0.06 \tag{1}$$

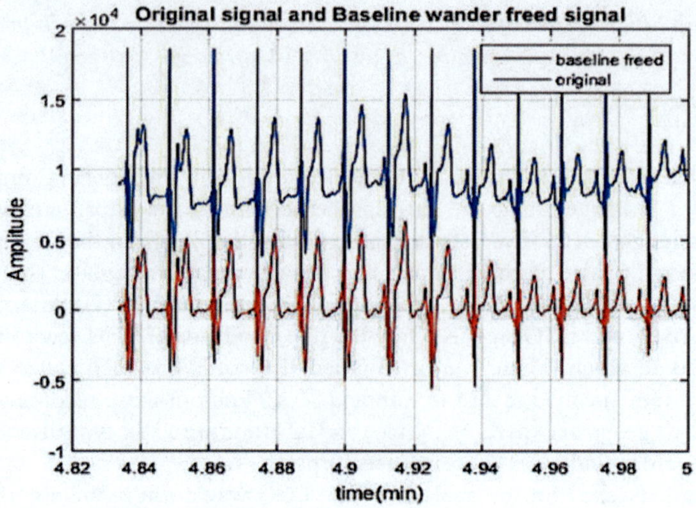

Fig. 1 The image of merged signal F45 (channel 1) of before (original) and after (baseline freed) the baseline wander removal

Initially, the samples were bipolar normalized. Then, based on (1), each of the consecutive ECG signals is checked and those with peak greater than 6% (0.06) are replaced with arbitrary mean value of the data. Figure 2a is the original ECG signal before and Fig. 2b shows the signal after the spiking removal.

In addition, trends are another common noise in ECG that exist due to fluctuation of the surrounding ambient temperature and disruption of the electrode surface electrochemical activity at the point of contact. Low-pass Chebyshev filter (CLPF) [12] and high-pass Butterworth filter (BHPF) were experimented to identify most effective method to remove trends in this project. The cutoff frequency for this filter is obtained from the spectrum domain plot as in Fig. 3 which was identified as 5 Hz with passband ripple of 3 dB and stopband attenuation of 20 dB at the 0.1 Hz. The marked points are the part with the trend and chosen as the cutoff frequency.

The red circles in Fig. 4 indicate that the peak deterioration is more visible in Fig. 4c compared to that of Fig. 4b relative to the original signal Fig. 4a. CLPH method has greater standard deviation thus preserving the peaks better. The filter with greater standard deviation value, CLPF, was chosen as it proves better preservation of peaks of the ECG signal.

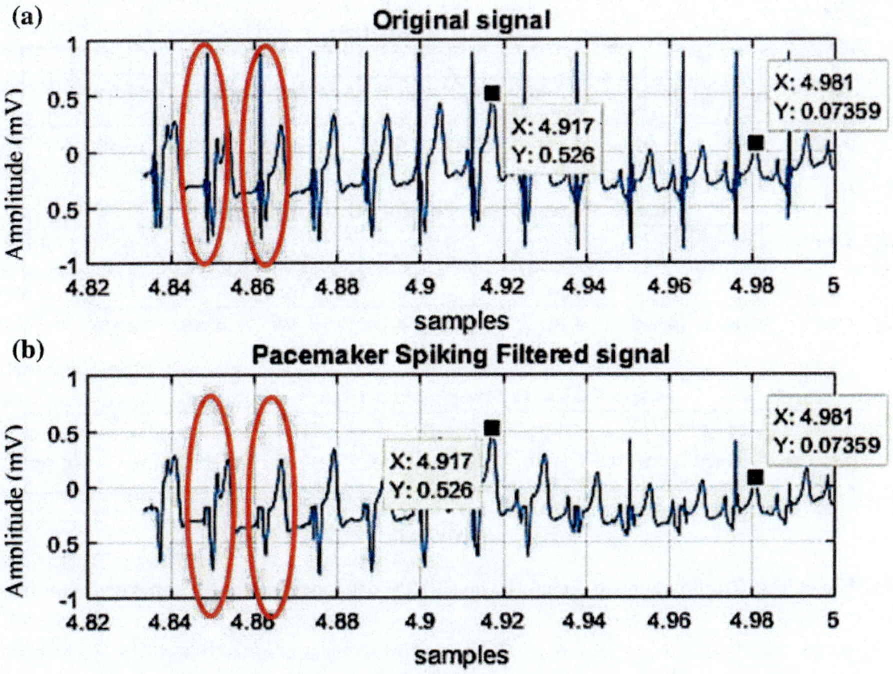

Fig. 2 **a**, **b** The pacemaker spiking in F45 (lead 1) were effectively removed (indicated by red circles) as shown on the before (original signal) and after (pacemaker spiking removed signal) filtered signal above

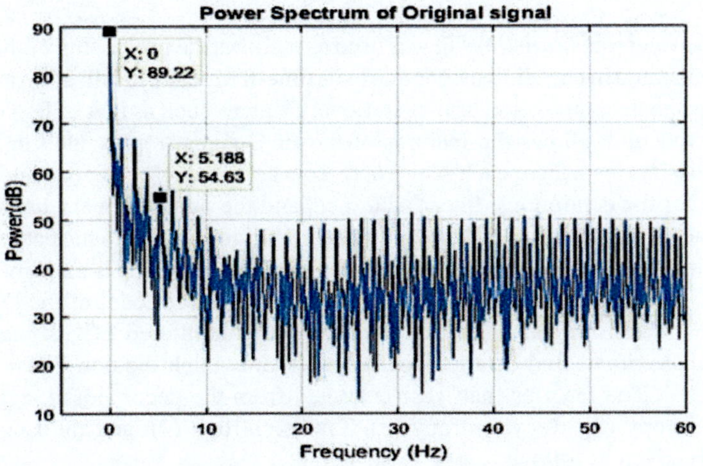

Fig. 3 The spectrum domain plot representing power spectrum of signal F45 (lead 1) in dB

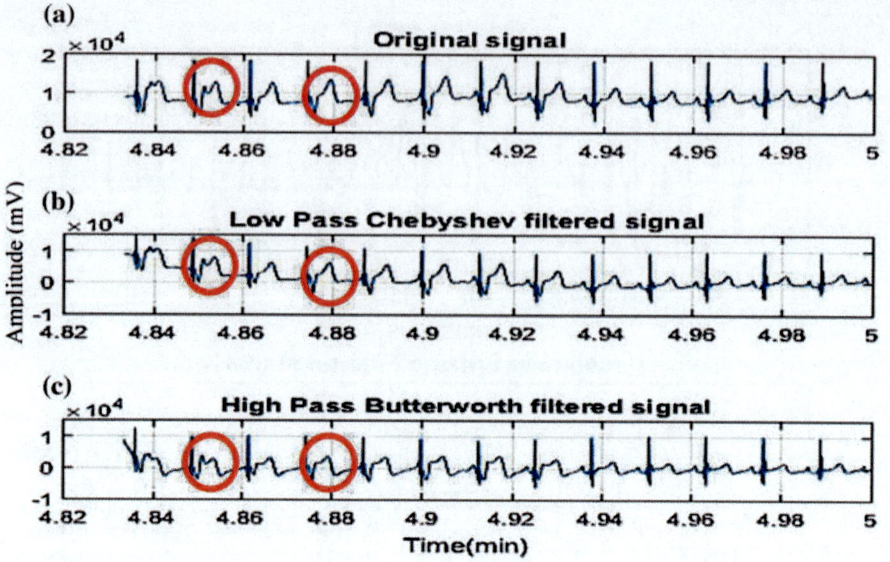

Fig. 4 a–c The CLPF detrended signal (**b**) and BHPF detrended signal (**c**) compared with the original signal F45 (lead 1) (**a**)

4 Feature Extraction

4.1 Non-linear Features

Discrete Wavelet transform (DWT) was used as it allows the processing of the signals to be implemented in new dimensions due to its time-frequency localization technique which suits the decomposition of non-stationary signals such as that of ECG signals. The decomposition allows the features of the ECG signal represented in multiple resolutions or scales, where each scale represents certain degree of coarseness of the signal. As for the continuous, linear patterned, and the non-stationary nature of the ECG signal, the wavelet model that would be suitable and similar to the pattern of this physiological signal would be the Daubechies 4 (Db4) [13]. This is majorly because of the QRS complexes of the ECG that is analogues to the model of the Db4. This Db4 wavelet transform has been used to decompose the filtered ECG signals into 4 levels of approximate and detailed coefficients from which the non-linear features are computed. The features that were extracted from the decomposed signal were Shannon entropy (2), Renyi entropy (3), Tsallis entropy (4), and the Log-Energy entropy (5), given as follows:

(i) Shannon entropy,

$$S_{en} = \sum_{i=1}^{n} p(x_i) \log_a \frac{1}{p(x_i)}, a > 1 \quad (2)$$

(ii) Renyi Entropy,

$$R_{en} = \frac{1}{(1-\alpha)} \ln_{i-1}^{n} p(x_i)^{\alpha} \quad (3)$$

(iii) Tsallis Entropy,

$$T_{en} = \frac{1}{\alpha-1}(1 - \sum_{i=1}^{n} p_i^{\alpha}) \quad (4)$$

(iv) Log-Energy Entropy,

$$L_{EEi} = \sum_{i=1}^{n} \log(s_i^2) \quad (5)$$

where S_{en} is the entropy of dataset x, $p(x_i)$ is the probability of acceptance of dataset X with values x_i, and α is the order of entropy. The α value of 3 was used for deriving the features from the wavelet coefficients.

The entropy features were derived from detailed coefficients of level 2, 3, 4, and the approximate coefficient of level 4 as these levels contain prominent ECG signal-related information [14].

4.2 Morphological Features

The morphological features utilized in this project were based on QRS peak detection, in which, RR amplitude and interval measures are determined. Based on the RR intervals, the heart beat rate was computed. However, the classification based on heart beat rate feature alone would not be suitable due to irregularity of the signal pattern of each channel even for the same subject, therefore, measures such as the standard deviation and mean of the RR interval and amplitude obtained were also obtained and used for classification. The QRS peak detection was accomplished using Pan and Tompkins algorithm [15]. The algorithm originally consist of a few parts such as assigning the fiducial marks, thresholding to distinguish noise and real peaks, search again the missed peaks along the signals, eliminate the possible multiply detected peaks within the refractory period, and finally discriminate the T-wave. As for the heart rate computation, heart rate is taken as one of the features to be used for the classification of the true and false alarms. The equation to compute the heart rate is given as in (6).

$$HeartRate\_(Beats/\min) = \frac{Number\_of\_RR\_interval}{Total\_time(minutes)} \quad (6)$$

4.3 Adaptive Synthetic Sampling Approach (ADASYN)

The 750 ECG dataset that had been used for this project consists of 456 sets of false alarm subjects and 294 sets of true alarm subjects in ratio of 3: 2, respectively. Therefore, it caused an imbalance in the dataset that might affect the learning algorithm of the classifiers. ADASYN is used to produce synthetic samples by sampling minority dataset based on the density distribution [16]. In addition, the weights of each of the minority examples were changed adaptively so that the resulting distribution of the synthetic minority samples is not skewed and generalized. Thus, the dataset has been synchronized by sampling the minority set of true alarm data up to be the same number as that of the majority set of data of false alarms. This algorithm has been performed on the non-linear and morphological features of both the channels in true alarmed signals as can be inferred from Figs. 5 and 6.

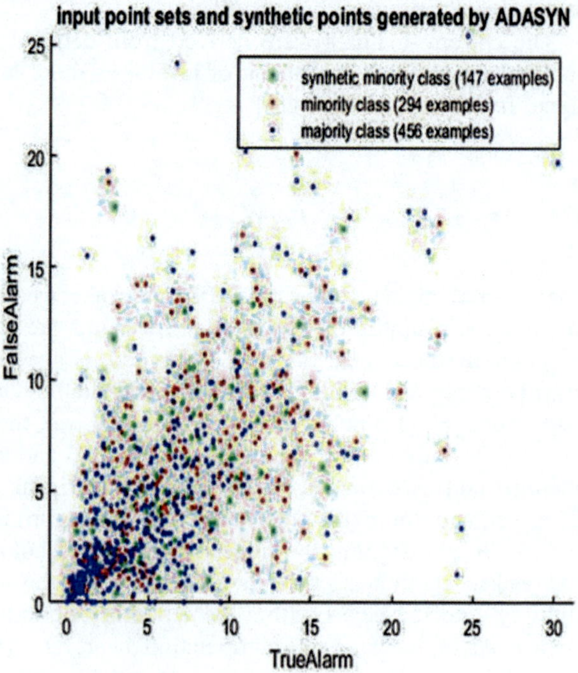

Fig. 5 The non-linear feature distribution for both false (majority dataset) and true alarm (minority dataset) after ADASYN

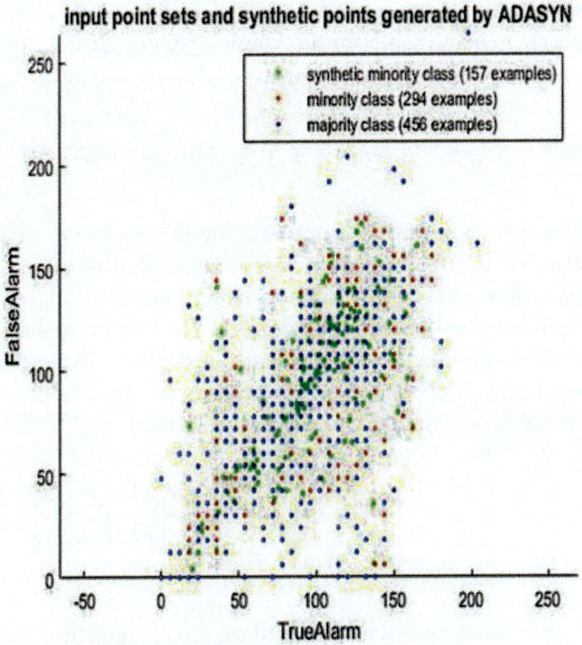

Fig. 6 The morphological feature distribution for both false (majority dataset) and true alarm (minority dataset) after ADASYN

4.4 Dimension Reduction Using Principle Component Analysis (PCA)

Statistical analysis is not adequate to identify the most significant features as it does not specify how much the features differ that eventually still results in large set of data [17]. Therefore, dimensionality reduction is used as it encompasses the process of mapping data with high dimension space into lower dimensions as a way to get rid of the data or feature set with uninformative variances for which Principle Component Analysis (PCA) was used where it finds the principle components with orthogonal linear combinations of the samples of each feature that is arranged in accordance with the largest variance [18]. The highest principle component containing sample features were determined based on Kaiser's rule, where those features with eigenvalue more than 1 are chosen as they have greater significance.

5 Classification

Fuzzy K-Nearest Neighbour (FKNN), Random Forest algorithm, and AdaBoost classifier are employed in this work to quantify the false arrhythmia alarm prediction. All the three classifiers are validated using ten-fold cross-validation method. The cross-validation method allows to determine the accuracy rate of the classifier for the multiple portion of input data in the training set with 90% confidence interval and is generalizable for the whole data [19, 20].

5.1 *Fuzzy K-Nearest Neighbour (FKNN)*

The FKNN is the advance and modified version of typical K-Nearest Neighbour (KNN). In FKNN, fuzzy class membership has been allocated to the input, which had allowed the classification being performed based on the fuzzy labels that had been assigned to the input vectors. The class membership is assigned to a vector of random sample instead of a particular class. The membership is assigned based on the function of the vector's distances in accordance with the K-nearest neighbours (Euclidean distance measure) as described in (7) [21, 22].

$$u_i(x) = \frac{\sum_{j=1}^{K} u_{ij}(1/\|x - x_j\|^{2/(m-1)})}{\sum_{j=1}^{K} (1/\|x - x_j\|^{2/(m-1)})} \tag{7}$$

where

k = samples obtained based on KNN algorithm.

u_{ij} = membership in the ith class and jth vector.

$u_i(x)$ = The assigned membership of the vector x.

m = the weightage of the distance between the neighbours that contribute to the membership.

The membership of the vector x is being influenced by the inverse of distance from nearest neighbour and their membership classes. The weights assigned to the vector's membership are greater if the vector distances are closer to each other or else lesser weights will be assigned to the membership if the vectors are far apart from each other due to the inverse relationship between the weightage assignment and the distance between the vector. The variable m on the other hand is the weightage assigned to the distance between the neighbours, when the contribution of each neighbour to the membership is computed. As the value of m increases, the neighbours are weighted evenly and their relative distance from the points classified have lesser effect. However when m approaches 1, it will result in the closer neighbours being weighted heavily relative to those farther, that will eventually lead to reduce in the number of points that contribute to the membership value of the point being classified. Thus $m = 2$ has been used in the application of FKNN in this study.

5.2 *Random Forest Classifier*

Random forest classifier is ensemble learning method that actually consists of numbers of tree-structured classifiers where every tree actually casts votes to the most dominant classes [9]. The trees in this algorithm have grown based on the random input combination during the training and 10 cross validation [6]. The Random forest classifier used in this project was tested using 100, 300, and 500 trees. When the number of trees increased from 100 to 300, the accuracy was observed to be increased as well. However, further increase in the number of trees beyond 500 had not given any significance on the accuracy of the classifier, other than making the computation more complex and hinders the development of real-time applications. Therefore, 300 trees were used for the execution of the algorithm for the study.

5.3 *AdaBoost Classifier*

This classifier is a boosting algorithm with ensemble learning method. This classifier basically functions by modifying the weights for voting on the training data on each iteration. The iteration is the number of times the classifier attempts to perform the classification process. AdaBoost assigns heavier weights to misclassified sets and reduced weights to correctly classified signals. In this project, the AdaBoost with iteration of 500 and the Learn rate of 0.1 was used.

6 Results and Discussion

Table 1 shows the true positive rate (TP), true negative rate (TN), and the percentage of accuracy of all the classifiers for non-linear feature of all the three categories of feature inputting.

Table 2 shows the true positive rate (TP), true negative rate (TN), and the percentage of accuracy of all the classifiers for morphological feature of all the three categories of feature inputting.

Table 3 shows the true positive rate (TP), true negative rate (TN), and the percentage of accuracy of all the classifier for both types of features of all the three categories of feature inputting. The highlighted region indicates the highest accuracy achieved by each classifier.

Table 1 Accuracy of all the classifiers for non-linear features

Non-linear features									
	Raw features			PCA			PCA and ADASYN		
Classifier	TN	TP	ACC	TN	TP	ACC	TN	TP	ACC
FKNN	73.46	55.10	**66.27**	86.67	56.67	**74.67**	82.61	83.33	**82.89**
Random forest (300)	88.10	54.57	**74.89**	89.20	60.38	**75.92**	85.50	80.14	**81.63**
Adaboost	86.34	56.51	**74.62**	82.91	64.91	**75.87**	83.42	80.40	**81.63**

Table 2 Accuracy of all the classifiers for morphological features

Non-linear features

	Raw features			PCA			PCA and ADASYN		
Classifier	TN	TP	ACC	TN	TP	ACC	TN	TP	ACC
FKNN	69.74	53.74	**63.47**	73.33	55.17	**66.22**	78.26	68.97	**74.67**
Random Forest (300)	83.57	40.63	**66.67**	71.86	55.41	**67.50**	74.72	64.43	**69.55**
Adaboost	77.68	41.45	**63.48**	65.39	46.79	**58.12**	63.83	64.71	**62.65**

Table 3 Accuracy of all the classifiers for combined feature

Non-linear features

	Raw features			PCA			PCA and ADASYN		
Classifier	TN	TP	ACC	TN	TP	ACC	TN	TP	ACC
FKNN	74.12	56.46	**67.20**	75.56	62.07	**70.27**	86.67	82.76	**85.14**
Random Forest (300)	87.46	57.43	**75.59**	85.77	71.37	**76.28**	77.64	84.25	**81.94**
Adaboost	87.89	57.47	**75.95**	84.12	60.93	**75.06**	79.04	82.37	**80.77**

From the tables, it can be observed that for FKNN, the highest accuracy of 85.14% was achieved using consolidation of both the non-linear and morphological features compared to 82.89% achieved using non-linear features and 74.67% of accuracy achieved using morphological features alone, respectively, in PCA and ADASYN combinations. Generally, the false arrhythmia prediction accuracy for the morphological features in Table 2 are seen lower than the predictive accuracies for non-linear features in Table 1 for any combination of methods. This observation elucidates that when examined independently, the non-linear entropies derived from DWT were able to effectively capture much of the information from the ECG signals compared to the morphological features. However, when both of these features are combined, increase of predictive rate is observed for most of the combinations of method tested across the three classifiers. This means that although morphological features provide lesser accuracies when used alone, they provide critical observation of the ECG signals that are lacking in the non-linear entropies method, and when used in unison, they can contribute to better performance of classifiers.

The use of PCA method to dimensionally reduce the features was observed to be well supported in FKNN and random forest classifiers. This is because, both the classifiers were observed to have improved in performance when the dominant principal components were used for the classifications. However, for AdaBoost classifier, the same combination is seen to reduce its effectiveness as can be seen from Tables 2 and 3 where the accuracies drop from 63.48 to 58.12% for morphological features and from 75.95 to 75.06% in consolidated feature combinations. This demonstrates that the AdaBoost classifier is not compatible with PCA-based approach for the features tested.

For the use of ADASYN method to overcome the imbalance dataset problem, the approach seems to improve the recognition rate of all the classifiers. This might be due to the fact that the use of ADASYN results in more synthetic true alarm data generated to balance the dataset, therefore, providing more training data to the classifiers. This in turn results in better training accuracy for the classifier and better predictive rate. Therefore, from the analysis, it can be observed that FKNN is the most suitable classifier out of the three classifiers tested, and the use of PCA and ADASYN strategies are able to boost the performance of the predictive algorithm.

The comparison of the current research findings are difficult to be compared to the existing studies due to the difference in nature of study, selection of data, and methods employed. In 2018, Lehman et al. studied the false arrhythmia alarms in ICU using the Physionet 2015 computing in cardiology challenge database [1]. A subset of the available data of 562 samples containing v-tach alarms was selected for their study. Utilizing Supervised Denoising Autoencoder (SDAE) to learn non-linear embeddings of beat-by-beat FFT-transformed ECG, they reported area under curve of 0.91 with 77.59% of challenge score. Studies conducted by Srivastava et al. in 2019 utilizing the same dataset were able to achieve overall accuracy of 83.96% using multi-modal signals approach [23]. They have utilized electrocardiogram (ECG), arterial blood pressure (ABP), and photo-plethysmogram (PPG) signals-based features for the analysis. Although these studies have reported good accuracies for the detection of false arrhythmia alarms, the proposed method utilizing all the 750 samples from the available training data was able to achieve 85.14% using parameters derived from only the ECG signals. The ability of the proposed method in identifying specific arrhythmia failures could be an interesting topic to be investigated in future.

7 Conclusion

From the result of the classification, it can be concluded that the feature being combined as in non-linear together with that of morphological features gives sufficiently high recognition rate. The high accuracy was also being achieved through the dimension reduction by PCA and balancing the dataset procedure of ADASYN. Therefore, the proposed method is able to effectively predict the false arrhythmia alarms in ICU and enables the medical care providers to concentrate on the truly critical patients. This can lead to better patient management and reduce the fatigue experienced by the caregivers due to inaccurate alarms in intensive care units. In future, advanced classification algorithms could also be tried for classification such as the AdaBoost, SVM, Bayer's Optimization classifier, and so on to ensure an accurate and at the same time a faster computation capability.

References

1. E.P. Lehman, R.G. Krishnan, X. Zhao, R.G. Mark, L.H. Lehman, Representation learning approaches to detect false arrhythmia alarms from ECG dynamics. Proc. Mach. Learn. Res. **85**, 571–586 (2018)
2. F. Michard, R. Bellomo, A. Taenzer, The rise of ward monitoring: opportunities and challenges for critical care specialists. Intensive Care Med. **45**(5), 671–673 (2018)
3. V. Kalidas, L.S. Tamil, Cardiac arrhythmia classification using multi-modal signal analysis. Physiol. Meas. 1253–1272 (2016)
4. S. Ansari, A. Belle, H. Ghanbari, M. Salamango, K. Najarian, Suppression of false arrhythmia alarms in the ICU: a machine learning approach. Physiol. Meas. 1186–1203 (2016)
5. R. Vullings, R.M. Aarts, L.M. Eerikäinen, J. Vanschoren, M.J. Rooijakkers, Reduction of false arrhythmia alarms using signal selection and machine learning. Physiol. Meas. **37**(8), 1204–1216 (2016)
6. R. Rodrigues, P. Couto, Detection of false arrhythmia alarms with emphasis on ventricular tachycardia. Physiol. Meas. **37**(8), 1326–1339 (2016)
7. A.B. Patwary, M.T.I. Chowdhury N. Mamun, Comparison among ECG filtering methods for non-linear noise, in *2018 International Conference on Advancement in Electrical and Electronic Engineering (ICAEEE)*, Gazipur, Bangladesh (2018), pp. 1–4. https://doi.org/10.1109/ICAEEE.2018.8642991.
8. N. Sadr, J. Huvanandana, D.T. Nguyen, C. Kalra, A. McEwan, P. de Chazal, Reducing false arrhythmia alarms in the ICU using multimodal signals and robust QRS detection. Physiol. Meas. **37**(8), 1340 (2016)
9. C.H. Antink, S. Leonhardt, M. Walter, Reducing false alarms in the ICU by quantifying self-similarity of multimodal biosignals. Physiol. Meas. **37**(8), 1233–1252 (2016)
10. S. Fallet et al., False arrhythmia alarms reduction in the intensive care unit: a multimodal approach. Physiol. Meas. 1217–1232 (2016)
11. S. Mousavi, A. Fotoohinasab, F. Afghah, Single-modal and multi-modal false arrhythmia alarm reduction using attention-based convolutional and recurrent neural networks. PLoS ONE **15**(1), e0226990 (2020)
12. X. Cao, *Detrending and Denoising of Impedance Cytometry Data* (2015), pp. 17–21
13. Y.V. Parkale, S.L. Nalbalwar, Investigation on Daubechies Wavelet-based compressed sensing matrices for ECG compression, in *Computing, Communication and Signal Processing* ed. by B. Iyer, S. Nalbalwar, N. Pathak. Advances in Intelligent Systems and Computing, vol. 810 (2019)
14. R.J. Martis, C. Chakraborty, A.K. Ray, Mach. Learn. Healthcare Inform. **56**, 1–23 (2014)
15. M.A.Z. Fariha, R. Ikeura, S. Hayakawa, S. Tsutsumi, Analysis of Pan-Tompkins algorithm performance with noisy ECG signals. J. Phys: Conf. Ser. **1532**, 012–022 (2020)
16. N. Naghavi, S. Borhani, E. Wade, Improving machine learning based detection of freezing of gait using data synthesis methods, in *18th IEEE International Conference On Machine Learning And Applications (ICMLA)*, Boca Raton, FL, USA (2019), pp. 1432–1437. https://doi.org/10.1109/ICMLA.2019.00234
17. K. Rajeswari, R. Garud, V. Vaithiyanathan, Improving efficiency of classification using PCA and apriori based attribute selection technique. Res. J. Appl. Sci. Eng. Technol. **6**(24), 4681–4684 (2013)
18. H. Telgaonkar Archana, Dimensionality reduction and classification through PCA and LDA. Int. J. Comput. Appl. **122**(17), 4–8 (2015)
19. V. Vijean, M. Hariharan, S. Yaacob, M.N.B. Sulaiman, A.H. Adom, Objective investigation of vision impairments using single trial pattern reversal visually evoked potentials. Comput. Electr. Eng. **39**(5), 1549–1560 (2013)
20. Q.W. Oung, S.N. Basah, H. Muthusamy, V. Vijean, H. Lee, W. Khairunizam, S.A. Bakar, Z.M. Razlan, Z. Ibrahim, Objective evaluation of freezing of gait in patients with Parkinson's disease through machine learning approaches, in *2018 International Conference on Computational Approach in Smart Systems Design and Applications, ICASSDA* (IEEE, 2018), pp. 1–7

21. Q.W. Oung, H. Muthusamy, S.N. Basah, H. Lee, V. Vijean, Empirical wavelet transform based features for classification of Parkinson's disease severity. J. Med. Syst. **42**(2), 29 (2018)
22. I. Ben Fredj K. Ouni, Comparison of crisp and fuzzy kNN in phoneme recognition, in *2017 International Conference on Advanced Systems and Electric Technologies (IC_ASET)*, Hammamet (2017), pp. 118–122. https://doi.org/10.1109/ASET.2017.7983676.
23. C. Srivastava, S. Sharma, A. Jalali, A novel algorithm for reducing false arrhythmia alarms in intensive care units, in *38th Annual International Conference of the IEEE Engineering in Medicine and Biology Society (EMBC)*, Orlando, FL (2016), pp. 2525–2528. https://doi.org/10.1109/EMBC.2016.7591244

Human Recognition Using Omni-Directional Camera for Service Robot Application

A. H. Ismail, M. N. Ayob, H. B. Hassrizal, M. S. Muhamad Azmi, M. S. M. Hashim, S. M. Othman, S. N. A. M. Kanafiah, M. K. Ali Hassan, and I. I. Ibrahim

Abstract Service robots have received increasing attention in mobile robotics field with the recent advancement in technology and computing power. One of the important aspects of service robots is the vision recognition system. This system is crucial for the robot to navigate through the surrounding and detect the objects such as obstacles and targets. Most of the recent service robots use costly sensors such as LIDAR and high-end vision system in order to recognize its environment. This paper proposes to use a cost-effective sensor which is an omni-directional camera to obtain a full 360° field of view (FOV) of the surrounding environment. The images from the camera were then used to identify the location of the human in any direction from the robot. From the results, the optimal method to unwrap the omni-directional image and recognition of human is promising. In addition, the direction and heading of the robot is also configurable in future works using the proposed method.

Keywords Omni-directional image · Service robot · Image processing · Mobile robot

1 Introduction

Human detection for service robot application has gained significant research attention over the past decades. Many works related to this area focus on, for instance, assistive service robots either in public spaces [1, 2] or industry [3], and surveillance robots [4, 5]. The critical feature enabling this function uses the camera, in which

A. H. Ismail (✉) · M. N. Ayob · H. B. Hassrizal · S. M. Othman · S. N. A. M. Kanafiah · M. K. Ali Hassan · I. I. Ibrahim
Intelligent Computing and System Informatics Research Group (ICSI), Faculty of Electrical Engineering Technology, Universiti Malaysia Perlis (UniMAP), Perlis, Malaysia
e-mail: ihalim@unimap.edu.my

M. S. Muhamad Azmi · M. S. M. Hashim
Faculty of Mechanical Engineering Technology, Universiti Malaysia Perlis (UniMAP), Perlis, Malaysia

F. Thakkar et al. (eds.), *Proceedings of the International e-Conference on Intelligent Systems and Signal Processing*, Advances in Intelligent Systems and Computing 1370,
https://doi.org/10.1007/978-981-16-2123-9_24

the robot employed image processing techniques to recognize the surrounding environment and related objects such as obstacles, neighboring objects as well as nearby humans. Human Recognition is quite essential in the case of service robots since the main goals of the service robot are to help and assists us human daily [6, 7]. This paper explores the use of the omni-directional camera with image processing algorithm for service robot application.

The omni-directional camera has more advantages where one omni-directional camera is similar to more than three traditional cameras, to cover an entire surveillance area with 360° field of view (FOV) [8, 9]. A patent described in [10] shows the architecture of an omni-directional camera. Omni-directional camera is a modification of a standard camera where a spherical mirror is put on top of the camera view as depicted in Fig. 1. A modification to the static camera (non-rotating) by adding a spherical mirror on top of the camera will result in a full 360° FOV which is very useful to image the environment, particularly for mobile robot use.

In this research, the omni-directional camera was put on top of the robot, which yields an image with 360° view environment in the robot's surrounding. Figure 2 shows the placement of the camera on top of the robot. The omni-directional image obtained from this camera showing the useful information of the environment such as the hallway, wall, and the object of interest in the works, i.e., a human.

Fig. 1 The architecture of an omni-directional camera [10]

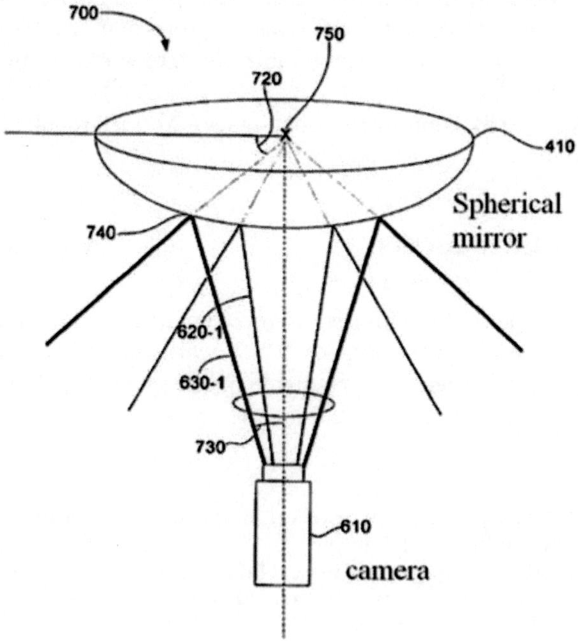

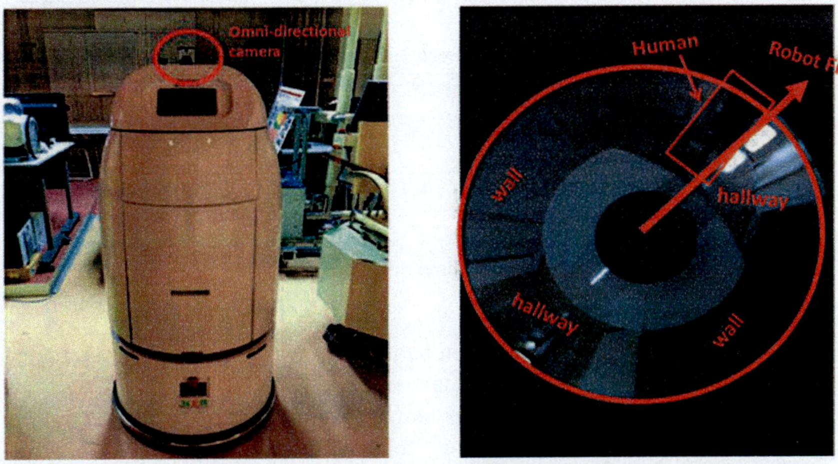

Fig. 2 Placement of omni-directional camera on top of the robot (left), omni-directional image, and possible surrounding environment of the robot obtained from the camera (right)

2 Methodology

Figure 3 shows the flowchart in this works. These preliminary works are done offline, i.e., all the images are sampled concerning the robot static position while the human move around the robot. After the omni-directional photos have been obtained, the image processing and analysis is then conducted. As the obtained omni-directional image is in polar form, it is essential to convert the omni-directional image into its corresponding panoramic format to extract any useful information.

Figure 4, on the other hand, depicts the conversion from polar into panoramic form to all parameters required for the conversion, known as image unwrapping. In this paper, we modified the image unwrapping technique in [11] for our use which is much faster and effective.

One of the important parameters in this works is the determination of the concentric circles. This is because the omni-directional image yield in the outer ring, denoted

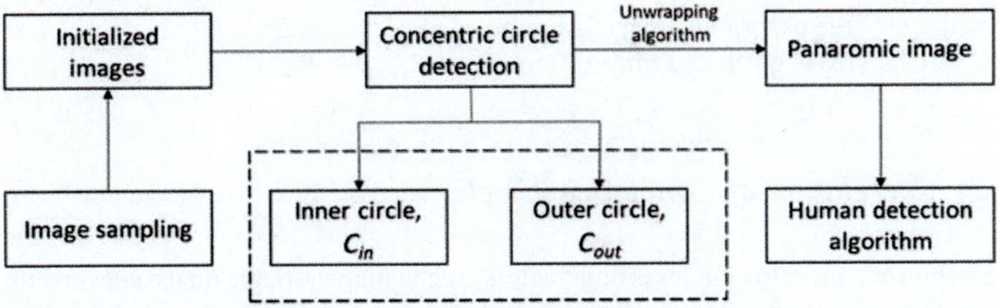

Fig. 3 The methodological flowchart in this works

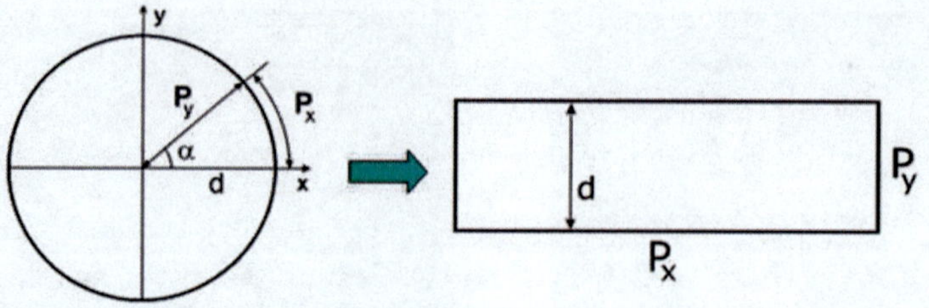

Fig. 4 Panoramic image conversion from omni-directional images [11]

as C_{out}. As the robot head is also a circle in nature, then we also have the parameter of inner circle indicated as C_{in}. The graphical view of C_{in} and C_{out} is shown in a later section. Since the image processing results in matric computation which proportion with image resolution, then the concentric circles will also change concerning the change of omni-directional image resolution. We analyze this change by using the linear correlation method.

Finally, the human in the robot FOV is detected by using typical image processing techniques such as color to grayscale conversion, thresholding, and image scene analysis.

2.1 *Image Unwrapping Algorithm*

Image unwrapping is a method to convert the omni-image into its panoramic equivalent image. This conversion is given by Eq. (1).

$$
\begin{aligned}
x_m &= x_c + \left[\cos\left(\frac{x}{2\pi\, R_{outer}} + \beta\right) \times (R_{inner} + y)\right] \\
y_m &= y_c + \left[\cos\left(\frac{x}{2\pi\, R_{outer}} + \beta\right) \times (R_{inner} + y)\right]
\end{aligned} \tag{1}
$$

where x_m and y_m are the panoramic pixel map, x_c and y_c are the center pixels R_{inner} and R_{outer} are the radius of concentric circles C_{in} and C_{out} and β is the bias offset. The horizontal size of the panoramic image is computed as perimeter $2\pi \times (R_{outer} - R_{inner})$, and vertical size is given by radius difference of $R_{outer} - R_{inner}$.

2.2 *Correlation of Concentric Circles*

Detection of concentric circles is nonetheless crucial for panoramic image conversion from omni-directional image. Figure 5 shows the concentric circles of the omni-directional image obtained in this works from the top of the mobile robot head.

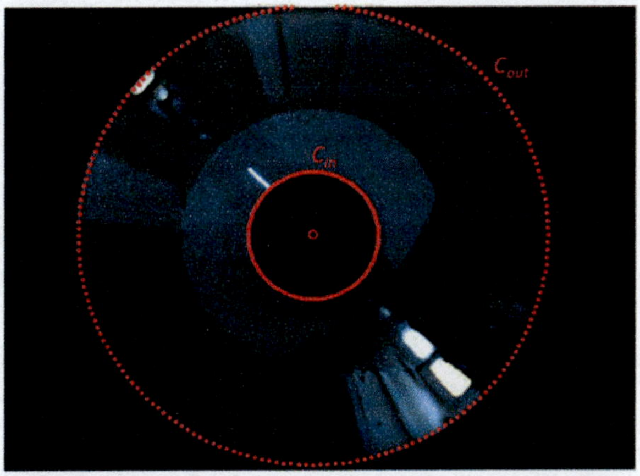

Fig. 5 The concentric circles in the omni-directional image

However, the larger the image resolution means more time for computation. Image processing algorithms such as background estimation and subtraction use a lot of computational memory and processing time. Thus, limiting the system in terms of real-time (or near real-time).

In this works, we employed a linear regression method to correlate the concentric circles with the image resolution. A curve fitting technique is used to develop a generalized linear function between the resizing factor corresponding to image resolution and the radii of the concentric circle.

2.3 *Human Detection Algorithm*

The human is detected by using a typical approach in image processing method. Figure 6 shows the algorithm of the human detection algorithm employed in this project, which is started with panoramic image conversion. Then, the background at image n is estimated by using a subtraction function from later image $n + 1$, as described in Eq. (2) below.

$$b_{est_n}(x, y, \text{dim}) = \max(im_n(x, y, dim), im_{n+1}(x, y, dim)) \quad (2)$$

The process continued with color to grayscale conversion using a weighted function of RGB color components before applying a histogram-based adaptive thresholding method delineated in [12] to obtain the binary image. A simple de-noising filter is then used to clean the image, resulting in an exact human feature particularly suitable for feature extraction and selection method. Finally, the location of the robot as the input for the mobile robot can be computed using trigonometric functions.

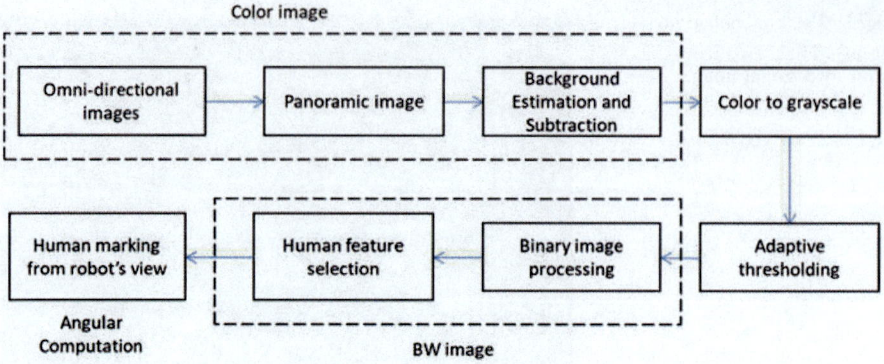

Fig. 6 Human detection algorithm proposed in this paper

3 Result and Discussion

Figure 7 shows the effects of image resolution and the fixed value of radii of the concentric circles. As one could be observed, the larger the image resolution for the resizing factor, the problematic the issues became. Therefore, a correlation between these parameters is required. A linear graph depicted in Fig. 8 shows the relation of the curve fitting technique. With 95% confident interval, the coefficients of determination

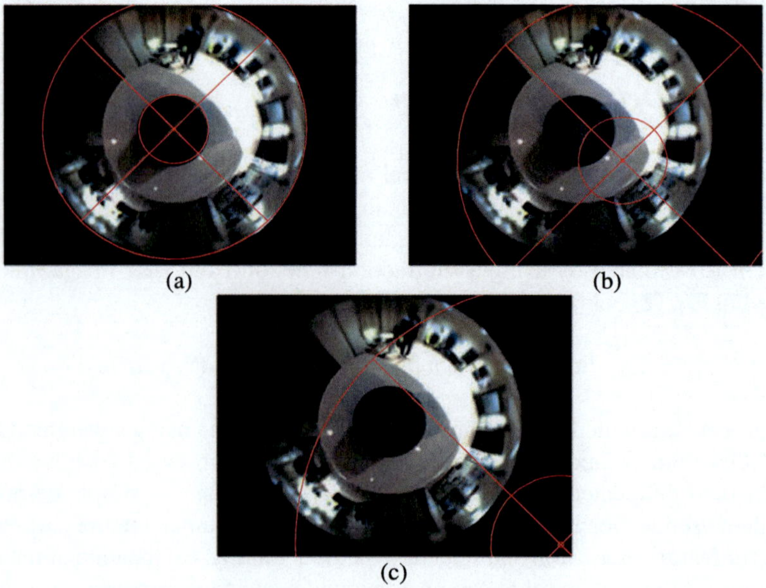

Fig. 7 The effect of resizing factor toward omni camera parameters, **a** Resizing factor = 1.0, **b** resizing factor = 0.8, and **c** resizing factor = 0.5

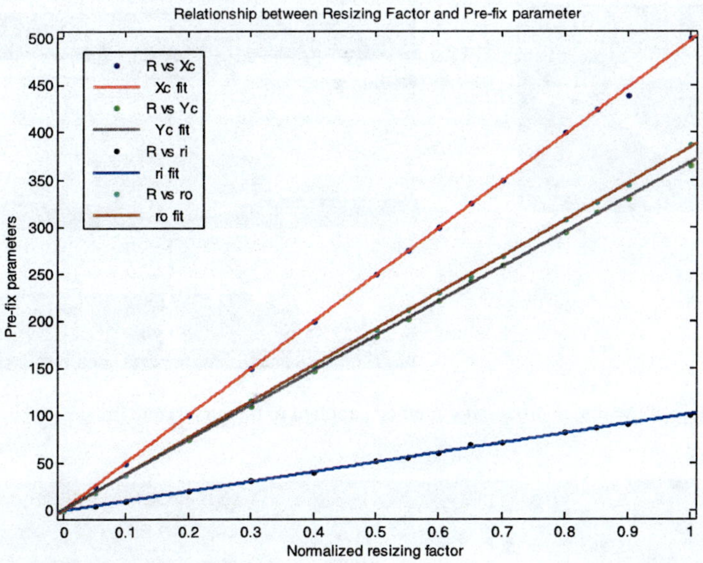

Fig. 8 Relationship between resizing factors and omni-directional image parameters

R^2 ranging from 99.91% to 99.98%. The resulting linear equation is shown in Eq. (3).

$$\begin{bmatrix} x_c \\ y_c \\ r_i \\ r_o \end{bmatrix} = \begin{bmatrix} 497.3 \\ 367.5 \\ 102.0 \\ 385.1 \end{bmatrix} R + \begin{bmatrix} 0.7084 \\ 0.9485 \\ 0.0377 \\ -0.3799 \end{bmatrix} \tag{3}$$

Figure 9 shows the image unwrapping result using the method discussed earlier. The concentric circles have been successfully identified even if a different image resolution is used. And more importantly, unnecessary information such as the 'black' robot head before C_{in} has been removed. Figure 10, on the other hand, depicts some of the image results obtained from the image processing procedures such as color conversion, image complement and the binary image after applying adaptive thresholding algorithm. Figure 11 shows the detected objects and the human of interest by using the simple recognition. With too many noises and unwanted items,

Fig. 9 The result of image unwrapping from omni-directional to a panoramic image

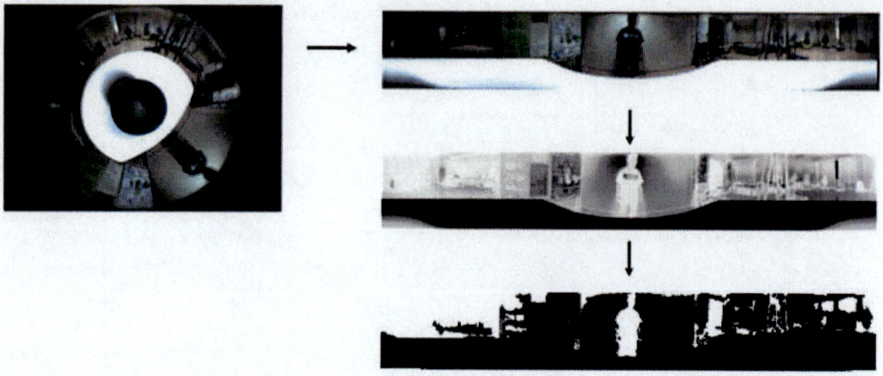

Fig. 10 Some of the image processing methods applied to the panoramic image

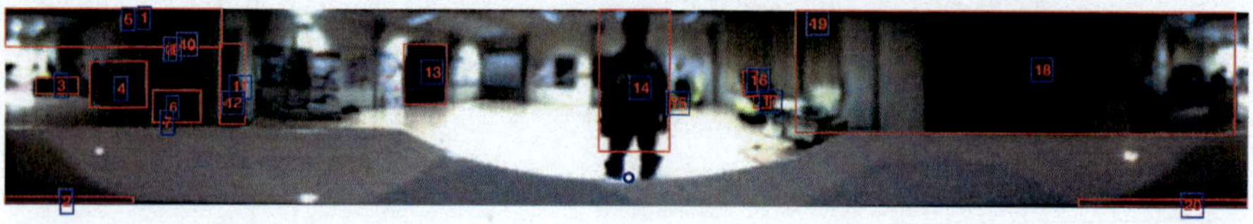

Fig. 11 Detected human and objects from the panoramic image

the human feature is needed for extraction. It is common to use the human body shape as a feature descriptor. By analyzing this descriptor, the human can be easily detected, which is shown in Fig. 12.

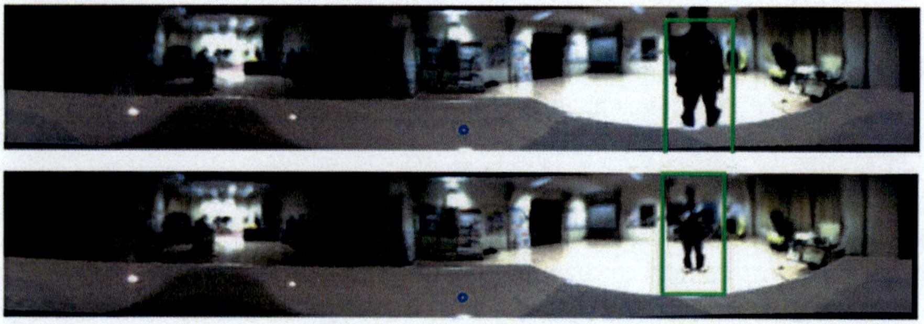

Fig. 12 Human detected in the image; closer distance (top) and far distance (bottom)

4 Conclusion and Future Works

This paper has proposed a new approach for human detection for a service robot application using an omni-directional camera resulting in 360° FOV of omni-directional images. The images were converted into panoramic images using a modified conversion function by introducing the concentric circles' approach, and the human detection using appropriate image processing method.

The human in the robot FOV was identified, and the noises surrounding has been suppressed by using a human feature descriptor. Next, we plan for the dynamic motion of the human and also the dynamic motion of the service robot itself and observe its detection rate.

Acknowledgements The authors would like to acknowledge the support from the Fundamental Research Grant Scheme (FRGS) under a grant number of FRGS/1/2018/ICT05/UNIMAP/02/4 from the Ministry of Education Malaysia.

References

1. F. Bahrmann, et al., Towards an all-day assignment of a mobile service Robot for elderly care homes. Am. J. Nursing **9**(5), 329–337 (2020)
2. C.J. Lai, C.-P. Tsai, Design of introducing service robot into catering services, in *Proceedings of the 2018 International Conference on Service Robotics Technologies* (2018)
3. S. Ivanov, C. Webster, What should robots do? A comparative analysis of industry professionals, educators and tourists, in *Information and Communication Technologies in Tourism 2019* (Springer, Cham, 2019), pp. 249–262
4. A.U. Bokade, V.R. Ratnaparkhe, Video surveillance robot control using smartphone and Raspberry pi, in 2016 International Conference on Communication and Signal Processing (ICCSP), IEEE (2016)
5. M. Ashokkumar, T. Thirumurugan. Integrated iot based design and android operated multipurpose field surveillance robot for military use, in *International Conference for Phoenixes on Emerging Current Trends in Engineering and Management (PECTEAM 2018)* (Atlantis Press, 2018)
6. H.K. Ragb, V.K. Asari, A feature fusion strategy for human detection in omnidirectional camera imagery. Electron. Imaging **375**(1) (2018)
7. T. Shimizu, K. Kobayashi, Development of a person-searching algorithm using an omnidirectional camera and LiDAR for the Tsukuba challenge, in *2018 57th Annual Conference of the Society of Instrument and Control Engineers of Japan (SICE)* (IEEE, 2018)
8. M. Hirabayashi, et al., Flying object detection system using an omnidirectional camera. Forensic Science International: Digital Investigation **35** (2020)
9. X. Gong, et al., High-precision calibration of omnidirectional camera using an iterative method. IEEE Access **7**, 152179–152186 (2019)
10. J.Z. Geng, Method and apparatus for omnidirectional three dimensional imaging, United States Patent 6744569 (2004)
11. I. Potucek, Omni-Directional image processing for human detection and tracking, PhD Thesis, Brno University of Technology, Czech Republic (2004)
12. A.H. Ismail, M.H. Marhaban, A simple approach to determine the best threshold value for automatic image thresholding, in *IEEE Inernational Conference on Signal and Image Processing Applications* (IEEE, 2009)

13. M. Tamura, S. Horiguchi, T. Murakami, Omnidirectional pedestrian detection by rotation invariant training, in *2019 IEEE Winter Conference on Applications of Computer Vision (WACV)* (IEEE, 2019)

Recent Trends in WDM Passive Optical Network for Smart IoT-Based Applications

Shailesh Khant and Atul Patel

Abstract Modern era is the era of Artificial intelligence and IoT applications. Almost every field involves the concept of IoT and AI. In this paper, we have presented a survey based on concept where optical network is used with wireless network to develop an IoT application. There are many types of access networks available but here passive optical network (PON) is used because it is having higher efficiency and reduced cost compared to active optical network (AON) due to simple configuration of its components, PON becomes very popular nowadays. Wavelength division multiplexing (WDM) technique with PON is discussed here and how it is connected with IoT is presented by various means. If optimization of various nodes is used in IoT over PON architecture, then it is also possible to reduce power consumption and optimization is achieved in all available resources.

Keywords Passive optical network · Wavelength division multiplexing (WDM) · Time-division multiplexing (TDM) · Access network · Internet of things (IoT) · Free space optics (FSO)

1 Introduction

1.1 Optical Network and IoT

Telecommunication industry now requires the features like higher bandwidth, excellent bitrate, lower BER, Higher Q-factor, and efficient power budget. The demand for bitrate is now shifted from several hundreds of Mbps to several Gbps. In conventional network, it is not possible to get these advantages. Using Optical Network, it

S. Khant (✉) · A. Patel
Smt. Chandaben Mohanbhai Patel Institute of Computer Applications, Charusat University, Changa, India
e-mail: shaileshkhant.mca@charusat.ac.in

A. Patel
e-mail: atulpatel.mca@charusat.ac.in

F. Thakkar et al. (eds.), *Proceedings of the International e-Conference on Intelligent Systems and Signal Processing*, Advances in Intelligent Systems and Computing 1370,
https://doi.org/10.1007/978-981-16-2123-9_25

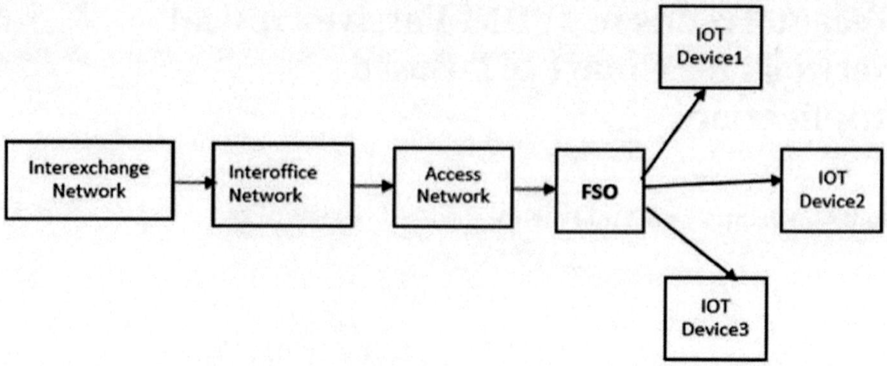

Fig. 1 Optical network with IoT devices

is possible to achieve lower signal attenuation, higher speed, higher bitrate, Lower BER, Higher quality factor, lower interference, and lower distortion.

An optical network can be classified into three major categories, Interexchange network, interoffice network, and Access network. The main type which is involved in IoT-based application is Access network. It is preferred for shorter distance. It can be made wireless by using component which are compatible with free space like free space optics (FSO). FSO should have lower power consumption and efficient conversion from O-E-O transfer. During this transfer, regeneration, reshaping, and retiming must be proper. Various O-E-O convertors are available with Optical Network. Free space communication is the major type which is used by IoT application based on optical network [1].

The basic structure of how IoT devices are connected in optical access network is shown in Fig. 1. Interexchange network is also termed as Long haul network which connects various towns or regions. Interoffice network is also termed as Metropolitan network which connects inter departments or inter offices in particular city. Optical Access network is the major area where recent research is focused. It creates direct connections between network and IoT devices through wire or wireless link. As a wireless link generally FSO is preferred if the communication is in light form else RF link is used to communicate between optical access network and IoT devices. Nowadays almost all researchers focus on access network to enhance the efficiency, power requirements, power budget, cost effectiveness, errors, and rate of data transfer [2].

How various IoT applications are connected with optical access network and major areas where IoT applications are growing nowadays is shown in Fig. 2. Almost each and every sectors use the IoT applications [3]. These sectors are shown above but it is not limited to these areas only. The major growth is in industrial applications and residential applications. It can also be used to provide safety and security in these areas.

IoT technology is based on four different layers which creates IoT stack. These four layers are IoT devices, IoT gateways, IoT Clouds, and IoT Analytics [3]. IoT

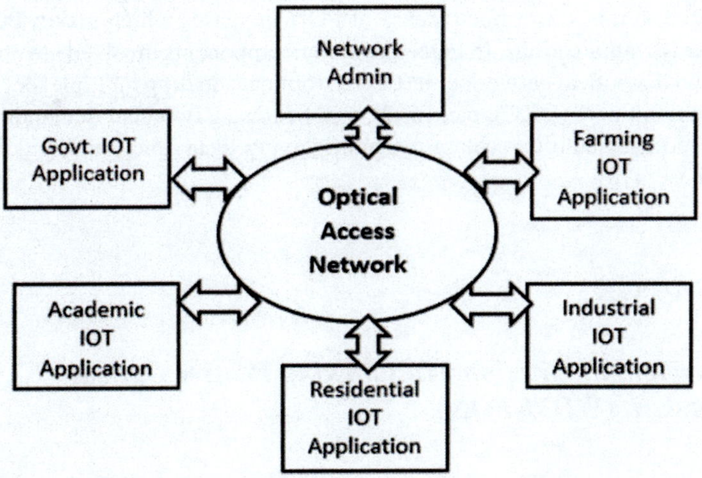

Fig. 2 Optical access network with IoT applications

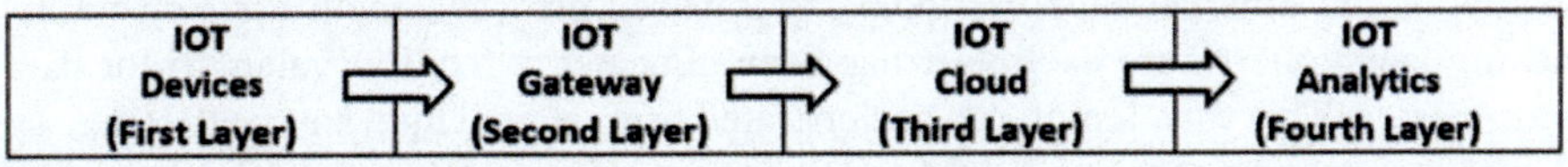

Fig. 3 IoT technology layers

devices layers include various sensing devices which can be power efficient. It can have capability to sense, compute, and communicate the sensed data. IoT gateway layers capture the sensed data and send it to connected network. In some literature, extra layer is mentioned named as IoT relay layer which can connect non-compatible device with gateway using relay devices. IoT cloud layer includes cloud-hosted server which can read the data from network, stores data for future use, and process the raw data for analysis purpose. IoT analytics layer collects data from cloud and converts this data into appropriate actions. Data analytics can be used in almost all sectors for processing the collected raw data and based on analysis it can perform the desired action. Figure 3 shows various layers involved in IoT. PON network concept can be introduced at IoT gateway layer where the device is connected with network [4, 5].

1.2 Passive Optical Network

To achieve the features like higher bandwidth, excellent bitrate, lower BER with cost effectiveness criteria, there are two types of optical networks which can be designed, active optical network (AON) and passive optical network (PON). AON includes active components like optical amplifiers while PON does not involve any active devices. Due to this, cost effectiveness of PON increases which makes PON more suitable for IoT applications. In PON, the major components involved are optical line terminal (OLT), optical network unit (ONU), (optical add drop multiplexer (OADM), optical cross connect (OXC), etc. [6, 7]. PON can satisfy recent demands in speed by its powerful capability, scalability, multiplexing techniques, and wide range of supported users [8].

2 WDM-PON

2.1 Wavelength Division Multiplexing Passive Optical Network (WDM-PON)

Number of types of PON networks are available based on techniques used like wavelength division multiplexing (WDM-PON), time-division multiplexing (TDM-PON), broadband PON (B-PON), A-PON (ATM PON), Gigabit Ethernet PON (G-PON), and ethernet PON (E-PON). It is based on wavelength division multiplexing technique where each user has been allocated a fixed wavelength for data transmission. This wavelength can be generated from optical light source [9] (Fig. 4).

The major component of WDM-PON system is its wavelength multiplexer and wavelength de-multiplexer. Wavelength multiplexer is the component which is used to combine all available wavelength in single optical wave. Fiber gratings or AWG is used for this purpose. It is located immediately after OLT and connects it with

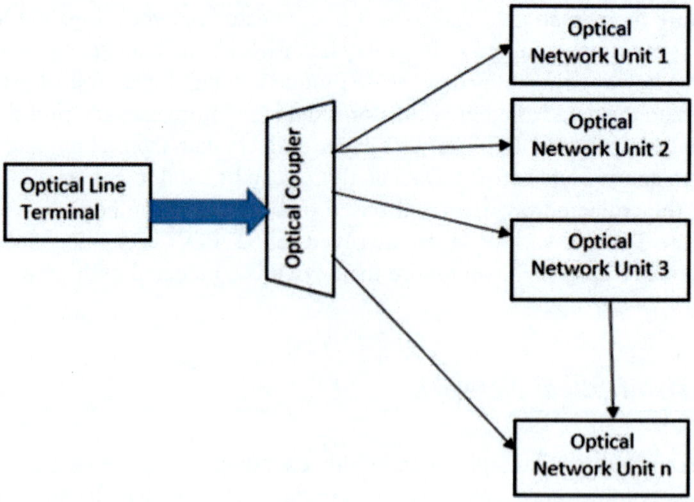

Fig. 4 PON block diagram

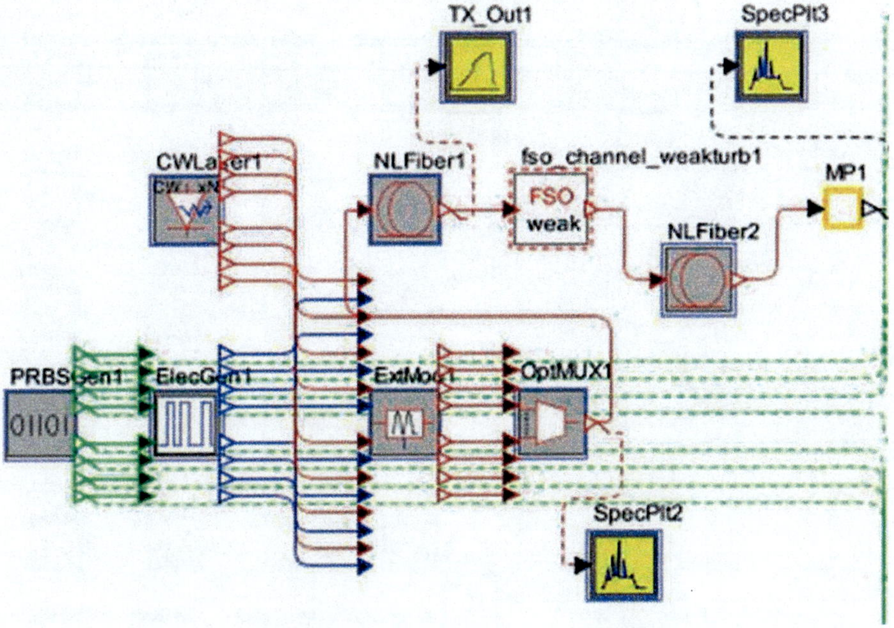

Fig. 5 WDM-PON OLT block diagram [11]

channel. Here, channel used can be fiber optics (wired) or FSO (wireless). Wavelength de-multiplexer is another important component of WDM-PON system [10].

WDM-PON OLT block diagram is shown in Fig. 5 where 8 IoT users are communicating with system. Data can be made bidirectional for both sides of communication in IoT system. Incoming data from central office is digital in network which is converted to electrical domain. This converted data is intensity modulated with LASER signal at specific wavelengths. For all users, different wavelength modulations are used. These incoming optical signals are multiplexed together using WDM technology where each IoT user has been assigned a separate wavelength for data communication. This multiplexed signal has been transmitted to Optical receiver through optical fiber channel and then to specific IoT user. IoT user response can be measured through various types of analysis where bit error rate can be measured with respect to optical received power, electrical power, bit rate or no of users, Signal can also be analyzed based on noise power and distortion. Eye diagram is also used to measure signal strength of digital signal and noise or distortion (Fig. 6).

WDM-PON wireless system is shown in Fig. 7. FSO is the basic building simulated block used for wireless channel simulation, and it can be characterized by the following equations [12].

$$\propto_{Geo} = 10\log\frac{\pi\theta^2 s^2}{4A_E} \tag{1}$$

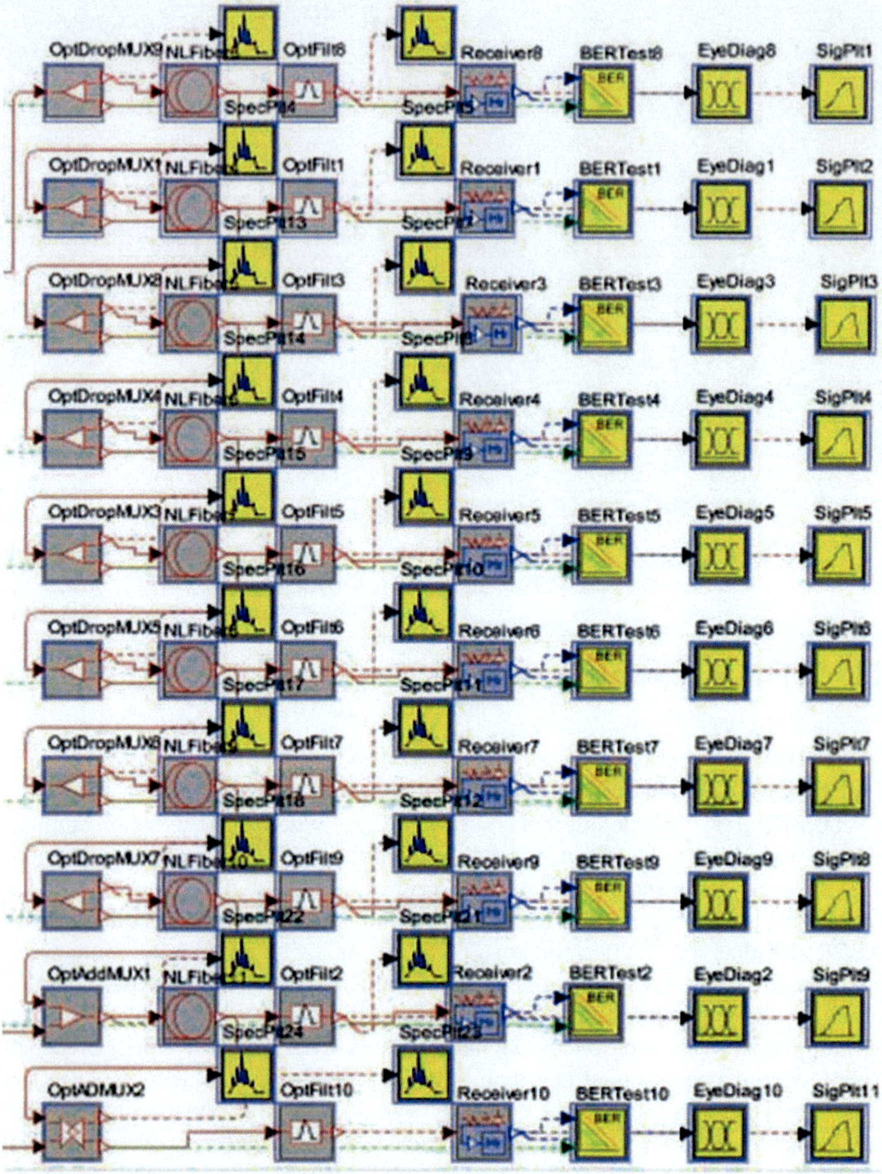

Fig. 6 WDM-PON ONU block diagram [11]

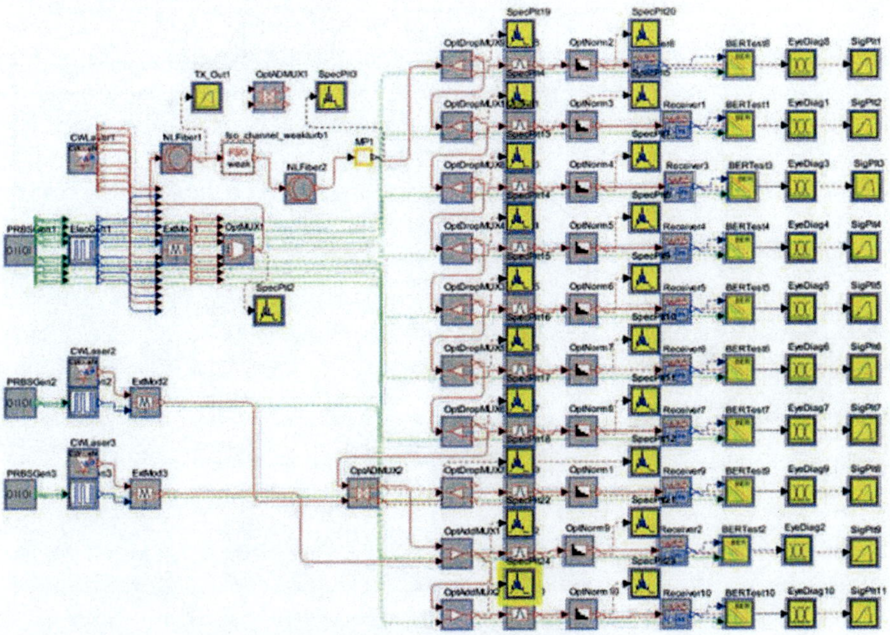

Fig. 7 WDM-PON with wireless channel (FSO) block diagram [11]

where $\propto_{Geo}$ = Geometrical Attenuation,

θ = Divergence Angle,

S = distance between OLT and ONU.

A_E = Receiving Area.

3 Recent IoT Access Network

3.1 Mixed Integer Linear Programming (MILP) in IoT PON Network

To increase the efficiency in IoT with PON network MILP can be used [3]. The PON is used here to connect two IoT networks using the concept of PON. PON is an energy efficient network. MILP is used for virtualization in IoT network. The Architecture is modeled and designed [3] for two IoT networks. IoT network is made up of four layers as shown in Fig. 8.

IoT network is made up of 4 layers. These layers are IoT object layer, IoT relay layer, IoT coordinator layer, and IoT gateway layer. Each IoT network has approximately 50 IoT Objects, 25 relays, one coordinator, and one gateway. These are connected with PON. Each ONU of PON is connected with one IoT network. If

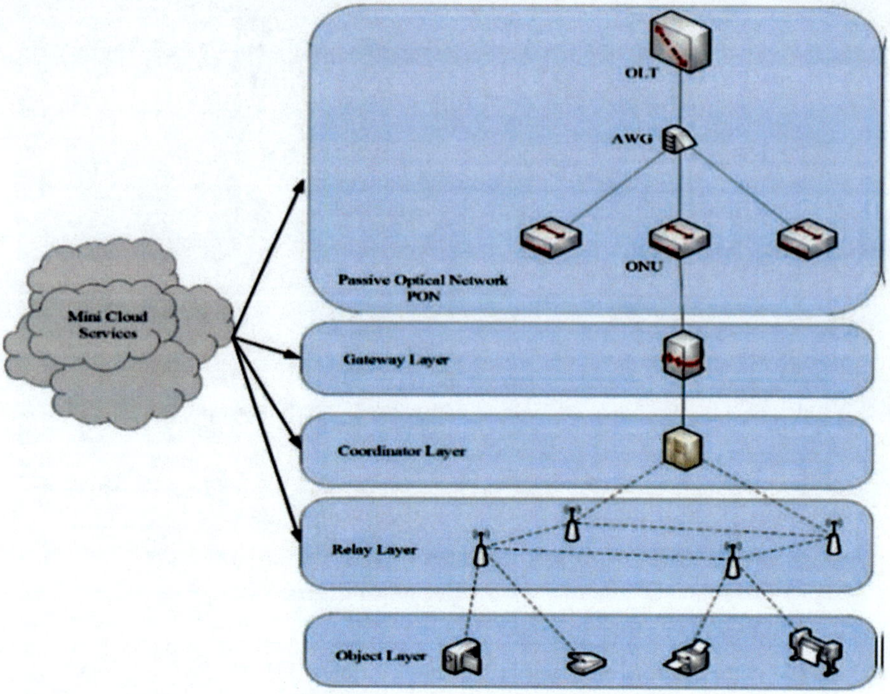

Fig. 8 WDM IoT network with PON [3]

there are n number of ONUs, then it can able to connect n number of IoT networks. MILP is successfully evaluated to measure the processing, traffic, and total power consumption. Energy efficiency is computed and concluded that 21% of total power is saved if proper utilization is there in OLTs [3].

3.2 Power Consumption of IoT Access Network

Power consumption for access network (PON) is modeled by various configurations [13]. It has been proved that Wi-Fi network with PON backhaul is the most efficient and modest network for IoT when there is large amount of traffic that occurs in the IoT network. We can use different frameworks where IoT can use four main task nodes: IoT gateway, IoT Customer Premises equipment, IoT Remote network node, and the IoT Edge node or central office. The schematic diagram of an IOT access network is shown in Fig. 9.

The analysis of an access network is made by four different methods. First is the use of PON access network. Second is very-high-bit-rate digital subscriber line (VDSL). Third is direct or point-to-point optical network and fourth is 4G LTE

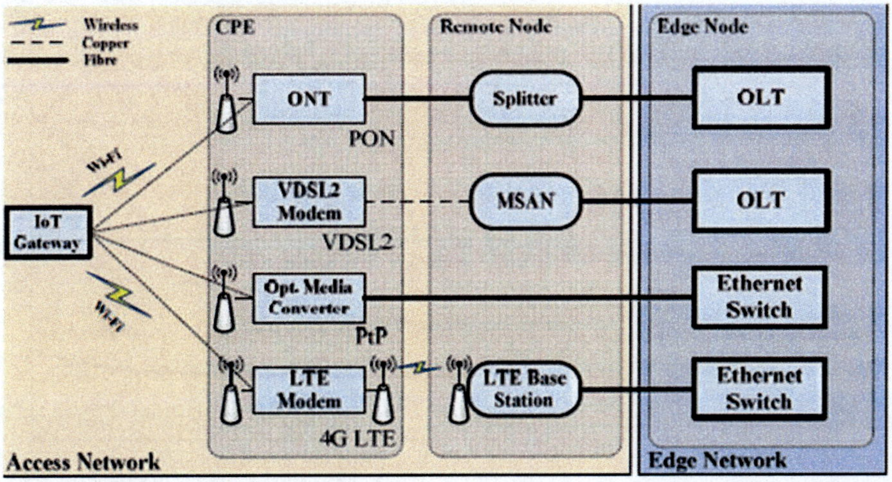

Fig. 9 IoT access network with PON [13]

cellular network. The total network power consumption of IoT gateway is calculated by following Power equation.

$$P = P_{IOT} + P_{CPE} + 1.5(P_{RN} + P_{EN}) \tag{2}$$

where P_{IOT} = Power consumption at IOR gateway,

P_{CPE} = Power consumption at CPE Modem,

P_{RN} = Power consumption at Remote Node,

P_{EN} = Power consumption at Edge Node.

P_{IOT}, P_{CPE}, P_{RN}, and P_{EN} indicate power consumption of various devices of IoT access network. Multiplier 1.5 in the equation is used as representative of overhead. IoT gateway is independent of which type of access network is used. Based on sharing the networks elements used in IoT, access network is classified into two categories. These are shared and unshared network devices. If we assume that total load in the network and power consumption are related using linear graph, then from slope of the graph we can find the equation of power for unshared and shared device. Power consumption of unshared device is given by the following equation.

$$P(C)_{Unshared} = P_{Idle} + \frac{P_{Max} - P_{Idle}}{C_{max}} C \tag{3}$$

where $P(C)_{Unshared}$ = Power consumption of unshared device at load C.

P_{Idle} = Power at no load.

P_{Max} = Power at maximum load.

C_{max} = Maximum Load.

C = Traffic load (bits/second).

Fig. 10 Comparison of various access network from power v/s IoT data access rate [13]

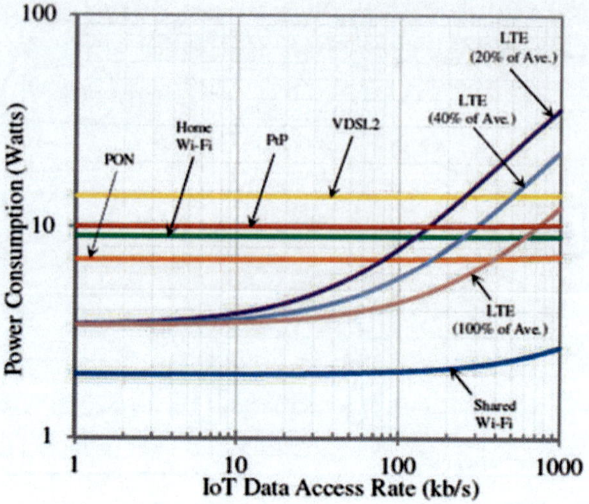

For shared network device, the total power equation is given as follows:

$$P(C)_{shared} = P_{Idle} + \frac{P_{Max} - P_{Idle}}{C_{max}}(C_{bgd} + C_{IoT}) \tag{4}$$

where $P(C)_{shared}$ = Power consumption of shared device at load C.

P_{Idle} = Power at no load.

P_{Max} = Power at maximum load.

C_{max} = Maximum Load.

C_{bgd} = Background Traffic load (bits/second).

C_{IoT} = Traffic load generated by IoT device (bits/second).

It has been found that PON is the most efficient in terms of power efficiency among all available wired optical IoT Network when the data rate is high. LTE is suitable when the data rate is lower. It has been observed using graph shown in Fig. 10.

3.3 Hybrid Long Rich Access Network (HYDRA)

HYDRA is the recent novel architecture for fiber access network which solve the problem arises in long-reach optical network and cellular network. The key idea is to introduce active remote node (ARN) [14]. It improves the cost as well as power consumption compared to long-reach fiber network. Due to these advantages, it is most suitable for IoT applications. Based on software defined networking (SDN) principle, it can improve the efficiency.

HYDRA introduces hybrid network architecture where any type of convergence is easily implemented. As shown in Fig. 11, it uses active remote node (ARN) at

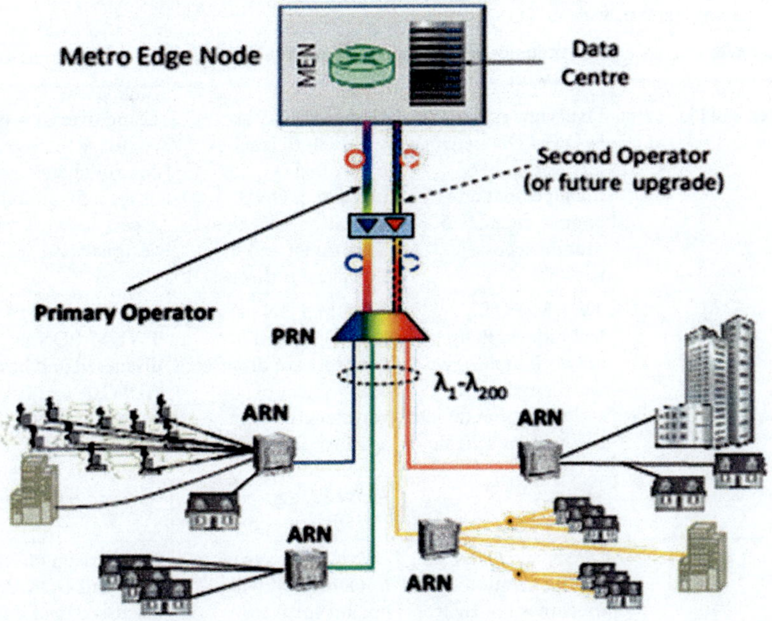

Fig. 11 Schematic of HYDRA access network with ARN [14]

first stage and passive remote node (PRN) at second stage. Service provider can define rules for collecting/distributing traffic from data center. Through virtualization techniques resources can be allocated dynamically. Here due to dynamic capability TDM is more preferred compared to WDM for designing PON. HYDRA can easily provide unlimited bandwidth, effective utilization and of programmable resources and control over distributed data for access network [14].

Following Table 1 shows comparison of various methods and applications implemented for access networks used for IoT. All methods and features are explained above like MILP programming techniques, power consumption with data rates.

4 Conclusion and Future Work

Recent trends of IoT access network is discussed where PON is used as basic access network. Using MILP programming and WDM-PON IoT access network, it is easy to increase the efficiency in terms of power. MILP is successfully evaluated to measure the processing, traffic, and total power consumption. Energy efficiency is computed and concluded that 21% of total power is saved if proper utilization is there in OLTs. Power consumption for access network (PON) is modeled by various configurations. It has been proved that Wi-Fi network with PON backhaul is the most efficient and modest network for IoT when there is large amount of traffic that occurs in the IoT

Table 1 Comparison of various PON frameworks for IoT

Author/Year	Theoretical framework	Analysis/Results	Conclusion
Honda et al. [15]	Overview of WDM-PON network, auxiliary management and control channel, 5G mobile access network	Using AMCC ho wavelength can be controlled is discussed, ONU activation, wavelength adjustment and its flowchart is discussed	Using overview of auxiliary management, control channel and its usage a 5G mobile access network can be designed and accessed
Butt et al. [16]	TWDM-PON techniques, PON network topology, wavelength switching, bandwidth assignment scheme,	TWDM-PON network topologies and concepts are discussed Assignment of wavelength using adaptive method is discussed, and ADWBA and FWDBA is compared	Performance of TWDM-PON is discussed and how ADWBA and FWDBA is used in next generation IoT network is presented
Arévalo et al. [17]	DPSK and OOK for next generation PON transmission, BER performance measurement,	DPSK PON system is implemented with optsim software. optsim matlab interface is used to get the results. BER versus received power relations were discussed for DPSK and OOK system	Comparison of DPSK PON and OOK PON is discussed, and it is proved that DPSK PON gives better results compared to OOK PON
Effenberger[18]	PON standards, TDMA and WDMA multiple access methods, next generation PON	Various PON standardization methods are discussed ad reviewed with future perspective. Block diagram of WDM-PON, TWDM-PON, and TDMA-PON is discussed for 5G Transport	For application like FTTH and other IoT applications, PON is most suitable. TDMA-PON and TWDM-PON are most suitable for these types of applications
Romero and Cerqueira [19]	Photonic assisted frequency convertor (PBRC), photonic assisted RF amplifier (PBRA), photonic assisted RF front end (PBRF), Implementation of PBRA	Three different systems PBRC PBRA and PBRF are discussed and implemented and compared n term of sensitivity and BER	PON and other networks to support RF technology or microwave is discussed and analyzed for next generation network

(continued)

Table 1 (continued)

Author/Year	Theoretical framework	Analysis/Results	Conclusion
Al-Shammari et al. [2]	Energy efficient IoT, smart road system, service embedding for IoT with PON	Energy efficient service embedding is used to find power consumption performance and compared with objectiveless service embedding	Energy efficient framework is most suitable for IoT based applications. It has been proved using MILP model

network. It has been found that PON is the most efficient in terms of power efficiency among all available wired optical IoT Network when the data rate is high. LTE is suitable when the data rate is lower. HYDRA is the recent novel architecture for fiber access network which solve the problem arises in long-reach optical network and cellular network. The key idea is to introduce ARN. It improves the cost as well as power consumption compared to long-reach fiber network. Various configurations can be designed and compared for other network parameters also.

Acknowledgements The author would like to thank Principal CMPICA and management Charusat University for their constant motivation and support.

References

1. J. Segarra, V. Sales, J. Prat, Versatile metro-access network integrating FTTH, enterprises, IoT and 5G services, in *2019 21st International Conference on Transparent Optical Networks (ICTON)* (2019)
2. H.Q. Al-Shammari, A. Lawey, T. El-Gorashi, J.M.H. Elmirghani, Energy efficient service embedding in IoT over PON, in *2019 21st International Conference on Transparent Optical Networks (ICTON)* (2019)
3. Z.T. Al-Azez, A.Q. Lawey, T.E.H. El-Gorashi, J.M.H. Elmirghani, Energy efficient IoT virtualization framework with passive optical access networks, in *2016 18th International Conference on Transparent Optical Networks (ICTON)* (2016)
4. B.A. Yosuf, M. Musa, T. Elgorashi, J.M. Elmirghani, Impact of distributed processing on power consumption for IoT based surveillance applications, in *2019 21st International Conference on Transparent Optical Networks (ICTON)* (2019)
5. J. Segarra, V. Sales, J. Prat, L. Kazovsky, Flexible coherent UDWDM-PON adapting dynamically to different bandwidths, in *2017* 19th *International Conference on Transparent Optical Networks* (ICTON) (2017)
6. H.S. Chung, High speed and low latency PON for 5G networks, in *2019 IEEE Photonics Society Summer Topical Meeting Series (SUM)* (2019)
7. I. Di Luch1, M. Ferrario, P. Boffi, G. Rizzelli, H. Wang, R. Gaudino, Demonstration of structural vibration sensing in a deployed Pon infrastructure, in *45th European Conference on Optical Communication (ECOC 2019)* (2019)
8. S. Ogawa, Y. Okumura, K. Fujii, Superposition of next generation PON to existing PON's OOK signals, in *2019 IEEE 8th Global Conference on Consumer Electronics (GCCE)* (2019)

9. G.V. Arévalo, M. Villavicencio, G. Flores, Experimental demonstration of a PAM-4 based 20 Gb/s PON using dispersion pre-compensation, in *2019 XVIII Workshop on Information Processing and Control (RPIC)* (2019)
10. K. Sun, Y. Danwen, L. Cao, X. Zhai, Y. Xing, Electricity information collection system based on WDM-RoF-PON, in *2019 IEEE 3rd Information Technology, Networking, Electronic and Automation Control Conference (ITNEC)* (2019)
11. S. Khant, A. Patel, Performance ratings of high speed WDM-PON and TDM-PON for wired and wireless environment. Ph.D. thesis (2019)
12. G. Hansel, E. Kube, Simulation in the design process of free space optical transmission systems, in *Proceedings 6th Workshop "Optics in Computing Technology"*, Paderborn, Germany (2001)
13. C. Gray, R. Ayre, K. Hinton, R.S. Tucker, Power consumption of IoT access network technologies, in *International Conference on Communication (ICCW)* (2015)
14. T.G. Orphanoudakis, C. Matrakidis, A. Stavdas, Next generation optical network architecture featuring distributed aggregation, network processing and information routing, in *European Conference on Networks and Communications* (2014)
15. K. Honda, H. Nakamura, K. Hara, K. Sone, Flexible management of WDM-PON using AMCC for 5G radio access networks, in *2019 24th Opto Electronics and Communications Conference (OECC) and 2019 International Conference on Photonics in Switching and Computing (PSC)* (2019)
16. R.A. Butt, M. Jawaid, M.W. Ashraf, M. Faheem, Load adaptive dynamic wavelength and bandwidth assignment for TWDM PON, in *2019 International Workshop on Fiber Optics in Access Networks (FOAN)* (2019)
17. G. Arévalo, D. Cárdenas, R. Cachimuel, L. Rivadeneira, 10 Gb/s non-coherent DPSK PON for next generation networks, BER performance, in *2019 IEEE Fourth Ecuador Technical Chapters Meeting (ETCM)* (2019)
18. F.J. Effenberger, PON standardisation status and future prospects, in *45th European Conference on Optical Communication (ECOC 2019)* (2019)
19. M.A. Romero, S.A. Cerqueira, Microwave-photonics and WDM-PON Fronthaul for 5G mobile systems: research activities in Brazil, in *2019 49th European Microwave Conference (EuMC)* (2019)

Applied Electronics and Emerging Technologies

Grid Synchronization for Three-Phase Grid-Tied Converter Using Decouple-Second-Order Generalized Integrator

Bhavik Brahmbhatt and Hina Chandwani

Abstract As universally well known, the phase angle of the grid voltage is a crucial parameter to the control performance of the grid-tied power converters. This paper emphases on robust grid synchronization technique for three-phase grid-tied photovoltaic system in order to detect fundamental frequency, phase, and amplitude. However, the accuracy of the estimated frequency depends on the type of abnormal grid voltages and the structure of the phase-locked loop or frequency-locked loop control scheme. The widely accepted phase-locked loop (PLL) algorithm has a complex architecture and requires a tedious tuning process to obtain a good stability margin. Decouple-second-order generalized integrator-based frequency-locked loop (SOGI-FLL) is proven to have the most promising performance by eliminating both the negative and harmonic distorted components under non-ideal grid voltage conditions. Moreover, even in the absence of any PI controller, when the grid voltage has characteristics such as dips, swells, harmonics, unbalance, and frequency changes, it can accurately track the frequency of the grid voltage. The simulation results show that the proposed DSOGI-FLL is effective under all abnormal grid voltage conditions. The results are verified through experiments to prove the superior performance of DSOGI-FLL under adverse conditions.

Keywords Frequency lock loop · Harmonics rejection · Negative sequence elimination · Phase-lock loop · Tan-arc angle computation

B. Brahmbhatt (✉)
EC Department, Government Engineering College, Modasa, Gujarat, India

H. Chandwani
The Maharaja Sayajirao University of Baroda, Vadodara, Gujarat, India

F. Thakkar et al. (eds.), *Proceedings of the International e-Conference on Intelligent Systems and Signal Processing*, Advances in Intelligent Systems and Computing 1370,
https://doi.org/10.1007/978-981-16-2123-9_26

1 Introduction

The power grid is a complex dynamic system that is affected by a variety of events, such as continuous variation in the connection of electrical loads, interference and resonance due to harmonic currents flowing through the line, faults caused by lightning strikes, and errors in the operation of electrical equipment. Therefore, when the power converter is connected to the power grid, the grid variable cannot be regarded as a constant amplitude and constant frequency, but it should be continuously monitored to ensure the desired performance of the grid-connected power converter. The basic amplitude, frequency, and phase-angle information become the key controllable parameters for the effective operation of the grid-connected converter (GCC) [1, 2]. In addition, precise and accurate phase angle and frequency measurement can ensure continuous active power and reactive power delivery to the grid through GCC. For this reason, the control algorithm based on the concept of grid synchronization is very important for estimating the grid voltage parameters required for GCC control, protection, and synchronization. The phase angle of the grid voltage is critical and essential parameter for the control system of grid-connected power converters, because popular control strategies (for example, direct power control, vector oriented control) strongly depend on the phase angle of the grid voltage to achieve Park transformation (dq reference frame conversion). The synchronous frame voltages (voltages at dq axis) are key parameter for the control system to decouple the current control loop under the dq axis. Therefore, the grid synchronization technology is a basic component of the grid-connected converter control strategy, and it is very important to ensure the satisfactory performance of the converter. Therefore, the grid synchronization technology is a basic part of the control strategy of the grid-connected converter, and ensures the satisfactory performance of the converter. In order to achieve precise and fast synchronization, many different grid synchronization technologies have been intensively studied and proposed [2, 3]. It should be noted that the grid voltage during grid synchronization may be under normal or abnormal conditions (i.e., 1-Φ phase grid voltage, 3-Φ phase unbalanced grid voltage, 3-Φ phase distortion grid voltage, phase-angle jump, frequency shift), and in the above method, PLL and FLL are crucial techniques to compute the utility grid parameters. However, interrelated feedback can be observed in these technologies, which may affect the tuning process as well as the stability margin and of the entire control scheme [4]. The conventional three-phase PLL, commonly known as synchronous reference frame (SRF), has excellent performance during the ideal grid situation (i.e., balanced grid voltage). However, if the measured grid voltage signal is unbalanced or distorted, it will give oscillatory response of computed fundamental frequency [5]. The decouple synchronous reference frame PLL (DDSRF-PLL) technique reported in the literature [6] improves the poor performance of the traditional SRF-based PLL. The DDSRF-PLL extracts positive and negative sequence components from the non-ideal grid voltages to compute the frequency. However, the performance of DDSRF-PLL will decrease during the highly distorted grid voltages. It can be found that under this distorted grid signal condition, the PD structure based on the generalized integrator

(GI) has better performance and reliability [7]. Thus, the second-order generalized integrator-based PLL (SOGI-PLL) is superior over the above-described PLLs and gives fast and accurate frequency and phase-angle estimation under the adverse grid conditions. Instead of the estimated frequency signal from the SOGI-PLL feed-backed to the SOGI block structure, this frequency can be computed adaptively by the frequency-locked loop (FLL) structure, added in SOGI structure and combination is known as SOGI-FLL, to obtain excellent performance over the SOGI-PLL. Instead of PLL, tan-arc phase-angle computation is used to compute the phase angle and the adaptive feature improves performance [8–11]. Two SOGI blocks connected in parallel form a decouple-SOGI-FLL (DSOGI-FLL) structure, which can provide good transient response even in the presence of grid abnormalities (i.e., harmonics distortions, voltage imbalances, frequency changes, voltage imbalances, etc.) [12–21]. The performance of the DSOGI-FLL is analyzed in the presence of different grid voltage abnormalities, and its superior performance is demonstrated through simulation and experimental results. The remaining part focuses on the detailed information of DSOGI-FLL and its performance under different conditions.

2 Decouple-Second-Order Generalized Integrator-Frequency Lock Loop (DSOGI-FLL) Based Grid Synchronization

Second-Order generalized integrator behaves as a resonator oscillating at the frequency $\omega' = 2\pi 50\,\text{rad/s}$ and used to generate direct-quadrature signal which can be further use in phase-angle detection of PLL. In three-phase grid-connected system, this two transfer functions in Eqs. 1 and 2 are looking as second-order band-pass filters, which can extract only AC signals at the fundamental angular frequency ω' while others AC sequences (likes 5th and 7th harmonic components) are discarded, but unfortunately unable to eliminate negative sequence in the v_α and v_β.

$$D(s) = \frac{v'}{v} = \frac{k * \omega' * s}{s^2 + k * \omega' * s + \omega'^2} \tag{1}$$

$$Q(s) = \frac{v'_q}{v} = \frac{\omega'}{s}\frac{v'}{v} = \frac{k * \omega'^2}{s^2 + k * \omega' * s + \omega'^2} \tag{2}$$

According to Eqs. (1) and (2), together with Fig. 1 by taking $k = 0.1$ and $\omega = 100\,\pi$rad/s, both the two transfer functions can be considered as second-order band-pass filters, which are able to extract only the AC signals at the fundamental angular speed ω, while the other AC sequences (for instance, 5th and 7th harmonic components) can be eliminated. However, it is worth mentioning that by using the single SOGI to either $u\alpha$ or $u\beta$, the harmonics can be eliminated, but the negative sequence still unfortunately remains. It can be observed from Figs. 1 and 2 that bode curve of

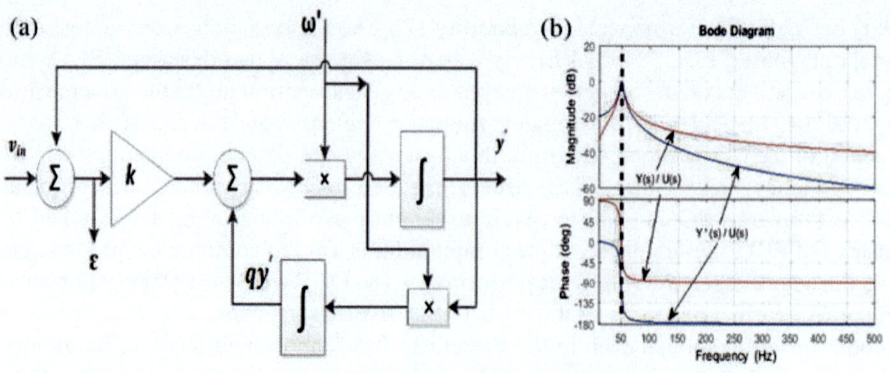

Fig. 1 **a** Representation of the SOGI, and **b** Bode plot of SOGI with k = 0.1 and ω = 100 πrad/s

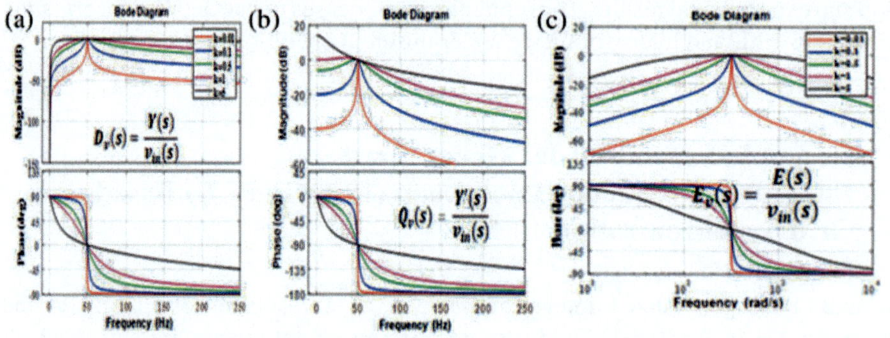

Fig. 2 Bode diagram of **a** D_v (s), **b** Q_v (s), and **c** E_v (s), with different values of k

transfer functions can maintain 0 dB at the fundamental frequency, i.e., 50 Hz and see significant amplitude attenuation at higher frequency, i.e., 250, 350 Hz, etc. In addition, from the Eq. (1) and Eq. (2), the phase responses of the two curves at 50 Hz are 0° and 90°, respectively, which show that the direct voltage y′ has a 90° phase lead compared with the quadrature component qy′. They can only extract AC signals at the grid frequency, while the other can eliminate the AC sequence (for example, the fifth and Seventh harmonic components). However, it is worth mentioning that by using a single SOGI for v_α andv_β, harmonics can be eliminated, but unfortunately, the negative sequence still exists.

In order to eliminate harmonics as well as negative sequence in the v_α and v_β, two SOGI are employed separately to v_α and v_β to generate direct/quadrature signal of v_α and v_β respectively, which is known as Decouple-Second-Order generalized integrator as shown in Fig. 3.

Figure 3 presents schematic diagram of decouple-second-order generalized integrator (DSOGI) along with frequency lock loop (FLL) and tangent-based phase-angle computation block. Figure 4 shows the four phasor diagram in which first and

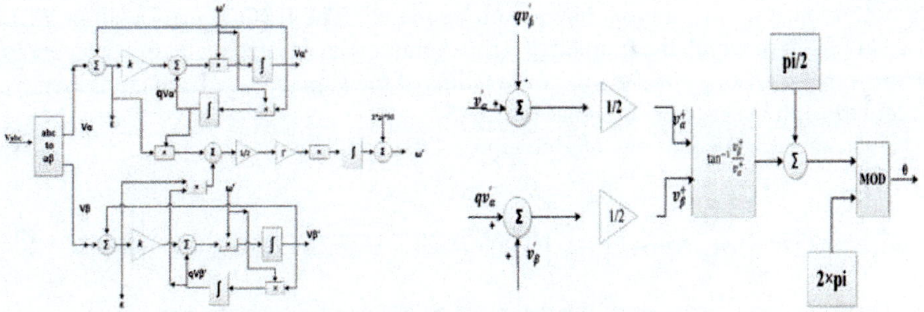

Fig. 3 Schematic diagram of the decouple-SOGI and phase-angle computation

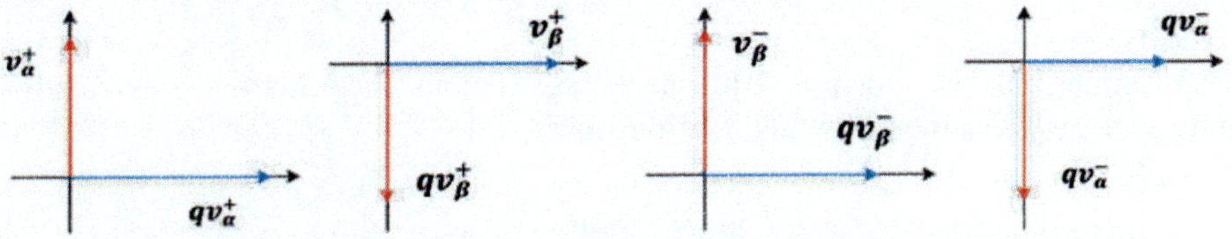

Fig. 4 Phasor diagram of direct/quadrature components of v_α and v_β grid voltage

second are positive direct/quadrature components of v_α and v_β, respectively, while third and fourth phasor diagram are negative direct/quadrature components of v_α and v_β, respectively.

It is pointed out from Figs. 1b and 2 that direct component is 90° phase leading with respect to quadrature components and negative component are rotating opposite direction of −50 Hz.

Hence, by observing Fig. 3 and complex phasor diagram Fig. 4 which gives $v'_{\alpha+} = (-qv'_{\beta+})$ and $v'_{\beta+} = (-qv'_{\alpha+})$, the output of DSOGI can be written as:

$$v'_\alpha - qv'_\beta = (v'_{\alpha+} + v'_{\alpha-}) - (qv'_{\beta+} + qv'_{\beta-}) = v'_{\alpha+} - qv'_{\beta+} = 2v'_{\alpha+} \tag{3}$$

$$qv'_\alpha + v'_\beta = (qv'_{\alpha+} + qv'_{\alpha-}) + (v'_{\beta+} + v'_{\beta-}) = qv'_{\alpha+} - qv'_{\beta+} = 2v'_\beta \tag{4}$$

$$\varepsilon_f = \frac{\varepsilon_{f(\alpha)} + \varepsilon_{f(\beta)}}{2} = \frac{1}{2}(\varepsilon_\alpha qv'_\alpha + \varepsilon_\beta qv'_\beta) \tag{5}$$

A frequency adaptive FLL is added in SOGI-QSG to achieve grid synchronization with frequency-changing environment. Moreover, normalized gain of FLL is derived by dividing with square of the amplitude of positive sequence components, i.e., $(v_\alpha^+)^2 + (v_\beta^+)^2$ in order to achieve linearize the response of the frequency lock loop. Although mathematically correct, using two independent FLLs in DSOGI-FLL may be conceptually strange because its two input signals have the same frequency. Due to both inputs, i.e., v_α and v_β have same frequency. The DSOGI uses a single FLL (see in Fig. 3) in which the frequency error signals are generated by taking an average error signal, i.e., $\varepsilon_{f(\alpha)}$ *and* $\varepsilon_{f(\beta)}$ of the QSGs of the signals have been combined by calculating an average error signal in Eq. (5).

The state-space equations of Decouple-SOGI can be written as

$$\begin{bmatrix} v_\alpha^+ \\ v_\beta^+ \end{bmatrix} = \frac{1}{2} * \begin{bmatrix} D(s) & -Q(s) \\ Q(s) & D(s) \end{bmatrix} \begin{bmatrix} v_\alpha \\ v_\beta \end{bmatrix} = \frac{1}{2} * \frac{k \times \omega'}{s^2 + k \times \omega' \times s + \omega'^2} \begin{bmatrix} s & -\omega' \\ -\omega' & s \end{bmatrix} \begin{bmatrix} v_\alpha \\ v_\beta \end{bmatrix} \tag{6}$$

The frequency response of the decouple-SOGI can be described as

$$\begin{bmatrix} v_\alpha^+ \\ v_\beta^+ \end{bmatrix} = \frac{1}{2} * \frac{k \times \omega'}{k \times \omega \times \omega' - j(\omega^2 + \omega'^2)} \begin{bmatrix} j\omega & -\omega' \\ -\omega' & j\omega \end{bmatrix} \begin{bmatrix} v_\alpha \\ v_\beta \end{bmatrix} \tag{7}$$

Assuming balanced positive sequence voltage vector at the grid frequency, steady-state analysis gives the following relationship:

$$v_\beta(\mathrm{s}) = -\mathrm{j}v_\alpha(\mathrm{s}) \tag{8}$$

Hence, the steady-state frequency response of the decouple-SOGI is re-written as

$$\begin{bmatrix} v_\alpha^+ \\ v_\beta^+ \end{bmatrix} = \frac{1}{2} * \frac{k \times \omega'(\omega' + \omega)}{k \times \omega \times \omega' - j(\omega^2 + \omega'^2)} \begin{bmatrix} v_\alpha \\ v_\beta \end{bmatrix} \tag{9}$$

The transfer functions give the relation between the amplitude of the positive sequence component detected by the DSOGI and the actual amplitude of a given positive sequence voltage vector applied to its input. The transfer function $\frac{|v_{\alpha\beta}^{+'}|}{|v_{\alpha\beta}^{-}|}$ is obtained by substituting ω by –ω. The bode plot of transfer function is shown in Fig. 5.

3 Simulation Results

The usefulness of DSOGI-FLL is validated through simulation results carried out from MATLAB/Simulink environment. The control variables used in simulation model are as k = 1, $\gamma = -2000$. The simulation results are obtained by considering four cases as (i) balance grid voltages (ii) unbalanced sag in grid voltage (iii) unbalanced swell in grid voltage (iv) harmonics distorted grid voltage. Figure 6a shows that the case (i) where balanced three-phase grid voltages are given to DSOGI-FLL up to t = 0.1 s. Afterward, voltage sag and swell are introduced in the phase A of grid voltage from t = 0.1 s to t = 0.15 s and from t = 0.15 s to t = 2.0 s, respectively. The grid synchronization block is essential and vital block in the design and

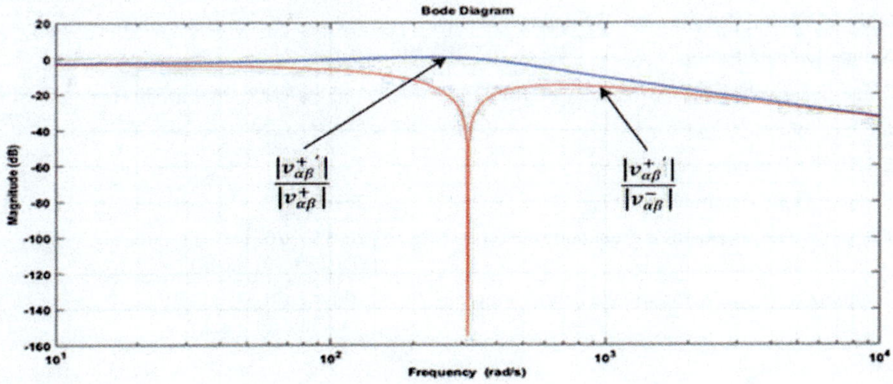

Fig. 5 Bode plot of transfer functions $\frac{|v_{\alpha\beta}^{+'}|}{|v_{\alpha\beta}^{+}|}$ and $\frac{|v_{\alpha\beta}^{+'}|}{|v_{\alpha\beta}^{-}|}$ in the Decouple-SOGI

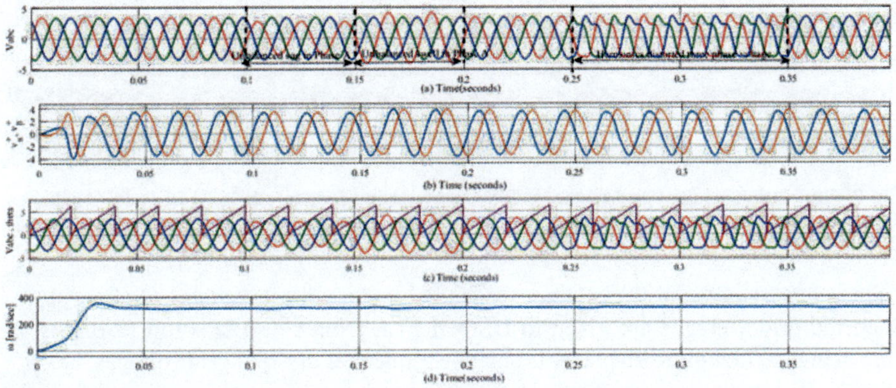

Fig. 6 Simulation result of grid-connected PV inverter using decouple-SOGI-FLL: **a** PCC voltages with grid abnormalities, **b** positive component of stationary frame voltages, **c** comparative results of phase angle and PCC voltages, and **d** frequency extraction

implementation of control system for grid-connected converter. The grid-connected system is simulated in the MATLAB/Simulink environment using simulation parameter described in Table 1. Simulation results of grid-connected system are shown in Figs. 6 and 7.

Figure 6 demonstrates the PCC voltages, grid currents, and inverter currents under grid abnormalities. The DSOGI-FLL is used to extract phase angle as well as positive sequences, i.e., v_{α}^{+}, v_{β}^{+} and both are used to generate v_d and v_q for the control system of grid-connected system. Due to superior performance of DSOGI-FLL, inverter currents $i_{inv,abc}$ are not affected by sag in phase A of v_{PCC} from t = 0.1 s to t = 1.5 s, swell in phase A of v_{PCC} from t = 0.15 s to t = 0.2 s, and harmonics distortion in a v_{PCC} from t = 0.25 s to t = 0.35 s as shown in Fig. 6c. In addition, the harmonics

Table 1 Control parameters and component values

Simulation parameters	Values
Three-phase voltages and frequency of grid	110 V (RMS, Ph–Ph), 50 Hz
DC bus voltage (V_{dc})	250 V
DSOGI-FLL parameter: k and γ	1, −2000
Sampling time for simulation (T_s)	2e−6
Sampling time, k_p, k_i for current control loop	10 * T_s, 70, 15 (Discrete-PI block)
Sampling time, k_p, k_i for voltage control loop	100 * T_s, 0.03, 0.0001 (Discrete-PI block)

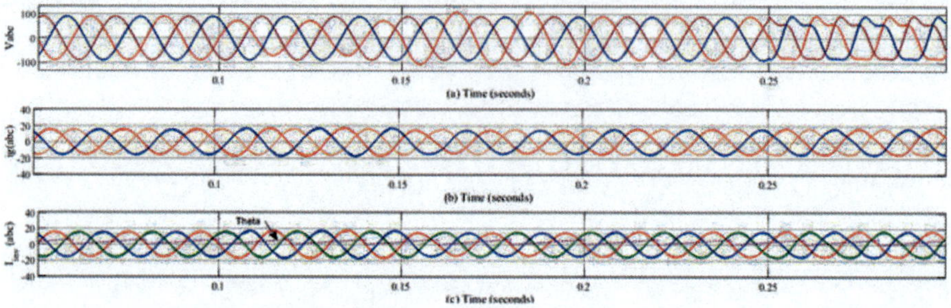

Fig. 7 Simulation result of grid-connected PV inverter using decouple-SOGI-FLL: **a** PCC voltages with grid abnormalities, **b** grid currents, and **c** comparative results of phase angle and inverter currents

distorted grid voltages are given to DSOGI-FLL from t = 0.25 s to t = 0.35 s as shown in Fig. 7a. Figure 7b presents positive sequences of $\alpha\beta$ of grid voltages, i.e., v_α^+ and v_β^+, FLL which are free from any effect of sag, swell, and harmonics distortion, and confirms the band-pass filtering capabilities of DSOGI-FLL. Figure 7c presents the synchronism of grid voltages and phase angle extracted using DSOGI-FLL by considering all simulation cases. From Fig. 7d, the transient response of estimated frequency in DSOGI-FLL does not indicate any fluctuation or minimum change at the instant in which either sag or swell encounter in the three-phase voltages and quickly settles down to the desired value.

Figure 8 gives the comparative analysis of frequency extraction of different grid synchronization techniques when subjected frequency shifted from 50 to 45 Hz at the t = 0.25 s. The dynamic parameters are chosen for analysis as follows: ω_n = 2π50 rad/s, ζ = 0.707. Table 2 presents settling time of frequency extraction, negative Sequence Components detection, and harmonics detection for the different grid synchronization techniques. Hence, DSOGI-FLL gives superior performance over the other grid synchronization techniques.

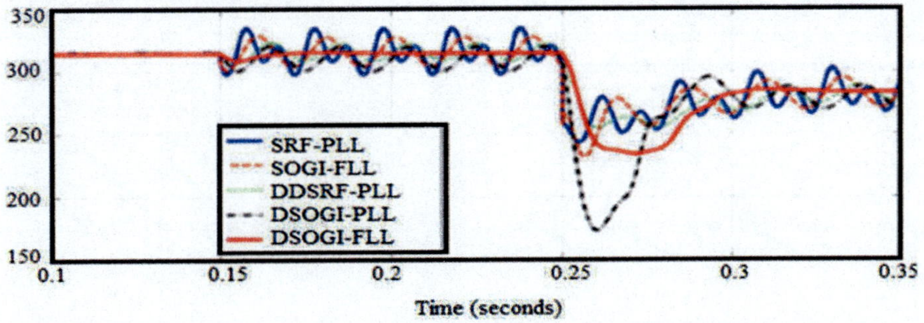

Fig. 8 Time responses of frequency extraction in different grid synchronization techniques

Table 2 Time response of frequency extraction and performance assessment of different grid synchronization techniques

Different grid synchronization techniques	Observed time (ms)	Negative sequence components detection	Harmonics detection
SRF-PLL	60	NO	NO
SOGI-FLL	54	NO	YES
DDSRF-PLL	50	YES	Required decoupling term
DSOGI-PLL	50	YES	YES
DSOGI-FLL	35	YES	YES

4 Experimental Results

In this section, the effectiveness of DSOGI-FLL is validated through experimental results carried out with different test cases. The simulation results are confirmed through experimental results using 32-Bit ARM-based STM32F407VGA microcontroller and waijung block-set in the MATLAB/Simulink environment. The DSOGI-FLL discrete model is made in the MATLAB/Simulink using waijung block-set and dumped into STM32F407VGA microcontroller using KEIL and ST-utility environments. The control variables for DSOGI-FLL are chosen as follows: $k = 1$ and $\gamma = -2000$. The sampling frequency and grid frequency are 10 kHz and 50 Hz, respectively. Hence, control variable chosen for experimentation and simulation are same for justifying DSOGI-FLL-based control system. Figure 9 displays three-phase grid voltages along with the phase angle extracted by DSOGI-FLL. The three-phase grid voltages, displayed in Fig. 9, undergoe a frequency shift of 10% (frequency change from 50 to 45 Hz) and phase shift of 45° to observe the frequency change. The frequency and phase shift in the three-phase grid voltages are measured by taking difference of cursor positions before and after frequency and phase shift, shown in Fig. 9a and b, respectively. The estimated frequency is settled down to the

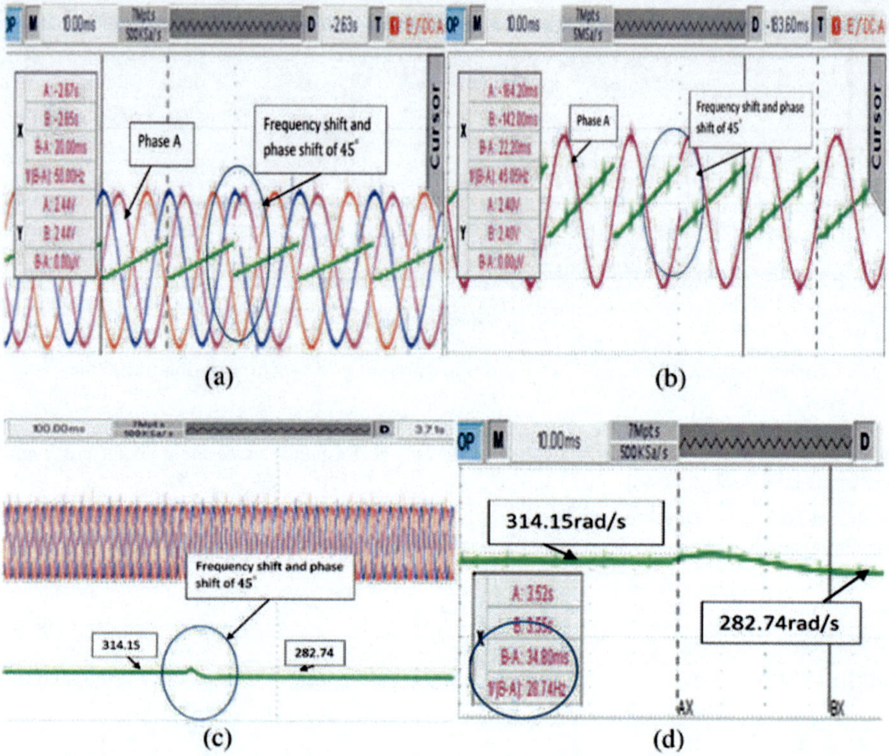

Fig. 9 Experimental results measured during 10% frequency and 45° phase shift in grid voltages

new frequency of 45 Hz within 35 ms, as shown in Fig. 9c, d. Figure 9 shows the three-phase grid voltages along with extracted phase angle from DSOGI-FLL.

It is noted that extracted phase angle from DSOGI-FLL is not affected much more when initially balanced grid voltage experiences the balanced sag of 50%, as shown in Fig. 10. Figure 10b shows the positive sequence of grid voltages, i.e., v_{α}^{+} and v_{β}^{+} and along with extracted phase angle from DSOGI-FLL. During the

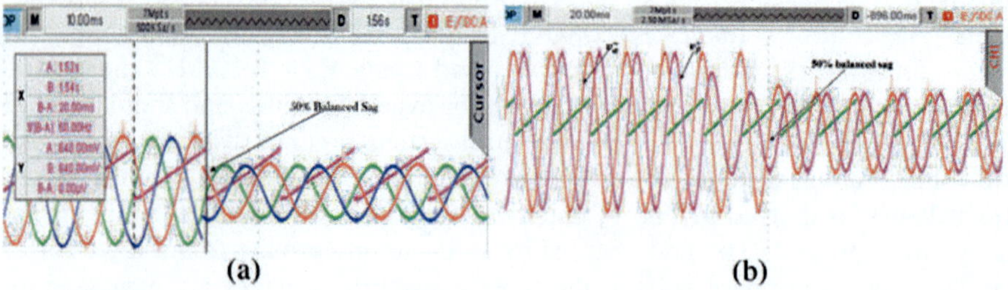

Fig. 10 Experimental results obtained during 50% balanced sag in the grid voltages

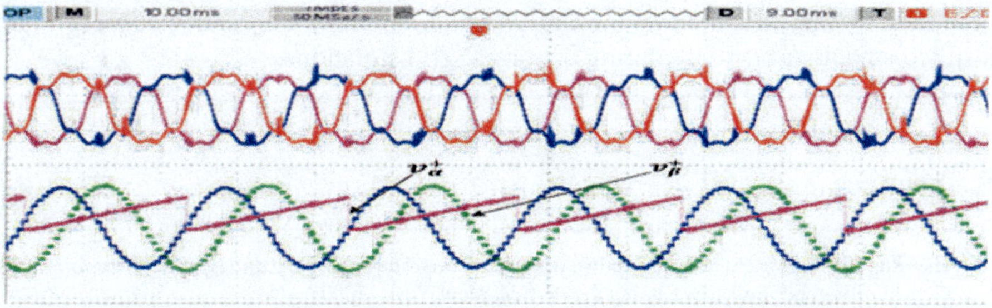

Fig. 11 Experimental results obtained during grid voltages affected by the harmonics

voltage sag, frequency and extracted phase angle are settled down within two cycles, i.e., around 35 ms, as shown in Fig. 10. The DSOGI-FLL behaves as second-order band-pass filter, which provides immunity toward highly distorted three-phase grid voltages. The three-phase grid voltages experience the presence of 5th, 7th, 11th order harmonics with amplitude proportional 20%, 15%, and 10% respectively, as shown in Fig. 11. It is noted that DSOGI-FLL is capable to eliminate the negative sequences and harmonics component from grid voltages. Figure 11 shows the highly distorted grid voltages, the positive sequence of grid voltages, i.e., v_α^+ and v_β^+ and along with extracted phase angel from DSOGI-FLL. The DSOGI-FLL gives superior performance when grid voltages experience multiple abnormalities at the same time. It is also observed that extracted phase angle is free from high frequency components due to the distorted grid voltages.

5 Conclusion

The SRF-PLL and SOGI-FLL detect phase angle, frequency, and better performance during ideal grid condition, i.e., balanced three-phase sinusoidal grid voltage. During non-ideal grid condition, they give disappointing performance in grid synchronization, and unfortunately addition of 100 Hz oscillating component in the phase angle of grid voltage. In this paper, the DSOGI control approach is used to generate positive sequence of grid voltage for the phase-angle computation, furthermore, eliminates both harmonics distortion and negative sequences of three-phase grid voltage in order to detect accurately phase angle of the grid voltage during abnormal grid voltage conditions. Meanwhile, frequency of the three-phase grid voltage is a more stable variable than the phase angle of grid voltage, FLL block used to track grid frequency adaptively. The tan-arc angle computation approach has minimized computation burden as observed in the second-order PLL techniques. Hence, the presented control approach has notable immunity toward frequency drifts and unbalance grid situations. DSOGI-FLL has demonstrated as appropriate control approach to be used in the design of stationary and synchronous controllers for grid-connected systems.

From Simulation and experimental results, the DSOGI-FLL control approach has justified smoother response during abnormal grid situations.

References

1. H. Akagi, E.H. Watanabe, M. Aredes, *Instantaneous Power Theory and Applications to Power Conditioning*, ed. by M.E. El-Hawari (Wiley, New York, 2007)
2. R.N. Beres, X. Wang, M. Liserre, F. Blaabjerg, C.L. Bak, A review of passive power filters for three-phase grid-connected voltage-source converters. IEEE. J. Emerg. Sel. Top. Power Electron. **4**(1), 54–69 (2016)
3. M.I.M. Montero, E.R. Cadaval, F.B. Gonzalez, Comparison of control strategies for shunt active power filters in three-phase four-wire systems. IEEE Trans. Power Electron. **22**(1), 229–236 (2007)
4. S.K. Chauhan, M.C. Shah, R.R. Tiwari, P.N. Tekwani, Analysis, design and digital implementation of a shunt active power filter with different schemes of reference current generation. IET Power Electron. **7**(3), 627–639 (2014)
5. R. Panigrahi, B. Subudhi, P.C. Panda, Model predictive-based shunt active power filter with a new reference current estimation strategy. IET Power Electron. **8**(2), 221–233 (2015)
6. C.H. da Silva, R.R. Pereira, L.E.B. da Silva, G. Lambert-Torres, B.K. Bose, S.U. Ahn, A digital PLL scheme for three-phase system using modified synchronous reference frame. IEEE Trans. Ind. Electron. **57**(11), 3814–3821 (2010)
7. F. Cupertino, E. Lavopa, P. Zanchetta, M. Sumner, L. Salvatore, Running DFT based PLL algorithm for frequency, phase, and amplitude tracking in aircraft electrical systems. IEEE Trans. Ind. Electron. **58**(3), 1027–1035 (2011)
8. B.C. Babu, K. Sridharan, E. Rosolowski, Z. Leonowicz, Analysis of SDFT based phase detection system for grid synchronization of distributed generation systems. Eng. Sci. Technol. Int. J. **17**(4), 270–278 (2014)
9. T. Thacker, D. Boroyevich, R. Burgos, F. Wang, Phase-locked loop noise reduction via phase detector implementation for single-phase systems. IEEE Trans. Ind. Electron. **58**(6), 2482–2490 (2011)
10. S. Golestan, M. Monfared, F.D. Freijedo, J.M. Guerrero, Design and tuning of a modified power-based PLL for single-phase grid-connected power conditioning systems. IEEE Trans. Power Electron. **27**(8), 3639–3650 (2012)
11. R. Teodorescu, M. Liserre, P. Rodriguez, A thesis on grid converters for photovoltaic and wind power systems (2011). ISBN 978-0-470-05751-3
12. M. El-Habrouk, M.K. Darwish, P. Mehta, Active power filters: a review. Proc. IEE Electr. Power Appl. **147**(5), 403–413 (2000)
13. H. Akagi, Active harmonic filters. Proc. IEEE **93**(12), 2128–2141 (2005)
14. B. Singh, J. Solanki, An implementation of an adaptive control algorithm for a three-phase shunt active filter. IEEE Trans. Ind. Electron. **56**(8), 2811–2820 (2009)
15. Q.N. Trinh, H.H. Lee, An advanced current control strategy for three-phase shunt active power filters. IEEE Trans. Ind. Electron. **60**(12), 5400–5410 (2013)
16. P. Kanjiya, V. Khadkikar, H.H. Zeineldin, A noniterative optimized algorithm for shunt active power filter under distorted and unbalanced supply voltages. IEEE Trans. Ind. Electron. **60**(12), 5376–5390 (2013)
17. V.M. Moreno, A.P. Lopez, R.I.D. Garcias, Reference current estimation under distorted line voltage for control of shunt active power filters. IEEE Trans. Power Electron. **19**(4), 988–994 (2004)
18. P. Rodrigues, A. Luna, R. Santiago, M. Aguilar, I.E. Otadui, R. Teodorescu, F. Blaabjerg, A stationary reference frame grid synchronization system for threephase grid-connected power converters under adverse grid onditions. IEEE Trans. Power Electron. **27**(1), 99–112 (2012)

19. Y. Han, M. Luo, X. Zhao, J.M. Guerrero, L. Xu, Comparative performance evaluation of orthogonal-signal-generators-based single-phase PLL algorithms—a survey. IEEE Trans. Power Electron. **31**(5), 3932–3944 (2016)
20. G. Buticchi, E. Lorenzani, G. Franceschini, A dc offset current compensation strategy in transformerless grid-connected power converters. IEEE Trans. Power Del. **26**(4), 2743–2751 (2011)
21. S. Golestan, J.M. Guerrero, G. Gharehpetian, Five approaches to deal with problem of DC offset in phase-locked loop algorithms: design considerations and performance evaluations. IEEE Trans. Power Electron. **31**(1), 648–660 (2016)

Regression-Based Model of Low Cost Air Quality Monitoring System

Mitul Kumar Ahirwal, Gautam Patidar, Neeraj, Sandeep Kumar, and Amit

Abstract In this paper, an attempt has been made to build a prediction model based on regression analysis of air quality data. The proposed prediction model will help in air quality monitoring and analysis with less number of sensor data. For this, linear and polynomial regression models have been created and tested. Air quality monitoring data has been used, which has been collected through air quality monitoring devices. The compositions of air have been also explored. The dataset consists of information about the composition (different gases) of air at different times. Among all the components of air, CO (Carbon Mono-Oxide) is a major air pollutant. Hence, the regression analysis-based model is used to predict the readings of CO sensors through specified sensors. This study contributes toward the modeling and development of low cost devices for monitoring air quality in the environment. In future devices may be developed based on the presented model that fits in budget and used widely in cities at different locations.

Keywords Sensors · Air quality · Air pollution · Regression model

1 Introduction

At present, air and environmental pollution is a major issue across the globe. People are facing many health issues due to air pollution. Some of the pollutants are highly responsible for it. Air Quality Index (AQI) is the term which is used by scientists to show the amount of pollution in the air. Risk of public health increase as the AQI rises. This index is generally used to check the quality of air. It is also used for reporting the severity levels of air pollution to the living things in that area [1, 2]. Major pollutants of air are Carbon Mono-Oxide (CO), Nitrogen Dioxide (NO2), Ozone (O3), Particulate Matter (PM10), and Sulfur Dioxide (SO2). These pollutants are mainly originated from motor vehicles, industries, fires, unpaved roads, construction

M. K. Ahirwal (✉) · G. Patidar · Neeraj · S. Kumar · Amit
Computer Science and Engineering Department, Maulana Azad National Institute of Technology, Bhopal 462003, India

F. Thakkar et al. (eds.), *Proceedings of the International e-Conference on Intelligent Systems and Signal Processing*, Advances in Intelligent Systems and Computing 1370,
https://doi.org/10.1007/978-981-16-2123-9_27

sites, volcano eruptions, and fossil fuel burning. These pollutants create health problems like throat irritation, coughing, airway inflammation, decreased lung function, asthma, irregular heartbeat, breathing difficulties, coughing, and chest problems. To overcome problems related to air pollution, the first step is to monitor the quality of air. Ratio of different gases which compose the air and our environment also needs to be observed. For this, many national and international agencies are continuously collecting the air quality data from different locations [1–3].

Agencies like United States Environmental Protection Agency (USEPA), American Association of State Highway Transportation Officials (AASHTO), Italian National Agency for New Technologies—Energy and Sustainable Economic Development (ENEA) are collecting the air quality data and provide their access to research community working in environmental studies, air quality monitoring, and other similar fields. Normally, air quality data consists of recordings of CO, NOx, Non-Metanic Hydrocarbons (NMHC), NO_2, and Benzene (C_6H_6). Many studies have been conducted over air quality analysis and monitoring [4–8]. In [4], street-level urban air quality measurement and modeling have been done through mobile device. Land Use Regression (LUR) model is used to develop low cost device. This study was performed over data taken from Open Street Map (OSM), South Korea. Prediction of air quality and gas concentration in coal mines has been presented in [5]. In this, modeling of relation between gas emission and coal falling fragmentation and mining speed was proposed. Auto-Regressive Moving Average (ARMA), Chaos Time Series (CHAOS) model, Encoder-Decoder (ED) framework are used. Integration of information from multiple sensors through network was proposed in [6]. This integration improves the spatial resolution of monitoring area. Air Box low cost sensors data is shared by satellites. In [7], low cost sensors and their field calibration has been explored. Raw and scaled sensor data is used with Linear Regression (LR), multiliniear regression, and Artificial Neural Networks (ANNs). Testing of IoT-based air quality monitoring setup is done in [8]. Low cost Arduino setup was found to be suitable for air quality monitoring system. Application of Gaussian Process Mixture (GPM) has been proposed in [9], to adopt hidden variables in characteristics of the gas concentration for air quality monitoring. The GPM is found to be more accurate than the other traditional models. In [10], investigation and analysis of effects of employment, network of road, status of traffic, and social Demographic features on air pollutant discharge has been observed. Results show that employment characteristics, status of traffic, and density of roads are significantly associated with the concentrations of PM10, CO, and NO_2 discharge. Regression analysis, machine learning, weighted regression, and interpolation methods are utilized for investigation.

For timely prediction of pollutant concentrations in air, a forecasting model of air pollution to take necessary actions was presented in [11]. This model guarantees good accuracy with improvements in the forecasted values over other state-of-the-art methods. Examination of spatial outlier's effect over estimation of air pollutant concentrations using regression methods was proposed in [12]. Spatial outlier's worse the quality of prediction. Spatial outliers detection is an option that can be used when observation-based prediction model is used. Forecasting of pollutant concentrations is presented in [13], data is pre-processed by logarithmic procedure and use of Immune Algorithm (IA) was done with Support Vector Regression (SVR). Short-term association and interaction of different pollutants of air and daily temperature level on specific emergency causes have been explored in [14]. The relation of temperature with NO_2 has been observed statistically significant. Multivariate calibration for multisensor device was proposed with neural celebration using ANN [15]. It is capable to estimate concentrations of gases in field successfully with multisensor device.

Semi-Supervised learning methods and artificial olfaction approach are utilized to enhance the performance of classifier and regression by reducing sensor drifts [16]. Neural regression is used in multivariate calibration for pollutants concentration estimation. The study explores that the distribution of concentrations leads to best possible performances and also satisfactorily insensitive to outliers [17].

From the above literature survey, it has been observed that use of regression and forecasting methods are in trend over air quality data. Continuing the same track of research, from the best of our knowledge, one air quality dataset (provided by ENEA) has been noticed and identified. Over this dataset regression analysis has not been applied with the aim to reduce the number of sensors. Therefore, in this study regression analysis over said dataset has been presented with the aim to overcome the need for large number of sensors, which leads to overall cost reduction of the system. Further study is arranged as; Sect. 2 includes details of the dataset and methods used. Results are provided in Sect. 3. Sections 4 and 5 present conclusions and future directions, respectively.

2 Methodology

2.1 Dataset

Data set provided by ENEA—National Agency for New Technologies, Energy and Sustainable Economic Development, consists of 15 attributes with 9358 samples. The source of the dataset is UCI machine learning repository [15, 18]. This dataset contains hourly measured averaged response of chemical sensors attached with the air quality sensor device. The dataset is of location of Italian city. The dataset contains the values of concentration of CO, Non-Methanic Hydrocarbons, Nitrogen Dioxide, Nitrogen Oxides, and Benzene. The missing values are represented by -200.

The dataset contains the following attributes:

(1) *Date* (DD/MM/YYYY): It is the parameter which gives the date when the data is measured by the respective sensors.
(2) *Time* (HH.MM.SS): It is the parameter which gives the time when the data is measured by the respective sensors.
(3) *Carbon Monoxide* (CO): This is colorless and odorless gas which forms when the incomplete combustion of carbon takes place. This parameter is measuring

the value of hourly average concentration of CO in microgram per cubic meters.

(4) *PT08.S1* (Tin Oxide): Nominally, targeted to CO, this parameter is the measurement of tin oxide which is calculated on hourly basis.

(5) *Non-Metanic Hydrocarbons* (NMHC): This parameter measures the concentration of hydrocarbons in microgram per cubic meters.

(6) *Benzene* (C_6H_6): This parameter is measuring hourly average of the quantity of benzene in microgram per cubic meters.

(7) *PT08.S2* (Titania): Nominally, targeted to NMHC, this parameter is the measurement of titania which is calculated on hourly basis.

(8) *Total Nitrogen Oxides* (Nox): This parameter measures the hourly average concentration of Nox in parts per billion.

(9) *PT08.S3* (Tungsten Oxide): Nominally, targeted to Nox, this parameter is the measurement of tungsten oxide which is calculated on hourly basis.

(10) *Nitrogen Dioxide* (NO_2): This parameter measures the hourly concentration of NO_2 in microgram per cubic meters.

(11) *PT08.S4* (Tungsten Oxide): Nominally, targeted to NO_2, this parameter is the measurement of tungsten oxide nominally targeted NO_2 on hourly basis.

(12) *PT08.S5* (Indium Oxide): Nominally, targeted to O3, this parameter is the measurement of indium oxide nominally targeted O_3 on hourly basis.

(13) *Temperature* (T): This parameter shows the temperature at the time when the readings of the other parameters were recorded. It is showing the values in degree Celsius (°C).

(14) *Relative Humidity* (RH%): This parameter gives the value of the relative humidity in percentage at the time at which the reading of the other parameters was taken.

(15) *Absolute Humidity* (AH): This parameter gives the value of the absolute humidity at the time at which the reading of the other parameters was taken.

Properties and effects of above-listed gases: Serious health effects can be caused by CO. This stops certain oxygen amount from reaching body parts like heart, brain, and tissues. This is also a major component responsible for air quality. Hydrocarbons are responsible for formation of ozone. They are toxic in nature. Benzene is light yellow colorless liquid. Benzene levels are generally high in the region where lump of waste materials is found. Total Nitrogen Oxides are highly reactive gases and are also poisonous in nature. Burning of large amount of fuel is responsible for the emission of Nitrogen Oxide. Nitrogen Dioxide is formed by emission from motor vehicles, stoves, etc. This is highly reactive gas and also a respiratory irritant [2, 3, 19].

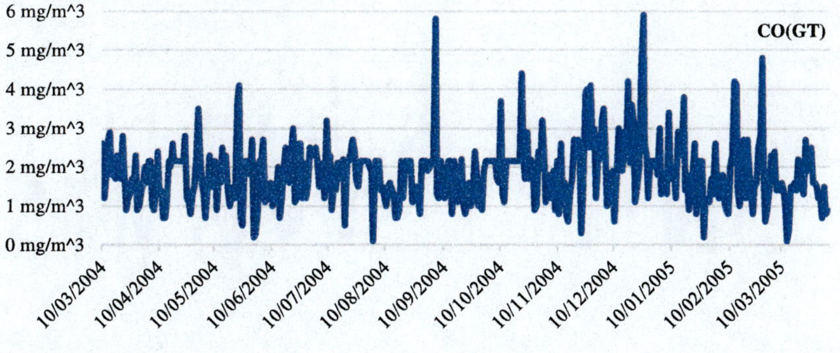

Fig. 1 Readings of CO

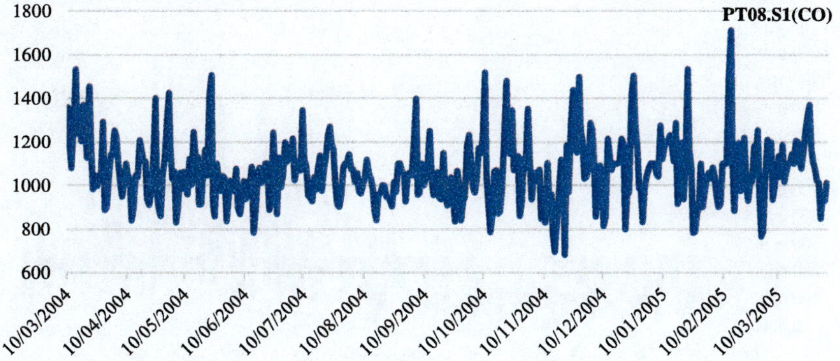

Fig. 2 Readings of PT08.S1

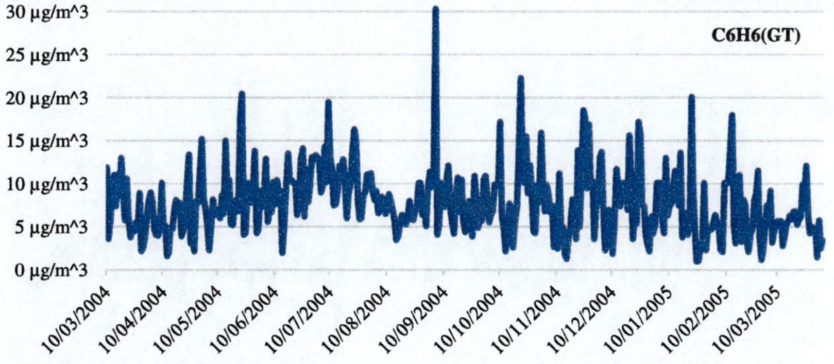

Fig. 3 Readings of C_6H_6

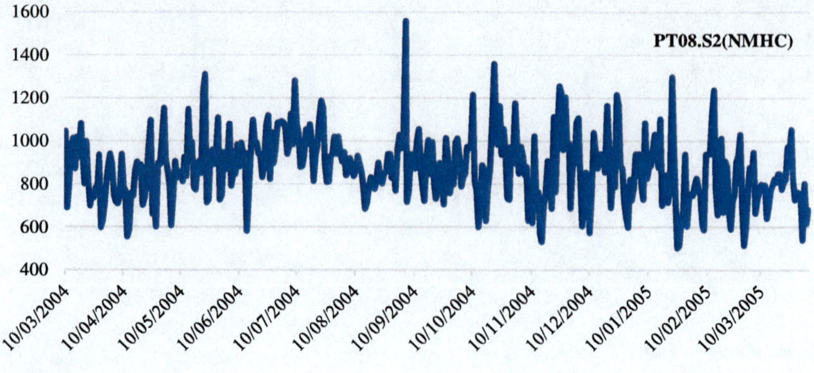

Fig. 4 Readings of PT08.S2

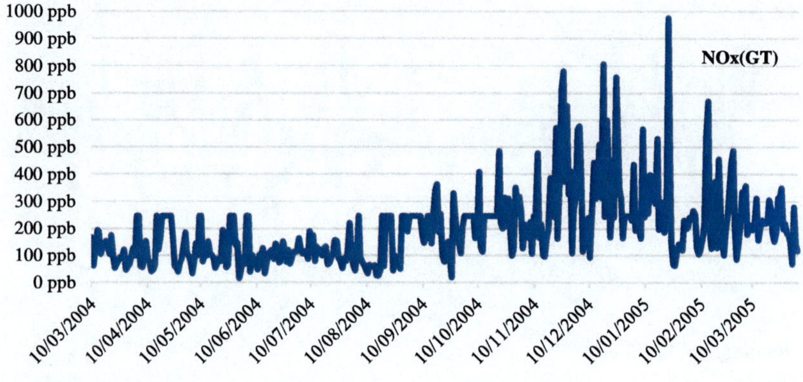

Fig. 5 Readings of NOx

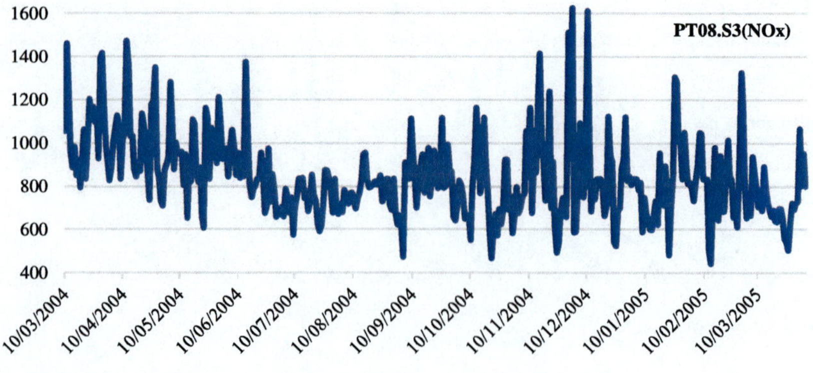

Fig. 6 Readings of PT08.S3

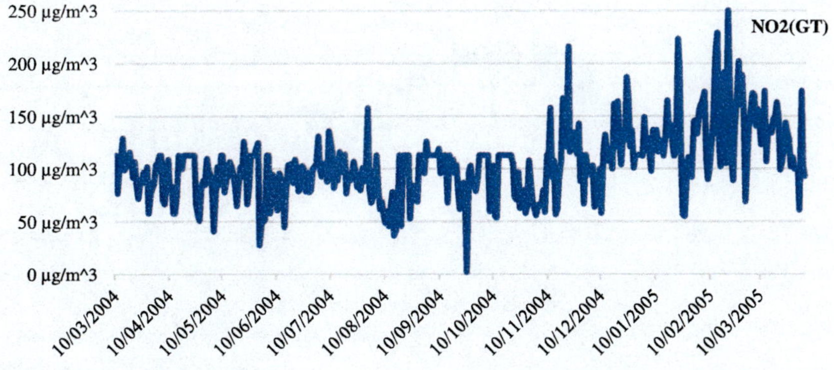

Fig. 7 Readings of NO_2

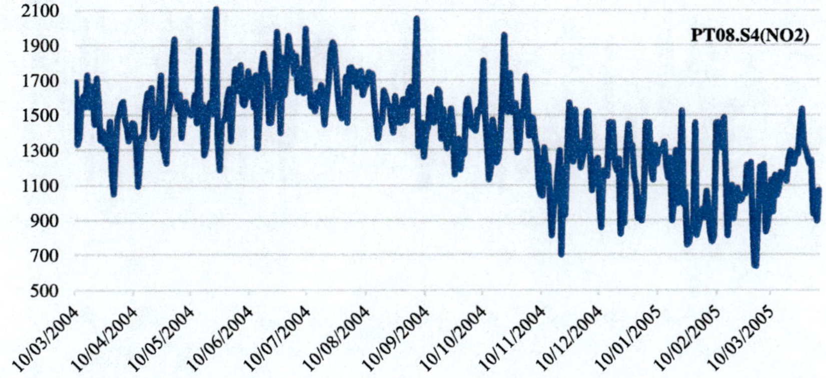

Fig. 8 Readings of PT08.S4

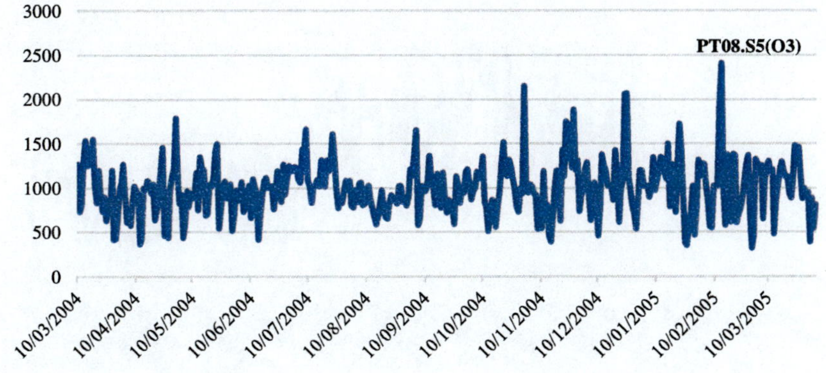

Fig. 9 Readings of PT08.S5

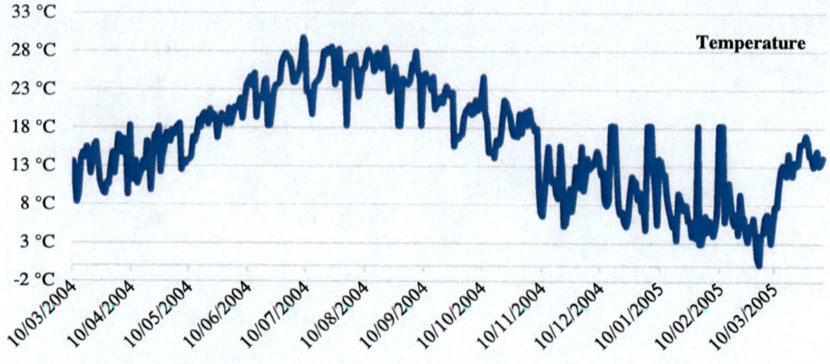

Fig. 10 Readings of temperature

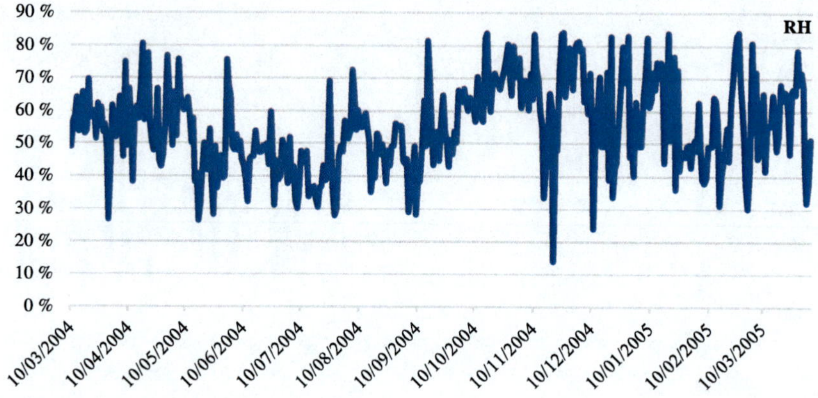

Fig. 11 Readings of RH

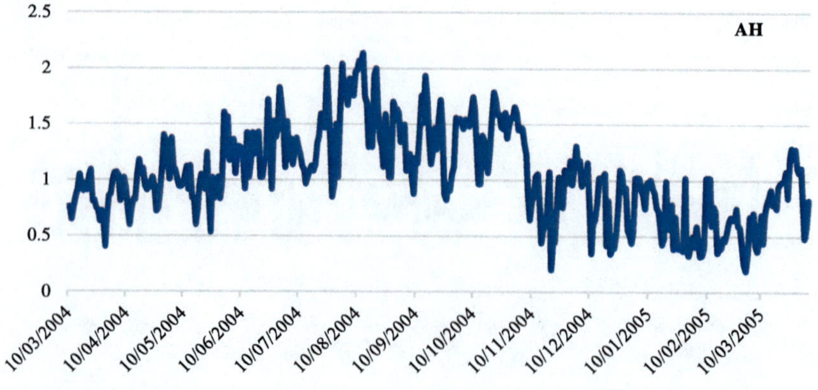

Fig. 12 Readings of AH

Table 1 Statistical analysis of all sensors readings

Sensors	Minimum value	Maximum value	Mean value	Standard deviation value
CO	0.1000	11.9000	2.15280	1.3081
PT08.S1	647.00	2040.00	1099.83	211.50
NMHC	7.0000	1189.00	218.811	63.484
C6H6	0.1000	63.7000	10.0831	7.2586
PT08.S2	383.00	2214.00	939.153	259.98
NOx	2.0000	1479.00	246.896	192.25
PT08.S3	322.00	2683.00	835.493	250.22
NO2	2.0000	340.000	113.091	43.655
PT08.S4	551.00	2775.00	1456.26	337.31
PT08.S5	221.00	2523.00	1022.90	388.25
T	−1.900	44.6000	18.3178	8.6054
RH	9.2000	88.7000	49.2342	16.872
AH	0.1847	2.23100	1.02550	0.3934

2.2 *Data Analysis*

In this section, all attributes recording are graphically represented, from Figs. 1, 2, 3, 4, 5, 6, 7, 8, 9, 10, 11, and 12. Statistical analysis of all attributes is also provided in Table 1.

Sensor selection as dependent variable: In the present study, focus is to reduce the cost of the sensing devices. CO is the main air pollutant and its value is required for calculating the air quality index. So, CO is selected as dependent variable to predict its value with the help of other independent parameters [19]. Carbon monoxide is generally formed in household surroundings, so it is important that sensing devices are effective in these regions.

Data contains samples which are not having values of all the parameters. Mean value of all the values of the attribute is used to replace the empty and missing values.

2.3 *Regression Model*

Regression analysis is a technique to establish the relation between two or more variables. In regression, there can be more than one independent variables; this type is known as multiple regression. The relation between dependent and independent variables can be obtained by linear regression, multi-linear regression, or polynomial regression [20, 21]. The mathematical function to model the data can be shown as $y = f(x)$. Equation (1) represents the modeling using *P*th order polynomial,

$$y = \beta_p x^p + \beta_{p-1} x^{p-1} + \cdots + \beta_1 x + \beta_0 \quad (1)$$

where the coefficients of the polynomial equations are represented as β and x is used to represent independent variables. y represents dependent variable [4].

Measures for Model performance: R^2 is coefficient of determination, and it is calculated by Eq. (2),

$$R = 1 - \frac{SSE}{SSM} \tag{2}$$

where *SSE* is sum of squares of the errors, which shows the measure of deviations of the predicted values from the observed values. *SSM* is sum of squares about the mean, which shows the measure of deviations of the mean from the observations. *SSE* and *SSM* are calculated by Eqs. (3) and (4),

$$SSE = \sum_{i=i}^{N} (y_i - \hat{y}_i)^2 \tag{3}$$

$$SSM = \sum_{i=1}^{N} (y_i - \overline{y})^2 \tag{4}$$

where actual value is represented by y and predicted value is shown by $\hat{y}$. Mean of observed N points is represented by $\overline{y}$.

To see the effect of extra variables, $\overline{R}^2$ is used, known as adjusted R^2 calculated by Eq. (5),

$$\overline{R}^2 = 1 - (1 - R^2)\frac{N - 1}{N - p - 1} \tag{5}$$

where sample size is represented by N and number of regressors is represented by p.

Mean Square Error (MSE) can be measured by Eq. (6),

$$MSE = \frac{1}{N} \sum_{i}^{N} (y_i - \hat{y}_i)^2 \tag{6}$$

Training data: Training data consists of 6000 samples. Models are trained with training data after that testing data is used for testing. Training data contains values of parameters in range (upper & lower limits) as mentioned in Table 2. CO is selected as dependent variable and other attributes as independent variables. NMHC data is removed from independent variables because it has too many empty values.

Interpolation Testing: For interpolation testing, sample values which lie inside the range of training dataset are used. About 30% samples of dataset are separated out for this testing.

Table 2 Range of independent variable for training of model

Sensor	Lower limit	Upper limit
PT08.S1	900	1200
C6H6	5	11
PT08.S2	750	1000
NOx	100	400
PT08.S3	700	1000
NO2	70	100
PT08.S4	1400	1600
PT08.S5	800	1100
T	13	22
RH	30	45
AH	0.7	1.2

Extrapolation Testing: For extrapolation testing, sample values which lie outside the range of training dataset are used. There are 11 independent variables in model and for all the variables selection of samples outside training range is difficult. After statistical analysis on data, only 280 samples are selected for this testing. Equation (7) shows the regression model with 11 sensors to predict the value of CO.

$$\begin{aligned} CO = {} & \beta_{11} \times AH + \beta_{10} \times RH + \beta_9 \times T + \\ & \beta_8 \times PT08.S5 + \beta_7 \times PT08.S4 + \\ & \beta_6 \times NO2 + \beta_5 \times PT08.S3 + \\ & \beta_4 \times NOx + \beta_3 \times PT08.S2 + \\ & \beta_2 \times C6H6 + \beta_1 \times PT08.S1 + \beta_0 \end{aligned} \tag{7}$$

After training the above model, values of β_0 to β_{11} are obtained as 5.12621946e-16, 0.13565925, 0.31641755, −0.15527467, 0.48651143, 0.0310526, 0.11056275, 0.41313658, −0.11145817, −0.07326808, −0.05334946 and −0.09505575, respectively.

3 Results

Regression model of linear and polynomial nature has been used. To predict value of CO, linear model (degree 1), polynomial model of degree 2 and degree 3 are utilized and performance of models is listed in Tables 3 and 4. Table 3 consists of interpolation testing results. In Table 4, extrapolation testing results are listed.

From above results, it has been observed that on average interpolation results are better than extrapolation results. One of the reasons behind this is less number of

Table 3 Interpolation testing results

Model degree	$\overline{R}^2$	R^2	SSE	SSM	MSE
Degree 1	0.791987	0.792231	982.731	4729.92	0.324762
Degree 2	0.81524	0.81678	866.615	4729.92	0.28639
Degree 3	0.720527	0.731399	1270.46	4729.92	0.419849

Table 4 Extrapolation testing results

Model degree	$\overline{R}^2$	R^2	SSE	SSM	MSE
Degree 1	0.821097	0.821308	231.932	1297.94	0.825382
Degree 2	0.800105	0.801771	257.29	1297.94	0.915621
Degree 3	-2.80293	-2.65499	4743.98	1297.94	16.8825

samples in testing for extrapolation. But liner model is working more accurately for extrapolation sample.

4 Conclusion

The models are fitted over the air quality data for prediction of CO gas, achieved up to 82% extrapolation testing accuracy for linear models. In case of interpolation testing, a polynomial model of degree 2 is suitable with 81% accuracy. The developed models can also be used in real devices to save the cost of sensors whose value is predicted by the regression model. In the present study, CO gas sensor is selected because it is a major component of air pollution and treated as a dependent variable for prediction. Results show that CO sensors can be eliminated, which leads to developing low cost devices. As an overall conclusion study presents, regression-based model development for low cost devices monitors air quality by eliminating the need for specific sensors.

5 Future Work

On the basis of the presented results, further advanced methods for modeling can be applied over the same data. This model can also be used with other datasets also. This study is also considered as a prototype for low cost air quality sensing devices, which can be implemented as real devices.

References

1. Z. Khan, R. Srivastava, Air pollution and its impact on human health in India: a review. J. Biol. Sci. Med. **5**(4), 9–12 (2019)
2. K. Kanchan, A.K. Gorai, P. Goyal,A review on air quality indexing system. Asian J. Atmos. Environ. **9**(2), 101–113 (2015)
3. R. Jain, K. Palwa, Air pollution and health, in *Discussion Paper, The Energy and Resources Institute (TERI)* (New Delhi, 2015). https://www.teriin.org/
4. C.C. Lim, H. Kim, M.J.R. Vilcassim, G.D. Thurston, T. Gordon, L.C. Chen, K. Lee, M. Heimbinder, S.Y. Kim, Mapping urban air quality using mobile sampling with low-cost sensors and machine learning in Seoul, South Korea. Environ. Int. **131**, 105022 (2019)
5. P. Lyu, N. Chen, S. Mao, M. Li, LSTM based encoder-decoder for short-term predictions of gas concentration using multi-sensor fusion. Process Saf. Environ. Prot. **137**, 93–105 (2020)
6. J. Li, H. Zhang, C.Y. Chao, C.H. Chien, C.Y. Wu, C.H. Luo, L.J. Chen, P. Biswas, Integrating low-cost air quality sensor networks with fixed and satellite monitoring systems to study ground-level PM2. 5. Atmos. Environ. **223**, 117293 (2020)
7. L. Spinelle, M. Gerboles, M.G. Villani, M. Aleixandre, F. Bonavitacola, Field calibration of a cluster of low-cost commercially available sensors for air quality monitoring. Part B: NO, CO and CO_2. Sens. Actuators B: Chem. **238**, 706–715 (2017)
8. K.B.K. Sai, S. Mukherjee, H.P. Sultana, Low cost IoT based air quality monitoring setup using arduino and MQ series sensors with dataset analysis. Procedia Comput. Sci. **165**, 322–327 (2019)
9. Y. Zhou, X. Zhao, K.-P. Lin, C.-H. Wang, L. Li, A Gaussian process mixture model-based hard-cut iterative learning algorithm for air quality prediction. Appl. Soft Comput. **85**, 105789 (2019)
10. C. Xu, J. Zhao, P. Liu, A geographically weighted regression approach to investigate the effects of traffic conditions and road characteristics on air pollutant emissions. J. Clean. Prod. **239**, 118084 (2019)
11. S. Masmoudi, H. Elghazel, D. Taieb, O. Yazar, A. Kallel, A machine-learning framework for predicting multiple air pollutants' concentrations via multi-target regression and feature selection. Sci. Total Environ. **715**, 136991 (2020)
12. S. Araki, H. Shimadera, K. Yamamoto, A. Kondo, Effect of spatial outliers on the regression modelling of air pollutant concentrations: a case study in Japan. Atmos. Environ. **153**, 83–93 (2017)
13. K.-P. Lin, P.-F. Pai, S.-L. Yang, Forecasting concentrations of air pollutants by logarithm support vector regression with immune algorithms. Appl. Math. Comput. **217**(12), 5318–5327 (2011)
14. Y. Chen, M. Zheng, J. Lv, T. Shi, P. Liu, Wu. Yan, W. Feng, W. He, Pi. Guo, Interactions between ambient air pollutants and temperature on emergency department visits: analysis of varying-coefficient model in Guangzhou, China. Sci. Total Environ. **668**, 825–834 (2019)
15. S. De Vito, E. Massera, M. Piga, L. Martinotto, G. Di Francia.On field calibration of an electronic nose for benzene estimation in an urban pollution monitoring scenario. Sens. Actuators B: Chem. **129**(2), 750–757 (2008)
16. S. De Vito, G. Fattoruso, M. Pardo, F. Tortorella, G. Di Francia, Semi-supervised learning techniques in artificial olfaction: a novel approach to classification problems and drift counteraction. IEEE Sens. J. **12**(11), 3215–3224 (2012)
17. S. De Vito, M. Piga, L. Martinotto, G. Di Francia, CO, NO_2 and NOx urban pollution monitoring with on-field calibrated electronic nose by automatic bayesian regularization. Sens. Actuators B: Chem. **143**(1), 182–191 (2009)
18. D. Dua, C. Graff, *UCI Machine Learning Repository* (University of California, School of Information and Computer Science, Irvine, CA 2019). http://archive.ics.uci.edu/ml
19. T. Nandy, R.A. Coutu, C. Ababei, Carbon monoxide sensing technologies for next-generation cyber-physical systems. Sensors **18**(10), 3443 (2018)

20. M.K. Ahirwal, A. Kumar, G.K. Singh, N.D. Londhe, Performance prediction of adaptive filters for EEG signal. IET Sci., Meas. Technol. **11**(5), 525–531 (2017)
21. D.C. Montgomery, E.A. Peck, G.G. Vining, *Introduction to Linear Regression Analysis*, vol. 821 (Wiley, 2012)

Initial Results on High Sensitivity Magnetic Induction Spectroscopy Circuit for Metal Detection in Human Body

Gin Haur Khor, Zulkarnay Zakaria, Jaysuman Pusppanathan, Anas Mohd Noor, Ahmad Nasrul Norali, Yen Fook Chong, Asyraf Hakimi Abu Bakar, Muhamad Khairul Ali Hassan, Muhammad Juhairi Aziz Safar, and Ahmad Faizal Salleh

Abstract Bone fracture healing has a relative close connection with the fixation stability. A dislocation of the bone screw may cause the screw loosening and future lead to screw defect. Early detection of the position shifting of the metal (bone screw) is the key for faster healing. Therefore, a high sensitivity Magnetic Induction Spectroscopy (MIS) is proposed to identify the relative position of the implanted metal in human body. The purpose of this study is to investigate the potentials of MIS for metal detection in human body. In this project, the MIS circuit is designed, simulated using NI Multisim Software and hence fabricated on the PCB. This study has found that a magnitude measurement technique of MIS system is sufficiently sensitive for metal detection in human body, due to the conductivity of the biological tissues is million times lower than metallic object. The study also proven that the developed MIS circuit is capable in detecting the position shift of metal in human body at operating frequency of 500 kHz.

Keywords Magnetic induction spectroscopy · Metal detection · Biological tissue

---

G. H. Khor · Z. Zakaria (✉) · A. M. Noor · A. N. Norali · Y. F. Chong · A. H. A. Bakar · A. F. Salleh
Biomedical Electronic Engineering, Faculty of Electronic Engineering Technology, Universiti Malaysia Perlis, Arau, Perlis 02600, Malaysia
e-mail: zulkarnay@unimap.edu.my

Z. Zakaria · A. M. Noor · A. N. Norali · Y. F. Chong · A. H. A. Bakar · M. K. A. Hassan · M. J. A. Safar · A. F. Salleh
Medical Device & Life Sciences Cluster, Sports Engineering Research Centre (SERC), Universiti Malaysia Perlis, Arau, Perlis 02600, Malaysia

J. Pusppanathan
Sport Innovation & Technology Centre (SiTC), Institute of Human Centered Engineering (IHumen), Universiti Teknologi Malaysia, Skudai, Johor 81310, Malaysia

F. Thakkar et al. (eds.), *Proceedings of the International e-Conference on Intelligent Systems and Signal Processing*, Advances in Intelligent Systems and Computing 1370,
https://doi.org/10.1007/978-981-16-2123-9_28

1 Introduction

Bone fractures are most commonly seen in patients with traumatic injuries [1]. Generally, for the bones to become fractured is due to trauma, for example, an automobile accident, a fall, a sport injury, and so on. As a bone become fracture, it requires an external support to stabilize and restore the bone anatomy of the fracture site [2]. Therefore, a bone fracture treatment known as internal fixation is applied.

During the internal fixation treatment, the metal plate and screw act as supportive device for the injure bones by providing mechanical support for physiological load at the fracture site to enable fast healing of the injured bone [3]. The fracture healing has a relative close connection with the fixation stability [4]. A dislocated screw in the bone fixation may cause the screw loosening to occur and future leads to the screw crack and screw defect [5]. Thus, it is important to identify the relative position of the implanted bone screw after the internal fixation surgery [6]. Hence, Magnetic Induction Spectroscope (MIS) which is a non-invasive and ionizing radiation-free technique for early detection of the metal in human body is proposed.

In general, Magnetic Induction Spectroscopy (MIS) for metal detection in human body is started with the electromagnetic field generated at the transmitter, which propagates through the material (object under investigation), and hence eddy current is induced within the object due to passive electrical properties of in the object itself. This eddy current then generates its own field known as secondary fields that will

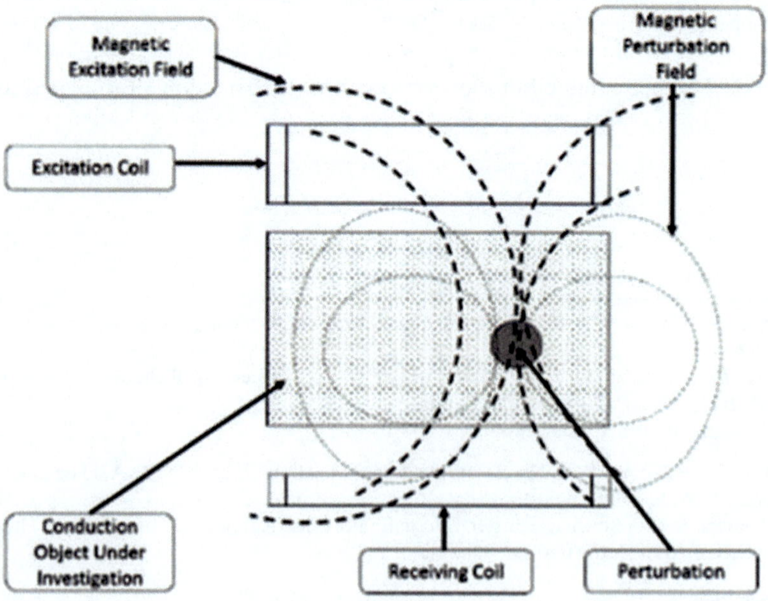

Fig. 1 Principle of magnetic induction spectroscopy [9]

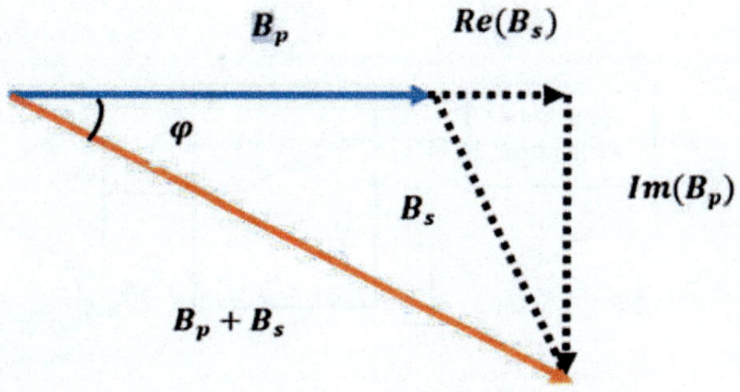

Fig. 2 Vector diagram of magnetic induction spectroscopy principle

be received or sensed by the receiver coils[7, 8]. Figure 1 shows the principle of the magnetic induction spectroscopy.

The principle of the signal received by the receiver can be explained and represented by a vector diagram, as shown in Fig. 2. In general, two signal will be sensed by the receiver which are $\boldsymbol{V_p}$ and $\boldsymbol{V_s}$, where $\boldsymbol{V_p}$ represented the signal induce from the excitation field or known as primary fields, $\boldsymbol{B_{p.}}$ Whereas the $\boldsymbol{V_s}$ is the signal that derived from the eddy currents. In addition, these eddy current will induce perturbation of the magnetic fields, which will change the strength and spatial distribution of the original excitation magnetic field by a phase angle of φ[10]. Basically, the signal $\boldsymbol{V_s}$ can be divided into two parts, a real part $\boldsymbol{Re(V_s)}$ and an imaginary part $\boldsymbol{Im(V_s)}$. The real part represents permittivity, while imaginary represents conductivity of the object under investigations. Essentially, the $\boldsymbol{V_s}$ signal is the solution for detection of metal in human body, where the $\boldsymbol{V_s}$ signal carries information of the object of interest (bone screw) [11].

2 Methodology

2.1 Hardware Design

Basically, the magnetic induction spectroscopy hardware design can be divided into three parts which are the transmitter (Tx), receiver (Rx), and controller unit.

2.1.1 Transmitter and Receiver

The arrangement of the MIS transmitter and receiver circuit is shown in Fig. 3 and Fig. 4, respectively.

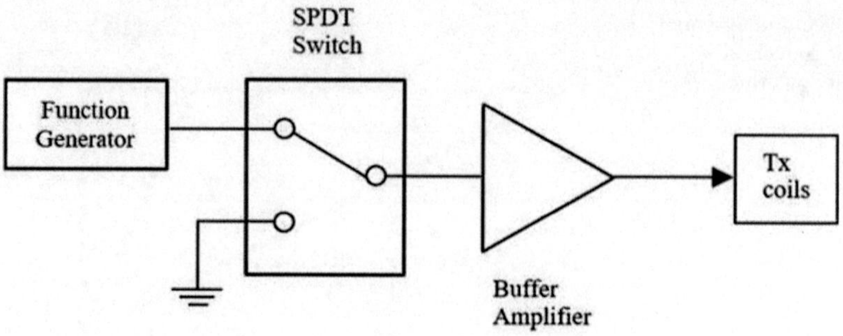

Fig. 3 Arrangement of the MIS transmitter circuit

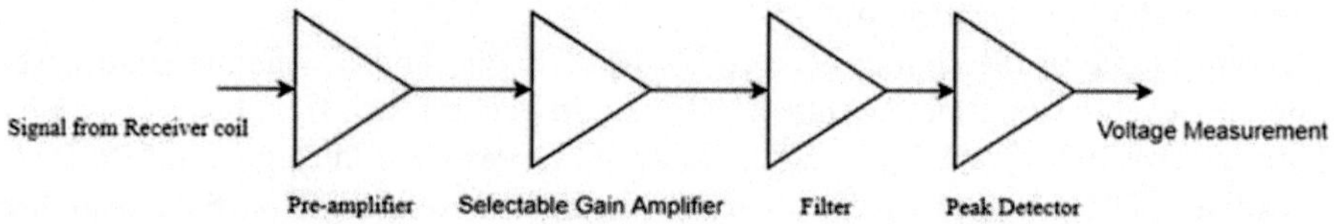

Fig. 4 Arrangement of the MIS receiver circuit

2.1.2 Controller Unit

Arduino Mega 2560 is used as a controller unit to control the MIS system. Where the controller is responsible to control the pre-amplifier, programmable gain amplifier, and reset for the peak detector. The advantage of Arduino Mega 2560 controller is that itconsists of 54 digital pin and it also consists of Universal Asynchronous Receiver-Transmitter (UART) module, which allows the MIS system to interface with the computer Hardware Setup and Data Collection.

2.2 Experimental Setup

The setup is shown in Fig. 5. The sensor jig is made of acrylic and surrounded by electromagnetic screen to avoid interference from outside. The diameter of the Region Of Interest (ROI) is 48 cm.

In this experiment, steel rod of cylindrical types with three different sizes (1.5, 2.0, and 2.5 inch) as in Fig. 6 has been tested. The steel rod will be immersed a container filled with water and placed in the ROI. The water medium is to mimic the human biological body, where average of 60% of human biological body is built from water. The steel rod moves along the Tx and Rx pathway in a straight line starting from the position of −20 cm to +20 cm with an increment of 2 cm with the reading at each point is recorded. The experiment was done at frequency range from 300 to 700 kHz as these frequencies are within Beta-dispersion range.

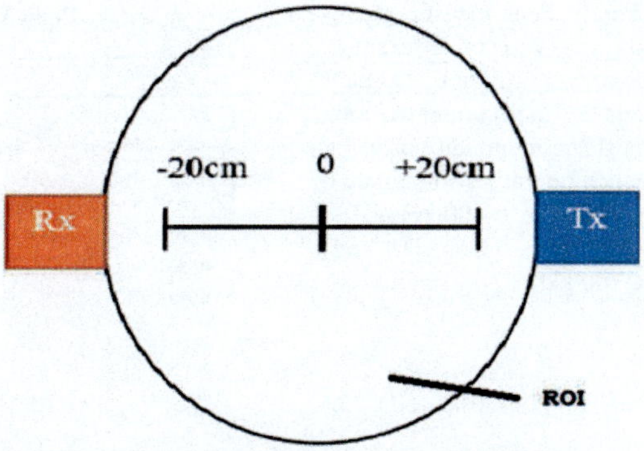

Fig. 5 Measurement pathway between Tx and Rx

Fig. 6 Cylinder steel rod with different sizes; 1.5 inch, 2.0 inch, and 2.5 inch diameter

3 Results and Discussions

The results in Fig. 7 show the peak voltage reading at every position with 2 cm increment starting from point −20 cm to +20 cm. In general, we can see that the magnitude of the peak voltage is increasing with the applied frequency which 700 kHz records the highest at all positions. All graphs show the similar pattern with higher reading at both ends, whereas at the center of ROI the readings are the lowest. The higher reading at both ends is due to the nearest location of both ends to the transmitter/receiver as the signal is relatively stronger compared to the center of ROI.

Signal at the center of ROI is the weakest as it is the farthest point to both receiver and transmitter. On the other hand, the voltage measurement at the position toward receiver sensor coils (Rx) side is higher than the transmitter coils (Tx) side since the number of turn for Rx coils is larger than the Tx coils. The experiments show that only frequency of 400 and 500 kHz produce the better results for all three different sizes as the readings are clearly separated (not overlapped) compared to the readings at other applied frequencies. However, 500 kHz is the best as the recorded signal is stronger than that is measured at 400 kHz.

Fig. 7 Peak measurement of the steel rods of different diameter (1.5 inch, 2.0 inch, and 2.5 inch) under different position in human body when operating frequency of **a** 300 kHz, **b** 400 kHz, **c** 500 kHz, **d** 600 kHz, and **e** 700 kHz

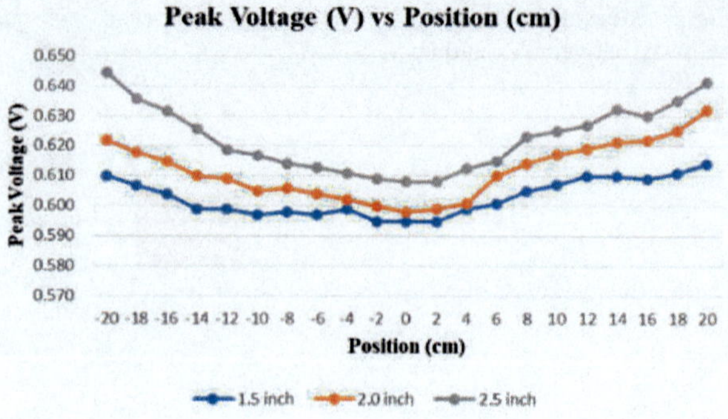

(a)

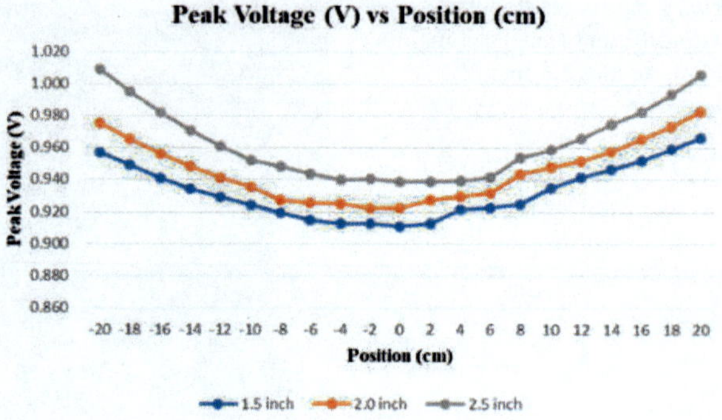

b)

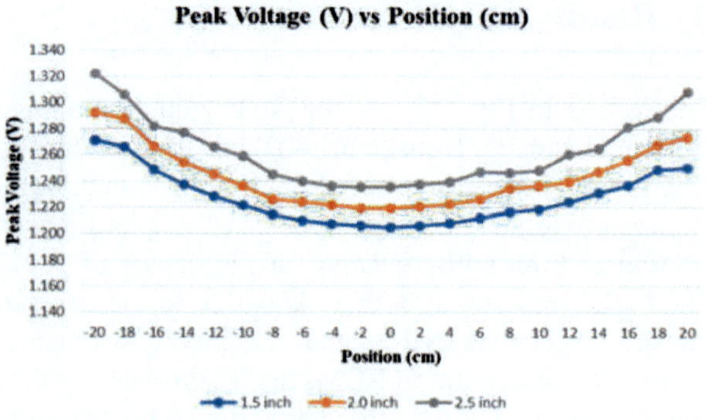

(c)

Fig. 7 (continued)

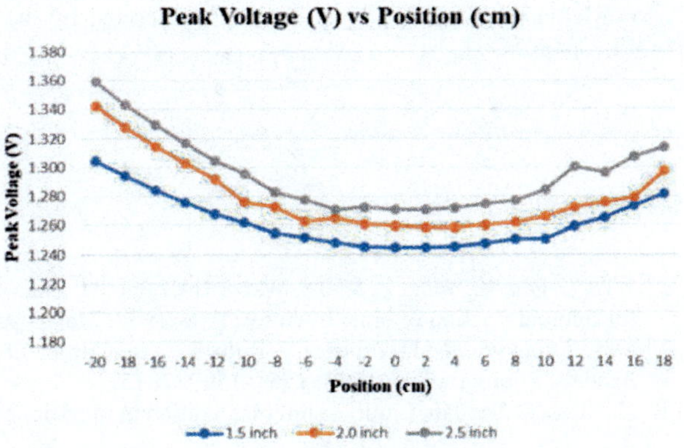

(d)

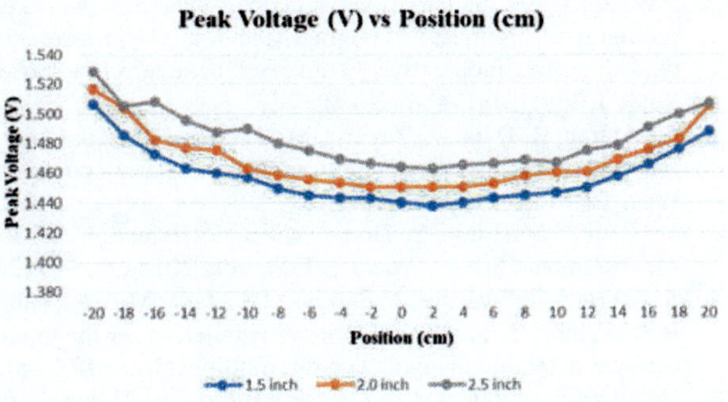

(e)

Generally, the coil used in this research project is a ferrite core coils. The nature properties the ferrite core coils is low losses at high frequency and due to this characteristics, ferrite coils tend to absorb noise from the surrounding environment.

4 Conclusions

Magnetic Induction Spectroscopy (MIS) is a non-invasive, contactless, and electrodeless technique for metal detection in human body based on the Passive Electrical Properties (PEP) of the object under investigation where in this research the dominant property is conductivity. The outcome of this project concludes that the designed MIS circuit has the high sensitivity to detect the metal size of 1.5 inch in diameter in human body with the best frequency of 500 kHz. The success of the MIS system will provide opportunity for moving toward to a more advance non-invasive measurement in the future.

Acknowledgements This project has been supported by the grant UTMSHINE 09G18, TDR 06G17, CRG 05G04.

References

1. J. Li, L. Qin, K. Yang, Z. Ma, Y. Wang, L. Cheng, D. Zhao, Materials evolution of bone plates for internal fixation of bone fractures: a review. J. Mater. Sci. Technol. **36**, 190–208 (2020)
2. M.S. Taljanovic, M.D. Jones, J.T. Ruth, J.B. Benjamin, J.E. Sheppard, T.B. Hunter, Fracture fixation. Radiographics **23**(6), 1569–1590 (2015)
3. H. Nieto, C. Baroan, Limits of internal fixation in long-bone fracture. Orthop. Traumatol. Surg. Res. **103**(1), S61–S66 (2017)
4. X. Wu, J. Shi, J. Wu, Y. Cheng, K. Peng, J. Chen, H. Jiang, Pedicle screw loosening: the value of radiological imagings and the identification of risk factors assessed by extraction torque during screw removal surgery. J. Orthop. Surg. Res. **14**(1), 1–9 (2019)
5. J.M. Spirig, R. Sutter, T. Götschi, N.A. Farshad-Amacker, M. Farshad, Value of standard radiographs, computed tomography, and magnetic resonance imaging of the lumbar spine in detection of intraoperatively confirmed pedicle screw loosening—a prospective clinical trial. Spine J. **19**(3), 461–468 (2019)
6. M. Amran, R. Daud, Z. Zakaria, M.K.A. Hassan, M.I. Omar, F. Mat, Effect of electromagnetic induction spectroscopy of screw crack implant on electromagnetic signal strength. Mater. Today Proc. **16**, 2153–2159 (2019)
7. J. Xiang, H. Xiong, Y. Dong, Magnetic Induction Sensing with a Gradiometer Coil and Measurement Circuit. Proc. IEEE Sensors 2018 (1), 1–3 (2018)
8. S. Sarkawi, Z. Zakaria, I. Balkhis, J.A. Jalil, M.A.A. Rahim, M.H.F. Rahiman, N. Mustafa, R.A. Rahim, Z. Shaffie, 3D model simulation on magnetic induction spectroscopy for fetal acidosis detection using COMSOL multiphysics. AIP Conf. Proc. **7**, 1774 (2016)
9. H. Scharfetter, R. Casañas, J. Rosell, Biological Tissue characterization by magnetic induction spectroscopy (MIS): requirements and limitations. IEEE Trans. Biomed. Eng. **50**(7), 870–880 (2003)
10. A.J. Jurimah, Z. Zakaria, B. Ibrahim, S. Shazwani, A.R.M. Alif, A.R. Ruzairi, M.A. K.A. Nasir, Development of magnetic induction spectroscopy system in biomedical application to detect jaundice, in AIP Conference Proceedings, vol. 1774, October, 2016
11. M.S.B. Mansor, Z. Zakaria, I. Balkhis, R.A. Rahim, M.F.A. Sahib, Y.M. Yunos, S. Sahlan, S. Bunyamin, K.H. Abas, M.H.I. Ishak, K.A. Danapalasingm, Magnetic induction tomography: a brief review. J. Teknol. **73**(3), 91–95 (2015)
12. P. De Tillieux, Y. Goussard, Biomedical magnetic induction tomography: an inhomogeneous Green's function approach. Appl. Comput. Electromagn. Soc. J. **34**(2), 360–362 (2019)

Additional Control Based on ANFIS Algorithm to Improve Transient Current of Converter-Side in HVDC Transmission System

I Made Ginarsa, I Made Ari Nrartha, Agung Budi Muljono, Sultan Sultan, and Sabar Nababan

Abstract Soft computing likes ANFIS now penetrating the control system, especially for power semi-conductor and power system controls. Control of trigger angle is one of an important method to deliver power through high-voltage direct current (HVDC) transmission system effectively. Higher transient of current is one problem of HVDC on start-period, for up-ramp at 20 pu/s or higher. To cove this problem, the ANFIS control scheme is proposed to suppress the transient current in this research. The ANFIS control is built by using data train that these data are obtained from the simulation of HVDC system equipped by PI control under various loads and up-ramp rates. Training of ANFIS controller is done by input–output in 5000 data points. After training stage, the ANFIS controller is embedded into HVDC converter through fuzzy logic controller tool. Simulation results show that the proposed control is more effective to reduce the transient current in all scenarios compared to conventional PI control. Where the peak overshoot of transient current is achieved at the value of 1.11 pu for the proposed control. While the peak overshoot of transient current for conventional PI control is obtained at the value of 1.148 pu.

Keywords ANFIS · Additional control · HVDC · Improvement · Transient time

1 Introduction

By growing-up of industrial and commercial load demands in power systems, and difficulty to develop power plant near load center makes transmission line is the only one choice to connect the power plant and the load center. So, load locations are commonly very far from power plant sites. In this situation, the transmission line has a vital role to deliver amount of power successfully in all condition operations. While high-voltage direct current (HVDC) transmission system is set up to substitute/replace high-voltage alternating current (HVAC) transmission one, especially for area that the HVAC is not be installed such as long-distance transmission

I. M. Ginarsa (✉) · I. M. A. Nrartha · A. B. Muljono · S. Sultan · S. Nababan
University of Mataram, NTB, Mataram 83125, Indonesia
e-mail: kadekgin@unram.ac.id

F. Thakkar et al. (eds.), *Proceedings of the International e-Conference on Intelligent Systems and Signal Processing*, Advances in Intelligent Systems and Computing 1370,
https://doi.org/10.1007/978-981-16-2123-9_29

line, long submarine-cable, and asynchronous interconnected [1, 2]. Deep discussion of ultra-high voltage application to deliver of power included renewable energy for supply Europe-Asia continents through long-distance overhead transmission line and submarine cable is done in [3]. Also, security-constrained (SC) optimization model for DC/AC transmission is developed to support the HVDC and HVAC optimal operation [4], for multi-area coordination [5] and a new corrective SC optimization is built for support AC grid-voltage source converter (VSC) multi-terminal DC [6]. The SC optimization economic dispatch for combination of wind power was combined with AC/DC grids controlled by HVDC system [7]. The SC unit commitment under penetration of high wind turbine is developed with distributed dispatch for bulk AC/DC meshed transmission system [8]. Starting strategy for HVDC-link using VSC-HVDC is introduced for supply passive network [9]. A new recovery control is developed in [10] to overcome real/reactive powers exchange between bus AC and rectifier converter during fault on AC-side. Robust stability power in transmission using UPFC controlled by adaptive neural network [11] and transient stability of AC-DC power systems is able to improve by HVDC controller based on neural network [12]. Some efforts to improve performance of HVDC are developed such as by using 24-pulse AC-DC-converter to mitigate current/voltage harmonics to improve power quality in [13] and some voltage source converter (VSC)-HVDC models to enhance their interaction to AC grids. In VSC-HVDC model, it is found that significant reduction in total sum-error for longer AC grid around a VSC station, especially emphasized on fast phase locked loop (PLL) speed control in dynamic model [14]. Simple dynamic model of HVDC included line commutated converter (LCC) and VSC models are built to avoid model complexity and to shorter simulation time in an English-Scottish power system. Also, small-signal stability model for critical inter-area oscillation is classified using mode shape quantification [15].

Development of artificial intelligent control (AIC) is an active topic research and spread into power system and power electronic engineering in recent years. Some of the AIC schemes are applied on power system stability and control such as ANFIS control is used to suppress chaos and voltage collapse in power systems [16, 17], fuzzy type-2 is applied on design power system stabilizer (PSS) to enhance small-signal stability of a large-scale power system [18], fuzzy type-2 algorithm combined by digital signal processing is used to control permanent magnet synchronous machine drive [19], and ANFIS-PSS based on feedback linearization is developed to improve stability of a non-linear model single-machine [20]. Moreover, fuzzy control is used to regulate a solar PV generation connected to grid system for control both active and reactive power on its inverter device during the grid fault [21], fuzzy and neural network control is implemented on distributed generation to reduce power ripple, harmonic distortion and to increase response time [22], and the fuzzy logic control is applied on distributed-statcom to enhance the power quality in distribution system [23]. The ANFIS control-scheme is implemented to control HVDC on starting-time and short-circuit faults [24], to suppress transient of direct current [25]. The ANFIS method is used on space vector pulse width modulation to voltage source converter HVDC [26], and the ANFIS control is built to regulate average value model of inverter [27] and rectifier-side detail model of HVDC system [28].

Rest of this paper is organized as follows: High voltage direct current transmission model is introducted in Sect. 2. Next, design of additional ANFIS control is explored in Sect. 3. Result and analysis are explained in Sect. 4. And finally, conclusion is summarized in Sect. 5.

2 High Voltage Direct Current Transmission Model

In order to simplify HVDC model, it can be developed by diagram blocks to represent the respective component such as Sending-end, HVDC transmission, receiving-end, and the HVDC transmission line. The sending-end (bus rectifier converter) is commonly built near power plant to supply the HVDC transmission system. Meanwhile, the receiving end (bus inverter converter) is commonly developed near load center, and the power is distributed directly to consumer. The transmission HVDC is developed over long-distance area to connect the sending-end and receiving-end. The HVDC system model in this research is provided by [29]. This model is shown in Fig. 1. The system is arranged by an AC ideal source (Bus 1) 500 kV, 5000 MVA equivalent, freq. 60 Hz, angle 80 degrees. Converter bus (Bus 2) consists of device rectifier device. The function of this rectifier is to convert the AC voltage/current into DC voltage/current forms. And the DC voltages/currents are transmitted to Bus 3 through the HVDC transmission line. Bus 3 consists of inverter device, where the function of inverter is to convert the DC voltage/current into AC voltage/current forms again. The AC voltages/currents are sent to Bus 4. The Bus 4 is modeled by 10,000 MVA equivalents, Voltage 345 kV, frequency 50 Hz, and angle 80 degrees. Assumed that the Bus 4 is functioned as a load bus. The power is delivered from Bus 1 to Bus 4 and transmitted via HVDC line in this scenario.

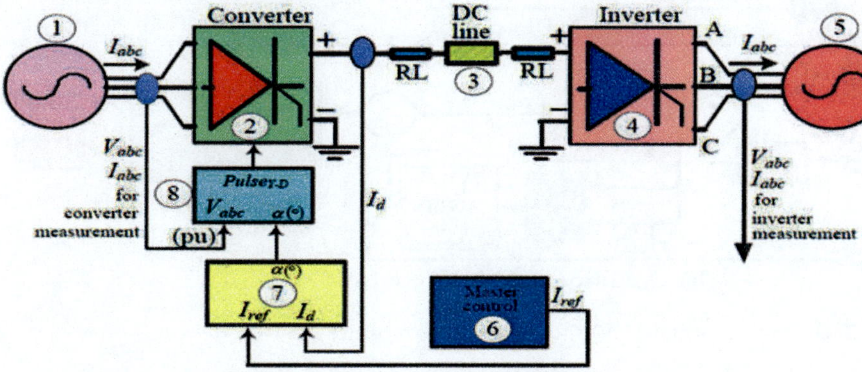

Fig. 1 HVDC and its control scheme represented in a simple model

3 Design of Additional ANFIS Control

This section is conducted to design additional ANFIS control. At first stage, row-data sets were collected from simulation of HVDC model [29] that was equipped by proportional–integral (PI) controller in various load. The PI controller diagram block that is used in this research is shown in Fig. 2a. Next stage, an input–output matrix of ANFIS control was developed on Matlab script by using I_{err}, dI_{err}, and α_{AN} for inputs and output, respectively. This input–output matrix consists of [2 Inputs + 1 Output] $\times$ $5000 = 3 \times 5000$ data points. The input–output matrix was formed into train-data set and was used to train the additional ANFIS control. Neuro-fuzzy designer (anfisedit tool) packet program was used in this learning on off-line train mode. In training stage, the train-data set was given as input into the anfisedit tool. Some ANFIS parameters are set as follows: Gaussian and linear membership function (MF) types were used for the inputs and output MFs, respectively. Every training session was taken on 20 epochs. Moreover, least squares estimate and backpropagation methods are applied to optimize the ANFIS parameter on forward and backward steps, respectively. Fuzzy inference system (FIS) was obtained at every training session. Values of the FIS parameter were generated and maintained automatically according to the train-data set that was used to learn the ANFIS control. At final stage, the FIS structure was saved into *.fis file format and was embedded to fuzzy control on Simulink model. The additional ANFIS control and its installed at rectifier of HVDC is illustrated by diagram block in Fig. 2b.

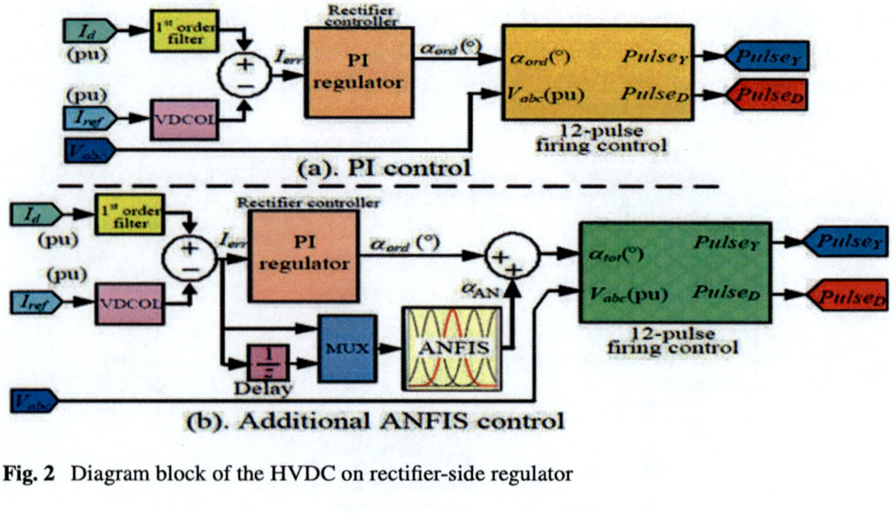

Fig. 2 Diagram block of the HVDC on rectifier-side regulator

4 Result and Analysis

Performance of additional ANFIS control is tested on an Intel Core i5-7400 processor, 6.0 MB cache, 3.0 GHz, and LGA 1151 personal computer. The Matlab/Simulink 7.0.9 (2013b) [30] packed program was implemented to simulate the HVDC model equipped by ANFIS control.

In this scheme control, the PI parameters were set at the values of 45, 4500, and 90° for the P, I, and initial alpha degree, respectively, on current regulator of rectifier-side. The results are also listed in Table 1, for peak overshoot (M_p) and settling time (t_{st}) of all control schemes.

It is shown that the peak overshoot was obtained at the values of 1.088 and 1.05 pu for PI and ANFIS controls, respectively, when the up-ramp rate at 21 pu/s. When the up-ramp rate was increased to 23 pu/s, the peak overshoot was observed at the values of 1.106 and 1.068 pu for the PI and ANFIS controls. Next, the up-ramp rate was taken on 25 pu/s, the peak overshoot was achieved at the values of 1.118 and 1.077 pu for the PI and ANFIS controls, respectively. Figure 3 shows simulation results of the proposed control compared to conventional (PI) control for the up-ramp

Table 1 Performance of the proposed control examine on various up-ramp rate

Up-ramp rate (pu/s)	PI control		Additional ANFIS	
	M_p(pu)	t_{st}(s)	M_p(pu)	t_{st}(s)
21	1.088	0.589	1.05	0.572
23	1.106	0.594	1.068	0.586
25	1.118	0.605	1.077	0.597
27	1.136	0.613	1.097	0.601
29	1.148	0.618	1.11	0.609

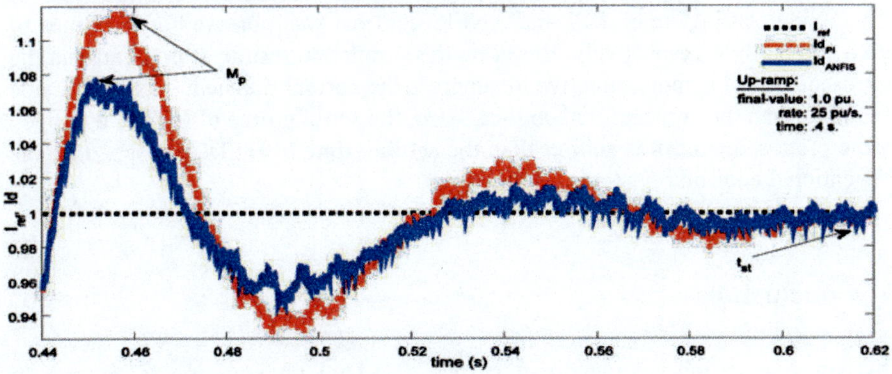

Fig. 3 Transient current response improved by the proposed control

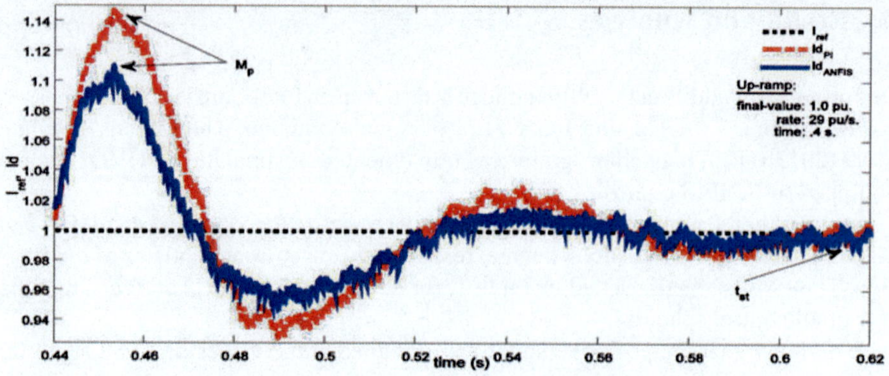

Fig. 4 Improvement of transient current response for up-ramp rate at 29.0 pu/s

rate, up-ramp final value, and up-ramp time were set at 25 pu/s, 1.0 pu, and 0.4 s, respectively.

Next, the up-ramp rate was increased again to 27 pu/s, the peak overshoot was at the values of 1.136 and 1.097 pu. Finally, the up-ramp rate was increased again to 29 pu/s, the peak overshoot was obtained at the values of 1.148 and 1.11 pu for the PI and ANFIS controls, respectively. Furthermore, the proposed control performed is compared to PI control are shown in Fig. 4, for the up-ramp rate, up-ramp final value, and up-ramp time set at 29 pu/s, 1.0 pu, and 0.4 s.

The settling time of the PI and ANFIS controls was obtained at times of 0.589 and 0.572 s for the up-ramp rate at 21 pu/s. Moreover, for the up-ramp rate increased to 23 pu/s, the settling time of the PI and ANFIS controls was achieved at the times of 0.594 and 0.586 s. For the up-ramp rate at 25 pu/s, the settling time of the PI and ANFIS controls was at 0.605 and 0.597 s. Next, for the up-ramp rate increased again to 27 pu/s, the settling time of the PI and ANFIS controls was achieved at the times of 0.613 and 0.601 s. Finally, the up-ramp rate was increased to the value of 29 pu/s, the settling time of PI and ANFIS controls was observed at the times of 0.618 and 0.609 s, respectively. Based on the simulation results, it is found that the proposed control is more effective to suppress the current transient on rectifier side of HVDC than the conventional control. Also, the settling time of HVDC equipped by the proposed control is shorter than the settling time of HVDC equipped by the conventional control.

5 Conclusion

Thyristor-based high-voltage direct current (HVDC) transmission is commonly used on interconnected high-voltage alternating current (HVAC) transmission as the backbone of modern power system. In order to support properly operation of HVDC/HVAC transmission, regulation rectifier of HVDC is very important to control sending power as the consumers needed at receiving end. In thyristor-based HVDC system, it is found that transient current too high when up-ramp rate is set more than 20 pu/s. To solve this problem, an additional control based on ANFIS algorithm is proposed in this research to reduce transient current response of HVDC system. The proposed control is able to maintain the transient current response of converter-side in HVDC model. This result has been compared to the PI conventional control in order to validate the simulation result. The peak overshoot is achieved at the value of 1.11 pu for the proposed control. Meanwhile, the peak overshoot is achieved at the value of 1.148 pu for conventional control. The next research is to suppress the transient current by using fuzzy type-2 or other control schemes.

Acknowledgements The Authors would like thank to the Lembaga Penelitian dan Pengabdian Masyarakat University of Mataram to support this research through PNBP 2020 Program.

References

1. J. Arrillaga, Y.H. Liu, N.R. Watson, *Flexible Power Transmission: The HVDC Options*. (Wiley, 2007), pp. 529–551
2. V.F. Pires, J. Fialho, J.F. Silva, HVDC transmission system using multilevel power converters based on dual three-phase two-level inverters. IJEPES J. **65**, 191–200 (2015)
3. T.J. Hammons et al., State of the art in ultrahigh-voltage transmission. Procs. IEEE **100**(2), 360–384 (2012)
4. M. Giuntoli, S. Schmitt, Security analysis of embedded HVDC in transmission grids, in, *Advanced in Energy Optimization, Trends in Mathematics*, ed. by V. Bertsch, et al. (2020) pp. 13–25
5. N. Huebner et al., Multi-area coordination of security-constrained dynamic optimal power flow in AC-DC grids with energy storage, in *Advanced in Energy Optimization, Trends in Mathematics* ed. by V. Bertsch, et al. (2020), pp. 26–40
6. Y. Li, Y. Li, Security-constrained multi-objective optimal power flow for a hybrid AC/VSC-MTDC system with lasso-based contingency filtering. IEEE Access **8**, 6801–6811 (2020)
7. H. Huang, et al., Exploiting the operational flexibility of wind integrated hybrid AC/DC power systems. IEEE Trans. Power Syst. **36**(1) (2021).
8. M. Zhou et al., Distributed dispatch approach for bulk AC/DC hybrid systems with high wind power penetration. IEEE Trans. Power Syst. **33**(3) (2018).
9. C. Guo, C. Zhao, Supply of an entirely AC network through a double-infeed HVDC system. IEEE Trans. Power Electron. **24**(11), 2835–2840 (2020)
10. J. Wang et al., A new strategy of HVDC system during AC faults. IEEE Trans. Power Del. **34**(2), 486–495 (2019)
11. B. Abdelkrim, Y. Merzong, Robust stability power in transmission line with the use of UPFC system and neural controllers based adaptive control. IJPEDS J. **10**(3), 1281–1296 (2019)
12. I. Kulkarni, S. Dash, Development of neural network based HVDC controller for transient stability enhancement of AC/DC system, in *IEEE Student's Conference on Electronics and Computer Science 2012* (2012).
13. C.E.G. Ogoulola et al., An alternative proposal for HVDC transmission systems using 24-pulse AC/DC converters based on three-winding non-conventional transformers. EPSR J. **182** (2020)
14. G. Grdenic, M. Delimar, J. Beerten, Assessment of AC modeling impact on small-signal stability of AC system with VSC HVDC converters. IJEPES J. **119**, 105897 (2020)

15. S. Asvapoositkul, R. Preece, Impact of HVDC dynamic modelling on power system small signal stability assessment. IJEPES J. **123** (2020).
16. I.M. Ginarsa et al., Improvement of transient voltage response using an additional PID-loop on ANFIS-based composite controller-SVC (CC-SVC) to control chaos and voltage collapse in power systems. IEEJ Trans. Power Energy **131**(10), 836–848 (2011)
17. I.M. Ginarsa, A. Soeprijanto, M.H. Purnomo, Controlling chaos and voltage collapse of transient voltage response using an ANFIS-based composite controller-static var compensator in power systems. IJEPES J. **46**, 79–88 (2013)
18. A.B. Muljono et al., Coordination of adaptive neuro-fuzzy inference system (ANFIS) and type-2 fuzzy logic power system stabilizer (T2FLS) to improve a large-scale power system stability. IJECE J. **8**(1), 76–86 (2018)
19. S. Wahsh, Y. Ahmed, E.A. Elzahab, Implementation of type-2 fuzzy logic controller in PMSM drives using DSP. IJPEDS J. **9**(3), 1098–1105 (2018)
20. I.M. Ginarsa, O. Zebua, Improvement of stability of single machine nonlinear model using ANFIS-power system stabilizer based on feedback linearization. TELKOMNIKA J. **12**(2), 315–324 (2014)
21. J.P. Roselyn et al., Design and Implementation of fuzzy logic based modified real-reactive power control of inverter for low voltage ride through enhancement in grid connected solar PV syatem, Control Eng. Practice **101**, 104494 (2020)
22. M. Jamma et al., Advanced direct power control for grid-connected distribution generation system based on fuzzy logic and artificial neural networks techniques. IJPEDS J. **8**(3), 979–989 (2017)
23. B.S. Kumar et al., Improvement of power quality using fuzzy controlled D-statcom in distribution system. IJ-AI J. **7**(2), (2018)
24. I.M. Ginarsa et al., Regulation of 12-pulse rectifier using ANFIS-based controller in a HVDC transmission system, in ICSTAR Book Chapter, vol. 1 (UNILA, Indonesia, 2015), pp. 44–53
25. I.M. Ginarsa, A.B. Muljono, I.M.A. Nrartha: Simulation of ANFIS controller to line commutation based on current source converter high voltage direct current system, in *Proceedings of IEEE CENCON* (Yogyakarta, Indonesia, 2019)
26. M.A. Kumar, N.V. Srikanth, An adaptive neuro fuzzy controlled space vector pulse width modulation based HVDC light transmission system under fault AC conditions. Cent. Eur. J. Eng. **4**(1) (2014)
27. I.M. Ginarsa et al., ANFIS-based controller to regulate firing angle of inverter in average value model-high voltage direct current transmission system. In: Proceedings of ICSGTEIS, (Bali, Indonesia, 2018), pp. 1–6
28. I.M. Ginarsa et al., Transient response improvement of direct current using supplementary control based on ANFIS for rectifier in HVDC. IJPEDS J. 11(4), 2107–2115 (2020)
29. S. Casario, Thyristor-based HVDC transmission system, in Matlab Demo (Hydro Quebec, Canada, 2013)
30. Matlab/Simulink, Matlab V. 7.9.0 (The Matworks Inch., 2013)

Comparative Performance of CMOS Active Inductor

Darshak B. Marvania, Dhruvi S. Parikh, and Dhara P. Patel

Abstract In this work, a Gyrator-C-based CMOS active inductor is designed using 180 nm and 90 nm CMOS technology. Moving from 180 to 90 nm technology will diminish the cost of the system and provide more functionalities. The nominated circuit is designed using a three-stage Common Source (CS) configuration, and the resemblance is made with the recent work. From the simulation results, it is monitored that the maximum quality factor (Q_{max}) lifts, and the area sinks as we wane the technology. The consequences depict the Q_{max} of 311 at 478.63 MHz and an area of 216.04 μm^2 for 180 nm technology, whereas the ensue for 90 nm technology gives the Q_{max} of 659 at 10.96 GHz and an area of 68.73 μm^2.

Keywords Active inductor (AI) · Area · CMOS · Gyrator-C · Quality factor (q)

1 Introduction

Inductors are the heart of wireless communication and RF integrated systems [1, 2]. It plays an integral role in designing Voltage Control Oscillators (VCO), Low-Noise Amplifiers (LNA), impedance matching and gain boosting in wireless transceivers, bandwidth improvement in broadband data communication over the wire and optical channels. Generally, passive inductors are off-chip discrete components. However, the off-chip inductors have convinced obstructions of providing the limited bandwidth, reduce the reliability, and increase the outlay of the system by consuming more silicon area [3]. In the meantime, the need for a large silicon area to fabricate spiral inductors has also provoked great interest in and intensive research on the

D. B. Marvania (✉)
Charotar University of Science and Technology, Changa, Anand, India

D. S. Parikh
Charotar University of Science and Technology, Changa, Nadiad, India

D. P. Patel
Charotar University of Science and Technology, Changa, Vadodara, India
e-mail: dharapatel.ec@charusat.ac.in

F. Thakkar et al. (eds.), *Proceedings of the International e-Conference on Intelligent Systems and Signal Processing*, Advances in Intelligent Systems and Computing 1370,
https://doi.org/10.1007/978-981-16-2123-9_30

synthesis of inductors using active devices. The indispensable recognitions of CMOS Active Inductors (AIs) are less silicon area consumption, high-quality factor, large and tunable inductance value. The most fundamental capability of AIs is compact and broad tuning range [3]. Out of all this, there are few stumbling blocks of AI: poor linearity, higher noise, and less stability [4, 5]. CMOS AIs are comprehended using varied methodologies in the industry. Mainly, there are two methods to realize the AI circuits: op-amp-based realization deals with MHz frequency, whereas; gyrator-C-based realization deals with GHz frequency. Quality factor and size are compelling parameters. Common gate (CG)–Common Source (CS) gyrator-C-based AI is discussed in Ref. [6]. Less number of transistors results in the small chip area and low power consumption of the AI circuit, but the higher value of parasitic series resistance and lower value of parallel resistance provide the lower value of the quality factor [5, 7, 8].

Nowadays, the challenging task is to shrink the channel length of CMOS active components. In the proposed work, 180 and 90 nm conventional AI circuits are considered as shown in Fig. 1. The proposed work shows the analogy of AI in 180 and 90 nm technology. The quality factor upsurges as we reduce the CMOS technology in the frequency range of 10–100 GHz. The peak quality factor for granted work is 311 at 478.63 MHz for 180 nm and 659 at 10.96 GHz for 90 nm. Moreover, the area for 180 nm layout is 216.04 $\mu m^{2,}$ whereas for 90 nm layout is 68.73 μm^2.

Two peculiar technologies are being implemented in the paper. In the first one, the AI circuit is being implemented by 180 nm technology and the second same AI circuit is being implemented by 90 nm technology. Both the technologies are being

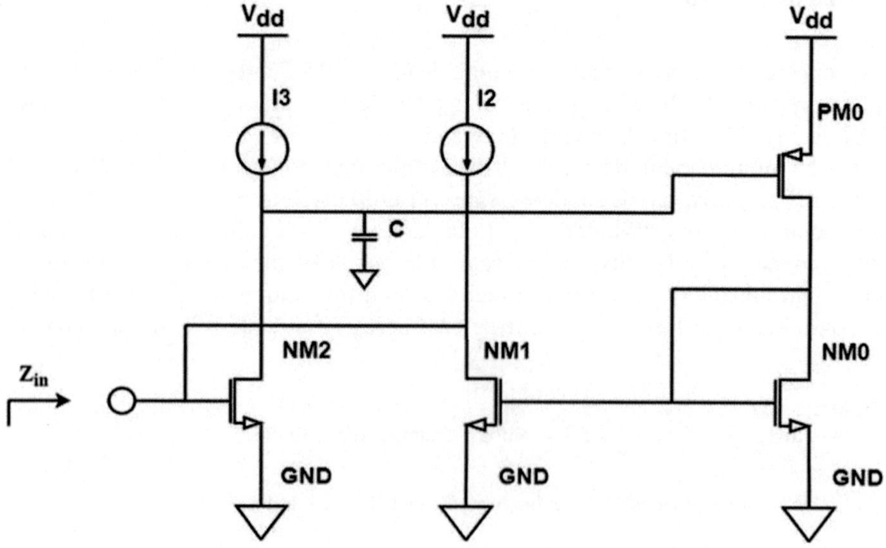

Fig. 1 Conventional active inductor circuit

compared based on inductance value (L), quality factor (Q), power dissipation (P_{diss}), silicon area, Monte Carlo analysis, and PVT analysis [9, 10].

2 Small-Signal Model

The simplified small-signal model of Fig. 1 is shown in Fig. 2.

RF Integrated circuit is being realized using the small-signal model in order to know which parameters will affect which component during the simulation in high frequency. The inductance behavior can be seen in the frequency range where the phase is between +90° and −90°. Gyrator-C topology is being shown in Fig. 3 and its equivalent passive model is shown in Fig. 4. During the inductor behavior, all the NMOS and PMOS are working in the saturation region. For small-signal model analysis, one assumption is being taken, and that is gate-source capacitance (C_{gs}) is much higher than the gate-drain capacitance (C_{gd}). The equation gives the input impedance Zin at the input port:

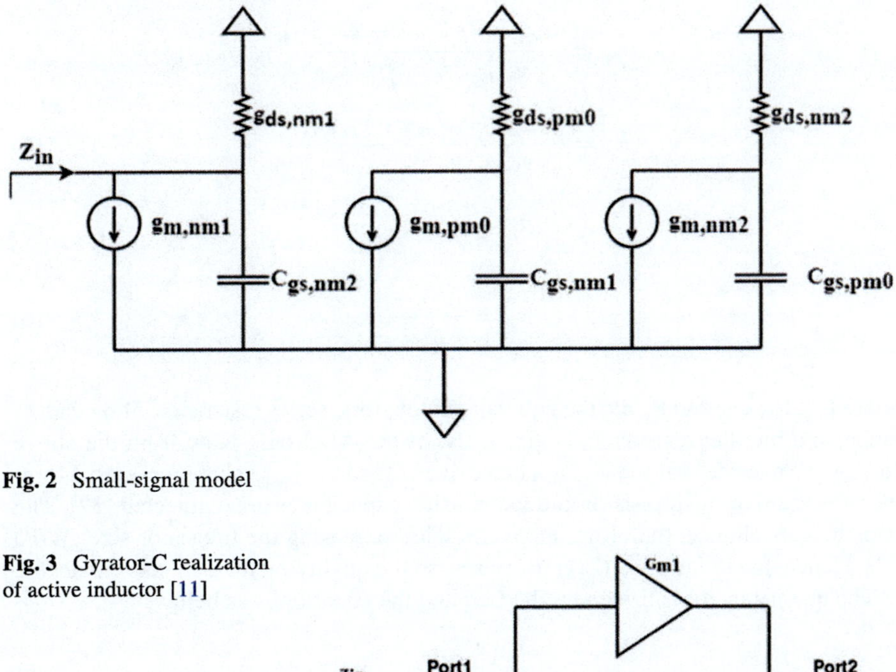

Fig. 2 Small-signal model

Fig. 3 Gyrator-C realization of active inductor [11]

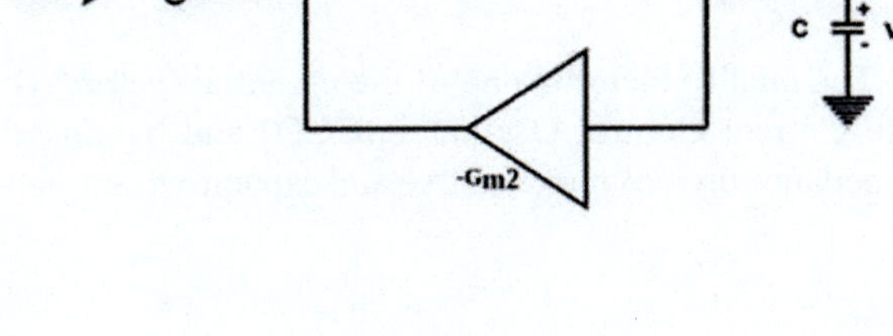

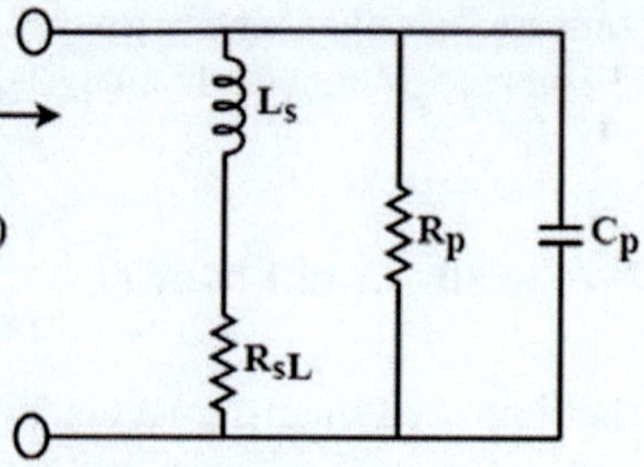

Fig. 4 Equivalent passive model [11]

$$Z_{in} = \frac{1}{sC_{gs,nm2}} + \frac{1}{g_{ds,nm1}} + \frac{(sC_{gs,nm1} + g_{ds,pm0})(sC_{gs,pm0} + g_{ds,nm2})}{g_{m,nm1}g_{m,nm2}g_{m,pm0}} \quad (1)$$

$$Z'_{in} = \frac{sC_{gs,nm1} + g_{ds,pm0}(sC_{gs,pm0} + g_{ds,nm2})}{g_{m,nm1}g_{m,nm2}g_{m,pm0}} \quad (2)$$

Now, by considering the values $C_{gs,\,nm1} <<< C_{gs,\,pm0} + C_{gs,\,nm2,}$ the other components value can be derived as follows:

$$L_{eq} = \frac{C_{gs,pm0}(g_{m,pm0} + g_{m,nm2})}{g_{m,nm1}g_{m,nm2}g_{m,pm0}} \quad (3)$$

$$R_s = \frac{g_{ds,nm2}(g_{m,pm0} + g_{m,nm2})}{g_{m,nm1}g_{m,nm2}g_{m,pm0}} \quad (4)$$

$$C_d = \frac{1}{C_{gs,nm2}} \quad (5)$$

$$R_p = \frac{1}{g_{ds,nm1}} \quad (6)$$

where L_{eq}, R_s, C_d, and R_p are the equivalent inductors, series resistance, drain capacitance, and parallel resistance, respectively, of the AI circuit. Now from the above Eq. (3), we can see that the $C_{gs,\,pm0}$ can drive L_{eq}, so the L_{eq} value increases. Moreover, the scale of the transistor and its parasitic capacitance are connected [12]. The parasitic capacitance, therefore, increases with increasing the transistor size (W/L) [13]. Hence, the inductance (L_{eq}) increases, so the quality factor also gets enhanced.

The quality factor is driven by the Leq and the equation is as below:

$$Q = \frac{wL_{eq}}{R_s} \quad (7)$$

The quality factor is one of the essential factors when we deal with applications like Current Control Oscillator (CCO) and bandpass filter. The CMOS transistor impedance divides into resistive and capacitive components due to channel resistance and gate ability. In the present work, the quality factor is being enhanced by reducing the effect of the series resistance (R_s) by a pure three-stage CS circuit.

3 Performance of AI

The conventional AI circuit (Fig. 1) is being designed in two different technologies, i.e., 180 and 90 nm CMOS technologies. In the proposed circuit, two stages are being considered where one stage gives negative transconductance and another stage provides positive transconductance as shown in Fig. 5 for 180 and 90 nm CMOS technologies.

The data shown in Table 1 is chosen in such a manner that all the transistors should be in the saturation region.

The need for migrating from 180 to 90 nm technology is to lessen the feature size of the transistor [14]. The minimal feature size means that during the fabrication process of a transistor, how closely can the transistors be placed on a chip to be used

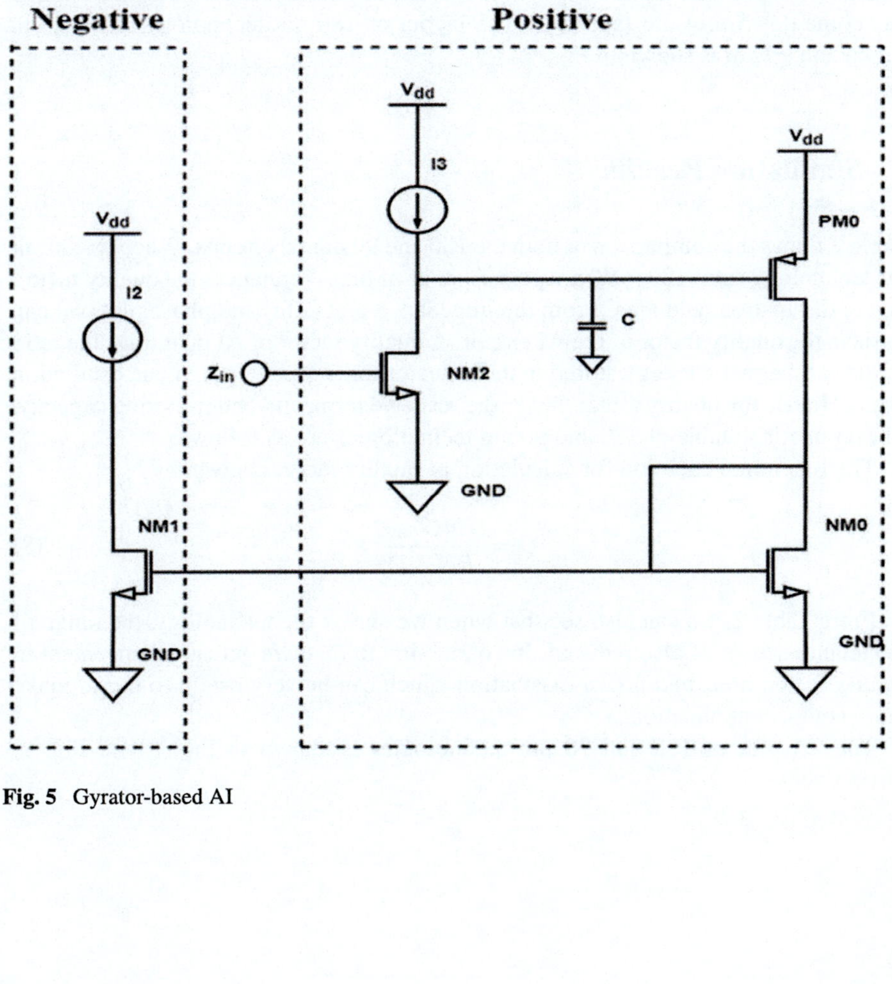

Fig. 5 Gyrator-based AI

Table 1 Proposed circuit transistor data for 180 nm and 90 nm CMOS technologies

Parameters	180 nm				90 nm			
V_{dd} (V)	1.8				1.1			
Length (L)	180 nm				90 nm			
Width (w)	nm0	nm1	nm2	pm0	nm0	nm1	nm2	pm0
	2.8u	2.8u	2.8u	5u	4u	4u	4u	4u
Capacitor (c)	100f F				500a F			
I_2	200u A				150u A			
I_3	25u A				100u A			

for various purposes. The paltrier the size is, the enormous number of transistors can be assembled on the chip. The numbers 180 and 90 nm serve the minimum channel length that will be used in fabrication. Also, these numbers are not aimlessly authorized but decided by dividing the previous number by the square root of 2 (2, because it is neither too small nor too big). Different technologies are being used today and the transistor size is shrinking day-by-day to lower the cost of production of a chip as smaller the chip, the cheaper is to make it. In the year 2020, the technology has come down to 4 nm (approx.). The layout of 180 nm technology is shown in Fig. 6 and 90 nm is shown in Fig. 7.

4 Simulation Results

Table 2 shows the comparison of both the 180 and 90 nm AI circuits. When we shrink the technology from 180 to 90 nm performance of the AI enhances the quality factor, power dissipation, and area. From the impedance plot (Zin) and phase plot, we can retrieve the quality factor of the AI circuit. A quality factor of AI quantifies the ratio of the net magnetic energy stored in the inductor to its ohmic loss in one oscillation cycle. Higher the quality factor, lower the resistive term, and better-storing capacity. The comparison table of 180 and 90 nm technologies are as follows:

The functional equation for calculation of quality factor is given by

$$Q = \frac{Im(Z_{in})}{Re(Z_{in})} \tag{8}$$

From Table 2, we can also see that when we shrink the technology, the range of inductance we get is also reduced. In contraction to that, we get an enhancement in quality factor, area, and power dissipation which can be very useful to use to make more compact applications.

The Z_{in} plot of 180 and 90 nm technologies is shown in Fig. 8 and Fig. 9, respectively.

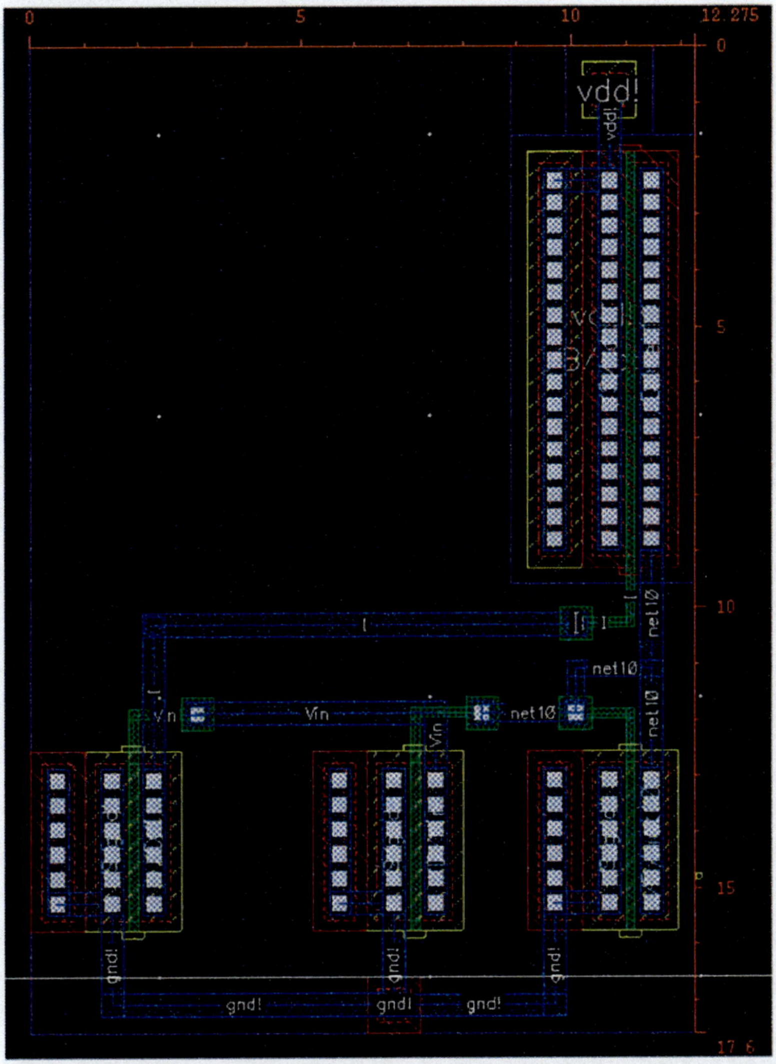

Fig. 6 Layout of AI for 180 nm CMOS technology (216.04 μm^2)

5 Conclusion

The proposed circuit gives us the highest quality factor with lower V_{dd}, high L, low P_{diss}, and more excellent inductive frequency range.

The comparison with Ref. [15–17] is shown in Table 3 for 90 nm technology.

The present work with a better-quality factor is being used in the application like temperature sensor where the sub-threshold proportional-to-absolute temperature current element strives the CCO.

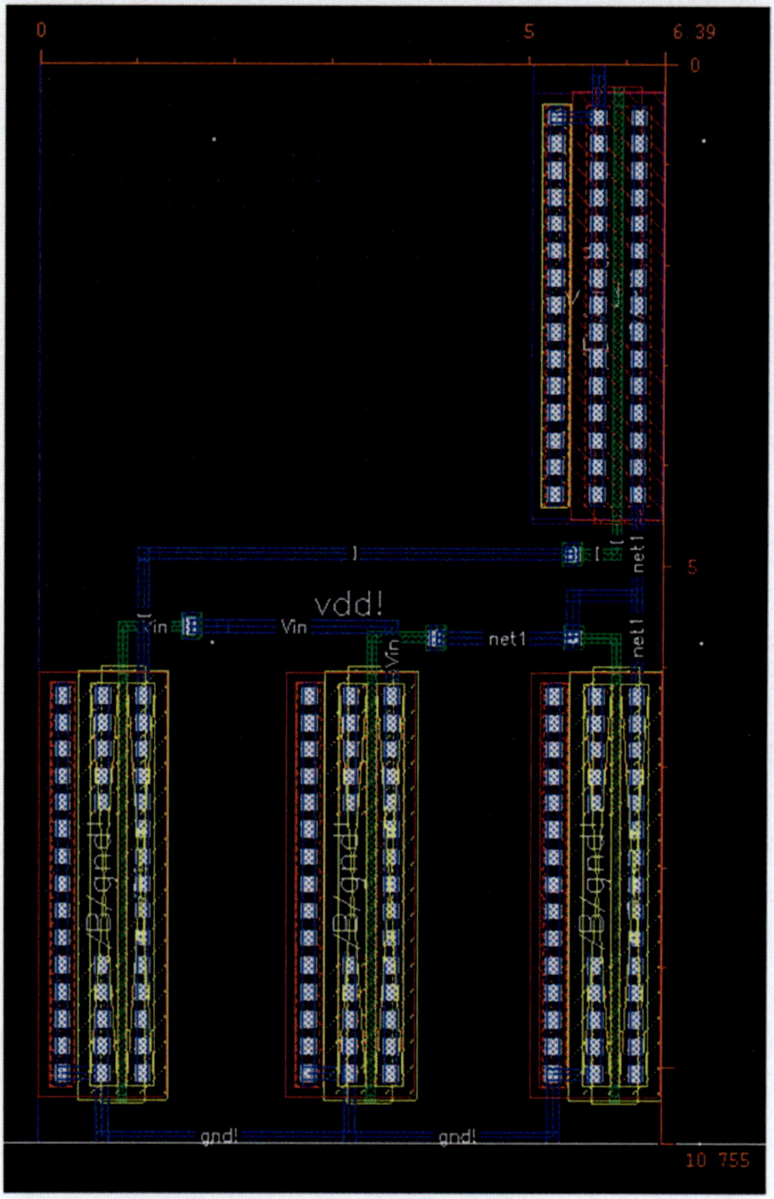

Fig. 7 Layout of AI for 90 nm CMOS technology (68.73 μm^2)

Table 2 Comparison of AI circuit for 180 nm and 90 nm technologies

Parameters	180 nm		90 nm	
L (nH)	348–3051		40–347	
Q_{max}	311 at 478.63 MHz		659 at 10.96 GHz	
P_{diss} (mW)	0.784		0.283	
Area (μm^2)	216.04		68.73	
Monte Carlo	50 iterations Min: Z_{in}: 7140Ω Max: Z_{in}: 9450Ω		50 iterations Min: Z_{in}: 7000Ω Max: Z_{in}: 12100Ω	
PVT	Typical Values (P: NN, V: 1.8 V, T: 27 °C)		Typical Values (P: NN, V: 1.1 V, T: 27 °C)	
	Process	**Z_{in}(Ω)**	**Process**	**Z_{in}(Ω)**
	FF	51,600	FF	4030
	NN	70,300	NN	9510
	SS	32,500	SS	19,400
	Voltage (V)	**Z_{in}(Ω)**	**Voltage (V)**	**Z_{in}(Ω)**
	1.7	59,700	1.7	8710
	1.8	70,300	1.8	9510
	1.9	81,400	1.9	6330
	Temp. (°C)	**Z_{in}(Ω)**	**Temp. (°C)**	**Z_{in}(Ω)**
	−20	34,400	−20	5050
	27	70,300	27	9510
	74	70,300	74	32,100

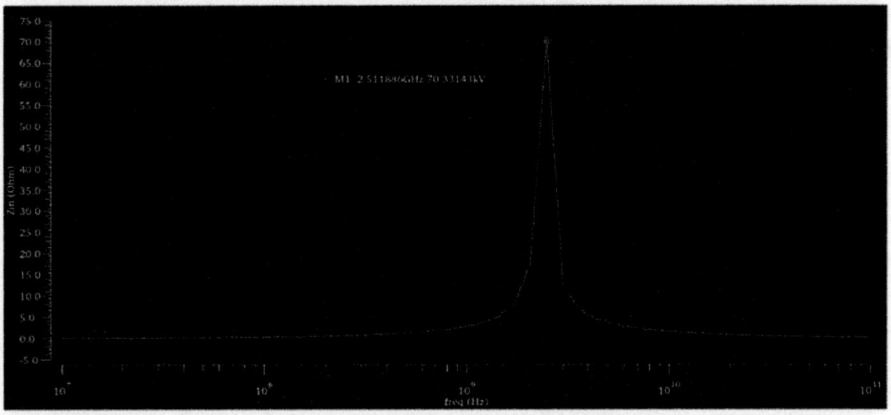

Fig. 8 Amplitude plot of AI circuit for 180 nm CMOS technology

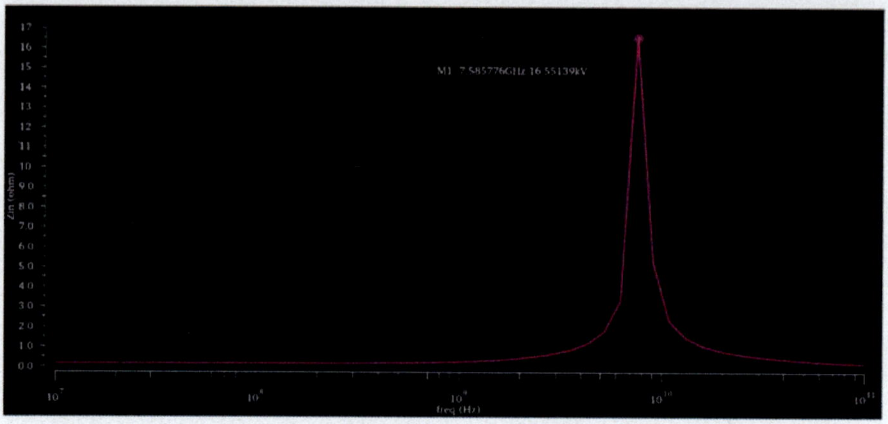

Fig. 9 Amplitude plot of AI circuit for 90 nm CMOS technology

Table 3 Comparison of active inductor

Parameters	Ref. [15]	Ref. [16]	Ref. [17]	This work
L(nH)	6.83–11.7	1.9	25.7–180	40–347
V_{dd} (V)	1.2	1.2	1.8	1.1
P_{diss}(mW)	22.5	6.4	0.99	0.283
Q_{max}	240	38.8	341	659
Freq. (GHz)	2.4–5.35	3.5	0.79–2.69	1–7.58
Area (μm^2)	–	–	–	68.73

References

1. P.H. Woerlee, RF-CMOS Performance Trends. IEEE Trans. Electron Devices **48**(8) (2001)
2. P.I. Mak, R.P. Martins, High-/Mixed-Voltage RF and analog CMOS circuits come of age. IEEE Circuits Syst. Mag. **10**(4) (2010)
3. D.P. Patel, S. Oza-Rahurkar, CMOS active inductor/resonator based voltage controlled oscillator. Recent. Adv. Electr. Electron. Eng. (Formerly Recent Patents on Electrical Electronic Engineering) **12**(6) (2019)
4. S. Saad, M. Mhiri, A.B. Hammadia, K. Besbes, A new low-power, high-Q, wide tunable CMOS active inductor for RF applications. IETE J. Res. **62**(2) (2016)
5. D.P. Patel, S. Oza-Rahurkar, Tunable CMOS active inductor using widlar current source. J. Circuits, Syst. Comput. **28**(2) (2019)
6. M. Konal, F. Kacar, MOS only grounded active inductor circuits and their filter applications. J. Circuits, Syst., Comput. **26**(6) (2017)
7. D.P. Patel, CMOS active inductor and its applications, in *Electromechanical Devices and Machines* (Intech Open, 2019)
8. D. Akella Kamakshi, A. Shrivastava, B.H. Calhoun, A 0.2 V, 23 nW CMOS temperature sensor for ultra-low-power IoT applications. J. Low Power Electron. Appl. (2016). https://doi.org/10.3390/jlpea6020010

9. D. Sachan, M. Goswami, P.K. Misra, A high-Q floating active inductor using 130 nm BiCMOS technology and its application in IF band pass filter. Analog Integr. Circuits Sig. Process. (2018).https://doi.org/10.1007/s10470-018-1196-3
10. O. Faruqe, Md. M.M. Saikat, Md. A.K. Bulbul, Md. T. Amin, Comparative analysis and simulation of active inductors for RF applications in 90 nm CMOS, in *3rd International Conference on Electrical Information and Communication Technology (EICT)* (2017). https://doi.org/10.1109/EICT.2017.8275233
11. F. Yuan, *CMOS Active Inductors and Transformers Principle, Implementation and Applications* (Springer Science and Business Media, New York, 2008)
12. R. Barnett, S. Lazar, J. Liu, Design of multistage rectifiers with low-cost impedance matching for passive RFID tags, in *IEEE Radio Frequency Integrated Circuits Symposium* (2006), p. 257e60
13. M. Cansiz, D. Altinel, G.K. Kurt, Efficiency in RF energy harvesting systems: a comprehensive review. Energy **174** (2019). https://doi.org/10.1016/j.energy.2019.02.100
14. D.P. Patel, S. Oza, CMOS active inductor: a technical review. Int. J. Appl. Eng. Res. **13**(11) (2011)
15. M.A.Y. Abdalla, K. Phang, G.V. Eleftheriades, Printed and integrated CMOS positive/negative refractive-index phase shifters using tunable active inductors. IEEE Trans. Microw. Theory Tech. **55**(8) (2007)
16. A. Ahmed, J. Wight, 6.7 GHz high-Q active inductor design using parasitic cancellation with process variation control. Electron. Lett. **46**(7) (2010)
17. L. Bharath, D. Anila, C.N. Ajay, B. Shravani, A. Jain, A wide band, low-power grounded active inductor with high Q factor for RF applications, in *International Conference on Communication, Computing and Electronics Systems* (2020). https://doi.org/10.1007/978-981-15-2612-1-52

Continuous Monitoring of Banana Plantations

Soham Gandhi and Anup Shah

Abstract The growth and production of fruits from crops are influenced by many factors. Light, temperature, nutrients, and water are the four major factors that influence the crop's growth (Environmental factors affecting plant growth, [1]). The main factor left to be controlled is water and nutrients. An array of electroconductive sensors are used to figure out the soil moisture throughout the farm. Each moisture sensor is paired with a stem diameter sensor to measure plant health and growth. This pair of sensors are used to measure the growth of crops over time. This allows for continuous monitoring of the crops to help the farmer understand how the crops are progressing. Continuous monitoring and history of information provided by these sensors help the farmer make changes to the watering schedule and nutrients in the soil to ensure optimal crop growth.

Keywords Drip irrigation automation · Continuous crop monitoring

1 Introduction

1.1 Background of Automation in Agriculture

India is the top producer of bananas throughout the world. Each year India produces roughly 28 million tons of bananas [2] with Gujarat being the second-highest producing state producing roughly 14% of India's bananas [3]. Banana is an economically important crop for India with roughly 20% of all farmland being used for it [4]. However, banana farming is difficult as it requires accuracy in irrigation and precision in fertigation [5]. Excess water increases the possibility of fungal and viral infection, while insufficient water causes poor plant growth to high -mortality rates in plants due to the hot climate, both of which lead to poor yield quality and quantity

S. Gandhi · A. Shah (✉)
InSignEx, F-14 Kishor Plaza, Station Road, Anand 388001, India
e-mail: Anup@insignex.com

F. Thakkar et al. (eds.), *Proceedings of the International e-Conference on Intelligent Systems and Signal Processing*, Advances in Intelligent Systems and Computing 1370,
https://doi.org/10.1007/978-981-16-2123-9_31

[6]. This heavy dependence on the proper balance for water and nutrients makes banana farming a difficult task for many farmers.

One innovation recently gaining popularity is the introduction of automation in agriculture. Farming for a long time has been done based on instinct, however, in recent years, the need for such instinct has slowly been dwindling with the introduction of automation. More and more farmers are making a move toward automation to help increase crop growth and crop output. These systems have taken a lot of the guesswork out of the hands of the farmers and replaced them with numbers from sensors throughout the farm. A system similar to this has already been proven to the work by Professor Zhaojiang Meng and his team. Professor Zhaojiang Meng used tomato crops (Lycopersicon esculentum L.). They were able to find a statistically significant difference between the crops that had 70–80% field water capacity compared to those that under 70% field water capacity. They also found that crops with a sufficient amount of water had a larger stem diameter as well as better resistance to solar radiation, vapor pressure deficit, and humidity [7].

1.2 Banana Crop Schedule

Banana crops are unique compared to other crops due to one major factor. Banana crops have a 9–13 month cycle with five stages: sucker growth, inflorescence, flowering, bunch development, and maturity [5]. During each stage, the banana crop requires specific amounts of water as well as certain nutrients as one of example shown in Fig. 1, to help produce a healthy crop. This makes each cycle more important since each cycle requires a long duration of time to produce, and maintain. This is both beneficial, and at the same time can be detrimental. The longer cycle gives farmers the ability to change nutrients, and watering schedules are being given to the

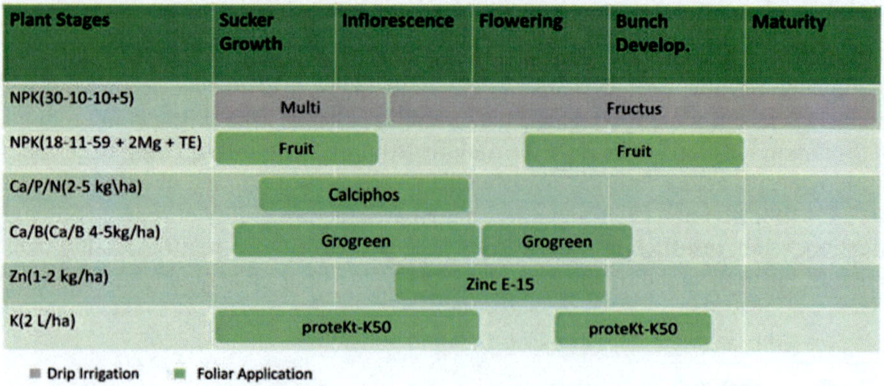

Fig. 1 This diagram illustrates the different minerals that are required for each stage of the banana crop as well as what method is required to distribute the nutrient [5]

crops based on how the banana crops are progressing. However, at the same time if the banana crops go unnoticed it could lead to a lower and poorer yield.

1.3 Factors of Influence

The growth and production of fruits are heavily influenced by the factors influencing their growth. The four major factors that play a role in the growth of crops are light temperature, nutrients, and water. Two of the variables, light and temperature, are not able to be controlled since these are based on the location of the farm itself. This leaves two variables that a farmer can influence to try and increase the quality and quantity of their yields.

Water. One of the biggest problems when it comes to the amount of water absorbed by the roots of the plants is that it is heavily dependent on other factors such as soil, sunlight, and heat. If able to regulate water flow one can optimize the watering schedule to best work around these factors that are not able to be changed on a farm by watering more on days when the temperature is higher to counteract evaporation or by watering less over a long duration to ensure percolation in soil. Soil is important, but other factors like percolation, water amount, and water evaporation play a role in the root zone absorption. One of the most common systems used on farms in India is drip irrigation due to the affordability as well as the practicality of it. However, there is one large downside to such systems; they are highly prone to cuts, and breakage. This leads to problems in pressure which in turn means crops being water improperly.

Nutrients. The second biggest factor influencing plant growth is nutrients. During each particular stage, the crop needs different nutrients to help it develop into a mature crop. Each stage is dependent on the previous, making it more crucial the banana crop growth is monitored to figure out if there are nutrients that are lacking. Banana crops have a benefit in this area since they can recover if care is taken at the proper time. However, if left neglected or unmonitored the crop can sustain permanent damage leading to poorer yields.

2 Banana Farm System

2.1 Irrigation and Fertigation

To successfully spread nutrients needed by the banana crops, one of the most effective systems is to incorporate them through the drip lines [8]. Example line diagram of such system is shown in Fig. 2. This allows farmers to easily know what nutrients are put into the soil as well as what days which nutrients are put. This system makes it easier as well as more accurate since it removes human error. This fertigation system

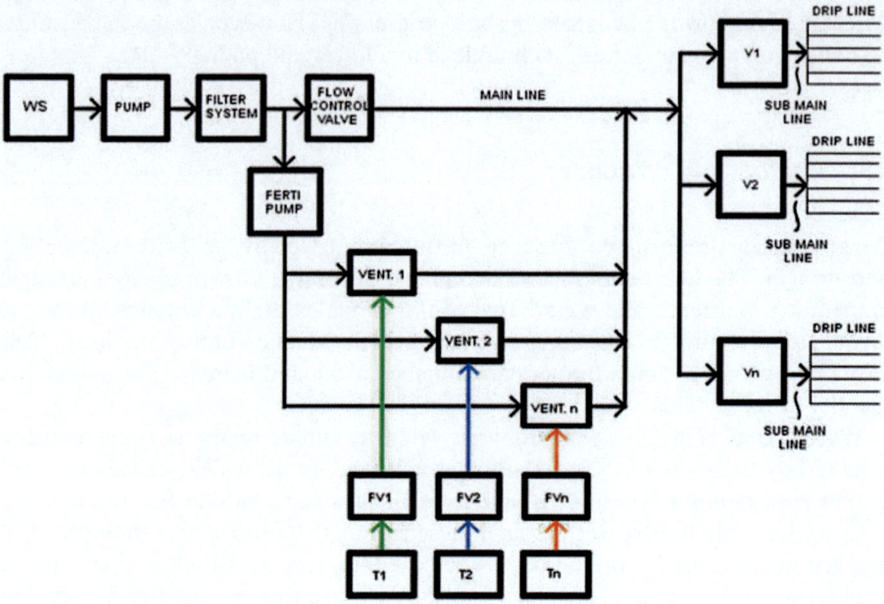

Fig. 2 The diagram illustrates how the fertigation system incorporates the nutrients into the water which is then spread through the drip lines laid out through the farm

also integrates with the sensors to help let the farmer know the amount of nutrients that were put into the soil in each region as well as what kinds of nutrients were put in. This level of customization helps farmers increase yields and quality of crops.

2.2 System

The banana farm system shall use an array of sensors strategically spread out to measure moisture as well as crop growth through the circumference of the crop as depicted in Fig. 4. These arrays of sensors will then report data back through sensor node network shown in Fig. 3 and have the data displayed on a simple to use app interface so farmers can take action based on what is reported back as depicted in Figs. 5 and 6.

2.3 Sensors

The system consists of two kinds of sensors each of which will be paired to one plant to help determine the effects of an independent variable is changed.

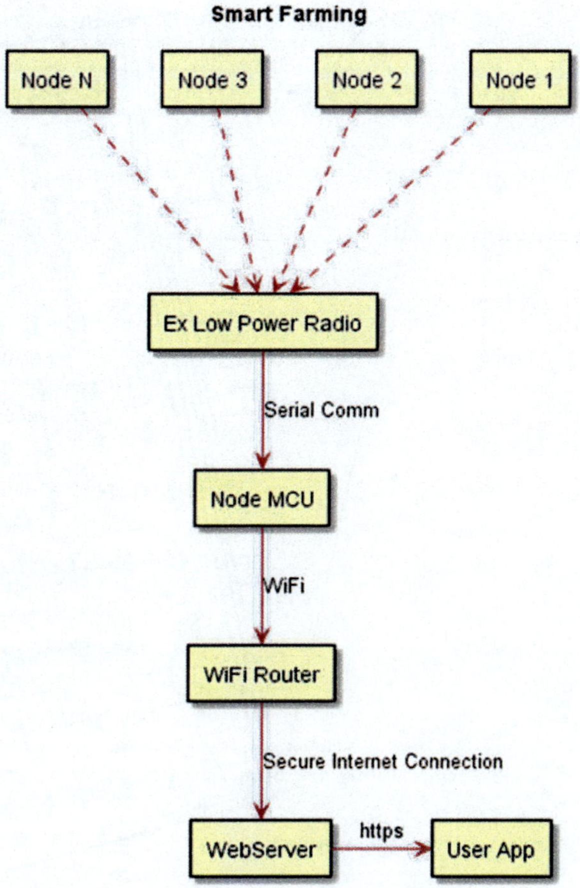

Fig. 3 Setup for relaying data back to the farmer. The goal behind this is to avoid keeping high powered devices constantly on if they are running on battery

Water Moisture Sensors. The first kind of sensor is a water moisture sensor. There are three specific groups of sensors in this category: Volumetric sensors, tensiometers, and solid-state sensors.

Volumetric sensors. These sensors are the most accurate. They measure the dielectric constant in the soil to determine soil moisture content. These sensors consume the most power and are the most expensive making them not practical for large-scale applications.

Tensiometer sensors. These sensors measure the tension between the soil particles and water molecules. The tension between the molecules helps to determine the soil moisture content. These sensors are quite accurate, however, they require heavy maintenance, and are useless in drier conditions.

Solid-State Sensors. These sensors are the least expensive and work by measuring the resistivity in the ground using two electrodes. The more water in the soil the easier the current flows through. The biggest drawback with these sensors is they do not

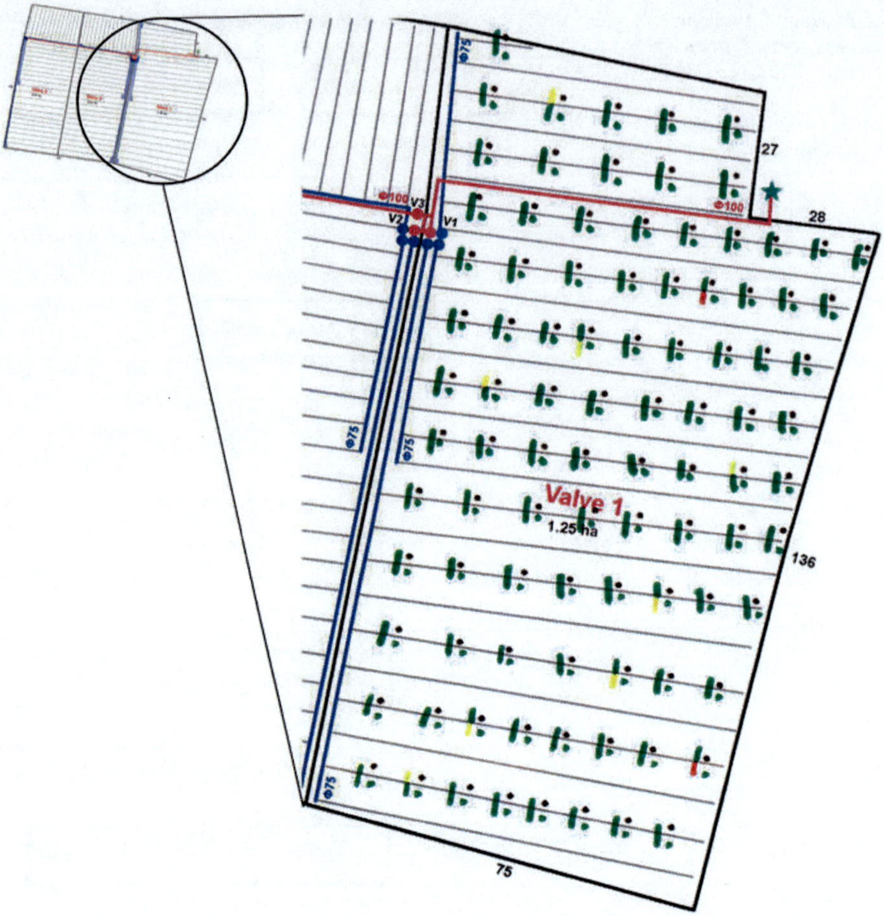

Fig. 4 Sample farm layout with sensors placed throughout the farm. The bars indicate the status for each sensor and the dot indicates what kind of sensor it is

work in sandy regions and regions with a high salinity value. However, for a large-scale application like this, they are perfect due to a decent accuracy and low cost [9].

Stem Diameter Sensors. The second kind of sensor is a stem diameter sensor. This sensor will be a way for farmers to track the progress of the crops and determine whether the crops are positively or negatively reacting to the current conditions. These sensors will be strapped around banana crops once the circumference has exceeded 4–6 inches to ensure a proper mount as well as reliable readings.

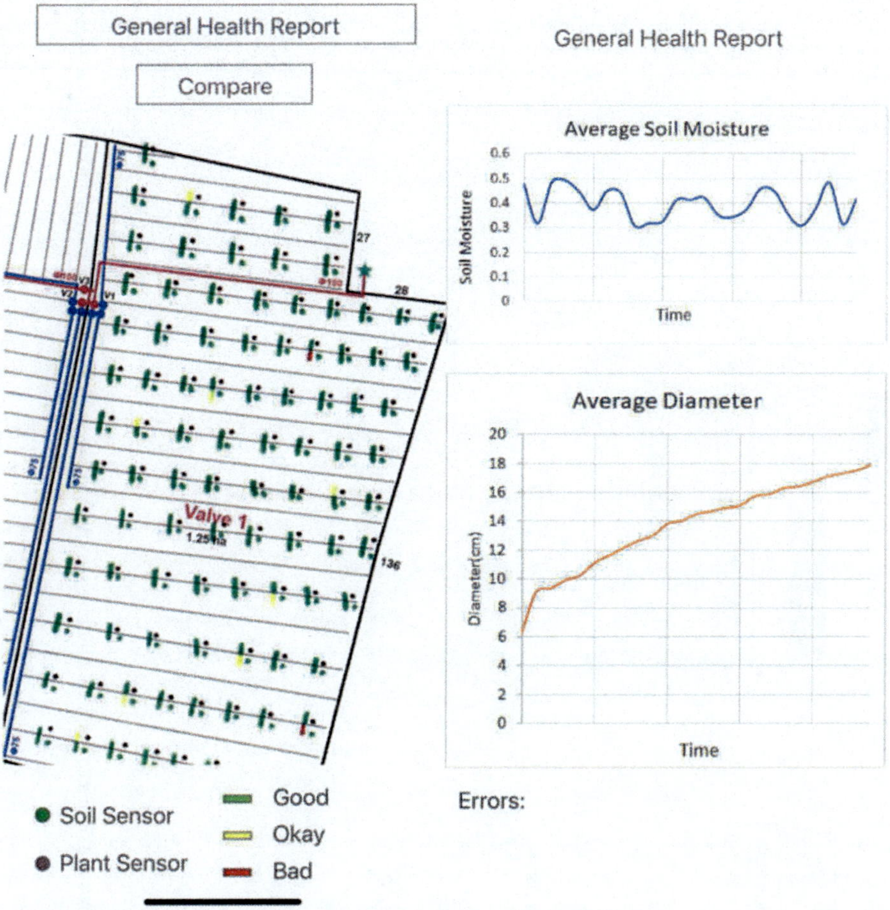

Fig. 5 Sample app layout for farmers. The left side displays the main screen, and the right side provides a general report of all the sensors on the farm

2.4 Network Coordinator

The network coordinator is a device capable of operating with a low-power draw to help extend the life of the battery. The main functionality of the network coordinator is its ability to connect with WiFi. This gives it the ability to report data back to the web server with a low power consumption. The network coordinator fetches data from the sensors using a radio attached to all the sensors. Through this, the data is sent to the network coordinator where it is formatted and then sent using WiFi to a central web server.

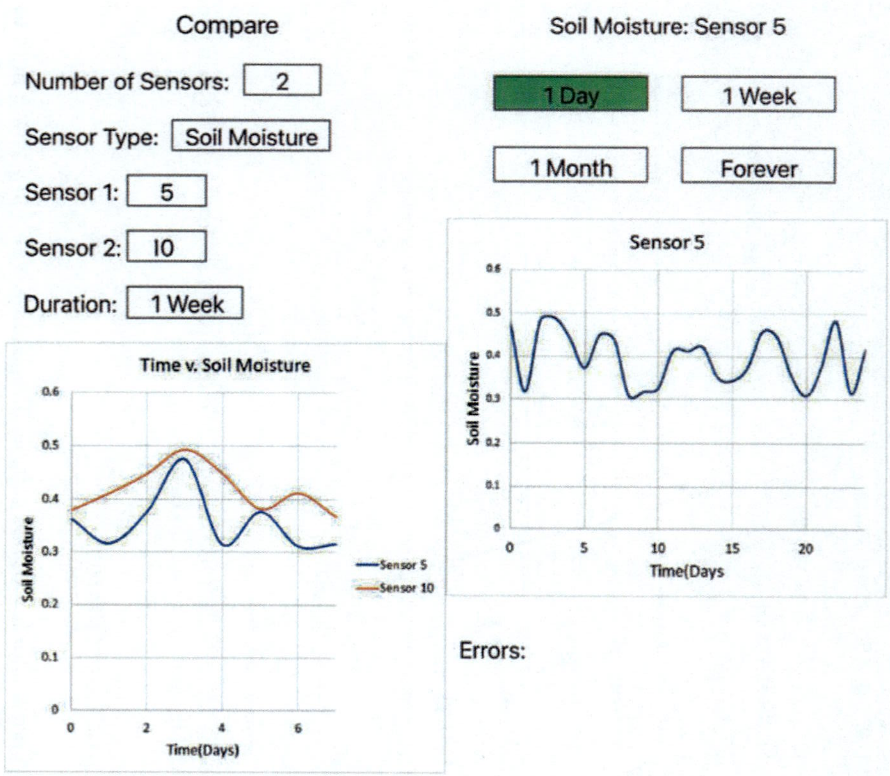

Fig. 6 Sample app layout for farmers. The left side allows the farmer to compare different sensors, and the right side displays the information for a specific sensor

2.5 Web Server

An important aspect of the system is the webserver. The web server stores the data from each sensor. This allows farmers to go back and see past sensor data to determine if there are patterns or a consistent issue in a section. The web server also generates visualizations.

2.6 Interface

One of the most crucial aspects of any system is the portrayal of information to the farmer in an easy-to-understand format. The user interface design has a main page that provides the user with a general overview of the sensors on the farm and buttons to guide the user to other pages to see more specific information. On the subsidiary pages, the user can compare data, see how the farm as a whole is doing or look more in depth into a specific sensor.

Main Screen. The main screen shows the farm and all the sensors with indicators. If a sensor is green the data is within two standard deviations of the mean, and the plant is growing well. If a sensor is marked as yellow the data being reported is two standard deviations below the mean indicating to the farmer that there may potentially be an issue. Finally, if the sensor is marked as red the data is at least three standard deviations away from mean or the sensor is having errors. This indicates to the farmer that there is a severe issue and immediate action is required.

Sensor Screen. If the farmer would like to look into more details for a particular sensor they can click on the sensor from the page which will guide them to the page for the sensor that provides the farmer with information on the sensor data as well as any errors that may have occurred. The farmer on this page can select the duration for which they would like to see the data for helping the farmer get more choice, and understand the data more easily.

General Health Screen. From the main page, the farmer has another option to open up a general report of the farm. This provides average data of all the sensors and presents the data through graphs. The graphs help the farmer understand if there is a need to change the amount of water being put into the soil if the nutrients need to be changed based on the soil moisture and plant diameter.

Compare Screen. The final screen is a page that allows the farmer to compare different sensors to better understand if a region is not growing as expected or if there are issues that are occurring such as cuts in drip lines, or clogs. The compare screen provides information through graphs and allows the farmers to select which sensor, how many sensors, as well as how far back.

3 Conclusion and Recommendations

3.1 Benefits

The system being proposed offers numerous benefits to the farmers. It gives an early warning to farmers before it is too late to react to any issues that the crops are facing. It helps farmers identify breakage in drip lines or clogs that could be preventing water from reaching the plants. It reduces the probability of an area being neglected. It also gives leeway for the farmer to experiment with different nutrients and watering schedules to see what is best for the crops. The system being proposed helps farmers get an early edge to any problems that could lead to permanent effects on the crop yield.

3.2 Recommendations

Future considerations may include making the system expandable as the farm size increases. Using the system to work not only with banana crops but also other crops in the region such as wheat, cotton, and rice [10]. It may also be a good idea to incorporate a predictive algorithm in the system to automatically adjust watering and nutrients based on the information provided by the sensors rather than having the farmer compute a logical explanation based on the data. This would help to reduce errors.

Acknowledgments The algorithm and its application described in this paper were developed as part of dissertation work of Soham Gandhi and Anup Shah at InSignEx.

All IP rights for "algorithms, described applications, all code, software, notes, analyses, compilations, studies or other documents, whether prepared by Soham or other employee of InSignEx (and whether in printed form or electronically stored in digital or analog form, on electronic or magnetic media, or otherwise), which contain or otherwise reflect such information" is fully perpetually assigned to InSignEx by all the authors of this publications.

The authors are greatly thankful to Jignesh Gandhi, Nutan Gandhi, and the InSignEx Company.

References

1. Environmental factors affecting plant growth, https://extension.oregonstate.edu/gardening/techniques/environmental-factors-affecting-plant-growth. Accessed 19 Oct 2020
2. Map of top 10 banana producing states in India, https://www.mapsofindia.com/indiaagriculture/fruits-map/banana-producing-states.html. Accessed 19 Oct 2020
3. Top banana producing countries in the world, https://www.worldatlas.com/articles/top-banana-producing-countries-in-the-world.html, Accessed 2020/10/19.
4. Banana farming; painting; care; harvesting guide, https://www.agrifarming.in/banana-farming. Accessed 19 Oct 2020

5. Crop stages, https://www.lima-europe.eu/wp-content/uploads/2017/04/Captura-de-pantalla2017-04-11-a-las-10.58.48.png. Accessed 19 Oct 2020
6. Banana tree plant profile, https://www.thespruce.com/banana-tree-growing-profile-3269353. Accessed 19 Oct 2020
7. Z. Meng, Suitable indicators using stem diameter variation-derived indices to monitor the water status of greenhouse tomato plants. PLoS One **12**(2), e0171423 (2017)
8. The Indian drip irrigation Jugaad, https://www.forbesindia.com/article/breakpoint/the-indian-drip-irrigation-jugaad/17302/1. Accessed 19 Oct 2020
9. 3 types of soil moisture sensors—which is best for you?, https://mytrellis.com/blog/smstypes. Accessed 19 Oct 2020
10. Gujarat—region profile, http://sap.ipni.net/article/gujarat. Accessed 19 Oct 2020

An Experimental Study to Evolutes Performance of PV Module at Workplace

Sohankumar G. Prajapati and Sanjay R. Vyas

Abstract This paper investigates the performance of a polycrystalline solar module at Government Engineering College, Patan, Gujarat, India. The experiments were carried out during September month on a 10-W PV module under variable climate conditions. The open-circuit voltage and short-circuit current observed. The PV module has the facility to change tilt angle and direction manually. The performance graph was designed based on monthly average solar irradiation and ambient temperature, variable tilt angle, and panel orientation. Results show that the PV module performs better when it is inclined and placed at a 24-degree tilt angle and panel orientation the in south direction. It is also observed that PV module generated 78% power of its rated maximum on average irradiation of 695.6 W/m^2/day and average temperature of 30.6 °C. Experimental results compared with the HOMER Pro microgrid software data for the same location and received good agreement between them. Similar system design can be applied for future large-scale projects for same location.

Keywords Solar module performance · PV output analysis · Experiment on solar PV

1 Introduction

To meet the higher demand for energy all over the world in the past decade, more attention has been paid recently to solar PV power generation by most countries due to its universal, non-polluting, vast, noiseless, safe, and inexhaustible features. In addition, traditional power generation has a number of shortcomings, including consumption of coal, oil, and natural gas resources, serious environmental pollution, and low utilization rate. On the other hand, cost of solar PV system has dropped

S. G. Prajapati (✉)
Kadi Sarva Vishwavidyalaya, Gandhinagar, Gujarat, India

S. R. Vyas
LDRP Institute of Technology and Research, Gandhinagar, Gujarat, India
e-mail: srvyas_ee@ldrp.ac.in

F. Thakkar et al. (eds.), *Proceedings of the International e-Conference on Intelligent Systems and Signal Processing*, Advances in Intelligent Systems and Computing 1370,
https://doi.org/10.1007/978-981-16-2123-9_32

significantly in recent years, which encourage to set up large-scale solar plants. However, relatively high space required to produce a large amount of energy, storage devices, and low efficiencies has formed the demerits of solar PV system. The growth of India's energy consumption will be the fastest among all significant economies by 2040, with coal meeting most of this demand followed by renewable energy. Renewable became the second most significant source of domestic power production, overtaking gas and then oil, by 2020 [1]. Renewable energy can supply two-thirds of the total global energy demand, and contribute to the bulk of the greenhouse gas emissions reduction that is needed between now and 2050 for limiting average global surface temperature increase below 2 °C [2]. Recently, India achieved the fifth global position in solar power deployment by surpassing Italy. Solar power capacity has increased by more than 11 times in the last 5 years from 2.6 GW in March 2014 to 30 GW in July 2019. Presently, solar tariff in India is very competitive and has achieved grid parity [3].

The Patan, city is located in northern Gujarat state, west-central India, (latitude 23.75° N and longitude 72.23° E). The city covers a total land area of 5,792 km^2 with an estimated population of 1, 33,737. The average annual irradiation and temperature are about 540 W/m^2 and 27.53 °C, respectively [4] and the city has good light conditions, making this region suitable to build a certain capacity of PV power stations. Near Patan City, Gujarat Power Corporation Limited (GPCL) installs a 590-MW solar park. It is the Nodal Agency for the development of "Solar Park" in Gujarat.

The performance of the solar photovoltaic (PV) panel depends mainly on two environmental factors, namely temperature and solar irradiance. In this paper, the output characteristics of solar two different type solar modules under various irradiance, temperature, shading patterns, various tilt angle, panel direction has been studied to check its performance at workplace.

These factors affect the PV output parameters such as maximum power; short-circuit current, and open-circuit voltage. The real-time measurement data collected from a small-scale experimental test setup. The results demonstrate the relationships between the environmental conditions, system configuration, and power generation by solar modules and can be extended to large-scale photovoltaic power plants. The experimental results of the study Provide reliable and useful information regarding the behavior of solar PV systems, which can be considered for model validation and system design purposes for further research Projects.

The tilt angle of a solar panel is also an important consideration. The geographical latitude determines the angle that a solar panel should be set at to produce the most energy in a given year. A general rule for optimal annual energy production is to set the solar panel tilt angle equal to the geographical latitude. The performance of a solar radiation conversion system is affected by the tilt angle with the horizontal plane. Thus, a photovoltaic array needs to be tilted at the correct angle to maximize its performance [5]. For example, if the location of the solar array is at 40° latitude, the optimum tilt angle is also 40°. Essentially, the closer a solar panel is located to the equator the more the panel should be pointing straight up. The closer the panel is to the poles, the more they should tilt toward the equator.

The voltage dependence on temperature is given by a simplified relation expressing V_{oc}—(T) as a function of V_{oc} at S_{TC}, the open-circuit voltage temperature coefficient β, and the module operating temperature T_c.

$$V_{oc}(T) = V_{oc} 25\,^{\circ}\mathrm{C}[1 + b(T_C - 25)] \tag{1}$$

Whenever the temperature of the solar panel increases above the operating temperature of 25 °C, the bandgap of the semiconductor material in solar cell decreases, and the open-circuit voltage drops, and as the temperature of the solar cell decreases, there is an increase in open-circuit voltage.

The effect of temperature and radiation on the parameters of photovoltaic modules evaluate by Perraki and Kounavis [6] from the analysis found that the impact of irradiance on PV module's properties, the generated current is proportional to the total incident irradiance at fixed temperatures with a dependence given by

$$I_{sc}(T) = I_{sc}(25\,^{\circ}\mathrm{C})[1 + \alpha(T - 25)]\frac{G}{1000\ \mathrm{W/m^2}} \tag{2}$$

where I_{sc} (25 °C) stands for the short circuit current at 25 °C, G for the global solar radiation incident on the surface of PV cells/modules (W/m^2), a for the short circuit current correction coefficient for the temperature (°C^1) considered constant, and T for the temperature of the solar cell. According to this relation, the photogenerated current is directly related to the irradiation incident (G), while its precise value depends on the spectrum of the irradiation (global or direct). Moreover, the measured Isc is adjusted to STC by multiplying the measured values by the ratio of the reference irradiance at STC to the measured irradiance G, as measured during the experiments. With reference to the V_{oc}, the logarithm dependence of this parameter on the irradiance is considered constant, as long as there is typically less than 5% change in the voltage coefficients over the change in irradiance from 100 to 1000 W/m^2.

2 Solar PV Modeling

Modeling is very important to understand the behavior and characteristics of solar PV under varying atmospheric conditions.

The PV module manufacturers Provided basic parameter on the nameplate at standard test conditions (usually 1000 W/m^2 radiation, Temperature 25 °C, 1.5 Air mass Index), the parameters are; rated power or maximum power (Wp), voltage at the maximum power point (*Vmp*), the open-circuit voltage (*Voc*), the short-circuit current (*Isc*), the current at maximum power point (*Imp*), temperature coefficient for the short-circuit current (%/°C), and temperature coefficient for the open-circuit voltage (%/°C).

Fig. 1 Equivalent circuit of a single-diode model of PV cell

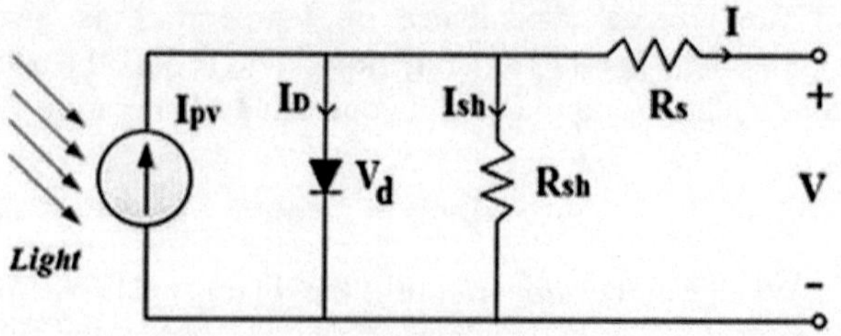

The PV output power is calculated with the help of a single diode model. The model consists of a cell photocurrent IPV, a diode D, series resistance Rs, and a shunt resistance Rsh as shown in Fig. 1.

The output current of PV primarily depends on temperature (T) and solar irradiance intensity (G). Following parameters use in mathematics equations.

STC—Standard test condition (usually 25 °C and 1000 W/m^2 radiation, Temperature of the cell—25 °C),

The temperature of the solar cell itself, not the temperature of the surrounding.

Solar Irradiance—1000 Watts per square meter. This number refers to the amount of light energy falling on a given area at a given time.

Mass of the air—1.5. This number is somewhat misleading as it refers to the amount of light that has to pass through Earth's atmosphere before it can hit Earth's surface, and has to do mostly with the angle of the sun relative to a reference point on the earth.

$\mathbf{I_{PV}}$—current generated by the incident light (it is directly proportional to the Sun irradiation),

$\mathbf{I_d}$—diode current

$\mathbf{R_{sh}}$—Shunt resistance in ohm

$\mathbf{R_s}$—Series resistance in ohm

The output terminal current I is an evaluate by applying KCL in the equivalent circuit of solar cell;

$$I = I_{PV} - I_d - I_{sh} \tag{3}$$

where

- $I =$ output current (ampere)
- $I_{PV} =$ photo generated current (ampere)
- $I_d =$ diode current (ampere)
- $I_{sh} =$ shunt current (ampere).

By the Shockley diode equation, the current diverted through the diode is

$$I_d = I_0\left\{\exp\left(\frac{V + IRs}{nV^T}\right)^{-1}\right\}$$

where

- I_0 = reverse saturation current (ampere)
- n = diode ideality factor (1 for an ideal diode)
- q = electron charge (1.602×10^{-19} C)
- k = Boltzmann's constant (1.380×10^{-23} J/K)
- T = absolute temperature of the p–n junction
- $V_T = KT/q$, the thermal voltage. At 25 °C, $VT \approx 0.0259$ V.

Note: The ideality factor of a diode is a measure of how closely the diode follows the ideal diode equation. The derivation of the simple diode equation uses certain assumptions about the cell.

Now, by applying Ohm's law, the current diverted through the shunt resistor is

$$I_{sh} = \frac{V + IRs}{R_{sh}}$$

Substituting these into the first equation produces the characteristic equation of a solar cell, which relates solar cell parameters to the output current and voltage:

$$I = I_{PV} - I_0 \left\{ \exp\left(\frac{V + IRs}{nV_T}\right)^{-1} \right\} - \frac{V + IRs}{R_{sh}} \tag{4}$$

By substituting VT values in Eq. (2), we get

$$I = I_{PV} - I_0 \left\{ \exp\left(\frac{q(V + IRs)}{nKT}\right)^{-1} \right\} - \frac{V + IRs}{R_{sh}} \tag{5}$$

The values of IPV (light-generated current) of the PV cell depend linearly on the solar irradiation and are also influenced by the temperature according to the following equation:

$$I_{PV} = \left[I_{SC\_STC} + K_i (T - T)_{STC} \frac{G}{G_{STC}} \right] \tag{6}$$

where

I_{sc_STC} = short-circuit current at STC

K_i = short-circuit temperature coefficient. T = surface temperature of the module, T_{STC} = Temperature at STC.

And the diode saturation current I_0 and its dependence on the temperature are expressed as

$$I_0 = \frac{I_{PV}}{\left[\exp\left(\frac{V_{OC\_STC} + Kv(T - T_{STC})}{nV_T}\right)^{-1} \right]} \tag{7}$$

where

V_{OC_STC} = Open-circuit voltage at STC, Kv = open-circuit temperature coefficient.

3 Experiment System

The system consists of a polycrystalline solar module of 10-Wp power capacity mounted on a stand that can be adjusted for optimum tilt and exposure to sunlight. A solar power meter (Make: Thrive Energy Technology Pvt Ltd, Hyderabad) was used to measure solar irradiance. Using an ammeter and voltmeter Voc and Isc of the solar panel is measured over the single solar panel. The experiment is repeated at the outdoor common space of the Government Engineering College, Patan, Gujarat, India. The module properties and other experimental parameters are given in Table 1. The system setup is as shown in Fig. 1 (Fig. 2).

Duration of test: 1st September–30 September 2020, Equipments/Instruments used for test:

1. Solar PV module (10 Wp)
2. Solar power meter for radiation measurement in W/m^2
3. Infrared thermometer for temperature measurement in °C
4. Angle Pro mobile app for tilt angle measurement
5. Digital multimeter.

Table 1 Specifications of 10 W solar module employed for the experiment

Solar module capacity	10 Wp
Type	Multi & mono crystalline
Manufacturer	Thrive Energy Technology Pvt Ltd Hyderabad
Vmp	16.4 V
Imp	0.61 A
Voc	21 V
Isc	0.70 A

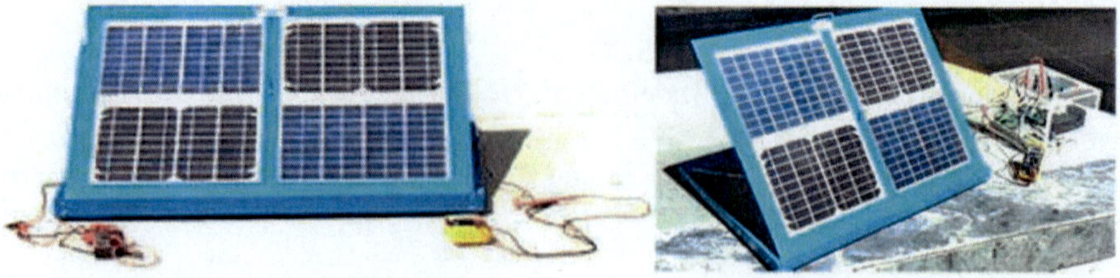

Fig. 2 Experiment setup of solar panel

4 Results and Discussion

The Proposed model presented above to some actual datasets as obtained through experiments in Patan, Gujarat, India, which is located at latitude 23.75° N and longitude 72.23° E. The test was performed from 10:30 a.m. to 5:30 p.m. during September month, 2020, and the data measured included total solar irradiation, ambient temperature, open-circuit voltage, and short-circuit current. The PV output power performance is presented based on hourly total solar radiation measured for various tilt angles, panel direction to calculate available energy from the photovoltaic system. September month average Global Horizontal Irradiance (GHI) and ambient temperature data are taken from HOMER Pro software for same location and latitude, and from a comparison of experiment data and software data, it is found that difference in average radiation and temperature is 0.89 KWh/m^2/day and 1.83 °C, respectively.

The following observation is found during the experiment.

The observed data from experiment is graphically presented here (Fig. 3).

From the above graph, it can be seen that, at 24° tilt angle, PV module generates an average 13.5 W power from 14.7 W maximum-rated, so it generates 91% of its rated maximum power at 24° tilt angle on module orientation from 10° to 30° variable angle.

The second exercise was conducted to evaluate the PV module placement facing for optimum power generation.

Graphical representation of PV Module facing versus output power is presented here (Fig. 4).

From the PV module placement analysis, it is observed that it delivers an average of 12.35 W from the 14.7 maximum-rated powers. Hence, PV module generates 84% of its rated maximum power when it places in the south direction. On the other hand,

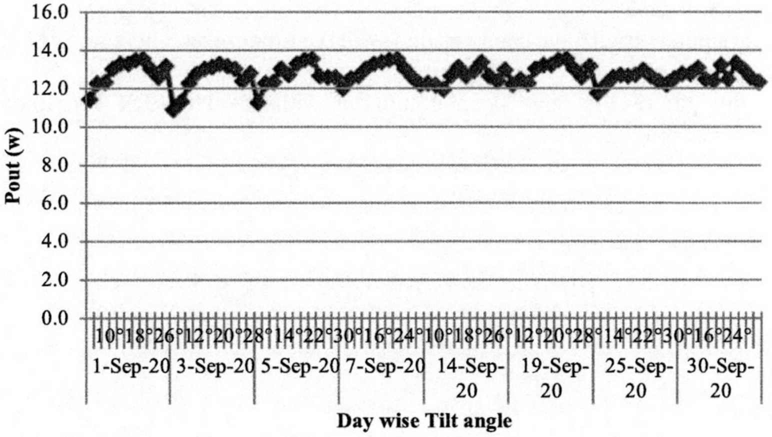

Fig. 3 Day-wise tilt angle versus output power

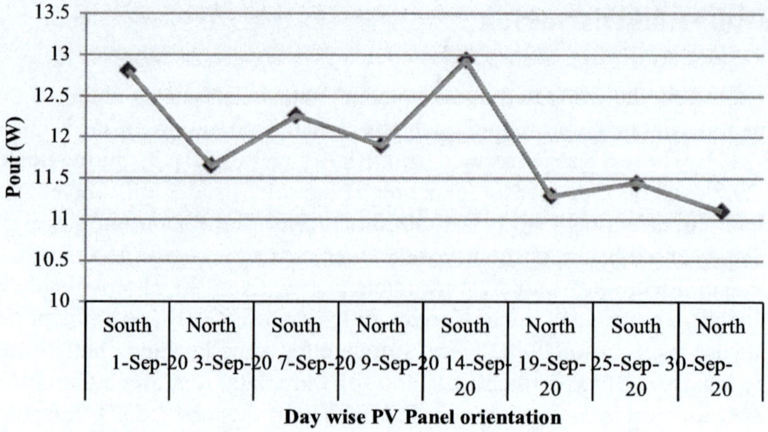

Fig. 4 PV module orientation versus output power

on north direction placement of PV module, it generated an average of 11.48 W that is 78% of its rated maximum power.

The experimental study carried on September moth day between 10:30 a.m. to 5:30 p.m. by mounting PV module orientation at a fixed 24° tilt angle and module face the south direction. The PV module performance was observed. Finally, average power output was calculated based on September month irradiation and temperature.

The graphical representation of September moth average solar irradiation, ambient temperature is present here (Table 2 and Fig. 5).

From the above graph, it is clear that average radiation of September month observed was 695.6 W/m^2 and PV module output average power observed was 11.6 W from 14.7 W rated maximum, so the module generates 79% of its rated power (Figs. 6, 7, 8, 9, and 10).

The parameters used have been verified with the other researchers work [7, 8] and received good agreement.

Experimental results have been compared with the HOMER Pro microgrid software [9] (Table 3).

Table 2 Monthly average data of radiation, temperature, Voc, Isc, and power

Day	Average Ambiant temperature (°C)/day	Average radiation (W/m^2/day)	Average irradiation (KWh/m^2/day)	Voc (V)	Isc (A)	P (W)
1-Sep-20	31.1	784.0	5.5	20.1	0.6	12.1
3-Sep-20	31.1	648.0	4.5	18.9	0.6	11.3
5-Sep-20	30.6	698.1	4.9	19.3	0.6	11.6
7-Sep-20	31.3	730.0	5.1	20.1	0.6	12.0
9-Sep-20	31.3	663.6	4.6	19.7	0.6	11.8
14-Sep-20	30.5	733.1	5.1	19.3	0.6	11.6
16-Sep-20	30.0	649.3	4.5	18.4	0.6	11.1
19-Sep-20	30.7	702.0	4.9	19.3	0.6	11.6
21-Sep-20	31.5	640.6	4.5	19.6	0.6	11.7
22-Sep-20	31.3	671.6	4.7	19.2	0.6	11.5
24-Sep-20	30.3	673.5	4.7	19.2	0.6	11.5
25-Sep-20	30.4	663.8	4.6	19.1	0.6	11.4
28-Sep-20	29.0	738.0	5.2	19.7	0.6	11.8
30-Sep-20	29.4	742.4	5.2	19.6	0.6	11.7
Monthly average	**30.6**	**695.6**	**4.9**	**19.4**	**0.6**	**11.6**

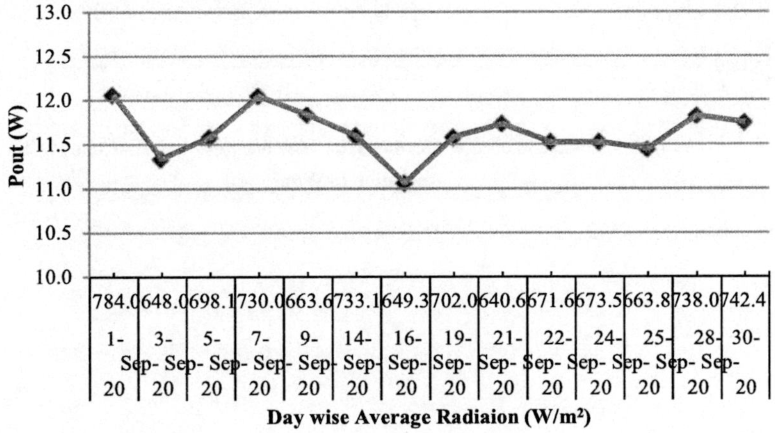

Fig. 5 Radiation versus pout

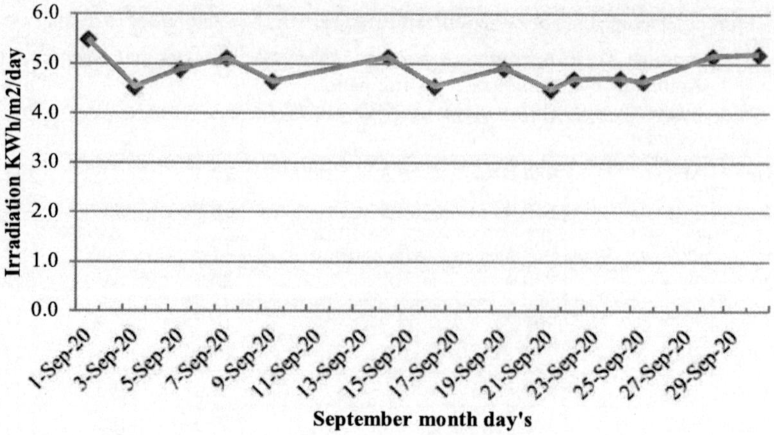

Fig. 6 Day versus irradiation

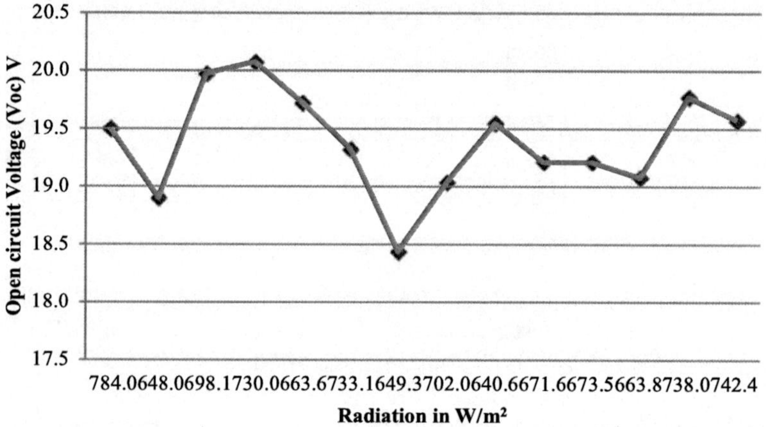

Fig. 7 Radiation versus Voc

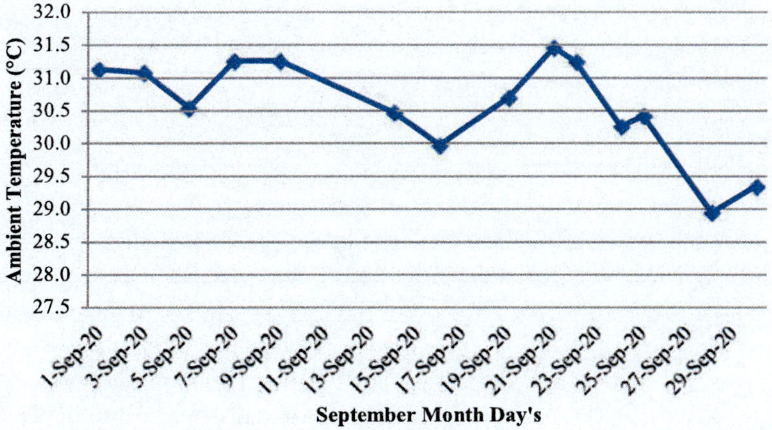

Fig. 8 September month days versus ambient temperature

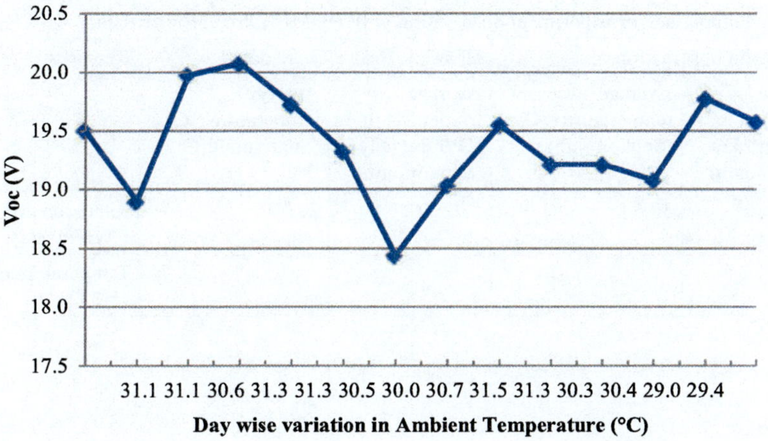

Fig. 9 Ambient temperature versus Voc

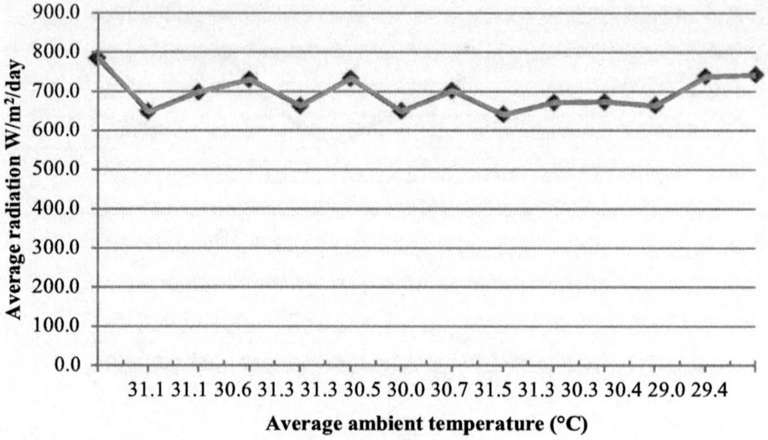

Fig. 10 September month days average ambient temperature versus radiation

Table 3 Comparison of experiment observation with HOMER Pro software data

Experiment observation		HOMER Pro software data		Remarks
Average irradiation in KWh/m^2/day (Sept. month)	Average temperature (°C) (Sept. month)	Average irradiation in KWh/m^2/day (Sept. month)	Average temperature (°C) (Sept. month)	
4.9	30.6	5.72	28.77	Radiation diff. 0.89 KWh/m^2/day
				Temperature diff. 18.3 C

5 Conclusion

From the experimental investigations performed on solar PV panel in climate condition of Patan, the following points are concluded:

1. At 24° tilt angle, PV module generates an average of 91% of its rated maximum power from the panel orientation range of 10°–30°.
2. At the south direction placement of PV module, it generates an average of 84% of its rated maximum, while at north direction facing, it generates average 78% of its rated maximum. Hence, 6% more average power generates when it is mounted on south face compare to north. Similar results are also obtained by many researchers.
3. The average irradiation of September month found 695.6 W/m^2/day in which PV module generates average 11.6 W from total maximum 14.7. Hence, it is remarkable that Patan site is preferable to set up a large capacity plant.
4. The average ambient temperature (°C)/day of September month found 30.6 which is near to STC 25 °C. So, the performance of solar module at workplace found near to panel rating.
5. The result obtained from experiments also compared with HOMER Pro microgrid software and reached a good agreement. Actually, there are many new techniques available, but the present paper contains the experimental study on PV panel.

For future work, a solar PV module with tracking is suggested to enhance the performance.

References

1. J. Charles Rajesh Kumar, M.A. Majid, Renewable energy for sustainable development in India: current status, future prospectus, challenges, employment, and investment opportunities. Energy, Sustain Soc **10**. Article number: 2 (2020)
2. D. Gielen, F. Boshella, D. Saygin, M.D. Bazilian, N.W. Ricardo, Gorini, The role of renewable energy in the global energy transformation. Elsevier J. Energy Strat. 2211–467X (2019)
3. Ministry of New and Renewable Energy (MNRE), Government of India, Solar Energy Overview. https://mnre.gov.in/solar/current-status. Updated On: 05-Dec2020
4. Government of Gujarat official. https://patan.nic.in/about-district/. Patan, Developed and Hosted by National Informatics Centre, Ministry of Electronics & Information Technology, Government of India
5. Z.R. Dogaheh, V. Puig, Tilt Angle Optimization of Photovoltaic Panels, in *IEEE Systems Sponsored 6th International Conference on Control, Decision and Information Technologies*. https://doi.org/10.1109/CoDIT.2019.8820549 (2019)
6. V. Perraki, P. Kounavis, Effect of temperature and radiation on the parameters of photovoltaic modules. J. Renew. Sustain. Energy **8**, 013102 (2016)
7. S. WanQuan, E. Niringiyimana, V. Ndayishimiye, Analysis of electrical characteristics and performance of poly-crystalline solar PV module by I-V tester under temperature and solar, in *IEEE Conference*. https://doi.org/10.1109/PowerAfrica49420.2020.9219930 (2020)

8. S. Chandra, P. Bachan, Experimental investigation to emphasize on true potential of solar PV module, in *2020 International Conference on Power Electronics & IoT Applications in Renewable Energy and its Control (PARC)*, GLA University, Mathura, UP, India, Feb 28–29, 2020
9. HOMER Pro microgrid software by HOMER Energy. http://www.homerenergy.com, HOMER Pro Version 3.14.2 (August 10th, 2020)

Automatic Capacitor Switching Method for Power Factor Improvement with HMI Interface and Cloud Data Logger

Risfendra and Gheri Febri Ananda

Abstract This research aims to improve the power factor in a power grid. The large inductive load of power factor is less than 1 which results in the high value of the electric current and the installed power cannot be used optimally. The correction system is designed using an automatic switching method of four capacitors equipped with solid-state relay drivers to produce 14 combinations of capacitor values. The interface is designed to use a TFT HMI display and a cloud data logger to be able to display the value of the power factor in real time. The test sample used two 36-W fluorescent lamps or tube lamps and one 1-phase motor with 1/4 HP of power that had an average cos phi of 0.40. From the tests that have been done with some variation of load, the results show that the system has been able to make improvements with the final power factor value above 0.80. Furthermore, the results also show that HMI display TFT and Cloud systems are able to display data from power factors in rea ltime.

Keywords Power factor · Capacitor banks · IoT · Interface · Cloud

1 Introduction

The electrical power factor is the comparative value between active power and apparent power. Active power is the power used to do the actual energy. While apparent power is the power generated by multiplication between voltage and current in a network. The power factor is said to be good when it has a value close to one [1]. But the value of the power factor can decrease or is commonly referred to as the low power factor. This is because the load used on the house is inductive, such as tube lamps (TL), air conditioners (AC), fans, water pumps, and others [2, 3]. Inductive loads result in decreased power factors and reactive power. Reactive power is the amount of power required for the formation of magnetic fields and from the formation of such magnetic fields will form a flux of magnetic fields [4]. Thus, the low power

Risfendra (✉) · G. F. Ananda
Universitas Negeri Padang, Padang, Indonesia
e-mail: risfendra@ft.unp.ac.id

F. Thakkar et al. (eds.), *Proceedings of the International e-Conference on Intelligent Systems and Signal Processing*, Advances in Intelligent Systems and Computing 1370,
https://doi.org/10.1007/978-981-16-2123-9_33

factor leads to less effective use of electrical energy due to the many power losses generated. In addition, low power factors can make the installed power capacity not optimally used [5]. In addition, the power provider only allows the lowest limit of power factor is 0.85 if the power factor is less than that limit then there will be a fine for excess use of KVAR power which causes high electricity payment rates [6].

The effort that can be done to increase the value of the declining power factor is to regulate the value of reactive power by using the reactive power compensation method applied to the bank capacitor [7]. Capacitor components as reactive power generation serve to balance the reactive power needs of inductive loads. In this study, capacitor switching method was applied to perform power factor improvement to overcome changes in conditions of load; in addition, this study also implemented an interface with HMI (Human–Machine Interface) display TFT (Thin-Film Transistor) and cloud data logger.

HMI is a useful system for connecting humans who act as operators with machine technology. HMI will display information data on the computer screen to meet the needs of the user [8]. HMI can record data, collect, and process information obtained from controlled machines into information that is easy for humans to understand. HMI can be in the form of real-time computer visualization and control [9, 10]. Meanwhile, through the Cloud data logger system, data can be stored and monitored in real time.

1.1 Active Power (Watt)

Active power is the actual electrical power used in electrical equipment [11, 12]. The unit of active power is Watts and can be calculated by Eq. (1).

$$P = V.I.\cos\varphi \tag{1}$$

1.2 Apparent Power (Volt Ampere)

Apparent power is the power generated by the multiplication of the RMS voltage (root mean square) and the RMS current in a network or power which is the trigonometric sum of active power and reactive power [13] and can be calculated by Eq. (2).

$$P = V.I \tag{2}$$

1.3 Reactive Power (Volt Ampere Reactive)

Reactive power is the power that arises from the formation of a magnetic field that does not generate work. Reactive power is absorbed by loads which have inductive or capacitive reactances [4]. This power has units of reactive Volt Ampere (VAR) and can be calculated by Eq. (3).

$$Q = V.I.\sin\varphi \tag{3}$$

1.4 Power Factor

Power factor is the ratio between active power (W) and real power (VA) or it can also be called the cosine of the angle between active power and real power [14]. The power factor value can be calculated by Eq. (4).

$$PF = \frac{P}{S} = \frac{V.I.\cos\varphi}{V.I} = \cos\varphi \tag{4}$$

1.5 Power Factor Improvement

If the power factor is less than 0.85, it is necessary to improve the power factor. This is done by installing a capacitor bank. The principle of power factor improvement can be shown in Fig. 1.

To correct the low power factor value, the capacitor value of the bank must have the same Qc power as the reactive power of the system to be fixed by the power factor [15]. If this condition is met, the capacitor of the bank will increase the power factor to maximum or value 1. The amount of reactive power required to change the power factor from cos φ1 to cos φ2 can be calculated with Eq. (5).

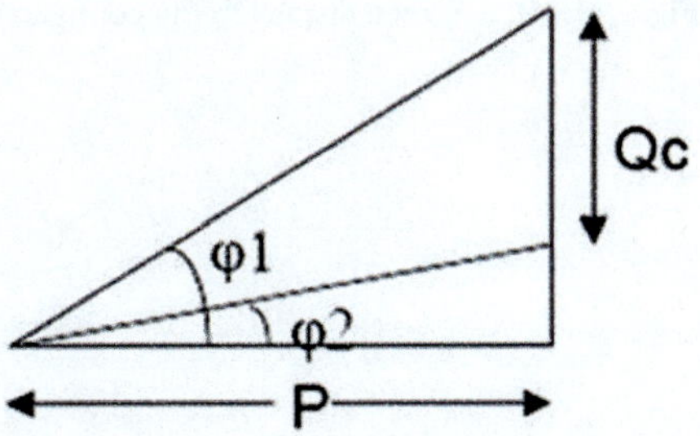

Fig. 1 The principle of power factor improvement

$$Qc = P(\tan\varphi 1 - \tan\varphi 2) \tag{5}$$

$$C = Qc/2.\pi.f.V^2 \tag{6}$$

1.6 Human–Machine Interface (HMI) with Display TFT

HMI Nextion is a Liquid Crystal Display (LCD) consisting of a series of TFTs in the form of touch screens with resistive touchscreen technology. This nextion-type HMI is equipped with a UART (Universal Asynchronous Receiver Transmitter) system with an Rx, Tx pin to communicate with other devices [16] (Fig. 2).

1.7 Cloud Data Logger

Data logger is a device that can be used to read different types of input signals and store that data within a certain period of time [17, 18]. Compared to other data acquisition devices, logger data has the advantage of being able to operate separately with a computer and has many variations according to the needs of the user [19, 20]. The logger data interface used is the Cloud that stores, collects, and analyzes data that is the ThingSpeak platform.

Fig. 2 HMI; **a** Screen display, **b** Pin configuration [16]

2 Method

2.1 System Function Diagram

In the block diagram of the system starting from the HMI display to log into the internet network on page 2 of the HMI display, after the system is connected to the internet, the system starts to read data by the PZEM-004T sensor, the data from the sensor is processed by the microcontroller to determine the value of the power factor. If the power factor value is smaller than 0.80, the system will calculate the C value requirement of the power grid and will activate the Solid-State Relay (SSR) to connect the capacitor according to the electricity grid requirements. Apart from that the microcontroller also processes data to be sent to the cloud via the ESP8266 so that it can be monitored in real time on the monitor device. The block diagram image and its physical form are shown in Figs. 3 and 4, respectively, below.

2.2 Sensor Design

Sensors serve to obtain data parameters from the power grid for processing on microcontrollers. In this study, the sensors used were PZEM-004T modules that can be directly connected to the power grid because the wiring of this module has two parts,

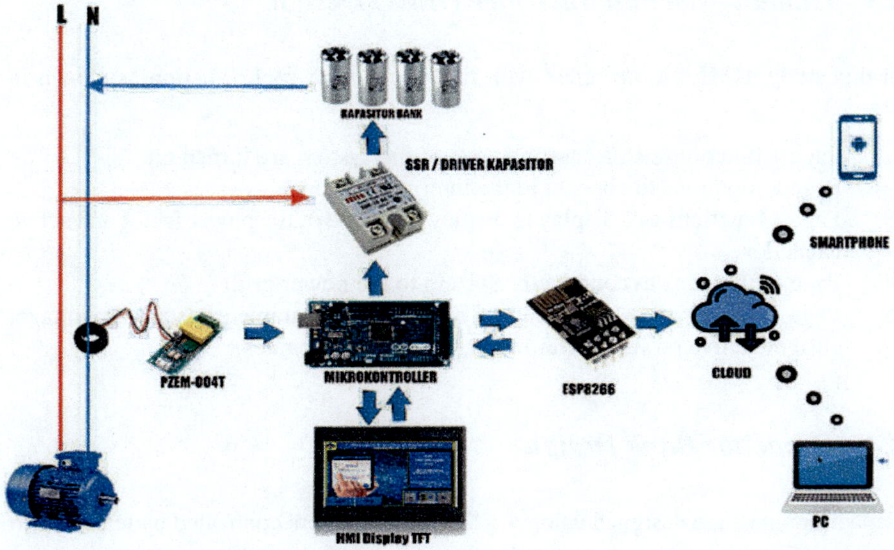

Fig. 3 System block diagram

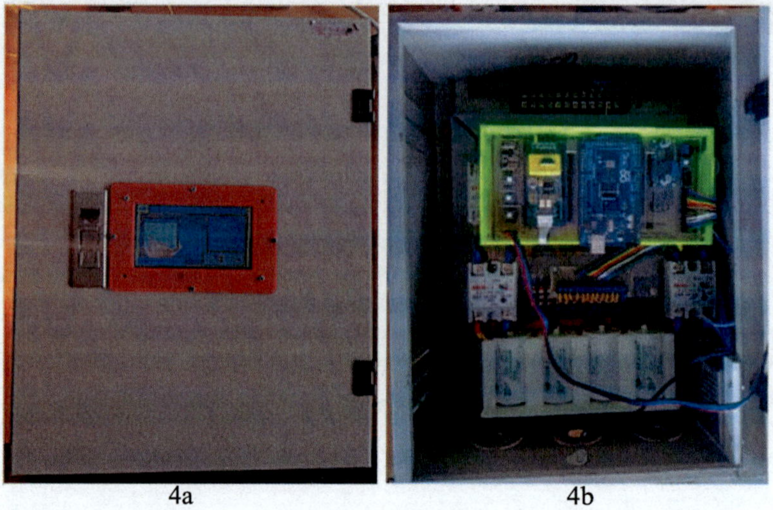

Fig. 4 Real system design; **a** outside view, **b** inside view

namely the wiring of voltage and current input terminals (CT), as well as serial communication wiring (Rx, Tx) as a communication line with microcontrollers.

2.3 Human–Machine Interface (HMI) Design

In this study, HMI was designed with five pages with each function as shown in Fig. 5.

(1) Page 1 functions as the starting page of the system are turned on
(2) Page 2 functions to connect to the internet network
(3) Page 3 functions as a display to display the value of the power factor and active capacitor
(4) Page 4 functions to connect the system to the power grid
(5) Page 5 functions as a display of all electrical parameters such as voltage, current, active power, power factor, and electricity rates.

2.4 Capacitor Bank Design

Capacitor banks are designed using 4 × 20-μF capacitors controlled by ten SSRs to produce 14 combinations of capacitor values. This design aims to meet the variety of capacitor value needs on an electrical grid. The following is a picture of the capacitor control circuit wiring (Fig. 6).

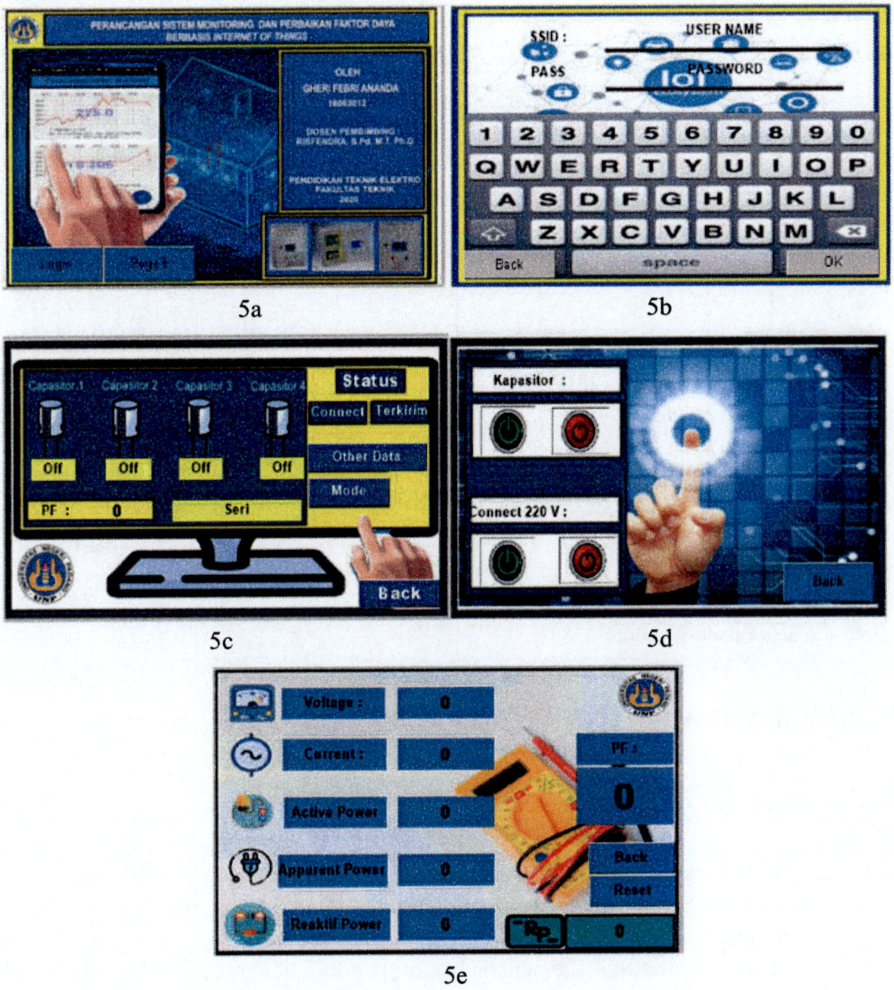

Fig. 5 The HMI design; **a** *Page* 1, **b** *Page* 2, **c** *Page* 3, **d** *Page* 4, **e** *Page* 5

From the control circuit above, 14 variations of capacitor values are obtained as in the configuration in Table 1 below.

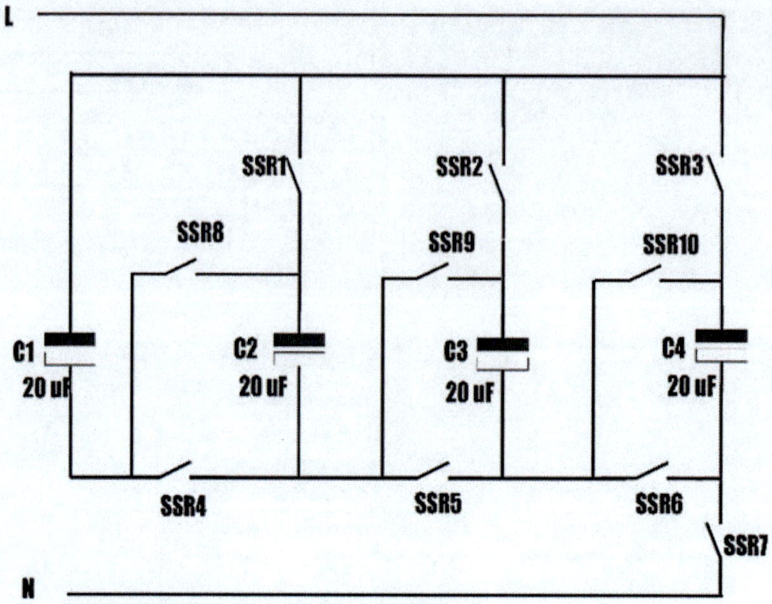

Fig. 6 Capacitor control circuit

3 Result and Analysis

3.1 Sensor Testing

Following are the test results of the PZEM-004T sensor readings by comparing with the Cos phi meter measuring instrument. Tests are carried out with several load variations as shown in Table 2.

The test results show that the sensor has an average accuracy level of 93.96 in power factor measurements. The following is a graph of the readings of sensors and measuring instruments with four different load tests (Figs. 7 and 8).

3.2 Human–Machine Interface (HMI) Testing

HMI displays communication with Arduino controller via serial communication (Rx, Tx). The following are the test results for each page on the HMI that has been designed (Table 3).

Table 1 Capacitor circuit configuration

Combination series	Capacitance value (μF)	Active SSR
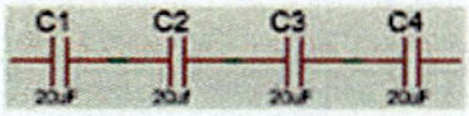	5	7,8,9,10
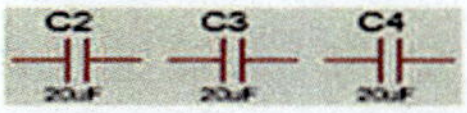	6.67	1,7,9,10
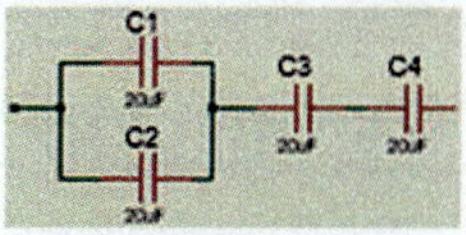	8	1,4,7,9,10
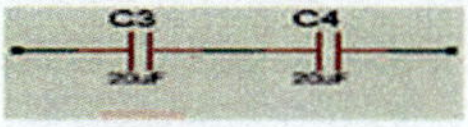	10	2,7,10
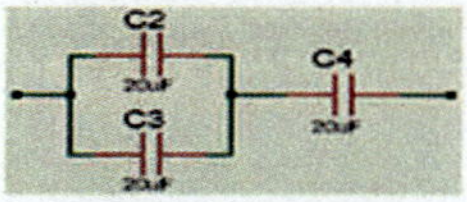	13.3	1,2,5,7,10
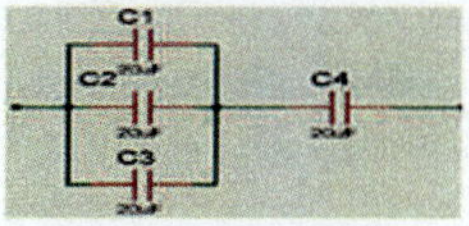	15	1,2,4,5,7,10
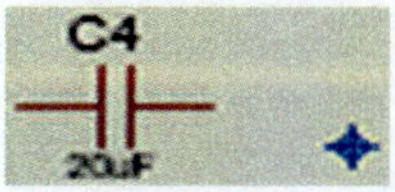	20	3,7
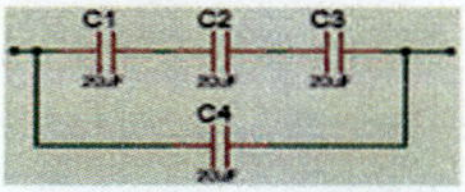	26.67	3,6,7,8,9
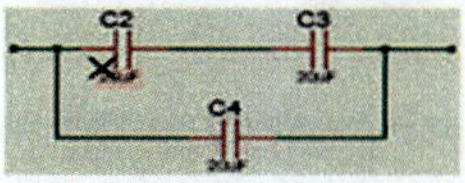	30	1,3,6,7,9
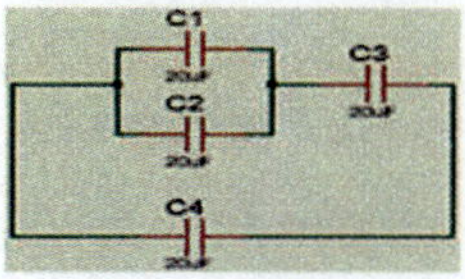	33.3	1,3,4,6,7,9
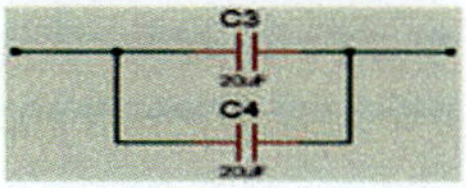	40	2,3,6,7

(continued)

Table 1 (continued)

Combination series	Capacitance value (μF)	Active SSR
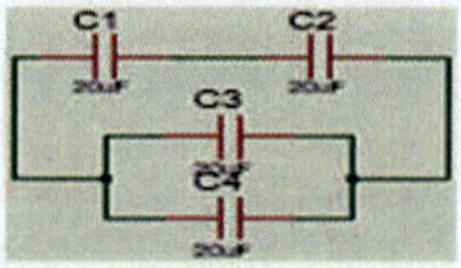	50	2,3,5,6,7,8
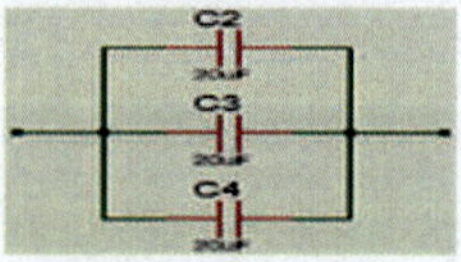	60	1,2,3,5,6,7
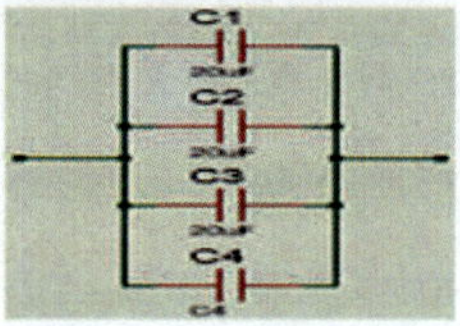	80	1,2,3,4,5,6,7

Table 2 Sensor test data results for power factor readings

Load	Measurement results measuring instrument	Sensor reading results	Accuracy (%)
No Load	0	0	100
1 × TL 36 W	0.44	0.40	88.8
2 × TL 36 W	0.42	0.39	92.8
1 × Induction motor ¼ HP	0.34	0.36	94.1

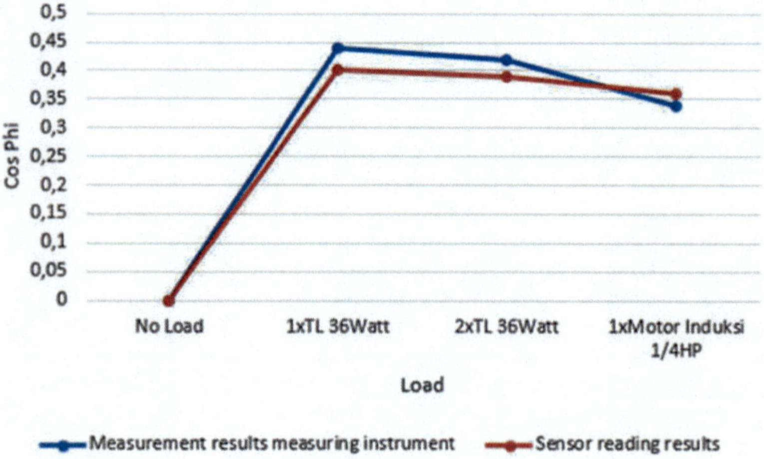

Fig. 7 Graph of the reading of Cos phi by sensors and measuring instruments

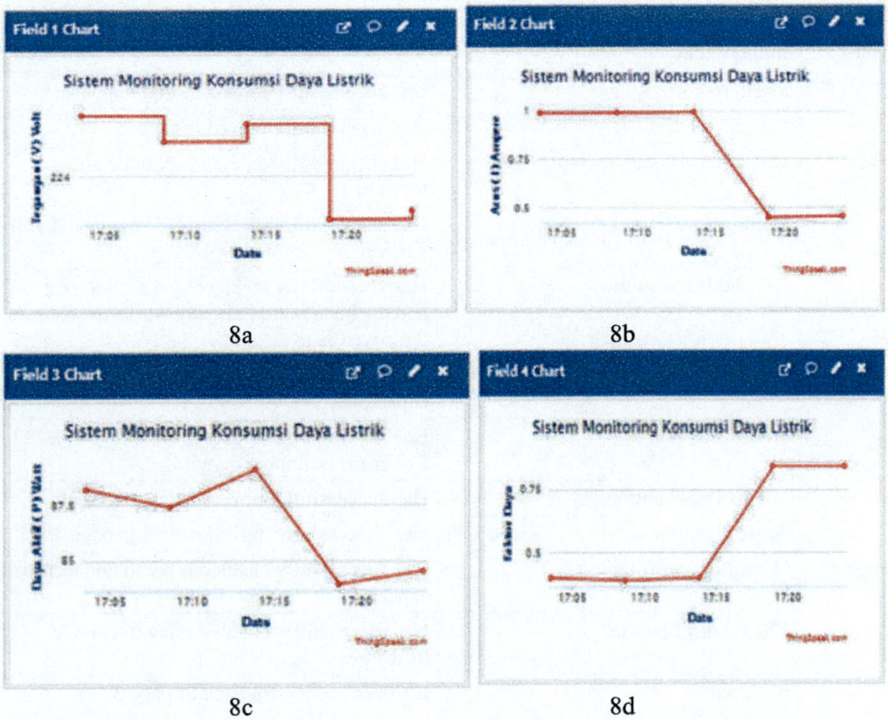

Fig. 8 Display on ThingSpeak: **a** Voltage, **b** Current, **c** Active power, **d** Power factor

3.3 Cloud Testing

Testing aims to see whether the system is able to transmit reading data from the sensor so that it can be monitored remotely. In this test, the data is set or the settings will be sent every 5 min. The data sent is in the form of voltage (V), current (A), active power (W), and the value of the power factor (Tables 4, 5).

3.4 Testing Tools for Power Factor Improvement

The test results of the automatic power factor repair system are shown in Table 5.

Analysis:
Load 1 × TL 36 W
$P = 46$ W
$\text{Cos } \varphi 1 = 0.40$
$\text{Cos } \varphi 2 = 0.85$ (sett point)
$\text{Cos } \varphi 1 = 0.40$ changed to angle value: $\text{Cos}^{-1}(0.40) = 66.42$□

Table 3 The test results for each page on the HMI

Page	Input/output	Result
Page 1	Login button	Has successfully moved to login page
	"Page3" button	Has successfully moved to page 3
Page 2	USERNAME and Password button	Has successfully entered username and password
	Number and letter keys	Has successfully entered SSID name and wifi password
	OK and Back button	Has successfully connected to the internet and the back button already works to return to the cover page
Page 3	Capacitor Indicator	Has successfully displayed the active capacitor
	Status Indicator	Has successfully displayed Connect/disconnect on wifi
	"Other Data" button	Has successfully moved page 3 to page 5
	"Mode" button	Has successfully moved page 3 to page 4
Page 4	Capacitor on/off button	Has successfully enabled relay to connect to capacitor
	220 V on/off button	Has successfully enabled relay to connect to load
	Back button	It's been successful back to page 3
Page 5	Sensor data indicator	Has successfully displayed all sensor data
	Reset button	Has successfully reset the value of electricity bills
	Back button	Already successfully moved back to page 3

Table 4 Data results of testing data sending on cloud ThingSpeak Table

Time	V	I	W	Cos φ
17.03	225	0.99	88	0.40
17.08	224	0.99	87.5	0.39
17.13	224.5	0.99	89	0.40
17.18	223.5	0.44	84	0.84
17.23	223.5	0.44	84.5	0.84
Desc	Success	Success	Success	Success

Table 5 Results of test data for automatic power factor improvement

Load	Power (Watt)	Cos phi before repair	Cos phi after repair
1 × TL 36 W	46	0.40	0.9
2 × TL 36 W	92	0.39	0.91
1 × Induction Motor ¼ HP	153	0.36	0.83

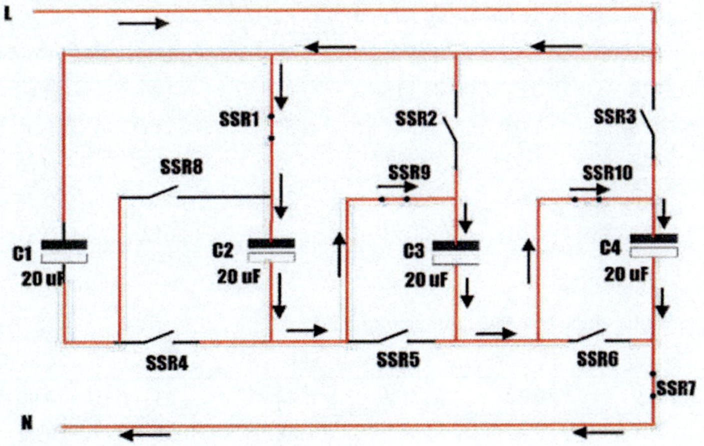

Fig. 9 SSR configuration that is active on mode 2

Cos φ2 = 0.85 changed to angle value: Cos-1(0.85) = 31.79□
Qc = P (tan φ1 – tan φ2)
= 46 (tan (66.42□) – tan (31.79□)
= 46 (2.29 – 0.62)
= 46 (1.67)
= 76.89

$$C = \frac{Qc}{2\pi f v^2}$$

$$C = \frac{76,89}{2.3, 14.50.220^2}$$

$$C = \frac{76,89}{15197600}$$

So at a load of 1 × TL 36 watts, the system requires a capacitor value of 5.05 μF so that the Cos phi value is 0.80. Thus, the live mode is SSR2 mode with SSR active 1, 7, 9, and 10 which has a capacitance value of 6.67 μF (Fig. 9).

3.5 *Analysis Before and After Using APFC*

This test aims to see a comparison of the value results of A values, W, and electrical power factors before and after using Automatic Power Factor Correction (APFC). In this test, it is done with some load variations as shown in Tables 6 and 7 below.

Load A = 1 × TL 36 W

Table 6 Results of test data before using APFC

Load	V	I	S	P	Q	Cos phi
A	228	0.51	116	46	106	0.40
B	228	1.03	234	92	215	0.39
C	228	1.92	437	153	409	0.35
D	229	2.43	556	202	518	0.36
E	227	2.91	660	246	612	0.37

Table 7 The results of the test data after using APFC

Load	V	I	S	P	Q	Cos phi
A	229	0.22	50	44	23	0.90
B	228	0.42	95	88	35	0.91
C	227	0.75	170	141	94	0.83
D	227	0.94	213	188	100	0.88
E	225	1.19	267	230	135	0.85

Load B = 2 × TL 36 W
Load C = 1 × Induction Motor ¼ HP
Load D = 1 × TL 36 W + 1 × Induction Motor ¼ HP
Load E = 2 × TL 36 W + 1 × Induction Motor ¼ HP

Based on Tables 5 and 6 above, it is obtained that testing with the highest load of a load of 2 × TL 36 Watts + 1 × Motor Induction 1/4 HP, before using APFC, the current value reaches up to 2.91 A, while after using APFC with the same load, the current value comes down to 1.19 A. Figure 10 is a graph of the relationship of current value and power factor and active power value before and after the installation of APFC (Fig.11).

From the graph above, it can be seen that the value of active power (Watt) after repair is lower than before making the repair. This proves that improving the power factor can save electricity consumption.

3.6 *Effect of APFC Installation on MCB Trip*

Load A:

2 × TL 36 W + Induction Motor ¼ HP + 75 W Incandescent lamp + 35 W Fan

Load B:

2 × TL 36 W + Induction Motor ¼ HP + 75 W Incandescent lamp + 35 W Fan

Load C:

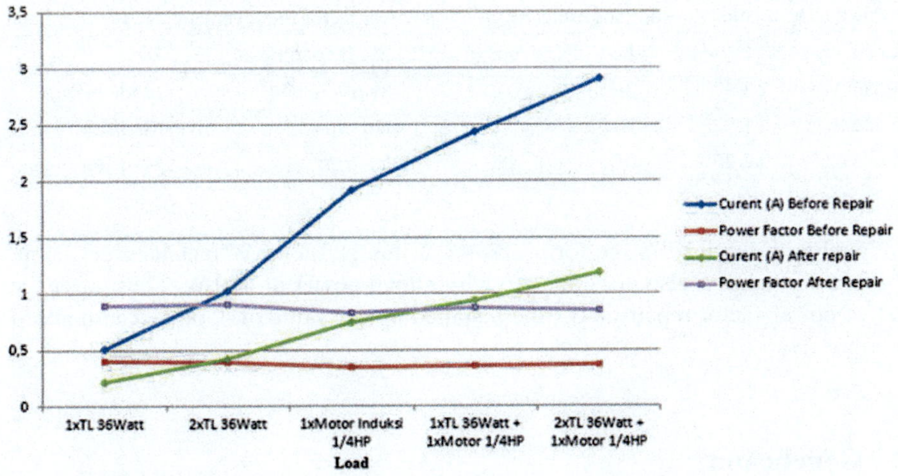

Fig. 10 The relationship between current values and power factor before and after repair

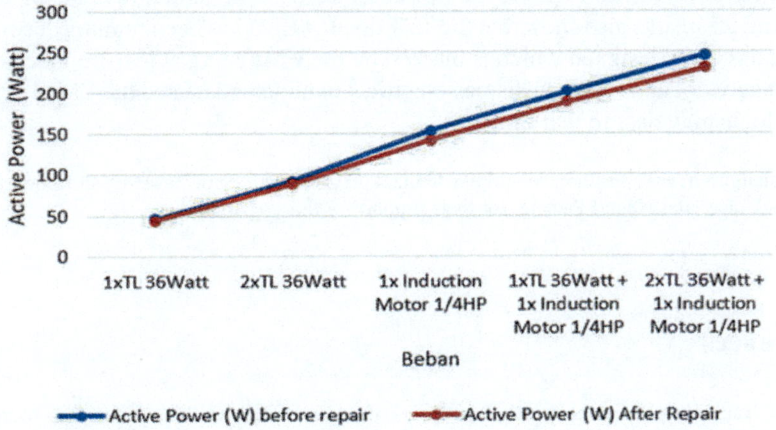

Fig. 11 Graph of active power values before and after repair

2 × TL 36 W + Induction Motor ¼ HP + 75 W Incandescent lamp + 35 W fan + 45 W fan + 75 W Incandescent Lamp.

From Table 8 above, it can be seen that when the power factor is fixed before repairing the power that can be used is only 333 W because, when testing with a load of 2 × 36 W TL + Induction motor ¼ HP + 75 W incandescent lamp, the MCB condition is still not trip with 333 watts of power, after additional load in the form of a 35 W fan on testing point 1, the MCB condition becomes a trip. This indicates that the usable power is only 333 W when before the repair mode is active. Unlike after the repair mode is active, with the same load on the test point 2, the MCB condition does not trip with a power capacity of 357 W. And then the test at point 3 when

Table 8 The results of testing the effect of MCB on power factor improvement

Load	I	P	cos φ	MCB conditions	Desc
Load A	3.37	368	0.47	Trip	Mode off
Load B	1.77	357	0.87	No trip	Mode on
Load C	2.25	0.91	480	No trip	Mode on

coupled with the load in the form of a 45 W fan and a 75 W incandescent lamp, the MCB condition also does not trip with a total power of 480 W. Thus, after the active power factor repair mode, the installed power value of 450 VA can be used completely.

4 Conclusion

Based on the results of the tests that have been done, it can be concluded that the system has been able to improve the power factor with the final power factor above 0.80. And the results also show that the TFT display HMI has been running according to what has been designed which is marked by the HMI being able to connect to the internet network and is able to display electrical parameter data and the Cloud system is able to display data in real time.

Acknowledgments The authors would like to thank LP2M (Institute of Research and Community Service), Universitas Negeri Padang, for their support for this research.

References

1. M.F. Hakim, Analisis Kebutuhan Capacitor Bank Beserta Implementasinya Untuk Memperbaiki Faktor Daya Listrik Di Politeknik Kota Malang. Jurnal Eltek **12**(1), 105–118 (2017)
2. S. Noor, N. Saputera, Efisiensi Pemakaian Daya Listrik Menggunakan Kapasitor Bank. Jurnal Poros Teknik **6**(2), 73–79 (2014)
3. M.B. Khan, M. Owais, Automatic power factor correction unit, in *2016 International Conference on Computing, Electronic and Electrical Engineering (ICE Cube)* (IEEE, 2016), pp. 283–288
4. D.P. Alfauzi, A.G. Permana, A. Novianti, Rancang Bangun Alat Perbaikan Faktor Daya Listrik Satu Fasa Berbasis Mikrokontroler. Proc. Appl. Sci. **5**(3), (2019).
5. T.N. Hartono, M. Shidiq, H. Santoso, Perancangan Alat Perbaikan Faktor Daya Beban Rumah Tangga dengan Menggunakan Switching Kapasitor dan Induktor Otomatis. Jurnal Mahasiswa TEUB **2**(1) (2014).
6. F.E. Nasution, T. Juara Mangapul, ST, M.: Pengaruh Faktor Daya Pelanggan Industri Terhadap Rugi-Rugi Pada Jaringan Sisi Sekunder Transformator Distribusi Pt. Pln (Persero) Area Serpong. Prosiding Semnastek, **1**(1), (2019)
7. A. Yani, Pemasangan Kapasitor Bank untuk Perbaikan Faktor Daya. JET (J. Electr. Technol.) **2**(3), 31–34 (2017)

8. H. Haryanto, S. Hidayat, Perancangan HMI (Human Machine Interface) Untuk Pengendalian Kecepatan Motor DC. Setrum: Sistem Kendali-Tenaga-elektronika-telekomunikasi-komputer **1**(2), 58–65 (2016)
9. A. Amin, R. Risfendra, Human machine interface for automation system of handling station. Motivection: J. Mech. Electr. Indus. Eng. **1**(3), 13–20 (2019)
10. N.P. Gireesh, Designing HMI using RX series RX63N microcontroller, in *2017 2nd IEEE International Conference on Recent Trends in Electronics, Information & Communication Technology (RTEICT)* (IEEE, 2017), pp. 1049–1053
11. P.A. Juliantara, I.W.A. Wijaya, C.G.I. Partha, Rancang Bangun Kapasitor Bank Otomatis Berbasis Mikrokontroler ATmega 328P Untuk Perbaikan Faktor Daya. Jurnal SPEKTRUM **5**(1), 157–163 (2018)
12. A. Taye, Design and simulation of automatic power factor correction for industry application. Int. J. Eng. Technol. Manage. Res. **5**(2), 10–21 (2018)
13. Y. Yuniarto, E. Ariyanto, Korektor Faktor Daya Otomatis pada Instalasi Listrik Rumah Tangga. Gema Teknologi **19**(4), 24–30 (2018)
14. N. Fartino, T. Tarmizi, M. Syukri, Kajian Perancangan Alat Perbaikan Faktor Daya Otomatis. Jurnal Karya Ilmiah Teknik Elektro **5**(1) (2020)
15. Y. Kabir, Y.M.Mohsin, M.M. Khan, Automated power factor correction and energy monitoring system, in *2017 Second International Conference on Electrical, Computer and Communication Technologies (ICECCT)* (IEEE, 2017), pp. 1–5
16. A. Supriyadi, A. Setyawan, J.E. Suseno, Rancang Bangun Sistem Kendali Unit Pengolahan Air Bersih Berbasis Arduino Uno R3 Dan Nextion. Berkala Fisika **22**(2), 42–55 (2019)
17. F. Hadiatna, R. Susana, Implementasi Metode Multiplexing pada Sistem Akuisisi Data di Perangkat Data Logger, in *Prosiding Seminar Nasional Energi Telekomunikasi dan Otomasi (SNETO) 2017, Jurusan Teknik Elektro, Institut Teknologi Nasional (Itenas) Bandung* (2017)
18. M.S. Hadi, IoT Cloud Data Logger Untuk SIstem Pendeteksi Dini Bencana Banjir Pada Pemukiman Penduduk Terintegrasi Media Sosial. Jurnal Edukasi Elektro **1**(2) (2017).
19. R. Rismawati, M. Sadli, Desain Data Logger Sensor Suhu Berbasis Mikrokontroler ATMega16 dengan Empat Kanal Input. Jurnal Litek: Jurnal Listrik Telekomunikasi Elektronika **17**(1), 19–22 (2020)
20. A. López-Vargas, M. Fuentes, M. Vivar, IoT application for real-time monitoring of solar home systems based on Arduino^{TM} with 3G connectivity. IEEE Sens. J. **19**(2), 679–691 (2018)

Analysis of Novel 15Gbps WDM–FSO System Employing Hybrid Amplifier Under Various Weather Conditions

Preet Kaur, Maninder Singh, Manish Sharma, and Rajeev Kumar

Abstract This paper presents high-capacity WDM (Wavelength Division Multiplexing)-based FSO system which utilizes two EDFAs (Erbium-Doped Fiber Amplifiers).It is used for long-distance transmission as a booster amplifier and SOA (Semiconductor Optical Amplifier) as a pre-amplifier. The proposed system supports 15Gbps transmission over optimized link length on various weather conditions, i.e., haze, light rain, medium rain, and heavy rain. This novel system provides Q-factor and BER of 17.55, 2.5e-69, respectively, for heavy rain conditions, whereas the conventional system provides Q-factor and BER of 9.14, 1.55e-20, respectively, for same weather condition. This paper increases Q-factor and BER, respectively, for heavy rain condition.

Keywords EDFA · FSO · MZM · SOA · WDM

1 Introduction

The rapidly increasing high bandwidth wireless access demand requires the development of new technology [1]. Thus, FSO becomes the prominent solution for accomplishing wireless broadband demands as this technology utilizes the optical signal as carrier wave for the transmission of radio frequency signal [2, 16, 17]. But, the main limitation of the FSO system is the dependence of transmission performance on free space which is affected by the weather conditions, i.e., snow, haze, rain, fog, etc. [3]. To compensate for the limitations of FSO system various techniques has been analyzed to enlarge the transmission coverage [4]. The analysis of different modulation techniques along with the employment of an optical amplifier has been presented to increase the spectral efficiency of FSO link. The hybrid modulation techniques have

P. Kaur (✉)
J C Bose University of Science & Technology, YMCA, Faridabad, India

M. Singh · M. Sharma · R. Kumar
Institute of Engineering and Technology, Chitkara University, Chandigarh, Punjab, India
e-mail: Maninder.s@chitkara.edu.in

F. Thakkar et al. (eds.), *Proceedings of the International e-Conference on Intelligent Systems and Signal Processing*, Advances in Intelligent Systems and Computing 1370,
https://doi.org/10.1007/978-981-16-2123-9_34

been employed in the transmitter section to reduce BER [5]. The analytical model utilizing DF (decode and forward) and SC-QAM (subcarrier-Quadrature Amplitude modulation) techniques has been presented to improve symbol-error rate (SER) [6]. The employment of PPM (Pulse Position Modulation) and ASK (Amplitude Shift Keying) modulation techniques improves FSO performance [7]. The performance evaluation of various modulation techniques namely BPSK-SIM (binary phase shift keying-Subcarrier Intensity Modulation), QAM-SIM, and DPSK-SIM (Differential PSK-SIM) by adjusting the turbulence strength, bit rate, and link distance has been presented to reduce the BER at receiver output [8]. DDM (Dual-Diffuser Modulation) technique has been presented to improve the transmission capacity in the FSO system [9]. The transmission link range has been improved by the utilization of EDFA [10]. The FSO link performance in terms of Q-factor and BER has been improved by the employment of various optical amplifiers in conjunction with photodiode arrays [11]. The hybrid combination of EDFA and SOA has been described for WDM-based 2.5Gbps FSO system [12].

Based on the literature survey, it is concluded that the link distance can be improved by utilizing modulation technique in conjunction with optical amplifiers. Therefore, this paper presents a novel 15Gbps WDM-based FSO system utilizing two EDFAs before FSO channel act as booster amplifier and one SOA after FSO channel work as pre-amplifier to support the long-distance high bit rate transmission.

2 Impact of Different Weather Condition

The main challenge faced by FSO communication system is the attenuation due to weather condition. The different weather condition affects the performance of FSO links which limits the long-distance transmission. The strength of optical signals gets reduced due to attenuation caused by weather condition.

Attenuation caused by fog is due to Mie scattering [13] and the attenuation coefficient value for fog weather condition is described by Kim model [14]. The attenuation caused by fog depends on visibility (V), operating wavelength (λ), and distribution of diffusing particle size (q). Attenuation caused by fog is given as [15]

$$\alpha_{fog} = \frac{3.91}{V}\left(\frac{\lambda}{550nm}\right)^{-q} \tag{1}$$

where

$$q = \begin{cases} 1.6 \quad for\ V > 50\,\text{km} \\ 1.3 \quad for\ 6\text{km} < V < 50\,\text{km} \\ 0.16 \quad for\ 1\text{km} < V < 6\,\text{km} \\ V - 0 \quad for\ 0.5\,\text{km} < V < 1\,\text{km} \\ 0 \quad for\ 0.5\,\text{km} < V \end{cases}$$

Raindrop geometrical scatters laser beam. Attenuation caused by the water particles of rain cause major distortion in the transmission link and the attenuation caused by rain is variable in nature, i.e., the attenuation decreased with the decrease in rainfall rate. Carbonneau model [15] describes the value of attenuation in terms of precipitation intensity R(mm/hr) is expressed as

$$\alpha_{rain} = 1.076R^{0.67} \quad (2)$$

3 Hybrid Amplifier-Based WDM–FSO Setup

Figure 1 depicts the simulation setup of the proposed 15Gbps WDM-based FSO system. The simulation parameters employed in the proposed setup are described in Table 1. Typical devices involved in this setup are laser transmitters, hybrid booster amplifiers, multiplexers, pre-amplifier, demultiplexers, and receivers. The internal view of transmitter subsystem is shown in Fig. 2a, which consists of PRBS (Pseudo-Random Bit Sequence) generator, NRZ pulse generator, Continuous Wave Laser, and MZM (Mach–Zehnder Modulator). PRBS generates a bit rate of 5Gbps, CW Laser generates carrier signal of varying frequency from 193.1 to 193.3THz and this signal is modulated by the PRBS generated signal using MZM. The modulated signal is fed to the 3:1 Multiplexer, thereby the overall system capacity of the proposed system is 15Gbps, i.e., 5Gbps (PRBS bit rate) × 3 (no. of channels). The multiplexed signal is applied to the combination of two EDFA to provide the maximum output power

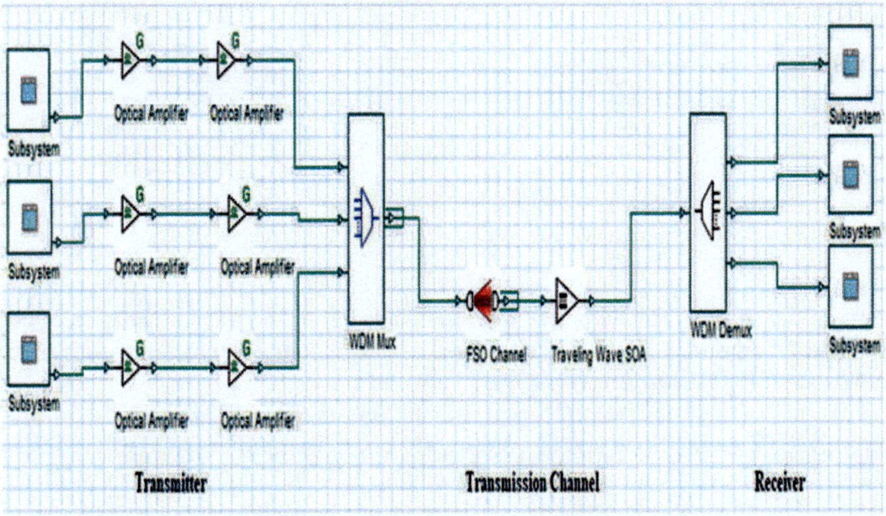

Fig. 1 Simulation setup of 15Gbps WDM–FSO system employing hybrid amplifier

Table 1 Simulation parameters

Layout parameters	Value
Bit rate	5Gbps
Sequence length	128bits
Transmitter parameters	Value
PRBS bit rate	5Gbps
CW laser power	30dBm
CW laser frequency	193.1–193.3THz
FSO parameters	Value
Receiver aperture	30 cm
Beam divergence	2.5mrad
WDM parameters	Value
Number of channels	3
Optical amplifier	Value
Gain	25–50 dB
Injection current	0.15A
Receiver parameters	Value
PIN responsivity	1A/W
Dark current	10nA
Low-pass bessel filter cut off frequency	0.75 * Bit Rate Hz

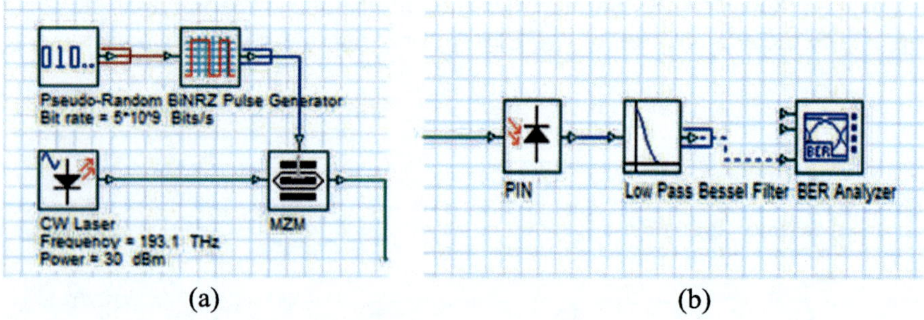

Fig. 2 **a** Internal view of transmitter subsystem **b** Internal view of receiver subsystem

from the transmitter. The amplifier which is placed after the transmitter is referred to as booster amplifier. EDFA has been used as booster amplifier as it provides high saturation power as compared to SOA.

The gain of first and second booster amplifiers is set as 50dBm and 25dBm, respectively. The amplified signal from hybrid booster amplifier is fed to the FSO. FSO comprises of transmitting antenna, receiving antenna, and atmosphere work as channel. The transmission performance depends on atmosphere condition, i.e., rain, fog, haze, turbulence, snow, etc. The amplified modulated optical signal is fed to the channel through a transmitting antenna and propagated signal from atmosphere is captured by the receiving antenna. The received optical signal is amplified by SOA which works as pre-amplifier (amplifier placed immediately in front of optical receiver). The fundamental role of pre-amplifier is to amplify the low-strength optical signal to improve receiver sensitivity. The SOA is compact thereby employed as pre-amplifier in the proposed system. The received signal after SOA is fed to demultiplexer which splits the multi-wavelength signal into distinct wavelength signal. The amplified signal is applied to receiver subsystem and its internal view is illustrated in Fig. 2b which consists of photodiode, low-pass Bessel filter, and BER analyzer. The PIN photodiode converts the received signal into an electrical signal. The recovered signal is filtered by low-pass Bessel filter and applied to BER analyzer. BER analyzer measures the system performance in terms of Q-factor, BER, etc.

4 Results and Discussion

Figure 3a depicts the eye diagram obtained for haze condition (attenuation 2.37 dB/km) at optimized length, i.e., 40 km. The proposed system provides the Q-factor of 17.14 with BER of 3.11e-66, whereas the conventional system provides the Q-factor of 16.28 with BER of 3.04e-60, respectively. The eye diagram for light

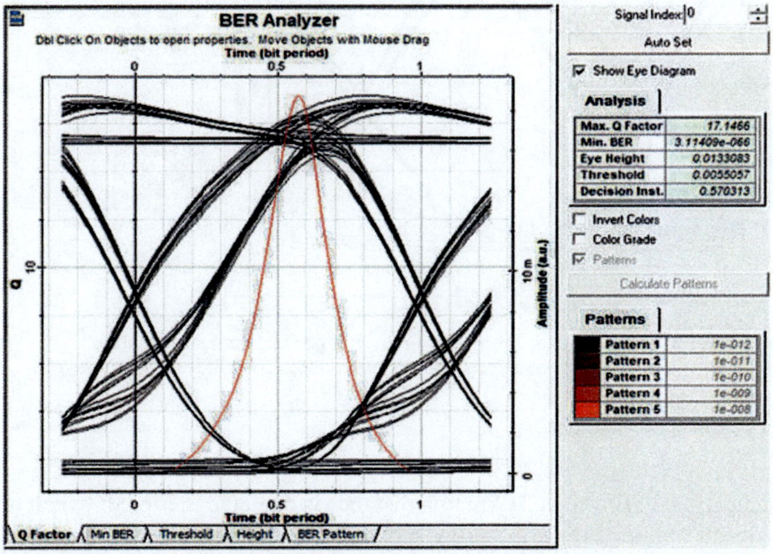

(a) Simulated result using haze condition Range (40 km) ,Attenuation (2.37 db/km) . The proposed system provides Q-factor of 17.14 with BER of 3.11e-66.

Fig. 3 Eye Diagram of 15Gbps WDM–FSO system with **a** Haze condition, **b** Light rain condition, **c** Medium rain, **d** Heavy rain

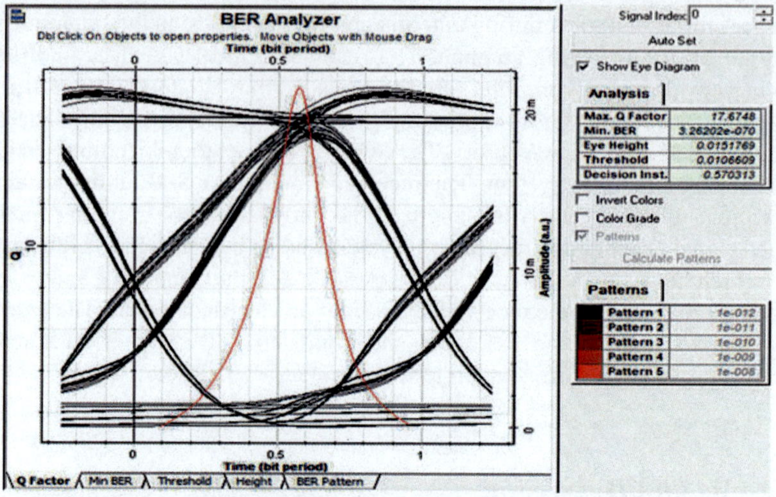

(b) Simulating result for light rain Range (31km), Attenuation (3dB/km).The proposed system enhances the system capacity as Q-factor of 17.64.

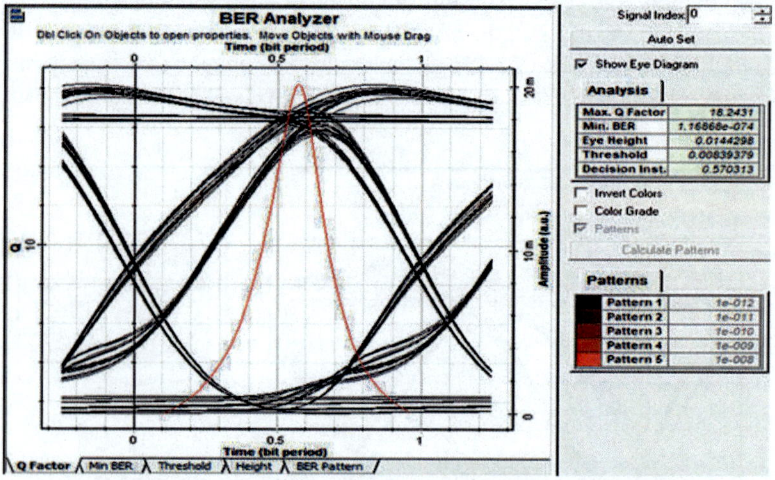

(c) Simulating result for medium rain Range (11.5 km), Attenuation (9 dB/km) which reveals that proposed system reduces BER 4.5e-33 to 1.16e-74.

Fig. 3 (continued)

rain (attenuation 3 dB/km) at 31 km link distance is depicted in Fig. 3b which reveals that the result obtained from the proposed system enhances the system capacity of FSO system as Q-factor of 17.64 is obtained, whereas conventional system provides Q-factor of 9.27. Figure 3c and d shows the eye diagram for medium rain (attenuation 9 dB/km) and heavy rain (attenuation 20 dB/km) at optimized link distance of 11.5 km and 5.4 km, respectively, which reveals that proposed system reduces BER

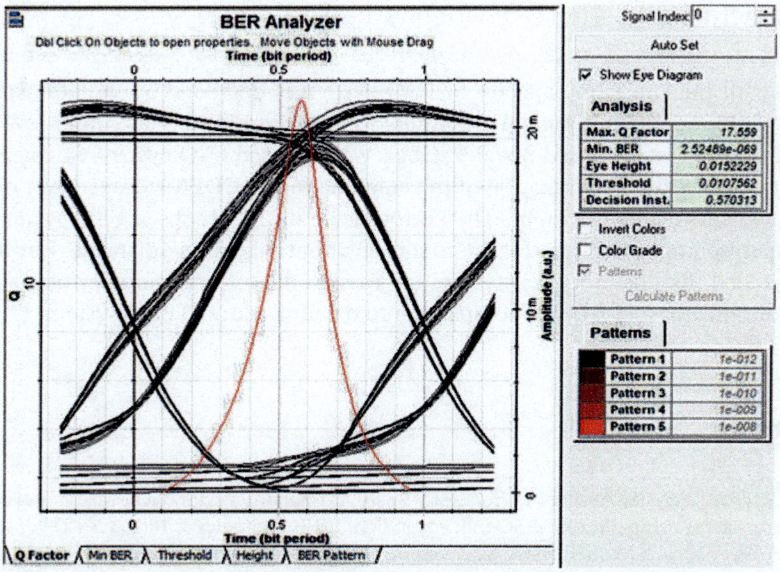

(d) Simulating result for heavy rain Range (5.4km) , Attenuation (20dB/km) which reveals that proposed system reduces BER 1.55e-20 to 2.5e-69.

Fig. 3 (continued)

from 4.5e-33 to 1.16e-74 and 1.55e-20 to 2.5e-69 for medium rain and heavy rain, respectively. Table 2 shows the comparison between the conventional and proposed systems at different weather conditions for optimized transmission range. Table 2 reveals that the proposed system improves the transmission system performance by increasing Q-factor and reduces BER.

Table 2 Performance of 15Gbps WDM–FSO System

Weather condition	Attenuation (dB/km)	Range (km)	Conventional		Proposed	
			Q-factor	BER	Q-factor	BER
Haze	2.37	40	16.28	$3.04e^{-60}$	17.14	$3.11e^{-66}$
Light rain	3	31	9.27	$4.31e^{-20}$	17.67	$3.27e^{-70}$
Medium rain	9	11.5	11.56	$4.5e^{-33}$	18.24	$1.16e^{-74}$
Heavy rain	20	5.4	9.14	$1.55e^{-20}$	17.55	$2.5e^{-69}$

5 Conclusion

The overall transmission capacity of FSO system is degraded by the atmospheric effect but by employing optical amplifiers in FSO system, the distortion level can be decreased. The proposed novel 15Gbps WDM-based FSO system enhances the system capacity by the utilization of the combination of EDFA at transmitter output and SOA before receiver input. The performance of the proposed FSO system has been analyzed on different weather condition at optimized link distance. Further, it is concluded that the proposed system improvises the transmission performance in terms of Q-factor and BER in comparison to the conventional FSO system.

References

1. B. Bag, A. Das, I.S. Ansari, A. Prokes, C. Bose, A. Chandra, Performance analysis of hybrid FSO systems using FSO/RF-FSO link adaptation. IEEE Photonics J. **10**(3) (2018)
2. Y.F. Al-eryani, A.M. Salhab, S.A. Zummo, M. Alouini, Protocol design and performance analysis of multiuser mixed RF and hybrid FSO/RF relaying with buffers. J. Opt. Commun. Netw. **10**(4), 309–321 (2018)
3. M.T. Dabiri, S. Mohammad, S. Sadough, Performance analysis of all-optical amplify and forward relaying over log-normal FSO channels. J. Opt. Commun. Netw. **10**(2), 79–89 (2018)
4. Y.F. Al-eryani, A.M. Salhab, S.A. Zummo, M. Alouini, Two-way multiuser mixed RF/FSO relaying: performance analysis and power allocation. J. Opt. Commun. Netw. **10**(4), 396–408 (2018)
5. A.M. Shukla, M.K. Singh, V. Kapoor, Performance of FSO system using PPM and MSK sub-carrier intensity modulation scheme, in *Proceedings of the 2017 International Conference Wireless Communications Signal Process*, vol. 2. (Networking, WiSPNET 2017, 2018), pp. 2475–2479
6. B.T. Vu, T.C. Thang, A.T. Pham, Selective relay decode-and-forward QAM/FSO systems over atmospheric turbulence channels, in *2014 9th International Symposium on Communicaton Systems, Networks and Digital Signal Processing CSNDSP 2014* (2014), pp. 407–410
7. S. Das, M. Chakraborty, ASK and PPM modulation based FSO system under varying weather conditions, in *2016 IEEE 7th Annual Ubiquitous Computing, Electronics & Mobile Communication Conference UEMCON 2016* (2016)
8. A. Mishra, P.S. Pati, R.K. Giri, Performance analysis of SIM based FSO using various modulation techniques with APD receiver over turbulent channel, in *IEEE WiSPNET 2017 Conference* (2017), pp. 2730–2734
9. K.R. Ummul, M.S. Anuar, A.K. Rahman, C.B.M. Rashidi, High Speed transmission in Giga bps performance for FSO Communication using new modulation technique, in *2016 IEEE 6th International Conference on Photonics (ICP)* (2016), pp. 7–9
10. A. Hatziefremidis, K.E. Zarganis, K.E. Zoiros, Flexible high power free space optical system for aircraft to ground communications. ICTON **2014**, 3–6 (2014)
11. A. Guptab, Performance analysis of FSO link under effect of fog with array of receivers and amplifiers, in *Proceedings of 2015 RAECS* (2015)
12. G. Sharma, L. Tharani, Performance evaluation of WDM-FSO based hybrid optical amplifier using bessel filter, in *2018 International Conference on Communication and Signal Processing* (2018), pp. 653–656
13. A. Prokes, V. Skorpil, Estimation of free space optics systems availability based on meteorological visibility, in *LATINCOM'09. IEEE Latin-American Conference* (2009), pp. 1–4

14. Propagation data required for the design of terrestrial free-space optical links, in *ITU-R Recommendation* (Geneva 2012), p. 1817
15. S.A. Zabidi, W. Khateeb, M.R. Islam, A.W. Naji, The effect of weather on free space optics communication (FSO) under tropical weather conditions and a proposed setup for measurement, in *International Conference on Computer and Communication Engineering (ICCCE)* (2010), pp. 1–5
16. S. Mahajan, D. Prakesh, H. Singh, Performance analysis of free space optical system under different weather conditions, in *2019 6th International Conference on Signal Processing and Integrated Networks (SPIN)* (2019). https://doi.org/10.1109/spin.2019.8711687
17. S.K. Modalavalasa, R. Miglani, Performance analysis of FSO system for different amplification strategies, in *2020 International Conference on Emerging Smart Computing and Informatics (ESCI)* (2020). https://doi.org/10.1109/esci48226.2020.9167517

Optimized Molecular Structure, Vibrational Spectra, and Frontier Molecular Orbitals of 1,4-Benzene Diamine with Palladium Electrodes as a Molecular Switch—A Computational Analysis

T. Rafsa Koyadeen, A. R. Abdul Rajak, and Vilas H. Gaidhane

Abstract Molecular electronics focuses on shrinking the electronic circuit with functional single-molecular devices. Molecular logic circuits have the potential to replace conventional semiconductor logic circuits very shortly. This emerging field aims at investigating molecules with exciting properties and discovering strategies to acquire reliable and replicable contacts between the molecular backbone and the nanoelectrodes. Here, molecular structure and vibrational spectroscopic studies of 1,4-benzene diamine (BDA) with palladium nanoelectrode have been investigated as a molecular electronic switch using the density functional theory (DFT) and Hartree–Fock (HF) method with BPV86 (Becke Perdew 1983 functional modified by Vosko) and basis set Stuttgart/Dresden (SDD). The noticed vibrational spectra and frontier molecular orbitals have been resolved and analyzed in detail for the comparison of both the OFF and ON state of electrical conductivity. This paper provides salient information regarding enhancing the conductivity of benzene-based single-molecular switches with metal nanoelectrodes by maintaining the stability and switching property. The current flow in 1,4-benzene diamine with the palladium nanoelectrode has been analyzed using theoretical simulations and it is extensively reconcilable with the recent experimental results.

Keywords Molecular switch · DFT and HF method · Vibrational spectra · Frontier molecular orbitals · Ionization energy · Potential energy

1 Introduction

The current flowing molecular backbones have immeasurable possibilities for applications in molecular electronics and nanotechnology. The emerging implementations of single molecule-based electronics require an elementary knowledge of how the

---

T. R. Koyadeen (✉) · A. R. A. Rajak · V. H. Gaidhane
Birla Institute of Technology and Science Pilani, Dubai Campus, Dubai 345055, UAE

F. Thakkar et al. (eds.), *Proceedings of the International e-Conference on Intelligent Systems and Signal Processing*, Advances in Intelligent Systems and Computing 1370,
https://doi.org/10.1007/978-981-16-2123-9_35

intrinsic properties of molecular backbones and nanoelectrodes interpret their interfacial electronic structure [1]. Unfold the new molecular backbones and molecule–electrode cluster combination for desired applications and technology dispenses a motivation to develop a profound understanding of the molecular electronic structure and bonding [2]. Like traditional semiconductor logic devices, molecular electronics also possess functional molecular devices such as molecular transistors, molecular wires, molecular diodes, molecular memory, molecular switches, etc. Among these devices, molecular switches are analogous to conventional complementary metal oxide semiconductor (CMOS) switches in the VLSI (Very Large-Scale Integration) circuits. Although a very simple device, a suitable molecular switch has the potential to become the fundamental unit of complicated molecular machines [3, 4].

Principally, a molecular switch comprises a single-molecular backbone connected with nanoelectrodes on either side; sometimes referred to as a molecular backbone sandwiched between two nanoelectrodes. The central molecular backbone is also known as the switching molecule since it can switch controllably between two stable states, i.e., ON and OFF. An electrical signal, light energy, change in absolute temperature, or change in the chemical environment may be the reason to shift from ON to OFF states or vice versa [5]. One of the mature conjugated switching molecules to design an efficient single molecular switch is Oligo phenylene ethynylenes (OPEs). This molecule has been systematically inspected by Tour et al. 20 years back [6]. OPEs are benzene ring-dominated chains with huge electronic conjugation between phenyl rings and also when connecting with nanoelectrodes. It shows two different stable geometries with two different conductivity [7]. BDA has been considered as one of the paradigm molecules for both theoretical and experimental studies in molecular electronics [8]. It is a molecule with benzene (C_6H_6) as the switching molecule and amine($-NH_2$) as the anchoring group [9]. Here, in this paper, the BDA molecule is connected between two Palladium (Pd) electrodes with four atoms each to get an electronic molecular switch design.

The analysis of relative and absolute vibrational spectroscopy (Raman intensity and IR intensity) is a very powerful tool to go deeper into the study of nanomaterials and nanostructures [10]. Moreover, comprehension of the physics and chemistry of organic compounds demands the formulation of molecular orbital properties and spectra. Regardless of a large number of already synthesized and investigated molecular geometrical structures, there remains a rich structural variety to explore. The structure–property correlations of the model compounds report a big challenge between theory and experiment [11]. Computational chemistry has become a convenient tool for the interpretation of the foundations of chemical phenomena and inspection of molecular structural properties. Furthermore, it is preferable to avoid a profligate experimental screening through rapid computational assessment. The computational methods based on ab initio or first principle approaches are very good in understanding the molecular structural properties [12]. Nowadays density functional theory (DFT) methods are very popular as it gives more accurate quality computer simulation results comparing to Hartree–Fock (HF) since it neglects correlations in the motion between two electrons with anti-parallel spins [13, 14].

The motivation behind the present work is to investigate the molecular structures, vibrational spectra (with IR and Raman intensity), and frontier molecular orbital analysis of BDA-based single-molecular electronic switch with Pd electrode using DFT incorporated with the basis set BPV86/SDD andHF theory with the same basis set. The stability and conductivity of the optimized molecular switches have been discussed based on the calculations and findings of our work. Discussed the effect of the number of aromatic rings in the zero-bias molecule.

2 Model and Computational Details

The first principle calculations have performed using density functional theory and HF method at the Gaussian 09 program to calculate bond length, bond angles, and vibrational frequencies with the IR intensities and Raman scattering activities [15]. For the calculations of the molecular structure properties, the Stuttgart/Dresden (SDD) basis set and the BPV86 (Becke Perdew 1983 functional modified by Vosko) functional was chosen because it is computationally less demanding than the B3LYP (Becke three-parameter Lee–Yang–Parr) functional. The count of SCF cycles is increased and this is very frequently required with transition metal electrodes. To obtain the Raman spectra, Gaussian calculated activities are converted using the empirical relation of Raman scattering theory with corresponding Raman intensities [16]. The structural and electrical properties of all the molecular structures have been explored and compared with each other.

3 Results and Discussion

3.1 Geometric Structure

The geometrically optimized molecular structures of Pd–BDA–Pd for both OFF and ON state by DFT method have been shown in Fig. 1a, b. The contact geometry between molecule–electrode plays the key role in determining the "ON" and "OFF" state of a benzene ring-based aromatic molecular switch. The same molecular–metal junction can display different conductance values due to the different molecule–metal bonding forms. This is due to the different coupling strengths with different contact geometry. Among these, top-top geometry shows low conductance and it corresponds to the "OFF" state of a molecular switch. It is a molecular junction where both ends of switching molecules sit on top sites of both electrodes. Whereas one headgroup is sitting on the three-fold hollow site of the electrode and the other end is connected to the top sites is denoted as top-hollow, and it shows high conductance and it corresponds to the "ON" state of a molecular switch. [7, 12].

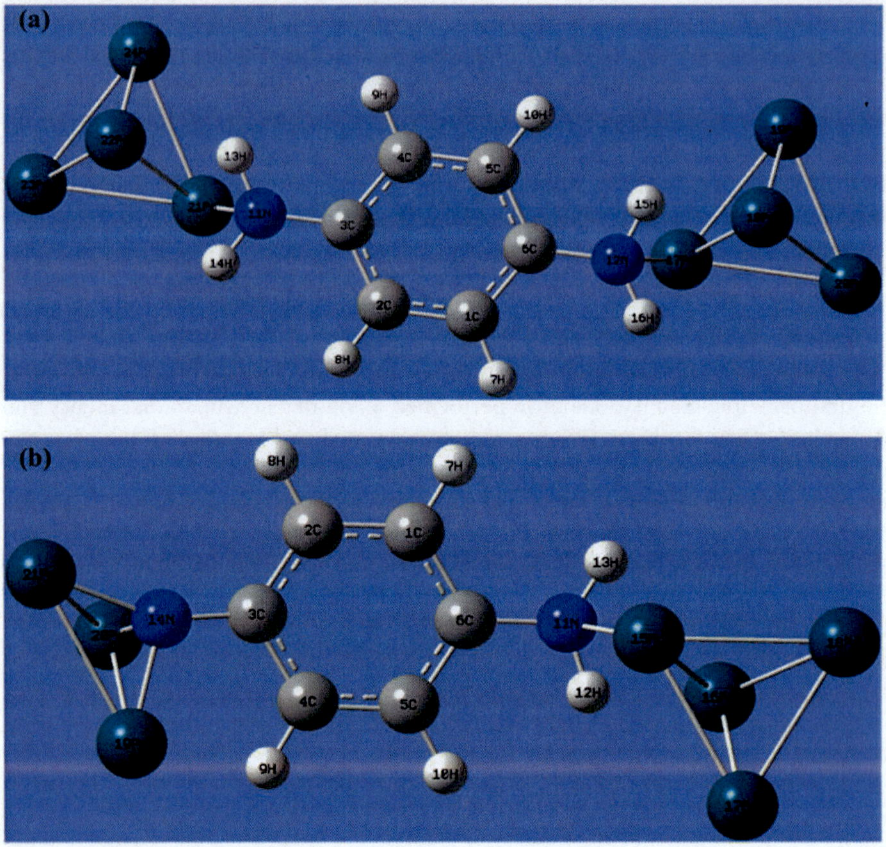

Fig. 1 **a** Optimized electronic structure of the Pd–BDA–Pd OFF state and. **b** Pd–BDA–Pd ON state molecules using the DFT method

To realize and design new types of molecular switches with different anchoring groups and nanoelectrode clusters Fig. 1a, b is vital. Optimized bond lengths, bond angles, and dihedral angles at the DFT/BPV86/SDD level and HF/BPV86/SDD level of both states of molecules have been given in Tables 1 and 2. The molecular structural values are marginally different for both DFT and HF method electron–electron interaction term is not included in HF calculation. Hence, here discussed only the DFT results. Transparently, the ON state molecule shows one Pd atom less compared to the OFF state molecule in its optimized structure. It is observed that the ON state molecule shows shorter values of bond length and bond angle at the left side electrode, while the OFF state molecule has slightly larger values. The anchoring group amine (–NH_2) and nanoelectrode Pd bonding at both left and right sides of the switching molecule determine these critical geometrical parameters to design a molecular switch. Moreover, the computed data (Tables 1 and 2) has marginally different values of bond length and bond angle in the two chosen first principle studies

Table 1 Optimized geometrical parameters of Pd–BDA–Pd for OFF state using the DFT and HF methods

Pd–BDA–Pd OFF State					
Parameters	DFT	HF	Parameters	DFT	
Bond length in (Å)			Bond angle in degrees (°)		
C1–C2	1.391	1.392	C1–C2–C3	121.60	121.62
C2–C3	1.402	1.400	C2–C3–C4	116.81	116.72
C3–C4	1.406	1.401	C3–C4–C5	121.59	121.99
C4–C5	1.391	1.389	C4–C5–C6	121.55	121.52
C5–C6	1.407	1.399	C5–C6–C1	116.80	116.83
C6–C1	1.403	1.402	C6–C1–C2	121.57	121.54
C3–N11	1.388	1.362	C3–N11–Pd21	120.82	120.81
N11–Pd21	2.090	1.98	N11–Pd21–Pd24	106.85	106.81
Pd21–Pd22	2.680	2.673	Pd21–Pd24–Pd23	60.0001	60.12
Pd21–Pd23	2.680	2.673	Pd24–Pd23–Pd22	60.00	60.11
Pd21–Pd24	2.680	2.672	Pd23–Pd22–Pd21	60.0002	60.00
Pd22–Pd23	2.680	2.672	C6–N12–Pd17	116.60	116.32
Pd22–Pd24	2.680	2.680	N12–Pd17–Pd19	112.70	112.67
Pd23–Pd24	2.680	2.679	Pd17–Pd19–Pd20	60.14	60.00
C6–N12	1.388	1.386	Pd19–Pd20–Pd18	60.00	60.02
N12–Pd17	1.916	1.905	Pd20–Pd18–Pd17	62.04	62.12
Pd17–Pd18	2.506	2.501	N11–Pd21–Pd22	64.33	63.00
Pd17–Pd19	2.662	2.661	N12–Pd17–Pd18	4.34	64.13
Pd17–Pd20	2.677	2.673			
Pd18–Pd19	2.680	2.678			
Pd18–Pd20	2.680	2.677			
Pd19–Pd20	2.680	2.655			

in this paper. These contrast values may be due to the hydrogen bonding inside the single-molecule switch.

3.2 *Vibrational Spectra*

The IR and Raman intensities for harmonic vibrational frequencies are calculated for the Pd–BDA–Pd molecule for both ON and OFF states separately with BPV86, SDD, and functions of polarization and diffuse, given in Tables 3 and 4. The findings of this paper reveal the effect of neglecting a harmonicity in the real system [17, 18]. Nevertheless, the decrement of the frequency value is because of electron correlation to a defined range in DFT. Here, it has been observed only a slight difference from the

Table 2 Optimized geometrical parameters of Pd–BDA–Pd for ON state using the DFT and HF methods

Pd–BDA–Pd ON state					
Parameters	DFT	HF	Parameters	DFT	HF
Bond length in (Å)			Bond angle in degrees (°)		
C1–C2	1.390	1.391	C1–C2–C3	121.60	121.54
C2–C3	1.401	1.400	C2–C3–C4	116.81	116.23
C3–C4	1.403	1.401	C3–C4–C5	121.59	123.11
C4–C5	1.389	1.382	C4–C5–C6	121.55	121.76
C5–C6	1.402	1.349	C5–C6–C1	116.80	116.21
C6–C1	1.407	1.402	C6–C1–C2	121.57	121.34
C3–N14	1.382	1.312	C3–N14–Pd28	94.57	95.13
N14–Pd19	1.835	1.821	N14–Pd20–Pd21	44.66	45.61
N14–Pd20	1.916	1.872	Pd20–Pd21–Pd19	60.00	61.22
N14–Pd21	1.884	1.883	Pd21–Pd19–Pd20	59.99	60.11
Pd19–Pd20	2.681	2.672	C3–N14–Pd19	109.59	110.11
Pd19–Pd21	2.682	2.679	C6–N11–Pd15	116.59	119.32
Pd21–Pd20	2.680	2.682	N11–Pd15–Pd17	112.70	111.67
C6–N11	1.389	1.386	Pd15–Pd17–Pd18	60.13	60.02
N11–Pd15	1.918	1.915	Pd17–Pd18–Pd15	59.59	60.12
Pd15–Pd16	2.505	2.501	Pd15–Pd16–Pd17	61.66	63.10
Pd15–Pd17	2.661	2.641	Pd17–Pd16–Pd18	61.00	62.13
Pd15–Pd18	2.676	2.671			
Pd16–Pd18	2.680	2.638			
Pd16–Pd17	2.680	2.677			
Pd17–Pd18	2.599	2.605			

calculated values of harmonic vibrations. The selection of the basis set in the DFT calculation has a crucial role in it. However, it is exclusively conventional to cut back the computed harmonic frequencies. Tables 3 and 4 show the computed vibrational frequencies. The majority of the normal modes of potential energy distributions (PEDs) are overlapping of various dissimilar harmonic vibrations. The computations of IR and Raman spectra values are given in Figs. 2 and 3.

3.3 *Vibrational Frequencies in 4000–1500 cm^{-1} Range*

For the aromatic molecule, the characteristic region for the identification of C–H stretching vibrations is 3100–3000 cm^{-1} [18]. Distinctly for both ON and OFF states of BDA sandwiched between Pd nanoelectrode molecular structure show aromatic

Table 3 Computed vibrational spectral frequencies (cm^{-1}) of the Pd–BDA–Pd OFF state molecule using the DFT/BPV86/SDD level

Frequency	IR	Raman	Assignment (PEDs)
3234	10.8	314.67	rPdN
3123	1.1	567.00	Arom rCH
3067	3.26	31.90	Arom rCH
2909	10.49	64.50	Arom rCH
2900	23.27	14.89	Arom rCH
1562	51.07	54.08	Arom rCH
1559	78.30	211.86	rNH_2 asymmetric
1556	12.79	270.65	rNH_2 asymmetric
1496	16.45	345.98	rNH_2 symmetric
1484	31.78	21.54	rNH_2 symmetric
1483	0.81	34.56	rNH
1480	94.67	28.95	Arom rCC
1478	2.79	4.16	Arom rCC + iprPdPd
1469	8.69	94.90	ρNH_2
1388	7.93	32.38	– ωNH_2
1371	112.24	23.15	βPdN + rNH + μNH_2
1399	7.35	54.59	Arom rCC + βCH + ωNH_2
1272	98.82	43.03	rNH + β CH + βPdN
1269	423.75	83.29	βCH
1256	89.72	4.12	βCH
1241	3.79	3.02	βCH + rCC2 + ωNH_2
1226	41.63	1.01	Arom ϕ_a(CCC)
1216	9.87	51.27	ωNH_2 + rCC2
1214	10.42	87.13	γPdN + μNH_2 + γCH
1166	9.76	98.19	PdN + Arom ϕ_a(CCC) + rCC1
1129	39.76	21.43	μNH_2 + γCH
1111	49.11	0.98	γCH + μNH_2 + τCC1
1106	25.11	6.90	γCH + γPdN + Φ_{c1}(CNPd)
1104	32.98	58.78	Arom ϕ_a(CCC)
1099	45.32	7.62	Arom ϕ_a(CCC) + γPdN + ϕ_{c1}(CNPd)
1088	32.56	0.12	γPdN
937	12.00	0.18	τCC2 + μNH_2
911	6.81	3.92	Φ_{c2}(CCC)
836	9.18	10.76	μNH_2 + ϕ_{c2}(CCC)
812	81.38	2.98	τCC2 + Arom ϕ_a(CCC)
749	4.12	1.69	Arom ϕ_a(CCC)

(continued)

Table 3 (continued)

Frequency	IR	Raman	Assignment (PEDs)
640	7.26	1.98	μNH_2

C–H stretching vibrations at around 3000 cm^{-1}. Furthermore, they are a pure mode of vibrations. Precisely, it is 2909, 3067, and 3123 cm^{-1} for OFF state and the range of band is observed for ON molecule as 2900–3110 cm^{-1}. Generally, semi-circle stretching is a C–C aromatic stretch. Here, it is noticed at 1562 for both ON and OFF state molecules. For the appropriate PED spectra, another C–H in-plane bending vibration has been explored. Moreover, both the ON and OFF state geometries show C–H scissoring vibrations too.

3.4 *Vibrational Frequencies in 1500–1000 cm^{-1} Range*

Great modes of vibrations have been observed for both the Pd–BDA–Pd molecules in the range of 1500–1000 Cm^{-1}. From the computations, it is investigated that the transition of an electron from HOMO to LUMO is highly dependent on excitation energies, wavelengths, and oscillator strengths. For OFF state molecule, it is investigated wagging $\omega(NH_2)$ connected with benzene C–C stretching vibration at 1166 cm^{-1}. Also in-plane blending of –β(CH), ring breathing mode coupled with an in-plane blending of –β(CH), and in-plane bending –β(PdN) coupled with in-plane bending-β (CH) are observed. In-plane blending of –β(CH) at 1226, 1214, 1256, 1269, 1088 cm^{-1} and ring breathing mode coupled with in-plane blending at 1241 cm^{-1} and in-plane bending –β(PdN) at 1106, and 1104 cm^{-1} are investigated, respectively. At 1216 cm^{-1}, the in-plane stretching-*ips* (NPd_3) vibrational frequency has occurred, whereas at frequency 1272 cm^{-1}, out-of-plane stretching-*ops* (NPd_3) coupled with other modes of vibrations have been observed. At the frequencies, 1011 and 1129 cm^{-1} benzene angle bending-ϕ_a (CCC) vibration has been noticed. Similarly, in ON state molecule, the vibrational frequencies are investigated at 1399 cm^{-1}, 1499 cm^{-1}, 1099 cm^{-1}, and 1111 cm^{-1} for in-plane bending-β(NH) coupled with scissoring-$\rho(NH_2)$, in-plane bending-β(NH) coupled with other modes of vibrations, in-plane bending –β(NH) coupled with twisting-$\tau(NH_2)$, and stretching-r(CN) coupled with twisting-$\tau(NH_2)$, respectively.

3.5 *Vibrational Frequencies Below 1000 cm^{-1}*

For Pd–BDA–Pd OFF state geometry molecule, three frequencies are allotted at 937, 836, and 749 cm^{-1} concerning the three different modes: out-of-plane bending-γ (NH) coupled with other modes of vibrations, rocking-μ (NH_2) coupled with

Table 4 Computed vibrational spectral frequencies (cm^{-1}) of Pd–BDA–Pd ON state molecule using the DFT/BPV86/SDD level

Frequency	IR	Raman	Assignment (PEDs)
3234	9.8	380.17	rPdN
3123	4.1	647.90	Arom rCH
3067	3.16	31.12	Arom rCH
2909	9.49	44.59	Arom rCH
2900	26.17	13.49	Arom rCH
1862	50.07	52.08	Arom rCH
1861	23.30	210.85	rNH_2 asymmetric
1855	342.79	23.34	rNH_2 asymmetric
1799	126.45	456.18	βNH
1786	131.78	45.56	rNH_2 symmetric
1781	10.81	67.45	rNH
1634	194.67	34.78	Arom rCC
1612	22.79	3.56	Arom rCC + iprPdPd
1611	28.69	100.12	ρNH_2
1581	0.99	45.76	ωNH_2
1499	16.24	45.12	βNH
1389	37.85	67.56	Arom rCC + βCH + ωNH_2
1372	196.82	67.34	rNH + β CH + βPdN
1399	413.75	19.09	βNH
1290	989.72	24.12	βCH
1281	63.80	83.02	βCH + rCC2 + ωNH_2
1269	141.74	50.26	Arom ϕa(CCC)
1256	59.88	0.98	ωNH_2 + rCC2
1244	610.31	7.92	γPdN + μNH_2 + γCH
1156	79.76	8.00	PdN + Arom ϕa(CCC) + rCC1
1129	739.77	21.33	μNH_2 + γCH
1121	19.11	0.28	γCH + μNH_2 + τCC1
1116	225.29	16.90	γCH + γPdN + Φc1(CNPd)
1114	2.18	9.18	Arom ϕa(CCC)
1091	201.00	14.72	βCH + τNH_2
1089	8.92	23.41	βCH + rCC2
1016	42.15	2.17	rNH + βCH
1013	20.19	4.86	rNH + τCH_2
1012	10.67	3.03	βNH
1005	12.86	34.23	βCH + Arom ϕa(CCC)

(continued)

Table 4 (continued)

Frequency	IR	Raman	Assignment (PEDs)
999	30.19	30.05	γCH
989	50.07	20.16	γCH
918	90.11	0.31	γPdN + μNH_2
888	76.20	50.58	γCH + γPdN
869	1.99	5.72	γCH + μNH_2
844	65.54	3.69	Arom ϕa(CCC) + τCC2
834	0.32	5.62	Arom ϕa(CCC) + γPdN + ϕc1(CNPd)
823	2.16	8.92	γPdN
817	324.06	86.98	τCC2 + μNH_2
816	63.91	38.92	Φc2(CCC)
815	39.18	50.76	μNH_2 + ϕc2(CCC)
812	91.38	12.98	τCC2 + Arom ϕa(CCC)
749	4.82	91.69	Arom ϕa(CCC)
640	86.26	61.98	μNH_2

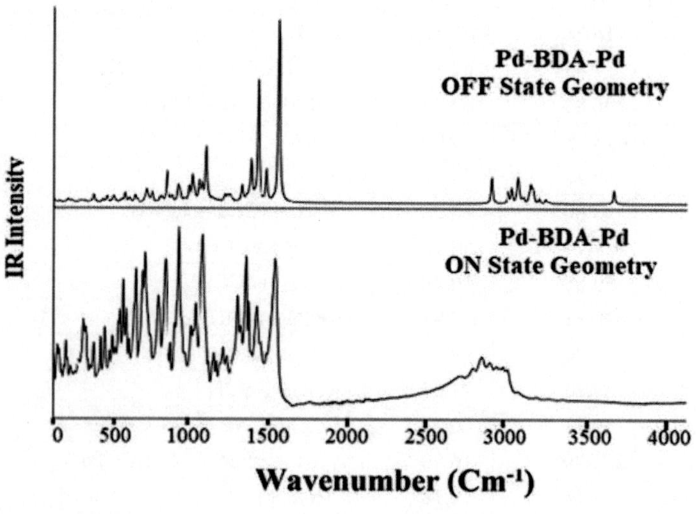

Fig. 2 Theoretical IR spectra of Pd–BDA–Pd OFF state and Pd–BDA–Pd ON state molecules using the DFT method

out-of-plane bending γ(CH), and aromatic angle bending-ϕa (C–C–C), respectively. Likewise, the other frequency bands have been allocated for Pd–BDA–Pd ON state geometry molecule at the frequency range 800–100 cm^{-1}. This includes out-of-plane bending-γ(CH) coupled with out-of-plane bending- γ(NH), rocking-$\mu(NH_2)$

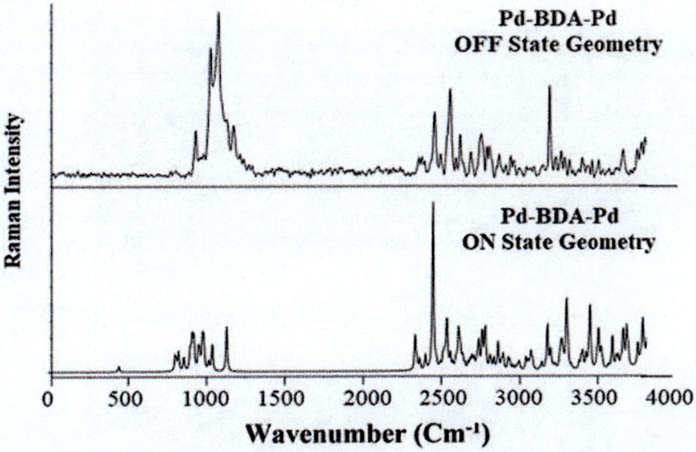

Fig. 3 Theoretical Raman spectra of Pd–BDA–Pd OFF state and Pd–BDA–Pd ON state molecules using the DFT method

coupled with other modes of vibrations, and rocking-$\mu(NH_2)$ coupled with the other two modes of vibrations, respectively.

3.6 Frontier Molecular Orbital Analysis

The narrow energy gap between the lowest unoccupied molecular orbital (LUMO) and the highest occupied molecular orbital (HOMO) is appropriate for injection and tunneling of charged carriers in a molecular switch between switching molecule and nanoelectrode [19]. The electrons in the tunneling mechanism of molecular junctions are transported through the closest frontier molecular orbitals (HOMO or LUMO) of the switching molecule and the Fermi energy level (EF) of the nanoelectrode. Amine-terminated molecules in which tunneling is primarily mediated by the highest occupied molecular orbital (HOMO) [6]. In this paper, the coupling of the benzene molecule and NH_2 anchoring group with the Pd metal nanoelectrodes results in the broadening and shifting of the discrete energy levels of the switching molecule. Hence broadening can be determined by the coupling interaction between the BDA switching molecule with its two Pd metal nanoelectrodes on both sides. Figure 4a, b shows the molecular orbital detailing of both the OFF and ON state of the Pd–BDA–Pd molecule.

The bandgap energy of the molecular device is reduced by connecting the molecular backbone with the two metal nanoelectrodes in a sandwich manner [5]. In an experimental setup, it can be calculated the electric current across a molecular device since the conduction happens purely through either HOMO or LUMO depending on the applied bias voltage across the nanoelectrodes [20]. To design a molecular switch,

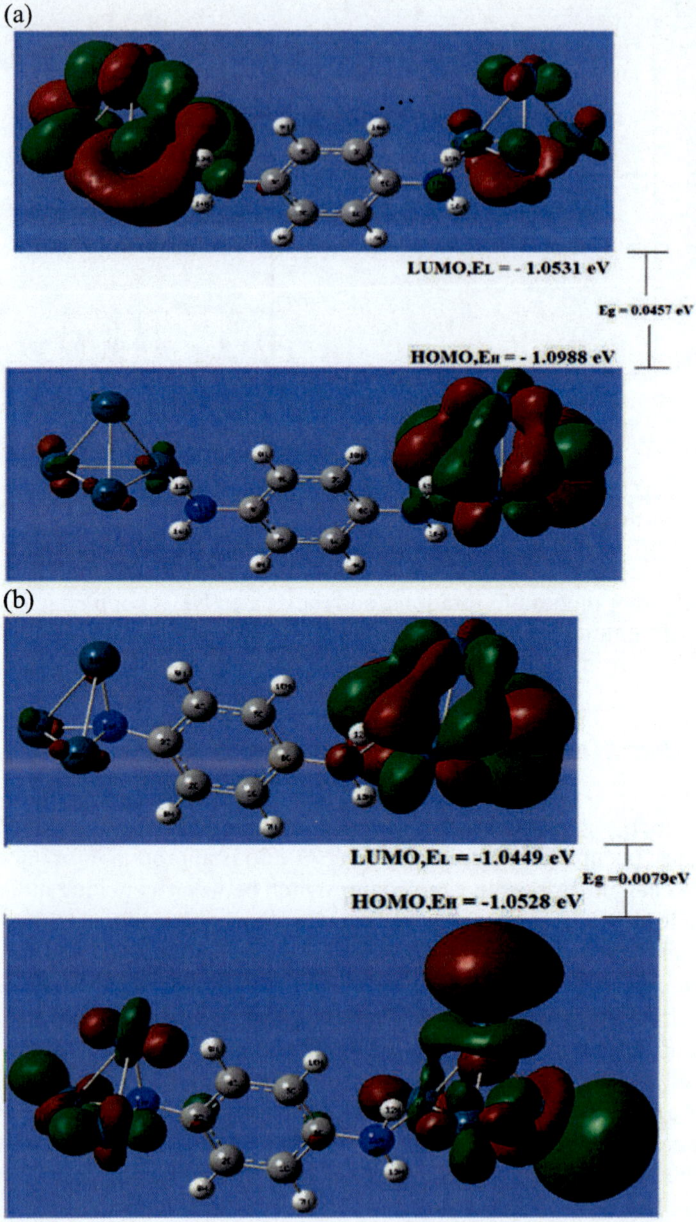

Fig. 4 **a** HOMO and LUMO pattern of Pd–BDA–Pd OFF state molecule. **b** HOMO and LUMO pattern of Pd–BDA–Pd ON state molecule

Table 5 Calculated values of HOMO energy, LUMO energy, and Bandgap for BDA, Pd–BDA–Pd OFF state, and ON state molecule

Molecule	HOMO energy	LUMO energy	Bandgap $E_g = E_{LUMO} - E_{HOMO}$
BDA	4.14 eV	0.48 eV	4.62 eV
Pd–BDA–Pd OFF; Top-Top Geometry	−1.0988 eV	−1.0531 eV	0.0457 eV
Pd–BDA–Pd ON; Top-Hollow Geometry	−1.0528 eV	−1.0449 eV	0.0079 eV

the HOMO–LUMO gap is crucial. It mainly depends on the optimized geometrical structure of a molecular switch. Also, it is affected by changing the contact geometry.

The HOMO–LUMO gap is increased for the switching molecule without the electrode [21]. The bandgap is the fundamental criterion for determining the electrical conductivity of a solid material [22]. From Table 5, it is seen that the ON state geometry has smaller bandgap energy compared to the OFF state geometry and therefore has a higher conductivity. BDA without the Pd electrode will act as a pure insulator compares with BDA sandwiched between two Pd electrodes. In a molecular switch design, increasing the number of phenyl rings in the switching molecule between the nanoelectrodes will lead to the decrement of the HOMO–LUMO gap [23]. Thereby, the conductance of molecular switch is strictly related to the HOMO–LUMO gap of molecule–electrode sandwich structure [24].

When explaining charge transport through a molecule–electrode junction one more factor must take into account that the number of electrons on this molecular system. By the Koopmans' theorem [25, 26], the ionization energy also called ionization potential (I) and electron affinity (A) of any molecule can be calculated using the relations, $E_{HOMO} = -I$ and $E_{LUMO} = -A$. Therefore, by applying Koopmans' theorem, the ionization potential and electron affinity are seen to be the negative value of HOMO energy and LUMO energy of the molecular device, respectively. Theoretically, whenever the ionization potential of a molecule decreases the conductivity of that molecule increases. Similarly, atoms in molecules possess potential energy since they are attracted to each other. And, whenever the potential energy of a molecule increases, the stability of that molecule decreases obviously [5]. Therefore, by checking ionization potential and potential energy one can check the conductivity and stability of a specific molecular device to an extent. From Table 6, it is clear that the ON state has higher conductivity than the OFF state with low ionization potential or ionization energy.

Also, in this paper using Koopmans' theorem, quantum chemical descriptors of both the ON and OFF states of BDA molecule with the Pd electrode have been investigated. The hardness of the molecule $\eta = (I-A)/2$, the electronic chemical potential of the molecule $\mu = -(I + A)/2$, the softness of the molecule $S = 1/2\eta$, the electronegativity of the molecule, $\chi = (I + A)/2$, and electrophilicity index of the molecule $\omega = \mu 2S$ has been computed for both the ON and OFF states of BDA molecule with Pd electrode on both sides. The calculations of quantum

Table 6 Calculated values of ionization potential (I), electron affinity (A), electronegativity (χ), electronic chemical potential (μ), chemical hardness (η), softness (S), electrophilicity index (ω). (Units for all the below-calculated values-eV)

Molecule	I	A	χ	μ	η	S	ω
Pd–BDA–Pd OFF State	1.0988	1.0531	1.076	−1.076	0.0223	22.422	25.959
Pd–BDA–Pd ON State	1.0528	1.0449	1.049	−1.049	0.004	126.58	139.29

Table 7 Calculated values of Potential Energy (P.E) for BDA, Pd–BDA–Pd OFF state and ON state molecules

Molecule	Potential energy (P.E)
BDA	0.0849×10^{-14} J
Pd–BDA–Pd OFF state (top-top geometry)	0.998×10^{-14} J
Pd–BDA–Pd ON state (top-hollow geometry)	10^{-14}J

chemical descriptors have been disclosed in Table 6 using their respective formulae. The bandgap energy for OFF and ON state molecules is established to be 0.0457 eV and 0.0079 eV, respectively. BDA molecule without the Pd electrode shows an energy band gap of 4.62 eV. Additional, enhancement in the number of the benzene ring, increases the thermal vibration and reduces the HOMO–LUMO gap. This must lead to the low packing efficiency of the molecular crystal geometry with a significant boost in electrical conductivity [30].

The values of potential energy for each state have been computed in Table 7 and it can be concluded that the ON State geometry (or Top-Hollow geometry) is more stable than the OFF-State geometry (or Top-Top geometry) since the ON state shows low potential energy compared to OFF state. Hence, these observations on the stability of molecular switch give discernment on designing and modeling of stable BDA-based molecular switch with Pd electrode.

4 Conclusions

The wide range of the first-principle study of the Pd–BDA–Pd molecule concludes that DFT and HF computational values are slightly different as HF calculations do not include electron–electron interaction term. The HOMO–LUMO gap gets drastically reduced with the attachment of the nanoelectrode on both sides of a switching molecule. A comparison of the investigated vibrational frequencies stipulates the overestimation of the values due to the neglect of anharmonicity in the real system. Nevertheless, the insertion of electron correlation in DFT to a certain extent makes the frequency value smaller. It is inspected that the left side electrode makes shorter bond lengths and bond angles in top-hollow geometry (ON state). This may be due to the intramolecular hydrogen bonding due to the absence of one Pd atom in the left Pd cluster. The length of the molecular backbone depends on the addition of phenyl rings and that leads to the radical control on the energy gap and the electrical conductivity of switching molecules for the design of an electronic molecular switch. BDA with the Pd electrode on both sides shows a high switching band gap ratio between ON and OFF states, proving Pd–BDA–Pd as a suitable candidate for the design of a molecular switch. Geometrically optimized structure of top-hollow geometry (ON state) shows slightly higher conductivity and better stability.

References

1. M. Nasrollahzadeh, S. Sajadi, M. Sajjadi, Z. Issaabadi, An introduction to nanotechnology. Interface Sci. Technol. **1**(28), 1–27 (2019)
2. D.W. Chang, H.J. Choi, J.B. Baek, Wet-chemical nitrogen-doping of graphene nanoplatelets as electrocatalysts for the oxygen reduction reaction. J. Mater. Chem. A **3**(14), 7659–7665 (2015)
3. T. Masateru, Paving the way to single-molecule chemistry through molecular electronics . Phys. Chem. Chem. Phys. **21**(19), 9641–9650 (2019)
4. M. Bravo-Sanchez, T.J. Simmons, M.A. Vidal, Liquid crystal behavior of single wall carbon nanotubes. Carbon **48**(12), 3531–3542 (2010)
5. S. Devisree, A. Kumar, R.T. Koyadeen, A methodology to simulate and analyse molecules as molecular switches. Int. J. Simul.-Syst. Sci. Technol. **17**, 22–31 (2016)
6. J.M. Tour, Molecular electronics. Synthesis and testing of components. Acc. Chem. Res **33**(11), 791–804 (2000)
7. D. Xiang, X. Wang, C. Jia, T. Lee, X. Guo, Molecular-scale electronics: from concept to function. Chem. Rev. **116**(7), 4318–4440 (2016)
8. Y. Sugita, A. Taninaka, S. Yoshida, O. Takeuchi, H. Shigekawa, The effect of nitrogen lone-pair interaction on the conduction in a single-molecule junction with amine-Au bonding. Sci. Rep. **8**(1), 1–8 (2018)
9. M. Kiguchi, S. Fujii, Governing the metal–molecule interface: towards new functionality in single-molecule junctions. Bull. Chem. Soc. Jpn. **90**(1), 1–11 (2017)
10. Z. Zhang, H.J. Guo, Transferability of coarse-grained force field for n CB liquid crystal systems. J. Phys. Chem. B **118**(17), 4647–4660 (2014)
11. M.V. Gorkunov, M.A. Osipov, A. Kocot, J.K. Vij, Molecular model of biaxial ordering in nematic liquid crystals composed of flat molecules with four mesogenic groups. Phys. Rev. E **81**(6), 061702 (2010)
12. K.T. Rafsa, K. Anand, R. Abdul, H.G. Vilas, Theoretical investigation of design methodology, optimized molecular geometries, and electronic properties of benzene-based single molecular switch with metal nanoelectrodes. J. Nanomater. 6260735 (2020).
13. S. Singh, H. Singh, T. Karthick, P. Tandon, D.H. Dethe, R.D. Erande, Conformational study and vibrational spectroscopic (FT-IR and FT-Raman) analysis of an alkaloid–borreverine derivative. Anal. Sci. **33**(1), 99–104 (2017)
14. W. Koch, M.C. Holthansen, *A Chemist's Guide to Density Functional Theory*, 2nd edn. (Wiley-VCH, Weinheim, 2001)
15. M.J.E.A. Frisch, G.W. Trucks, H.B. Schlegel, G.E. Scuseria, M.A. Robb, J.R. Cheeseman, G. Scalmani, V. Barone, B. Mennucci, G. Petersson, H. Nakatsuji, *Gaussian 09, Revision d. 01*, vol. 201 (Gaussian. Inc., Wallingford CT, 2009)
16. G. Keresztury, J.M. Chalmers, P.R. Griffith, *Raman Spectroscopy Theory in Handbook of Vibrational Spectroscopy*, vol. 1 (Wiley & Sons Ltd., New York, 2002)
17. J.B. Foresman, A. Frisch, *Exploring Chemistry with Electronic Structure Methods* (Gaussian Inc, Pittsburg, USA, 1996).

18. R. Walesa, T. Kupka, M.A. Broda, Density functional theory (DFT) prediction of structural and spectroscopic parameters of cytosine using harmonic and anharmonic approximations . Struct. Chem. **26**(4), 1083–1093 (2015)
19. V. Krishnakumar, R.J. Xavier, Normal coordinate analysis of 2-mercapto and 4, 6-dihydroxy-2-mercapto pyrimidines. Indian J. Pure Appl. Phys. **41**, 597–601 (2003)
20. S. Seshadri, M.P. Rasheed, R. Sangeetha, Vibrational spectroscopic (FT-IR and FT-Raman) studies, HOMO LUMO analysis and electrostatic potential surface of 2-amino-4, 5-dimethyl-3-furancarbonitrile. IOSR J. Appl. Chem. **15**, 87–100 (2015)
21. V. Nagarajan, R. Chandiramouli, First-principles studies on electronic properties of Oligo-p-phenylene molecular device. Solid State Commun. **269**, 50–57 (2018)
22. R.P. Kaur, R. Preet, R.S. Sawhney, D. Engles, Quantum tunneling through aromatic molecular junctions for molecular devices: a review. Chin. J. Phys. **56**(5), 2226–2234 (2018)
23. P.L. Praveen, D.P. Ojha, Photosensitivity, substituent and solvent-induced shifts in UV-visible absorption bands of naphthyl-ester liquid crystals: a comparative theoretical approach. Liq. Cryst. **41**(6), 872–882 (2014)
24. W.R. Yan, X. Jin, S.Y. Guan, X.G. Zhang, R. Pang, Z.Q. Tian, D.Y. Wu, B.W. Mao, Theoretical study of quantum conductance of conjugated and nonconjugated molecular wire junctions. J. Phys. Chem. C **120**(22), 11820–11830 (2016)
25. M.A. Abbas, F.H. Hanoon, L.F. Al-Badry, Possibility designing XNOR and NAND molecular logic gates by using single benzene ring. Solid State Commun. **263**, 42–49 (2017)
26. V.M. Geskin, C. Lamber, J.L. Bredas, Origin of high second-and third-order nonlinear optical response in ammonio/borato diphenylpolyene zwitterions: the remarkable role of polarized aromatic groups. J. Am. Chem. Soc. **125**(50), 15651–15658 (2003)
27. D.A. Kleinman, Nonlinear dielectric polarization in optical media. Phys. Rev. **126**(6), 1977 (1962)
28. S. Prasad, D.P. Ojha, Optimized molecular geometries, vibrational spectra, and frontier molecular orbital of fluorinated liquid crystals-a comparative quantum chemical calculations. Mol. Cryst. Liq. Cryst. **669**(1), 12–26 (2018)
29. V. Mukherjee, T. Yadav, Conformational study of neutral histamine monomer and their vibrational spectra. Spectrochim. Acta Part A Mol. Biomol. Spectrosc. **165**, 167–175 (2016)
30. S. Prasad, D.P. Ojha, Molecular structure, vibrational spectroscopic and chemical reactivity of nematogenic pn-alkylbenzoic acids: a comparative computational analysis. Mol. Cryst. Liq. Cryst. **665**(1), 52–63 (2018)
31. S. Prasad, D.P. Ojha, Study of group charges, molecular conformations, and electrochemical properties of nematogens: a DFT approach. Am. J. Mater. Sci. **7**(2), 41–45 (2017)

Computer Vision and Artificial Intelligence

Bag of Visual Words Methodology in Remote Sensing—A Review

Jyoti S. Shukla, Kriti Rastogi, Hetal Patel, Gaurav Jain, and Shashikant Sharma

Abstract With the increasing popularity and use of geographical data processing, more and more research efforts have been placed on geographical scene classification. Not only in the field of satellite imagery, but considering the recent overall, the overwhelming increase in the amount of visual information such as digital images present, there's an urgent need of developing a model that aids in complete representation of an image and its features and also its retrieval. To obtain a complete representation of satellite images from a high spatial resolution satellite, bag of visual words (BoVW) model is emerging as a significant tool with quite promising results in the fields of image classification and image retrievals. This paper gives a detailed review of the introduction of BoVW methodology in the field of satellite imagery, its implementation, results, and future scope of BoVW method in remote sensing science and technology.

Keywords Bag of visual words (BoVW) · Classification · Remote sensing (RS) · Region of interest (ROI) · Bag of words (BOW) · Satellite imagery · Image representation · Image retrieval · Reverse image search (RIS)

J. S. Shukla (✉)
Junior Research Fellow, ISRO RESPOND Project, A D Patel Institute of Technology, V.V.Nagar, Anand, Gujarat, India
e-mail: jyoti.shukla037@gmail.com

K. Rastogi · G. Jain · S. Sharma
Space Applications Center, Indian Space Research Organization (ISRO), Ahmedabad, India
e-mail: kritirastogi@sac.isro.gov.in

G. Jain
e-mail: gvj@sac.isro.gov.in

S. Sharma
e-mail: sasharma@sac.isro.gov.in

H. Patel
Electronics and Communications, A D Patel Institute of Technology, V.V.Nagar, Anand, Gujarat, India
e-mail: ec.hetal.patel@adit.ac.in

F. Thakkar et al. (eds.), *Proceedings of the International e-Conference on Intelligent Systems and Signal Processing*, Advances in Intelligent Systems and Computing 1370,
https://doi.org/10.1007/978-981-16-2123-9_36

1 Introduction

With the increasing data processing technology in the field of remote sensing and image processing, there has been a significant rise in the usage of bag of visual words (BoVW) model. BoVW is being visualized as a tool for image feature representation and its subsequent classification along with an efficient image retrieval mechanism for both non-satellite (aerial images) and satellite images. BoVW method is found to perform well in mid-level feature representation and thus works as a bridging model between low-level features (such as pixel intensity, gradient, variance, color, etc.) and their contrary high-level features (such as shape, object-based, edges, MBI, etc.). Nowadays, the classification of high-resolution images has become an open research problem. BoVW method helps in the classification process by providing the feature distribution throughout the image. With this knowledge of versatile features, one can even develop an algorithm on the basis of which interclass and intraclass variations can also be achieved. This upgradation will further help in a more detailed classification up to levels 2 and 3. With the increase in high-resolution remote sensing imagery, BoVW narrows down the gap between the visual features and the semantic information present with the images to better interpret the image for wide use in further many applications [1].

Taking analogy from the bag of words model used in NLP (Natural language processing), BoVW's model has been developed wherein the words are replaced with visual words or the image patches that define the key features of an image. It in a way jumbles all the features removing the limitation of spatial information and creates a list of the features comprising of the standout features of an image. The list is further clustered to form a visual dictionary that will serve as a reference for a query image supplied to categorize the features of the query image into the already known clusters of the parent image.

Sometimes, there is a need of segmenting the large RS image and working individually on patch level for operations like, either to classify the landmass present in the image, or to extract information about the landscape, or to categorize the objects present in the image. In all the stated scenarios, there's a need of defining an ROI (region of interest) and do the processing on the ROI/class. However, there's also the need for proper retrievals of these images back after processing so that if any test image is passed on it can be demonstrated properly as to which ROI it belongs to. There are many methodologies, for example, as can be seen in the now-in-use machine learning algorithms and neural network methodologies present which work in this layout. However, there's a need to develop new methodologies that can better perform the task and provides more smooth services and prototype systems which not only excel in performance compared to their counterparts but also is computationally efficient. BoVW model has been experimented by many researchers to fulfill criteria like the ones mentioned and has shown results with a good visual selection that assists in image information mining [2].

This paper briefs about the usage of BoVW in the image processing/classification community. The remainder of this paper is organized as Sect. 2 outlines the history and evolution of BoVW methodology in the field of image processing and its emergence as a tool in the Remote Sensing community. Section 3 details the methodology and implementation steps of BoVW. Section 4 describes the various applications created and the experiments done by various researchers using BoVW along with the inferences obtained by them. Section 5 sums up the overall concept of the BoVW methodology. It is summarized based on the literature survey done. It also marks the progress made by BoVW so far in the world of image processing and what it can be best utilized and modified for. Section 6 concludes this paper by remarking on the future scope of BoVW and its possible potential for easing out the large data handling and processing mechanism.

2 History and Evolution of Bag of Visual Words

The BoVW model was first introduced by Sivic and Zeserman [3] in 2003 for object matching in videos. They described the method as an analogy to the bag of words method used for text retrieval. The bag of words model which is the most rigorously used method in NLP and text data analysis was developed far ago in around the year 1954. Taking analogy from the methodology used in the BOW of creating a frequency histogram for the words present in the article irrespective of their order, BoVW was developed by replacing the words with features present in an image. The word frequencies were considered as the signature of the document in BOW and then used for classification. A similar layout is attempted in BoVW [4].

2.1 Extension of BoVW

A similar extension and early introduction to BoVW was done by Csurka in 2004. The term *bag of keypoints* was coined for the method and was explained as a vector quantization process of invariant descriptors of the image patches [5, 6]. Being one of the earliest researches to discover the use of the collection of features in an unordered way in image analysis and further applications, a review on the advantages and disadvantages of the method was calibrated. BoVW was found as one of the potential solutions to the problem of generic visual categorization and was claimed that BoVW was one such method that can provide a generic model to cope with many object types simultaneously and can be readily extended to new object types [6]. A novel research done during the upliftment of BoVW method was by Li Fei Fei in 2005, in which an approach was proposed to learn and recognize natural scene categories [7]. The image of a scene was represented as a collection of local regions, denoted as codewords obtained by unsupervised learning, and then later categorized the themes or the local regions [7].

2.2 *Extension of BoVW to Satellite Imagery*

Since 2004, there have been many experiments and research work done in the field of image processing using BoVW method. From around 2007, the usage of BoVW method began in the field of remote sensing and the RS data was used to implement the method. Fortunately, the method gave good results on the RS data as well for applications like scene categorization [8], classification, feature representation, etc. Considering the potential of BoVW, a research was done using BoVW for detecting urban zones in satellite images [9]. A normalization algorithm was applied in the procedure, as a result of which the method was robust to atmospheric disturbances during the acquisition time. Good results on the classification of urban zones were acquired using BoVW on IKONOS image dataset which has a resolution of 4 m (multispectral) and 1 m (panchromatic) [9]. However, with RS data, the non-spatial characteristic of BoVW was creating erroneous and less accurate results, thus the spatial extensions for BoVW were developed to implement it on the satellite data. An early research work was carried out with the use of BoVW model along with spatial extensions for land-use classification in which firstly the non-spatial representation of BoVW was implemented and then later the results were compared by combining it individually with two spatial extensions—SPMK (Spatial Pyramid Match Kernel), and a novel approach proposed–SCK (Spatial co-occurrence Kernel). The data used for experimentation was a manually labeled ground truth data which consisted of 21 land-use classes and had a pixel resolution of 1 foot [10]. Better results were obtained by coalescing the spatial extensions and the traditional method [10].

3 Bag of Visual Words—Methodology

In this section, the layout of the methodology is explained. BoVW can be defined as a method to represent an image as a set of features. The features are marked by keypoints and descriptors. Keypoints are the chief, discriminative, or "stand-out" points in an image and the descriptors are the descriptions of the keypoints [11]. This method helps in an unordered collection of the defining points of an image and then quantizing only those features to form the feature vectors. These vectors will be then further used to develop the frequency histogram denoting the number of occurrences of particular image patterns in an image and also to train the classifier model. The chief advantage of BoVW is its computational efficiency and also it is a very simple generic model for visual categorization [6]. From the analysis of the literature survey done, the method can be broken down into four chief steps [6, 11] (Fig. 1).

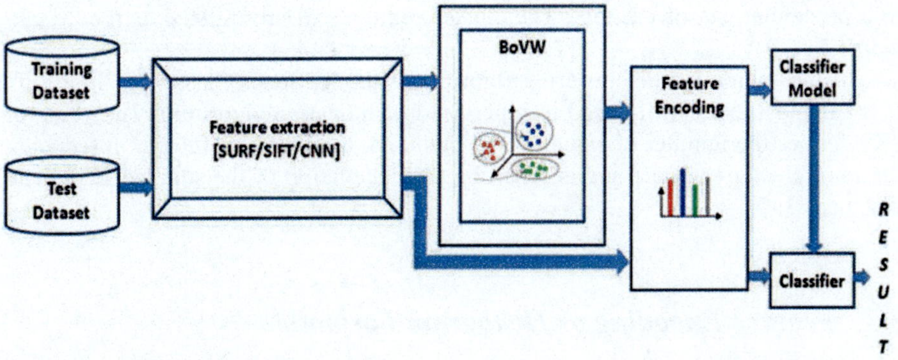

Fig. 1 Block diagram of BoVW methodology

3.1 Local Feature Detection, Description, and Feature Extraction

The first step comprises of the splitting of an image into a collection of local overlapping or non-overlapping regions/patches/sub-images [6, 7] and then detecting the keypoints or descriptive points present in the patches that define the patches judiciously [12]. After that, these keypoints are extracted by using any feature extraction algorithm, and then we will have a list of keypoints of all the patches and hence of the entire image. For a large dataset, *feature selection* is also required; it leads to the selection of the most distinctive and significant features from the pool of features [2].

There are numerous feature extraction methods being implemented based on the data which is being processed and the type of features required. Some of the feature extraction methods are [13]: BRISK (Binary Robust Scalable Keypoints), FAST(Features from accelerated segment test), SIFT(Scale Invariant Feature Transform), SURF (Speeded-up Robust Features), HOG (Histogram of Gradients), LBP (Local Binary Patterns), PCA (Principal Component Analysis), LDA (Linear Discriminant Analysis), etc. The most widely used feature extraction method among these was found to be SIFT and SURF. The high usage of these techniques can be reasoned to their property of being scale and rotation invariant which is required for unordered collection of features [2, 6, 9, 14].

3.2 Codebook/Visual—Vocabulary Construction

The features extracted are used to construct a visual vocabulary which can be later referenced during categorizing a query image on the basis of the features of the query image. The features are clustered into groups or clusters for vector quantization [6, 9]. The construction of a visual vocabulary requires the assignment of the descriptors to a predefined set of clusters. The cluster centers will are utilized as the *"visual words."*

The most widely used clustering algorithm is the K-Means Clustering algorithm. It is the most extensively used unsupervised quantization algorithm. The value of "K" defines the number of visual words obtained. K-Means is efficient in reducing the intra-cluster variance and is based on the calculation of the squared Euclidean distance.

3.3 Feature Encoding or Histogram Formation

This is the final step toward the creation of the visual codebook. A histogram is constructed which demonstrates the number of patches assigned to each cluster. A histogram is constructed for each image patch denoting the frequency of occurrence of features in each patch. This histogram serves as a reference for any "new feature descriptors" in query images and relates the new features with the already existing cluster centers. The histogram is created both for the test and train data. The histograms of the test data can then be used to match with the pre-computed frequencies of visual words and can thus be categorized. Histogram computation serves as the final step towards converting it from a generic model to a multi-class classifier model. The cluster centers are used as vectors hereafter [6].

3.4 Classifier Model

The feature vectors or the cluster centers calculated are then used to classify the image. This model now converts into a multi-class classifier model. The training data is used to train the classifier model and it categorizes the image patches into various classes. The feature vectors of a query image calculated are then supplied to the classifier model which is already trained and hence the classification of query image is done and the accuracy is calculated.

The most extensively used classifier in BoVW is the SVM classifier [6, 10, 15, 16]. The support vector machine works by creating a hyperplane between two separate class data. There are many other classifier models [17] used along with BoVW such as Naïve Bayes classifier, Random Forest Classifier [18, 19], etc.

4 Review of Work Done with Bag of Visual Words Method

In this section, the overview of the diverse research works carried out in the field of remote sensing and image processing using BoVW are discussed. Most of the applications of BoVW are related to the classification of regions or objects; however, there are few other applications in which they have been used.

4.1 Early Development

The early development in BoVW was done in 2003 in which the method was proposed for object matching in videos [3]. A similar initiation was done in 2004 [6], in which the objective was to curb the problem of generic visual categorization and the need of having a more generic method with flexibility to extend in the future for new data types and objects. Hence, BoVW was introduced for image processing. The paper describes the complete layout of how to bag the features and then use them for classification. Two classifiers SVM and Naïve Bayes Classifier were used for classification. As per the results obtained, SVM outperformed Naïve bayes classifier [6]. In 2005, a novel approach for natural scene categorization was done [7]. In the project, around 13 different categories were categorized using bag of keypoints method and got satisfactory results. The method was further modified to use for the application of effective facial expressive image classification [4], wherein the spatial relationship was preserved by computing the relative conjunction matrix of the data by coding the relationships between the visual words. An overall good model was developed with the help of SVM classifier [4]. A novel approach was proposed for fingerprint classification using BoVW along with an image compressing mechanism for reducing storage. SVD—singular value decomposition was used as the compressing algorithm. The results incurred through the model show that the method led to the occupation of less storage space and computational time [20]. The usage of BoVW was extended by using the BoVW approach for reverse image search and user-supplied image searching application which is of use in e-commerce websites etc. [21]. It also integrated the spatial pyramid matching technique with the traditional BoVW method and got good results. Many more such applications in the field of non-satellite imagery have been implemented that show reliable results. Some of them are illustrated, as for example, in a research, the method was used primarily for image retrieval and recognition [22], in another research, the method was proposed to construct the codebook using predictive clustering trees in order to enhance the efficiency of generated codebook [23]. On comparing the results, they found that the codebook generated through trees was more efficient than the traditional method.

4.2 Development in RS Field

There has also been an extensive research work done in the field of remote sensing using BoVW from around 2008. In 2008, the method was extended to the RS data for the detection of urban zones from the images of a high-resolution satellite overhead imagery [9]. The improved method was tested on IKONOS images as well and satisfactory results were availed. A more advanced step was taken of creating a multi-BoVW for the classification of RS images as illustrated in one of the research papers [1]. In the paper, the argument put was that a single BoVW constricts the possible more efficient features and leads to a restricted set of significant features. Thus, multi-bag creation will lead to fairly selecting the features and a good set of features can be processed [1]. In another research, BoVW was implemented by considering both the local and global feature sets for the creation of a codebook [24]. It was shown that the combination of local and global features leads to good performance as compared to the state-of-the-art scene classification method. Local features consider the features like mean, variance, etc., and texture, color, etc., are considered as the global features [24]. A step forward was taken in enhancing the incorporation of spatial extensions to BoVW by introducing a 2D wavelet decomposition method, which exploits the textural characteristics of an image and also involves color information transformation into grayscale [25]. The method is found to have not only worked well for textural feature extractions but also for the multi-resolution representation of the image [25]. A similar study was carried out under which different methods of texture feature analysis were mutually compared to test their efficiency on satellite imagery based land-use classification models [26]. In another research, the feature extraction method was enhanced by involving CNN for the feature extraction so that they could get better visual vocabularies which will ultimately define the performance of BoVW as a geoprocessing tool for classification. The model was tested on three datasets with diverse scene categories and satisfactory results were obtained. The BoVW method was also used for SAR image classification [14]. The method was modified by introducing a de-noising algorithm which will help reduce the multiplicative speckles present in SAR images and thus enhance the performance of BoVW method. Similar work was done on SAR data for classification purposes in which they used the Gabor-based features for training the SAR images and then used those Gabor-based features to form a discriminative codebook [16]. Another research using satellite images and BoVW was done on PolSAR images that play an important role in land cover management [27]. In the project, the pixel-based low-level features were used to form the codebook, and also segment-based decision-making technique was utilized. On comparison with the traditional method, highly good results were obtained through the modification. The BoVW method was also used to process interferometric SAR images [11] (InSAR images). Gabor and fractional Fourier transform were used for feature extraction of low-level features. It was found that for each baseline, BoVW along with FrFT gave good and improved accuracies [11].

A lot of updation is being done with the traditional method of BoVW such as integrating it with machine learning or with new algorithms for feature extraction and clustering mechanisms. One such advancement was done by using the analogy of BoVW method to form a bag of convolutional features for scene classification [28]. The method differed in the features extracted, by use of handcrafted feature descriptors in traditional method to the generation of visual words through deep convolutional features using the CNN technique. The results obtained, show that the model proves to be more effective than the authentic method on remote sensing data for classification [28]. Apart from that, a novel multi-sample, multi-tree approach for the construction of the visual dictionary was also introduced [29]. The model initiated with the use of random forest and k-d trees for feature extraction and also sampled multiple image patches out of an image region. Following it, a multi-tree approach for quantizing the descriptors with each tree representing a particular codebook is implemented. In this way, packets of visual words for each region were formed and then later combined into a single good efficient codebook for further use. There are many such other applications in which BoVW has given considerable and satisfactory results like, it was used for automatic cloud detection [30], for aerial image classification using the attention-based local descriptors proposed [15], etc.

5 Understanding BoVW

From the extensive literature survey done, BoVW is found to be an efficient method for representing an image in terms of its features. The BoVW is a robust and simple method for visual categorization and image feature representation. It is also computationally efficient. It has also proven to handle large databases very effectively and in shorter times than its counterparts. From the literature survey done, it has been found that the base or the fundamental factor on which the performance of the model depends is its visual vocabulary; the better the vocabulary, the better the model. Not only the choice of features but also the dictionary size considerably affects the performance of the methodology. A good and diverse training feature set leads to the construction of a good visual codebook. However, the dictionary sizes are also supposed to be judiciously chosen as they, in a way, define the number of categories to be generated on the final classification. The studies show that the best tradeoffs of accuracy and computational efficiency are obtained with intermediate sizes of clustering.

BoVW is a good alternative for mid-level feature representation and fills the semantic gap. The spatial extensions provided to BoVW method in various researches has strengthened the model for its extension to various RS-based applications where spatial information is primary for usage. Incorporating the method along with the new techniques and algorithms being researched can further be helpful to make it more efficient as was seen, for example, in the case of a bag of convolutional features [28].

The advantages of BoVW are inevitable, but still, there are some lacks present with the method, such as inaccurate classification for the not-so-diverse training set. Also, it does not excel the image classification for SAR images but does well after some modification [14]. It was found that overall it is a very good technique for feature representation but performs mediocre in classifications of some datasets. The lack can be owed to inefficient feature collection due to nearby and indistinctive pixel-based features present in RS data of a low-resolution satellite or a homogeneous collection of images for a pre-set database. These gaps can always be filled as are being done in various studies and the advantage of the method can be taken.

BoVW method can be used on all types of dataset and the power of the method lies in its unordered collection of features thus breaking the spatial boundaries. Its features being invariant to rotation also adds to the pros of the model. From the understanding acquired about BoVW, it can be stated that it shall be implemented in projects where good feature distribution is needed and needs to be visualized which is possible through histogram computation for each region. The feature vectors or the visual words then obtained, can be used further for categorization.

6 Conclusion and Future Scope

In this paper, a review of the literature survey carried out on the "*Bag of Visual Words methodology*" is demonstrated. The paper describes the evolution of the method and how it was broadcasted to the vast areas of remote sensing and also to various corners of the world of image processing and visual data handling systems. The use of BoVW can be extended to many more remote sensing-based applications, such as analyzing the urban sprawl, landmass categorization, segment-based classification, etc., and a plenty of image processing applications. The modifications that are being made day-to-day by integrating various other algorithms of machine learning or digital signal processing with BoVW, are leading to further enhancement of such methods. The advantages present with the use of BoVW can be availed through these integrations and the loopholes of BoVW can also be filled at the same time.

Acknowledgements The work presented in this paper is carried out under the funded research project of Indian Space Research Organization (ISRO), RESPOND (OGP198).

References

1. L. Zhao, P. Tang, L. Huo, Feature significance-based multibag of-visual-words model for remote sensing image scene classification. J. Appl. Remote Sens. **10**(3), 035004 (2016)
2. J. Yang, J. Liu, Q. Dai, An improved bag-of-words framework for remote sensing image retrieval in large-scale image databases. Int. J. Digital Earth **8**(4), 273–292 (2014)
3. Sivic, Zisserman, Video Google: a text retrieval approach to object matching in videos, in *Proceedings Ninth IEEE International Conference on Computer Vision*, Nice, France, 2003
4. D.A. Chanti, A. Caplier, Improving bag-of-visual-words towards effective facial expressive image classification, in *Proceedings of the 13th International Joint Conference on Computer Vision, Imaging and Computer Graphics Theory and Applications*, Funchal, Madeira, Portugal, 2018
5. K. Vyas, Bag of visual words model for image classification and recognition (13 July 2016) [Online]. https://kushalvyas.github.io/BOV.html
6. G. Csurka, C. Dance, L. Fan, J. Willamowski, C. Bray, Visual categorization with bags of keypoints, in *Workshop on Statistical Learning in Computer Vision, ECCV* (2004)
7. L. Fei-Fei, P. Perona, A Bayesian hierarchical model for learning natural scene categories,in *IEEE Computer Society Conference on Computer Vision and Pattern Recognition (CVPR '05)* (2005)
8. J. Feng, Y. Liu, L. Wu, Bag of visual words model with deep spatial features for geographical scene classification. Hindawi Comput. Intell. Neurosci. **2017**(2), 1–14 (2017)

9. L. Weizman, J. Goldberger, Detection of urban zones in satellite images using visual words, in *IEEE International Geoscience and Remote Sensing Symposium—IGARSS* (2008)
10. Y. Yang, S. Newsam, Bag-of-visual-words and spatial extensions for land-use classification, in *Proceedings of the 18th SIGSPATIAL International Conference on Advances in Geographic Information Systems—GIS '10*. (California, USA, 2010)
11. N.D. Cagatay, M. Datcu, Bag-of-visual-words model for classification of interferometric SAR images, in *Proceedings of EUSAR 2016: 11th European Conference on Synthetic Aperture Radar-IEEE*, Hamburg, Germany, Germany, 2016
12. B. Davida, Bag of visual words in a nutshell (3 July 2018) [Online]. https://towardsdatascience.com/bag-of-visual-words-in-a-nutshell-9ceea97ce0fb
13. S. Karim, Y. Zhang, M.R. Asif, S. Ali, Comparative analysis of feature extraction methods in satellite imagery. J. Appl. Remote Sens. **11**(4), 042168 (2017)
14. J. Feng, L.C. Jiao, X. Zhang, R. Niu, An effective bag-of-visual-words framework for SAR image classification, in *MIPPR 2011: Remote Sensing Image Processing, Geographic Information Systems, and Other Applications. Proceedings of SPIE* (2011)
15. S. Xu, T. Fang, H. Huo, D. Li, A novel method of aerial image classification based on attention-based local descriptors, in*The 6th International Conference on Mining Science & Technology*, ELSEVIER, 2009
16. M. Amrani, S. Chaib, I. Omara, F. Jiang, Bag-of-visual-words based feature extraction for SAR target classification, in*Ninth International Conference on Digital Image Processing (ICDIP 2017)* (2017)
17. D. Rajyalakshmi, K.K. Raju, G.S. Varma, Taxonomy of satellite image and validation using statistical inference, in *2016 IEEE 6th International Conference on Advanced Computing (IACC)*, Bhimavaram, India, 2016
18. W. Koehrsen, An implementation and explanation of the random forest in python (30 August 2018) [Online]. https://towardsdatascience.com/an-implementationand-explanation-of-the-random-forest-in-python-77bf308a9b76
19. N. Donges, A complete guide to the random forest algorithm (16 June 2019) [Online]. https://builtin.com/data-science/random-forest-algorithm?utm_campaign=News&utm_medium=Community&utm_source=DataCamp.com. Accessed Sept 2020
20. P.N. Andono, C. Supriyanto, S. Nugroho, Image compression based on SVD for BoVW model in fingerprint classification. J. Intell. Fuzzy Syst. **34**(4), 2513–2519 (2018)
21. K. Vyas, Y. Vora, R. Vastani, Using bag of visual words and spatial pyramid matching for object classification along with applications for RIS. Procedia Comput. Sci. ScienceDirect **89**, 457–464 (2016)
22. H. Kato, T. Harada, Image reconstruction from bag-of-visual-words, in *2014 IEEE Conference on Computer Vision and Pattern Recognition*, Columbus, OH, USA (2014)
23. I. Dimitrovski, D. Kocev, S. Loskovska, S. Džeroski, Improving bag-of-visual-words image retrieval with predictive clustering trees. Inf. Sci.—Elsevier-ScienceDirect **329**, 851–865 (2016)
24. Q. Zhu, Y. Zhong, B. Zhao, G.-S. Xia, L. Zhang, Bag-of-visual-words scene classifier with local and global features for high spatial resolution remote sensing imagery. IEEE Geosci. Remote Sens. Lett. **13**(6), 747–751 (2016)
25. L. Zhao, P. Tang, L. Huo, A 2-D wavelet decomposition-based bag-of-visual-words model for land-use scene classification. Int. J. Remote Sens. **35**(6), 2296–2310 (2014)
26. P. Kupidura, The comparison of different methods of texture analysis for their efficacy for land use classification in satellite imagery. MDPI-Remote Sens. **11**(10), 1233 (2019)
27. R. Mohammadi, M.R. Sahebi, M. Omati, M. Vahidi, Synthetic aperture radar remote sensing classification using the bag of visual words model to land cover studies. Int. J. Geol. Environ. Eng. **12**(9), 588–591 (2018)
28. G. Cheng, Z. Li, X. Yao, L. Guo, Z. Wei, Remote sensing image scene classification using bag of convolutional features. IEEE Geosci. Remote Sens. Lett. **14**(10), 1735–1739 (2017)

29. Z. Wu, Q. Ke, J. Sun, H.-Y. Shum, A multi-sample, multi-tree approach to bag-of-words image representation for image retrieval, in *2009 IEEE 12th International Conference on Computer Vision*, Kyoto, Japan, 2010
30. Y. Yuan, X. Hu, Bag-of-words and object-based classification for cloud extraction from satellite imagery. IEEE J. Select. Top. Appl. Earth Observ. Remote Sens. **8**(8), 4197–4205 (2015)

Automated Segmentation and Classification of Psoriasis Hand Thermal Images Using Machine Learning Algorithm

Snekhalatha Umapathy, Meghna Sampath, Nelufer, and Sakshi Srivastava

Abstract Psoriasis is an autoimmune disorder which affects the skin and causes inflammation in the skin regions which leads to lesions in hand, leg, shank, and upper back regions. The objectives of the study are to compare the performance of a hybrid automated segmentation algorithm with fuzzy c-means algorithm for segmenting the hand region of psoriasis and normal subjects and to extract the handcrafted features from the segmented image and implement the classification process using SVM classifier. In this specific study, hybrid algorithms of k-means clustering and region growing have been used for segmentation. The GLCM features are extracted and classification of psoriasis and normal hand region was performed using machine learning classifier. The percentage difference of 7.9% is achieved between the normal and psoriasis subjects. The mean average temperature difference of 2.5 °C was found between the normal and psoriasis subjects. The features such as contrast and entropy exhibit the highest percentage difference of 12.8% and 9.05%, respectively, between the normal and psoriatic hand. The sensitivity, specificity, and accuracy of the classifier were found to be 83.3%, 73.3%, and 78.3%, respectively. The Proposed hybrid algorithm provides better segmentation results compared to fuzzy c-means algorithm in segmenting psoriasis hand thermal image. The thermal imaging technique can be used as screening diagnostic tool for analyzing and detecting psoriatic lesions.

Keywords Psoriasis · Segmentation · k-means · Region growing · GLCM features · Support vector machine

1 Introduction

Psoriasis is an autoimmune skin disorder which produces inflammation in the skin and causes red patches with white scale in various body regions such as forearm, knees, elbow, and upper back regions. Worldwide, about 2–3% population were

S. Umapathy (✉) · M. Sampath · Nelufer · S. Srivastava
Biomedical Engineering Department, College of Engineering and Technology, SRMIST, Kattankulathur, Chennai, Tamil Nadu, India
e-mail: snehalau@srmist.edu.in

F. Thakkar et al. (eds.), *Proceedings of the International e-Conference on Intelligent Systems and Signal Processing*, Advances in Intelligent Systems and Computing 1370,
https://doi.org/10.1007/978-981-16-2123-9_37

affected with psoriatic disorder [1]. The prevalence of Psoriasis is high in Western Europe (3.46%) compared to America (2.96%), Australia (2%), and Asia (0.4%) [2]. In India, the prevalence rate of psoriasis varies from 0.4% to 2.8% in which men were severely affected compared to women [3]. The causes for psoriasis is may be due to genetic conditions.

Current diagnostic methods available for assessing the severity of Psoriasis are physical examination, biopsy, and digital imaging. Infrared thermography is one of the research methods used for diagnosing psoriatic disorder. The inflammation in the skin region can be characterized by heat signatures produced by the thermal imaging technique. Thermal imaging has been known to be one of the most efficient non-contact methods for diagnosing abnormalities. A condition like psoriasis usually produces significant heat signatures. A thermal imaging system can be used as a preliminary diagnostic tool in visualizing the presence of these heat signatures. Several studies are conducted to evaluate the severity of psoriasis based on thermal imaging techniques [4, 5]. Thermal imaging method is considered as a novel screening method to evaluate psoriatic disorder [5].

Image segmentation algorithms play a vital role in the segmentation of various medical images. Puneeth et al. implemented a semi-supervised scaling segmentation algorithm in digital images for automated segmentation of Psoriasis and classified the lesions using a Support vector machine (SVM) [6]. George et al. used unsupervised methods such as a combination of superpixels and k-means algorithm for automated detection of the lesion. They performed fusion of multi-scale segmentation result based on the majority of voting of decision rule to obtain final segmented image [7]. Our earlier work focused on automated segmentation of psoriatic hand images using fuzzy-c means algorithm [8]. Manoranjan et al. implemented a complete automatic segmentation of skin lesions in the psoriasis patients by using a modified fully convolutional network U-Net [9].To the best of our knowledge, automated segmentation of hand psoriasis thermal images is limited. In the proposed work, a hybrid algorithm is formulated based on k-means algorithm and region growing method for automated segmentation of hand thermal image and feature extracted using GLCM method and further classified using SVM classifier.

The aim and objective of the study were as follows: (i) To compare the performance of hybrid automated segmentation algorithm with fuzzy c-means algorithm for segmenting the hand region of psoriasis and normal subjects and extract the hand-crafted features from the segmented image and implement the classification process using SVM classifier.

2 Materials and Methods

2.1 *Study Population*

A total of 30 normal subjects and 30 patients with active psoriasis are included in the study. The Inclusion criterion of the study involves patients who have above 18 years of age with active psoriasis with a duration of disease existed at least for 6 months. Subjects with pustular, erythrodermic form of psoriasis were excluded from the study [10]. The diagnostic criteria for Psoriasis include subject with more than 10% body surface area affected by psoriasis. Approval from the institutional ethics committee and participants' signed consent form was obtained for the study.

2.2 *Image Acquisition Procedure*

The thermal imaging procedure was carried out by the standard protocol specified by the International Academy of Clinical Thermology. During the thermal imaging procedure, any jewelry or other ornaments worn by the participants is taken off temporarily during the time span of capturing the image because again this can interfere with the image acquisition process. The subjects were seated in an air-conditioned room having a temperature of 20 °C for a duration of 15 min at a humidity rate of 35–40%. The sensitivity and accuracy of the thermal camera were found to be 0.05 °C and 2 °C, respectively. Normal subjects and selected patients who were either recently diagnosed with psoriasis or been administered medications for quite a long period of time are made to stand in front of the thermal camera lens. It is made sure that the background using which the images taken is completely dark because interference of light can alter the thermal changes and heat signatures captured by the thermal camera. The forearm thermal images were captured using the FLIR SC 305 version. It is composed of 320 × 240 resolutions along with 8× wide zoom lens with lens size is 18 mm with a manual and automatic focus. The data's were analyzed using the FLIR software installed in computer. The minimum, maximum and average temperature was obtained on the temperature gradient for the forearm region.

2.3 *Hand Image Segmentation Method*

The image segmentation was performed in hand forearm region using hybrid algorithm (combined k means and regions growing) for the total population studied. The hybrid algorithm employs the functions of the k-means clustering and region growing. This makes the entire system more efficient since initially the similar regions are clustered and then further these similar regions are cultivated over each other. The high-temperature region in the psoriasis images were segmented using this hybrid algorithm.

(1) Centroids for the k clusters are created randomly based on a predefined value of k.
(2) Every input point of data is allocated to its centroid which is in its closest vicinity by the segmentation algorithm. This means that a point is part of a cluster if it is nearest to that cluster's centroid.
(3) The centroids are relocated by the k-means by taking into consideration; all data points' mean which have been given by the algorithm to the centroid of that particular cluster. This greatly reduces the intra-cluster variance.
(4) The image is separated into initial seed regions $R_i^{(0)}$ (e.g., splitting an image into 7×7 regions).
(5) A planar model is fit into each region of the seed. If $E(R_i^{(0)}, a, m)$ is small enough to kick start the algorithm, $R_i^{(0)}$ is accepted along with its model; otherwise $R_i^{(0)}$ is rejected.
(6) For every divided region, all points that happen to be compatible with that region are found out, by considering the neighbors of that region (this is done to avoid outliers). $C_i^{(k)} = [(x, y): (g(x, y) - f(x, y, a, m))^2 <$. Given that (x, y) is a 4 and is a neighbor of $R_i^{(k)}]$
(7) If there are no compatible points, then $m = m + 1$. This is under the condition that if $m > M$, $R_i^{(k)}$ does not grow further; otherwise, step 3 is repeated.
(8) Form the new region $R_i^{(k+1)} = R_i^{(k)} \cup C_i^{(k)}$, the model is refit into $R_i^{(k+1)}$, and $E(R_i^{(k+1)}, a, m)$ is computed.
(9) The error in difference is computed, $(k) = E(R_i^{(k+1)}, a, m) - E(R_i^{(k)}, a, m)$

The algorithms for Fuzzy c-means are described as follows:

1. The numbers of clusters to be selected are determined initial stage.
2. Each pixel co-efficient in the images are randomly assigned to the clusters.
3. Find the cluster centroid.
4. Find the Euclidean distance between the pixel point and cluster centroid and assign to the appropriate clusters based on minimum distance.
5. Continue the process until the convergence level is reached with maximum iteration.

2.4 *Statistical Feature Extraction*

The statistical features of the segmented images are obtained by performing the Grey-level co-occurrence matrix method. The GLCM method gives a high level of information regarding the inter-pixel correlation of an investigated texture and provides statistical features of the neighbouring pixels in an image. The features such as energy, entropy, contrast, homogeneity and correlation were obtained from the thermal images extracted by the following GLCM method.

(a) Energy—for every square inch of the image, the square of every pixel value present within is defined as the energy. This parameter mainly focus on calculating the orderliness of the image (b) Entropy—The intensity of randomness or the magnitude of non-uniformity of grey levels in an image. (c) Contrast—defined as the element of the image with a difference of order 2 of the moments. (d) Homogeneity—Measures the closeness of the distribution of elements present in the image. (e) Correlation—The differences between two consecutively processed images in the same order is defined as correlation.

2.5 Support Vector Machine Classifier

The SVM trains a given set of data by categorizing them into two groups in the study normal and psoriasis. Upon implementing the SVM algorithm on our data set, it was able to provide us with a clear classification of which images are normal and which ones are psoriasis affected. Support Vectors are simply the co-ordinates of individual observation. Support Vector Machine is a frontier which best segregates the two classes (hyper-plane/line). Support vectors are the data points that lie closest to the decision surface (or hyperplane). SVM selects the hyper plane which classifies the classes accurately prior to maximizing margin.

2.6 Performance Measures

The Confusion matrix was formulated to determine the performance measure of the classifier. (i) True Positives (TP): The subjects with actual psoriasis predicted as Psoriasis. (ii) True Negatives (TN): Healthy subjects classified as normal subjects. (iii) False Positives (FP): Normal subjects misclassified as Psoriasis affected subjects. (iv) False Negatives (FN): Psoriasis affected subjects misclassified as normal. Furthermore, the characteristics of the classification system are obtained as follows:

Sensitivity = TP/TP + FN (OP = Observed Positive; TP + FN).

Specificity = TN/TN + FP (ON = Observed Negative; TN + FP).

Accuracy= (TP + TN)/Total Observations.

The ROC curve is a graphical plot which is plotted as sensitivity vs specificity which illustrates the diagnostic accuracy of the classifier system.

3 Result and Discussions

The proposed study focused on skin temperature measurements of normal and psoriasis subjects in hand region and further automated segmentation of high-temperature region of hand thermal imaging using Hybrid algorithm. The statistical features

Table 1 Temperature parameter values for psoriasis and normal subjects

	Max. temp	Min. temp	Average
Normal (N = 30)	32.74 ± 0.9	34.75 ± 0.6	28.8 ± 0.6
Psoriasis (N = 30)	25.02 ± 0.4	28.03 ± 0.7	31.3 ± 0.6
% Difference	5.7%	10.7%	7.9%

were derived from the segmented image and classified using SVM classifier. Table 1 depicts the temperature measurement of normal and psoriasis hand region for the total population studied. The percentage difference of 7.9% is achieved between the normal and psoriasis subjects. The mean average temperature difference of 2.5 °C was found between the normal and psoriasis subjects.

Figure 1 depicts the segmentation process of normal and psoriatic hand image using hybrid algorithm. Figure 1a represents the thermal image of normal hand and Fig. 1b illustrates the implementation of hybrid algorithm i.e. combination of region growing and k-means clustering method. Figure 1c shows the segmented image of normal hand. Figure 1d displays the thermal image of Psoriatic hand, Fig. 1e illustrates the implementation of hybrid algorithm in psoriatic hand. Figure 1f shows the segmented image of psoriatic hand depicting the hot spot regions.

Figure 2 illustrates the segmentation process of fuzzy c-means algorithm for Normal and Psoriatic hand thermal image. Figure 2a depicts the normal hand thermal image; Fig. 2b is the segmented normal thermal image; Fig. 2c shows the psoriatic hand thermal image; Fig. 2d represents the segmented psoriatic thermal image.

Table 2 describes the feature extracted parameters from the segmented image of normal and psoriatic hand. The features such as contrast and entropy exhibit the

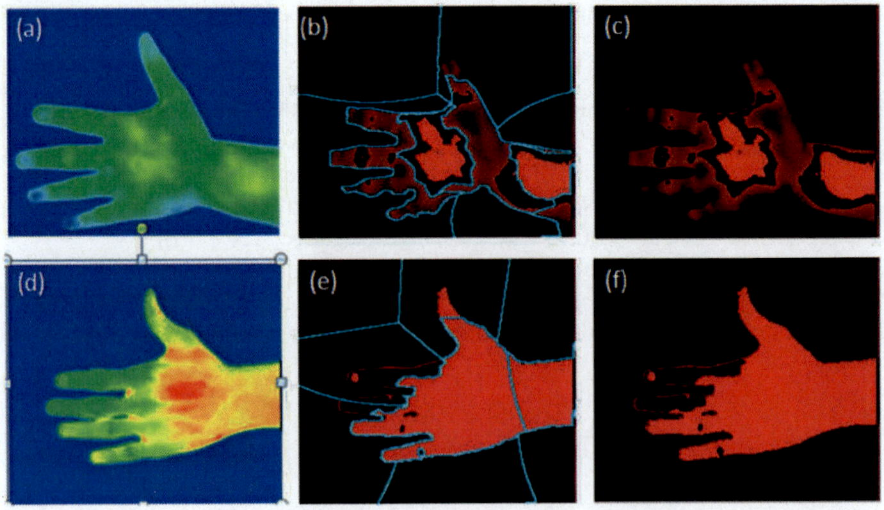

Fig. 1 **a–f** Segmentation process based on hybrid algorithm for normal and Psoriasis hand

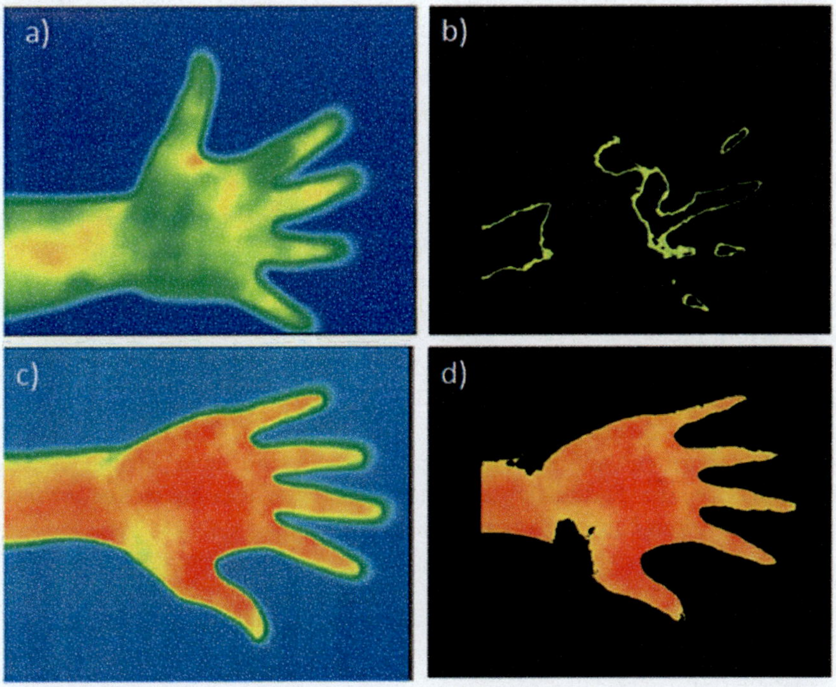

Fig. 2 **a–d** Segmentation process based on Fuzzy c-means algorithm for normal and Psoriasis hand

Table 2 (a) Feature extracted parameter for psoriasis and normal subjects using hybrid algorithm. (b) Feature extracted parameter for psoriasis and normal subjects using fuzzy c means algorithm

Features	Normal	Psoriasis	Difference %	Significance p value
(a)				
Energy	0.206 ± 0.08	0.28 ± 0.06	26.42	0.02
Entropy	4.07 ± 0.6	4.38 ± 0.5	7.07	0.01
Contrast	64.06 ± 16.8	65.81 ± 15.8	2.65	0.02
Homogeneity	0.789 ± 0.05	0.878 ± 0.02	10.13	0.01
Correlation	0.930 ± 0.003	0.967 ± 0.005	3.82	0.04
(b)				
Energy	0.18 ± 0.08	0.19 ± 0.06	5.26	0.05
Entropy	4.32 ± 0.4	4.75 ± 0.6	9.05	0.02
Contrast	16.7 ± 3.7	19.16 ± 3.3	12.8	0.01
Homogeneity	0.78 ± 0.05	0.8 ± 0.04	2.5	0.04
Correlation	0.98 ± 0.004	0.99 ± 0.007	1.01	0.05

highest percentage difference of 12.8% and 9.05%, respectively, between the normal and psoriatic hand.

The five features are given as the input features to the SVM classifier for the classification of normal and psoriasis hand. The confusion matrix illustrating the performance measures were given in Table 3.

The sensitivity, specificity, and accuracy of the classifier based on hybrid algorithm were found to be 83.3%, 73.3%, and 78.3%, respectively. Figure 3 depicts the ROC curve illustrating the sensitivity vs. specificity; Fig. 3a represents the ROC curve of the SVM classifier based on features extracted using hybrid algorithm and Fig. 3b indicates the ROC curve of the SVM classifier based on feature extracted using fuzzy c-means algorithm.

The sensitivity, specificity, and accuracy of the SVM classifier based on fuzzy c-means algorithm were found to be 69.7%, 74%, and 71.6%, respectively.

In analyzing the thermal images of psoriatic disorder and normal subjects, the temperature parameter was found to be higher in psoriasis-affected region compared to the normal region. The feature extraction parameter was higher for psoriasis-affected hand regions compared to normal.

Table 3 (a) Confusion matrix for SVM classifier based on features extracted from Hybrid algorithm. (b) Confusion matrix for SVM classifier based on features extracted Fuzzy c-means algorithm

	Abnormal skin	Normal skin
(a)		
Abnormal skin	25 (TP)	5 (FP)
Normal skin	8 (FN)	22 (TN)
(b)		
Abnormal skin	23 (TP)	7 (FP)
Normal skin	10 (FN)	20 (TN)

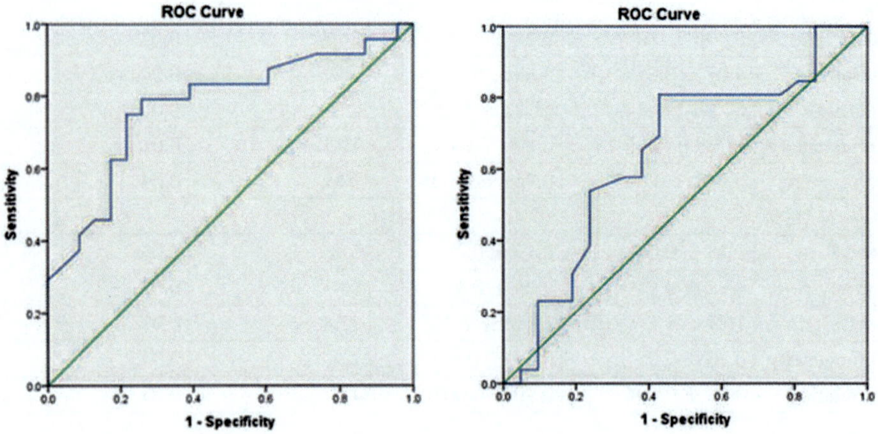

Fig. 3 Roc curve of the SVM classifier **a** hybrid algorithm **b** fuzzy c-means algorithm

Taur et al. classified the normal and psoriasis vulgaris images using orthogonal subspace classifier and fuzzy feature spectrum. They adopted manual method to pick up training regions in order to achieve good segmentation results [11]. Raaghavi et al. segmented the psoriasis image using k-Means clustering method and further classified using SVM classifier and Markov Random field. They estimated the growth rate of disease affected area and identified the types of psoriasis [12]. Wei et al. performed automated segmentation of psoriasis images using marker-controlled watershed segmentation method. They extracted the texture and color features using GLCM method and further classified the different types of psoriasis using SVM classifier [13]. Lu et al. applied semi-supervised scaling segmentation algorithm for segmentation of psoriasis images and extracted image color and texture features and further classified using SVM classifier [14]. In the proposed study, hybrid segmentation algorithm is applied for the automated segmentation of hand psoriasis thermal image and extracted features using GLCM method. Then the classification of psoriasis hand images was performed using SVM classifier. The accuracy of the SVM classifier was found to be 8.3%. The ROC curve depicts the detailed efficiency of the entire training system with respect to sensitivity and specificity. The limitations of the study include obtaining the amount of data was a considerable task because of the availability of fewer psoriasis patients.

4 Conclusion

The proposed hybrid algorithm provides better segmentation results compared to the fuzzy c-means algorithm in the segmentation of psoriasis hand thermal image. The thermal imaging technique can be used as a screening diagnostic tool for analyzing and detecting the psoriatic lesions. In future, multi-class classification of psoriasis thermal images can be computed using deep learning techniques.

References

1. C. Huerta, E. Rivero, L.A.G. Rodríguez, Incidence and risk factors for psoriasis in the general population. Arch. Dermatol. **143**(12), 1559–1565 (2007)
2. R. Parisi, I.Y.K. Iskandar, E. Kontopantelis, M. Augustin, C.E.M. Griffiths, D.M. Ashcroft, National, regional, and worldwide epidemiology of psoriasis: systematic analysis and modelling study. BMJ **369**, m1590 (2020)
3. S. Dogra, S. Yadav, Psoriasis in India: prevalence and pattern. Indian J. Dermatol. Venereol. Leprol. **76**, 595–601 (2010)
4. A. Zalewska, G. Gralewicz, G. Owczarek, B. Wiecek, Thermography in psoriasis vulgaris evaluation. Conf. Proc. IEEE Eng. Med. Biol. Soc. **1**, 627–630 (2005)
5. M. Strakowska, A. Kaszuba, B. Wiecek, M. Strzelecki, G. De Mey, System and software for thermal images screening in medicine-application to psoriasis. 127–136 (2015). https://doi.org/10.1080/17686733.2015.1035532

6. G. Puneetha, H. Girisha, Automatic segmentation of scaling in 2-D psoriasis skin images using a semi-supervised algorithm. Int. J. Mod. Eng. Res. **4**(7) (2014)
7. Y. George, M. Aldeen, R. Garnavi, Automatic psoriasis lesion segmentation in twodimensional skin images using multiscale superpixel clustering. J. Med. Imag. **4**(4), 044004 (2017). https://doi.org/10.1117/1.JMI.4.4.044004.
8. R. Meghna Sampath, U. Snekhalatha, K. Nelufer, S. Sakshi, N. Murali, Thermal imaging method in the evaluation of psoriasis in upper limb region. IOP Conf. Ser.: Mater. Sci. Eng. **912**, 062026 (2020)
9. M. Dash, N.D. Londhe, S. Ghosh, A. Semwal, R.S. Sonawane, PsLSNet: automated psoriasis skin lesion segmentation using modified U-Net-based fully convolutional network. Biomed. Signal Process. Control **52**, 226–237 (2019). https://doi.org/10.1016/j.bspc.2019.04.002
10. S. Lakshmi, A.K. Nath, C. Udayashankar, Metabolic syndrome in patients with Psoriasis: a comparative study. Indian Mermatol. Online J. **5**, 132–137 (2014)
11. J.S. Taur, C.W. Tao, C.C. Chen, C.W. Yang, Segmentation of psoriasis vulgaris images using orthogonal subspace techniques. IEEE Trans. Cybern. (2006)
12. S. Raaghavi, M. Ragini, Psoriasis severity assessment of 2D psoriasis skin images. Adv. Eng. Res. **142**, 56–59 (2018)
13. L.S. Wei, Q. Gan, T. Ji, Skin disease recognition method based on image color and texture features. Comput. Math. Methods Med. ID 8145713 (2018)
14. J. Lu, E. Kazmierczak, J.H. Manton, R. Sinclair, Automatic segmentation of scaling in 2-D psoriasis skin images. IEEE Trans. Med. Imaging **32**(4), 719–730 (2013). https://doi.org/10.1109/TMI.2012.2236349

Thermal Imaging Analysis in Detection of Childhood Obesity in Cervical Region Using Machine Learning Classifiers

Richa Rashmi and U. Snekhalatha

Abstract Thermal imaging which uses infrared radiation can be used to quantify obesity in children. The aim of this study is to determine the mean temperature in three body regions and to classify normal and obese children using machine learning algorithm. Participants were 7–10 years (8.17 ± 1.58) boys (N = 30) and girls (N = 30). All children underwent thermal analysis at the anterior cervical, anterior fingerbed, and posterior thigh region. Fingerbed region showed the highest percentage temperature difference (13.559%) between normal (30.69 ± 1.1) and obese (35.15 ± 0.7) children. The SVM classifier showed the best performance accuracy (96.67%) for both anterior cervical and anterior fingerbed region. Thermal imaging along with machine learning approach can be utilized as a screening diagnostic tool for evaluating and classifying obesity in children.

Keywords Childhood obesity · Thermal imaging · Machine learning algorithm · Feature extraction · Support vector machine · Naive Bayes

1 Introduction

Childhood obesity is a severe medical condition that develops when a child intakes improper nutrition [1]. It has reached its extensive level with the highest prevalence rate in developing countries. In 2016, over 340 million children and adolescents of age 5–19 years were reported as overweight and obese globally [2]. Child obesity is associated with many complications such as cardiovascular illnesses, endocrine disorders, musculoskeletal diseases, renal, gastrointestinal, pulmonary, and psychosocial disorders [3]. There are several existing diagnostic methods that can be accessed for the assessment of child obesity. The frequently used methods are anthropometric parameters, bioelectrical impedance examination (BIA), magnetic resonance imaging (MRI), isotope dilution, ultrasound strategy, regional thermal

R. Rashmi · U. Snekhalatha (✉)
Department of Biomedical Engineering, College of Engineering and Technology, SRMIST, Kattankulathur, Chennai 603203, Tamil Nadu, India
e-mail: snehalau@srmist.edu.in

F. Thakkar et al. (eds.), *Proceedings of the International e-Conference on Intelligent Systems and Signal Processing*, Advances in Intelligent Systems and Computing 1370,
https://doi.org/10.1007/978-981-16-2123-9_38

imaging, etc. [4]. The non-invasive nature of infrared thermal (IRT) imaging and its non-ionizing radiation nature made this technique highly acceptable for young children and adolescents [5].

Thermal camera is used to measure thermal energy released by the area of interest using infrared imaging [6]. Sangamithirai et al. conducted a study in which temperature variations in various parts of the body were measured in 11 healthy normal and 10 healthy obese individuals, ranging from 20 to 29 years [7]. Jalil et al. examined the function heat dissipation change in five healthy obese women and five healthy normal women at the abdominal and fingerbed regions, and a temperature map was generated as the result of the study [8]. Hartwig et al. described the temperature difference associated with brown adipose tissue in adults with varying BMIs [9]. Simmonds et al. studied the prediction of adult obesity using the BMI parameter from childhood obesity. In their study, they followed DerSimonian Laird random effect meta-analysis study for predicting adult obesity. They investigated the follow-up study of 200,777 subjects from age 7 to 30 years and reported that 55% of obese children develop into obese adolescents and about 80% of obese adolescents continue to be obese in their adulthood [10]. Evensen et al. studied the tracking of obesity from childhood to adolescents for 523 (279 boys and 253 girls) individuals using the BMI parameter. They conducted a follow-up study for 2 years–17 years of age. They reported that children which were identified as obese at the age of 5–7 years were found to be obese at 15–17 years of age in comparison with normal children (crude odds ratio: 11.1, 95% confidence interval: 6.4–19.2). They also reported that six out of ten children who were recognized as obese at 5–7 years were found to be obese at 15–17 years of age [11]. Numerous classifiers have been utilized previously using machine learning to predict obesity [12]. Zhang et al. reported that SVM was the best classifier for predicting childhood obesity [13]. Naïve Bayes classifier was considered as a useful probabilistic model and is utilized in various medical studies [14].

Therefore, the objective of the study is as follows: (i) to obtain the mean temperature in three different anatomical regions and to determine the statistical aspect in thermograms of healthy normal and obese children from the area of interest and (ii) to evaluate the machine learning algorithm to classify normal and obese children.

2 Methodology

2.1 Participant

A total of 60 children (30 healthy normal and 30 obese), age varies between 7 and 10 years (8.17 ± 1.58) were selected for this study out of which 30 were boys and 30 were girls. The Ethical Committee of SRM Hospital, Chennai, India, granted permission for this study. The study camp was organized at Sivananda Gurukulam School, Kattankulatur, Chennai, Tamil Nadu, India, during the month of February

2020. None of the children included in this study were taking medications that would affect their body's heat regulation. Moreover one day prior to the study, written consent was obtained from the parents and the school management.

2.2 *Experimental Setup*

The study room temperature was controlled at 21 °C and comfortable light clothes were provided and children were asked to change the school dress to light clothing. Firstly, with the help of the digital and physical scales, weight and height of the subject were computed. The children were divided into two groups, namely, normal and obese according to the calculated BMI values. A crossmark was made on the floor with a marker and the thermal camera was fixed at that point having a focal length of 600 mm. Another crossmark was made at the focal length of 600 mm from the thermal camera for the fixed position of the participants the assessment to be carried out. The children were requested to stand in an anatomical posture and expose anterior cervical, anterior fingerbed, and posterior thigh to capture the thermal images (thermograms) of the three study regions for the analysis.

2.3 *Measurement of Regional Body Temperature*

FLIR camera (FLIR A325$_{SC}$, Wilsonville, U.S.) was used for the acquisition of thermal images and further processed by using Matlab (MATLAB R2018, U.S.) software. During the thermogram imaging, the children were asked to sit in order to prevent the potential transfer of objects. Meal intake can induce heat dissipation due to the production of energy which is also referred to as diet-induced thermogenesis (DIT). Heat energy dissipation occurs through brown adipose tissue in humans [15]. To maintain the heat balance throughout our study, the thermograms were collected between 2 h post breakfast and before lunchtime. During the imaging of the anterior cervical and posterior thigh region, participants were asked to lift clothes to reveal the two study regions. Then the children were requested to show the palm area to the thermal camera from which the thermal distribution of the anterior finger bed was calculated. The temperature values for each region were attained by positioning the circular ROI of 25 × 25 mm in the specific region of thermogram. For the regions of the anterior cervical, anterior fingerbed, and posterior thigh, the circular ROI of 25 × 25 was positioned at the center of each regional thermogram.

2.4 Feature Parameters

The statistical features were derived from the anterior cervical, anterior fingerbed, and posterior thigh regional thermogram using MATLAB (MATLAB R2015, Math-Works, California) software. A total of five features were derived, namely, mean, total variance (TV), total standard deviation (TSD), entropy, and contrast.

2.5 Machine Learning Algorithm

Machine learning is a technique in which the computer is trained to make decisions for similar cases. Artificial neural network (ANN) is an inspired biological model that replicates the human brain system. ANN can extract and learn different information from input data which are spatially associated with each other. The classifier models are trained by training the computer through data samples and then classifying by identifying the pattern of hidden data. In this study, support vector machine (SVM) and Naïve Bayes were compared for their performance.

The SVM classifier is the most reliable classifier because of its ability to solve linear and nonlinear problems by acquiring a hyperplane that separates the different classes present in the problem. The main objective of this supervised learning method is to segregate the data until a minimum distance boundary is reached.

Naïve Bayes is a Bayes theorem-based probabilistic classifier. In this model, the features are independent of their class in order to detect some association between them. The main aim of this classifier is to create a relationship between the function and the class to give rise to a conditioned probability of the relationship between the predictive value of the attribute and the class.

Among the entire sample population ($N = 60$), 70% of the data was used as a training set, 10% of the data was used as a test set, and the rest 20% of the data was used for validation. In the present study, data from obese subjects were considered to be positive, and data from average subjects were considered to be negative. The confusion matrix was produced and the efficiency of each classifier was assessed and compared. The receiver operating characteristics (ROC) curve was plotted by using Weka software for each classifier (version 3.8.4).

3 Result and Discussion

The mean temperature was determined in regions such as the anterior cervical, anterior finger bed, and posterior thigh area for $N = 60$ children. Figure 1a, b represents the average temperature of overweight and obese children in the anterior cervical region, respectively. The mean temperature at the anterior cervical region in both healthy normal and healthy obese children were observed to be 36.54 ± 0.5 °C and

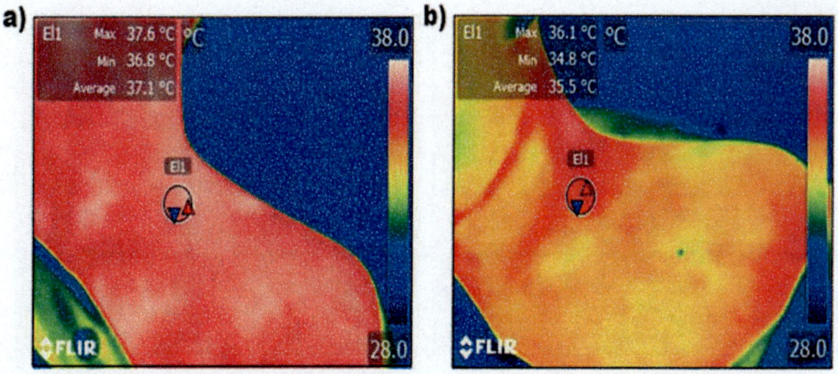

Fig. 1 Thermogram of a **a** healthy normal and an **b** obese child at anterior cervical region

32.40 ± 0.8 °C, respectively, as shown in Table 1. Figure 2 illustrates the temperature measured at the anterior fingerbed region. The average temperature derived at anterior finger bed area of healthy and healthy obese children is 30.6 ± 1.01 °C and 35.15 ± 0.7 °C, respectively (Table 1). Figure 3a, b depicts the temperature evaluated at the posterior thigh region of healthy and obese children, respectively. A percentage difference of −5.799 exists between the normal and obese children for posterior thigh region as the mean regional temperature of obese (35.203 ± 0.5 °C)

Table 1 Representation of mean average temperature (°C) and percentage difference of three study regions in children

Anatomical body region	Anterior cervical	Anterior fingerbed	Posterior thigh
Normal subjects	36.542 ± 0.589	30.691 ± 1.019	33.219 ± 0.688
Obese subjects	32.406 ± 0.866	35.155 ± 0.785	35.203 ± 0.592
% difference	11.997	−13.559	−5.799

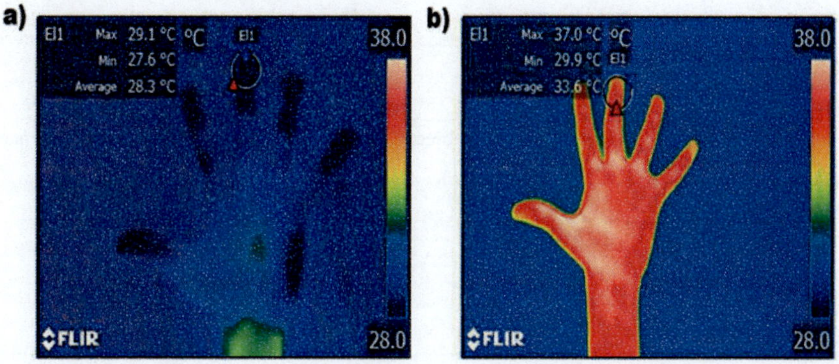

Fig. 2 Thermogram of a **a** healthy normal and an **b** obese child at anterior fingerbed region

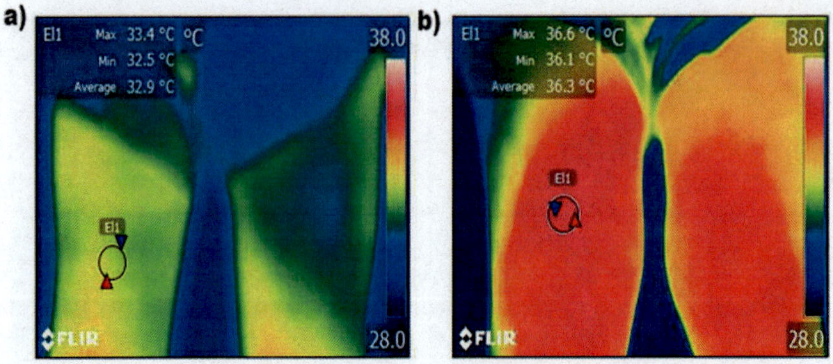

Fig. 3 Thermogram of a **a** healthy normal and an **b** obese child at posterior thigh region

children is found to be greater than that of normal (33.21 ± 0.6 °C) children. Table 2 represents the difference in skin surface temperature between both the study groups.

Table 3 displays the features derived from anterior cervical region in obese and normal children. It is observed that the feature mean (38.63) displays the highest significant difference between normal (231.53 ± 13.8) and obese (156.56 ± 10.9) participants. This corresponds to the elevated skin temperature measured at the cervical region of normal children. Table 4 displays the features derived from the anterior fingerbed area of normal and obese people. The contrast was reported to be

Table 2 Variation of temperature (°C) at different regions in normal and obese children

Anatomical body region	Overall normal healthy subjects (N = 30) regional temperature	Overall obese healthy subjects (N = 30) regional temperature
Anterior cervical	Higher	Lower
Anterior fingerbed	Lower	Higher
Posterior thigh	Lower	Higher

Table 3 Representation of the five features extracted for normal and obese subjects at the anterior cervical region

Feature extracted parameters	Anterior cervical		
	Healthy normal subjects (N = 30)	Obese subjects (N = 30)	% difference
Mean	231.532 ± 13.849	156.563 ± 10.973	−38.634
Total variance (TV)	29.617 ± 4.838	54.681 ± 7.797	59.465
Total standard deviation (TSD)	5.638 ± 1.146	7.512 ± 0.872	28.502
Entropy	4.965 ± 0.509	5.569 ± 0.437	11.467
Contrast	0.057 ± 0.057	0.074 ± 0.092	25.954

Table 4 Representation of the five features extracted for normal and obese subjects at the anterior fingerbed region

Feature extracted parameters	Anterior fingerbed		
	Healthy normal subjects (N = 30)	Obese subjects (N = 30)	% difference
Mean	115.082 ± 6.176	231.086 ± 11.881	67.022
Total variance (TV)	42.952 ± 5.481	51.614 ± 10.988	18.319
Total standard deviation (TSD	7.631 ± 0.948	6.379 ± 0.968	−17.873
Entropy	5.086 ± 0.501	5.017 ± 0.441	−1.365
Contrast	0.241 ± 0.119	0.078 ± 0.049	−102.194

Table 5 Representation of the five features extracted for normal and obese subjects at the posterior thigh region

Feature extracted parameter	Posterior thigh		
	Healthy normal subjects (N = 30)	Obese subjects (N = 30)	% difference
Mean	167.665 ± 13.546	228.246 ± 17.298	30.603
Total variance (TV)	32.507 ± 6.298	57.365 ± 8.836	55.318
Total standard deviation (TSD)	4.113 ± 1.062	7.686 ± 0.776	60.564
Entropy	4.291 ± 0.667	4.703 ± 0.502	9.162
Contrast	0.046 ± 0.071	0.032 ± 0.005	−35.897

highly significant between normal (0.241 ± 0.1) and obese (0.078 ± 0.04) subjects. Table 5 displays the characteristics of the posterior thigh area derived from normal and obese children. The feature contrast (35.89) was observed to provide the highest significance difference between obese (0.032 ± 0.005) and normal (0.046 ± 0.07) children. The findings from ANOVA demonstrate that greater variations in temperature found in regions such as anterior cervical [F(5, 174) = 3473,833, $p < 0.05$], anterior finger bed [F(5, 174) = 5363,907, $p < 0.05$], and posterior thigh [F(5, 174) = 3633,301, p < 0.05] regions for the total population studied (Table 6).

Six attributes including regional temperature, mean, TV, TSD, entropy, and contrast for both normal and obese children were given as a data set to the two excellent machine learning classifiers having tenfolds of cross-validation technique. Tables 7, 8, and 9 represent the confusion matrix of SVM and Naive Bayes (NB) for the anterior cervical, anterior fingerbed, and posterior thigh regions, respectively. The SVM outperformed NB classifier with an accuracy of 96.67% in anterior cervical and anterior fingerbed.

Figure 4, 5, and 6 illustrate the receiver operating characteristic (ROC) curve of SVM and NB classifier for anterior cervical, anterior fingerbed, and posterior thigh regions, respectively. The result depicted that SVM classifier exhibits area under the

Table 6 Analysis of Variance (ANOVA) test of three study regional temperatures performed between healthy normal and obese children

Analysis of variance						
Region	Group (Normal and Obese)	Sum of squares	df	Mean square	F score	Sig (P value)
Anterior cervical temperature	Between group	529510.426	5	105902.085	3473.833	0.002
	Within groups	5304.505	174	30.486		
	Total	534814.931	179			
Anterior fingerbed temperature	Between groups	1178693.270	5	235738.654	5363.907	0.000
	Within groups	7647.135	174	43.949		
	Total	1186340.406	179			
Posterior thigh temperature	Between groups	145968.681	5	229,193.736	3633.301	0.001
	Within groups	10976.168	174	63.081		
	Total	1156944.850	179			

Table 7 Confusion matrix of anterior cervical region for SVM and Naive Bayes machine learning classifiers

SVM classifier			
	Obese subjects (Predicted)	Normal subjects (Predicted)	Accuracy (%)
Obese subjects (Predicted)	29 (TP)	1 (FN)	96.67 (Sensitivity)
Normal subjects (Predicted)	1 (FP)	29 (TN)	96.67 (Specificity)
Overall accuracy (%)	96.67 (PPV)	96.67 (NPV)	96.67
Naive Bayes classifier			
Obese subjects (Predicted)	29 (TP)	1 (FN)	93.55 (Sensitivity)
Normal subjects (Predicted)	2 (FP)	28 (TN)	96.55 (Specificity)
Overall accuracy (%)	96.67 (PPV)	93.33 (NPV)	95

Table 8 Confusion matrix of anterior fingerbed region for SVM and Naive Bayes machine learning classifiers

SVM classifier			
	Obese subjects (Predicted)	Normal subjects (Predicted)	Accuracy (%)
Obese subjects (Predicted)	29 (TP)	1 (FN)	96.67 (Sensitivity)
Normal subjects (Predicted)	1 (FP)	29 (TN)	96.67 (Specificity)
Overall accuracy (%)	96.67 (PPV)	96.67 (NPV)	96.67
Naive Bayes classifier			
Obese subjects (Predicted)	29 (TP)	1 (FN)	93.55 (Sensitivity)
Normal subjects (Predicted)	2 (FP)	28 (TN)	96.55 (Specificity)
Overall accuracy (%)	96.67 (PPV)	93.33 (NPV)	95

Table 9 Confusion matrix of posterior thigh region for SVM and Naive Bayes machine learning classifiers

SVM classifier			
	Obese subjects (Predicted)	Normal subjects (Predicted)	Accuracy (%)
Obese subjects (Predicted)	29 (TP)	1 (FN)	90.62 (Sensitivity)
Normal subjects (Predicted)	3 (FP)	27 (TN)	96.43 (Specificity)
Overall accuracy (%)	96.67 (PPV)	90.00 (NPV)	93.33
Naive Bayes classifier			
Obese subjects (Predicted)	29 (TP)	1 (FN)	96.67 (Sensitivity)
Normal subjects (Predicted)	1 (FP)	29 (TN)	96.67 (Specificity)
Overall accuracy (%)	96.67 (PPV)	96.67 (NPV)	96.67

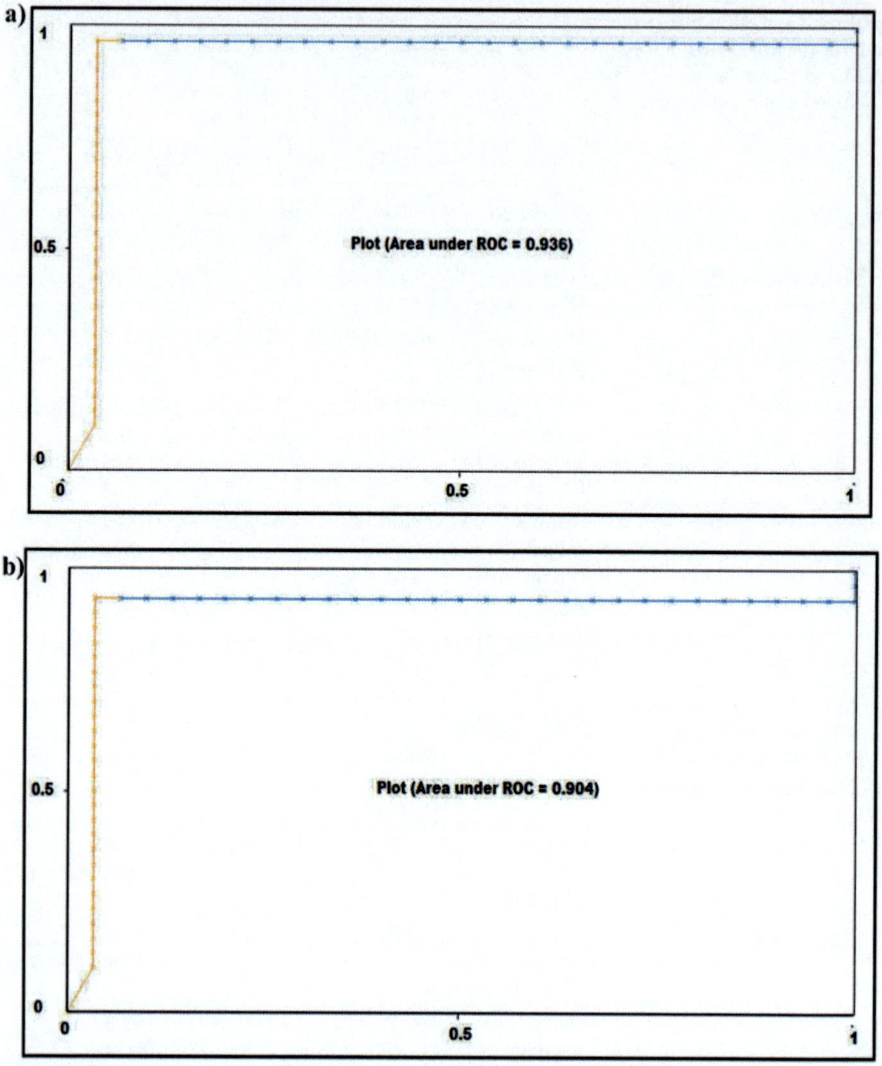

Fig. 4 ROC (Receiver operating characteristic) acquired by conducting machine learning using **a** SVM and **b** Naive Bayes for anterior cervical region

curve (AUC) as 0.936 which was higher than that of NB (0.904) for anterior cervical region. Moreover, the area under the curve for SVM and NB for anterior fingerbed region was computed as 0.948 and 0.936, respectively. However, for posterior thigh region, the area under the curve for SVM and NB was found to be 0.901 and 0.977, respectively.

Sangamithirai et al. attained mean skin temperature measured at the thigh region for normal (32.65 ± 1.37 °C) and obese (32.61 ± 1.39 °C) children [7]. In this

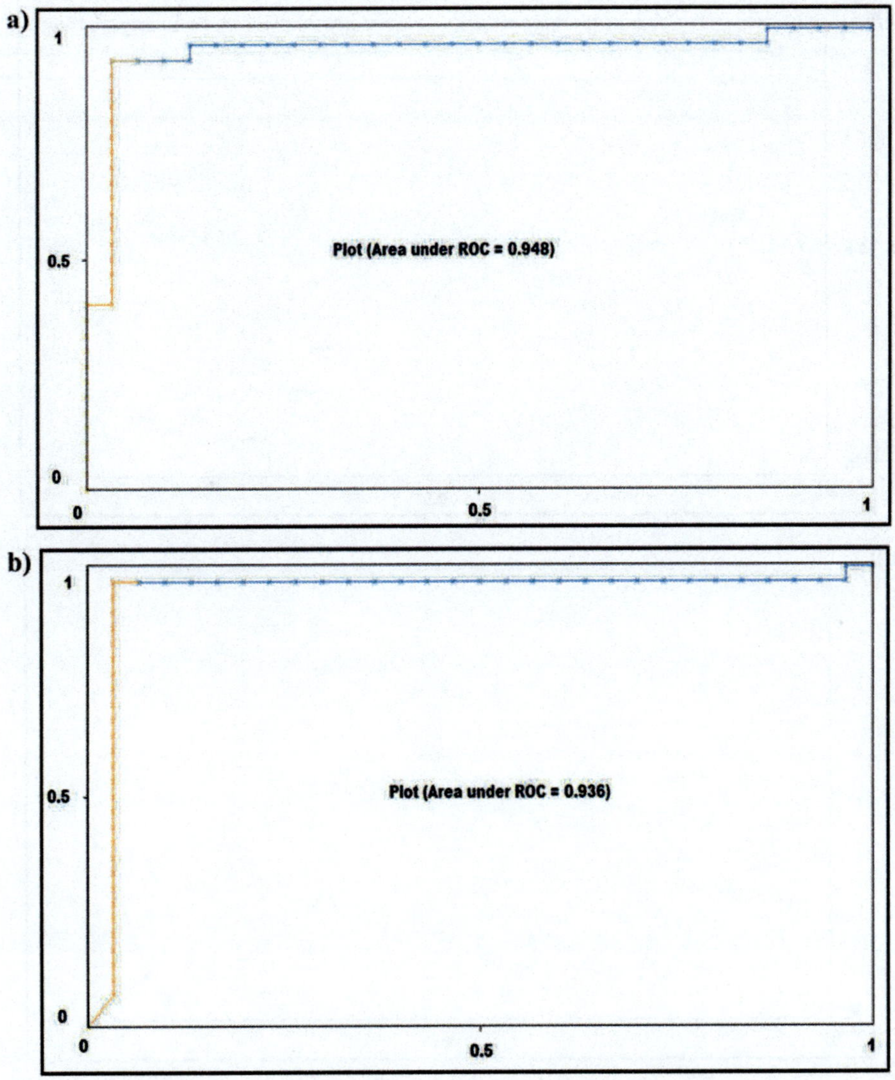

Fig. 5 ROC (Receiver operating characteristic) attained by conducting machine learning using **a** SVM and **b** Naive Bayes for anterior fingerbed region

present study, we acquired equivalent outcomes for thigh region (33.219 ± 0.688 °C for normal and 35.203 ± 0.592 °C for obese). Jalil et al. assessed that mean hand temperature was notably elevated in obese (31.87 ± 3.06 °C) than normal (28.22 ± 3.11 °C) subject [8]. Identical outcomes were attained in this current study which is in agreement with Jalil et al. (Table 1). Hartwig et al. detected the activation of brown adipose tissue in women and measured the neck region temperature for lean and obese women. They measured that the neck region temperature is greater in

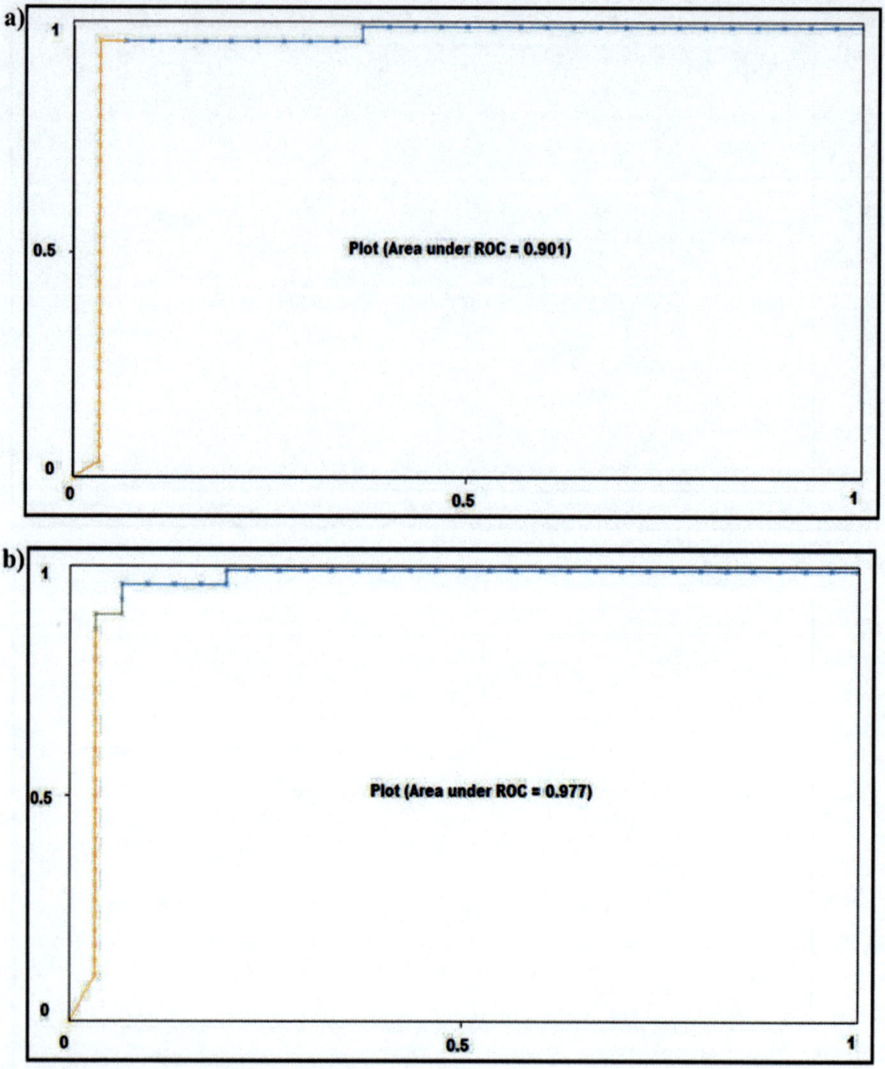

Fig. 6 ROC (Receiver operating characteristic) attained by conducting machine learning using **a** SVM and **b** Naive Bayes for posterior thigh region

the case of healthy normal and lower in healthy obese subjects [9]. A similar result is obtained in this current study illustrated in Fig. 1. The drawback of the current study involves the manual fixing of the circular ROI for analysis of thermograms in different areas, which may have affected the precision of the results.

4 Conclusions

The mean skin surface temperature of the anterior cervical, anterior fingerbed, and posterior thigh was positively associated with the derived features. Out of the three regions surveyed, the anterior fingerbed area displays a higher temperature difference among the sample population followed by the anterior cervical and posterior thigh regions. The SVM provided excellent performance with an accuracy(96.67%), sensitivity (96.67%), and specificity (96.67%) for both anterior cervical and anterior fingerbed regions. In future, computer-aided diagnosis tools will be used for automated analysis of ROI.

References

1. K. Sahoo, B. Sahoo, A.K. Choudhury, N.Y. Sofi, R. Kumar, A.S. Bhadoria, Childhood obesity: causes and consequences. J. Family Med. Primary Care **4**(2), 187–192 (2015). https://doi.org/10.4103/2249-4863.154628
2. B. Armoon, M. Karimy, Epidemiology of childhood overweight, obesity and their related factors in a sample of preschool children from Central Iran. BMC Paediatrics **19**(1), 159 (2019). https://doi.org/10.1186/s12887-019-1540-5
3. D. Kinlen, D. Cody, D.O. Shea, Complications of obesity. QJM Int. J. Med. **111**(7), 437–443 (2018). https://doi.org/10.1093/qjmed/hcx152
4. R. Rashmi, S. Umapathy, Evaluation of body composition parameters using various diagnostic methods: a meta analysis study. Obes. Med. **16**, 100150 (2019). https://doi.org/10.1016/j.obmed.2019.100150
5. J. Law, J. Chalmers, D.E. Morris, L. Robinson, H. Budge, M.E. Symonds, The use of infrared thermography in the measurement and characterization of brown adipose tissue activation. Temperature (Austin, Tex.) **5**(2), 147–161 (2018). https://doi.org/10.1080/23328940.2017.1397085
6. S. Bagavathiappan, T. Saravanan, J. Philip, T. Jayakumar, B. Raj, R. Karunanithi, T.M. Panicker, M.P. Korath, K. Jagadeesan, Infrared thermal imaging for detection of peripheral vascular disorders. J. Med. Phys. **34**(1), 43–47 (2009). https://doi.org/10.4103/0971-6203.48720
7. S. Sangamithirai, S. Umapathy, R. Sanjeena, L.S.U. Alla, Thermal imaging of abdomen in evaluation of obesity: a comparison with body composition analyser—a preliminary study, in *Proceedings of International conference on ISMAC in computational vision and Bio-Engineering 2018*. Lecture Notes in Computational Vision and Biomechanics (Springer Nature Switzerland AG, 2019), pp. 79–88
8. B. Jalil, V. Hartwig, D. Moroni, O. Salvetti, A. Benassi, Z. Jalil, L. Pistoia, T. Minutoli Tegrimi, A. Quinones-Galvan, G. Lervasi, A. L'Abbate, L. Guiducci, A pilot study of infrared thermography based assessment of local skin temperature response in overweight and lean women during oral glucose tolerance test. J. Clin. Med. **8**(2), 260 (2019). https://doi.org/10.3390/jcm8020260
9. V. Hartwig, L. Guiducci, M. Marinelli, L. Pistoia, T.M. Tegrimi, G. Iervasi, A. Quinones-Galvan, A. L'Abbate, Multimodal imaging for the detection of brown adipose tissue activation in women: a pilot study using NIRS and infrared thermography. J. Healthc. Eng. 5986452 (2017). https://doi.org/10.1155/2017/5986452
10. M. Simmonds, A. Llewellyn, C.G. Owen, N. Woolacott, Predicting adult obesity from childhood obesity: a systematic review and meta–analysis. Obes. Rev. **12**(2), 95–107 (2016). https://doi.org/10.1111/obr.12334

11. E. Evensen, T. Wilsgaard, A.S. Furberg, G. Skeie, Tracking of overweight and obesity from early childhood to adolescence in a population-based cohort—the Tromsø study, fit futures. BMC Pediatr. **16**(64), 1–11 (2016). https://doi.org/10.1186/s12887-016-0599-5
12. T.M. Dugan, S. Mukhopadhyay, A. Carroll, S. Downs, Machine learning techniques for prediction of early childhood obesity. Appl. Clin. Inform. **6**(3), 506–520 (2015). https://doi.org/10.4338/ACI-2015-03-RA-0036
13. S. Zhang, C. Tjortjis, X. Zeng, H. Qiao, I. Buchan, J. Keane, Comparing data mining methods with logistic regression in childhood obesity prediction. Inf. Syst. Front. **11**(4), 449–460 (2009). https://doi.org/10.1007/s10796-009-9157-0
14. K.M. Al-Aidaroo, A.A. Bakar, Z. Othman, medical data classification with Naive Bayes approach. Inf. Technol. J. **11**, 1166–1174 (2012). https://doi.org/10.3923/itj.2012.1166.1174
15. K.K.Y. Ho, Diet-induced thermogenesis: fake friend or foe? J. Endocrinol. **283**(3), 185–191 (2018). https://doi.org/10.1530/JOE-18-0240

Real-Time Tomato Detection, Classification, and Counting System Using Deep Learning and Embedded Systems

Stavan Ruparelia, Monil Jethva, and Ruchi Gajjar

Abstract Detection of healthy tomatoes and discrimination of infected ones is quite desirable for yield estimation. Automating this process can ease up the cultivation and harvest process for the farmers. Real-time in-filed conditions like soil and background, varying illuminations, occlusions due to leaves or other tomatoes pose as a challenge for tomato detection and counting. In this paper, we propose a deep learning-based tomato detection system that discriminates healthy, ripe, and unripe tomatoes from the infected ones and also provides a count for each of them. Different versions of the YOLO architectures are used for detection and a comparison of their performance in real-time is presented. From the experimental results, it is observed that YOLOv4 achieved high mean average precision, while YOLOv3 Tiny gave a better performance in terms of fps. These models were deployed on an embedded platform, NVIDIA Jetson TX1 for real-time testing in a tomato farm.

Keywords Deep learning · Real-time embedded systems · Object detection · Precision agriculture · Tomato detection

1 Introduction

Tomato is the second most-consumed horticulture vegetable in the world [1] and also is important due to its huge production of around 182 million tonnes around the world [2]. Tomato has both, commercial as well as dietary benefits. Its a rich source of nutrient contents like minerals, vitamin A, and vitamin C. Tomato cultivation is desirable since it gives a high yield within a short growth cycle [3].

Nowadays, because of global warming and climate changes, crop yield has been affected heavily which resulted in an imbalance between supply and demand and also caused loss to farmers. To ensure good crop yield, for harvesting, it is necessary

S. Ruparelia (✉) · M. Jethva · R. Gajjar
Department of Electronics and Communication Engineering, Institute of Technology, Nirma University, Ahmedabad, Gujarat, India
e-mail: stavanrupareliya7878@gmail.com

F. Thakkar et al. (eds.), *Proceedings of the International e-Conference on Intelligent Systems and Signal Processing*, Advances in Intelligent Systems and Computing 1370,
https://doi.org/10.1007/978-981-16-2123-9_39

to have proper climatic conditions, water supply, and irrigation methods [4] and [6]. Lowering the damage from crop diseases and pest control [5] is necessary for better growth and production of the crop. Detection of the tomatoes on a plant based on location, discrimination among ripe and unripe tomato, and identification of infected tomatoes are highly desired for better tomato cultivation and higher yield [7] and [8].

With rapidly growing technology, the involvement of artificial intelligence (AI) in agriculture and vegetation can contribute to sustainable development, and also efficient uses of the resources are required for the same. It is a difficult task to classify ripe tomatoes easily and accurately on the farm, which replaces manual labor. A system for the identification of ripe tomatoes that integrates deep learning with edge contour detection has been suggested in paper [10]. The proposed method is capable of identifying ripe tomatoes whether or not the ripe tomato fruits overlap. This proposed method achieved 80.00% Average Precision (AP) where it's tested in the greenhouse. Authors in paper [11] have proposed one real-time embedded system for Apple detection. In this proposed system, they have used the modified YOLOv3-tiny for detection. This model achieved 83.64% average accuracy as well as a frame rate of up to 30 fps, detecting and counting the Apples as well as measuring their size in real time. In [9], YOLOv3-based tomato detection system was proposed. This system replaced the rectangular bounding box for tomato localization with a circular bounding box and achieved higher prediction accuracy comparatively.

The main objective of this paper is to propose an automatic tomato detection and classification system which captures the video using a camera attached to a handheld embedded device and identifies the type of tomatoes present on the plant from the real-time images separated as a frame from the captured video. As shown in Fig. 1, the proposed system is capable of automatic detection, classification, and counting of tomatoes in real-time on a particular plant in the farm.

This paper's contributions are as follows:

- Deep convolutional neural network architecture-based robust system is proposed to detect and classify tomatoes in three different categories, i.e., Ripe Tomato,

Fig. 1 Detection and identification of tomatoes on the plant in real-time

Unripe or Green Tomato, and Infected Tomato and also provide an automatic counting of the same.

- The proposed robust system is deployed on an embedded platform for real-time inference.
- A Tomato detection, classification, and counting system that can identify and count all three different classes very efficiently in real time under different illumination conditions, background area, size, lighting, and orientation.

The remaining paper is organized as follows: Section 2 describes the dataset details as well as the hardware description. In Sect. 3, the proposed Tomato detection and classification system are discussed. The experimental results and discussion are discussed in Sect. 4, followed by the conclusion in Sect. 5.

2 Materials and Data

2.1 Data Collection

This work aims to detect and identify tomatoes on the plant as well as discriminate them into three different classes: ripe tomatoes, unripe, and infected tomatoes, using a convolutional neural network (CNN). For training the model, images of tomatoes for all the three classes were collectively gathered from internet sources as well as a tomato farm. The in-field images from a tomato farm nearby were manually collected by us. These images were captured with complex backgrounds having leaves and branches covering or occluding the tomatoes. The photographs taken included either individually hanging green, red, and infected tomatoes or numerous tomatoes grown on the same bunch with different conditions, such as completely exposed from the front or back to the sun, and shadows of leaves/branches or other tomatoes, as seen in Fig. 2.

2.2 Data Pre-processing, Augmentation, and Annotation

Pre-processing is the most important step for preparing a training dataset. The images collected for our dataset were of varying sizes. They were reshaped to $512 \times 512 \times 3$ before feeding them to the CNN model for training and testing.

The efficiency of deep CNN-based systems is higher than conventional approaches, but overfitting is a downside of this system [12]. Overfitting depends on the numbers of images and hyperparameters used for the training process [13]. When the number of images is not enough and if the model is much complex, the chances of overfitting or memorizing the training dataset increases. So to avoid overfitting, data augmentation techniques have been adopted to increase the number of images. We applied shift, rotate, and zooming operations on the images in the training dataset.

Fig. 2 Sample images of our tomato dataset (**a**) to (**c**) images taken from the internet; (**d**) to (**f**) images captured from the farm

Table 1 Details of healthy and infected tomato images

Class label	Class name	No. of images
0	Ripe (Red) tomato	900
1	Unripe (Green) tomato	600
2	Infected tomato	500
Total images		2000

Table 1 consists of the tomato dataset image class names, the class label assigned to each type, and the number of images present in each class.

For the data annotation process, we labeled the class and location of the tomato in the images of the dataset as well as generated a .xml file using labelImg software [15] for the training and testing of the CNN model.

2.3 Hardware Description

To deploy the trained CNN tomato detection and identification model, NVIDIA Jetson TX1 is used as an inference engine. It features an NVIDIA Maxwell GPU with 256 NVIDIA CUDA Cores, Quad-core ARM processor, 4GB RAM, and other peripherals [16].

3 Proposed Methodology

3.1 Proposed System Overview

The main aim of this work is to detect, classify, and count all three mentioned classes of tomatoes in real-time using deep learning. Figure 3 presents the block diagram of the proposed deep learning-based approach for Tomato detection and counting.

As our dataset was very small to train from scratch, we took pre-trained models and decided to go with transfer learning to get the optimal performance. The trained CNN model is deployed on an embedded platform for inference. Additionally, one camera module is also connected to the embedded hardware, which feeds input from the field to the trained CNN model for real-time detection, classification, and counting of tomatoes. The count of tomatoes is displayed class-wise on the top-left corner of the output.

3.2 Tomato Detection Model

Object detection has been a very challenging task for researchers, in the field of computer vision, as it is a combination of object classification and localization. Several object detection models were available earlier, but there was always a trade-off between accuracy and real-time inference speed. Also, the models were not that accurate when it comes to detecting objects at a different scale a from single image.

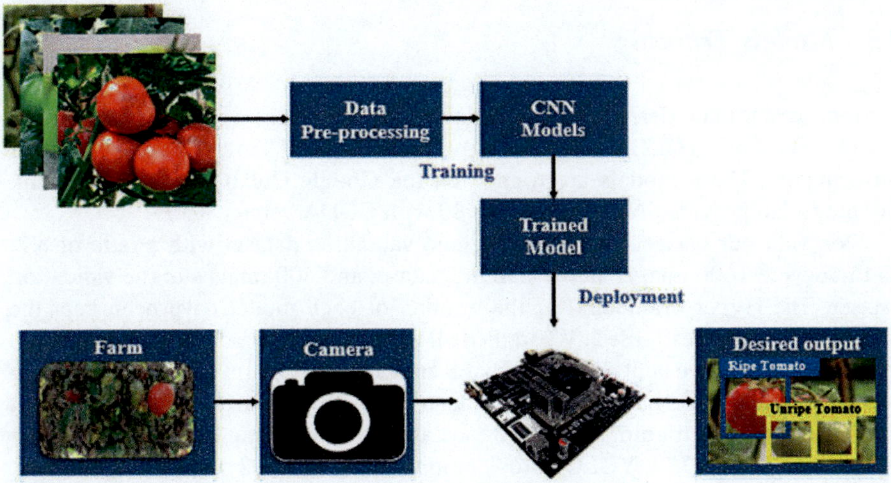

Fig. 3 Proposed system block diagram of the tomato detection system

To overcome all these problems, YOLO (You Only Look Once) was introduced with a different strategy for object detection. YOLO outperformed other models in terms of accuracy, inference time as well as detecting objects at a different scale. The first version of YOLO [17] was introduced in 2016 and after that other different versions of YOLO [18–20] were introduced. YOLO is a fully deep CNN as it only uses the convolutional layers for object detection. YOLO architecture can be divided into mainly two blocks: feature extraction which is responsible for feature extraction from the input images. The extracted features are used as input for the second block, detection layers which are responsible to predict the class and the location of the objects, from extracted features, present in the input images.

YOLOv3 [19] uses Darknet-53 in the first block for feature extraction which has 53 convolutional layers along with a batch normalization [21] layer and Leaky ReLU [22] activation and the second block also uses 53 convolutional layers for detection, making YOLOv3 [19] a 106 layers fully deep convolutional architecture. YOLOv3 Tiny is a compressed version of YOLOv3 [19] which uses Darknet-19 for feature extraction and also uses Max Pooling layers to make the model less computational complexity. YOLOv4 [20] uses Cross Stage Partial (CSP) [25] Darknet-53 having Mish [24] as an activation function stacked with Spatial Pyramid Pooling (SPP) [26] and to aggregate different first block levels, to the different detection layers of the second detection block, it uses Path Aggregation Network (PANet) [27]. YOLOv4 Tiny is a compressed version of YOLOv4 [20] which uses Leaky ReLU [22] as an activation function instead of Mish [24] used in YOLOv4 [20]. Instead of SPP [26] used in YOLOv4 [20], YOLOv4 Tiny uses detection layers directly to decrease the processing cost.

3.3 Models Training

For real-time tomato detection, classification, and counting, we used four different models, YOLOv3, YOLOv3 Tiny, YOLOv4, and YOLOv4 Tiny, and compared their performance. These models are trained on the Google Colab GPU environment. Google Colab provided Nvidia's Tesla K80 with CUDA version 10.1.

We split our dataset into a training and validation dataset with a ratio of 8:2. So there were 1600 images in the training dataset and 400 images in the validation dataset. The Hyperparameters for the training of each model have been kept the same which are as per Table 2. We trained all models for 25 epochs, each having 200 steps per epoch as we kept a batch size of 8 and we had 1600 images for the training dataset. As each model had different architectures and complexity, they took different amounts of time for training. YOLOv3 took around 01 h 43 min 07 s, YOLOv3 Tiny took around 35 min 35 s, YOLOv4 took around 01 h 54 min 02 s, and YOLOv4 took around 57 min 41 s for training.

Table 2 Hyperparameters of detection models for training

Hyperparameters	Values
Learning rate initial	1e−4
Learning rate end	1e−5
Max BBox per scale	100
Anchor per scale	3
Optimizer	Adam
Batch size	8
Number of steps	200
Input size	416
IOU loss threshold	0.5
Warmup epochs	2
Epochs	25

4 Experimental Results and Discussion

In this section, the validation results and validation loss versus epochs graphs are presented. The performance evaluation results are discussed for all four models in brief. At last, real-time testing results of detection and counting tomatoes on the field by deploying the detection model on an embedded platform is shown.

4.1 Validation Results

As mentioned earlier in the hyperparameter table, we trained each model for 25 epochs, having 200 steps per epoch. To avoid overfitting as well as to keep eye on model performance, we plotted the graphs of GIoU Validation loss [23] versus Epochs, Confidence Validation loss Vs. Epochs, Probability Validation loss Vs. Epochs and Total Validation loss versus Epochs for each model. Figure 4 refers to the graphs of Validation Losses versus Epochs for each model, YOLOv3, YOLOv3 Tiny, YOLOv4, and YOLOv4 Tiny, (a), (b), (c), and (d), respectively. As we can observe from Fig. 4, for each model, total loss and confidence loss were much higher in the first few epochs but over time the model improved the parameters through backpropagation and reduced the loss.

4.2 Performance Evaluation

For performance evaluation of the object detection model, we need to evaluate the performance for both terms, i.e. for object classification as well as object localization.

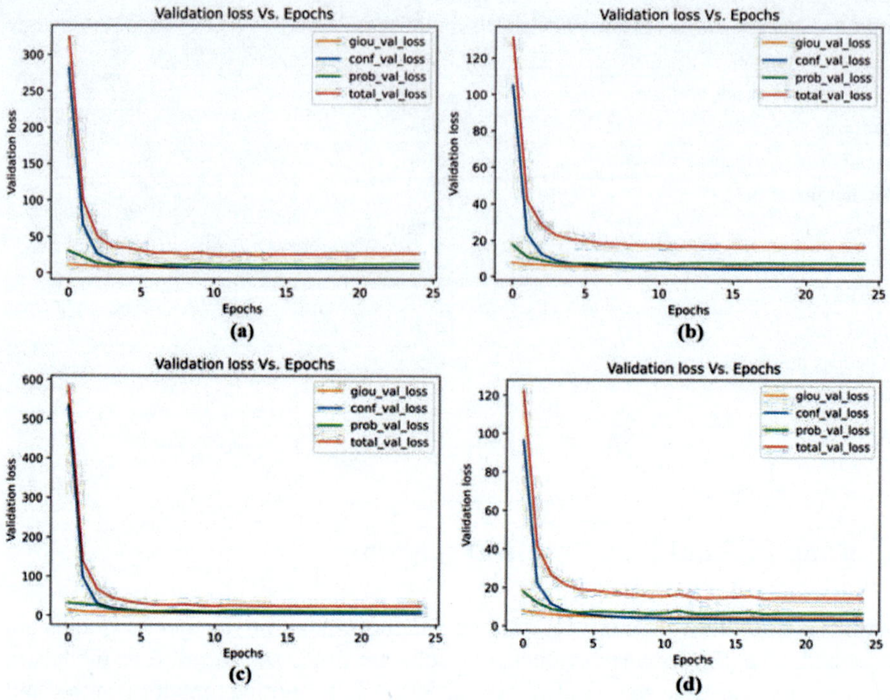

Fig. 4 Validation results

To evaluate the object detection model, the concept of Intersection over Union (IoU) is being used. Intersection over Union (IoU) computes the intersection over union between the bounding box of ground truth and the predicted bounding box.

$$Precision = \frac{\text{True Positive}}{\text{True Positive} + \text{False Positive}} \tag{1}$$

$$Recall = \frac{True\ Positive}{True\ Positive + False\ Negative} \tag{2}$$

$$f1\ Score = \frac{2 * (Recall * Precision)}{(Recall + Precision)} \tag{3}$$

$$mAP = \frac{1}{N}\sum_{i=1}^{N} AP_i \tag{4}$$

To evaluate the models, we kept the IoU threshold of 0.5. Usually, it is 0.5 or above. If the IoU value is equal or above the threshold value, then it is considered as True Positive (TP). If the IoU value is less than the threshold value then it is considered as

Table 3 Detection models performance comparison

	mAP	Precision	Recall	f1 score	fps
YOLOv3	78.49	0.90	0.49	0.63	16.56
YOLOv3-Tiny	72.15	0.86	0.49	0.62	26.85
YOLOv4	81.28	0.90	0.52	0.66	14.68
YOLOv4-Tiny	77.25	0.89	0.5	0.64	24.91

False Positive (FP). If a model fails to detect an object even if ground truth is present, then it is considered as False Negative (FN). Based on these values, we calculate the precision using Eq. (1) and recall using Eq. (2) for the object detection model. Along with the precision and recall, we also calculated the f1 score using Eq. (3) as FP and FN can affect the real-time performance. mAP [12] precision is a mean of AP of all classes, whereas fps shows the inference time a model takes. To calculate mAP, different 11 confidence thresholds are chosen such that the recall at those confidence values is 0, 0.1, 0.2, 0.3, ... , 0.9, and 1.0. The Average Precision (AP) is defined as the mean of the precision values at these chosen 11 recall values. We calculate AP for each object class the same way and as per the Eq. (4), the mean of all these AP is called mean Average Precision (mAP).

Table 3 shows the evaluation results of all four object detection models, which are considered for performance evaluation. We can observe that YOLOv3 and YOLOv4 performed very well in terms of accuracy but as these, both models have too much computational complexity and fps achieved were low. On the contrary, the Tiny versions of YOLOv3 and YOLOv4 have good fps but the accuracy was relatively low compared to their original versions.

4.3 *Real-Time Testing Results*

The detection models were deployed on NVIDIA Jetson TX1 and tested in real-time on a Tomato farm. Figure 5, presents these real-time in-field inference results. It can be observed that the proposed system is efficiently able to detect tomatoes on the plant and accurately count them, class, wise at the same time. In Fig. 5, results (a)–(c), (d)–(f), (g)–(i), and (j)–(l) are the real-time inferences using YOLOv3, YOLOv3 Tiny, YOLOv4 and, YOLOv4 Tiny, respectively. Comparing the real-time inference results of YOLOv3 and YOLOv3 Tiny, if we look at the image (a) and image (d), YOLOv3 was able to detect the tomatoes on the left side which were camouflaged by leaves whereas YOLOv3 Tiny missed to detecting those camouflaged tomatoes. Comparing image (b) and (e), both, YOLOv3 and YOLOv3 Tiny, performed the same. Looking at the image (c) and (f), YOLOv3 detected all tomatoes whereas YOLOv3 Tiny missed the tomato which had been camouflaged by the front three tomatoes.

Fig. 5 Real-time in-field test results **a–c** YOLOv3 results **d–f** YOLOv3-tiny **g–i** YOLOv4 and **j–l** YOLOv4-tiny

Looking at the real-time inference results of YOLOv4 and YOLOv4 Tiny, if we compare the image (g) and (j), the results are almost identical except that one tomato in the top right of the image where YOLOv4 Tiny predicted it like a ripe tomato as well as infected tomato also. Inference results on the image (h) and (k) are the same. Comparing real-time inference on the image (i) and (l), YOLOv4 detected all tomatoes accurately where YOLOv4 Tiny missed the tomato which is camouflaged by the front three tomatoes. So overall if we look at all the real-time inferences, YOLOv4 performed better compared to YOLOv3, YOLOv3 Tiny, YOLOv4, and YOLOv4 Tiny in terms of accuracy and localization.

In real-time in-field inference results, on the top-left corner, we have displayed the total counts of the tomatoes for each class which represents how many tomatoes have been detected in each class. The yellow bounding box represents the detected ripe tomato, the blue bounding box represents the detected green tomato and the red bounding box represents the detected infected tomato. As you can see in Fig. 5(h), there are three yellow bounding boxes and two blue bounding boxes which represent three ripe and two green tomatoes respectively and the same counts have been shown on the top-left corner of the image. In Fig. 5(i), there are four red bounding boxes which represent four infected tomatoes and the counts of the same have been shown on the top-left corner of the image [14].

5 Conclusion

A system to detect, classify, and count tomatoes in real time in a field using deep learning was proposed in this paper. Four deep learning models; YOLOv3, YOLOv3 Tiny, YOLOv4, and YOLOv4 Tiny; were trained on a Tomato dataset collected by us. Due to the heavy hardware specification requirements of all four models, the training is done on Google Colab with GPU as a hardware accelerator. After testing on several images, results revealed that, for the detection, classification, and counting, YOLOv3 produced mAP-fps of 78.49% – 16.56, YOLOv3 Tiny with 72.15% – 26.85, YOLOv4 with 81.28% – 14.68, and YOLOv4 Tiny with 77.25% – 24.91. The detection models were trained on RGB color space features that are a comparable representation to the human eye's behavior in observing the tomato fruit detection, classification, and counting with different backgrounds, lighting conditions, and occlusion effects. The experimental results show that YOLOv4 gives the best performance in terms of mAP, while YOLOv3 Tiny is better in terms of fps. The proposed models were also deployed on embedded hardware for real-time detection and count in a real field.

References

1. J.V.M. Peixoto, C.M. Neto, L.F. Campos, W.D.S. Dourado, A.P. Nogueira, A.D. Nascimento, Industrial tomato lines: morphological properties and productivity. Genet. Mol. Res **16**(2), 1–15 (2017)
2. Food and Agriculture Organization of the United Nations. FAOSTAT, http://www.fao.org/faostat/en/#data/QC

3. Y. Li, H. Wang, Y. Zhang, C. Martin, Can the world's favorite fruit, tomato, provide an effective biosynthetic chassis for high-value metabolites? Plant cell reports **37**(10), 1443–1450 (2018)
4. Food and Agriculture Organization of the United Nations. Tomato—Land water, http://www.fao.org/land-water/databases-and-software/crop-information/tomato/en/
5. R. Srinivasan, Safer tomato production techniques: a field guide for soil fertility and pest management. **10**(740). AVRDC-WorldVegetableCenter (2010)
6. A.D. Rutledge, Commercial greenhouse tomato production, https://extension.tennessee.edu/publications/Documents/pb1609.pdf
7. A. Koirala, K.B. Walsh, Z. Wang, C. McCarthy, Deep learning-Method overview and review of use for fruit detection and yield estimation. Computers and Electronics in Agriculture **162**, 219–234 (2019)
8. S. Nicola, G. Tibaldi, E. Fontana, A.V. Crops, A. Plants, Tomato production systems and their application to the tropics. Proc. IS on tomato in the tropics. Acta Hort. **821**, 27–33 (2009)
9. G. Liu, J.C. Nouaze, P.L. Touko Mbouembe, J.H. Kim, YOLO-Tomato: A Robust Algorithm for Tomato Detection Based on YOLOv3. Sensors **20**(7), 2145 (2020)
10. C. Hu, X. Liu, Z. Pan, P. Li, Automatic detection of single ripe tomato on plant combining faster R-CNN and intuitionistic fuzzy set. IEEE Access **7**, 154683–154696 (2019)
11. V. Mazzia, A. Khaliq, F. Salvetti, M. Chiaberge, Real-Time Apple Detection System Using Embedded Systems With Hardware Accelerators: An Edge AI Application. IEEE Access **8**, 9102–9114 (2020)
12. A. Géron, *Hands-on Machine Learning with Scikit-Learn, Keras, and TensorFlow: Concepts, Tools, and Techniques to Build Intelligent Systems* (O'Reilly Media, 2019)
13. A. Fuentes, S. Yoon, S.C. Kim, D.S. Park, A robust deep-learning-based detector for real-time tomato plant diseases and pests recognition. Sensors **17**(9), 2022 (2017)
14. Jetson TX1 system on module datasheet, http://images.nvidia.com/content/tegra/embedded-systems/pdf/JTX1-Module-Product-sheet.pdf
15. Tzutalin, Labelimg. Free software: MIT license (2015), https://github.com/tzutalin/labelImg
16. NVIDIA Developer, Jetson TX1 Module. https://developer.nvidia.com/embedded/jetson-tx1
17. J. Redmon, S. Divvala, R. Girshick, A. Farhadi, You only look once: unified, real-time object detection, in *Proceedings of the IEEE Conference on Computer Vision and Pattern Recognition* (2016), pp. 779-788
18. J. Redmon, A. Farhadi, YOLO9000: better, faster, stronger, in *Proceedings IEEE Conference on Computer Vision and Pattern Recognition (CVPR)*, July 2017, pp. 7263–7271
19. J. Redmon, A. Farhadi, YOLOv3: an incremental improvement (2018), arXiv:1804.0276
20. J. Redmon, A. Farhadi, YOLOv4: optimal speed and accuracy of object detection (2020), arXiv:2004.10934
21. S. Ioffe, C. Szegedy, Batch normalization: accelerating deep network training by reducing internal covariate shift. arXiv preprint (2015), arXiv:1502.03167
22. B. Xu, N. Wang, T. Chen, M. Li, Empirical evaluation of rectified activations in convolutional network. arXiv preprint (2015), arXiv:1505.00853
23. H. Rezatofighi, N. Tsoi, J. Gwak, A. Sadeghian, I. Reid, S. Savarese, Generalized intersection over union: a metric and a loss for bounding box regression, in *Proceedings of the IEEE Conference on Computer Vision and Pattern Recognition* (2019), pp. 658–666
24. D. Misra, Mish: a self regularized non-monotonic activation function. arXiv preprint (2019), arXiv:1908.08681
25. C.Y. Wang, H.Y. Mark Liao, Y.H. Wu, P.Y. Chen, J.W. Hsieh, I.H. Yeh, CSPNet: a new backbone that can enhance learning capability of CNN, in *Proceedings of the IEEE/CVF Conference on Computer Vision and Pattern Recognition Workshops* (2020), pp. 390–391
26. K. He, X. Zhang, S. Ren, J. Sun, Spatial pyramid pooling in deep convolutional networks for visual recognition. IEEE transactions on pattern analysis and machine intelligence **37**(9), 1904–1916 (2015)
27. S. Liu, L. Qi, H. Qin, J. Shi, J. Jia, Path aggregation network for instance segmentation, in *Proceedings of the IEEE Conference on Computer Vision and Pattern Recognition* (2018), pp. 8759–8768

A Standard Proposed Voice-Based Application for Virtual Assistance on Cloud Service Platforms

Jeel Patel and Jiya Patel

Abstract Virtual assistants associate the clients utilizing natural language processing with the human language. Implementation of custom skill applications can analyze client's choices and return intended responses. Subsequently, virtual assistance is becoming an integral of customer services. The main motivation for analyzing the customer's order data is that the consumer can pass commands on the fly and in response get the details for the ordered products. In this research, the order-related data is taken from the CimCommerce Platform and analyzed. This paper is an implantation of an intelligent virtual assistant in finding the order data on mobile application. This virtual-based application would gather the user preferences and model collective user knowledge base, which will improve the user-to-machine communication.

Keywords Alexa · Amazon web services · Machine learning

1 Introduction

Amazon Alexa is otherwise called Alexa, a menial helper which consistently tunes in for your voice, gauges your every ideal data, and performs activities when essential. Thanks to artificial intelligence voice colleagues that richness is currently available. They (voice collaborators) arrive in small bundles and can do an assortment of tasks following your commands. Users can provide a command to put an order, track order, switch off–on lights, and a lot more AI-based exercise. Voice associates are not only menial helpers but also individuals who work easily and handle a wide range of operations, wherein voice partners are tech-based. Alexa is based on Natural Processing Language (NPL), a methodology of changing discourse into articulation, plans, and sounds. Advanced assistance, for example, Alexa, Siri, Cortana, or the

J. Patel (✉)
Department of Computer Science, Birla Institute of Technology and Science, Pilani, Pilani, India

J. Patel
Shri I.B. Patel Angel Senior Secondary School, Anand, India

F. Thakkar et al. (eds.), *Proceedings of the International e-Conference on Intelligent Systems and Signal Processing*, Advances in Intelligent Systems and Computing 1370,
https://doi.org/10.1007/978-981-16-2123-9_40

Google Assistant are the absolute best instances of standard appropriation of Artificial Intelligence (AI) advances. These aides are getting increasingly pervasive and handling new space explicit errands which makes the upkeep of their basic AI, especially testing. The customary way to deal with fabricating computerized associates has been founded on NLP and Automated Speech Recognition(ASR) strategies which depended on comments on datasets. As of late, the Amazon Alexa group distributed a paper proposing a self-learning technique to permit Alexa right mix-ups while collaborating with clients. At the point when you approach Alexa for something, it triggers an Alexa Skills Kit, which is frequently connected to AWS Lambda to execute the code. So, before we find a good pace code, we must get an Amazon Developer Account and set up our cloud.

2 Literature Review

Argal [1], the fast development of language and speech AI techniques has made the guarantee of advanced associates a reality. These AI strategies have become a typical segment of any profound learning structure permitting any engineer to assemble genuinely modern conversational operators. In any case, the difficulties are totally different while working at the size of a computerized associate like Alexa. Normally, the precision of the AI models in these conversational operators is improved by physically interpreting and explaining information. Sun et al. [2], presented that it is becoming unimaginably costly and tedious when you must actualize it across various areas and errands. Subsequently, AI scientists have begun investigating systems, for example, semi-regulated learning or fortification figuring out how to improve the retraining procedures of conversational specialists. The Alexa group chose to go on a marginally extraordinary course by utilizing a not-very notable profound learning discipline.

To plan and actualize characteristics and natural connection modalities is an essential research field in the Human–Computer Collaboration Domain. Frameworks that can communicate with the client in their characteristic language are being investigated overwhelmingly at present. Voice-empowered chatbots are turning out to be increasingly mainstream with the appearance of gadgets and innovations like Google Home, Amazon Echo, NLP, ML, AI, and so on. A chatbot is an assistance that can begin, proceed, and handle complex collaborations with human accomplices in their regular language. Voice-empowered chatbots, today, are considered as old-style yet inventive interfaces for characteristic language association with machines. The chatbot simulates a certifiable trip specialist that accomplishes some outcome by speaking with the machine in a dialogic design utilizing normal language. The front line of voice-driven computerized collaborators is from the large four: Amazon's Alexa, Google's new right hand, Apple's Siri, and Microsoft's Cortana. The MIT Innovation Review records conversational interfaces as one of the ten leap forward advancements of 2016. Wei et al. [3], presents content-based chatbots that are utilized generally in business web and entertainment setups. What we propose is a voice-based chatbot which collaborates with clients in their regular language through discourse for empowering clients to scan for inquiries in regards to travel areas. Fruitful usage of our framework will comprehend what you are stating, examine it, and give you a reasonable reaction if the question is identified with the movement area. The client interacts with the Amazon Echo gadget which triggers the Alexa ability that forms the solicitation and builds an appropriate reaction from the information recovered from the information base. The solicitation is handled by the backend server which is facilitated by Amazon Web Services (AWS). The backend database is fused into an information base which has been prepared to utilize client evaluations. The client inquiry is dissected, and the reaction is produced utilizing the proposals from our profound neural arrange (DNN) motor which executes Restricted Boltzmann Machine for community sifting dependent on RBM utilized for tally information. Dale [4–6], presents the requirement for intelligent agents is demanding with the boundless utilization of individual machines and the longing of their producers to give common language interfaces. The proposed framework will change how to suggest travel agenda by fusing individual inclinations, recalling its clients' expectations and travel history, and producing reactions utilizing synergistic sifting. Georgiev [7], have prepared a model both client and thing connections in a bound together half and half non-IID structure. The authors have further utilize genuine qualities in the noticeable layer rather than multinomial factors, along these lines exploiting the regular request between client thing evaluations. At long last, they investigated the capability of consolidating the first preparing information with information produced by the RBM-based model itself in a bootstrapping style. Reshmi [8], Chatbots are getting a ton of consideration from the business local area right now as they can save costs in client support focuses and can deal with numerous customers all at once. This paper reports the execution of a curious chatbot, which tracks down the missing information in inquiry and tests the inquiries to clients to gather the information that is needed to answer the question. Through this execution, the degree of intelligence between the client and the chatbot is improved. Sun [9, 10], exhibits a conversational items proposal specialist. This framework shows how we consolidate research in customized proposal frameworks with research in exchange frameworks to construct a virtual deals specialist. In light of the new profound learning innovations we created, the virtual specialist is fit for figuring out how to collaborate with clients, how to respond to client questions, what is the following inquiry to pose, and what to suggest while visiting with a human client. Typically a plunge conversational specialist for a specific area requires a huge number of hand-marked conversational information or manually written standards. This is a significant obstruction when dispatching a discussion specialist for another space. Zhou [11, 12, 13] authors have shown observationally that the exhibition of ALS-WR (as far as root mean squared mistake (RMSE)) monotonically improves with both the number of highlights and the quantity of ALS emphasizes. We applied the ALS-WR calculation for a huge scope CF issue, the Netflix Challenge, with 1000 secret highlights and got an RMSE score of 0.8985, which is perhaps the best outcome dependent on an unadulterated strategy. Furthermore, consolidating with the equal variant of other known strategies, we accomplished an exhibition improvement of 5.91% over Netflix's CineMatch suggestion framework. Our strategy is straightforward and scales well to exceptionally huge datasets. Many proposal frameworks recommend things to clients by using the methods of community sifting (CF) in light of chronicled records of things that the clients have seen, bought, or evaluated. Two significant issues that most CF approaches need to battle with are versatility and scantiness of the client profiles. Al-Zubaide [14, 15], proposed approach conquers various conventional chatbots disadvantages including the need to learn and utilize chatbot explicit languages like AIML, high botmaster obstruction, and the utilization of non-developed innovation. OntBot has the extra force of simple client connections utilizing their normal language, and the consistent help of various application spaces. This gives the proposed approach various novel versatility and interoperability properties that will be assessed in future periods of this exploration project.

3 Architecture

Figure 1 depicts the proposed architecture of the application. The front end of the framework is an Amazon Alexa device and the in the backend our application is hosted on the cloud server. The client collaborates with Alexa by a voice acknowledgment programming which changes over discourse to text and the other way around. The mobile application is utilized for showing Echo-created cards and client enlistment pages of the administration. The Alexa custom skill is the interface for correspondence between the Alexa device and the server. User's fly commands are received by Alexa and send the request to the AWS Lambda function that triggers the procedure to get the intended response of the client query. At the point when an appropriate response is created, the Alexa skill communicates with the response and sends it to the Alexa device which speaks with the client. The skill encodes demand into a JSON format and hits the skill. The response is produced by the skill which is again in the JSON format. This response is given to the skill and then after to Alexa. Alexa interprets it into articulation and gives the information to the client. Skill interface is the Alexa development stage where we make our communication model.

4 User Requirements

The basic component of the framework is that the Alexa gadget must have the option to deal with complex discussions with the clients regarding their movement questions. The information base ought to give exact answers to the client's demands; hence, it must be prepared appropriately to anticipate the answer. A database would be required to store and deal with the information base and store client data. Additionally, it is fundamental to convey an Application server that will speak with the gadget and

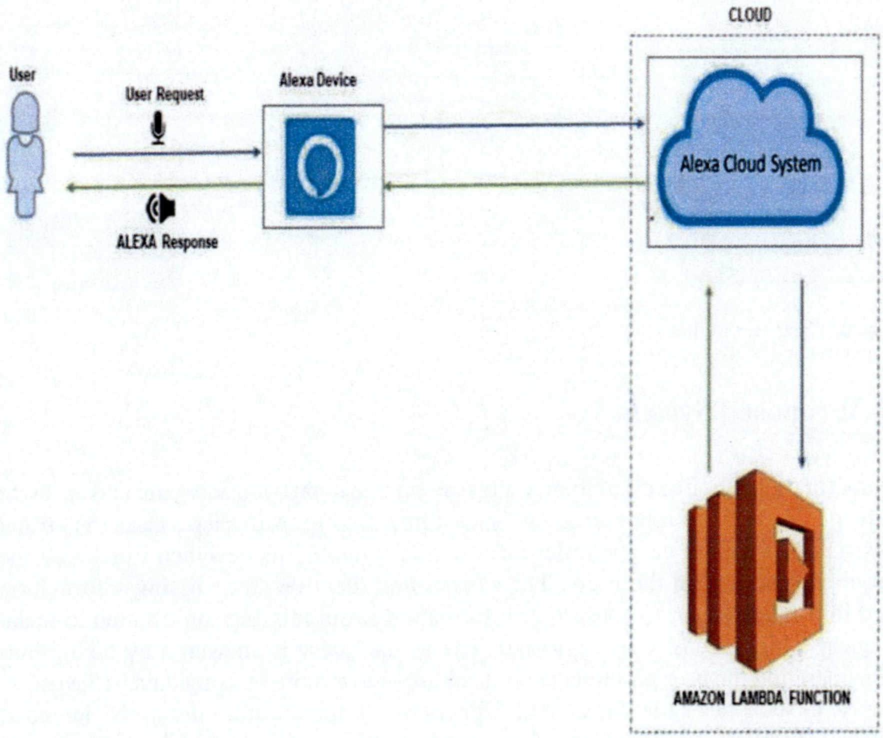

Fig. 1 Proposed architecture

database to support client demands. In this manner, the necessities are as per the following:

1. Install Alexa App on Device
2. Database: CimCommerce
3. Cloud Hosting
4. Web Client
5. Machine Learning Algorithms.

Figure 2 depicts the Orders are created when a customer completes the checkout process, and orders are visible by Admin and Shop Managers or Shopkeepers users only. For every customer, there is a unique customer id. They can track their order for latest, previous, line items products, etc. Each order has is given a unique ordered and has a status. Each order statuses let us know how far along the order is, i.e., starting from pending and ending with completed or delivered to the customer.

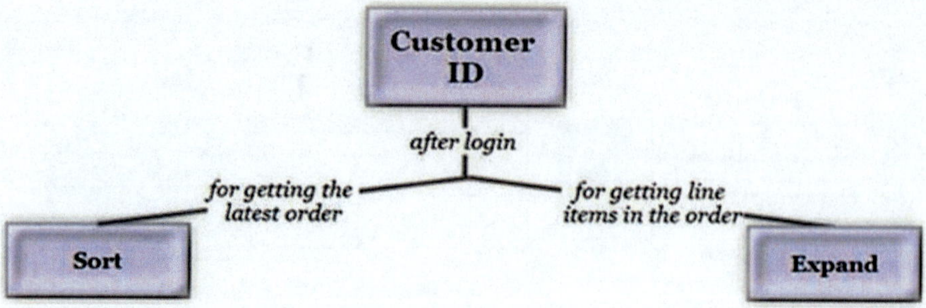

Fig. 2 Customer id flow

5 Proposed System

Sometimes, the online e-commerce sites are unable to provide adequate and accurate information for the order placed by the customer. In Fig. 3, in such a case, the virtual assistant is a key to get the order information. The skill is designed to take all the required inputs from the users. The system first identifies the missing information and informs the user, to get further information about missing information to make a query which needs to be answered. The actual query is answered by taking into consideration the user preferences. If there are one or more missing data in the users' query, in such cases the Alexa skill triggers to ask the question the users' intended query is fulfilled. After getting all the required information as the query is sent as a request to the handler function to analyze what is the user asking and in response the algorithms to find the appropriate status of the ordered product.

Given from Fig. 3, users query gets to pass on the backend framework which is facilitated on the cloud and has a server. Cimpress Assets custom APIs, and a database. The handler work is facilitated on AWS Lambda which collaborates with

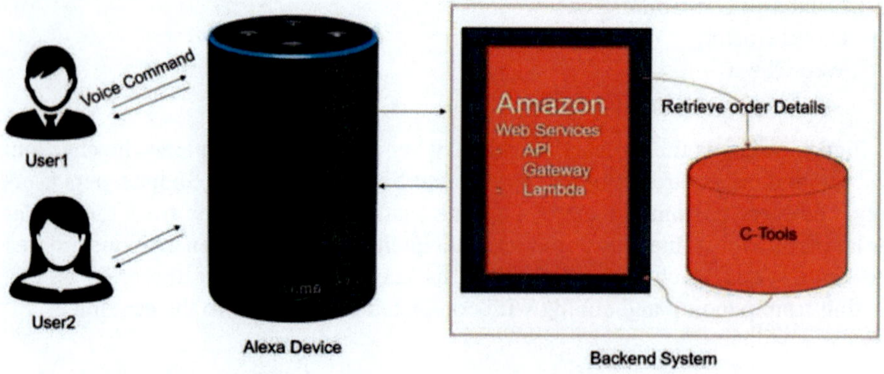

Fig. 3 Customer and virtual assistance architecture flow

the custom APIs to create client appropriate responses. The NodeJS server, facilitated on our private server, controls the whole framework.

The system is developed for various independent modules integrated together to provide all the necessary services. The modules covered in this are:

Order Module: The module will handle requests related to orders, which include order searches, order booking, etc. Orders are placed by the consumers on e-commerce sites. It should be able to handle all types of conversations related to orders.

Track Module: The tracking module to be designed and determine the current track of the order and line items of orders. The module would help the user to track their products.

Making the project more efficient, safe, and less time-consuming, there is a need to make a front-end web app to get the user's personal details, confidential information like, credit/debit card details, login credentials in advance which helps user to directly manage their products.

AWS Endpoint

An endpoint is a remote processing gadget that imparts to and for with a system to which it is connected. To get the endpoints for an Alexa Skill, we must make a Lambda Function with the assistance of Amazon Web Services (AWS). AWS Lambda is a serverless figure administration that runs your code considering occasions and consequently deals with the basic register assets for you. To interface the backend and front end, we need a help-type endpoint. We can have our ability's administration endpoint. An endpoint is a remote figuring gadget that imparts back and fronts with a system to which it is associated. There are two or three different ways by which we can get the endpoints for an Alexa Skill, make a Lambda Function, serverless arrangement with the assistance of AWS CLI. Requests that the Alexa service sends to your web service are transmitted over the internet.

Much of the time, when you make or change an AWS Lambda work utilized by your ability, you should utilize the normally suggested area that appeared in the accompanying table. We have hosted our skills in the Ireland region. We are utilizing the Alexa Skills Kit SDK for Node or Python; you can utilize a layout given in the AWS Serverless App Repository to make your capacity. This consequently makes the assets your requirement for the ability to work with Alexa. To interface the required with our skill, we must refresh the endpoint field for our expertise to the AWS Lambda ARN. After the ARN is generated, we host our custom skill and AWS lambda serverless. The lambda functions your code only when you need it and scales with your usage, so there is no provision or continuously running servers. Our skill encrypts its communications with Lambda utilizing TLS. We can manually test a Lambda function in the Lambda console by sending sample JSON events formatted in the same way as requests sent by Alexa. Sample events are provided for testing within the console. You can use these events as a starting point and modify them to represent requests that the Alexa service would send to your own function.

6 Results and Experiments

For users to interact with Alexa, we need to set up front-end functionality, which enables the user to interact with its virtual assistants. We would be using the Alexa Developer Console, for creating a custom skill application. Alexa gives a lot of inherent abilities, alluded to as aptitudes. Alexa's Software Development Kits (SDKs) dependencies and libraries installed in our skill. The orders can be set through any business unit. Their status can be followed up with Alexa. Clients to communicate with Alexa, their intents to work-in request to our Cimpress Cloud-Based Server.

The endpoint for reordering is hit in the same way as that for the iteration of tracking order. The recent shipped order is fetched if there is any. Or the last order is fetched with the following details last shipment state, shipment date, last revised arrival date, last promised arrival date, last line item name, and last line item number. A customer can track his/her last order placed.

In Fig. 4, the customer needs to enable product skill on the application. Further, users can start interacting with Alexa by giving utterances to Alexa. The customer can track his order by providing order details like Order Number, Product Name, etc. Customers usually track his order by providing Product Name.

User: 'Alexa, launch Tradeprint'.
Wake word: Users enable 'Alexa' to listen to mode.

Figure 5a depicts the status of the products. Consumer can get the current status of product by sending the name of the product in the slot value. With that, the customers can also get the status of the individual line items of the entire order and also with

Fig. 4 Enable product skill on application

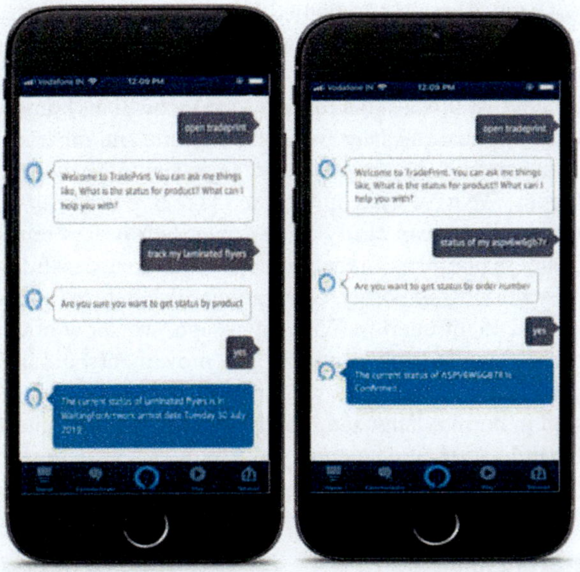

Fig. 5 **a** Getting status by product name. **b** Track by order number

the date of arrival of the products. In Fig. 5b. the customers can find the order status by order number. They need to pass the order number in slot value. This module will tell the status of the entire order. For instance,

User: Track my Laminated Flyers.
Alexa: The status of laminated flyers is in WaitingForArtWork arrival date Tuesday 30 July 2019.

User: Status of my ASPV6W6GB7R
Alexa: The current status of ASPV6W6GB7R is Confirmed.

If a user passes an order and if they have not ordered, then the customer gets the response as the order is not placed by them.

User: Alexa, get the status of my cricket balls
Alexa: You did not order this product.

Alexa fetches the intended results based on the user choices. Track ordered products and other order-related information which comes from our database server.

7 Conclusion

The savvy home space has plenty of intriguing difficulties to be understood. One significant issue that we attempted to address in this project is that of non-savvy. We utilized one of the most sweltering brilliant gadgets accessible today in the market,

Amazon Alexa, and coupled it with the webservices. The module we utilized for performing home robotization assignments functioned admirably. Further, on testing the application on Amazon Alexa progressively, we got promising outcomes. We accept that this a stage toward a savvy brilliant home. We mean to fabricate comparable applications later for Google Home and other individual collaborators that can rotate around utilizing basic voice orders to give a practical answer for non-brilliant homes. We have tried our best to make live business world environment applications where clients can easily track down their orders remotely without going onto their business platform. In the project, we have successfully completed all the functionalities working smoothly. Our framework can turn into a center point of all movement-related client queries. With this usage, we show how a shared proposal framework can be applied to support in the movement booking area to deliver increasingly customized and exact scan expectations for a client while animating a wise discussion in normal language. The absence of a visual interface will be managed through the utilization of data cards gave by Alexa also, the side web application.

References

1. A. Argal, S. Gupta, Intelligent travel chatbot for a predictive recommendation in echo platform, in *Conference 2016*, LNCS, vol. 9999. (Springer, Heidelberg, 2016), pp. 1–13
2. Y. Sun, Y. Zhang, Y. Chen, R. Jin, Conversational recommendation system with unsupervised learning, in *Proceedings of the 10th ACM Conference on Recommender Systems* (2016, September), pp. 397–398
3. J. Wei, J. He, K. Chen, Y. Zhou, Z. Tang, Collaborative filtering and deep learning based recommendation system for cold start items. Expert Syst. Appl. **69**, 29–39 (2017)
4. R. Dale, The return of the chatbots. Nat. Lang. Eng. **22**(5), 811–817 (2016)
5. M. Fischer, M. Lam, From books to bots: using medical literature to create a chatbot, in *Proceedings of the First Workshop on IoT-Enabled Healthcare and Wellness Technologies and Systems* (ACM, 2016, June), pp. 23–28. https://doi.org/10.1145/2933566.2933573
6. B.A. Shawar, E. Atwell, Chatbots: are they really useful? in *LDV Forum*, vol. 22, no. 1 (2007), pp. 29–49
7. K. Georgiev, P. Nakov, A non-iid framework for collaborative filtering with restricted Boltzmann machines, in *Proceedings of the 30th International Conference on Machine Learning* (2013), pp. 1148–1156
8. S. Reshmi, K. Balakrishnan, Implementation of an inquisitive chatbot for database supported knowledge bases. Sādhanā **41**(10), 1173–1178 (2016)
9. Y. Sun, Y. Zhang, Y. Chen, R. Jin, Conversational recommendation system with unsupervised learning, in *Proceedings of the 10th ACM Conference on Recommender Systems*. ACM, pp. 397–398 (2016, September). https://doi.org/10.1145/2959100.2959114.
10. P. Victor, C. Cornelis, A.M. Teredesai, M. De Cock, Whom should I trust?: the impact of key figures on cold-start recommendations, in *Proceedings of the 2008 ACM Symposium on Applied Computing* (2008), pp. 2014–2018
11. P.G. Campos, F. Dez, I. Cantador, Time-aware recommender systems: a comprehensive survey and analysis of existing evaluation protocols. User Model. User-Adapt. Interact. **24**(1–2), 67–119 (2014)
12. H. Wang, N. Wang, D.Y. Yeung, Collaborative deep learning for recommender systems, in *Proceedings of the 21st ACM SIGKDD International Conference on Knowledge Discovery and Data Mining* (2015), pp. 1235–1244

13. Y. Zhou, D. Wilkinson, R. Schreiber, R. Pan, Large-scale parallel collaborative filtering for the netflix prize, in *Algorithmic Aspects in Information and Management* (2008), pp. 337–348
14. E. Vysniauskas, L. Nemuraite, Transforming ontology representation from OWL to relational database. Inf. Technol. Control **35**(3A) (2006)
15. H. Al-Zubaide, A.A. Issa, Ontbot: ontology-based chatbot, in *2011 Fourth International Symposium on Innovation in Information & Communication Technology (ISIICT)*, pp. 7–12

COVID-19 Detection Using CNN and Decision Tree

Uttaran Roychowdhury and Mansi Subhedar

Abstract Currently, the outbreak of COVID-19 is a global challenge to be addressed. To address this issue, this paper presents a novel method to determine pneumonia and COVID-19 infection in lungs caused by the virus. This method is assisted by a decision tree algorithm based on the data collected during the initial breakout in china. The database of patient symptoms is used primarily to detect pneumonia infection and uses a sequential convolutional neural network to further identify COVID-19 infection. The proposed system is implemented using open-source resources and is optimal for use. Due to this, the execution of the proposed work can be achieved with much ease and high speed. The proposed convolutional neural network model, initiated after the decision tree model offered detection accuracy of 92.81%.

Keywords COVID-19 · CNN · Deep learning · Decision tree

1 Introduction

Amidst this global crisis when everyone has come together to fight against a very contagious virus, we need better solutions to identify and cure the infected candidates. Most of the affected countries lack medical personnel and test kits to run fast and efficient setup for diagnosis. The viral infection could be identified at the earlier stages and increase mortality of the person significantly. An accurate and efficient diagnosis is paramount in these times as many people fall prey to the virus, so we must turn to the modern imaging and detection techniques along with the collected data from the patients. This paper presents a data-based solution to identify the infection based on the patient's symptoms and lung X-ray images. World Health Organization (WHO) had released a set of symptoms which are visible in the infected patients. Since January 2020, a lot of research has gone into the computerized detection of COVID-19. The results are quite promising for the field of deep learning and image

U. Roychowdhury (✉) · M. Subhedar
Department of E & TC Engineering, Pillai HOC College of Engineering and Technology, Rasayani, Maharashtra, India

F. Thakkar et al. (eds.), *Proceedings of the International e-Conference on Intelligent Systems and Signal Processing*, Advances in Intelligent Systems and Computing 1370,
https://doi.org/10.1007/978-981-16-2123-9_41

processing. There were also many studies on the viability of X-ray and CT scans as proper diagnostic study tools [1]. A study comparing the radiology and CT scan images and differences can be found in [2, 3]. The study showed that the X-ray images lacked ground-glass and consolidative opacities compared to the CT scans which showed much more consistent of these features. But in [4], it was observed that lung consolidations, reticular nodular opacities and ground-glass opacities were visible. The studies conducted in [5] suggest that approximately 50% patients with COVID-19 infected have normal CT scans in 0–2 days period and patients' lungs abnormalities such as crazy paving and consolidation are more dominant in the CT scans around the 9–13 days mark after infection and slowly clear up in a month or more.

Work done by Hassanien et al. was conducted using X-ray images with classification using support vector machines and had excellent results with an average accuracy of 97.48%. A detailed study with different types of deep learning models trained of X-ray or CT scan images has been conducted in [6]. The model has achieved an average accuracy of 98.08%. A CNN-based method that uses a convolutional neural network with four convolutional layers and two dense layers is presented to determine if the input image of lung X-ray has pneumonia or not [7]. The method achieved an accuracy of 95.31% on the training set and 93.73% on the validation set. The work presented by Stephen et al. shows that the image size impacts the accuracy of the model and the loss function across the 100 epochs range [7]. It concluded that 200×200 is the most optimal size of input image as it gains the most accuracy amongst others that were tested.

The transfer learning method is presented in [8]. The use of already existing Inception $V3$ kind of models is a quick way to get detections up and running with a highly complex feature set and high accuracy. An accuracy of 96% is obtained. This viral infection has spread too rapidly, so there is a requirement of a simple and easy to deploy apparatus for the detection of this virus so it can be contained up in the primary stages and minimize the affected individuals. Most of the research conducted in this field is by using transfer learning, but as mentioned in [7], the models designed from scratch enable it to extend further as the need arises. Using the proposed method, such extension is easily possible.

The proposed method incorporates patient data as symptoms for a preliminary screening. Both, the decision tree and the CNN models, serve individual purpose of detection of the virus based on the patient's symptoms and the X-ray images. Using these symptoms as the criteria for identification, a binary decision is taken. Another layer of reassurance is added by using a CNN-based classifier which detects if a patient has pneumonia. Pneumonia is detected in patients in later stages of the infection but other symptoms might arise a bit earlier, hence the use of the decision tree classifier is made, due to the primitive nature of the decision tree algorithm. The decision tree has been paired with a convolutional neural network model for detecting images which have been more accurate and efficient nowadays due to the introduction of technologies like TensorFlow [9], PyTorch, etc. The proposed model can be changed and updated as per the infrastructure to provide higher accuracy and speed compared to the current designs.

2 Materials and Methodology

The decision tree uses the sample data based on [10] and the dataset used for training and testing the neural network model was based on X-ray images in [11]. The convolutional neural network was built on Keras API with TensorFlow backend [9]. The whole experiment is done on a standard laptop PC with Nvidia GeForce GTX 1650 card of 4 GB graphics card and Intel Core i5-9300H CPU, CUDA v11 and Tensorflow 2.1.

2.1 Dataset

The dataset used for the decision tree is based on the patient information available in [10]. The mentioned dataset has 14,937 entries out of which 486 were selected for this work as these entries only had all the required information needed for the proposed work. The choice of selection of entries was made on the basis of Age, Gender and Symptoms provided in the file. If one of the criteria is missing, it was not considered. In addition, 58 dummy data were added representing information about people without COVID-19. This was done by hand considering the features that may not be a definitive symptom of the viral infection and might be some other underlying disease or effect. The exact dataset file contained information of the patients who travelled to Wuhan, China, and are not residing within the city during detection which is an ideal case cause the majority of infections were outside the city and more across the globe giving a much more general understanding. The X-ray dataset originally has 5856 chest X-rays split into train and test sets which are further split into normal and pneumonia sets used to generate the labels for the network [11]. The training set consists of a total of 5232 images with 3883 images of infected lung X-rays and 1349 of normal lung X-rays. The testing dataset has 624 total images with 390 images of infected lung X-rays and 234 normal lung X-rays. The images shown in Fig. 1 are as first two rows of normal lungs and the rest are infected lungs either by bacterial or viral pneumonia.

2.2 Decision Tree

The decision tree is implemented using the scikit-learn python library and is used to predict the outcome based on the symptoms of the virus. The predictions are made based on the dataset provided in [10]. Only the major symptoms specified by the World Health Organization (WHO) and age and gender of the individual are considered. WHO has published a global generalized symptom list and the symptoms that correspond with those on the dataset were considered. Age and gender were derived directly from the dataset and are not related to WHO guidelines for COVID-

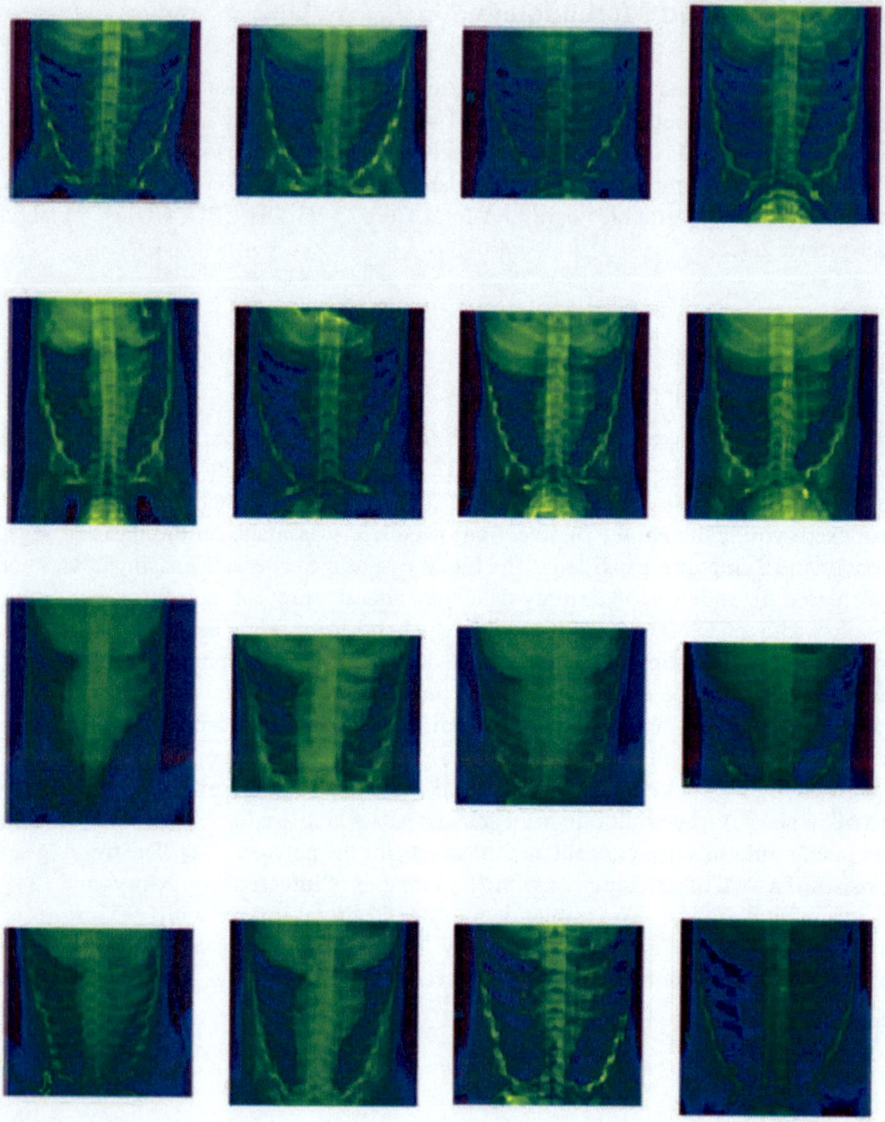

Fig. 1 Chest X-ray samples

19. Age, however, is a feature considered for COVID-19 classification as the number of articles mentioned that a greater age being a higher factor of risk for the viral infection. As per the dataset, elderly people are only mostly infected so using age as one of the parameters is a valid criterion. Though the results may be based on the age of the patient, the major symptoms such as respiratory distress trigger the deep learning model if selected as yes. This decision tree has root node as fever with 381 samples. A total of 95 nodes are present in this generated tree including the root node, internal node and leaf nodes (Fig. 2).

2.3 *Preprocessing and Augmentation*

The reason to do the augmentations is to provide the neural network with more amounts of variation of input data so it can detect much better results even if the input image shape is not in standard orientation. All the images are rescaled to 1/255. The first layer can also be considered as pre-processing to transform and prepare the images for the training purpose. Here, the images are sliced into 200 × 200 and split into three separate channels of the RGB colour spectrum. By trial and error, 200 × 200 is the size that worked best due to the memory limitations of only 8 GB.

2.4 *Model*

The proposed architecture is a triple dense layer design with five convolution layers stacked above it and four max pooling layers in between them. This model is exhibited in Fig. 3 and shows the input layer and conv2D 3 × 3, 32; with input image size of 200 × 200 pixel dimensions. The next layer is a max pooling2D layer which is connected to another conv2D 3 × 3, 64; max pooling2D layer; conv 3 × 3, 128; max pooling2D layer; then two conv 3 × 3, 128 and a maxpool2D. The flatten layer takes the input from the previous layers as pixel information that is in a matrix, this matrix is then sequenced in a continuous manner as pixel values in a single line. There is one drop-out layer with 0.25 tolerance, the first dense layer is 512 neurons in size. Another dropout layer with 0.25 tolerance; the second dense layer has 256 neurons which gives output to the final dense layer with a single neuron with a sigmoid activation function. The trainable parameters of the model are shown in Fig. 4. All the other layers use ReLU as an activation function except the last dense layers. The model is trained on a batch size of 96 and a validation batch size is 32. The number of steps per epoch for training and validation is 10 and 8, respectively. The smaller batch size is due to the memory limitations on the GPU. More dense layers allow to have more parameters for the classification part of the neural network. The whole model and the pipeline are built on the Keras API with TensorFlow backend.

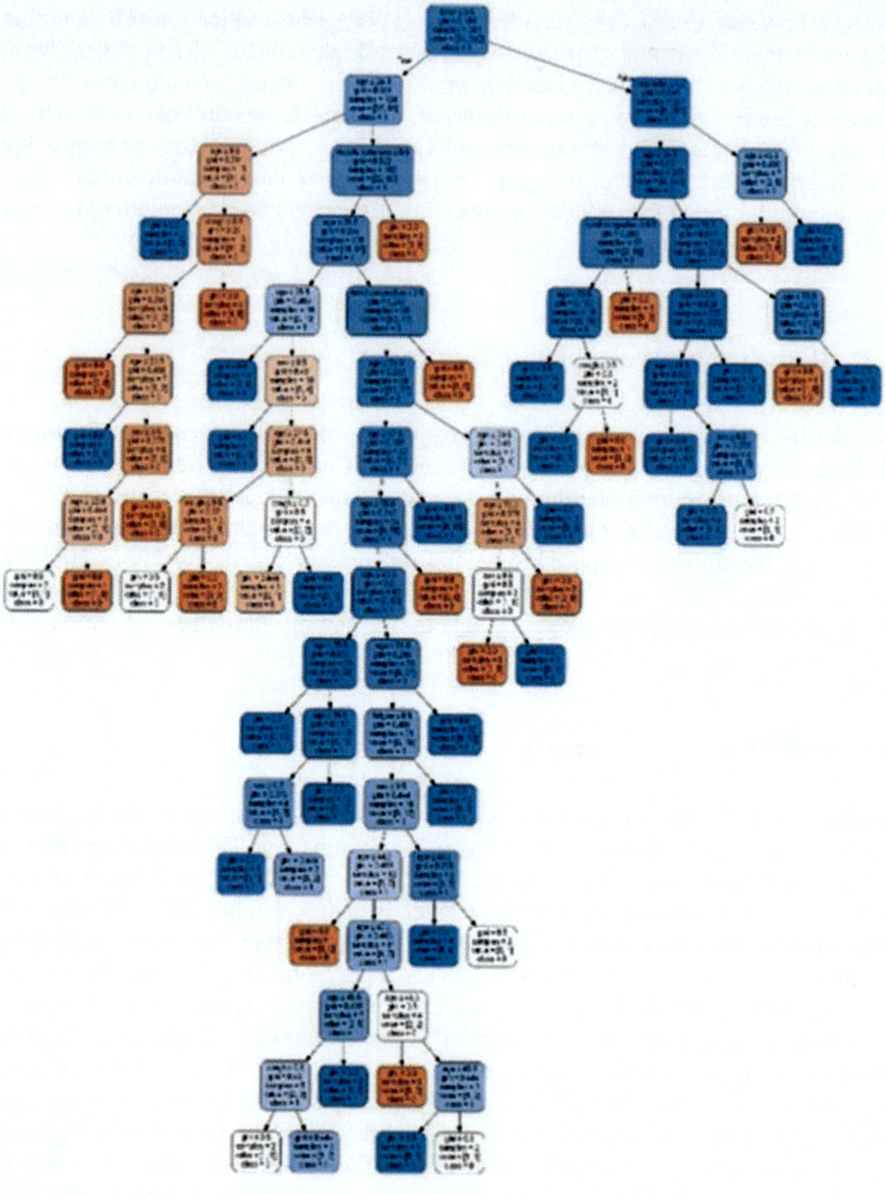

Fig. 2 Decision tree algorithm

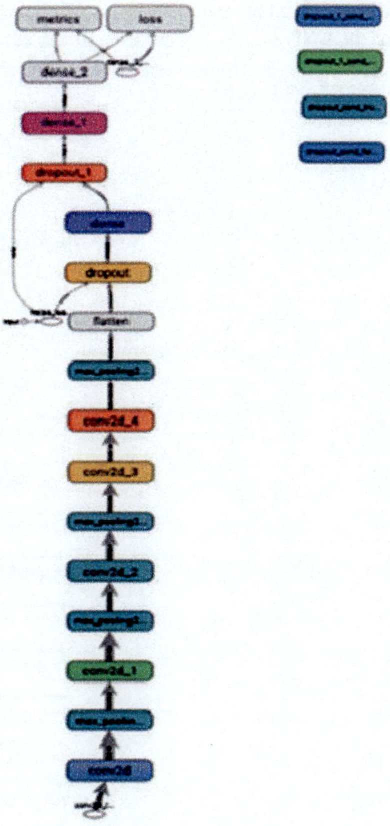

Fig. 3 Convolutional neural network structure

2.5 *Overall Working*

The CNN model as mentioned before is trained on the dataset of X-ray images [12]. The CNN model is only initiated by the output of the decision tree or by the entered symptoms list at the beginning. Age, gender and symptoms can be provided by the user. The user must enter the details as instructed. Then the decision tree takes those symptoms as input variables and provides the output as a binary classification of positive or negative. If respiratory symptoms are checked in the input criteria the CNN model is initiated and asks the user to input the X-ray image of the patient, regardless of the decision tree's output. It will then provide a definitive diagnosis if the lungs are infected with pneumonia which happens to be one of the major symptoms of nCoV-19 as it is a respiratory infection. If respiratory symptoms are not checked it goes through the standard process.

Fig. 4 Trainable parameters of the CNN

```
Model: "sequential"
_________________________________________________________________
Layer (type)                 Output Shape              Param #
=================================================================
conv2d (Conv2D)              (None, 198, 198, 32)      896
_________________________________________________________________
max_pooling2d (MaxPooling2D) (None, 99, 99, 32)        0
_________________________________________________________________
conv2d_1 (Conv2D)            (None, 97, 97, 64)        18496
_________________________________________________________________
max_pooling2d_1 (MaxPooling2 (None, 48, 48, 64)        0
_________________________________________________________________
conv2d_2 (Conv2D)            (None, 46, 46, 128)       73856
_________________________________________________________________
max_pooling2d_2 (MaxPooling2 (None, 23, 23, 128)       0
_________________________________________________________________
conv2d_3 (Conv2D)            (None, 21, 21, 128)       147584
_________________________________________________________________
conv2d_4 (Conv2D)            (None, 19, 19, 128)       147584
_________________________________________________________________
max_pooling2d_3 (MaxPooling2 (None, 9, 9, 128)         0
_________________________________________________________________
flatten (Flatten)            (None, 10368)             0
_________________________________________________________________
dropout (Dropout)            (None, 10368)             0
_________________________________________________________________
dense (Dense)                (None, 512)               5308928
_________________________________________________________________
dropout_1 (Dropout)          (None, 512)               0
_________________________________________________________________
dense_1 (Dense)              (None, 256)               131328
_________________________________________________________________
dense_2 (Dense)              (None, 1)                 257
=================================================================
Total params: 5,828,929
Trainable params: 5,828,929
Non-trainable params: 0
_________________________________________________________________
```

3 Results and Discussions

Out of the total sample set, 75% samples were used for training. The decision tree model showed a generic accuracy score of 86.13% with a 75% train and 25% test dataset split. The F1 metric in scikit-learn library test where the weighted average of precision and recall is represented with the highest score being 1 and the lowest 0, resulting in the score of 0.98 on training set and 0.91 on the test set. The F1 score metric had 75% training data and 25% testings data split with a random state of 1. This accuracy can be driven upwards by fine-tuning parameters such as the Gini index values. As can be seen in Fig. 5 and Fig. 6, respectively, the convolutional neural network model that is initiated after execution of the decision tree model has achieved an accuracy of 92.81% on training data and 90.63% on validation data at epoch 87 out of 100 with a loss of 0.2094 and 0.2747, respectively. The image classification model was able to pick out enough feature differences from the dataset mentioned

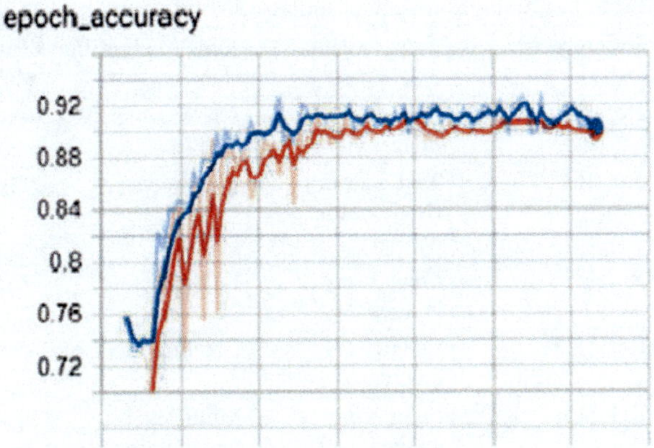

Fig. 5 Accuracy over 100 epochs

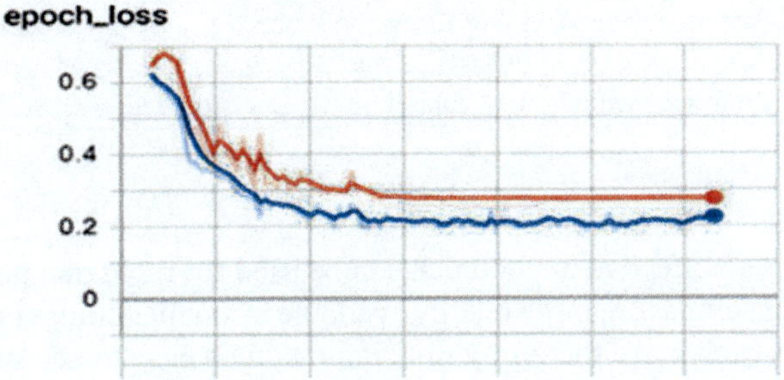

Fig. 6 Loss over 100 epochs

in [12] between pneumonia-infected lungs and normal lungs. Table 1 shows the comparison of results with existing works in literature. Both proposed systems run together in a sequential manner, the lower accuracy can be somewhat remedied by the fact that both symptoms, as well as lung X-ray imagery, are considered together. Considering a patient's symptoms can be a proper definitive way of determining if they are positive or not and adding another layer to reassure the fact by radiological imagery gives much more accurate diagnosis and helps the desk personnel who might not be a medical professional taking decisions based on both.

4 Conclusion

The proposed model provides a preliminary screening of the patients by using a decision tree. If the decision tree deems that the person might be infected and the CNN model doesn't show any kind of lung infection, it indicates that the virus is in the primary stage of its cycle. The proposed method provides a quick and accurate solution to detect the infected host till other more definitive tests results are obtained. A deep learning model with 92% accuracy and smaller size enables it to run very efficiently even on lower range computers. Both the systems running in a sequence

Table 1 Comparison with existing works

References	Type of data	Type of algorithm	Training accuracy (%)	Validation accuracy
Ref. [7]	Lung X-ray images	CNN	95.31	93.73%
Ref. [6]	Text dataset (Different text dataset)	Decision tree	92.5	–
Ref. [8]	Lung X-ray images	DCNN	97	93%
Ref. [12]	Lung X-ray images	CNN multiclass	97.9	–
Ref. [13]	Lung X-ray images	Multi-level thresholding + SVM	97.48	–
Proposed method	Lung X-ray images	CNN	92.81	90.63%
Proposed method	Text dataset	Decision tree	86.13	–

is an effective method that can be used for rapid diagnosis of the patients. The X-ray machines are more readily available in smaller clinics compared to CT scan machines which make the X-ray imaging methods easy to access at cheaper costs. This could be a very effective measure during the pandemic situation and even later in remote parts where sufficient medical aid has not reached yet and the virus persists.

References

1. J. Lei, J. Li, X. Li, X. Qi, CT imaging of the 2019 novel coronavirus (2019-nCoV) pneumonia. Radiology **295**(1), 200–236 (2020). https://doi.org/10.1148/radiol.2020200236
2. N. Ming-Yen, E.Y.P. Lee, J. Yang, F. Yang, X. Li, H. Wang, M.M. Lui et al., Imaging profile of the COVID-19 infection: radiologic findings and literature review. Radiol. Cardiothorac. Imaging **2**(1), e200034 (2020)
3. M.Z. Alom, M.M. Shaifur Rahman, M.S. Nasrin, T.M. Taha, V.K. Asari, COVID MTNet COVID-19 detection with multi-task deep learning approaches (2020), arXiv:2004.03747
4. D. Cozzi, M. Albanesi, E. Cavigli, Chest X-ray in new coronavirus disease 2019 (COVID-19) infection: findings and correlation with clinical outcome. Radiol. Med. **125**(8), 730–737 (2020). https://doi.org/10.1007/s11547-020-01232-9
5. J.P. Kanne, B.P. Little, J.H. Chung, B.M. Elicker, L.H. Ketai, Essentials for radiologists on COVID-19: an update radiology scientific expert panel. RSNA (2020). https://doi.org/10.1148/radiol.2020200527
6.)A.M.U.D. Khanday, S.T. Rabani, Q.R. Khan et al., Machine learning based approaches for detecting COVID-19 using clinical text data. Int. J. Inf. Tecnol. **12**, 731–739 (2020)
7. O. Stephen, M. Sain, U.J. Maduh, D.-U. Jeong, *An Efficient Deep Learning Approach to Pneumonia Classification in Healthcare* (Hindawi, 2019). https://doi.org/10.1155/2019/4180949

8. S. Asif, Y. Wenhui, H. Jin, Y. Tao, S. Jinhai, *Classification of COVID-19 from Chest X-ray images using Deep Convolutional Neural Networks* (2020). https://doi.org/10.1101/2020.05.01.20088211
9. M. Abadi, P. Barham, J. Chen, Z. Chen, A. Davis, Tensorflow: a system for large-scale machine learning, in *12th Symposium on Operating Systems Design and Implementation* (2016), pp. 265–283
10. Open COVID-19 Data Working Group, *Detailed Epidemiological Data from the COVID-19 Outbreak*. Accessed 08 April 2020
11. *Large Dataset of Labeled Optical Coherence Tomography (OCT) and Chest X-Ray Images*, http://dx.doi.org/10.17632/rscbjbr9sj.3file-4356bbc1-92da-4738-8e27-b0ba604f07f4
12. T. Ozturk, M. Talo, E.A. Yildirim, U.B. Baloglu, O. Yildirim, U. Rajendra Acharya, Automated detection of COVID-19 cases using deep neural networks with X-ray images. **121**, 103792 (2020). https://doi.org/10.1016/j.compbiomed.2020.103792
13. A.E. Hassanien, L.N. Mahdy, K.A. Ezzat, H.H. Elmousalami, H.A. Ella, Automatic X-ray Covid-19 lung image classification system based on multi-level thresholding and support vector machine (2020). https://doi.org/10.1101/2020.03.30.20047787

A Novel Approach for Spoken Language Identification and Performance Comparison Using Machine Learning-Based Classifiers and Neural Network

Vishal Tank, Manthan Manavadaria, and Krupal Dudhat

Abstract Spoken Language Identification (SLI) is the process of capturing a type of language of a speaker. In this research paper, the used database is created in three different languages, Gujarati, Hindi, and English. Language classification is performed using features like MFCC (Mel Frequency Cepstral Coefficients), Pitch, and average energy. Accuracy values of the created database are evaluated and compared using various pattern classifiers, namely, Fine Tree, Linear Discriminant, Gaussian Naïve Bayes, Linear SVM, Fine KNN, and feed-forward neural network in MATLAB 2019. Performance using individual speech features and hybrid features are compared. Training time of all the classifiers is also evaluated to decide the best among all classifiers.

Keywords Spoken Language Identification · Speech features · Classifier · Machine learning · Neural network

1 Introduction

Every language has a trademark structure which makes that language sound not quite the same as different dialects. The language structure is characterized by its telephone stock, syllable development, phonotactic rules, and prosodic examples [1]. For any speech processing system, Spoken Language Identification (SLI) is an important first step in several applications. Nowadays, lots of control of equipment are governed by voice only. Also with the increase of IoT and related development numbers will be a growing number which has voice-based assistants [2]. In this scenario, SLI has emerged as a popularly ongoing research field. Google Assistant, Alexa, and echo from Amazon, LG Vacuum Cleaner, Apple Home Kit are the best examples of such applications [3]. Secondly, here (Gujarati) language is preferred for experiment purpose as it is the mother tongue of authors as well as it has significant

V. Tank (✉) · M. Manavadaria · K. Dudhat
V. T. Patel Department of Electronics & Communication Engineering, C. S. Patel Institute of Technology, Charotar University of Science and Technology, Anand, Gujarat, India
e-mail: vishaltank.ec@charusat.ac.in

F. Thakkar et al. (eds.), *Proceedings of the International e-Conference on Intelligent Systems and Signal Processing*, Advances in Intelligent Systems and Computing 1370,
https://doi.org/10.1007/978-981-16-2123-9_42

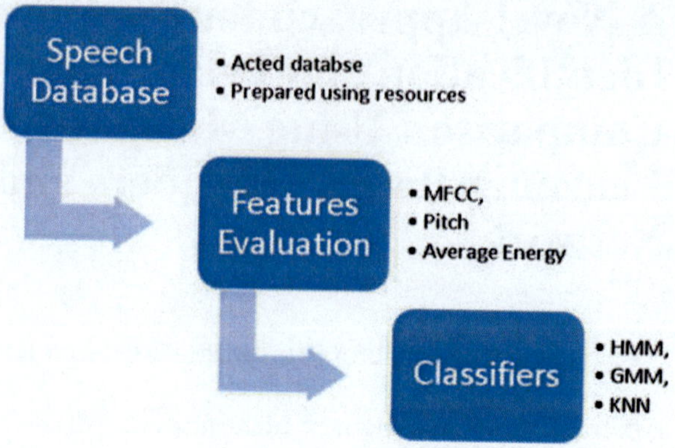

Fig. 1 Framework of Spoken Language Identification (SLI) system used in this research work

importance worldwide. The Gujarati language is majorly spoken in Gujarat state in India. It is ranked sixth in India spoken by 55.5 million speakers, which is 4.5% of the total Indian population as of 2011. As per 2007 data, worldwide Gujarati ranked 26th with the most number of speakers [4]. Gujarati is spoken in India, North America, the United States, Canada, Kenya, Uganda, Tanzania, Zambia, and South Africa. In Europe, Gujaratis form the second largest of the British South Asian speech communities, and Gujarati ranked with the fourth spoken language in London.

The framework of SLI system used for this research work is described in Fig. 1 [5, 6]. As per Fig. 1, the first step is to create a database of speech samples spoken in different languages [7, 8]. After the creation of database, speech features are extracted from speech samples. Finally, various types of classifiers are used to classify the language-based speech [9]. Classifiers available from different domains like machine learning, deep learning, and neural network can be used [10, 11].

The paper is organized as follows: Introduction, Related Work, and Creation of the proposed database, Speech Features, Classifiers, Experiment Setup, Results analysis, Conclusion, Future Work, and References.

2 Literature Survey

Many researchers have contributed SLI in foreign languages like Russian, German, and many more [12, 13], in addition to that many Indian researchers have contributed in Indian languages [14]. Literature survey of SLI can be summarized as follows: In 2017, Jothilakshmi et al. have classified nine Indian languages using GMM, HMM, and ANN with MFCC, delta, acceleration coefficients, and shifted delta cepstral (SDC) coefficients [15]. Madhu et al., in 2017, have classified seven Indian languages using ANN classifier. Ravi Kumar et al., in 2016, have implemented a system of language identification using cepstral features [16]. They used MFCC as a cepstral feature and GMM-UBM as a classifier. In 2012, Koolgudi et al. classified 15 Indian languages with MFCC, and language identification is mainly done using pattern classifiers, namely, GMM, SVM, ANN, and HMM [17]. Rania et al., in 2012, developed a language identification system using SVM Classifier which classified six languages like German, English, Arabic, Spanish, French, and Italian. Abhijeet Sangwanet et al., in 2010, identified five closely related Indian languages [18].

3 Creation of Database

There are various ways to create or develop a database for spoken language identification systems like acted database and natural database [19]. The proposed database is created using various dialogues, captured from online websites like YouTube, and other contents. The proposed database is created to classify Gujarati, Hindi, and English language. It contains 60 samples separated by 20 samples of each three languages for language identification. Gujarati and Hindi are the official languages as per the constitution of India and are majorly spoken across the globe. Creation and arrangement of each sample are saved in.wav format and saved using the following format:

01–03.wav, 01 indicates Gujarati, 03 indicates 3rd sample.

02–20.wav, 02 indicates Hindi, 20 indicates 20th sample.

03–09.wav, 03 indicates English, 09 indicates 9th sample.

Features: for the process of language identification various speech features like spectral features, prosodic features, and wavelet-based features are used [20]. In this research work, MFCC is used as a spectral feature, Pitch and average energy features are used as prosodic features. MFCC (Mel frequency cepstral coefficients): speech sound is produced with air passing through the vocal tract including tongue, teeth, etc. The shape of the vocal tract is represented by MFCC coefficients [21]. Pitch: pitch is a fundamental period of a speech signal. Pitch is Correlates with the fundamental frequency [22]. It represents the vibration frequency of the vocal cords during the sound productions. Average energy: energy values of each speech sample are evaluated as the mean square of the amplitude.

4 Classifiers

4.1 Fine Tree

Tree classifier is well known for its simple structure in classification. At each node based on searching conditions, each class is identified into the bottom approach. At each node, properties like entropy and gain are useful to decide the path of a tree [23].

4.2 Linear Discriminant

Linear discriminant analysis is concocted by Ronald Aylmer Fisher; hence, it is also called the Fisher method. It is typically known as dimension reduction technique for classification since, for the most part, it changes over high-dimensional datasets like video, picture, speech into low-dimensional space, still remains a robust technique [24].

4.3 Gaussian Naïve Byes

Gaussian Navïe Byes is based on the probabilities of each class. It is based on the likelihood of each class. In case of any unavailability of any missing attributes, it simply omits the respective probabilities [25].

4.4 Linear SVM

Support vector machine was identified by Vapnik and is well known for its supreme performance in space of limited data. By using various kernel functions, nonlinear high inseparable data into high-dimensional space is changed to separable. SVM identifies the distance between future and hyperplane. On each side of hyperplane, two parallel planes are created to separate data [26].

4.5 Fine KNN

KNN stands for K-nearest neighbor classifier. In KNN, only distance metric is used to classify a particular class, hence, it is called lazy learner among all the classifiers [27].

4.6 Feed-Forward Neural Network

A neural network is a progression of calculations that tries to perceive hidden connections in a lot of information through a process that copies the manner in which the neurons work. In this sense, neural networks allude to frameworks of neurons. Neural network can predict the most ideal output based on inputs by processing through various hidden layers. Various kinds of neural networks like feed-forward, backpropagation, and RNN becomes well known for classification and regression problems depending on the types of applications used in the present era [28].

5 Experiment Setup

The MATLAB 2019 software is used for performing all the experiments and observations are taken using its specific tools like Classifier Learner App and Neural Network Tool (nntool).

After creating the proposed database, speech features are evaluated for each sample. In this experiment, both spectral and prosodic features have been used in the identification of language. MFCC, which is a basic spectral feature along with Pitch and Energy as the prosodic features are used. Total 16 features are evaluated bifurcated by 14 coefficients of MFCC, 1 for Pitch, and 1 for Average Energy. Each sample is labeled with appropriate language numbered. Finally, .mat is created with speech features values and labeling.

Classification for machine learning algorithms: classification is performed using five classifier algorithms namely Fine Tree, Linear Discriminant, Gaussian Naive Bayes, Linear SVM, and Fine KNN. Cross-validation of 5% is kept during evaluation using mention classifiers.

Classification for feed-forward neural network: mention ANN is created using 10 neurons. Neural network is set up with 70% samples for training, 15% for testing, and the remaining 15% for validation.

6 Results and Discussion

In Tables 1 and 2, the accuracy/performance values and training time taken are evaluated and compared for each of the five classifiers, namely, (1) Fine Tree, (2) Linear Discriminant, (3) Gaussian Naïve Bayes, (4) Linear SVM, and (5) Fine KNN.

Performance values and training time taken by each classifier are compared using seven features, namely, Hybrid features like (1) MFCC + Pitch + Energy, (2) MFCC + Pitch, (3) Pitch + Energy, (4) MFCC + Energy, and individual features like (5) MFCC, (6) Pitch, and (7) Energy.

Table 1 Comparative performance analysis (%) of various classifiers using speech features

Sr no	Speech features	Fine tree	Linear discriminant	Gaussian Naïve Bayes	Linear SVM	Fine KNN
1	MFCC + Pitch + Energy	88.3	100	88.3	91.7	95
2	MFCC + Pitch	85	96.7	88.3	91.7	91.7
3	Pitch + Energy	66.7	58.3	50	55	63.3
4	MFCC + Energy	85	100	93.3	93.3	93.3
5	MFCC	81.7	93.3	88.3	90	90
6	Pitch	36.7	60	55	51.7	41.7
7	Energy	46.7	41.7	38.3	33.3	53.3

Table 2 Comparative training time (in seconds) analysis of various classifiers using speech features

Sr No	Speech features	Fine tree	Linear discriminant	Gaussian Naïve Bayes	Linear SVM	Fine KNN
1	MFCC + Pitch + Energy	1.69	3.29	3.86	6.11	8.85
2	MFCC + Pitch	1.07	1.3	1.13	1.37	3.2
3	Pitch + Energy	1.14	0.55	1.36	1.14	2.72
4	MFCC + Energy	1.24	0.59	1.29	1.9	4.81
5	MFCC	1.02	2.62	2.47	3.37	5.08
6	Pitch	1.10	1.45	1.22	1.96	5.64
7	Energy	1.19	0.57	1.31	1.86	8.22

As shown in Table 1, the highest accuracy of 100% with hybrid features MFCC + Pitch + Energy is given by linear discriminant and the lowest values of 88.3% are noted for Fine Tree and Gaussian Naive Bayes classifiers. The highest performance of 96.7% is given by linear discriminant and 85% by Fine Tree, 66.7% accuracy is noted by Fine Tree and a minimum of 50% for Gaussian Naive Bayes using Pitch and Energy. Super performance of 100% is given by linear discriminant and least performance of 85% is given by Fine Tree. For individual features, best accuracy values are provided by linear discriminant using MFCC and Pitch with respective values of 93.3 and 60%. Finally, 53.3% accuracy is given by the Fine KNN classifier using the Energy feature.

A conclusive remark of Table 1 is that hybrid features always give better performance than individuals. MFCC, which is a spectral feature, is more efficient in the process of classification of languages. Overall, linear discriminant has provided the best performances in language identification.

Less Training time will definitely help to have a speedy classifier. In real time, we always need a high-speed classifier so with this aspect training time is compared.

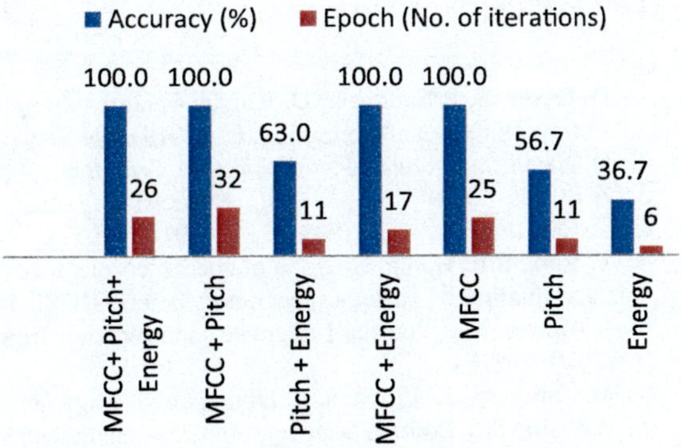

Fig. 2 Comparative of accuracy and no. of iterations using speech features in neural network

As shown in Table 2, using hybrid features, minimum and maximum training times taken by classifiers are noted as (1) MFCC + Pitch + Energy: 1.69 s for Fine Tree and 8.85 s for Fine KNN, (2) MFCC + Pitch: 1.07 s for Fine Tree and 3.2 s for Fine KNN, (3) Pitch + Energy: 0.55 for linear discriminant and 2.72 s for Fine KNN, (4) MFCC + Energy: 0.59 for linear discriminant and 4.81 s for Fine KNN. Training time taken by each classifier based on individual features is (5) MFCC: 1.02 s for Fine Tree and 5.08 s for Fine KNN, (6) Pitch: 1.1 s for Fine Tree and 5.64 s for Fine KNN, and finally, (7) Energy: 1.19 s for Fine Tree and 8.22 s for Fine KNN.

As shown in Fig. 2, 100% of performance is achieved by ANN in case of MFCC + Pitch + Energy, MFCC + Pitch, MFCC + Energy, and MFCC. Least accuracy is provided by the energy feature with the neural network. The highest number of iterations are taken while evaluating with MFCC + Pitch using ANN.

Conclusive remarks of Fig. 2 are again MFCC is the feature which has performed better than the other classifier and energy in the last place.

7 Conclusion and Future Work

At last, it can be concluded that each classifier performs better with combinations of hybrid features. Overall, neural Network gives better results than other classifiers in terms of accuracy. Spectral features are more efficient in the process of classification of languages. Fine KNN takes more training time and Fine Tree takes less training time than others. Overall, the performance of linear discriminant is noticeable than others.

In future, the number of samples in the created database can be increased to make it more effective and generalized. Also, our intention is to add more languages to this database, which will give a broad and creative way of understanding spoken language classification.

References

1. D. Deshwal, P. Sangwan, D. Kumar, Feature extraction methods in language identification: a survey. Wirel. Pers. Commun. **107** (2019)
2. D. Nandi, *Language Identification—A Brief Review* (2015)
3. A.B. Nassif, I. Shahin, I. Attilim, Speech recognition using deep neural networks: a systematic review. IEEE Access **7**(99), 19143–19165 (2019)
4. V. Tank, S.K. Hadia, Creation of speech corpus for emotion analysis in Gujarati language and its evaluation by various speech parameters. IJECE **10**(5), 4752–4758 (2020)
5. J. Anjana, S.S. Poorna, Language identification from speech features using SVM and LDA (2018), pp. 1–4
6. M. Zampieri, A. Ciobanu, L. Dinu, Native language identification on text and speech (2017)
7. P. Athira, S.S. Poorna, Deep learning based language identification system from speech (2019), pp. 1094–1097
8. S. Ishimitsu, Speech recognition method and speech recognition apparatus. J. Acoust. Soc. Am. **94**(109), 3538 (2018)
9. Z. Song, English speech recognition based on deep learning with multiple features. Computing **102**, 663–682 (2020)
10. H. Mukherjee, A. Dhar, S.K. Obaidullah, K.C. Santosh, S. Phadikar, K. Roy, A recurrent neural network-based approach to automatic language identification from speech (2020)
11. I. Lopez-Moreno, J. Gonzalez-Dominguez, D. Martinez, O. Plchot, J. Gonzalez-Rodriguez, P. Moreno, On the use of deep feedforward neural networks for automatic language identification. Comput. Speech Lang. **40** (2016)
12. C. Chandak, Z. Raeesy, A.L. Rastrow, H. Yuzong, W. Xiangyang, J. Dong, M. Roland, Streaming language identification using combination of acoustic representations and ASR hypotheses (2020)
13. R. Latypov, R. Nigmatullin, E. Stolov, Automatic spoken language identification by digital signal processing methods. Tatar Russ. Lang. (2017)
14. J. Chakraborty, S. Nath, S. Nirmala, K. Samudravijaya, Language identification of Assamese, Bengali and English Speech (2018) pp. 177–181
15. S. Jothilakshmi, V. Ramalingam, Sengottayan, A hierarchical language identification system for Indian languages. Digit. Signal Process. **22**, 544–553 (2012)
16. C. Madhu, A. George, L. Mary, Automatic language identification for seven Indian languages using higher level features, in *IEEE International Conference on Signal Processing, Informatics, Communication and Energy Systems (SPICES)* (2017), pp. 1–6
17. S. Koolagudi, D. Rastogi, Spoken language identification using spectral features. Commun. Comput. Inf. Sci. **306**, 496–497 (2012)
18. A. Sangwan, M. Mehrabani, J.H.L. Ansen, Automatic language analysis and identification based on speech production knowledge, in *IEEE International Conference on Acoustics, Speech and Signal Processing* (2010), pp. 5006–5009
19. S. Maity, A.K. Vuppala, K.S. Rao, D. Nandi, IITKGP-MLILSC speech database for language identification, in *IEEE National Conference on Communications (NCC)* (2012)
20. H. Ali, S.N. Tran, E. Benetos, Speaker recognition with hybrid features from a deep belief network. Neural Comput. Appl. **29**(6), 13–19 (2018)
21. K. Hwamin, P. Jeong-Sik, Automatic language identification using speech rhythm features for multi-lingual speech recognition. Appl. Sci. **10**, 2225 (2020)
22. M. Leen, B. Yegnanarayana, Prosodic features for language identification (2008), pp. 57–62
23. J. Fan, N. Zhou, J. Peng, L. Gao, Hierarchical learning of tree classifiers for large-scale plant species identification. IEEE Trans. Image Process. **24**(11), 4172–4184 (2015). https://doi.org/10.1109/TIP.2015.2457337
24. A.N. Younis, F.M. Remo, Distinguish musical symbol printed using the linear discriminant analysis LDA and similarity scale. Int. J. Comput. Appl. **179**(47), 20–24 (2018)

25. I.H. Sarker, A.S.M. Kayes, P. Watters, Effectiveness analysis of machine learning classification models for predicting personalized context aware smart phone usage. J. Big Data **6**(57), 01–28 (2019)
26. Y. Zhang, Support vector machine classification algorithm and its application, in *Information Computing and Applications. ICICA 2012.*, ed. by C. Liu, L. Wang, A. Yang. Communications in Computer and Information Science, vol. 308 (Springer, Berlin, Heidelberg, 2012). https://doi.org/10.1007/978-3-642-34041-3_27
27. S. Taneja, C. Gupta, K. Goyal, D. Gureja, An enhanced K-nearest neighbor algorithm using information gain and clustering, in *2014 Fourth International Conference on Advanced Computing & Communication Technologies*, Rohtak (2014), pp. 325–329. https://doi.org/10.1109/ACCT.2014.22.
28. A. Jamatia, A. Das, B. Gambäck, Deep learning-based language identification in English-Hindi-Bengali code-mixed social media corpora. J. Intell. Syst. **28**(3), 399–408 (2019). https://doi.org/10.1515/jisys-2017-0440

A Systematic Literature Review on Health Recommender Systems

Jagruti Prajapati and Keyur N. Brahmbhatt

Abstract In the present computerized world, medical care is one of the significant regions of the clinical space. A medical services framework is needed to break down a lot of patient information which assists with determining experiences and helps the prediction of diseases and treatments. Subsequently, this framework ought to be astute enough so as to anticipate a medical issue by dissecting a patient's way of life, physical wellbeing records, and social activities. Hence, health recommender systems (HRS) have developed as a driver for giving patient-driven customized medical care services. The principle target of recommender frameworks is to give recommendations dependent on recorded data on the clients' inclinations. In the case of HRS, rich health information is gathered through wearable gadgets and different sensors to survey the physical and mental conditions of the patient. Hence, to comprehend the best-in-class advancements in the medical care area, this paper gives an extensive overview of HRSs. This study additionally gives a comprehensive survey on the HRSs and the types of HRSs. At last, the open issues and challenges associated with the HRS are discussed.

Keywords Recommendation systems · Collaborative filtering · Health recommender systems · Deep learning

1 Introduction

Recommender systems (RSs) gather data on the preferences of their users for a lot of things (e.g., movies, songs, books, jokes, gadgets, applications, websites, travel destinations, and e-learning material). The data can be acquired explicitly (typically by collecting users' ratings) or implicitly (typically by observing users' behavior, for example, songs heard, applications downloaded, websites visited, and books read) [1, 2]. RS may use demographic features of users (such as age, nationality, gender). Social information, such as followers, followed, tweets, and posts, is commonly used

J. Prajapati (✉) · K. N. Brahmbhatt
Gujarat Technological University, Ahemdabad, India

F. Thakkar et al. (eds.), *Proceedings of the International e-Conference on Intelligent Systems and Signal Processing*, Advances in Intelligent Systems and Computing 1370,
https://doi.org/10.1007/978-981-16-2123-9_43

Table 1 Classification of recommender systems [5]

Strategies	Task	Output
Content-based filtering	General	Rating prediction
Collaborative filtering	Temporal	Item ranking
Hybrid methods	Sequential	

in recommendation techniques [4]. This part gives an outline of RSs from alternate points of view. As shown in Table 1 [5], the classification of RS mainly depends on recommendation strategies, tasks, and their respective output [1–5].

1.1 Classification by Strategies

Recommendation strategies can be classified into three categories: (1) Content-based filtering, (2) Collaborative filtering, and (3) Hybrid methods. More detailed descriptions of each category are as follows.

The main aim of the content-based filtering technique is to analyze features and attributes of products to generate predictions. Content-based filtering is usually utilized in document recommendations. In this filtering technique, user content determines recommendations. The user content deals with various attributes of items alongside the user's past purchasing history. Users give their preferences in terms of ratings, which are positive, negative, or neutral in nature. In this technique, the system recommends top-rated items to the user [1, 2]. Rather than thinking about features and attributes of items to decide their comparability, collaborative filtering (CF) utilizes user-based ratings to discover the similarity between items. Subsequent to gathering all user ratings, the system compares these ratings with other users with the help of a utility matrix and recommends top-rated items to the user. It uses various distance measures, for example, Jaccard's distance, cosine distance, Pearson's coefficient, and so forth to find a user's degree of similarity [1, 2]. This filtering method is normally used in e-commerce websites to recommend items based on users' ratings.

Generally, there are two sorts of CF-based strategies: Memory-based and Model-based. The previous one exploits original item interaction data (e.g., rating matrices) to predict unobserved ratings by accumulating the preferences of similar users or items. The latter assumes that the user preference or the item characteristics can be represented by low-dimensional latent vector. All the more precisely, model-based methods learn the feature of latent vectors of users and items from the user–item matrices and predict the recommendations by computing the dot product of latent vectors of the users and items [1, 2]. Observational investigations have demonstrated that model-based methods do better than memory-based ones much of the time. At last, hybrid technique consolidates the two strategies to reinforce the presentation and precision of the recommendation system. Hybrid filtering technique can be achieved by using any of the following ways: building a recommendation system that combines both of the above two techniques, applying some collaborative filtering in a content-based approach, and utilizing some content-based filtering in the collaborative approach [1–3].

1.2 Classification by Tasks

RSs are also classified based on task, for example, whether to consider time information (e.g., the order of historical interactions), RSs can be categorized by general, temporal, and sequential recommendation tasks. In case of general recommendation, it normally uses global user–item interaction data to recommend the top-N items for users. There are many algorithms, such as matrix factorization and its inferred models can successfully show user preferences, thus providing a static list of recommendations reflecting the long-term interests of each user [2, 5]. While temporal recommendation usually captures user preferences given a timestamp or a time-frame. More specifically, specific methods like TimeSVD++, which splits time into several segments, and model the user–item interactions in each segment to build an effective temporal RS [5]. The key is to model the dynamics of user preferences that exhibit significant (short- or long-term) temporal drift (e.g., "what users would like to have for Dinner" or "which places users would like to visit on weekends?"). On the other hand, Sequential recommendation (or next-item recommendation) predicts users' next preferences based on their most recent activities. In other words, sequential recommendation tries to display sequential patterns among successive items and generate well-timed recommendations for users.

1.3 Classification by Output

A further category is based on the form of outputs. Generally, there are two types of outputs of RS that is either predicted rating or ranked items. Based on this, it can be classified into two categories: rating-based and ranking-based item recommendation [1, 5]. Rating-based recommendation (rating prediction) predicts users' explicit preference scores towards items, which is usually considered as a regression task [2, 5]. Conversely, ranking-based recommendation (item ranking) focuses on the (relative) ranking positions of items and usually generates a top-N item recommendation list for each user.

2 Health Recommender System

Health information systems are turning into a significant stage for medical care administrations. In this specific situation, health recommender systems (HRSs) are introduced as complementary tools in decision-making processes in medical care administrations [6]. HRSs increase the usability of technologies and reduce information overload in processes. In this paper, a writing review was directed by following a survey strategy. Significant methodologies in HRS were plot and discoveries were examined. The paper presents current improvements in the market, challenges, and opportunities regarding HRS and emerging approaches.

2.1 The Main Determinants of Health

Wellbeing starts where we live, work, learn, and play some time before we visit the specialist's office or stroll in our nearby drug stores. Our health is largely determined by the social, economic, cultural, and physical environments we live in everything from where we work and live to our level of education and our access to healthy food and water [7, 8]. These conditions are known as social determinants of health: the social and financial assets, for example, lodging, instruction, food access, and transportation that help wellbeing and can decide length and quality of life. Over and over again, admittance to these wellbeing-deciding assets stays outside of people's control. Numerous health inequities experienced by communities of color, low-income individuals, and other vulnerable populations are due to the lack of appropriate distribution of social resources, rooted in a history of discrimination at the individual, institutional, and structural levels These social and financial elements significantly affect individual and community-level health. Fig. 1 shows how those factors impact health [8].

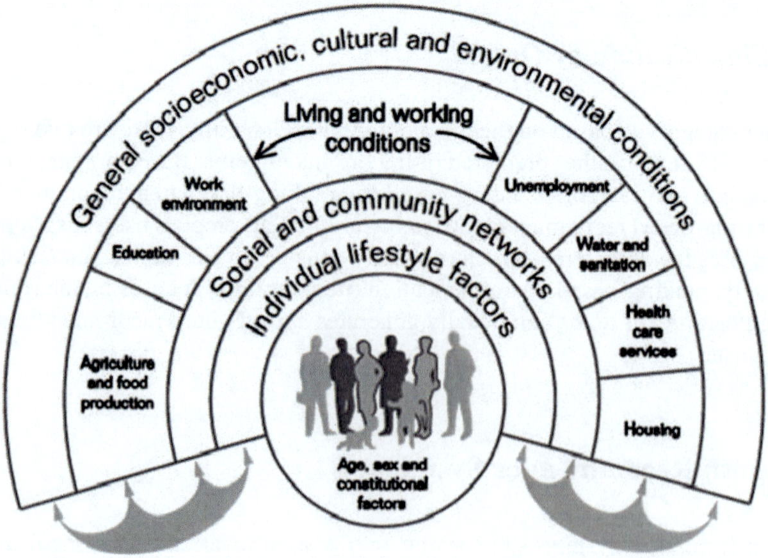

Fig. 1 The main determinants of health [8]

2.2 Architecture of HRS

The generalized architecture of healthcare RS is shown in Fig. 2 [6]. In this framework, various recommendations are produced for the users on the basis of diverse health factors and crucial clinical parameters. This design involves a medical care specialist organization, a physical director, a patient, a system administrator, and a user. The information created by these elements is kept up by a centralized server. This server also maintains and processes the recorded data in order to generate the recommendation. These recommendations are provided to the users through web portals, e-healthcare websites, healthcare applications, or mobile-based applications. The function of the system administrator is to develop the recommendation logics for generating the most relevant recommendation for the users. The web portal system is used by the users to transfers the health-related substance to the centralized server so that more reliable and accurate recommendations could be generated in the future [6].

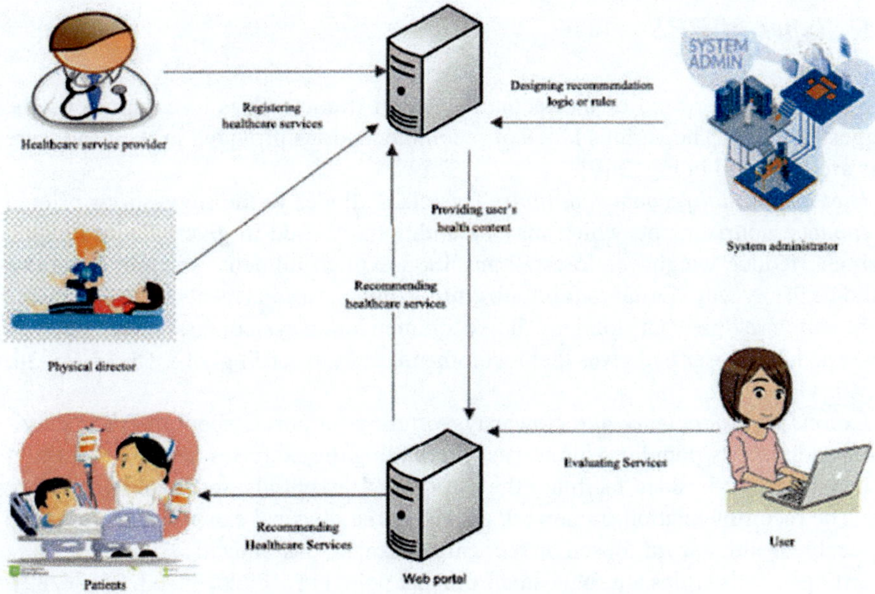

Fig. 2 Architecture of a generalized healthcare recommendation system [6]

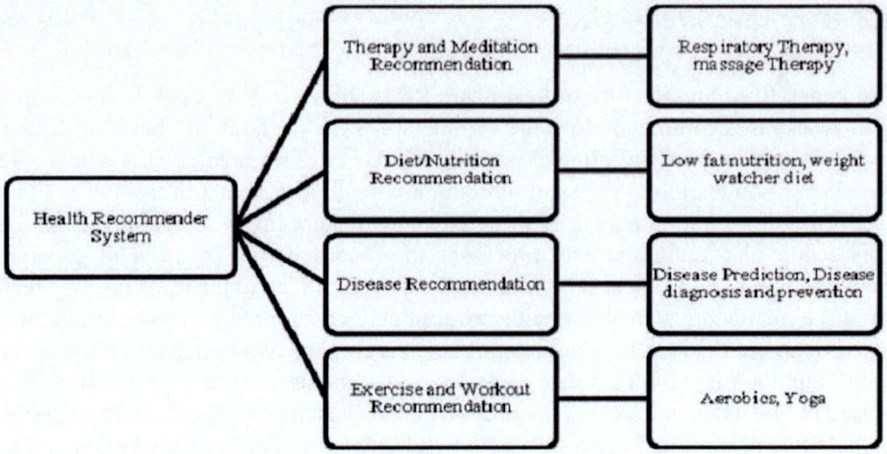

Fig. 3 Types of healthcare recommendation [9]

2.3 Types of HRS

The fundamental errand of the recommendation framework is to comprehend the suggested items. The various kinds of recommendations provided in the healthcare RSs are appeared in Fig. 3 [9].

Diet/nutritional recommendations: This class alludes to the suggestions offered to enhance nourishment, which may be additionally used to give balanced eating regimen, reduce weight, or forestall any food-explicit ailment. Recommendations could be in any way similar to food substitution items, various meals, or supplements to the eating regimen. Customized diet recommendation system dissects the day-by-day schedule of user and gives diet recommendations accordingly [6, 15, 17, 25, 34, 36, 40, 41].

Exercise/physical work out: Nowadays, offering proposals about physical activities is additionally popular. This assists with finding the exercises that are interesting or inspiring and besides facilitate the customers' essentials and requirements [6, 10]. The recommendation framework examines the physical exercise to foresee the applicable health hazard. Based on recognized danger, customized services for users regarding exercise rules are suggested from that point [11, 17, 25, 32, 33, 35, 36, 39].

Disease prediction/recommendations: In the earlier decade, the center has moved toward disease prediction, diagnosis, and prevention. For this reason, disease recommendations guarantee efficient decision-making [6, 11]. For representation, a smart recommender framework conjecture and gauge the malady hazard for cardiovascular breakdown patients [19, 24, 26, 28, 31, 38, 42].

Therapy/medication Recommendation: Providing recommendations with respect to the collection of possible choices that may be significant are likewise practical. A recommender structure could attest either persistent or potentially a specialist in looking through a patient-explicit treatment. This gives the best strategies to coordinate the client's wellbeing suggestions with a potential health treatment through reliable recommendations dependent on people's decision and wellbeing conditions [6, 11, 18, 22–24, 26, 30, 31, 38, 40, 42].

In light of the above conversation, the various sorts of healthcare recommendation frameworks have been examined in the current work that can be extensively arranged into content-based, collaborative filtering-based, and hybrid systems. These classified healthcare RSs are examined in the ensuing sections.

3 Related Work

We have considered the distinctive sort of healthcare recommendations in the above section. In this part, we will dissect the various methods used to distinguish medical problems and difficulties in an intelligent HRS in detail. This segment presents the key viable advancements, ideas, and the most famous strategies utilized in HRSs.

To begin with, Ezin and Kim [10] introduced 'Wellness that Fits', a model for suggesting physical exercise videos upon labeled video data from the Youtube-8M dataset. Likewise, they have coordinated essential content-based and collaborative filtering systems, in the proposed recommender model dependent on client profile and ongoing survey conduct. Moreover, an iterative substitution methodology motivated by neighborhood-collaborative filtering is acquainted with advance broadened recommendation records for users to improve with various sorts of fitness exercises as appeared in Fig. 4 [10].

Person's health equally depends on mental health. Therefore, Alqahtani and Al Khalifah [11] have examined the preliminary list of 437 mobile applications (258 applications from App Store and 179 applications from Google Play). For the examination, they have included applications whose primary objective as per the application's depiction and the demo of the application show that they are focused on psychological wellbeing, and applications that have more than five reviews (comments) altogether. Different Mobile applications were examined, which have demonstrated promising outcomes in individuals with emotional wellbeing issues to receive healthy lifestyles utilizing different convincing techniques. The majority of the mental health applications focused on a blend of emotional wellbeing issues which make it difficult to tell which powerful methodologies are more viable for a particular psychological health issue [11]. For example, Robin De Croon and Francisco Gutiérrez [28] have discussed the problem of stress at workplace and gave four solution, namely (1) personalized, in-time stress management therapies, (2) the application of HRSs to tailor interventions to employees, (3) the involvement of professional coaches to enable employees to explore recommendations outside therapy sessions, and (4) personalized gamification mechanics to increase employees' adherence to both recommended therapies and tracking data. They have proved with reasoned argument, that the combination of HRSs, the involvement of professional coaches, and the use of gamification have the potential to alleviate work-induced stress. In any

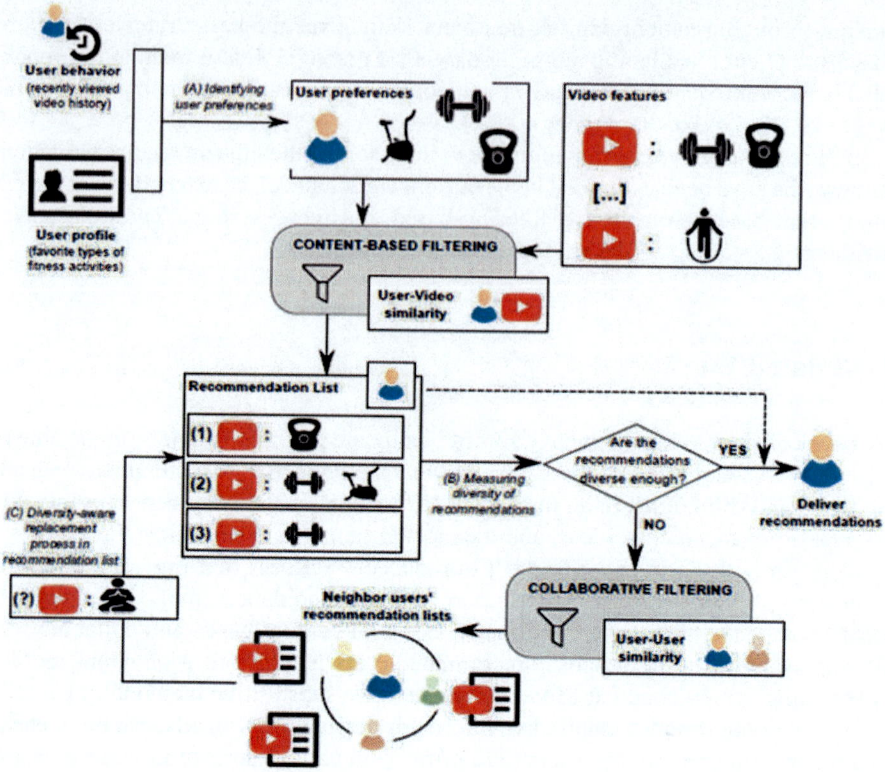

Fig. 4 Workout video recommendation [7]

case, personalization, self-observing, and reminder remain the most utilized enticing procedures in different mental health issues. Anxiety, stress, sadness, and general psychological wellness issues were the most issues the applications in this review target [11, 28].

M-Health applications are primarily intended for smart gadgets to give health-related administrations. Naseer and Jeon [12] have broken down different M-Health Applications, both Android and iOS-based, and proposed a complete M-Health model incorporated with a protected android application, secure cloud modes, and AI-based prescient examination for additional diagnosis. The information are gathered from sensor nodes and sent to nearby datasets through new innovations that empower cellular organizations and afterward store the information in the distributed cloud storage system. The information are gathered from cloud or clinical centers, for additional analysis [12]. Additionally, AI procedures are utilized for their exactness in the forecast of illness examination and have likewise been received for characterization.

One later and exhaustive investigation about the flow status of exploration in health recommender frameworks drops by Pincay and Teran [9]. In this work, the creators carried out a survey of significant HRSs distributions and characterized a multidisciplinary scientific categorization to evaluate four angles: domain, philosophy and methodology, wellbeing advancement techniques, and specialized perspectives. An aggregate of 890 publications was gathered, and after the assessment process, 19 research articles were chosen. With respect to field area, the creators found that the greater part of the investigations focused on nutrition and way of life.

Privacy is the major issue in HRS. Therefore, Sahoo and Pradhan [13], have compared different privacy-preserving collaborative filtering methods, along with the deep learning method and found that integration of an RBM (restricted Boltzmann machine) with CNN (convolutional neural network) in a deep learning environment provides a better recommendation quality for hospital recommendation to a patient. They have demonstrated that the proposed RBM–CNN has better accuracy of the HRS as compared to other profound methods. Deepa and Pandiaraja [16] have build up a Hybrid Context-Aware Recommendation System for E-Health Care (HCARS-EHC) using soft computing techniques. The proposed model has a new privacy-preserving phase to increase the security of the patient's reports as well as their ratings. Additionally, the model is efficient based on privacy preservation, recommendation, and ranking with less computation and communication complexity.

Another research article addresses privacy issue, which is proposed by Calero Valdez and Ziefle [14]. They have reviewed work done in the privacy-aware recommendation and personalized recommendation and found that despite the accessibility of technologies such as k-anonymity, differential privacy, privacy-aware recommendation techniques, and personalized privacy trade-offs, little exploration has been conducted on the users' willingness to share health data for utilization in such frameworks. In two conjoint-choice examinations (test size $n = 521$), they have explored the significance and utility of privacy-preserving techniques identified with sharing of individual wellbeing information utilizing k-anonymity and differential privacy techniques.

The large use of social networks provides opportunities for people to enhance their health knowledge and share their health-related information with others. Hence, Forouzandeh and Aghdam [15] developed the HRS for users of a Facebook social network site to evaluate their health status. Their methodology includes two phases, in the first phase, they have collected 1428 patients' data from 4 hospitals in Tehran and, in the second phase, RS provides some suggestions for users based on their current health status. The results show the effectiveness of suggested HRS in predicting the health status of Facebook users. Additionally, customizing healthy status and lifestyle for users using HRS can help them to stay in a healthy condition, reducing the costs of medicine. The outcome of this research can be beneficial for health organizations to enhance individuals' health knowledge in order to prevent diseases.

Because of the ongoing fashion trend of wearable sensors and smart gadgets with admittance to web-based media, migrating health services from the customary focus-based wellbeing framework to personal health care is unavoidable. Regarding this reality, Çelik Ertuğrul and Elçi [17] have reviewed many research articles and discovered that different AI, machine learning, data mining, information retrieval, graph theory, heuristics, and probabilistic approaches utilized in HRSs. Various

Personal HRSs were evaluated, examined, looked at, and modeled on clustering, classifying and regression, graph theory, knowledge-based, context-aware based, similarity matching, and matrix factorization techniques.

With the advancement of internet business, intelligent recommender framework has carried new chances to customized wellbeing observing for the clients with emotional distress. Therefore, Yang and Zhou [18] set forward the emHealth framework, which is a clever health recommendation system with depression prediction for emotion health. Framework design of emHealth is appeared in Fig. 5. They have planned customized cell phone Apps to gather emotional data of users with tendentious depressive mood and found the five principle external characteristics of depression by Pearson correlation analysis. They have partitioned 1047 volunteers' information into training set and test set and built a prescient model of predictive model of depression using decision tree and support vector machine algorithms. emHealth framework, which is an intelligent HRS for patients with emotional distress. The framework utilizes smart cell phones to gather and investigate client information whenever and anyplace, and anticipate emotional ailments, at that point give successful intercession measures to the treatment and medical services of patients.

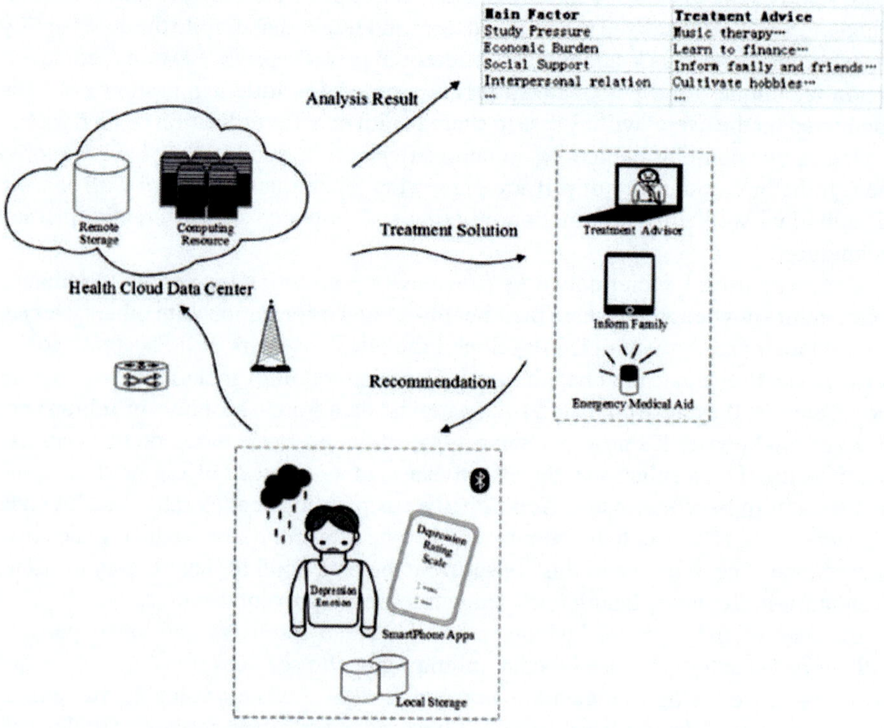

Fig. 5 System architecture of emHealth [18]

With the considerable advancement in Health Domain drug and the clinical test is also recommended by HRS. Komal Kumar and Vigneswari [22], have developed drug recommendation recommendations for patients suffering from various diseases. The recommendation model was developed using different supervised machine learning approaches such as Support Vector Machine (SVM), Random Forest, Decision Tree, and K-nearest neighbors. Out of all methods, Random forest achieved good recommendation accuracy. The experiment was carried out on sample dataset which was created for testing purpose only.

As there is group recommendation in e-commerce application similar way there is also group recommendation in health domain, For example, Maria Stratigi and Haridimos Kondylakis [20] have proposed a novel semantic similarity function, which uses health information of patients and additional information such as the education level, the health literacy, and the psycho-emotional status. The collaborative filtering method has been used to build the multi-dimensional group recommendation model. The model is able to generate fair recommendations to a small group of patients.

Most of the investigations utilized content-based techniques and similarity analysis to assemble their recommendation engine, which is somewhat common while implementing RSs [25, 29]. Nevertheless, a noteworthy level of the activities utilized artificial intelligence, fuzzy logic, and machine learning techniques as a method of accomplishing more significant levels of exactness and performance of the frameworks [26, 27].

In summary, there can be different ways to categorize HRSs from various perspectives; one of the ways is as shown in Table 2.

Table 2 Summary of health recommender systems

Recommendation type	Method used
Diagnosis and medication (treatment, drug, clinical test, decision, medical report)	Random forest [22], deep Learning (DL) [23], machine Learning(ML) [24, 40], fuzzy C-means clustering [26], Hybrid RS + ML + (Natural Language Processing) NLP [30], NLP + knowledge-based RS [31], cognitive computing [37], collaborative filtering [38, 42]
Healthcare services (personalized healthcare services, hospital, health apps)	Particle agent swam optimization [32], Partially penalized regression [33], ML + DL [35], Recurrent neural network (RNN) [39]
Wellness (diet, food recommendation, yoga, meditation, exercise recommendation)	Content-based RS + ML [25, 34], reinforcement learning [36], classification methods [40], RNN + (convolution neural network) CNN[41]

4 Deep Learning for HRS

There are numerous reasons for applying deep learning strategies to health recommender frameworks. It is clear from Table 2 that various deep RSs have been proposed in a short span of several years. The field is for sure clamoring with development. Additionally, deep neural networks are likewise composite as numerous neural structures can be formed into a single differentiable function and trained end to end. The key favorable position here is when managing content-based suggestions. This is unavoidable when modeling users/items on the web, where multi-modular information is basic spot. For example, when managing textual information (reviews, tweets, and so forth.), image data (social posts, item pictures), CNNs/RNNs become key neural structure blocks [17, 21]. Here, the conventional option turns out to be fundamentally less appealing, and thus, the recommender framework can't exploit joint (start to finish) representation learning [21, 29]. In some sense, improvements in the field of recommender frameworks are likewise firmly combined with propelling research in related modalities, (for example, vision or language networks). For instance, to deal with reviews, one would need to perform exorbitant preprocessing (e.g., key expression extraction, theme displaying, and so forth.) while more up-to-date deep learning-based approaches can ingest all textual data start to finish.

All things considered, the capacities of deep learning in this viewpoint can be viewed as outlook changing and the ability to speak to pictures, text, and communications in a bound together joint system is unimaginable without these ongoing advances. The scientific classification of deep learning strategies utilized in HRS is appeared in Fig. 6.

5 Open Issues and Challenges

In this comprehensive study, we reviewed recent developments in healthcare recommendation. Notwithstanding all the advancements, there are numerous challenges that ought to be taken care of in a viable way for the effective design of healthcare system. A portion of the proposed open issues and challenges are recorded as underneath.

Integration of electronic clinical records: The information identified with the clinical records is gathered from various sources like patients, sensors, wearable gadgets, implantable types of gear, IoT gadgets, and some more. The mix of the information gathered from these different sources into a typical electronic wellbeing record is a challengeable issue. The whole exhibition of any medical services arrangement relies carefully upon the information.

Information assortment and monitoring frameworks: The most troublesome assignment is the data collection, observing, and feedback systems on the off chance that these are ineffectively planned and actualized, at that point it will be extremely difficult to analyze the problem.

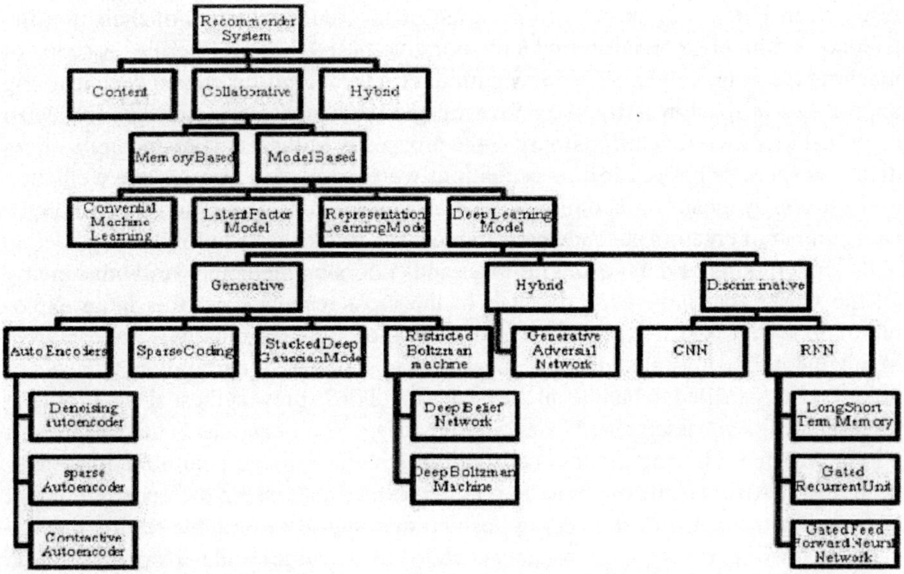

Fig. 6 Deep learning techniques used in health recommendation systems

Convincing patients that there is an issue and the arrangement picked is correct: It is a challenge to persuade the patients about the genuine issue and their answers utilizing a recommender framework. This is because of the absence of trust or mindfulness about the applicability of health care systems.

Moral difficulties: A serious deal is the moral/ethical difficulties in healthcare domain. To assemble the agreeableness of a brilliant medical care framework, initially, fundamental methodology ought to be straightforward and trustworthy. Medical delegates ought to comprehend that the advancement of any association legitimately depends on individual moral conduct.

Accuracy of health recommendations: Accuracy of results in health recommender framework is the most significant viewpoint than anything since it is answerable for tolerant' life. It is significant in the medical domain that outcomes ought to be extensively exact and near-real conditions and this can be conceivable if the colossal informational collection of data will be accessible to foresee precise recommendations.

6 Conclusion and Future Work

This paper adds to the present state-of-art review on RSs used in the health domain, also known as HRSs. Different research articles have been assessed by examining significant ideas and methods applied in HRSs. Based on this survey of recent research in HRS, it is classified into three main groups in terms of their filtering techniques namely, content-based, collaborative-based, hybrid filtering. A scope of machine learning, data mining, information retrieval, cloud-based and deep learning approaches utilized in HRSs were investigated under these principle groups. Most examinations were focused to improve the prosperity of the users, by recommending diets and exercise plans. Most investigations were targeted to improve the wellbeing of the users, by means of recommending diets and exercise plans. However, a significant number of creators put endeavors in planning frameworks fit for helping doctors in the undertaking of diagnosing illnesses and endorsing medicine. Notwithstanding all the works that have been directed in the most recent years, this is an exploration zone that is still at its outset and more thorough examinations are required. Managing wellbeing-related issues is perplexing and there is certainly not an exceptional strategy skilled to handle all the issues that HRSs present these days. From the research works considered and from our point of view, we can reason that we are still a long way from having a completely utilitarian and reliable health recommender framework. Also, efforts ought to be pointed toward addressing the uncertainty that making health-related decisions implies and defining more suitable testing mechanisms that will permit to appropriately evaluate the accuracy and performance of the implementations.

References

1. C.C. Aggarwal, *Recommender Systems: The Textbook* (IBM T.J. Watson Research Center: Springer, 2016). https://doi.org/10.1007/978-3-319-29659-3.
2. L.R. Francesco Ricci, *Recommender Systems Handbook* (2nd edn). (Springer Science+Business Media, New York, 2015). https://doi.org/10.1007/978-1-4899-7637-6.
3. A. Taneja, A. Arora, Recommendation research trends: review, approaches and open issues. Int. J. Web Eng. Technol. **13**(2), 123–186 (2018). https://doi.org/10.1504/IJWET.2018.092831
4. M. Eirinaki, J. Gao, I. Varlamis, K. Tserpes, Recommender systems for large-scale social networks: a review of challenges and solutions. Future Gener. Comput. Syst. **78**(1), 413–418 (2018). ISSN 0167–739X, https://doi.org/10.1016/j.future.2017.09.015.
5. Z. Sun et al., Research commentary on recommendations with side information: a survey and research directions. Electron. Commer. Res. Appl. **37**, 100879 (2019). https://doi.org/10.1016/j.elerap.2019.100879
6. D. Sharma, G. Singh Aujla, R. Bajaj, Evolution from ancient medication to human-centered Healthcare 4.0: A review on health care recommender systems. Int. J. Commun. Syst. 1–40 (2019). https://doi.org/10.1002/dac.4058.
7. S. Hooshmand, Factors influencing health outcomes across the least, average and healthiest states in America. Diss. Abstr. Int. Sect. A Humanit. Soc. Sci. **79**(4), (2017). https://doi.org/10.25777/sqdq-xm77.
8. B. Loring, A. Robertson, Obesity and inequities: Guidance for addressing inequities in overweight and obesity, World Heal. Organ. Eur. 1–6 (2014). ISBN 978 92 890 5046 3.
9. J. Pincay, L. Teran, E. Portmann, Health recommender systems: A state-of-the-art review, in *2019 6th International Conference eDemocracy eGovernment, ICEDEG 2019* (2019), pp. 47–55. https://doi.org/10.1109/ICEDEG.2019.8734362.
10. E. Ezin, E. Kim, I. Palomares, 'Fitness that Fits': a prototype model for workout video recommendation. CEUR Workshop Proc. **2216**, 40–45 (2018)

11. F. Alqahtani, G. Al Khalifah, O. Oyebode, R. Orji, Apps for mental health: an evaluation of behavior change strategies and recommendations for future development. Front. Artif. Intell. **2**, 1–11 (2019). https://doi.org/10.3389/frai.2019.00030.
12. K. Naseer Qureshi, S. Din, G. Jeon, F. Piccialli, An accurate and dynamic predictive model for a smart M-Health system using machine learning. Inf. Sci. (Ny). **538**, 486–502 (2020). https://doi.org/10.1016/j.ins.2020.06.025.
13. A.K. Sahoo, C. Pradhan, R.K. Barik, H. Dubey, DeepReco: deep learning based health recommender system using collaborative filtering. Computation **7**(2), (2019). https://doi.org/10.3390/computation7020025.
14. A. Calero Valdez, M. Ziefle, The users' perspective on the privacy-utility trade-offs in health recommender systems. Int. J. Hum. Comput. Stud. **121**, 108–121 (2019). https://doi.org/10.1016/j.ijhcs.2018.04.003.
15. S. Forouzandeh, A.R. Aghdam, Health recommender system in social networks: a case of Facebook. Webology **16**(1), 40–54 (2019). https://doi.org/10.14704/web/v16i1/a178
16. N. Deepa, P. Pandiaraja, Hybrid context aware recommendation system for E-Health care by Merkle hash tree from cloud using evolutionary algorithm. Soft Comput. **24**(10), 7149–7161 (2020). https://doi.org/10.1007/s00500-019-04322-7
17. D. Çelik Ertuğrul, A. Elçi, A survey on semanticized and personalized health recommender systems. Expert Syst., 1–23 (2019). https://doi.org/10.1111/exsy.12519
18. S. Yang, P. Zhou, K. Duan, M.S. Hossain, M.F. Alhamid, emHealth: towards emotion health through depression prediction and intelligent S. Forouzandeh and A. R. Aghdam, health recommender system in social networks: a case of Facebook. Webology **16**(1), 40–54 (2019). https://doi.org/10.14704/web/v16i1/a178
19. Health Evidence Network (HEN), What are the main factors that influence the implementation of disease prevention and health promotion programmes in children and adolescents? (2005) pp. 1–18.
20. M. Stratigi, H. Kondylakis, K. Stefanidis, Multidimensional group recommendations in the health domain. Algorithms **13**(3), 1–21 (2020). https://doi.org/10.3390/a13030054
21. S. Zhang, L. Yao, A. Sun, Y. Tay, Deep learning based recommender system: a survey and new perspectives. ACM Comput. Surv. **52**(1), 1–35 (2019). https://doi.org/10.1145/3285029
22. N. Komal Kumar, D. Vigneswari, A drug recommendation system for multi-disease in health care using machine learning, in *Advances in Communication and Computational Technology. Lecture Notes in Electrical Engineering*, vol 668, ed. by G. Hura, A. Singh, L. Siong Hoe (Springer, Singapore, 2021). https://doi.org/10.1007/978-981-15-5341-7_1
23. M.M. Islam, H.-C. Yang, T.N. Poly, Y.-C.J. Li, Development of an artificial intelligence-based automated recommendation system for clinical laboratory tests: retrospective analysis of the national health insurance database, JMIR Med. Inform. **8**(11), e24163 (2020). https://doi.org/10.2196/24163.
24. P. Scott, E. Andrikopoulou, H. Nakkas, P. Roderick, Exploring the Role of the Care and Health Information Exchange (CHIE) in Clinical Decision-Making : A Realist Evaluation (2020), pp. 1–20.
25. S. P. Rana, M. Dey, J. Prieto, S. Dudley, Content-based health recommender systems. Recomm. Syst. Mach. Learn. Artif. Intell., 215–236 (2020). https://doi.org/10.1002/9781119711582.ch11.
26. U. Mittal, N. Singla, G. Gupta, Part 5 HEALTHCARE RECOMMENDER Case Study 1 : Health Care Recommender Systems. https://doi.org/10.1002/9781119711582.ch17
27. A. Myren, P.S. Neto, *Evaluation of Machine Learning Algorithms in Recommender Systems* (2017), pp. 1–81.
28. R. De Croon, F. Gutiérrez, K. Verbert, Opportunities for recommended mental health strategies to reduce stress at work, in *Human Computer Interaction and Emerging Technologies: Adjunct Proceedings from the INTERACT 2019 Workshops*, ed. by F. Loizides, M. Winckler, U. Chatterjee, J. Abdelnour-Nocera, A. Parmaxi (Cardiff: Cardiff University Press, 2020), pp. 157–162. https://doi.org/10.18573/book3.t

29. J. Su, Y. Guan, Y. Li, W. Chen, H. Lv, Y. Yan, Do recommender systems function in the health domain : a system review. 92
30. Quintana-Wong, C., García, L., Rabanillo, A., Guillot, J., Amable, Z., *Recommender System in a Transactional Analytical Solution for Health Care and Health Promotion*, vol. 1 (2020), pp. 96–107
31. Wang Z, Huang H, Cui L, Chen J, An J, Duan H, Ge H, Deng N, Using natural language processing techniques to provide personalized educational materials for chronic disease patients in China: development and assessment of a knowledge-based health recommender system. JMIR Med. Inform. **8**(4), e17642 (2020). https://doi.org/10.2196/17642
32. R. Manikandan, V. Saravanan, A novel approach on Particle Agent Swarm Optimization (PASO) in semantic mining for web page recommender system of multimedia data: a health care perspective. Multimed. Tools Appl. **79**(5–6), 3807–3829 (2020). https://doi.org/10.1007/s11042-018-7141-8
33. E.J. Oh, M. Qian, K. Cheung, D.C. Mohr, Building health application recommender system using partially penalized regression, in *Statistical Modeling in Biomedical Research. Emerging Topics in Statistics and Biostatistics*, ed. by Y. Zhao, D.G. Chen (Springer, Cham, 2020). https://doi.org/10.1007/978-3-030-33416-1_6
34. N. Bagaria, F. Laamarti, H.F. Badawi, A. Albraikan, R. Martinez Velazquez, A. El Saddik, Health 4.0: digital twins for health and well-being, in *Connected Health in Smart Cities*, ed. by A. El Saddik, M. Hossain, B. Kantarci (Springer, Cham, 2020). https://doi.org/10.1007/978-3-030-27844-1_7
35. J. Saha, C. Chowdhury, S. Biswas, Review of machine learning and deep learning based recommender systems for health informatics, in *Deep Learning Techniques for Biomedical and Health Informatics. Studies in Big Data*, vol. 68, ed. by S. Dash, B. Acharya, M. Mittal, A. Abraham, A. Kelemen, (Springer, Cham, 2020). https://doi.org/10.1007/978-3-030-33966-1_6
36. J. Mulani, S. Heda, K. Tumdi, J. Patel, H. Chhinkaniwala, J. Patel, Deep reinforcement learning based personalized health recommendations, in *Deep Learning Techniques for Biomedical and Health Informatics. Studies in Big Data*, vol. 68, ed. by S. Dash, B. Acharya, M. Mittal, A. Abraham, A. Kelemen (Springer, Cham, 2020). https://doi.org/10.1007/978-3-030-33966-1_12
37. L. Terán, J. Pincay, D. Pacheco, M. Štěpnička, D. Simancas-Racines, Health recommendation system framework for the optimization of medical decisions, in Cognitive Computing. Edition Informatik Spektrum, ed. by E. Portmann, S. D'Onofrio (Springer Vieweg, Wiesbaden, 2020). https://doi.org/10.1007/978-3-658-27941-7_11
38. V. Frittelli, M.J. Diván, An architecture for e-health recommender systems based on similarity of patients' symptoms, in *Blockchain Technology for Smart Cities. Blockchain Technologies*, ed. by D. Singh, N. Rajput, (Springer, Singapore, 2020). https://doi.org/10.1007/978-981-15-2205-5_8
39. D.C. Edara, V. Sistla, V.K. Kishore Kolli, Health app recommendation system using ensemble multimodel deep learning. J. Eng. Sci. Technol. Rev. **13**(5), 7–19 (2020). ISSN: 1791–2377. https://doi.org/10.25103/jestr.135.03
40. N. Kim, S. Kim, Adaptive recommendation system for health screening based on machine learning **9**(2), 1–7 (2020)
41. R.Y. Toledo, L. Martinez, A health-awareness nutrition recommender system, in *Proceedings of the IEEE 14th International Conference on Intelligent and System Knowledge Engineering ISKE 2019*, (2019), pp. 36–42. https://doi.org/10.1109/ISKE47853.2019.9170432.
42. H. Jiang, L. Nie, W. Wang, L. Y. Duan, M. Liu, C. Xu, Market2Dish: a health-aware food recommendation system, in *MM 2019—Proceedings of the 27th ACM International Conference on Multimedia* (2019), pp. 2188–2190. https://doi.org/10.1145/3343031.3350594.

DNA Cryptography-Based Speech Security System

Shaikh Akib Shahriyar, Shovan Bhowmik, and Mahedi Hasan

Abstract Speech is one of the most vital forms of communication in today's age. We communicate with others through speech in the form of audio calls over the internet or mobile networks. While communicating with others, it is imperative to transfer speech securely. This paper proposes a novel Speech Security System (SSS) which leverages DNA cryptography to provide security to the transmitted speech signal. Our proposed method encrypts the speech signal into a DNA strand which works as a camouflage to the original speech signal. The DNA strand is decrypted at the receiver end and the original speech signal is retrieved. The proposed SSS is significantly robust against any cryptanalytic, statistical, and brute force attacks by an intruder and thus improves the security of the speech signals.

Keywords Speech encryption · DNA sequence · DNA cryptography · Mobile speech · Speech security system

1 Introduction

One of the commonly used means of communicating is speech communication. Yet even with the massive availability, the risk has increased making it more vulnerable. So, the importance of taking the appropriate measures to provide high-level protection to speech signal has increased. In recent days, it is very easy to interrupt/eavesdrop on any GSM-based voice call. It can be done by the controlling authority or hackers with very little effort. There are several security algorithms for GSM networks, such as A3, A5, and A8 [1]. The A3/A8 algorithm is applied in the SIM cards and the GSM network. The A5 encryption algorithm is used to scramble the speech of user and data traffic between the handset and the base station [1]. This algorithm is implemented to establish privacy on both the phone and the base station subsystem. There are also other methods for voice encryption, too. But there lies a danger of eavesdropping when it is all about speech transmission. In this paper, therefore, a

S. A. Shahriyar (✉) · S. Bhowmik · M. Hasan
Khulna University of Engineering & Technology, Khulna 9203, Bangladesh
e-mail: akib.shahriyar@cse.kuet.ac.bd

F. Thakkar et al. (eds.), *Proceedings of the International e-Conference on Intelligent Systems and Signal Processing*, Advances in Intelligent Systems and Computing 1370,
https://doi.org/10.1007/978-981-16-2123-9_44

reliable way of transferring speech into a secured form is proposed with the help of DNA cryptography. At first, it senses the signal and transforms it into a digital signal. Then, using DNA strands, this signal is encrypted and secured.

Speech enhancement measures are also taken prior to transferring the signal over any network. Shahriyar et al. used deep learning network to enhance the speech quality [2]. Deep learning is a subset of machine learning and artificial intelligence that is applied in multiple kinds of applications such as computer vision, cryptography, and computer security [3, 4].

Having some levels of cryptography to hide and secure handy information is a cherished task. Roy et al. provided a secure way to transmit text messages using a chaotic map and DNA encoding in [5]. Recently, a new field of DNA computing has been initiated. While the DNA-based algorithm remains primitive, it has created lots of prospects. The background of DNA computing is not so old but in this early stage, it has conquered huge elevation in technology. This DNA algorithm was first pioneered in the 1990s [6]. The concept to resolve the directed Hamiltonian Path Problem (Travelling Salesman Problem [7]) of size n in Q(n) by utilizing DNA molecules was developed by Adelman [6]. The practice of enzymes for emulation of easy computations and the coding of nodes and edges which hold information in DNA bundles follow the fundamentals of Adelman [8].

DNA computing is quite difficult to apply effectively in the field of Information Security [9–12]. It provides a significant complexity and arbitrary nature. As a result, it offers substantial apprehension that assists to encode data into DNA format superior to other cryptographic techniques. This area is still in emerging development. Different applications concerning this method are yet to perceive. The works related to DNA cryptography are in their beginnings and still there many issues needed to be answered. Since it delivers broad symmetry, good energy cost-effectiveness, and increased knowledge density found in DNA molecules, DNA cryptography has a unique benefit over conventional types of cryptography.

In this paper, a secured encryption and decryption method based on DNA cryptography is suggested. An arbitrary DNA segment will be taken as source sequence P which can be obtainable from a public database. This DNA strand will be familiar to the transmitter and the receiver. The source sequence P is integrated with the transmitter's hidden information M and then reconstructs into a new encrypted signal P'. This signal P' along with the received string is inspected and resolved. Eventually, the hidden information M is retrieved from the encrypted signal P'.

2 DNA Encoding

The abbreviation of DNA is Deoxyribonucleic Acid. DNA is the genetic structure that carries the genetic information of every living creature on the earth. It coheres to every compulsory "message" related to alive cells. Each distinct alive thing involving human holds an entire collection of DNAs. Yet, it varies wholly in carriers for their special attributes. Most notably, the DNA is a particular form of plastic consisting of several monomers known as deoxyribonucleotides. These nucleotides in addition are composed of deoxyribose sugar, a phosphate group, and a nitrogenous base. These nitrogenous bases are separated into two vast categories, namely, purins (A, G) and pyrimidines (C, T). Purin encompasses Adenine and Guanine. On the other hand, pyrimidine includes Cytosine and Thymine. A, G, C, and T are used to denote these four bases. DNA composition is devised by tying up A with T and G with C [8]. There have been various kinds of processes that consider DNA string into account and fabricate electrophoresis which means a fixed type of chain polymerase reaction.

DNA string has an exceptional characteristic. One cannot differentiate a real DNA string from a dupe one. This exceptional characteristic has been used and drawn the best of it in this research. Till now, countless amount of DNA strings has been gauged approximately 55 million [13] which can be openly accessed.

3 DNA Cryptography-Based Speech Security System

The proposed speech encryption process takes each speech signal and encodes the signal into a DNA sequence. For proper encryption, using DNA cryptography, every signal is transformed into a binary data format. The proposed system block diagram is given in Fig. 1. The system is split into two stages: Encryption and Decryption.

In this work, two confidential architectures are considered between two groups, a transmitter, and a receiver. In the beginning, a basic binary-coded statement is employed. This statement encoded the alphabetic A, C, G, and T into corresponding binary digits. As an instance, for this proposed system: 00, 01, 10, 11 commensurate to A, C, T, and G, respectively. In this case, two digits can be regarded for running the process. Next, for doing the reverse operation, a pairing regulation has been utilized. Each notion from A, C, G, and T is bounded with their reverse in the mentioned architecture represented as an (x) where $\alpha(\alpha(x)) \neq x$. The aforementioned regulation

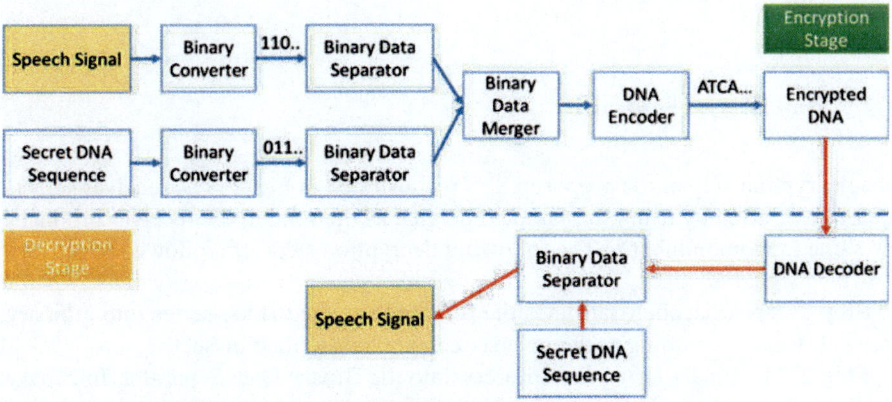

Fig. 1 Block diagram of speech security system using DNA cryptography

can be labeled as (A $\rightarrow$ C), (C $\rightarrow$ G), (G $\rightarrow$ T), and (T $\rightarrow$ A). Users can alter or improve it based on their requirements. It is also presumed that the secret message M is a binary string of the speech signal accepted over GSM.

3.1 Speech Encryption

The proposed Speech Security System (SSS) in Fig. 1 can encrypt any speech signal into an encrypted DNA sequence. From an available 55 million DNA sequences, a secret DNA sequence P is chosen that works as a secret key for both the receiver and the sender. The secret key of both the receiver and sender has a rich random number generator as a part where both know the seed value. The value of segment length "n" is produced by a random number generator. This length is used by Binary Data Separator. For encrypting a speech signal, the below steps are sequentially followed.

Step 1. Binary Converter transforms the selected speech signal M in a binary sequence that we need to send to the receiver (using the pyAudioAnalysis tool [14]).

Step 2. According to the DNA Encoding system described in Sect. 3, the secret DNA sequence P is transformed into a binary sequence.

Step 3. The binary signal generated from M is split into a fixed set of segments where each segment length is n. Then, to the start of every segment, one '0' is added. If the last segment length becomes less than n, k number of zeroes are padded to that segment. If no zero is added, then $k = 0$.

Step 4. The binary signal generated from P is split into a fixed set of segments. Each segment length is n.

Step 5. The Binary Data Merger helps to merge each ith segment of P after each ith segment of M. As a result, a mixed binary sequence is generated.

Step 6. The binary sequence is encoded into DNA series by the DNA encoder. So encrypted DNA sequence is formed.

Step 7. Then the receiver receives an encrypted DNA sequence and the secret key is (P, n, k).

3.2 Speech Decryption

The decryption step of the proposed SSS is illustrated in Fig. 1. The receiver selects the same P. Then by using the same seed value as the sender, the receiver produces the same random number n. The following decryption steps are followed to retrieve the voice signal M.

Step 1. DNA decoder translates the received encoded DNA series into a binary series. It works according to the reverse technique described in Sect. 3.

Step 2. The binary string is reinforced into the Binary Data Separator. In Binary Data Separator, the inputs are the confidential DNA string P and arbitrary number n. It transforms this string into binary series and disparate into an equal number of groups as happened in the encryption portion. If $k \neq 0$, then k number of zeroes are truncated from the end of the last segment of the data.

Step 3. The segments are the size of $2n$. For each ith segment of the sequence, the first n bits are truncated which denotes the ith part of M and the last n bits are matched with the ith segment of P. The method is followed iteratively to eliminate the segments of P from the whole chain and to obtain segments of M.

Step 4. From each segment of M, '0' is truncated from the beginning. After truncating and concatenating all the segments, the expected speech signal M is obtained.

4 A Simplified Example

Let's consider a speech signal of length 1 s from the National Anthem of Bangladesh "**Amar Shonar Bangla**" [15] where the sample rate of the audio is 8 kHz. "pyAudio-Analysis" tool is used to extract the binary form of the speech signal. The resultant binary sequence is stored in M. Multiple colors are used in this example for easier demonstration.

4.1 Encryption Stage

The encryption procedure of the binary string is given below.

Step 1.

M = **1110011110010110000001011000001110010001**

and let's consider

P = TCCT AAAC GCTG CAAG ATTC

Step 2.

After converting P into Binary form,

P = 10010110 00000001 11011011 01000011 00101001

A random number n is generated, where $n = 4$. As the length of M is 40, the number of segments would be 10. So, k = 0.

Step 3.

The segments of M are

01110	00111	01001	00110	00000
00101	01000	00011	01001	00001

Step 4.

Segments of *P*

1001	0110	0000	0001	1101
1011	0100	0011	0010	1001

Step 5.

Merged Binary Data =
01110100100111011001001000000110000100000110100101101101000010000011001101001001000001 1001

Step 6.

The encrypted DNA sequence is = CGCA TCGC TCAT AACT ACAA CTTC CTGC AATA AGAG CATC AACTC

.

4.2 Decryption Stage

The decryption procedure of the encrypted DNA string is given below.

Step 1.

The binary form of the encrypted DNA string is

0111010010011101100100100000011000010000011010010110110100001000 0011001101001001000001 1001

Step 2.

Segments of *P*

1001	0110	0000	0001	1101
1011	0100	0011	0010	1001

Step 3.
The overall segments are

011101001	001110110	010010000	001100001	000001101
001011011	010000100	000110011	010010010	000011001

The retrieved segments of *M*

01110	00111	01001	00110	00000
00101	01000	00011	01001	00001

Step 4.
After removing '0' from each segment

$M =$ 11100111100101100000010110000011100 10001

which matches exactly to our input speech signal.

5 Experimental Results

The proposed SSS encrypts any form of the speech signal into an encrypted DNA sequence. The SSS has been implemented in Python 3.6 in a Linux environment (Ubuntu 16.10) with 8 GB of RAM. The proposed SSS has been rigorously tested with different kinds of audio signals that are usually transferred over the cellular network. Any encryption method should be robust enough to withstand different kinds of cryptanalytic attacks. Thus, thorough cryptanalysis of SSS is performed in this section.

5.1 Key Space Analysis

The significance of any cryptographic encryption algorithm depends largely on the vastness of the keyspace. As the DNA sequence needed for encryption of any audio signal can be sampled from 55 million samples, the search space is quite huge. Also, arbitrary DNA sequences can be constructed based on the preference of the sender and receivers. Thus, it is hard for an attacker to retrieve the DNA sequence used for encryption by brute force.

5.2 *Histogram Analysis*

Speech signals can be analyzed to understand the underlying distribution of samples at different time steps. Thus, it is also necessary to evaluate how much the encrypted signal is different from the original signal. In Fig. 2, we can observe the histograms of the original input signal (from Sect. 4), the encrypted signal by our proposed SSS, the decrypted signals. As Fig. 2 suggests, the histogram of the encrypted signal is significantly different than the original input signal. The encryption done by the proposed SSS has completely changed the input signal to a degree that no statistical attack can be performed on the proposed SSS.

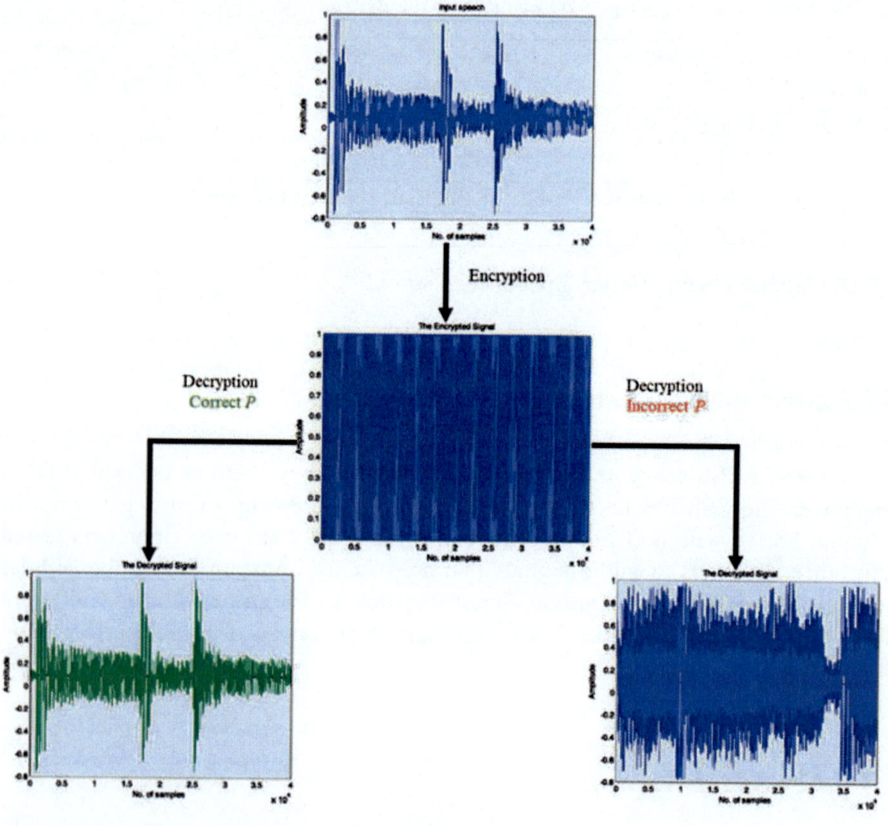

Fig. 2 Key sensitivity of the proposed speech security system

5.3 Key Sensitivity Analysis

Having a high key sensitivity is a core requirement of any cryptography-based security system. It refers to the process that the encrypted signal cannot be decrypted accurately even if there is a small difference between the encryption and decryption keys. Our proposed SSS is thoroughly tested with DNA sequences (*P*) which are different from the correct DNA sequence only by a single character. Figure 2 shows the key sensitivity of our proposed system where it is evident that our proposed SSS is tolerant of these slightly changed DNA sequences (incorrect *P*) and the original input speech signal could not be retrieved by any of them. The original input signal can be retrieved only by the correct *P*, which is depicted as a green histogram in Fig. 2.

5.4 Limitations and Future Scope

The DNA encoding scheme used in this paper is simple, yet, it has some limitations. As the random number generator depends on a specific seed value, it has to be shared as prior knowledge to both sender and the receiver. This can be a bottleneck to our proposed system. Although, it can be remedied by using public-key cryptography systems to transfer that seed value from sender to receiver.

Our proposed speech encryption system performs relatively well in accordance with the security requirements of a cryptographic system. Thus, it can create a great impact on securing the transmitted voice-over mobile networks or other forms of speech signal sharing platform.

6 Conclusion

Providing security to speech signals is a hefty task. The proposed SSS provides a robust method to secure speech signals. The DNA cryptographic scheme plays a vital role in the proposed SSS. It is a simple, yet effective method. Our proposed SSS has a set of significant features which helps it to stand out among other speech encryption system. Through rigorous cryptanalysis, the proposed SSS stands strong against different cryptanalytic, statistical, and brute force attacks. Our proposed SSS can be incorporated into current security solutions available in the cellular signal transfer domain, thus, further strengthening the security of the overall system. Thus, the proposed SSS can influence the use of speech security system in a more resilient manner.

References

1. Slashdot, NSA Able To Crack A5/1 Cellphone Crypto. https://yro.slashdot.org/story/13/12/14/0148251/nsa-able-to-crack-a51-cellphone-crypto. Accessed 24 June 2020
2. S.A. Shahriyar, M.A.H. Akhand, N. Siddique, T. Shimamura, Speech enhancement using convolutional denoising autoencoder, in *2019 International Conference on Electrical, Computer and Communication Engineering (ECCE)* (2019), pp. 1–5
3. S.A. Shahriyar, K.M.R. Alam, S.S. Roy, Y. Morimoto, An approach for multi label image classification using single label convolutional neural network, in *2018 21st international conference of computer and information technology (ICCIT)* (2018), pp. 1–6
4. M.I.R. Shuvo, S.A. Shahriyar, M.A.H. Akhand, Bangla Numeral Recognition from Speech Signal Using Convolutional Neural Network, in *2019 International Conference on Bangla Speech and Language Processing (ICBSLP)* (2019), pp. 1–4
5. S.S. Roy, S.A. Shahriyar, M. Asaf-Uddowla, K.M.R. Alam, Y. Morimoto, A novel encryption model for text messages using delayed chaotic neural network and DNA cryptography, in *2017 20th International Conference of Computer and Information Technology (ICCIT)*, Dhaka (2017), pp. 1–6
6. A. Agrawal, A. Bhopale, J. Sharma, M.S. Ali, D. Gautam, Implementation of DNA algorithm for secure voice communication. Int. J. Scient. Eng. Res. **3**(6). ISSN 2229-5518 (2012)
7. S.I. Ayon, M.A.H. Akhand, S.A. Shahriyar, N. Siddique, Spider monkey optimization to solve traveling salesman problem, in *2019 International Conference on Electrical, Computer and Communication Engineering (ECCE)* (2019), pp. 1–5
8. G. Jacob, A. Murugan, DNA based cryptography: an overview and analysis. Int. J. Emer. Sci. **3**(1), 27–36 (2013)
9. H.Z. Hsu, R.C.T. Lee, DNA based encryption methods, in *The 23rd Workshop on Combinatorial Mathematics and Computation Theory* (2012)
10. A. Gehani, T.H. LaBean, J.H. Reif, DNA-based Cryptography, in *5th Annual DIMACS Meeting on DNA Based Computers (DNA 5)*, MIT, Cambridge, MA (1999)
11. P.P. Singh, B. Singh, S.P. Ahuja, Need of secure voice encryption and its methods. Int. J. Adv. Res. Comput. Sci. Softw. Eng. **2**(1) (2012)
12. D. Ambika, V. Radha, Secure speech communication—a review. Int. J. Eng. Res. Appl. **2**(5), 1044–1049 (2012)
13. European Bioinformatics Institute. http://www.ebi.ac.uk/. Accessed 29 May 2020
14. T. Giannakopoulos, pyAudioAnalysis: an open-source Python library for audio signal analysis. PLoS ONE **10**(12), e0144610 (2015)
15. The National Anthem of Bangladesh "Amar Shonar Bangla". https://en.wikipedia.org/wiki/Amar_Shonar_Bangla. Accessed 21 May 2020

Obesity Detection in Thermal Imaging Using Convolution Neural Network: A Comparison with Machine Learning Models

U. Snekhalatha, K. Palani Thanaraj, and K. Sangamithrai

Abstract The purpose of the proposed study is to classify the thermal images of abdomen, forearm, and shank regions in obesity and normal subjects using the Speeded Up Robust Feature (SURF)-based machine learning classifiers compared to the convolutional neural network. Total populations of 60 subjects, among which 30 normal and 30 obese subjects, were recruited for the proposed study. The thermal images were acquired in abdomen, forearm, and shank regions using FLIR thermal camera (FLIR A300). SURF-based feature extraction method is used along with SVM classifiers for binary classification of adult obesity and normal subjects. After data augmentation process, convolution neural network VGG19 is used for the classification of obesity and normal subjects, and its performance is compared with SVM classifiers. The SURF-based SVM classifier provides an accuracy of 93% compared to VGG19 net. The machine learning classifier provided better results compared to the pre-trained models in CNN in the thermal image obesity analysis.

Keywords Thermal imaging · Speeded up robust feature · Support vector machine · Convolutional neural network

1 Introduction

Obesity develops in young adults gradually due to the accumulation of excess fat that may hinder their health. WHO estimated that the prevalence of obesity in children was dramatically increased from 4 to 18% during the period 1975–2016 [1]. The main causes of obesity are disproportion in energy between calories intake and calories dissipated. Body mass index (BMI) is the traditional method used to evaluate obesity

U. Snekhalatha (✉) · K. Palani Thanaraj
Department of Biomedical Engineering, College of Engineering and Technology, SRM Institute of Science and Technology, Kattankulathur, Chennai, Tamil Nadu 603203, India
e-mail: snehalau@srmist.edu.in

K. Sangamithrai
Department of Electronics and Instrumentation Engineering, St. Joseph College of Engineering, Anna Universtiy, Chennai, India

F. Thakkar et al. (eds.), *Proceedings of the International e-Conference on Intelligent Systems and Signal Processing*, Advances in Intelligent Systems and Computing 1370,
https://doi.org/10.1007/978-981-16-2123-9_45

in adults. But the major drawback is that it lacks in the measurement of body fat percentage and less accurate for different age groups.

The laboratory method used to evaluate the body composition are air displacement plethysmography, DEXA [2], etc. But these methods are time-consuming and require skilled technicians. The other diagnostic methods are bioelectric impedance analyzer [2], body composition analyzer, etc., which are found to be less precise than laboratory methods. Quantification of abdominal fat and body fat was assessed by ultrasound [3] and MRI [4] in obese subjects. Hence, the infrared thermography was used as the pre-screening method to assess obesity [5].

The recent trend in artificial intelligence was the development of automated applications using deep learning techniques. Machine learning and deep learning play a vital role in medical imaging, especially for medical image classifications. Extracting the hand-crafted features and classification using machine learning classifiers are common in many medical imaging applications. But deep learning techniques automated the process of segmentation, feature extraction, and classification using different defined pre-trained models. The main drawback of CNN is that more number of parameters needs to be fine-tuned to achieve optimum accuracy.

Farooq performed the real-time object recognition by extracting Speeded Up Robust Feature (SURF) features and classified the object using machine learning algorithms such as SVM and k-NN classifiers. They obtained the accuracy of 70.5% and 79% for SVM and k-NN, respectively [6]. Sumalatha et al. estimated the performance of face recognition in thermal images based on extracting the SURF features and classified using SVM classifier [7]. Ejindu designed a semi-automated system for estimation of the facial temperature and used the state-of-the-art techniques to detect the features such as SURF and classified using SVM classifier [8].

Numerous researchers have implemented deep convolution neural network for the classification of medical images [9–13]. Nanni et al. have utilized different medical image datasets and compared the performance of different deep learning networks such as Alex net, Google net, VGG16, VGG19, ResNet50, and Inception [14]. Yamashita et al. classified the benign and malignant tumors of primary and secondary lung cancers using the 3D inception CNN model. They performed the image segmentation using image patches in deep learning network [15]. Few researchers have used deep learning networks for the classification of thermal images for various applications such as human action recognition [16], breast cancer detection [17], cell classification [18], and face detection [19]. VGG19 is a benchmark image classification system based on deep learning architecture such as CNN. It is one of the widely used systems in general image classification extensively validated in the literature. Hence, we have used VGG19 for the classification of thermal images. The proposed approach involves comparing the performance of machine learning classifiers with the pre-trained CNN in obesity detection. The feature vector extracted using SURFsfrom thermal images were given as the input to the machine learning algorithms such as SVM and k-NN classifiers and compared with VGG19 model.

The study aims at the following: to classify the thermal images of abdomen, forearm, and shank regions in obesity and normal subjects, using SURF-based machine learning classifiers compared to convolutional neural network.

2 Methodology

2.1 *Study Participants*

Total population of 60 subjects, among which 30 normal and 30 obese subjects were recruited for the proposed study. The participants agreed to participate and filled the informed consent form. The approval of institutional ethical committee was obtained for the study. The inclusion criteria comprise of participants who are above the age group of 18 years and prone to the risk of obese and overweight.

2.2 *Thermal Image Analysis Procedure*

The thermal images were acquired in abdomen, forearm, and shank regions using FLIR thermal camera (FLIR A300). The participants are acclimated to a temperature-controlled environment with a room temperature of 20 °C for a period of 15 min prior to image acquisition process. The distance between the camera (FLIR A300) and the subject was standardized at 1 m. The images were analyzed using Python 3.7 version. Data augmentation of thermal images was carried out using translation, rotation, and shifting processes.

2.3 *SURF Feature Detection*

SURF used Determinant of Hessian Matrix (DOH) for interest point detection. The DOH is approximated using BOX filters. SURF characterizes the level of surface curvature in the image. Haar wavelet response is used for orientation assignment in both vertical and horizontal directions [20]. This wavelet response is computed in a circular region around the key points. After assigning the orientation, the descriptor is determined on the square region centered on the feature orientation. The neighborhood regions of key point are selected and divided into sub-regions [21]. Then the Haar wavelet is applied to each sub-region to obtain the SURF features.

2.4 SVM Classifier

SVM is a supervised machine learning algorithm for classification and regression processes. The hyperplane is identified to differentiate the two classes: Normal and obese in n-dimensional space. The distance between the nearest data point and the hyperplane should be maximum to choose the correct hyperplane margin.

2.5 VGG19 Net Architecture

VGG19 Net consists of input layer, five convolution layers, five maximum pooling layers, and two fully connected layers. The thermal images are resized and given as input to the convolution layer with size of 224 × 224 pixels. The input images were loaded in Python imaging library format and converted into NUMPy array. The input image size for CNN was determined after using preprocessing tools of Keras. The convolution layer convolves the input image with the fitter and provides the feature maps. The convolution layer consists of 3 × 3 kernel size with stride length 1 and has 64 output feature maps. The first convolution layer was followed by the first maximum pooling layer with kernel size 2 × 2 and the same output feature map as Conv net1. The max-pooling layer present on the CNN downscales the feature maps from the convolution layer and is used in VGG19 net. Each feature map size was minimized based on stride length and sub-sampling region. The maximum feature values were taken from the pool layer whose kernel length was considered as 2. In the proposed approach, five max pool layers were used in VGG19 net (Fig. 1).

The second convolution network is comprised of 128 feature maps as output with 3 × 3 kernel sizes. The next layer is the second maximum pooling layer with the same output feature map and kernel as in the first maximum pooling layer. The third convolution network contains 256 output feature maps with 3 × 3 kernels. The third maximum pooling layer has 2 × 2 kernel with 256 output feature maps. The fourth and fifth convolution layers have 512 feature maps with 3 × 3 kernel size. The fourth and fifth maximum pooling layers have 2 × 2 kernels with 512 output feature maps. The global average pooling layer layers attain the feature maps mean values obtained from downsampling the convolution layer. The main advantage of this layer

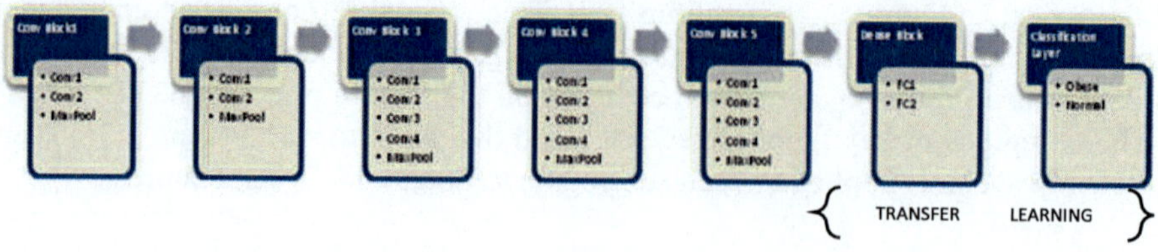

Fig. 1 Architecture of pre-trained model VGG19 net

was using a reduced number of learnable parameters and variable size input is given to enable the CNN.

The transfer learning process takes place in the fully connected layer of the VGG19 net. The initial weights were taken from the VGG19 net, and fine-tuning was performed on the dense layers. The VGG19 net consists of two dense layers (FC layers) in which the first layer perfroms based on RELU function. The second dense layer perfroms based on softmax activation function. FC1 provides has an output of 1024 feature maps followed by 2 feature maps as output in FC2. The output layer consists of two neurons which correspond to normal and obese. The final layer in the CNN is the classification layer which classifies the features into normal and obese. The target output was represented as $Z = \{z_0\}$, with $z_0 = -1$ for normal and $z_0 = 1$ for obese. The VGG19 Net architecture is illustrated in Table 1.

Table 1 VGG19 Net architecture for Obesity classification

Layer name kernel size, # filters	Input size	Output size	Stride	Parameters
Input Image	[224, 224, 3]	[224, 224, 3]	–	–
ConvNet-1 [3 × 3], 64	[224, 224, 3]	[222, 222, 64]	1	38, 720
Max Pool [2]	[222, 222, 64]	[112, 112, 64]	–	–
ConvNet-2 [3 × 3], 128	[112, 112, 64]	[112, 112, 128]	1	221, 440
Max Pool [2]	[112, 112, 128]	[112, 112, 64]	–	–
ConvNet-3 [3 × 3], 256	[112, 112, 64]	[56, 56, 256]	1	2, 065, 408
Max Pool [2]	[56, 56, 256]	[28, 28, 256]	–	–
ConvNet-4 [3 × 3], 512	[28, 28, 256]	[28, 28, 512]	1	8, 259, 584
Max Pool [2]	[28, 28, 512]	[14, 14, 512]	–	–
ConvNet-5 [3 × 3], 512	[14, 14, 512]	[14, 14, 512]	–	9, 439, 232
Global average pooling layer [2]	[14, 14, 512]	[7, 7, 512]	–	
Dense-1	[512,1]	[1024,1]	–	525, 312
Dense-2	[1024,1]	[1, 2]	–	2050
Total parameters				20, 551, 746

3 Results and Discussions

From the total 60 subjects, images were taken from three regions, hence, 180 images (60 subjects x 3 regions) of healthy and obese subjects were considered in the study. Hence, from these 180 images, 1000 images were generated using different transformation methods using data augmentation techniques. The training sets of 1000 images were used for training the network. Figure 2a–c represents abdomen, forearm and shank thermal imaging for normal subjects. Figure 2d–f indicates the thermal imaging of abdomen, forearm and shank of obese subjects.

The performance measures of SURF–SVM and VGG19 net for obesity detection are given in Table 2. The network performance was analyzed using the parameters such as precision, recall, and f1-score. The proposed SURF-SVM has n overall accuracy of 93%. Micro average for precision was calculated from individual true positive, true negative, false positive, and false negative of the total population. In Macro averaging, the performance of each individual class was calculated. The micro average and macro average of precision calculated for the two groups are 0.58 and 0.77, respectively, for VGG19. The weighted average accuracy of VGG19 net is obtained as 0.77 which was comparatively lower than SURF-SVM. Hence, the performance of machine learning approach is better than CNN in terms of accuracy (Fig. 2).

Figure 3a depicts the training and validation accuracy for VGG19 net for various epochs. Similarly, Fig. 3b shows the training and validation loss for VGG19 net. The overall classification accuracy for VGG19 net is found to be 77%.

The performance of the proposed net was computed based on sensitivity, specificity, and accuracy. The sensitivity was defined as correctly predicted true positive cases. The specificity is measured based on correctly predicting the normal cases. The ROC curve of VGG19 net and SURF-SVM are depicted in Fig. 4a and b, respectively. The ROC curve represents the plot of true positive versus true negative, hence for the proposed SURF-SVM, the achieved area under the curve is 1 and 0.58 for VGG19 net. Hence. from the comparative analysis of both SURF-SVM and VGG19 net, SURF-SVM provided a better performance in the classification of obesity and normal subjects.

Table 2 Performance measure of VGG19 net for obesity detection

VGG19	Precision	Recall	F1-score
Normal	0.54	1.00	0.70
Obese	1.00	0.16	0.28
Micro avg	0.58	0.58	0.58
Macro avg	0.77	0.58	0.49
Weighted avg	0.77	0.58	0.49
SURF-SVM			
Normal and obese	0.99	0.87	0.92

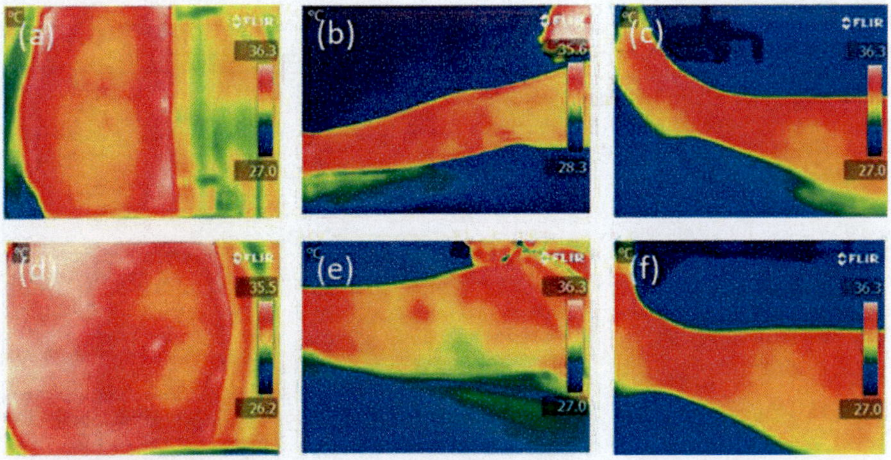

Fig. 2 **a–c** represents abdomen, forearm, and shank thermal imaging for normal subjects. **d–f** indicates the thermal imaging of abdomen, forearm, and shank of obese subjects

The limitation of the study was the need for a large amount of datasets to validate the study since deep learning could be performed on huge amount of data. Moreover, models are trained with only 1000 images which were generated by data augmentation techniques. An attempt to use CNN for thermal imaging was investigated in the paper. However, the CNN based on VGG19 architecture performed poorly compared to the machine learning approach. This is attributed to the low variability in the input images between the subject groups. More extensive data curation for the specific use case is needed for improving the accuracy of the deep learning approach, which we are hoping to do in future investigations.

4 Conclusion

In the proposed work, a pre-trained model, VGG19 net, is used for obesity detection and classification, and its performance is compared to the machine learning model. The SURF-based SVM classifier provides an accuracy of 93% compared to VGG19 net. The machine learning classifier provided better results compared to pre-trained models in CNN in the thermal image obesity analysis. Hence, it's better to design a customized network which could improve the performance and classification accuracy rather than using pre-trained networks.

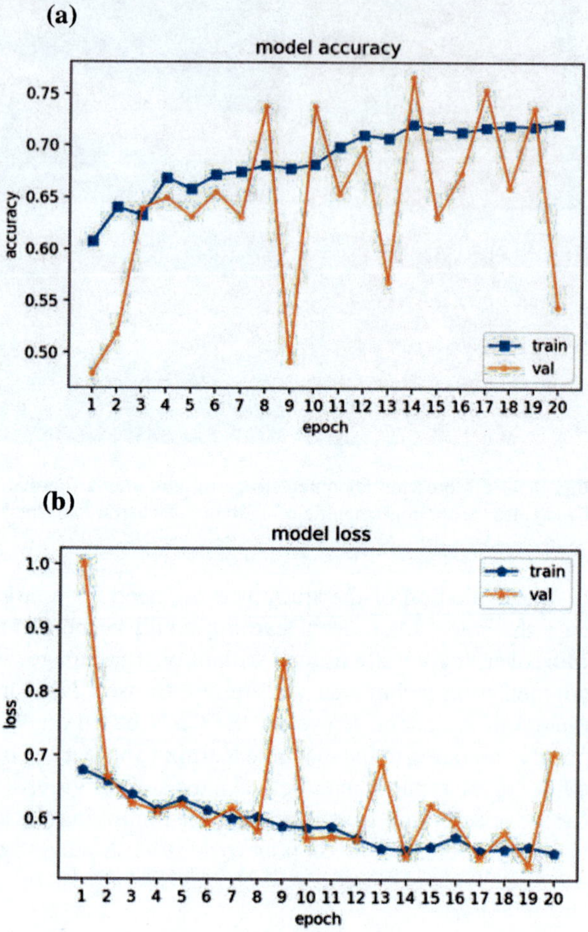

Fig. 3 **a** Training and validation accuracy of VGG19 for 20 epochs (blue color curve indicates the training accuracy and orange color curve indicates the validation accuracy. **b** Training and validation loss of VGG19 for 20 epochs (blue color curve indicates the training accuracy and orange color curve indicates the validation accuracy

Fig. 4 **a** ROC curve of VGG19 net. **b** ROC curve of SURF-SVM classifier

(a)

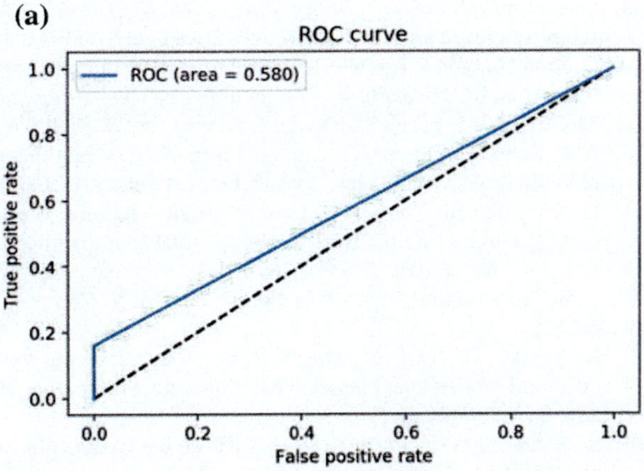

(b)

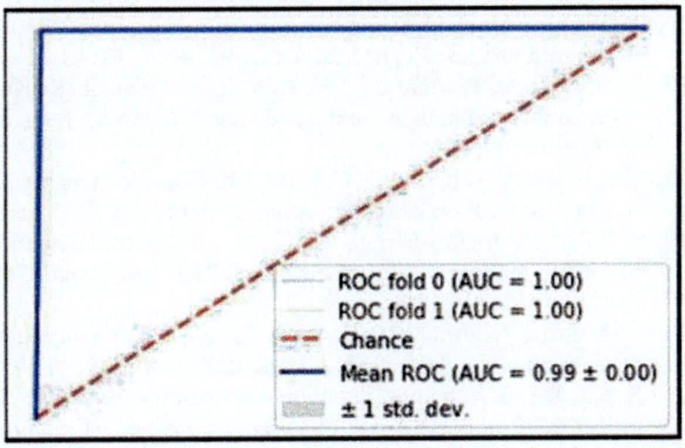

References

1. https://www.who.int/news-room/fact-sheets/detail/obesity-and-overweight
2. Yu. Ok-Kyeong, Y.-K. Rhee, T.-S. Park, Y.-S. Cha, Comparisons of obesity assessments in over-weight elementary students using anthropometry, BIA CT and DEXA. Nutr. Res. Prac. **4**(2), 128–135 (2014)
3. D.R. Wagner, Ultrasound as a tool to assess the body fat. J. Obes. **280713** (2013). https://doi.org/10.1155/2013/280713
4. J.C. Eloi, M. Epifanio, M.M. de Goncalves, A. Pellicioli, P.F.G. Vieira, H.B. Dias, et al., Quantification of abdominal fat in obese and healthy adolescents using 3 tesla magnetic resonance imaging and free software for image analysis. Plos one **12**(1), e0167625 (2017)
5. R. Heuberger, P. Kinnicutt, T. Domina, The relationship between thermal imaging and waist circumference in young adults. Health **4**(12A), 1485–1491 (2012)
6. J. Farooq, Object detection and identification using SURF and BoW model, in *2016 International Conference on Computing, Electronic and Electrical Engineering (ICE Cube)* (Quetta, 2016), pp. 318–323
7. R. Sumalatha, R. Varaprasada Rao, D. Rajasekhar, Thermal face recognition using bag of words. Int. J. Innov. Technol. Explor. Eng. (IJITEE). **8**(8) (2019). ISSN: 2278-3075

8. O.R. Ejindu, *Detecting Abnormalities in Thermal Pattern of Faces for Healthcare Applications*. (University of Calgary, Calgary, AB, 2019). http://hdl.handle.net/1880/110344
9. O. Ronneberger, P. Fischer, T. Brox, U-net: convolutional networks for biomedical image segementation, in *Proceedings of Medical Image Computing and Computer-Assisted Intervention-MICCAI 2015*, ed. by N. Navab, J. Hornegger, W. Wells, A. Frangi (2015)
10. A.A. Setio, F. Ciompi, G. Litjens, Pulmonary nodule detection in CT images: false positive reduction using multi-view convolution networks. IEEE Trans. Med. Imaging **35**, 1160–1169
11. G. Kang, K. Liu, B. Hou, N. Zhang, 3D multi-view convolution neural networks for lung nodule classification. PLoS ONE **12**, e0188290 (2017). https://doi.org/10.1371/Journal.pone.0188290
12. F. Lu, F. Wu, P. Hu, Z. Peng, D. Kong, Automatic 3D liver location and segmentation via convolution neural network and graph cut. Int. J. Comput. Assist. Radiol. Surg. **12**, 171–182 (2017)
13. K. Yasaka, H. Akai, O. Abe, S. Kiryu, Deep learning with convolution neural network for differentiation of liver masses at dynamic contrast-enhanced CT: a preliminary study. Radiology **286**, 887–896 (2018)
14. L. Nanni, S. Ghidoni, S. Brahnam, Ensemble of convolutional neural networks for bioimage classification. Appl. Comput. Inform. (2018), Article in press
15. R. Yamashita, M. Nishio, R.K. Gian Do, K. Togashi, Convolution neural networks: an overview and application in radiology. Insights Imaging **9**, 611–629 (2018)
16. R. Vigo, A. Akula, A.K. Shah, R. Ghosh, Deep learning approach for human action recognition in infrared images. Cogn. Syst. Res. **50**, 146–154 (2018)
17. S.J. Mambou, P. Maresova, O. Krejcar, A. Selamat, K. Kuca, Breast cancer detection using infrared thermal imaging and a deep learning model. Sensors **18**, 2799 (2018). https://doi.org/10.3390/s18092799
18. C.L. Chen, A. Mahjoubfar, C.T. Tai, I.K. Blaby, A. Huang, K.R. Niazi, B. Jalali, Deep learning in label-free cell classification scientific reports **6**, 21471 https://doi.org/10.1038/srep21471
19. S.S. Farfade, M.J. Saberian, L.J. Li, Multi-view face detection using deep convolutional neural networks, in *Proceedings of the 5th ACM International Conference on Multimedia Retrieval* (2015), pp. 643–650
20. T. Mouata, N. Aouf, D. Nam, S. Vidas, Performance evaluation of feature detectors and descriptors beyond the visible. J. Intell. Rob. Syst. **92**, 33–63 (2018)
21. S. Jain, B.L.S. Kumar, R. Shettigar, Comparative study on SIFT and SURF face feature descriptors, in *2017 International Conference on Inventive Communication and Computational Technologies (ICICCT)* (Coimbatore, 2017), pp. 200–205. https://doi.org/10.1109/ICICCT.2017.7975187

Non-invasive Technique for Detecting Neonatal Jaundice

S. Nihila, T. Rajalakshmi, Shradha Suman Panda, Nyelham Lhazay, and Ganga Devi Giri

Abstract Neonatal jaundice is a common ailment observed in infants. There is a lot of diagnostic approaches being made to detect it currently. When the infant's bilirubin level goes beyond 5 mg/L (85 mmol/L) it is indicative of jaundice. Bilirubin level increases often in infants just after birth till about a week. Almost 60 and 80% of full-term and premature babies, respectively, have jaundice. This paper aims for the early detection of jaundice in newly born infants. A total of 15 jaundice affected and 22 normal infants were considered for the purpose of this study, including babies of varied skin tones. A smartphone camera was used to capture images of the infant's skin. MATLAB was used to process the images that were taken, using algorithms for skin and facial detection. Color map transformations like RGB and YCbCr were used. Multiple quantitative features such as energy, entropy, mean, standard deviation and skewness were studied. PCA principle is the main technique that is used to reduce data redundancy. PCA uses energy and other features to come to a well-defined conclusion. It was noticed that the standard deviation and energy are much higher in jaundice-affected infants than normal ones. This research concludes that images can be used to identify the occurrence of neonatal jaundice.

Keywords Neonatal jaundice · Bilirubin · Face detection · Skin detection · Image processing · Color map transformation · PCA · Feature extraction

S. Nihila · S. S. Panda · N. Lhazay · G. D. Giri
Department of Biomedical Engineering, SRMIST, Kattankulathur, India
e-mail: nihila_sathish@srmuniv.edu.in

S. S. Panda
e-mail: shradha_kumar@srmuniv.edu.in

N. Lhazay
e-mail: nyelhamlhazay_@srmuniv.edu.in

G. D. Giri
e-mail: gangadevigiri_@srmuniv.edu.in

T. Rajalakshmi (✉)
Department of Electronics and Communication Engineering, SRMIST, Kattankulathur, India
e-mail: rajalakt@srmist.edu.in

F. Thakkar et al. (eds.), *Proceedings of the International e-Conference on Intelligent Systems and Signal Processing*, Advances in Intelligent Systems and Computing 1370,
https://doi.org/10.1007/978-981-16-2123-9_46

1 Introduction

Hyperbilirubinaemia also called jaundice indicates a high amount of bilirubin in the blood. If detected and treated early, it is not considered a disease. It is usually caused due to a lack of water in the body [1]. Almost 60% of infants in India are affected and need immediate care and attention [2]. The level of bilirubin can usually be determined by a transcutaneous bilirubinometer that measures serum bilirubin. The study relates the yellow color of skin to the bilirubin concentration [3]. The yellow color is visible usually after 24 h post birth when the concentration of bilirubin goes below 2 mg/dl [4]. A special therapy known as phototherapy is required when the bilirubin is more than 5 mg/dl and is dependent on the weight of the infant. Mainly there are three categories of jaundice, which are physiological jaundice, pathological jaundice and breast milk jaundice. Physiological jaundice is incapable of being detected post 14 days of the birth of the infant. The maximum intensity of the jaundice is usually at the 4–7th day of birth. With a peak of 17 mg/dl pathological jaundice onsets 24 h after birth, whereas breastmilk jaundice hits a peak of 10 mg/dl and persists after the third week. The cerebral lesion is very common at later stages of undetected jaundice [5]. More than adults generally the production of bilirubin is greater in babies and is usually 3–2 kg per body weight more [6]. The conventional transcutaneous bilirubinometer can be replaced by the non-invasive screening of jaundice usually done using a camera. The bilicam informs the parents on how severe the jaundice is and will help them take precautionary steps to reduce the severity and prevent serious cases that may cause cerebral palsy and hearing loss. It uses a color checker card and is based on calibration [7]. Comparison with actual skin color can be done using a color map transformation where there is a comparison of pixels between the current color and actual skin color [8].

The goal of this paper is to use PCA to detect jaundice in infants. It is more focused on the face, as it gives a more accurate estimation due to the fact that facial skin is more sensitive and its statistical feature extraction will reveal more information than the whole body region. There is an absence of a color checker or color calibration card, hereby making the device cost-efficient. IOS or android cameras may be used for capturing images of the infant.

Current research is usually limited to detecting jaundice for neonates. Therefore, the study that is proposed presents a non-invasive approach for jaundice detection in babies. This will classify infants into normal or jaundice-prone based on statistical facial features.

2 Materials and Methods

A total of 22 normal neonates and 15 jaundice-affected neonates were taken for examination in this proposed study. During the process, safety precaution steps were taken to ensure that no injuries were caused to the neonates. In this study, pictures of the baby were taken with a smartphone camera, and these data were loaded for processing into the MATLAB R2018a® environment. Secondly, using image processing methods to detect the face, these images were analyzed. The face was considered for identification because, relative to other sections, it is the thinnest and most vulnerable part of the skin. A skin detection algorithm based on pixel resemblance was performed after face detection. The Viola-Jones algorithm can be used to classify the faces, nose, eyes, mouth or upper body facial recognition of individuals. Skin detection can be achieved using the transformation of the color map, where the pixels are contrasted with the real skin color [8]. It is then accompanied by color map conversion to translate images into different color spaces. Finally, to differentiate between jaundice-affected and normal babies, the characteristics such as standard deviation, skewness, kurtosis, capacity, entropy, homogeneity and correlation were extracted.

Figure 1 shows the various steps and procedures involved to detect the presence of jaundice non-invasively. The methodology for the detection of jaundice is elaborated in detail in the following sections.

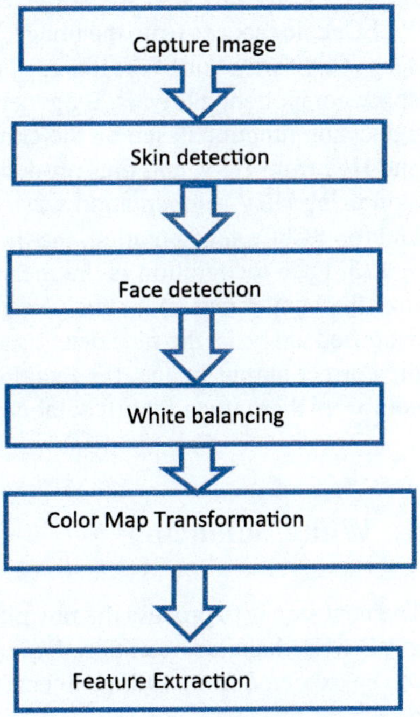

Fig. 1 Block diagram for detection of jaundice

3 Data Collection Phase

As the primary instrument for data collection, a smartphone camera was used. Neonatal jaundice, unlike the traditional approach in which blood samples are needed for analysis, was assessed by a direct method. The camera on the smartphone was used to capture a face-focused image of the newborn. These data (photos) have been loaded for processing in the MATLAB environment. The smartphone was positioned in a direction facing downward. After ensuring that the babies are in a quiet pose, the photos were taken.

4 Skin Detection and Face Detection

By detecting and labeling skin-like pixels within a given image, the skin detection algorithm is used to recognize the presence of a human face, hand or any other body part. Skin detection is implemented in this research, followed by a face detection algorithm to derive different statistical features from the images processed. By obtaining YCbCr color spaces from the image, skin detection is performed. Following the color space transformation of the image, the skin region was obtained based on unique color space component intervals. By generating a skin map, the resulting image undergoes a decision function based on the chroma components, i.e., Cb and Cr from YCbCbr and Hue from HSV, and thus produces an output image. The threshold for detecting skin using HSV segmentation was considered to be $h > 0.25$, $s \leq 0.15$ and $v > 0.9$; and for YCbCr segmentation, the threshold taken was $77 \geq Cb \geq 127$ and $133 \geq Cr \geq 173$. Face recognition is the mechanism by which photographs or videos recognize the human face(s) by analyzing the images and comparing the patterns. In this proposed analysis, the face detection of Viola-Jones is taken into account. To obtain the correct output image, the facial detection algorithm extracts facial features and compares them to an inbuilt database contained within the MATLAB program.

5 White Balancing

The next step is to process the raw pixel data obtained after face detection to perform a WB white balance operation. For each color channel, equivalent reflectivity values can be expected from a white object (for example, $R = G = B$). The involved process is:

1. The white entity is captured and the histogram is analyzed. The highest level color channel should be set as the goal average and the remaining two channels should be increased with a matching gain multiplier. For example, if the blue channel is considered to have the highest average, gain ‘x’ is applied to red and gain ‘y’ is applied to green ($B' = R'x = yG'$).

2. Depending on the light source (sunlight, fluorescent and tungsten) applied to the object and the sum of each color variable inside it, the white balance can differ with each color component. In the same way, a full-color natural scene may also be processed.
3. The white pixels are assumed to be the brightest in the image in order to find the objects that are actually white ($I = R + G + B$). Only pixels with the highest intensity are included in the mean measure, whereas pixels with a saturated channel are omitted.

6 Color Map Transformations

The white balance algorithm was implemented after the skin detection to eliminate the artificial colors from the picture after the skin detection. The color map transformation is implemented after white balancing. The method of transforming images from one color space to another color space is the transformation of color maps. In order to preserve the details, the transformed image looks identical to the original. The picture obtained after skin detection was translated to YCbCr and Lab color spaces in the RGB color model.

7 Feature Extraction

The required details for an image that can be used for analysis is a feature. In this stage, the images processed in the YCbCr and Lab color spaces, together with images in the RGB color space, were taken and nine features were extracted for each color channel.

The following features were calculated from each color space model:

The mean value determines the contribution of individual pixels to the image intensity.

$$\mu = \frac{1}{N}\sum_{i=0}^{N-1} x_i \tag{1}$$

where μ = mean, x_i = input, N = number of samples, i = is the index that runs from 0 to $N - 1$.

Standard deviation is a measure that quantifies the amount of variation or dispersion in a given set of data.

$$\sigma^2 = \frac{1}{N-1}\sum_{i=0}^{N-1} (x_i - \mu)^2 \tag{2}$$

where σ = standard deviation, N = number of samples, x_i = input sample, μ = mean and i = is the index that runs from 0 to N − 1.

Kurtosis is used to determine whether the data are peaked or flat relative to a normal distribution. High kurtosis values show low noise and low resolution.

$$K = n\frac{\sum_{i=1}^{n}\left(X_i - X_{avg}\right)^4}{\left(\sum_{i=1}^{n}\left(X_i - X_{avg}\right)^2\right)^2} - 3 \tag{3}$$

where K = kurtosis, $X_i - X_{avg}$ = deviation from the mean (difference between each value and arithmetic average of all values), n = number of samples and i = is the index that runs from 1 to n.

Skewness determines the symmetricity or absence of symmetricity from the given data set. A data set with symmetric distribution is equally divided from the center point.

$$b = \frac{\frac{1}{n}\sum_{i=1}^{n}\left(x_i - \overline{x}\right)^3}{\left[\frac{1}{n-1}\sum_{i=1}^{n}\left[x_i - \overline{x}\right]^2\right]^{3/2}} \tag{4}$$

where b = skewness, i = is the index that runs from 1 to n and $\overline{x}$ = mean and x_i= input.

Energy is used to extract information from an image by comparing the image pixels. In image processing, it is used to show how the gray levels are distributed. When the number of gray levels is low then energy is high.

$$E_x = \sum_{i=-\infty}^{\infty} |x_n|^2 \tag{5}$$

where E_x = energy, n = index that runs from $-\infty$ to ∞ and x_n = the input signal.

Entropy in image processing is the measure of the ability of pixels of an image to adapt to different intensity levels. It can be used to determine the texture of the input image.

$$Entropy = -\sum P_i log_2 P_i \tag{6}$$

P_i is the probability that the difference between two adjacent pixels is equal to i, and Log_2 is the base 2 logarithm.

Homogeneity is the measure of uniformity of distribution in an image.

$$\sum_{i,j}\frac{p(i, j)}{1 + |i - j|} \tag{7}$$

where matrix element P(i, j) is the relative frequency with two pixels, separated by a pixel distance within a given neighborhood; one with intensity i and the other with intensity j.

Correlation is the measure of how correlated or related a pixel is to its neighboring pixel of the whole image.

$$Correlation = \sum_{i=0}^{G-1}\sum_{j=0}^{G-1} \frac{\{i \times j\} \times P(i,j) - \{\mu_x \times \mu_y\}}{\sigma_x \times \sigma_y} \tag{8}$$

$$P_x(i) = \sum_{j=0}^{G-1} P(i,j) \tag{9a}$$

$$P_y(j) = \sum_{i=0}^{G-1} P(i,j) \tag{9b}$$

$$\mu_x = \sum_{i=0}^{G-1} i \sum_{j=0}^{G-1} P(i,j) = \sum_{i=0}^{G-1} i P_x(i) \tag{9c}$$

$$\mu_y = \sum_{i=0}^{G-1}\sum_{j=0}^{G-1} j P(i,j) = \sum_{j=0}^{G-1} j P_y(j) \tag{9d}$$

$$\sigma_x^2 = \sum_{i=0}^{G-1} (i - \mu_x)^2 \sum_{j=0}^{G-1} P(i,j) = \sum_{i=0}^{G-1} (P_x(i) - \mu_x(i))^2 \tag{9e}$$

$$\sigma_y^2 = \sum_{j=0}^{G-1} (i - \mu_y)^2 \sum_{i=0}^{G-1} P(i,j) = \sum_{j=0}^{G-1} (P_y(i) - \mu_y(j))^2 \tag{9f}$$

where G is the number of gray levels used, μ is the mean value of P (matrix), μ_x, μ_y, σ_x and σ_y are the means and standard deviations of P_x and P_y. $P_x(i)$ is the ith entry in the marginal probability matrix obtained by summing the rows of P(i, j). P(i, j) is the matrix element whose two pixels are separated with a pixel distance within their neighborhood.

8 Principle Component Analysis (PCA)

1. The original data is centered such that its average becomes zero.
2. Data normalization where each characteristic is separated by its standard deviation.

3. It is important to project the data onto the biggest eigenvectors. Let V be the matrix with columns containing the largest vectors of its own, and let D be the original data with different observations of its columns.

Then the projected data D′ is given by

$$D' = V^T D \tag{10}$$

One can use the V columns directly, or one can specify the amount of variation from the original data that is stored and delete one's own vectors. If only N eigenvectors are held, and e1….eN represents the corresponding eigenvalues, it is possible to calculate the quantity of variance remaining after projecting the original with d dimension as follows:

$$s = \frac{\sum_{i=0}^{N} e_i}{\sum_{j=0}^{d} e_j} \tag{11}$$

9 Results and Discussion

In this proposed study, from the images of 37 subjects, 22 were normal subjects and the remaining 15 showed symptoms of jaundice. These images are taken from Scientech and Pubmed databases. They are of unknown age and gender in specific. However, the age ranges from two days to 30 days. Figure 2 depicts the original input image before processing.

Figure 3 depicts the skin detected region of the original image. Figure 4 depicts the face detected region obtained by using Viola-Jones Algorithm after skin detection. Pre-processing (white balancing) is applied to the face detected (Fig. 5).

Figure 6a depicts the red channel extracted from pre-processed image. Figure 6b depicts the green channel extracted from pre-processed image. Figure 6c depicts the blue channel extracted from pre-processed image.

Fig. 2 Original input image

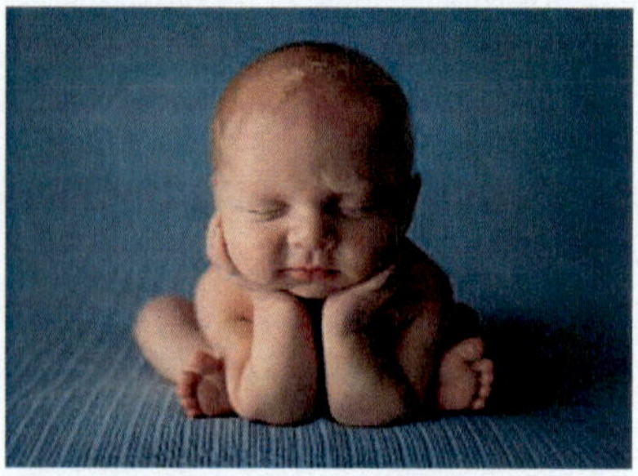

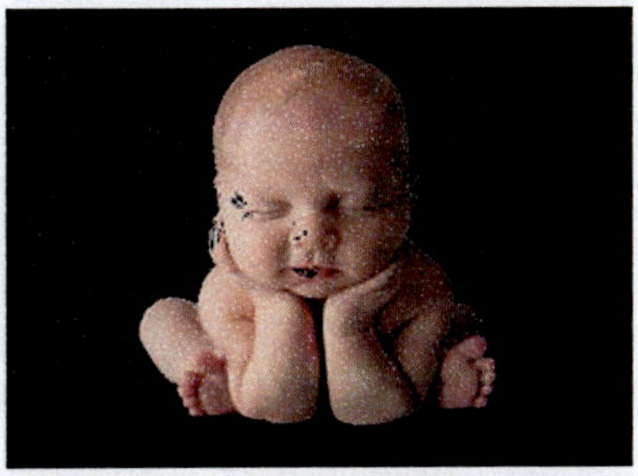

Fig. 3 Skin detected image

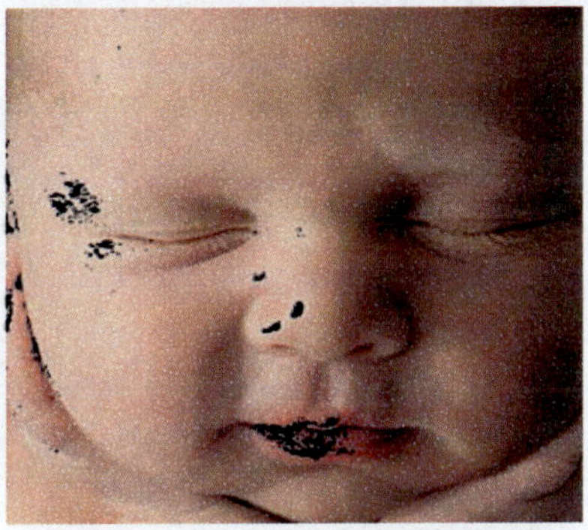

Fig. 4 Face detected image

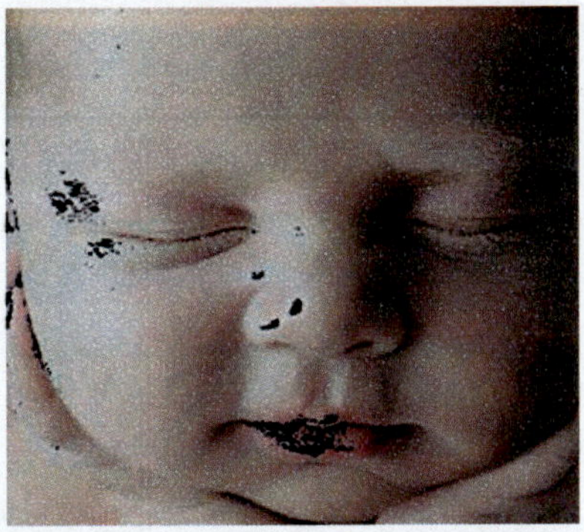

Fig. 5 Pre-processed (white balanced) image

Figure 7a depicts the Y (luminance) channel extracted from pre-processed image. Figure 7b depicts the Cb (Chrominance blue) channel extracted from pre-processed image. Figure 7c depicts the Cr (Chrominance red) channel extracted from pre-processed image.

Figure 8a depicts the L (light) channel extracted from pre-processed image. Figure 8b depicts the a (red green) channel extracted from pre-processed image. Figure 8c depicts the b (blue yellow) channel extracted from pre-processed image.

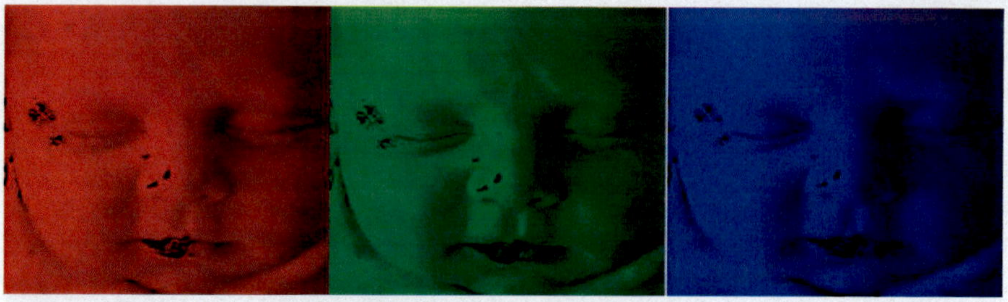

(a) Red Channel Image (b) Green Channel Image (c) Blue Channel Image

Fig. 6 RGB color space images

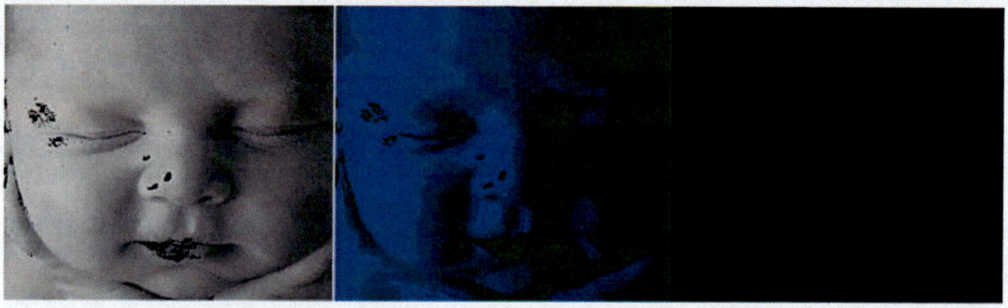

(a)Y Channel Image (b) Cb Channel Image (c) Cr Channel Image

Fig. 7 YCbCr color space images

(a) L Channel Image (b) a Channel Image (c) b Channel Image

Fig. 8 Lab colour space images

Table 1 depicts the statistical features extracted after performing principle component analysis and these include mean, entropy, standard deviation and skewness. It is observed that for mean the normal range is 17–30 approximately, and in the case of babies affected by jaundice it ranges from 30 to 51 approximately. It is observed

Table 1 Statistical features extracted to detect the presence of jaundice

Condition	Features			
	Mean	Standard deviation	Entropy	Skewness
Normal	**29.68**	**2.26**	0.99	−0.05
Normal	**28.39**	**2.88**	0.93	0.04
Normal	**25.49**	**2.3**	0.91	0
Normal	**26.78**	**2.73**	0.98	0.52
Normal	**21.62**	**1.69**	0.93	−0.04
Normal	**17.36**	**1.38**	0.93	0.79
Normal	**27.26**	**2.13**	0.93	−0.35
Normal	**26.58**	**2.67**	0.77	−0.52
Normal	**17.15**	**2.16**	1	0.32
Normal	**28.62**	**2.68**	0.97	0.02
Jaundice	**51.64**	**6.4**	0.57	−0.78
Jaundice	**37.88**	**5.68**	0.88	−0.29
Jaundice	**38.25**	**10.12**	0.86	−0.22
Jaundice	**40.9**	**8.15**	0.59	−1.12
Jaundice	**34.87**	**3.33**	0.99	0.52
Jaundice	**30.34**	**4.59**	0.75	−0.2
Jaundice	**30.25**	**5.95**	0.8	−0.28
Jaundice	**36.13**	**3.13**	0.87	0.14
Jaundice	**31.39**	**3.96**	0.99	0.46
Jaundice	**42.4**	**3.61**	0.85	−0.74

that for standard deviation, the normal range is 1 to 3 approximately and in the case of babies affected by jaundice it ranges from 3 to 11 approximately. Entropy showed a slight variation among normal and jaundice-affected babies (for normal babies, it is slightly higher). However, skewness did not exhibit any specific pattern and no accurate result can be inferred from this data.

10 Conclusion

We may interpret this non-invasive technique as a clear and usable method for self-assessing neonatal jaundice based on the experimental outcome. In diagnosing neonatal jaundice, the proposed method is reliable and can minimize the need for blood sampling. Of all the techniques tested during this research, facial detection seemed to be more promising. This approach could be used with all skin types and varying skin pigmentation populations. Significant brain damage is caused by hyper-bilirubinemia and non-invasive jaundice monitoring using a smartphone will pick up cases at an early age. Normal and jaundice-affected babies can be distinguished using the extracted features. Future research involves extracting more features from images and using 2–3 models of classifiers (machine learners) together to assess the level of bilirubin.

References

1. A. Arulmozhi, M. Ezhilarasi, Maximal information compression index (Mici) and PSO based detection of jaundice. **24**, 583–597 (2014)
2. A. Arulmozhi, M. Ezhilarsi, New born baby jaundice detection using time slot scheduling with GA based feature selection. Aust. J. Basic Appl. Sci. **8**(13), 105–114 (2014)
3. A. Knudsen, R. Brodersen, Skin colour and bilirubin in neonates. **64**, 605–609 (1989)
4. M.W. Abbas, T. Shamshad, M.A. Ashraf, R. Javaid, Jaundice: a basic review. Int. J. Res. Med. Sci. **4**, 1313–1319 (2016)
5. S. Mishra, R. Agarwal, A. Deorari, V.K. Paul, Jaundice in the newborns. Indian J. Pediatr. **75**, 157–163 (2008). https://doi.org/10.1007/s12098-008-0024-7
6. S. Werwath, Automatic coding for neonatal jaundice from free text data using ensemble methods (2018). arXiv:1805.01054v1[cs.CL]
7. L. De Greef, M. Goel et al., BiliCam: using mobile phones to monitor newborn jaundice, in *UBICOMP* (2014), pp. 331–342.
8. M.N. Mansor, S. Yaacob, Jaundice in newborn monitoring using color detection method, in *2012 International Workshop on Information and Electronic Engineering (IWIEE)*, vol. 29 (2012), pp. 1631–1635.

Augmented Reality Implementation on Physical Therapy Exercise

Ahmad Nasrul Norali, Anas Mohd Noor, Zulkarnay Zakaria, Wan Yi Khor, Yen Fook Chong, and Asyraf Hakimi Abu Bakar

Abstract Physical therapy exercise (PTE) is a planned physical activity that is enabled to enhance body functions, reduce risk and improve the overall health of a person. In this research, range of motion (ROM) for shoulder PTE was investigated. There are three types of shoulder PTE involved: shoulder horizontal flexion and extension, shoulder external rotation and shoulder flexion and extension. To compare the conventional shoulder PTE with the shoulder PTE that is implemented with current technology, an augmented reality (AR) application is designed and implemented to those shoulder PTE. In this research, it is found that ROM of shoulder horizontal flexion and ROM of shoulder flexion were classified well between the group with AR and the group without AR. This is due to their classification accuracy is quite high in the KNN classifier (accuracy >80%). Lastly, the group with AR and group without AR for each ROM of shoulder PTE are compared using paired sample *t*-test. The result showed that all the ROM of shoulder PTE has a high significant difference (when *p*-value <0.05) between the group with AR and group without AR as the *p*-value is extremely small.

Keywords Augmented reality · Human–computer interaction · Physical therapy exercise

1 Introduction

Physical therapy exercise (PTE) is a type of rehabilitation that is a combination of exercises and manipulation of the body parts to enhance body function, reduce and prevent the risk related to health [1]. The technique of PTE involved the range

A. N. Norali (✉) · A. M. Noor · Z. Zakaria · W. Y. Khor · Y. F. Chong · A. H. A. Bakar
Faculty of Electronic Engineering Technology, Universiti Malaysia Perlis, 02600 Arau, Perlis, Malaysia
e-mail: ahmadnasrul@unimap.edu.my

A. N. Norali · Z. Zakaria · Y. F. Chong
Sports Engineering Research Centre (SERC), Universiti Malaysia Perlis, 02600 Arau, Perlis, Malaysia

F. Thakkar et al. (eds.), *Proceedings of the International e-Conference on Intelligent Systems and Signal Processing*, Advances in Intelligent Systems and Computing 1370,
https://doi.org/10.1007/978-981-16-2123-9_47

of motion exercises, coordination and balance exercises and muscle strengthening exercises. A conventional PTE is usually done in rehabilitation centre and the PTE program is designed and monitored by the physical therapist.

To improve the PTE, new technology had been studied to implement on PTE. The technology used is augmented reality (AR). AR is a technology that has added computer-generated information such as 3D graphic objects or sound to a real-time video-based real-world environment [2]. The virtual objects are superimposed by sensing of reality physical objects. The information in the virtual situation can be used for entertainment purposes, daily task performance and so on.

Nowadays, duration of sitting, improper postures and psychosocial conditions at work are the factors that may cause problem or illness on the shoulder for the adults. It can be prevented or treated through a series of shoulder PTE. However, people nowadays feel bored to do the conventional PTE. Thus, this research is aimed to design an augmented reality application with PTE for the shoulder. The ROM of shoulder PTE is compared between the group with AR and the group without AR. Lastly, features extraction is analysed to determine suitable features for the classification of shoulder PTE between the group with AR and the group without AR.

2 Literature Review

There was some previous research related to AR. [3] had proposed a projection-based AR system called smARt.assembly. The system is able to project the digital-based guideline and assembly data of a user in the physical workplace without any other assistance tool. A location-based AR game proposed in [4] examines how playing games can improve the relationship between place and people (place attachment). The analyses in this research showed a positive result in place attachment.

AR software was developed and implemented in [5] to convert the languages from English to Tamil on smartphones. The smartphone with the AR software will sense the real words in the English language using the camera to augment the virtual Tamil language. The augmented virtual Tamil text provides the meaning of the captured English. A marker-based AR application is applied to an android mobile in [6] which allows the user to see the virtual objects in the real environment through the camera. There are several advantages of this application, which are it is able to display 3D objects and enable the object for virtual rotation compared to existing 2D applications.

In [7], AR was implemented for muscular improvement training and the results gave a significant improvement in muscle strength. The AR treatment in [8] was about the rehabilitation of the wrist and hand. Arm Motor Fugl-Meyer assessment was done and proved that AR helped to improve the rehabilitation. For [9], the AR system was used to improve the gait and posture of the patients. Lastly, AR games were created in [10] for upper limb rehabilitation. The design of AR included lifting, holding and moving.

3 Methodology

3.1 *Physical Therapy Exercise for Shoulder*

There are three types of PTE involved in this research, which are shoulder horizontal flexion, shoulder external rotation and shoulder flexion. The description of shoulder exercises is presented in Table 1.

Table 1 Descriptions of shoulder PTE in this research

Types of shoulder PTE	Description
Shoulder horizontal flexion	Shoulder horizontal flexion is shown in Fig. 1. Shoulder horizontal flexion is the movement of the arm in which the arm is held out to the side and then pulled across to the chest. The normal range of motion for shoulder horizontal flexion is 135°
Shoulder external rotation	Shoulder external rotation is shown in Fig. 2. The arm is placed at the side and the palm is faced to the body. Then, the elbow is bend 90°. Normal range of motion for shoulder external rotation is 90°
Shoulder flexion	Shoulder flexion is shown in Fig. 3 Shoulder flexion is the movement of arm where the arm is put perpendicular to the trunk and lifted 90° upward. Normal range of motion for shoulder flexion and extension is 90°

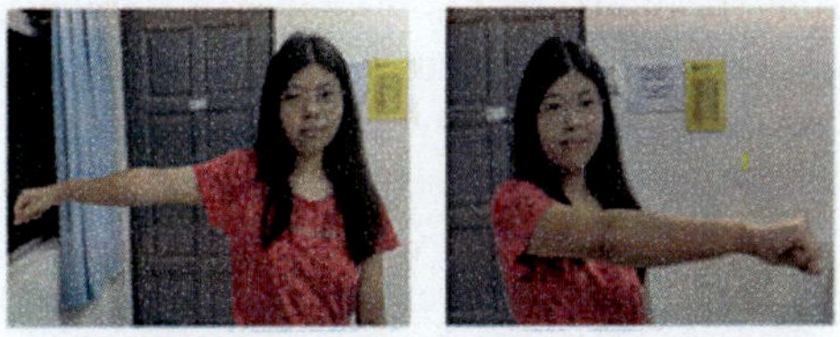

Fig. 1 Shoulder horizontal flexion

Fig. 2 Shoulder external rotation

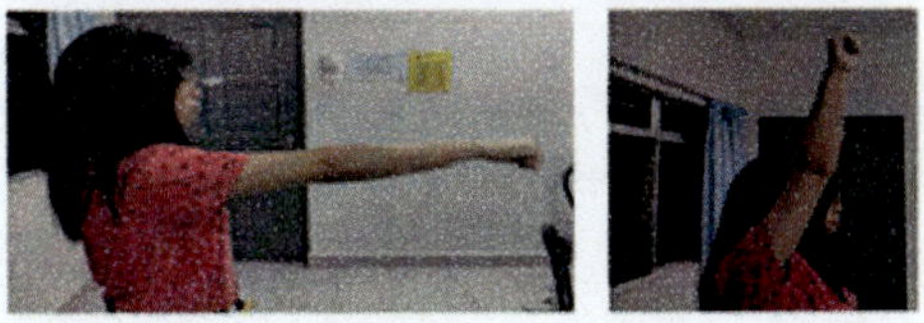

Fig. 3 Shoulder flexion

Fig. 4 "Hiro" marker

Fig. 5 Augmentation of the virtual sphere on the "Hiro" marker

3.2 Development of AR Application for Shoulder PTE

The AR application is designed using NyARToolkit for processing.EN (http://nyatla.jp/nyartoolkit/) that is written using Processing 3.5.3 (https://processing.org). This application is implemented to three different shoulder PTEs. Marker-based AR is chosen due to it is low cost and easier for AR coding nature. The type of marker used is shown in Fig. 4 and is provided together with the NyARToolkit. This marker will be detected by AR system through a webcam of the laptop to augment a virtual sphere on the computer screen as shown in Fig. 5.

Figure 6 showed the AR-based shoulder PTE. In the AR system, there will have 2 squares on the streaming video, one is the initial position and another square is the final position. The movement of the arm from one square to another square will represent the movement of the shoulder PTE. The marker is printed out on a manila card and attached to a strip as shown in Fig. 7.

3.3 Subject

In this research, 10 healthy subjects with age between 20 and 25 years old were involved to the experiment. Each subject will undergo two groups of shoulder PTE, which is shoulder PTE with AR and shoulder PTE without AR.

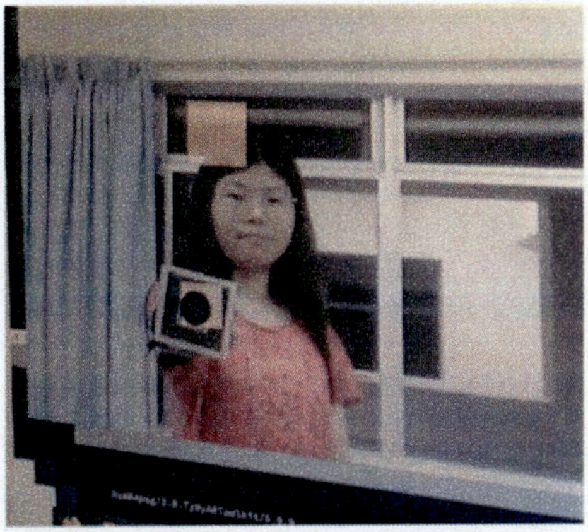

Fig. 6 AR-based shoulder PTE

Fig. 7 Marker placement

3.4 Data Collection Hardware Setup

The hardware setup consists of ADXL335 accelerometer sensor and Arduino Uno. ADXL335 accelerometer sensor is a tri-axial sensor. It is used to detect the acceleration from three different axes, which are x, y and z axes. The unit of the raw data collected by the accelerometer sensor is in gravity, g. PLX-DAQ is an add-on Microsoft Excel, which is used to interface with Arduino IDE to store the collected data from the accelerometer sensor in an excel file. Figure 8 shows the hardware setup. Figure 9 shows the placement of the sensor on the subject's arm. The accelerometer should be calibrated first before the experiment started.

Fig. 8 The setup of the Arduino hardware with accelerometer sensor

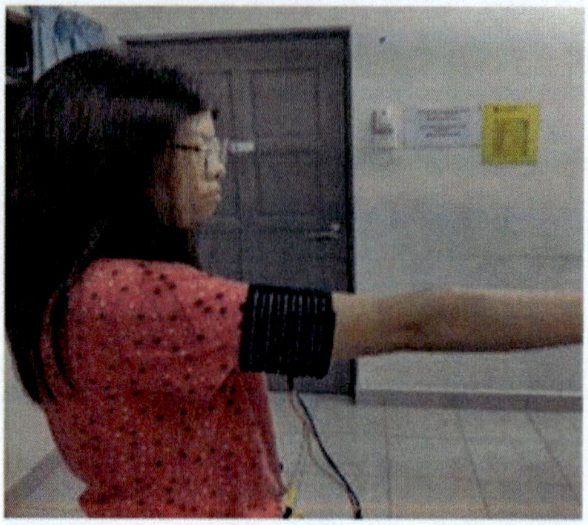

Fig. 9 The placement of the sensor on the arm

3.5 Experiment Protocol

A short briefing about the flow of the experiments is conducted to the subjects. The experiment is divided into two parts: with AR and without AR. The experiment is first conducted with shoulder PTE without AR. Each shoulder exercise is repeated for 10 times. The subjects will be rest for about 1 min each time a PTE is done. The experiment is further conducted on AR-based PTE. For this part, AR markers will be held by the subjects. The process is repeated the same as PTE without AR.

3.6 ROM Calculation

In this research, ROM of shoulder PTE is represented by the tilt angle of the accelerometer. Figure 10 shows how the inclination of the accelerometer represents the angle of ROM of shoulder PTE. Only the y-axis is used to analyse the tilt angle of the accelerometer. The tilt angle θ of the accelerometer is calculated as

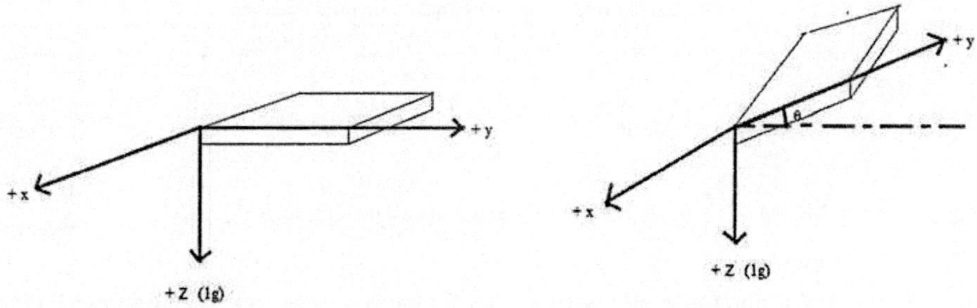

Fig. 10 Angle of inclination of accelerometer

$$\theta(°) = \text{atan}(\frac{a_y}{\sqrt{(a_x)^2 + (a_z)^2}}) \times \frac{180°}{\pi} \quad (1)$$

where a_x, a_y and a_z are the acceleration in g measured from the x, y and z axes of the accelerometer. When the accelerometer is placed horizontally, it gives the angle value of 0°. The motion of shoulder PTE will return a tilted angle based on the inclination of the accelerometer.

3.7 Segmentation and Preprocessing

The ROM signal was segmented to retain the range of ROM of shoulder PTE that was needed for further interpretation. Then a second-order low-pass Butterworth with 0.20 Hz cut-off frequency is applied to remove the high frequency of noise and thus retain the low frequency of pure data.

3.8 Feature Extraction

Feature extraction is the process to extract the important signal information from the input data. In this research, FFT is used to extract the information of the angle signal for all shoulder PTE with AR and without AR. There are five features extracted in this research, which are mean, maximum, minimum, variance and standard deviation.

3.9 Feature Selection and Classification

Feature selection is the process to determine which feature is best for classification. Cross-validation K-nearest neighbour (KNN) classifier was implemented to classify

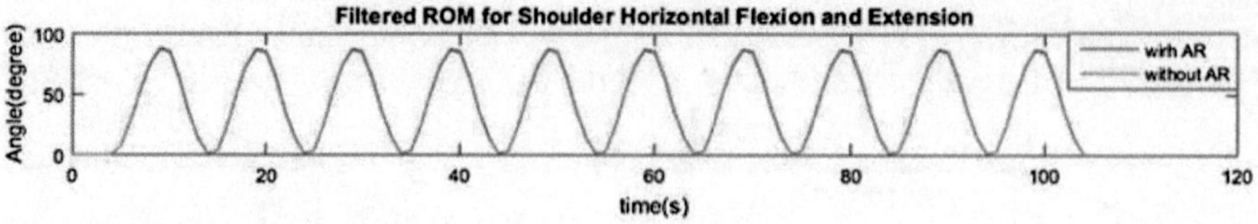

Fig. 11 Filtered ROM of shoulder horizontal flexion and extension for subject 2

the group with AR and the group without AR for each type of ROM of shoulder PTE. The Euclidean distance metric in the multidimensional feature spaces and number of neighbours, k, equals to 3 are implemented in KNN.

3.9.1 Statistical Analysis

Statistical analysis is performed by using Statistical Package for Social Science (SPSS) Statistics software. In this research, paired sample *t*-test is used to investigate the significant difference between the group with AR and the group without AR for each ROM of shoulder PTE. 95% confidence level is applied to the paired sample *t*-test. Thus, a *p*-value equals to 0.05 is important to determine the significant difference.

4 Results and Discussion

4.1 ROM Data Collection

Figure 11 showed the filtered graph after implementing the second-order low-pass Butterworth filter with cut-off frequency at 0.20 Hz.

4.2 Feature Extraction and Feature Selection

Time-domain features are extracted using Fast Fourier Transform (FFT) is applied as it is good to use for stationary ROM signals.

For feature selection, by referring to Table 2, the maximum feature is applied to shoulder horizontal flexion as it has the highest accuracy (90%). For ROM of shoulder external rotation, it has the highest accuracy in both variance (75%) and standard deviation (75%), the same as ROM of shoulder flexion (85%). Thus, the feature of standard deviation is chosen for these two ROM to compare between the group with AR and the group without AR.

Table 2 Classification accuracy on ROM for three shoulder PTEs

Classification accuracy on ROM for three shoulder PTE (%)					
	Mean	Minimum	Maximum	Variance	Standard deviation
Shoulder horizontal flexion	85	50	90	85	85
Shoulder external rotation	70	55	70	75	75
Shoulder flexion	80	55	70	85	85

Table 3 Classification accuracy on ROM for three shoulder PTEs

	Accuracy (%)		Sensitivity (%)		Specificity (%)	
	Selected features	All features	Selected features	All features	Selected features	All features
Shoulder horizontal flexion	90	85	100	90	80	80
Shoulder external rotation	75	60	80	60	70	60
Shoulder flexion	85	90	80	80	90	100

4.3 Classification

In general, the accuracy, sensitivity and specificity should be as high as possible to determine that the classification between the group with AR and group without AR for the ROM of shoulder PTE is performed well in the KNN classifier. In conclusion, the KNN classifier can classify well in ROM of shoulder horizontal flexion and ROM of shoulder flexion between the group with AR and the group without AR. This is due to both have a very high percentage in accuracy, sensitivity and specificity, which is greater than 80% based on Table 3. However, classification using KNN classifier in shoulder external rotation between the group with AR and the group without AR is not so accurate due to the low percentage of accuracy.

4.4 Statistical Analysis

In this research, three different shoulder PTEs are compared separately to determine their significant difference between with AR and without AR groups. Paired sample *t*-test is tested for each group of shoulder PTE by calculating the mean and standard deviation of ROM within 10 subjects.

Table 4 shows the pair samples test for three shoulder PTEs. Paired sample *t*-test is done for all shoulder PTE by using 95% confidence level and level of significance,

Table 4 Paired sample *t*-test

		Paired differences					t	Df	Sig. (2-tailed)
		Mean	Std. deviation	Std. error mean	95% Confidence interval				
					Lower	Upper			
Pair 1	Shoulder horizontal flexion	2.2687	1.0009	0.3165	1.5527	2.9847	7.16	9	0.00005
Pair 2	Shoulder external rotation	2.7872	2.0737	0.6557	1.3037	4.2707	4.25	9	0.00214
Pair 3	Shoulder flexion	2.6405	1.4638	0.4629	1.59332	3.68773	5.704	9	0.00029

p is equal to 0.05. If *p* is smaller than 0.05, it is meant that there is a significant difference between the 2 sample groups and vice versa. Overall, all the *p*-values for ROM of all shoulder PTE are smaller than 0.05, thus it is meant that there is a significant difference in ROM for shoulder PTE with AR and shoulder PTE without AR.

5 Conclusion

This research lays the foundation for the development of an AR application for different types of shoulder PTE. The developed AR-based shoulder PTE is used to compare with the conventional shoulder PTE, in which the shoulder PTE is performed without the aid of AR application. Thus, it can be said that this research is categorised into two different groups, which meant the group with AR and the group without AR. This research is limited to the scope of ROM of the shoulder PTE. The tilt angle of the accelerometer sensor is represented as the ROM of shoulder PTE. In this research, it can be concluded that the AR-based shoulder PTE is significantly different from the shoulder PTE as these two groups' samples can be classified well using the KNN classifier and paired sample *t*-test (p-value <0.05).

For future research, a more complicated AR application should be designed to implement shoulder rehabilitation. The type of AR used should be focussed on marker-less-based AR. Besides that, future research is aimed at individuals who have shoulder pain. Next, the data collection should focus on the shoulder electromyography measurements and determine whether shoulder rehabilitation could be performed well by using the implementation of AR application. Next, the number and type of feature sets should be increased for classification purpose. Lastly, further research can be done by implementing different types of classifiers to the features to figure out which type of classifier is suitable for the classification of each of the feature sets.

References

1. J. Carmigniani, B. Furht, M. Anisetti, P. Ceravolo, E. Damiani, M. Ivkovic, Augmented reality technologies, systems and applications. Multimed. Tools Appl. **51**(1), 341–377 (2011)
2. C. Kisner, L. Colby, *Therapeutic Exercise: Foundations and Techniques*, 7th edn. (F.A. Davis, Philadelphia, 2018)
3. O. Sand, S. Büttner, V. Paelke, C. Röcker, smart. assembly–projection-based augmented reality for supporting assembly workers, in *International Conference on Virtual, Augmented and Mixed Reality* (Springer, Cham, 2016), pp. 643–652
4. T. Oleksy, A. Wnuk, Catch them all and increase your place attachment! The role of location-based augmented reality games in changing people—place relations. Comput. Human Behav. **76**, 3–8 (2017)

5. R.F.J. Rose, G. Bhuvaneswari, Word recognition incorporating augmented reality for linguistic e-conversion, in *2016 International Conference on Electrical, Electronics, and Optimization Techniques (ICEEOT)* (IEEE, 2016), pp. 2106–2109
6. University Health Services, *Shoulder Rehabilitation: Treatment Includes Activity Modification, Stretching, and Strengthening the Affected Limb* (Princet. Univ. Athl. Med.), pp. 1–7
7. J. Lee, H.N. Yoo, B.H. Lee, Effects of augmented reality-based Otago exercise on balance, gait, and physical factors in elderly women to prevent falls: a randomized controlled trial. J. Phys. Ther. Sci. **29**(9), 1586–1589 (2017)
8. W. Ying, W. Aimin, Augmented reality based upper limb rehabilitation system, in *2017 13th IEEE International Conference on Electronic Measurement & Instruments (ICEMI)* (IEEE, 2017), pp. 426–430
9. C.H. Lee, Y. Kim, B.H. Lee, Augmented reality-based postural control training improves gait function in patients with stroke: randomized controlled trial. Hong Kong Physiother. J. **32**(2), 51–57 (2014)
10. H.L. Phan, J.P. Kim, K. Kim, C.H. Hwang, K.I. Koo, Wrist rehabilitation system using augmented reality for hemiplegic stroke patient rehabilitation: a feasibility study. Appl. Sci. **9**(14), 2892 (2019)

Deep Learning-Based Body Mass Index (BMI) Prediction Using Pre-trained CNN Models

Nur Alifah Megat Abd Mana, Chong Yen Fook, Lim Chee Chin, Vikneswaran Vijean, Saidatul Ardeenawatie, and Hariharan Muthusamy

Abstract Automatically predicting Body Mass Index (BMI) from face images is an interesting and challenging problem in the field of computer vision. Extracted facial features are the most important requirement for estimating BMI readings. The experiment was conducted on a visual BMI database containing 4206 face images. This work studies how to detect and crop face images through image processing methods for BMI prediction. The face detector used in the image pre-processing stage is Multi-Task Convolutional Neural Network (MTCNN). This method was based on Convolutional Neural Network (CNN) and Keras framework. Throughout this research, deep pre-trained CNN models were implemented to go through the training task and to evaluate the performance of those models for the BMI prediction system. Also, the aim of this research is to predict BMI scores and BMI classes from detected face images using three pre-trained CNN models. This research provides some deep insights which are image pre-processing method using MTCNN algorithm performed better than traditional methods such as Haar classifier or Adaboost algorithm and also pre-trained CNN models resulting in better accuracy with less computation time.

Keywords Deep learning · Haar · Adaboost · CNN · MTCNN · BMI

N. A. M. A. Mana · C. Y. Fook (✉) · L. C. Chin · V. Vijean · S. Ardeenawatie
Faculty of Electronic Engineering Technology, Universiti Malaysia Perlis, 02600 Arau, Perlis, Malaysia
e-mail: yenfook@unimap.edu.my

C. Y. Fook · L. C. Chin · V. Vijean · S. Ardeenawatie
Sports Engineering Research Centre (SERC), Universiti Malaysia Perlis, 02600 Arau, Perlis, Malaysia

H. Muthusamy
Department of Electronic Engineering, National Institute of Technology, Srinagar (Garhwal), Uttarakhand, India

F. Thakkar et al. (eds.), *Proceedings of the International e-Conference on Intelligent Systems and Signal Processing*, Advances in Intelligent Systems and Computing 1370,
https://doi.org/10.1007/978-981-16-2123-9_48

1 Introduction

Bodyweight status is becoming a major health issue and the subjects involved to be addressed nowadays are mainly individuals who are related to overweight issues. An increased risk of ten different types of cancers including oesophageal cancer, colorectum, liver, pancreas, gallbladder, breast, endometrium, ovarian, kidney, and prostate was causally linked to the excess body weight [1].

Many researchers have studied the association of excess body weight, cancer, and Body Mass Index (BMI). BMI is widely used as an indicator of total adiposity in the sector of medical and community healthcare [2]. Therefore, BMI is vital medical detail in diagnostic condition estimation and clinical treatment [3]. Generally, a person's weight was taken and divided by height in meters squared (kg/m^2) in order to compute BMI values.

There will be difficulty to retrieve BMI data from elderly folks, paralyzed, a bedridden, or diabetic patient who currently in a critical stage by using weighing and height scale. Plus, when doctors or medical personnel need to get their BMI readings regularly, they will feel uncomfortable and very frustrated. This research will be done to solve these problems by estimating BMI from face images. The suggested research can save time by calculating BMI manually and at the same time can help to minimize the usage of weighing and height scales.

Furthermore, previous studies have applied a geometrical-based approach in their BMI prediction system. But this method showed less accurate results. In order to avoid incorrect patient weight and height assessment, another approach using the image processing method was applied.

To overcome these difficulties, few objectives have been proposed throughout this project which are:

- To detect and crop face images through image processing method for BMI prediction.
- To apply three different pre-trained CNN models and to evaluate the performance of those models for the BMI prediction system
- To predict BMI score and BMI classes from detected face images using three pre-trained CNN models.

The database used is retrieved from Face-to-BMI [4] which consists of 4206 images. The overall process in this project was carried out using Python software since this open-source platform provides related code and also took less computation time. Implementation of the CNN framework in this project was applied to start from image pre-processing until the prediction stage.

2 Literature Review

2.1 Image Pre-processing

Image pre-processing is the most concerned aspect before further detection of the face. The reason is to obtain normalized intensity, uniform size, and shape of pure face images. To reduce noise, improve contrast, smooth, and sharpen the image, it requires basic operations which are image cropping, image resizing, image normalization, and advanced operation such as edge detection.

Face recognition through image processing technologies is used to enhance raw images obtained from cameras which are located on spacecraft or pictures taken for various applications of normal daily life. Image pre-processing techniques including image normalization, image resizing, image enhancement, image filtering, edge detection, etc. have been implemented many years before. Based on the research from authors Wen and Guo [5], normalization basically involves face translation, rotation, and scaling steps in order to create face alignment with typical eye coordinates. In developing a robust computing system, these image pre-processing steps are necessary.

This phase involves detecting a face in the given image data. It aims to remove all the unwanted things from the background picture and to keep only relevant information, the face, from the data. This phase employs various methodologies such as image processing techniques and curvature features. In face recognition technology, face detection is the most essential phase before starting to extract features and to carry out the evaluation task. According to authors Hemalatha et al. [6], face detection in a single image requires four approaches, namely knowledge-based invariants, facial invariants, fitting models, and appearance-based invariants.

Many researchers preferred the Viola–Jones algorithm in their face detection operation since the detection result gives fast detection processing time and higher accuracy. A large number of face detection algorithms are available, based on a large number of features and implementations. The classic Viola–Jones algorithm performed very well in which 100% of faces is detected correctly and thus incorporates it into the proposed algorithm [7]. The first step for a given facial image is to conduct a face detection process and the detection of two eyes will be applied in a face detection-like technique. The distinction is the templates used to train the classifier in AdaBoost. Facial templates are applied for face detection while for eye detection, the templates are replaced by the eyes region.

Then each face detected is standardized according to the detected eye coordinates. After detecting facial features, the face alignment is done on the basis of the eyes that were found in the same image row and this ensures that the faces are not rotated but scaled and positioned so that the eyes are in the same horizontal positions in all face images [8].

Besides, there is a top image pre-processor that was picked by most researchers other than Viola–Jones algorithm which is a Convolutional Neural Network (CNN) model-based face detector. Recently, CNN literally made impressive developments in facial recognition technology. Zhang et al. [9] proposed a new framework in their study to integrate face detection and face alignment through multi-task learning using unified cascaded CNNs. The proposed method consists of three major stages. It produces candidate windows quickly over a shallow CNN during the first stage. In the second stage, it refines windows into a more complex CNN by eliminating a large number of non-face windows. It uses a more effective CNN during the final stage to refine the result again and produce the location of five facial landmarks.

On the other hand, with good frontal or good light, the conventional method can detect the face but cannot detect the face with bad side or bad light, so this method is not appropriate for this field application. Once in the natural environment, the MTCNN algorithm is more robust to changes in light, angle, and facial expression, hence it showed better face detection results. MTCNN has improved performance compared to the conventional approach, can locate the face accurately and the speed is also faster, and can be applied in real-time scenario [10].

In advance, there is a system called FaceNet proposed by Schroff et al. [11] that learns a mapping of face images directly to a compact Euclidean space where distances correspond directly to a measure of face similarity. Using standard techniques with FaceNet embedding as feature vectors, face recognition, verification, and clustering tasks can be implemented easily once this space has been created. This proposed method implemented a deep convolutionary neural network trained to directly optimize the embedding itself, rather than an intermediate layer of a bottleneck as previous deep learning approaches did. They use triplets of roughly aligned matching or non-matching face patches that are generated using a novel online triplet mining method for training tasks. This proposed system achieved 99.63% accuracy.

2.2 *Feature Extraction*

Feature extraction is the method of separating facial features such as eyes, nose, mouth, and eyebrows from the image of the human face. Extraction of the facial feature is very important for initializing processing techniques such as facial tracking, recognition of facial expression, or face recognition. Singh et al. [12] stated that there were difficulties for localizing automatically the facial regions such as eyes, nose, eyebrows, mouth, and jawline, and those facial landmarks are placed on the face region. Hence, a number of exploitations with different methods have been directed to feature extraction from face images to obtain the best result for facial landmark detection.

Extraction of Region of Interests (ROIs) from the detected faces is proposed by Greche et al. [13] in their research, in order to extract information about the facial features by implementing the descriptor Histogram of Oriented Gradient (HOG). The number of features provided by the HOG descriptor can be reduced through Linear Discriminant Analysis (LDA) [14, 15]. The results of the process showed that the applications of HOG and LDA provide a higher recognition rate of 94.66%. ROIs including permanent features such as the eyes, nose, and mouth and transient features such as wrinkles, forehead lines, frown lines are extracted from each window face. This method helps to eliminate these irrelevant facial parts while selecting facial regions with permanent and transient features that are likely to be changed.

In advance, there is a new approach based on a deep CNN model that can be implemented as a feature extractor which is known as the pre-trained CNN model. The main principle of this model was based on a transfer learning basis. Transfer learning uses knowledge from one or more auxiliary domains to extract existing information from a specific domain. Transfer learning is widely used to derive information that exploits a series of carefully designed features. For this study, they use the principle of transfer learning to extract image features using a qualified CNN. In classification layers, the product of transfer learning can be retrained. One of the advantages of this method is not wasting time creating a model with the CNN being trained in large data [16].

According to Bengio et al. [17], VGG-Face is a CNN focused on the VGG16 architecture and is trained on a face dataset of over 2.5 million images and 2622 specific identities collected by the Visual Geometry Group which composed of 22 layers consisting of 13 convolution layers, 5 max-pooling layers, 3 fully connected layers, and a Softmax layer. The predicted input layer image resolution is 224 $\times$ 224. The dataset is chosen as the largest public face dataset in their scope for transfer research. Initially, the images in the dataset are transformed into the required input layers of the CNN. VGG-Face has already been trained in the face recognition task and may not require fine-tuning in the first place. After retraining the CNNs, features are extracted for facial recognition from images.

Singh et al. [12] studied the performance of three pre-trained CNN models, ResNet50, SENet50, and LightCNN-29, for the extraction of features. These models have been implemented in order to analyze the behavior of deep learning networks through face images in terms of their own age bias. The output of these pre-trained models will be used to test deep learning networks with Class Activation Maps (CAMs). CAMs are used to capture the most biased face image regions, focused on by the CNNs. The findings are then stated in terms of the accuracy of the verification which is Genuine Acceptance Rate (GAR) obtained at 1% False Acceptance Rate (FAR). Although feature visualizations are used to compare models beyond verification accuracy and evaluate the regions, the network is learning to extract features with the highest discriminative data across races.

2.3 Model Training and Evaluation

After accomplished with the facial feature extraction stage, a model evaluation task will be assigned to evaluate the performance of the machine learning-based classifier through certain parameters or metrics. Then, regression and classification tasks will be carried out during this stage in order to analyze and distribute the participants into BMI categories which are normal, overweight, and obese. The output of regression is performed in terms of numerical and continuous form, while classification output is showed in terms of categorical and discrete form. In order to provide a better performance in the BMI prediction system, various machine learning approaches in regression and classification were proposed by previous researchers.

For evaluation of the BMI classification task, Lee et al. [3] have implemented the Area Under the Curve (AUC) and kappa values as the main requirement in their research study. The result obtained was based on a tenfold- cross-validation method for statistical analysis of learning algorithms. The performance of the method with applying MDL has shown a higher accuracy in classification task compared to the method without applying MDL, by comparing AUC and kappa values.

The numerous variables, which usually exceed the number of cases, challenge the statistical estimation of BMI and WHR from facial structure and texture. As a result, the out-of-sample predictability is over-estimated within sample correlations. Cross-validation provides a method for calculating prediction accuracy more realistically. We were able to predict 25–27% of variation and 10% of variation in WHR by facial structure using multiple linear regression and leave-one-out cross-validation. Just around 3% of the variability in BMI and WHR was expected in facial texture [18]. Simonyan et al. [19] evaluated the classification performance using two measures which are top-1 and top-5 error. Multi-class classification evaluation is implemented including single-scale evaluation, multi-scale evaluation, multi-crop evaluation, and ConvNet fusion. In terms of the single-net performance, VGG models' structure has achieved the best performance results. On the other hand, the researchers [4] evaluated the three-age-based sub-groups using the pre-trained networks of ResNet50, SENet50, and LightCNN-29. These three pre-trained networks are used for evaluating the performance and visualizations. Also, the networks which are pre-trained on a large set of data capturing multiple variations, that is, ResNet50 and SENet50 obtained an accuracy of 77.9% and 75.5%, respectively, for the third group.

2.4 Research Gap

Most of the researchers studied the relationship between BMI and facial landmarks rather than using facial landmarks to predict BMI. Therefore, the major framework in this research is based on CNN structure since it provides less computation and results in high accuracy in face recognition applications. MTCNN algorithm can provide a shorter detection time and higher accuracy of face detection. Next, for the feature extraction stage, VGG-Face from deep learning is implemented because of the best performance results in terms of less execution time and higher accuracy compared to the conventional method or geometrical-based method. For the model evaluation task, different classifiers will be applied to analyze the best performance of BMI classification within the accuracy result that is appropriate and effective for BMI prediction.

3 Methodology

In this section, the details of the framework for the Body Mass Index (BMI) prediction system will be discussed. This research has three main stages which are image pre-processing, model training and evaluation, and then the final stage is prediction. In general, each stage including image pre-processing, model training and evaluation, and prediction were carried out in order to accomplish the three objectives of this research. Throughout this research, the deep CNN framework was implemented in every stage starting from image pre-processing to prediction task.

Figure 1 below shows the proposed framework of the BMI prediction system. This system begins by retrieving the visual BMI database from previous research [19]. At first, all images in the database undergoes image pre-processing using MTCNN face detector. During the image pre-processing stage, there are three main procedures to undergo including face detection, image cropping, and image resizing. MTCNN algorithm was implemented in this research because it promises better accuracy in face detection and is also prescribed with Keras and VGG-Face library.

3.1 Visual BMI Database Exploratory

This input data is retrieved from previous research Face-to-BMI which is studied by Kocabey et al. [4]. Visual BMI database contains 4206 face images and corresponding BMI data readings for each individual. Total images of each gender are 2438 and 1768 images for male and female, respectively. According to their BMI readings, there are four classes which are categorized into underweight, normal weight, overweight, and obese. Overall, there are 7 images in the underweight class, 680 images in the normal weight class, 1151 images in the overweight class, and 2368 images in the obese class. In this research, face images from the underweight class were eliminated due to insufficient amount of database, and also most of those images were in low-quality resolution. In order to obtain better accuracy of the system, this research will proceed with only three BMI classes which are normal, overweight, and obese.

3.2 Image Pre-processing

In this stage, MTCNN algorithm will be explained since there are a lot of challenges in detecting faces such as orientation, ratio, or scale of the face. This face detector algorithm was first proposed by Zhang et al. [9]. MTCNN algorithm consists of three main network models which are Proposal Network (P-Net), Refinement Network (R-Net), and Output Network (O-Net) in order to detect the faces extremely rapidly with high detection rates.

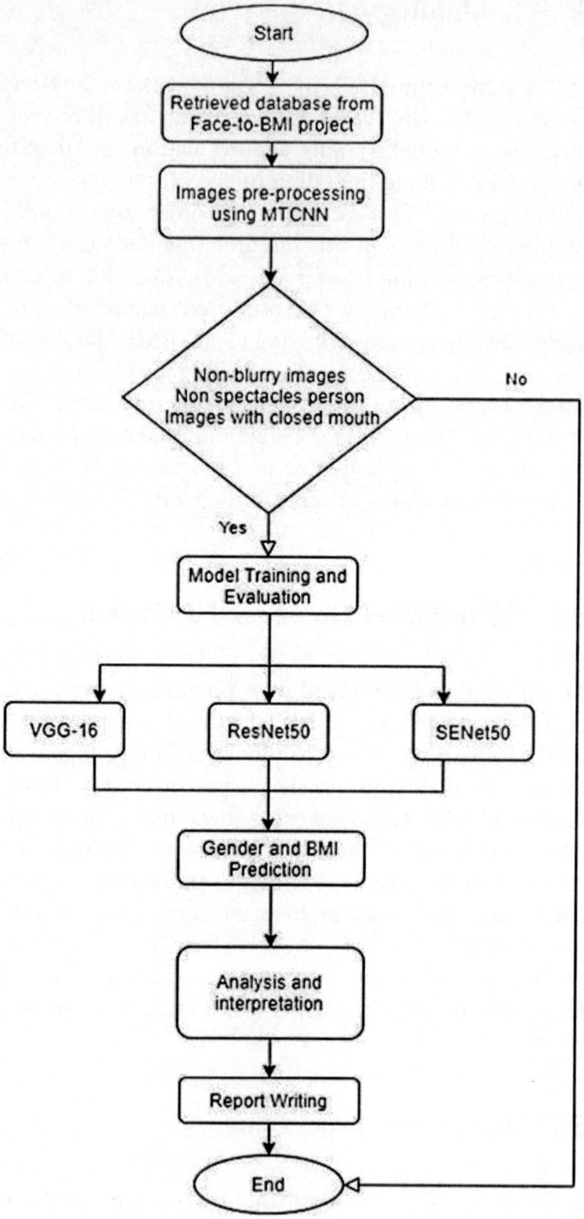

Fig. 1 Proposed framework throughout this BMI prediction project

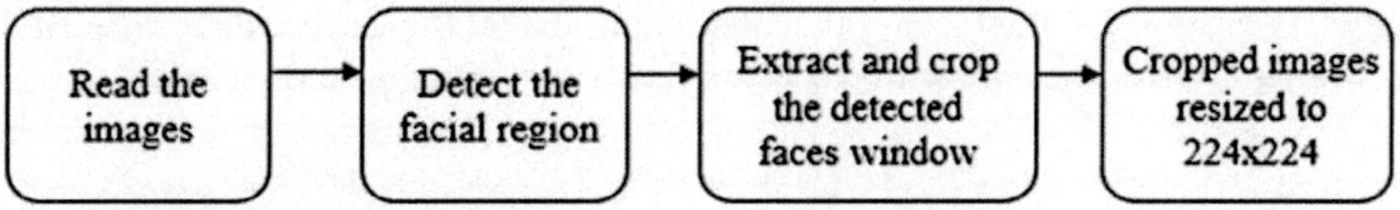

Fig. 2 Block diagram of the MTCNN algorithm framework in pre-processing the face images

During the image pre-processing stage, the main processes are carried out by MTCNN algorithm including reading the facial images, face detection, extracting the facial windows, cropping face images, and resizing detected face images into the specified size. In this research, MTCNN architecture was using Tensorflow, Keras, and VGG-Face framework. The key benefits of this research are firstly, it offers a top-performing pre-trained model and secondly, it can be built as a ready-to-use library of its own code. Figure 2 below shows the block diagram of the MTCNN algorithm framework in pre-processing the face images.

3.3 Model Training and Evaluation

After completing the pre-processing, the face images can feed into model training and evaluation. The main benefit of using pre-trained CNN models is these models can be applied for more than one task. Generally,pre-trained models in this research were implemented to extract the features, to evaluate the performance system, and to predict the gender and BMI score.

For model training and evaluation, the whole process was carried out using open-source libraries which are Scikit-learn and Scipy. These libraries are the most useful libraries when it comes to machine learning and statistical modeling including classification, regression, and so on. Also, they were implemented in most of the researches for many years till today since they can be applied in any kind of machine learning project. This machine learning library has been featured in various algorithms such as Linear Regression, Lasso Regression, Logistic Regression, Support Vector Machine, k-Neighbors, and Random Forest.

3.4 Prediction

After accomplished with pre-processing and model training and model performance evaluation, the final stage in this research is to predict the gender and BMI score from input face images. This was done by extracting the features from the testing set and creating gender and BMI models from loading the saved previous models which are three pre-trained CNN models. Finally, the overall results on gender and BMI score were compared among three different pre-trained CNN models. Since there is no classification for BMI score, thus the output for BMI classes will be manually converted from BMI score with respect to their corresponding BMI score.

4 Results and Discussions

4.1 Data Exploratory Analysis

Examples of original images before proceeding to the image pre-processing stage were shown in Fig. 3.

4.2 Results of Image Pre-processing

In the first step, all images from the visual BMI database which is 4206 images have undergone the process, and there were two output results from the process. There were 4179 images that succeed in the entire process while another 28 face images were failed in order to complete the image pre-processing process.

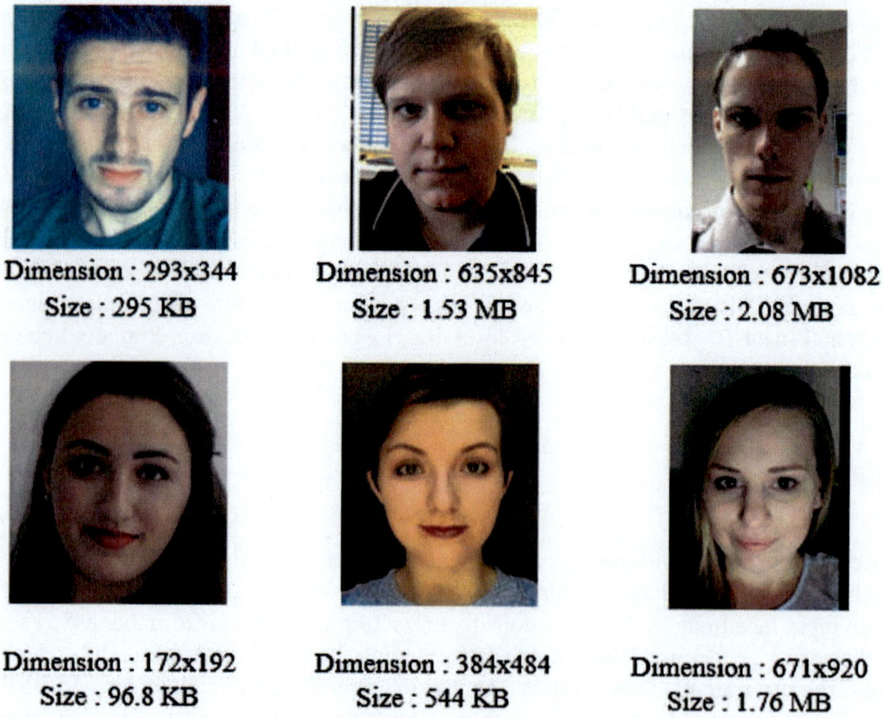

Fig. 3 Examples of original images in different dimensions and sizes

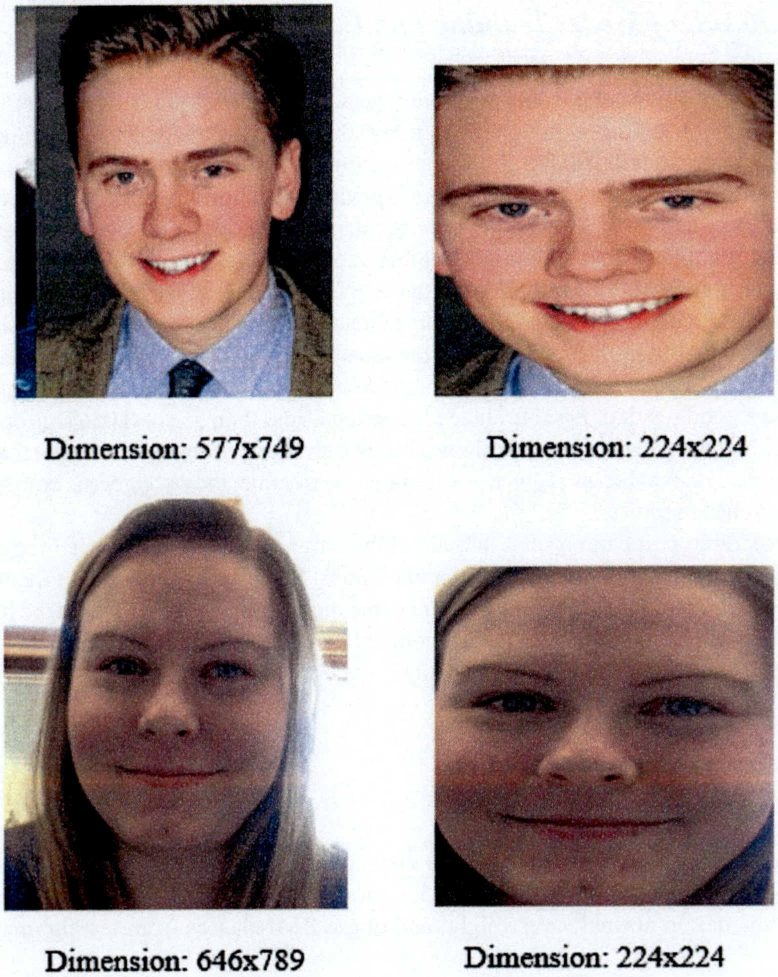

Fig. 4 Examples of face images that succeed to complete the image pre-processing stage. The left side of the figure is before undergoing the process while the right side is after done with the process

Figure 4 shows the result of image pre-processing based on MTCNN algorithm including before and after completing the process successfully. Before undergoing pre-processing, the dimension or input size of face images was different and after done with image pre-processing, all succeed images were performed in 224×224 input size.

4.3 Result of Model Training and Evaluation

Table 1 below shows the result for model evaluation among VGG16, ResNet50 and SENet50 when tuning alpha parameter to 0.001. The score for the area under the ROC curve graph is 1.000 for all three pre-trained models. This result explained that all facial features show a 100% relationship with the gender feature. While for BMI feature, the model that showed good performance is ResNet50 since the correlation coefficient is the lowest compared to other models. This is because the least value of the correlation coefficient indicates that the model showed the better performance of gender and BMI prediction. On the other hand, the least value of RMSE and MAE explained that the models have the least error which means they have a high probability to achieve a good performance. While all three pre-trained CNN models showed a good result of p-value since all p-value is less than 5% or 0.05. This optimal p-value indicates that the models show a better correlation between the facial features and gender and BMI score. Hence, the smaller the p-value, the stronger the correlation between those features.

From Table 1, it can be seen that the alpha parameter at 0.001 showed the better model evaluation result compared to other values. It can be explained that ResNet50 is a good model in the evaluation task since this model consistently showed the lowest value of p-value while the SENet50 model showed the lowest value of RMSE and MAE for the training and evaluation stage.

4.4 Result of Prediction

The final output of this research was to visualize the face images within the gender, BMI score, and also BMI class. Figure 5 shows the result of prediction using the pre-trained model on normal, overweight, and obese BMI classes from testing images.

Table 1 Result of performance metrics among three pre-trained models when tuning alpha parameter to 0.001

Features	Performance metrics	Pre-trained CNN models		
		VGG16	*ResNet50*	*SENet50*
Gender	ROC_AUC score	1.000	1.000	1.000
BMI	RMSE	6.308	5.618	5.214
	MAE	4.543	4.541	3.905
	Correlation	0.387	0.343	0.168
	P-value	0.007	0.003	0.005

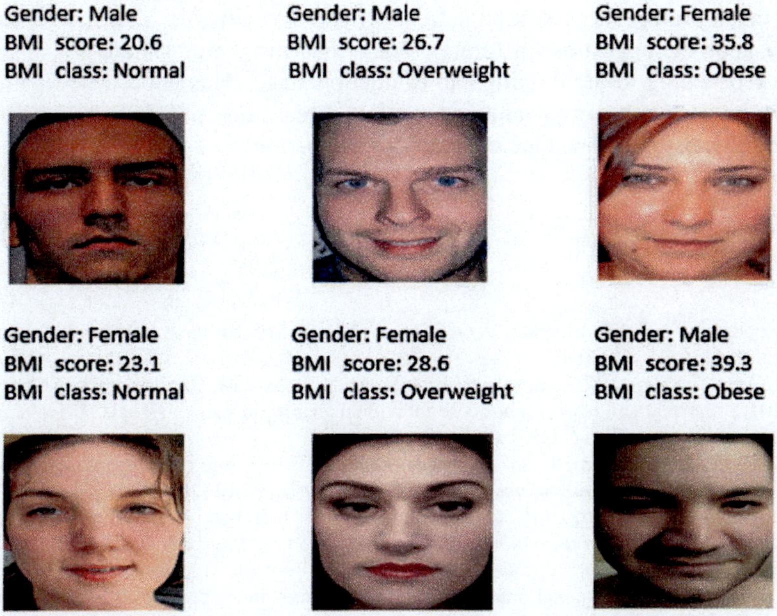

Fig. 5 Visualization of testing images on gender and BMI prediction

5 Conclusion and Future Work

The main objective of this research has been achieved to detect and crop face images through MTCNN algorithm for BMI prediction which was resulting in 4179 face images succeed and 28 face images were rejected during the process. Then, the best pre-trained model performance for model training and model evaluation tasks is the SENet50 model since this model showed the lowest value of RMSE and MAE. The lowest value of error produces good model performance. Also, this research predicts gender and BMI score from facial images more accurately using VGG16 and SENet50 models compared to the ResNet50 model. The accuracy is 60% for both models VGG16 and SENet50 while ResNet50 only showed 56% accuracy in the prediction result. It can be concluded that SENet50 is the best model to implement in the BMI prediction system.

First and foremost, the comparison between the performance of high-quality images and low-quality images should be done for the next research. This is due to developing a BMI prediction model that can predict BMI score and BMI class accurately even though the images are in low-quality condition.

Furthermore, face image samples from underweight BMI classes should be provided in sufficient amounts so that the BMI prediction system can help to predict BMI scores from face images accurately. Next, the application of GUI needs to be done at the end of the BMI prediction stage via Python software since there is still not so much project success through this software compared to MATLAB software. Finally, another suggestion for future work is that fine-tuning should be considered before proceeding to the training and evaluation stage. This is because fine-tuning helps in improving accuracy among the pre-trained models. This technique is also required in most real-time face recognition applications.

References

1. An update of the global impact, in *Obesity and Cancer*, ed. by M. Arnold et al. Cancer Epidemiol. **41**, 8–15 (2016)
2. C.C. Lim, C.Y. Fook, V. Vikneswaran, M.A.H. Abdul Rahim, M. Hariharan, Body Mass Index (BMI) of normal and overweight/obese individuals based on speech signals. J. Telecommun. Electron. Comput. Eng. (JTEC) **10**(1–16) (2018)
3. B.J. Lee, J. Do, J.Y. Kim, *A Classification Method of Normal and Overweight Females Based on Facial Features for Automated Medical Applications*, vol. 12 (2012)
4. E. Kocabey et al., Face-to-BMI: using computer vision to infer body mass index on social media, in *Proceedings of 11th International Conference Web Social Media, ICWSM 2017* (2017), pp. 572–575
5. L. Wen, G. Guo, A computational approach to body mass index prediction from face images. IMAVIS **31**(5), 392–400 (2013)
6. G. Hemalatha, C.P. Sumathi, *A Study of Techniques for Facial Detection*, vol. 5, no. 2 (2014), pp. 27–37
7. J. Salyut, C. Kurnaz, Profile face recognition using local binary patterns with artificial neural network, in *2018 International Conference Artificial Intelligence Data Process* (2018), pp. 1–4
8. R. Raisamo, *Evaluation of Gender Classification Methods with Automatically Detected and Aligned Faces*, vol. 30, no. 3 (2008), pp. 541–547
9. K. Zhang, Z. Zhang, Z. Li, S. Member, Y. Qiao, S. Member, (MTCNN) Multi-task cascaded convolutional networks. IEEE Signal Process. Lett. **23**(10), 1499–1503 (2016)
10. W. Qin, L. Wang, W. Luo, *Face Recognition Based on Gabor Local Feature and Convolutional Neural Network*, vol. 4, no. 1 (2017), pp. 37–42
11. F. Schroff, D. Kalenichenko, J. Philbin, FaceNet: a unified embedding for face recognition and clustering, in *Proceedings of IEEE Computer Social Conference Computer Vision and Pattern Recognition*, (2015), pp. 815–823. https://doi.org/10.1109/CVPR.2015.7298682
12. M. Singh, S. Nagpal, R. Singh, M. Vatsa, On recognizing face images with weight and age variations. IEEE Access **2**, 822–830 (2014)
13. L. Greche, M. Akil, R. Kachouri, N. Es-sbai, A new pipeline for the recognition of universal expressions of multiple faces in a video sequence. J. Real-Time Image Process. (2019)
14. C.Y. Fook, M. Hariharan, S. Yaacob, A.H. Adom, Malay speech recognition in normal and noise condition, in *2012 IEEE 8th International Colloquium on Signal Processing and its Applications*. Melaka (2012), pp. 409–412
15. S.K. Eng, H. Ali, A.Y. Cheah, Y.F. Chong, Facial expression recognition in JAFFE and KDEF datasets using histogram of oriented gradients and support vector machine. IOP Conf. Ser. Mater. Sci. Eng. **705**, 012031 (2019)
16. A. Phornchaicharoen, P. Padungweang, Face recognition using transferred deep learning for feature extraction, in *ECTI DAMT-NCON 2019—4th International Conference on Digital Arts, Media and Technology. 2nd ECTI Northern Section Conference on Electrical, Electronics, Computer and Telecommunications Engineering* (2019), pp. 304–s309
17. Y. Bengio, I.J. Goodfellow, A. Courville, Deep Learn. Nature **521**, 436–444 (2015)
18. C. Mayer, S. Windhager, K. Schaefer, P. Mitteroecker, *BMI and WHR Are Reflected in Female Facial Shape and Texture: A Geometric Morphometric Image Analysis* (2017), pp. 1–13

19. K. Simonyan, A. Zisserman, Very deep convolutional networks for large-scale image recognition, in *3rd International Conference on Learning Represent. ICLR 2015—Conference Track Proceedings* (2015), pp. 1–14

An Effective Approach to Classify White Blood Cell Using CNN

Kinjal A. Patel, Kinjal K. Gandhi, and Ankit S. Vyas

Abstract The white blood cell plays an important role in the human body, protecting the human body from being invaded by the WBC by removing parasites, fungi, bacteria, and viruses. There are many types of white blood cells, such as eosinophils, lymphocytes, monocytes, and neutrophils. Identification and classification of white blood cells is an important application in clinical trials. The WBC statistics provide information about the state of health of the human body. Many diseases and infections depend on white blood cells in the bloodstream, thus identifying and separating the different WBCs. Nowadays, blood cell diagnostics is performed by laboratory pathologists. Symptomatic testing, however, is a major time-consuming, expensive, and error-free test. An in-depth learning approach, especially the implementation of interactive neural networks, offers additional benefits in the medical field, where a large number of images are analyzed and processed. CNN is expected to acquire a number of semantic features, which will affect the accuracy of the sections to some degree. CNN automatically captures features from a single image and identifies specific types of white cells in different images. In this study, it is proposed to differentiate different types of WBC such as eosinophils, lymphocytes, monocytes, and neutrophils in confusing images of neural network formation. The test will take place on the WBC database with 12,436 images. The improved model was tested to give an accurate result with sectional accuracy of 97% in all classes.

Keywords Image processing · White blood cells · Deep learning · Convolutional neural network

K. A. Patel (✉)
Department of Computer Application and IT, Gujarat Law Society University, Ahmedabad, India

K. K. Gandhi
Department of Information Technology, Parul University, Vadodara, India

A. S. Vyas
Oracle HRMS Consultant, Evolutionary Systems Pvt.Ltd, Ahmedabad, India

F. Thakkar et al. (eds.), *Proceedings of the International e-Conference on Intelligent Systems and Signal Processing*, Advances in Intelligent Systems and Computing 1370,
https://doi.org/10.1007/978-981-16-2123-9_49

1 Introduction

White blood cells are part of the body's cells. In natural terms, red cells, white cells, and blood platelets are three types of cells. They are divided by color and size. White blood cells form in the bone marrow and lymphoid tissue [1]. In the blood, leukocytes play a vital role in the human immune system, so this cell is called an immune cell. Typically, hematologists use granular information in leukocytes to divide white blood cells into granule cells: neutrophils, eosinophils, basophils, and non-fertilized cells: monocyte and lymphocyte percentages of blood in these five different cell types among those without the disease [2]. To date, automated cell differentiation systems have not been able to meet the complexity of actual clinical needs [3]. Many research teams are focusing on the development of computer programs that can analyze a wide variety of medical images and gather useful information from medical professionals [3]. Doctors often use this data to diagnose the disease. Therefore, studying the differentiation of white blood cells is very important for medical professionals.

Neutrophils are very high and responsible for protection against bacterial or fungal infections. Eosinophils form 2% of the WBCs and act in response to allergies [4]. The lymphocytes act as specific signals for foreign agents and as markers and detectors in the host. Monocytes directly eliminate pathogenic cells and clean them from garbage and debris. The role of a signature in various sports is a clinical diagnosis and testing is about white blood cells. The body eliminates the hidden infection in the body and warns the hematologist [5]. A rarely growing symptom in WBC is called leukocytosis. It is helpful for nurses to monitor chemotherapy or radiation services in people with cancer.

There are many systems on the market for automatic pre-division of white blood cells. These analysts rely on a microscope to obtain images from blood smears and are a major algorithm for classifying blood cell images. These systems are more expensive due to the complexity of the image acquisition phase and the separation stages [6].

Eosinophils, lymphocytes, monocytes, and neutrophils: the main purpose of the CNN-based system for the automation of four blood classes.

2 Literature Review

Ownership of WBC quotes has been read and discovered by many founders and researchers on the implementation of the method. Image processing methods incorporate most of this process before using taxonomy algorithms, which separate the area of interest behind, the WBC nucleus (known as the nucleus), and the removal and deleting of sound in thumbnails. Do it again. To some extent, it is important to differentiate between the desired domain [7] and many other methods such as mergers have been modified [8]: restricted [9, 10], grammar operators [11], Gram-Schmidt orthogonization method [1 increase], margin marking [3], increased acreage [12], and water flow [13].

The first cellular division algorithms were mainly the KNN algorithm, the Bayesian division, the SVM division, and more. We have briefly reviewed this section.

The main idea of the KNN algorithm is that most models close to the feature area are within a certain range. Note that the models in this category have all the other model features as well. This method determines the model category to be classified based on the adjacent model category in determining the classification decision. The KNN method contains only a small number of neighboring models in class determination. They first used histogram limits to separate white blood cells and then separated them using distance division. Bakheda et al. used entropy-based and non-invasive methods for partial cells and were categorized by a detection rate of 90.14% [14].

Bayesian taxation is based on statistical tax calculations and uses its knowledge of potential statistics to classify data. In many categories, the naïve Bayes algorithm can be compared to a decisive tree and a neural network algorithm. Sinha and Ramakrishnan et al. used Bayes's economic tax to classify cells, and the detection rate was 82.3% [15]. Separating bone marrow images of the Ellis Fisher Cancer Center (only one cell per image) in the center of Mississippi, der-Yupenen and Dampongsa et al. found Beses tax rates and identification rates are 77% [16, 17]. Ghosh et al. Sect. 150 use a watershed algorithm for cell imagery and classifies it using Bayesian taxonomies with a detection rate of 3.2% [18].

The concept of SVM tax is essentially the same as the deferred LR taxonomic structure. This is to obtain a set of weight coefficients that can be separated after a direct representation. SVM first sets up a separate hyperplane and then flies to a split plane. The classic SVM algorithm is suitable for only two types of partition problems. After adjustment, SVM can also be used for most category problems. In the actual use of white blood cell isolation, it is often necessary to solve the problem. In many categories, for example, the problem of leukocyte classification has been studied and can be solved by combining multiple binary SVMs. Regatofigi et al. used Gram–Schmidt orthogonalization and snake algorithm in the 400 smears section and split it using SVM. Their level of recognition is 90% [16].

Recently, fine neural networks have been widely used in various fields of image classification. In particular, the conflict neural network (CONNETS) [14] has achieved unprecedented results in the 2012 ImageNet Large Scale Visual Recognition Challenge, where natural images are categorized by 1000 categories in the IMAGNET database [19]. CT imaging can greatly improve the performance of various medical imaging applications, such as pulmonary and pulmonary disease [20, 21], brain tissue fragmentation [22] (pixel separation), and MRI scans. According to vascular fundus images [23] and patient-level cervical images, multimodal data is used to identify cervical neoplasia (CIN, especially CIN2+) [24] In addition, the presented issues have shown excellent efficacy in image classification of cells, such as pleural cancer [23] and human epithelial cell images [25].

3 Proposed Work

3.1 Dataset Collection

Data is the most important parameter in any research activity. The data used in this research work was collected from [26]. The database contains four classes—eosinophil, lymphocyte, monocyte, and neutrophil. A total of 3120 images in the eosinophil phase, 3102 images in the lymphocyte phase, 3091 images in the monocyte phase, and 3123 images in the neutrophil phase were used. Figure 1 shows the sample images of each class.

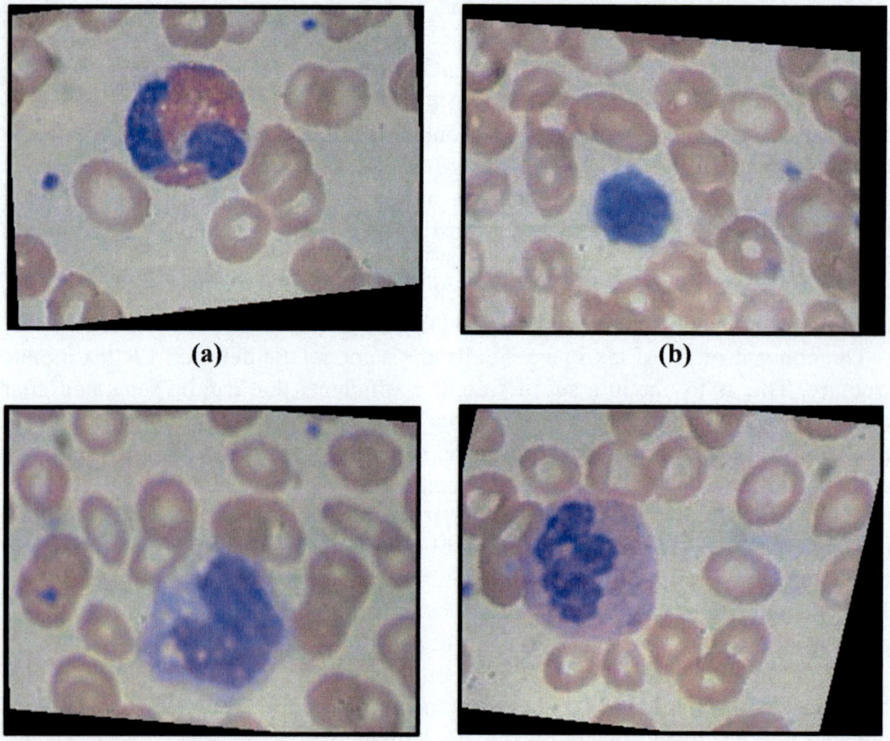

(a) (b) (c) (d)

Fig. 1 Dataset **a** Eosinophil image, **b** Lymphocyte image, **c** Monocyte image, **d** Neutrophil image

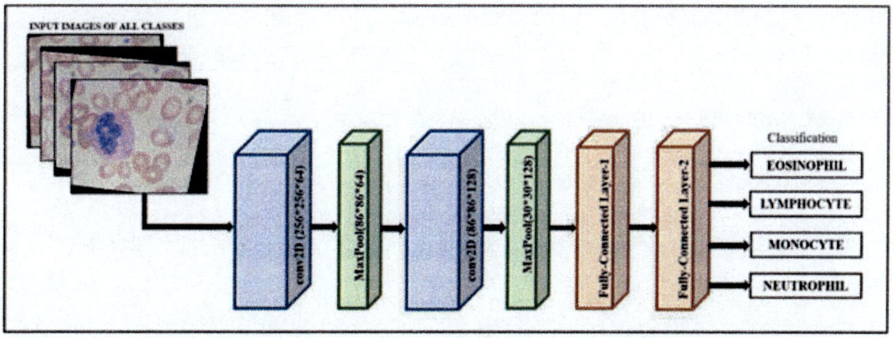

Fig. 2 Architecture of proposed LITENET model

3.2 Proposed LITENET Architecture

The convolutional neural network (CNN) is a leading model that is widely involved in image segregation activities. The CNN's main parameters such as convolutional layers, fully integrated layer neurons, batch, and kernel size, and pooling size are evident. Also, the other parameters such as depth, height, and width are tested throughout the CNN layer. The proposed CNN structure was linked to all of the above parameters after the investigation of a specific pre-training facility used by [1, 3, 4, 7, 27, 28]. Therefore, we designed the LITENET model from scratch with the appropriate parameters for this research work. The proposed LITENET model is shown in Fig. 2.

Earlier it was suggested that the LITENET model has two levels of decision-making. The Relu activation function is applied after each convolutional layer followed by a consolidation layer. The proposed LITENET model contains a total of two layers that are fully connected. The 0.4 value exit method was used behind the fully connected layers. It randomly sets the interface of the previous layer to overcome the problem of model overlap. The softmax activation function is the latest layer of output in this proposed model.

The size of the inserted image was 256 × 256. The image size was changed to 256 × 256. A 2 x 2 kernel size is used throughout this test. To avoid shrinkage, a 2 × 2 size of zero paddings is used in the test.

One of the most important limitations of the LITENET model is the size of the batch. A total of 64 batch sizes were used in this test. A larger batch size provides an improved gradient and slows closer, but too large batch size can have a memory problem. Some hyperparameters of CNN architecture are steps per epoch. This can be calculated by the size of the training sample divided by the batch size. In this trial, 8707 image samples were considered for training. Therefore, the steps per hour are 137 ending as 8707/64. LITENET model adjustments are recorded in Table 1.

The Adam tool was used to train the model with a reading level of 0.0001. Blood cell division testing using the LITENET model was achieved in Google Colaboratory

Table 1 Configuration details of the LITENET model

Layer	Processing unit	Input dimension	Output dimension	Kernel size	Batch size	Zero padding
1	Conv2D	256 × 256	256 × 256 × 64	2 × 2	64	2 × 2
2	MaxPool	256 × 256 × 64	86 × 86 × 64	2 × 2	64	–
3	Conv2D	86 × 86 × 64	86 × 86 × 128	2 × 2	128	2 × 2
4	MaxPool	86 × 86 × 128	30 × 30 × 128	2 × 2	128	–
5	Fully connected layer 1 with 1024 neurons followed by ReLu activation and a DropOut rate of 0.4					
6	Fully connected layer 2 with 4 neurons followed by ReLu activation and a DropOut rate of 0.4					
7	Output layer with softmax activation function					

with the Tesla K80 GPU configuration. The Tesla K80 GPU has been used by libraries such as Tensorflow and Cameras.

4 Performance Evaluation

Examination of the proposed LITENET model was taken from the database [26]. The database was divided into 70 and 30% for training and testing, respectively. Figures relating to the images used for training and assessment in each class are listed in Table 2. The LITENET structure is shown in Fig. 2. To avoid overstatement, 60 times have been used for model training. The time taken for each session is ~126 s during the training of the LITENET model. The performance test of the LITENET model was tested using a matrix for confusion and accuracy, memory, f1 points, and accuracy. The statistics of the results obtained for each class are listed in Table 2.

Table 2 Result statistics achieved by LITENET model

Class	Training data samples (70%)	Testing data samples (30%)	Precision	Recall	F1-score	Accuracy
Eosinophil	2184	936	0.91	0.97	0.94	0.97
Lymphocyte	2172	930	1.00	0.99	1.00	0.97
Monocyte	2164	927	1.00	1.00	1.00	0.97
Neutrophil	2187	936	0.96	0.91	0.93	0.97
Overall results	8707	3729	0.97	0.97	0.97	0.97

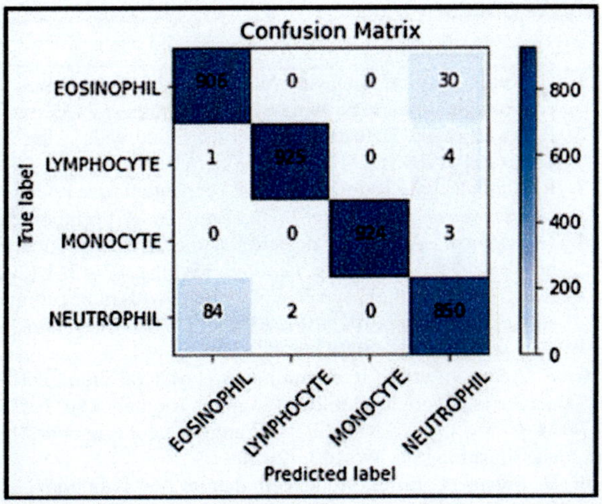

Fig. 3 Confusion matrix obtained on the test dataset

The result achieved by the proposed LITENET model from the dataset is verified in the confusion matrix of Fig. 3.

The result obtained in the eosinophil class is 91% accuracy, 97% memory, 94% f1 points, and 97% accuracy; class lymphocyte equals 100% accuracy, 99% memory, 100% f1 school, and 97% accuracy; class monocyte equals 100% accuracy, 100% memory, 100% f1 school, and 97% accuracy; the neutrophil class has 96% clarity, 91% memory, 93% f1 points, and 97% accuracy. The overall result was obtained as 97% with accuracy, recall, f1 points, and accuracy of all test data.

5 Conclusion

To improve the problem of white blood cell diagnostic method, the LITENET model was proposed, showing outstanding performance in white cell image classification, with high clarity, memory, f1 points, and accuracy. Our experiments show that the image-based approach to the convolutional network is highly promising for the separation of white blood cells. In the future, we hope that our proposed model can be used on mobile devices, in addition to the environment with inadequate health facilities and medical conditions. It will definitely increase the speed of identifying the type of white blood cells.

References

1. H. Kutlu, E. Avci, F. Özyurt, White blood cells detection and classification based on regional convolutional neural networks. Med. Hypotheses **135**, 109472 (2020)
2. P. Tiwari et al., Detection of subtype blood cells using deep learning. Cogn. Syst. Res. **52**, 1036–1044 (2018)
3. R. Adollah et al., Blood cell image segmentation: a review, in *4th Kuala Lumpur International Conference on Biomedical Engineering 2008* (Springer, Berlin, Heidelberg, 2008)
4. J. Zhao et al., Automatic detection and classification of leukocytes using convolutional neural networks. Med. Biol. Eng. Comput. **55**(8), 1287–1301 (2017)
5. R.R. Tobias, et al., Faster R-CNN model with momentum optimizer for RBC and WBC variants classification, in *2020 IEEE 2nd Global Conference on Life Sciences and Technologies (LifeTech)*. (IEEE, 2020)
6. A. Acevedo, et al., Recognition of peripheral blood cell images using convolutional neural networks. Comput. Methods Programs Biomed. **180**, 105020 (2019)
7. M.-C. Su, C.-Y. Cheng, P.-C. Wang, A neural-network-based approach to white blood cell classification. Sci. World J. (2014)
8. B. Sheng et al., A blood cell dataset for lymphoma classification using faster R-CNN. Biotechnol. Biotechnol. Equip. **34**(1), 413–420 (2020)
9. P. Yampri, C. Pintavirooj, S. Daochai, S. Teartulakarn, White blood cell classification based on the combination of Eigen cell and parametric feature detection, in *Industrial Electronics and Applications, 2006 1ST IEEE Conference on IEEE*, (2006), pp. 1–4
10. S.F. Bikhet, A.M. Darwish, H.A. Tolba, S.I. Shaheen, Segmentation and classification of white blood cells, in *Acoustics, Speech, and Signal Processing, 2000. ICASSP'00. Proceedings. 2000 IEEE International Conference*, vol. 4 (IEEE, 2000), pp. 2259–2261
11. H. Sheikh, B. Zhu, E. Micheli-Tzanakou, Blood cell identification using neural networks, in *Bioengineering Conference, 1996, Proceedings of the 1996 IEEE Twenty-Second Annual Northeast*, (IEEE, 1996), pp. 119–120
12. J.-M. Chassery, C. Garbay, An iterative segmentation method based on a contextual color and shape criterion. IEEE Trans. Pattern Anal. Mach. Intell. (6), 794–800 (1984)
13. K. Jiang, Q.-M. Liao, S.-Y. Dai, A novel white blood cell segmentation scheme using scale-space filtering and watershed clustering, in *Machine Learning and Cybernetics, 2003 International Conference on IEEE*, vol. 5 (IEEE, 2003), pp. 2820–2825
14. L.M. Soria-Morillo et al., Deep neural network for traffic sign recognition systems: an analysis of spatial transformers and stochastic optimisation methods. Neural Netw. **99**(12), 158–216 (2018)
15. N. Sinha, A.G. Ramakrishnan, Automation of differential blood count, in *Proceedings of TENCON Conference on Convergent Technologies for Asia-Pacific Region*, vol. 2 (2003), pp. 547–551
16. N. Theera-Umpon, P.D. Gader, System-level training of neural networks for counting white blood cells. IEEE Trans. Syst. Man Cybernet. Part C **32**(1), 48–53 (2002)
17. N. Theera-Umpon, Automatic white blood cell classification using biased-output neural networks with morphological features. Thammasat Int. J. Sci.Tech. **8**(1), 64–71 (2003)
18. M. Ghosh, D. Das, S. Mandal, et al., Statistical pattern analysis of white blood cell nuclei morphometry, in *Students Technology Symposium (TechSym)*, (2010), pp. 59–66
19. A. Krizhevsky, I. Sutskever, G.E. Hinton, ImageNet classification with deep convolutional neural networks. Adv. Neural Inform. Process. Syst. 1097–1105 (2012)
20. H.C. Shin et al., Deep convolutional neural networks for computer-aided detection: CNN architectures, dataset characteristics and transfer learning. IEEE Trans. Med. Imaging **35**(5), 1285–1298 (2016)
21. H.R. Roth, L. Lu et al., Improving computer-aided detection using convolutional neural networks and random view aggregation. IEEE Trans. Med. Imaging **35**(5), 1170–1181 (2016)
22. T. Xu, H. Zhang, C. Xin, E. Kim et al., Multi-feature based benchmark for cervical dysplasia classification evaluation. Pattern Recogn. **63**, 468–475 (2017)

23. P. Liskowski, K. Krawiec, Segmenting retinal blood vessels with deep neural networks. IEEE Trans. Med. Imaging **35**(11), 2369–2380 (2016)
24. P. Moeskops, M.A. Viergever et al., Automatic segmentation of MR brain images with a convolutional neural network. IEEE Trans. Med. Imaging **35**(5), 1252–1261 (2016)
25. P. Buyssens, A. Elmoataz, O. Lézoray, Multiscale convolutional neural networks for vision-based classification of cells. Asian Conf. Comput. Vis. **7725**, 342–352 (2012)
26. Kaggle [Online], https://www.kaggle.com/paultimothymooney/blood-cells. Accessed 10 Oct 2020
27. T.T.P. Thanh et al., Leukemia blood cell image classification using convolutional neural network. Int. J. Comp. Theory Eng. **10**(2), 54–58 (2018)
28. G. Liang, et al., Combining convolutional neural network with recursive neural network for blood cell image classification. IEEE Access **6**, 36188–36197 (2018)

Video Summarization for Multiple Sports Using Deep Learning

Chakradhar Guntuboina, Aditya Porwal, Preet Jain, and Hansa Shingrakhia

Abstract This paper proposes a computationally inexpensive method for automatic key event extraction and subsequent summarization of sports videos using scoreboard detection. A database consisting of 1300 images was used to train a supervised learning-based object detection algorithm, You Only Look Once (YOLO). Then, for each frame of the video, once the scoreboard was detected using YOLO, the scoreboard was cropped out of the image. After this, image processing techniques were applied on the cropped scoreboard to reduce noise and false positives. Following that, the processed image was passed through an optical character recognizer (OCR) to get the score. A rule-based algorithm was run on the output of the OCR to generate the timestamps of key events based on the game. Finally, clips of length 30 s around every detected event (15 s before and after the event) were extracted from the video file and stitched together to form the highlights video. The proposed method is best suited for people who want to analyze the games and want precise timestamps of the occurrence of important events. The performance of the proposed design was tested on videos of Bundesliga, English Premier League, ICC WC 2019, IPL 2019, and Pro Kabaddi League. An average F1 score of 0.979 was achieved during the simulations. The model was also tested on different video formats, and these were .mp4, .avi, .mov, and .mkv. The algorithm was trained on five different classes of three separate games (soccer, cricket, and kabaddi). The design was implemented using Python 3.7.

Keywords You Only Look Once (YOLO) · Optical Character Recognizer (OCR) · Intersection Over Union (IOU) · Region of Image (ROI) · True Positives (TP) · True Negatives (TN) · False Positives (FP) · False Negatives (FN) · Frames Per Second (FPS) · Mean Average Precision (MAP)

C. Guntuboina (✉) · A. Porwal · P. Jain · H. Shingrakhia
Electronics and Communication Engineering Department, IITE, Indus University, Ahmedabad, India
e-mail: guntuboinachakradhar.16.ec@iite.indusuni.ac.in

F. Thakkar et al. (eds.), *Proceedings of the International e-Conference on Intelligent Systems and Signal Processing*, Advances in Intelligent Systems and Computing 1370,
https://doi.org/10.1007/978-981-16-2123-9_50

1 Introduction

Sports is a major part of the entertainment industry, and with the rise in internet accessibility in recent years, its popularity and viewership have only increased.

A majority of those people find it difficult to devote time to watch full-length matches of their preferred sport in their busy lives and, as such, most individuals prefer to watch a shorter (summarized) version of the match. Manual summarization or highlight generation is a tedious and time-consuming process requiring professional video editing tools which puts an upper bound on the amount of video footage that can be summarized in a given amount of time. Moreover, video editing, in general, has a very steep learning curve meaning that not everyone—given access to a full-length video—can summarize a sports video without the necessary skills that are required for it. As a natural consequence of this, a lot of research work has been done to automatically summarize the videos into important video clips which are significantly shorter than full-length sports videos, and thus enable the viewers to enjoy the match without having to watch the entire match. Traditional works relied on image processing, audio signal processing, computer vision techniques, etc. for video summarization. But with the recent resurgence of neural networks owing to the exponential growth of computational power, a whole new way of extracting highlights using deep learning has emerged. The objective of this paper is to exploit the computational power by using image detection and classification techniques to summarize almost any sports video. The proposed methodology in brief uses image detection and localization techniques to determine the location of the scoreboard in a video frame, crop the scoreboard, run OCR over the cropped scoreboard, and compare the score with the previous score to detect important events in the video and combine video clips of 30 s around the events into a single video file.

2 Literature

Some of the previous works in this field include using oriented fast, rotated brief (ORB) to detect the bowler run-up sequence (BRS) in cricket and classify it as a highlight [1]. Pushkar Shukla et al. [2] have combined event-driven and excitement-based methods to generate highlights for cricket. In [3], they have used the intensity profile of the umpire's gestures to detect the occurrence of an event. Javed et al. [4] have used the fact that replays are generally preceded and followed by gradual transition effects to extract the replays and then pass them through an extreme learning machine (ELM) for the detection of key events (boundary, six, and wicket). The authors used three-dimensional convolutional neural networks (3D CNNs) and long short-term memory (LSTM) to extract features from a soccer video for summarization [5, 6]. Yu et al. [7] have designed an architecture that detects events and proposes temporal boundaries to the events to form what they call stories, i.e., the event clip and some additional context around it. And while these are all novel methods, they all focus on a single sport for summarization and would translate poorly to other sports. Then there are those that try to generalize the summarization for multiple sports like Kathirvel et al. [8] who have used referee's whistle sound as a cue for extraction of highlights. This would fail in games that don't require a whistle or when the whistle is inaudible due to background noise, etc. In [9], Antonio Tejero-de-Pablos et al. have proposed learning players' actions using deep neural networks and use that to detect important events applied to a game called Kendo (Japanese fencing). This would translate poorly to games where there are a lot of players like in cricket, soccer (football), etc. In this paper, we present a computationally inexpensive method to extract event-based highlights which could work on any game which has a scoreboard. The idea is to use a state-of-the-art object detection model called YOLO and train it to detect the scoreboard of the desired game. Then we run the model over images extracted from the video at the rate of one image per second and run an OCR over it to keep track of the score. If the score changes by some predefined delta, then an event is detected at that frame. The simplicity of our model is such that with only 250–300 images of the scoreboard for training, anyone can deploy this on any desired sports video with the domain knowledge of the sport to fix the predetermined delta. The rest of the paper is organized as follows. Section 3 gives information about YOLO and why we have chosen YOLO, Sect. 4 presents the methodology, Sect. 5 presents the results and performance of our model, and then finally, Sect. 6 presents concluding remarks and limitations/future work for the model.

3 You Only Look Once

3.1 A Brief Comparison of Object Detection Algorithms

There are a number of object detection algorithms such as R-CNN [10], faster R-CNN [11], YOLO. In this section, we discuss the reason(s) behind choosing YOLO as our go-to algorithm. This is done by briefing other algorithms, their limitations, and how YOLO counters/overcome those limitations. Region-based convolutional neural networks (R-CNN): Focus is on dividing the image into 2000 small regions and then finding the object in each region, an offset is also added to improve the accuracy. Limitations: Takes a huge amount of time to classify 2000 sections. Not practical for real-time detections as each image takes approximately 47 s to process (classify). Faster R-CNN: The network itself learns the region proposals. It is a lot faster than R-CNN and fast R-CNN [12] but still when it comes to real-time performance the algorithm is not suitable. It can be observed that almost every algorithm is based on three tasks to achieve object detection: (1) Use region proposal method to generate potential bounding boxes in an image. (2) Run the classifier on these boxes. (3) After classification, perform postprocessing to tighten the boundaries of the bounding boxes, remove duplicates. These tasks prove to be very complex and hard to optimize at the same time as each of the components is required to be trained separately. This is where YOLO comes in. YOLO: This algorithm is different from all the traditional object detection algorithms in the way that it considers object detection as a regression problem rather than a classification problem. A single convolutional neural network is used to predict the bounding box as well as the probabilities. A prediction having the threshold (maximum IOU) is considered as detection and the rest is discarded. The speed is of the order of 45 frames per second, the fastest available algorithm as of now. Limitation: The algorithm cannot detect small objects because of the spatial constraints of the algorithm. Another limitation of the algorithm is that it lacks accuracy compared to another algorithm such as faster R-CNN.

3.2 How YOLO Works

The algorithm divides the input image into an S S grid. If the center of an object falls into a grid cell, that grid cell is responsible for detecting that object. Each grid cell predicts a fixed number of boundary boxes and confidence scores for those boxes to locate the object.

Training data is a bounding box along with the class label(s). This is referred to as 'ground truth box',

$$b = \left[b_{\mathrm{x}}, b_{\mathrm{y}}, b_{\mathrm{h}}, b_{\mathrm{w}}, classname(or\ number)\right] \tag{1}$$

where b_x, b_y is the midpoint of annotated bounding box and b_h, b_w is height and width of box.

Output or prediction is bounding box b along with class c for an image i. Formally:

$$y = \left[p_{\mathrm{c}}, b_{\mathrm{x}}, b_{\mathrm{y}}, b_{\mathrm{h}}, b_{\mathrm{w}}, cn\right] \tag{2}$$

where b_x, b_y is the midpoint of annotated bounding box. b_h, b_w is height and width of box. p_c is the probability of having class(es) c in 'box' b.

YOLO uses a linear activation function for the final layer and the following leaky rectified linear activation for the rest of the layers:

$$\begin{aligned} \varphi(x) &= x \quad\ \ if\ x > 0 \\ \varphi(x) &= 0.1 * x\ if\ x \le 0 \end{aligned} \tag{3}$$

YOLO uses sum-squared error between the predictions and the ground truth to calculate the loss. The loss function comprises:

- The classification loss—If an object is detected, the classification loss at each cell is the squared error of the class conditional probabilities for each class.
- The localization loss—Measures the errors in the predicted boundary box locations and sizes. Only the box responsible for detecting the object is considered.

– The confidence loss—Computes the loss associated with the confidence score for each bounding box predictor.

The final loss adds localization, confidence, and classification losses together.

$$Loss = \left(\lambda_{coord} \sum_{i=0}^{S^2} \sum_{j=0}^{B} \Omega_{ij}^{obj} \left[\left(x_i - \hat{x}_i\right)^2 + \left(y_i - \hat{y}_i\right)2 \right] + \right.$$
$$\lambda_{coord} \sum_{i=0}^{S^2} \sum_{j=0}^{B} \Omega_{ij}^{obj} \left[\sqrt{(w_i} - \sqrt{\hat{w}_i})^2 + \left(\sqrt{(h_i} - \sqrt{\hat{h}_i})^2 \right) \right] +$$
$$\sum_{i=0}^{S^2} \sum_{j=0}^{B} \Omega_{ij}^{obj} \left[\left(C_i - \hat{C}_i \right)^2 \right] + \tag{4}$$
$$\lambda_{nobj} \sum_{i=0}^{S^2} \sum_{j=0}^{B} \Omega_{ij}^{nobj} \left[\left(C_i - \hat{C}_i \right)^2 \right] +$$
$$\left. \sum_{i=0}^{S^2} \Omega_i^{obj} \sum_{c \in classes} \left[\left(p_i(c) - \hat{p}_i(c) \right)^2 \right] \right)$$

where

$\Omega_t i^{obj}$ 1, if an object appears in cell i, otherwise 0.
$p\hat{}_i(c)$ denotes the conditional class probability for class c in cell i.
$\Omega_{ij}{}^{obj}$ 1, if the jth boundary box in cell i is responsible for detecting the object, otherwise 0. λ_{coord} increases the weight for the loss in the boundary box coordinates.

$C\hat{}_i$ is the box confidence score of the box j in cell i.

$\Omega ijnoobj$ is the compliment of $\Omega ijobj$.

λ_{noobj} weights down the loss when detecting the background.

For a detailed explanation of how the algorithm works refer to [13].

3.3 Advantages of YOLO

YOLO's convolutional network understands general object representation, allowing it to be trained on real-world images and predictions on artwork with fair accuracy. Unlike the traditional object detection algorithms that treat object detection as a classification problem, YOLO treats the same as a regression problem and performs all the tasks using a single convolutional neural network. Convolution is done on the entire image at once and only once. Hence, with a smaller architecture, fair results are achieved. YOLO predicts all the bounding boxes in one go, unlike other algorithms which have to go through the image multiple times, predicting bounding boxes and then again process them to remove duplicate ones. R-CNN is highly accurate but compromises with the speed, having the ability to process only 5 fps (frames per second). On the other hand, YOLO (V3-used in the project) is lightning fast having the ability to process 45 fps, faster than real time. YOLO makes fewer background mistakes, owing to the fact that it divides the image in S * S grid and then each cell is used to predict the bounding box, if the confidence score is above a threshold then the further process is conducted on the box. Additionally, YOLO is an open-source project, thus mending it into our project was an optimal choice.

4 Methodology

4.1 Training the YOLO Model

Approximately 250–300 images (for a single class, in total about 1300 images were used) containing scoreboards of varying scores were used to train the models [14]. In total, three models were trained, one for each sport (cricket, soccer, and kabaddi). The models were trained until the loss got down below 1. This is another advantage that YOLO provides in that it requires significantly fewer images to train on an object compared to other object detection algorithms. Since the only things that change in an image of a scoreboard are the digit fields, using 250 images for one type of scoreboard yields extremely good accuracy.

4.2 Feeding the Images to the Model

Using OpenCV we can extract one image per second (e.g., for a 30 fps video, 1 out of the 30 frames in a second) from the video sequence and feed it into the model. The model outputs whether there is a scoreboard in the image or not and, if there is, it will return the coordinates of the scoreboard on the image. The rate of feeding the images is kept at 1 image per second of video sequence so that the possible error in the output timestamp of an event and the actual timestamp of the occurrence of the event is at most 1 s. While a lower rate would reduce the computational complexity and increase performance even further, it would also increase this max possible error, for example, a rate of 0.2 images per second from the video sequence, i.e., 1 image per 5 s of video sequence would result in a max possible error of 5 s which is not acceptable.

4.3 Reading the Score from the Scoreboard

If the model finds the scoreboard on the image, the scoreboard will then be cropped out of the image. Then the following image processing operations will be performed on the resulting image of the scoreboard: convert the image to binary using thresholding, find contours on the image, and define a region of image (ROI) around each individual contour, crop out these regions from the image to create separate images each containing 1 digit or character. All of these regions will then be fed to a trained OCR [15] which will return the score in the current frame. There are many OCRs available online but none of them had satisfactory results in reading the scoreboard due to the fact that after cropping, the image of the scoreboard becomes very small, and as such its resolution also decreases. Hence, we trained a convolutional neural network (consisting of a single hidden layer) on the digits of the five classes that we have used in this project. Doing this meant that the classification accuracy of the OCR went from unsatisfactory to almost 100% (Fig. 1).

4.4 Steps to Minimize False Positive in OCR

1. Before processing the video, 10 frames (each taken 30 s apart) are extracted from the video and fed to YOLO. The output locations of these 10 frames are averaged to get a tighter bound on the scoreboard.
2. A threshold on the minimum size of individual contours is used to eliminate granular noise.
3. Regular expression pattern matching is used to eliminate random scores such as 12-3-0.

4.5 RegEx

A RegEx, or Regular Expression, is a special sequence of characters that helps you match or find other strings or sets of strings, using a specialized syntax held in a pattern [16]. RegEx can be used to check if a string contains the specified pattern of characters. Regular expressions are used extensively for pattern matching everywhere. For example, in online forms, a regular expression is generally used to verify that the email id entered by the user is a valid email id like so: r′([azA-Z0-9.]) + @(gmail|outlook|apple|live|hotmail|yahoo)(.com)’. This will match a string containing [a-zA-Z0-9.]—any combination of characters a to z (upper and lower case), digits, periods, and underscores followed by the @ character followed by (gmail|outlook|apple|live|hotmail|yahoo)—the email domains Gmail, Outlook, Apple, Live, Hotmail, yahoo followed by .com. If a match of the pattern is found in the string input by the user, then it is considered a valid email id. RegEx is

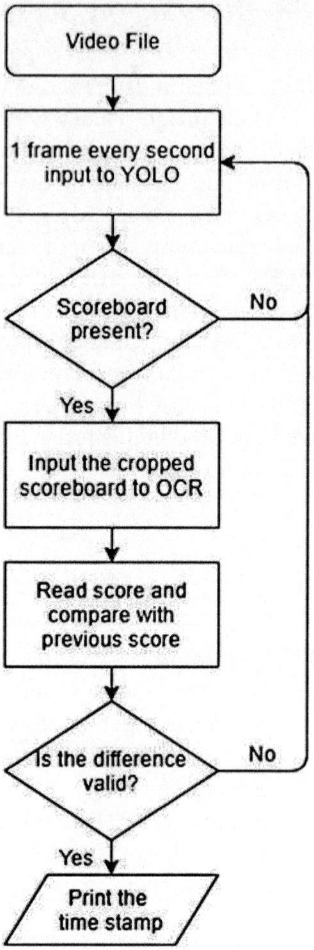

Fig. 1 Flowchart of the algorithm

also used to check the strength and validity of a password, for example, r′[a-zA-Z0-9@# &]{8,}′ can be used to only accept a password that is at least 8 characters long and contains only a–z (upper and lower case) digits and special characters @, #, and &. Similarly, a phone number can be verified using r′(\ + 91)?(\d{10}|\d{8})′. This will match a string that contains only digits, either 10 or 8 characters long, and has an optional +91 at the front.

4.6 Algorithm to Detect an Event

The score outputted by the OCR will be compared to the score from the previous frame and the difference of the scores is used to determine if there is an event at the current frame or not according to some predefined rules which are different for a different sport. For example, in cricket, if the difference in runs is among 6, 4, 3, 5, or if the difference in wickets is 1 then it is considered as an event and the timestamp of that particular frame is outputted along with the type of the event which has occurred.

4.7 Generating Highlights Video

Using MoviePy python module, short 30 s clips are generated around each and every event detected by the algorithm. The clips contain 15 s segments (before and after the event) from the video which are then combined into one video file and output as the highlights video.

5 Results

5.1 Performance of Trained Scoreboard Detection Models

The individual YOLO models (three models, one for each class of game) were tested on images of their respective class of game. These images were not part of the training dataset and were of varying resolution (between 360 and 1080p) to thoroughly examine the image detection capabilities of the models. The results are shown in Table 1.

5.2 Performance on Video Footage

The proposed method was tested on video footage of kabaddi, soccer, and cricket amounting to a total of 6 h and 35 min. The performance metrics used here are precision, recall, and F1 score.

Some terms are related to precision, recall, and accuracy (the following definitions have been derived from [17]):

Table 1 Performance of trained detection models

Model	Input images	Images with a detection	Best mAP
Kabaddi	647	639	1.00
Soccer	639	600	0.93
Cricket	341	341	0.98

Table 2 Confusion matrix

Class	Predicted: 'dog'	Predicted: 'not dog'
Ground truth: 'dog'	True positives	False negatives
Ground truth: 'not dog'	False positives	True negatives

- True Positives—These are the correctly predicted positive values, which means that the actual class (ground truth) is 'dog' and the predicted class is also 'dog' (Table 2).
- True Negatives—These are the correctly predicted negative values, which means that the actual class (ground truth) is 'not dog' and the predicted class is also 'not dog'.
- False Positives—These are the wrongly predicted positive values, which means that the actual class (ground truth) is 'no dog' and the predicted class is 'dog'.
- False Negatives—These are the wrongly predicted negative values, which means that the actual class (ground truth) is 'dog' but the predicted class is 'not dog'.

Precision is defined as the ratio of correctly predicted observations to the total predicted positive observations.

$$Precision = \frac{TP}{TP + FN} \tag{5}$$

Recall is defined as the ratio of correctly predicted positive observations to the total observations in the actual class.

$$Recall = \frac{TP}{TP + FN} \tag{6}$$

F1 score is defined as the weighted average of precision and recall. Therefore, this score takes both false positives and false negatives into account.

$$F1\,Score = \frac{2 * (Precision * Recall)}{Precision + Recall} \tag{7}$$

Since the model analyzes 1 frame from 1 s of the video sequence, the detected events are within 1 s of their actual occurrence. Figures 2, 3, 4, 5, 6, 7, 8, and 9 show the working of the model by depicting images at different stages of the algorithm (Tables 3 and 4).

Fig. 2 A frame from the video

Fig. 3 Output of the algorithm

Fig. 4 Input to the OCR

```
0:9 ~ Four
0:13 ~ 3 or 5 runs
0:16 ~ Wicket
0:38 ~ Six
0:45 ~ Four
0:56 ~ Wicket
0:56 ~ Six
1:1 ~ Four
```

Fig. 5 Output of the algorithm

Fig. 6 A frame from the video

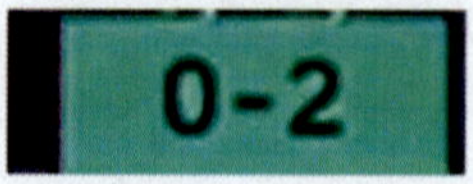

Fig. 7 Output of the algorithm

Fig. 8 Input to the OCR

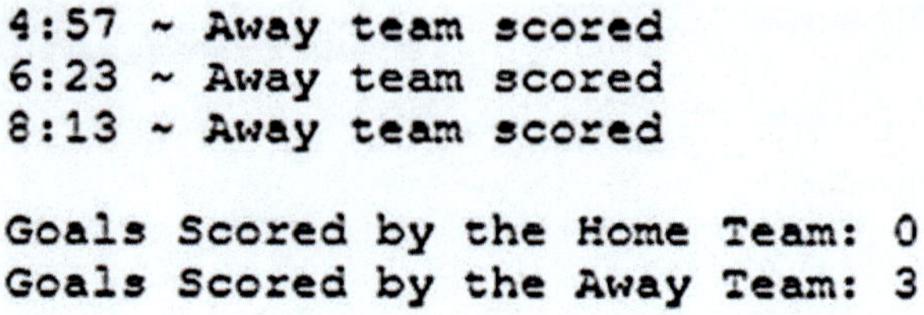

Fig. 9 Output of the algorithm

Table 3 Performance on video footage (A)

Game	Actual events	Detected events
Kabaddi	75	78
Soccer	11	11
Cricket	108	105

6 Conclusion, Limitations, and Future Work

We propose a deep learning model to summarize sports videos and output timestamps of important events. It uses YOLO for the detection of the scoreboard and then some python code for event detection. The model was implemented using Python 3.7 and TensorFlow 1.13 on a PC with 1.90 GHz, core i5-4460 T CPU with 4 GB of RAM running Windows 10. Google's open-source ML model training and deployment platform, Colaboratory [18], was used to train the models. Other software used for the model includes OpenCV, Keras API, and ImageAI modules for python. The output had an average (overall classes) F1 score of 0.979. The model can be trained and deployed for other classes (i.e., can be used to summarize any sports video). Below we describe the limitations of our current implementation and also layout the scope for future work:

Table 4 Performance on video footage (B)

Game	TP	FP	FN	Precision	Recall	F1 Score
Kabaddi	75	3	0	0.961	1	0.980
Soccer	11	0	0	1	1	1
Cricket	102	3	6	0.971	0.944	0.957

- A need for custom OCR emerged as the existing OCRs could not detect the text or were giving false output. The OCR used in the project works on very specific kinds of texts, and we would like to make it more generalized as part of a future project.
- We also plan to generalize the scoreboard detection model to detect any kind of scoreboard so as to mitigate the need to train the model on new scoreboard classes. This can be achieved by training the model on a very large image database.

Here we conclude our paper and express our heartfelt gratitude toward Professor Hansa Shingrakhia, who mentored us throughout our journey.

References

1. D. Ringis, A. Pooransingh, Automated highlight generation from cricket broadcasts using orb, in *2015 IEEE Pacific Rim Conference on Communications, Computers and Signal Processing (PACRIM)* (2015), pp. 58–63
2. P. Shukla, H. Sadana, A. Bansal, D. Verma, C. Elmadjian, B. Raman, M. Turk, Automatic cricket highlight generation using event-driven and excitement-based features, in *2018 IEEE/CVF Conference on Computer Vision and Pattern Recognition Workshops (CVPRW)* (2018), pp. 1881–18818
3. R. Hari, M. Wilscy, Event detection in cricket videos using intensity projection profile of umpire gestures, in *2014 Annual IEEE India Conference (INDICON)* (2014), pp. 1–6
4. A. Javed, A. Irtaza, Y. Khaliq, H. Malik, M. Mahmood, Replay and keyevents detection for sports video summarization using confined elliptical local ternary patterns and extreme learning machine. Appl. Intell. **02** (2019)
5. R. Agyeman, R. Muhammad, G.S. Choi, Soccer video summarization using deep learning, in *2019 IEEE Conference on Multimedia Information Processing and Retrieval (MIPR)* (2019), pp. 270–273
6. M.Z. Khan, S. Saleem, M.A. Hassan, M. Usman Ghanni Khan, Learning deep c3d features for soccer video event detection, in *2018 14th International Conference on Emerging Technologies (ICET)* (2018), pp. 1–6
7. Y. Junqing, L. Aiping, H. Yangliu, Soccer video event detection based on deep learning, in *MultiMedia Modeling*, (Cham) (Springer International Publishing, 2019), pp. 377–389
8. P. Kathirvel, M. Manikandan, K. Soman, Automated referee whistle sound detection for extraction of highlights from sports video. Int. J. Comput. Appl. **12**, 12 (2011)
9. A. Tejero, Y. Nakashima, T. Sato, N. Yokoya, M. Linna, E. Rahtu, Summarization of user-generated sports video by using deep action recognition features. IEEE Trans. Multimedia **20**(8), 2000–2011 (2018)
10. R.B. Girshick, J. Donahue, T. Darrell, J. Malik, Rich feature hierarchies for accurate object detection and semantic segmentation, *CoRR*, vol. abs/1311.2524 (2013)
11. S. Ren, K. He, R. Girshick, J. Sun, Faster r-cnn: towards real-time object detection with region proposal networks, in *Advances in Neural Information Processing Systems* (2015), pp. 91–99
12. R. Girshick, Fast r-cnn, in *Proceedings of the IEEE International Conference on Computer Vision* (2015), pp. 1440–1448
13. J. Redmon, A. Farhadi, Yolov3: An incremental improvement. *arXiv* (2018)
14. How to train a yolo model. https://medium.com/deepquestai/train-objectdetection-ai-with-6-lines-of-code-6d087063f6ff
15. Digit recognition using cnn. https://towardsdatascience.com/imageclassification-in-10-minutes-with-mnist-dataset-54c35b77a38d

16. Python—regular expressions. https://www.tutorialspoint.com/python/pythonregexpressions.htm.
17. Accuracy, precision, recall and f1 score: interpretation of performance measures. https://blog.exsilio.com/all/accuracy-precision-recall-f1-score-interpretation-ofperformance-measures/
18. "Google colaboratory." https://colab.research.google.com

Extraction and Recognition of Handwritten Gujarati Characters and Numerals from Images Using Deep Learning

Deepika Shukla and Apurva Desai

Abstract Automated recognition of Indian scripts from a document or image is a challenging problem. Gujarati is an Indian script meant for expressing the language Gujarati. Gujarati is the official language of the west Indian state of Gujarat and is spoken and written by nearly 60.3 million people. Gujarati script can be considered as more complicated for machine recognition owing to the presence of complex consonants, vowel modifiers compound character called 'Jodakshar' and absence of 'Shirorekha'. The problem becomes all the more challenging when the script to be recognized is handwritten rather than printed. Two basic approaches exist in the literature on the topic: Offline as well as online handwritten Gujarati script recognition. The online method takes the input in real time, whereas offline systems work on scanned images making the recognition task more lengthy and complicated. The main aim of this paper is twofold. First, a dataset of images of handwritten Gujarati characters and numerals is made available, and second, offline Gujarati character recognition from images of handwritten Gujarati text. Hence, the primary objective of the paper is to present an approach that could extract Gujarati characters from an image and then recognize them with deep learning. The accuracy of the system obtained in most of the cases is more than 98%. The best accuracy obtained when tested with unseen character images is 82.15%.

Keywords Handwritten Gujarati character recognition · Deep learning · Handwritten character recognition · Gujarati script · Gujarati numerals

D. Shukla (✉)
Computer Science and Engineering Department, Institute of Technology, Nirma University, Ahmedabad, India
e-mail: deepika.shukla@nirmauni.ac.in

A. Desai
Department of Computer Science and IT, Veer Narmad South Gujarat University, Surat, India

F. Thakkar et al. (eds.), *Proceedings of the International e-Conference on Intelligent Systems and Signal Processing*, Advances in Intelligent Systems and Computing 1370,
https://doi.org/10.1007/978-981-16-2123-9_51

1 Introduction and Motivation

The Gujarati language is from the Indo-Aryan family of languages; it's an official language of the west Indian state of Gujarat and is spoken by nearly 60.3 million people. Gujarati script is an adaptation of the traditional Devanagari. It's a multilevel script like Hindi or many other Indian languages. It is more complicated to recognize Gujarati script by machine due to the presence of complex consonants, vowel modifiers, and the compound character called 'Jodakshar'. The absence of Shirorekha on the top of the word like other North Indian languages makes it difficult to segment words from a sentence. The problem becomes all the more challenging when the script to be recognized is handwritten rather than printed. In the handwritten character recognition method, the input is from scanned images, documents etc. Several efforts are dedicated toward the development of an automated machine learning-based recognition system for recognition of such handwritten Gujarati script but can still be considered in its infancy as no system to date matches the efficiency of human beings. Two basic approaches exist in the literature on the topic: offline as well as online handwritten Gujarati script recognition. The online method takes the input in real time, whereas offline systems work on scanned images making the recognition task more lengthy and complicated.

1.1 Gujarati Script

The Gujarati language is from the Indo-Aryan family of languages, and the Gujarati script is an adaptation of the traditional Devanagari. It's a multilevel script like other Indian languages, the sole difference being that it does not have a headline or Shirorekha. The earliest known document within the Gujarati script may be a manuscript dating from 1592.

1.2 Properties of Gujarati Text

In the Gujarati script, there are 12 vowels, 36 consonants and 10 digits, as illustrated in Fig. 1. The consonants are often combined with the vowels and may form compound characters. Figure 2 shows the same. There are many possibilities for the conjunct consonants which are created using two or more consonants and vowels.

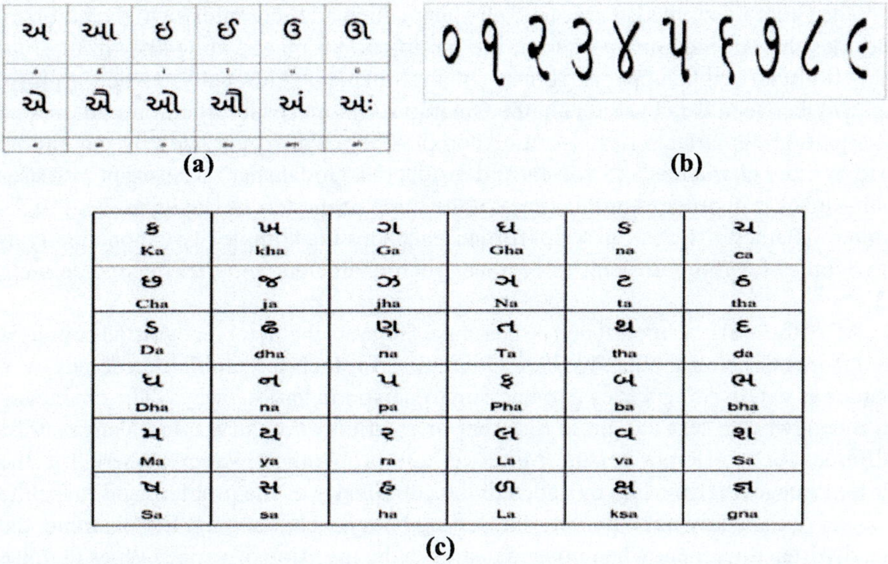

Fig. 1 **a** Gujarati vowels **b** Gujarati numerals **c** Gujarati consonants. Images taken from [1] and [2]

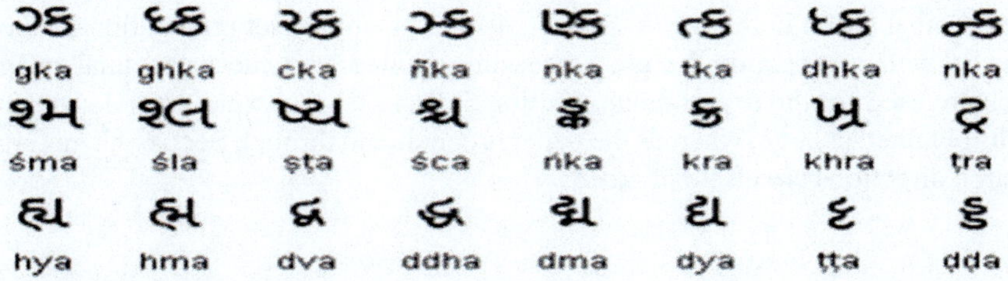

Fig. 2 Examples of half alphabets and compound characters in Gujarati script. Image taken from: https://omniglot.com/writing/gujarati.htm

1.3 Challenges in Gujarati Script Recognition

In this section, we explain the nature of complexity involved in extracting and then recognizing Gujarati characters and numerals from images where the text present in the image is handwritten rather than printed.

As cited earlier that most of the scripts similar to Gujarati script possess 'Shirorekha' in the script, the presence of Shirorekha helps to spot the boundaries of the word. The Shirorekha also facilitates extracting upper modifiers. The absence of Shirorekha in the Gujarati language makes it harder to segment text lines and thereafter extract words from the handwritten text.

The Gujarati script also uses half-alphabet which adds complexity to the problem. Besides this, in the Gujarati script, the modifiers are placed altogether in the four directions; left, right, upper and bottom of an alphabet. Additionally, the non-uniform spacing between the characters in the Gujarati script makes it difficult for automated recognition. Gujarati script also uses complex characters generated by combining two or more characters and vowels and results in a 'Jodakshar'. The script also uses half-alphabet in order to mimic tones of the same characters as shown in Fig. 2. This kind of characteristic of the script further enhances the complexity exponentially as it becomes far more difficult to segment such characters and then recognize them [3].

All of these above-mentioned issues make Gujarati character recognition complex, and if it is done on the handwritten Gujarati script then the problem becomes even harder as handwriting varies from person to person in various ways. The characters in a handwritten text can be in different orientations, the curves and slant may be different for a different person. Moreover, the problems which are inherent to the task of image analysis like this adds to the complexity of the problem and make the subject of the study not only very interesting but also challenging. For example, the handwritten document when imaged results in the insertion of various types of noise like lighting effects, background noise etc. in the resultant image.

After introducing the topic and motivation behind the study in detail in Sect. 1, the rest of this paper is organized as follows. Section 2 covers the discussion on related work existing in the field. Section 3 discusses briefly the technologies used for experimentation in this work. Section 4 focuses on dataset preparation methodology as well as explains the pre-processing applied to achieve the final images which are used for the experiments. Section 5 discusses the experimental setup and results obtained at large, whereas the paper is concluded through Sect. 6 where future research directions are also explained.

2 Related Works

Gujarati character recognition from images has been attempted by many researchers but most of the work is based on using traditional machine learning. Very few works have performed offline handwritten Gujarati character recognition using deep learning. This section discusses the major works existing on the topic, even if they have reported approaches based on traditional machine learning, and also deep learning-based instances. The section talks about such approaches, results obtained and limitations.

A lot of research work has been done previously and is still being done for character recognition. Extensive work has been reported for the widely used language English. Approaches also have been reported for many other languages like Hindi, Tamil, Chinese etc. but they are very few in number [4]. Moreover, relatively very less work exists for the Gujarati language. The set still reduces when it comes to the usage of deep learning for Gujarati characters and numeral recognition.

Many authors have reported separately their work on only Gujarati numeral recognition using various conventional machine learning approaches [5], where in such approaches different classifiers like a neural network [2] or support vector machine [6] or multiple classifiers K-NN, SVM or backpropagation neural network have been used [7]. Authors in [7] have also reported three different methods for feature extraction in spatial as well as frequency domain. However, in [8], the features used for the representation of numerals are chain code-based. In [9], the authors have proposed an approach for online Gujarati handwritten numeral recognition using SVM and achieved an accuracy of 92.01%.

Similarly, numerous authors have presented approaches, where they have worked on only Gujarati alphabet recognition [10–13]. These approaches have used traditional machine learning and differ in either the features used and/or the classifier chosen. Still, approaches exist which have reported both the recognition of alphabets and numerals but are dependent on conventional machine learning, where handcrafted features are provided to one or the other classifier.

The closest to this work is the work reported by [14] where authors have used deep learning for word-level recognition and have obtained 97.21% accuracy using CNN. The work uses 10,000 images from 250 people.

This study has presented an approach with the following characteristics. The approach is an offline process. It tackles the recognition task from images consisting of handwritten Gujarati characters as well as numerals. The total number of images taken for training and validating the model is 22704. The approach presented is deep learning-based using a convolutional neural network. The validation accuracy obtained in most of the cases is more than 98%.

3 Technologies Used

3.1 Deep Learning

Deep learning is a part of machine learning that is concerned with an algorithm inspired by the structure and function of the human brain [15]. It mimics the way a human brain works and thus becomes extremely useful in many real-world pattern-recognition applications. Traditional machine learning involves handcrafted feature engineering, whereas deep learning extracts features on its own, making it perform remarkably well where huge multidimensional data exist. Deep learning has wide applicability in image analysis, scene classification, text recognition, captioning and expressing the content of the image in natural language. Deep learning provides many learning frameworks, including TensorFlow, Keras, Caffe, Theano, PyTorch, Microsoft CNTK and many more. These frameworks also allow shifting back and forth between two or more. Deep neural networks provide many advanced architectures like AlexNet, VGG Net, GoolgeNet, ResNet, ResNeXt, RCNN, YOLO, SqueezeNet, segNet, GAN [10].

3.2 Convolutional Neural Network (CNN)

In this work, MATLAB is used to build a convolutional neural network (CNN), also known as ConvNet. ConvNet is a feed-forward multilayered hierarchical neural network used in various applications where machine intelligence is needed. Object detection and recognition, document analysis, recommendation systems, semantic web analysis, speech recognition are some of the areas where the use of CNN is already reported in the literature. CNN has got hierarchical feature extraction ability, i.e., learning high-level features from low and mid-level features as extracted from the image [16]. This characteristic of CNN makes it very near to the way the human brain works, making it popular for image analysis tasks. The work reported through this study involves analyzing visual images, so CNN is taken as the architecture for performing experimentation. A CNN ideally is composed of a series of layers, where each layer defines a specific computation. There are main four layers in the convolutional neural network:

(i) Convolutional Layer
(ii) Rectified Linear unit Layer
(iii) Max Pooling Layer and
(iv) Fully Connected Layer.

4 Dataset and Its Preparation Methodology

The availability of a dataset that captures variations encountered within the world may be a critical issue in any experimental research. One of the characteristics of the deep learning approach and a major limitation in applying the deep learning approach for any recognition problem is that it needs a huge number of images for training and validating the model. Own dataset was prepared for this study. Eleven individuals were asked to write every numeral and character of Gujarati script on a white sheet and then the images of those sheets were acquired with the use of a mobile camera phone. No special lighting arrangement was done while acquiring the images of these sheets. Also, an image of the printed instance of the character was taken that makes 12 unique instances of each character and numeral of the Gujarati script. The images so acquired were rotated by 2–42° clockwise and anticlockwise in a step size of 2° from their vertical position, thus the total number of rotations applied on each image becomes 43. The rationale behind this processing was that handwriting of different individuals usually holds some degree of slant. To capture this variability in a realistic manner, the images were rotated. This rotation at one end though introduces the required variability but at the same time makes the situation intriguing and complex. The idea of rotating the images also resulted in another benefit that very few individuals (11 in this case) were needed to create the dataset, and still needed variations are introduced in the dataset. The approach presented in [2] has also generated additional images for each of the patterns by rotating the numeral

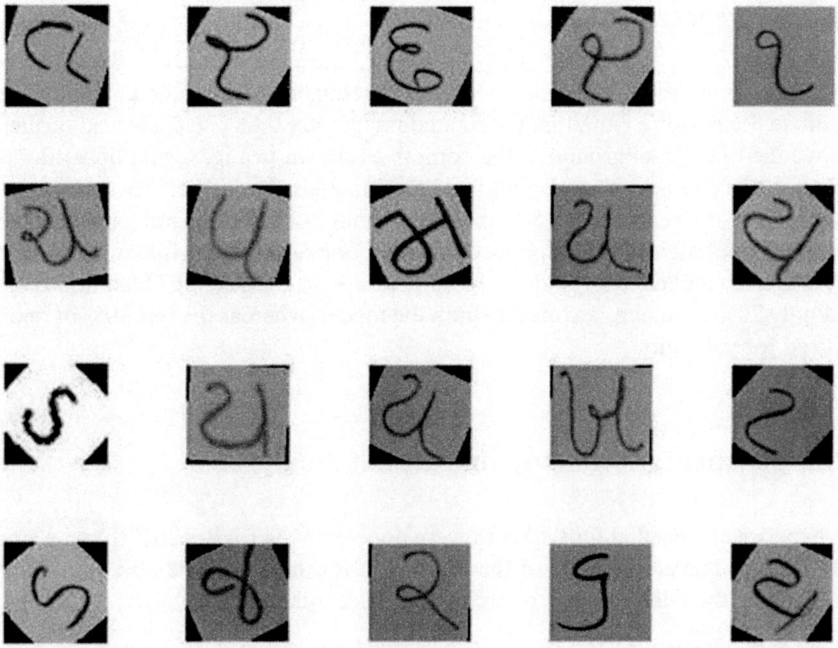

Fig. 3 Randomly selected snapshot of the dataset

up to 10° both clockwise and anti-clockwise direction to deal with the problem of skew, but here in this study, to include maximum slant that a handwritten character can have we have considered a total of 43 rotations, 21 in the clockwise direction as well as anti-clockwise in step size of 2°.

The entire procedure resulted in 43 images for each character and each numeral per user, and finally resulting in 516 images per character/numeral, and a **total of 22,704 images were generated of 44 classes**. Figure 3 shows some of the instances of characters and images as contained in the dataset. Additionally, another two individuals (other than earlier 11 individuals) were asked to write six random Gujarati characters and numerals in the same way on a white sheet of paper, and then images thus acquired were used for testing the system. These six characters were again rotated by 2–42° clockwise and anti-clockwise in a step size of 2° from their vertical position making 43 images for each character and each numeral and finally resulting in a total of 516 previously unseen testing images.

4.1 Sample Dataset

To give the idea of the dataset visually, Fig. 3 shows a snapshot of the dataset by randomly selecting few images from it.

4.2 Pre-processing

As the images are acquired without any special setup, in the most unconstrained setup to imitate the real-life situation, the rotated images are then pre-processed further to remove the black background at the corners as shown in Fig. 4, and bounding box enclosing the character is computed and the image is cropped as per the bounding box and then resized to 50 × 50 pixels. The images of each character/numeral are kept in a folder signifying its phonetic name. For example 'ka' folder for 'ક 'letter. This exercise resulted in 44 folders (10 numerals + 34 characters). Out of this dataset randomly 70% of images are used to train the model, whereas the rest 30% of images are used for validation.

5 Experimentation Setup and Result Analysis

The experiments needed for the work were done on MATLAB R2017(b) on a single CPU system. The architecture of the CNN used for this work and also the rationale behind using the configuration of architecture is explained as below:

- One input layer was fed by pre-processed 50 × 50 × 1 (grayscale images as resulted after pre-processing) images holding characters and numerals of Gujarati script.
- Convolutional layer: The convolutional layer in CNN is used to systematically apply filters to an input. The architecture of CNN used for this study uses three 2D convolutional layers (16 and 32 and 64 filters correspondingly with 3 × 3 window size with a padding of 1 pixel and stride of 1). As most of the features which are useful for an application are usually locally placed, so few local pixels should be taken to apply convolutions. Thus, a small filter size 3 × 3 is taken in this work. Also, the rationale behind taking stride of 1 as useful features may

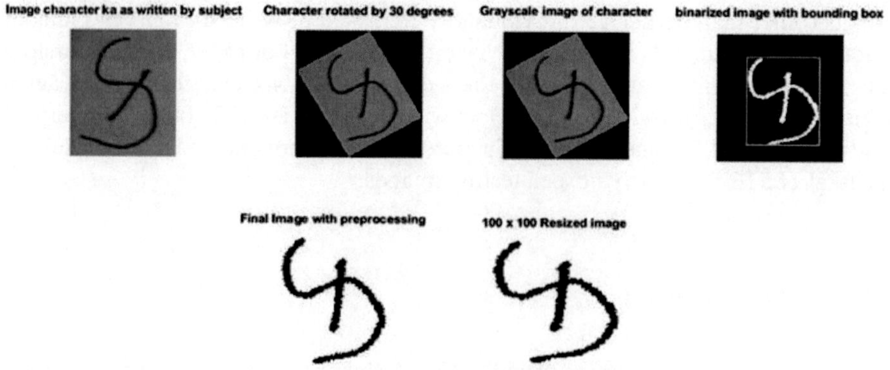

Fig. 4 Pre-processing applied on character images as acquired

exist in the image at more than one place, hence minimum sliding is taken so that these features can be extracted at the maximum.

- Additionally, the number of filters and number of convolutional layer is considered as hyper-parameter, and thus is to be tuned for the task at hand, but owing to the fact that as we go deeper into the architecture spatial resolution is reduced, at the same time the number of feature maps increases.
- Moreover, we know that CNN learns initially low-level features, then learned low-level features are combined to learn mid-level, and finally combining various mid-level features learns high-level features. So the number of filters if increased in deeper convolutional layers makes sense.
- Each convolutional layer was followed by a batch normalization layer to normalize the weights and error activation along the channel across a mini-batch. As a result of normalization, it speeds up the learning gradient, thus decreases sensitivity. The batch normalization layer was followed by ReLu (rectified linear unit).
- Two MaxPooling layers with a pooling region of 2×2 and a stride of 2 resulting in no overlap in the pooling regions.
- At the end of the pipeline, a fully connected layer was created followed by a softmax layer and classification layer with 44 nodes to correspond with 44 classes to be recognized.

5.1 Training and Validating the Model

The experiment was performed totally for nine times, with 3, 4 and 5 epochs, each for three times, respectively, where each epoch was of 695 iterations. Each iteration uses a batch size of 20. An initial learning rate of 0.01 was specified and the validation accuracy was computed after every 30 iterations. Table 1 shows the accuracy obtained as the complete experiment was executed for 3, 4 and 5 epochs, and Figs. 5, 6 and 7 show the training progress validation accuracies and load for best cases when the experiment was performed with 3, 4 and 5 epochs. From the graphs shown in Figs. 5, 6 and 7, it can be observed that initially for some iteration, the validation accuracy is low and the cost is higher. This can be explained initially as the network does not have optimal weights and biased. Gradually, optimal weights are achieved, so the accuracy increases speedily, and after the first epoch it almost becomes steady. The validating accuracy was noted. It can be easily observed that the average accuracy in most of the instances is above 98%. It can also be observed that the best results are obtained when the experiment is run with maximum 4 epochs.

5.2 Testing the Approach

The model so trained and validated was tested on 516 previously unseen test images, acquired as mentioned in the earlier section where dataset preparation methodology

Table 1 Accuracies obtained from the experiments

No. of epochs for which experiment was run	First instance of the experiment		Second instance of the experiment		Third instance of the experiment		Avg. validation accuracy	Avg. overall accuracy (%age)
	Accuracy (%age)	Overall accuracy (%age)	Accuracy (%age)	Overall accuracy (%age)	Accuracy (%age)	Overall accuracy (%age)		
3	98.38	98.32	**98.38**	**98.47**	98.08	98.08	98.28	98.29
4	98.38	98.47	**98.75**	**98.83**	98.35	98.34	98.49	98.50
5	98.16	98.20	96.75	96.84	**98.46**	**98.51**	97.79	97.85

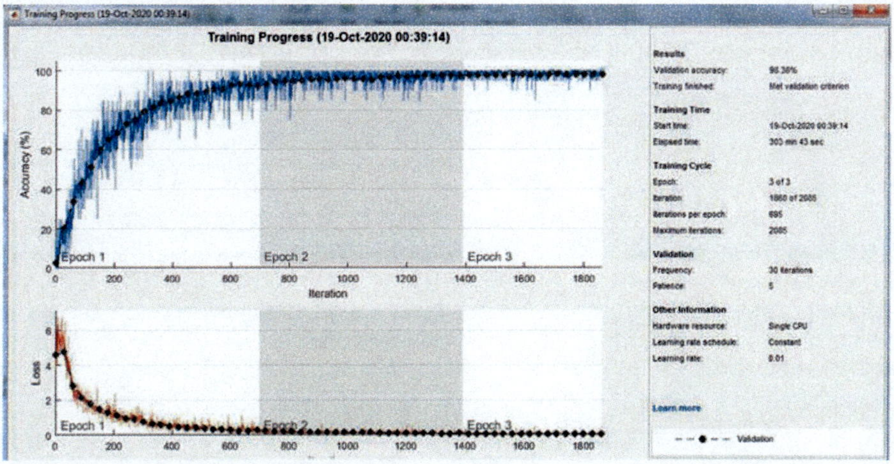

Fig. 5 Model performance for max epochs = 3

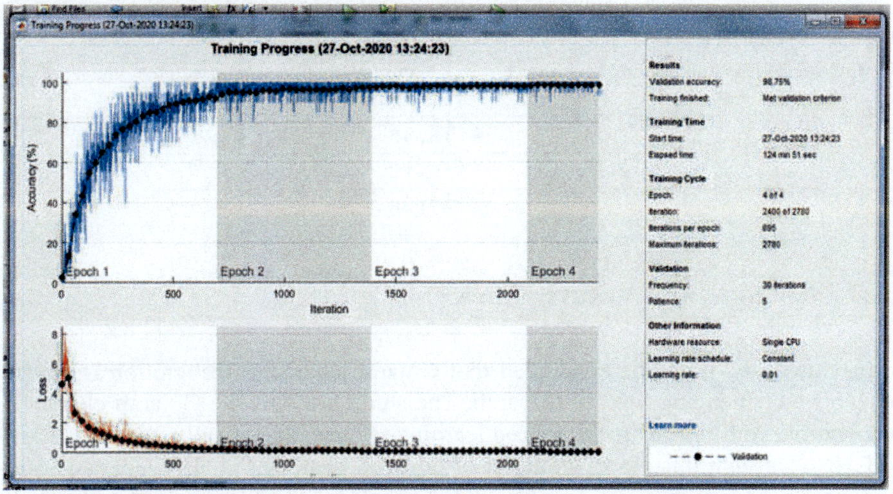

Fig. 6 Model performance for max epochs = 4

is discussed. The testing images were also pre-processed and then passed for classification. The three instances of the testing experiments were performed. Table 2 lists the accuracy of classification obtained on testing images. Rather than totally visually distinct characters, visually similar-looking characters like ‘gha’ and ‘dhaa’, ‘pa’ and ‘five’ were made part of test images so that the model classification ability can be tested to the fullest, thus resulting in relatively low accuracy.

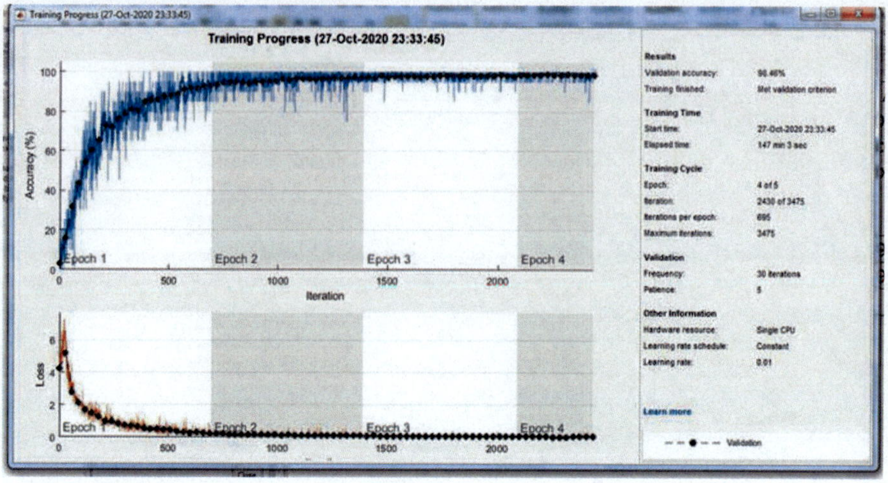

Fig. 7 Model performance for max epochs = 5

Table 2 Percentage accuracy obtained as a result of testing on unseen images

Testing model (max epoch) for which max accuracy was obtained	Accuracy obtained on unseen test images (%age)
3	82.15
4	68.82
5	71.48

6 Conclusion and Future Work

From this study, it can be concluded that Gujarati character recognition from the image is a challenging field of research. The study presents the Gujarati character recognition from images using a deep learning strategy specifically with the usage of CNN. Through this work, a dataset of Gujarati character is also generated and incorporate maximum variation in the acquired images through rotation.

It was observed that when the model was tested for previously unseen images of characters like ધ(Dha) and ઘ(Gha) or alphabet પ(pa) or numeral five (૫)or any such visually similar characters then the accuracy decreases. The best accuracy obtained was 98.83% for max epoch = 4, and the average validation accuracy in most of the instances is above 98%. The best testing accuracy on the unseen image dataset is 82.15%.

The future work can be for the recognition of characters when 'matras' (vowel modifiers) are also present along with the character and then segment and recognize the word in the image. Moreover, the work is to be extended for compound characters and half-alphabets occurring in the Gujarati script. The effect of other deep learning architectures can be explored, which can be yet another dimension along which the work can be extended.

References

1. https://omniglot.com/writing/gujarati.htm. Accessed 25 Oct 2020
2. A.A. Desai, Gujarati handwritten numeral optical character reorganization through neural network. Pattern Recogn. **43**(7), 2582–2589 (2010)
3. M. Parikh, A. Desai, Segmentation of frequently used handwritten gujarati conjunctive alphabet, in *2019 5th International Conference On Computing, Communication, Control And Automation (ICCUBEA)* (IEEE, 2019)
4. U. Pal, B.B. Chaudhuri, Indian script character recognition: a survey. Pattern Recogn. **37**(9), 1887–1899 (2004)
5. A.K. Sharma et al., Comparative analysis of zoning based methods for Gujarati handwritten numeral recognition, in *2015 5th Nirma University International Conference on Engineering (NUiCONE)* (IEEE, 2015)
6. M. Maloo, K.V. Kale, Support vector machine based Gujarati numeral recognition. Int. J. Comput. Sci. Eng. **3**(7), 2595–2600 (2011)
7. A.N. Vyas, M.M. Goswami, Classification of handwritten Gujarati numerals, in *2015 International Conference on Advances in Computing, Communications and Informatics (ICACCI)* (IEEE, 2015)
8. A. Sharma, D.M. Adhyaru, T.H. Zaveri, Chain code feature based recognition of handwritten Gujarati numerals. Int. J. Adv. Res. Comput. Sci. **8**(1) (2017)
9. V.A. Naik, A.A. Desai, Online handwritten Gujarati numeral recognition using support vector machine (2018)
10. https://www.analyticsvidhya.com/blog/2017/08/10-advanced-deep-learning-architectures-data-scientists/. Accessed 25 Oct 2020
11. A.A. Desai, Support vector machine for identification of handwritten Gujarati alphabets using hybrid feature space. CSI Trans. ICT **2**(4), 235–241 (2015)
12. M. Kayasth, B. Patel, Offline typed Gujarati character recognition. Nat. J. Syst. Inf. Technol. **2**(1), 73 (2009)
13. J. Dholakia, A. Negi, S. Rama Mohan, Zone identification in the printed Gujarati text, in *Eighth International Conference on Document Analysis and Recognition (ICDAR'05)* (IEEE, 2005)
14. J. Pareek et al., Gujarati handwritten character recognition from text images. Proc. Comput. Sci. **171**, 514–523 (2020)
15. N. Buduma, N. Locascio, *Fundamentals of Deep Learning: Designing Next-Generation Machine Intelligence Algorithms* (O'Reilly Media, Inc., 2017)
16. A. Khan et al., A survey of the recent architectures of deep convolutional neural networks. Artif. Intell. Rev. **53**(8), 5455–5516 (2020)

Application of Machine Learning for Estimating Empirical Parameters for Rectangular Microstrip Patch Antenna

Pankti Shah and Amit Patel

Abstract An antenna is a rudimentary element in communication technology. Combination with trending technology is apparent with a certain zenith for advancement in the domain. An approach of predicting the resonant frequency, return loss and input impedance by taking hundreds of combinations of length and width of a rectangular radiating patch is made by considering rectangular microstrip patch antenna. The design of this is made by using Ansys high-frequency structure simulator (HFSS) software. Design and optimization are made on 2.4 GHz patch antenna. Optimization of parameters is done by extracting the dataset from simulated design and optimization. Ridge regression, linear regression and decision tree regressor algorithms are used for training the model; moreover, the model works more efficiently with decision tree regressor. However, the model works very well for the decision tree by predicting practical resonant frequency, input impedance and return loss with 99.66%, 99.98% and 80.04% accuracy, respectively, on unseen data.

Keywords Microstrip patch antenna · Linear and ridge regression · Decision tree regressor · Return loss and impedance

1 Introduction

Antennas have many applications in the communications domain, such as in satellite communication, a global positioning system (GPS), digital cordless system (DCS), wireless local area network (WLAN), in the healthcare system as a sensor, spacecraft, aircraft and other private communication systems [1, 2]. Antenna designing using the various types of the patch is modified as per the requirements in communication. It is having some fascinating features such as the minimal cost of fabrication and material, lightweight, small in size and portable [3, 4]. In patch antennas, conductors are etched on the double side of a dielectric substrate, in which only one side

P. Shah (✉) · A. Patel
V. T. Patel Department of Electronics and Communication Engineering, Chandubhai S. Patel Institute of Science and Technology, Charotar University of Science and Technology (CHARUSAT), Anand, Gujarat, India

F. Thakkar et al. (eds.), *Proceedings of the International e-Conference on Intelligent Systems and Signal Processing*, Advances in Intelligent Systems and Computing 1370,
https://doi.org/10.1007/978-981-16-2123-9_52

behaves as a radiator, while the other side plays the role of a ground plane. This patch has de-merits in its designing such as low gain, narrow bandwidth and low power handling. Microstrip patch operates on narrow bandwidth, so an array of patches is given more preference. As per the studies in the literature, designs of the patch antennas are decided based on applications with eclectic patterns such as E-shaped, T-shaped monopole, A-shaped, rectangular, circular and stacked patched [1, 5–8]. The amalgamation of artificial intelligence domain with antenna is prevailing as per the study. Neural network algorithm is the common method used for prediction purposes in antenna because it consists of several numbers of hidden layers which divide the data into some parts, and at each neuron, some amount of feature extraction occurs. Algorithms such as artificial neural network (ANN), deep neural network (DNN), genetic algorithm, particle swarm optimization (PSO), lasso and k-nearest neighbour (kNN) are used [5–12].

In designing of E-shaped compact microstrip antenna (ECMA), data of 144 samples were used with an architecture of five hidden layers with DNN and ANN developed, which can anticipate resonant frequency [1, 7]. Estimating resonant frequency with Genetic algorithm and PSO approach with different patch designs such as stacked patch, circular patch, rectangular slot and T- shaped monopole [2, 5, 8, 9]. Predicting resonant frequency and return loss with all the above-mentioned methods is efficient. As per the analysis, the neural network model or its advanced architecture model is designed when they keep a very large number of epochs for getting precise outputs. Then, even with less data more training results can be obtained. Elliptical-shaped patch antenna which is circularly polarized is trained with the ANN model. It works efficiently for resonant frequency and the return loss with a very less error rate. Their data consists of generic parameters value for three dielectric substrates of the radiating patch such as Fr4 Epoxy and Roger3003 and RT Duroid [12]. In this paper, an approach to estimate resonant frequency, return loss and input impedance using linear regression, ridge regression and decision tree regressor is proposed. The regression-based algorithms work well on supervised methodology. We have obtained good accuracy on a small dataset with linear regression and ridge regression for resonant frequency and input impedance but they are incapable of predicting the values of return loss which is having a small dataset. So, an increase in data helps us to increase the prediction rate with the decision tree regression algorithm for all three desired outputs. The decision tree algorithm works well with all labelled analysed outputs. A brief explanation about the designed antenna is mentioned. These statistics are thoroughly described in Table 1. In Table 2, a detailed table of comparison with actual and predicted values is delineated for the most efficient algorithm that is decision tree regression with thousand varied combinations.

Table 1 Accuracy comparison table of all three algorithms

Parameters	Small dataset (%)	Large dataset (%)
Linear regression		
Practical resonant frequency	98.08	89.2
Input impedance	90.09	64.95
Return loss	45.26	48.86
Ridge regression		
Practical resonant frequency	97.71	89.2
Input impedance	91.125	64.95
Return loss	45	48.86
Decision tree regression		
Practical resonant frequency	97.44	99.66
Input impedance	97.09	99.98
Return loss	73.88	80.04

Table 2 Actual and predicted values comparison table for decision tree regression algorithm with 1000 combinations

Parameters (mm)		Practical resonant frequency (GHz)		Input impedance (Ω)		Return loss (dB)	
lp	wp	Actual	Predicted	Actual	Predicted	Actual	Predicted
12	28	4.2	4.2	109.26	102.55	−8.35	−12.15
25	23	2.8	2.8	471.19	471.19	−7.21	−7.21
17	25	3.9	3.9	195.34	195.34	−12.93	−12.93
15	36	4.5	4.3	76.77	76.77	−10.02	−10.22
23	38	2.9	2.9	152.56	152.56	−13.24	−15.12

2 Microstrip Patch Antenna with the Inset Feeding Technique

In this paper, the antenna is made in a rectangular shape with an inset feeding technique. Equations for designing purpose were referred from [13].

$$W = \frac{1}{2fr\sqrt{\mu_0 \in_0}}\sqrt{\frac{2}{\in r + 1}},\ W/h > 1 \tag{1}$$

$$\in_{reff} = \frac{\int r + 1}{2} + \frac{\int r - 1}{2}\left[1 + 12\frac{h}{W}\right]^{-1/2} \tag{2}$$

$$\Delta L = h(0.412)\frac{(\int_{reff} + 0.3)(W/h + 0.264)}{(\int_{reff} - 0.258)(W/h + 0.8)} \tag{3}$$

$$L = \frac{1}{2fr\sqrt{\mu_0 \in_0}\sqrt{\in_{reff}}} - 2\Delta L \tag{4}$$

where W is the width of the patch, L is the length of the patch, h is the height with respect to dielectric material, ϵ_r is a dielectric material and ϵ_{reff} is effective dielectric value.

By using these equations, we can obtain rudimentary values. The basic design of the antenna is made at 2.4 GHz resonant frequency with 29.5 mm length of patch and 38 mm width of a patch with FR-4 epoxy as its dielectric medium because FR-4 epoxy is low in price, easily available and performs well on narrow bandwidth with less conduction loss. Inset feeding technique was used to supply charge to the patch. Ideally, 50 Ω impedance is being there between the gap of patch and feed line which gives perfect return loss. This patch is mounted on a plain rectangle-shaped sheet with 60 mm length and width each and 1.6 mm height with copper as a substrate for the ground plane. In the inset feeding technique, a proper gap should be taken between the patch and the feed line for proper radiation and gain. In this technique, perfect return loss is obtained. With the help of HFSS design software, a rudimentary design was made. The basic diagram of antenna and design of microstrip patch using HFSS are shown in Fig. 1.

By design simulation using HFSS, we have obtained a good value of return loss at 2.4 GHz which is −15.78 dB, as shown in Fig. 2.

The design has a good value of return loss since it radiates properly, so this design was taken as a rudimentary design for optimization purpose for an ample amount of combinations using HFSS software. To use this software for optimization, a minimum requirement is to have at least 8 GB RAM because it takes too much time to optimize and dissipate much heat for some combinations. Datasheet which is generated by software consists of a complex structure; therefore much time is consumed to go through every element specified. Hence refurnished dataset is fed into the machine learning model, then the time consumption barrier becomes nugatory.

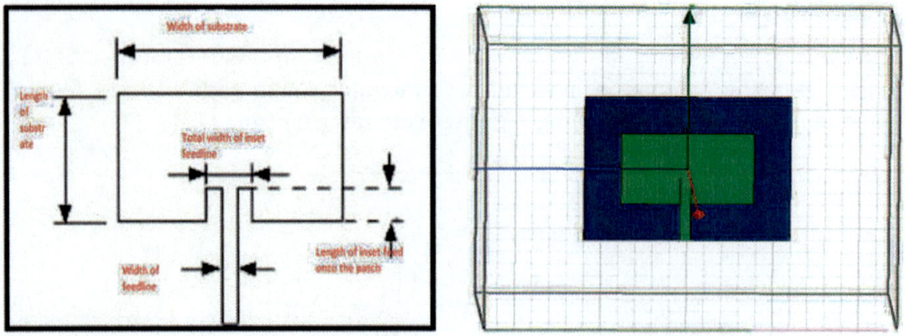

Fig. 1 **a** Diagram of rectangular microstrip patch antenna with the inset feeding and **b** Model design of antenna at 2.4 GHz using HFSS

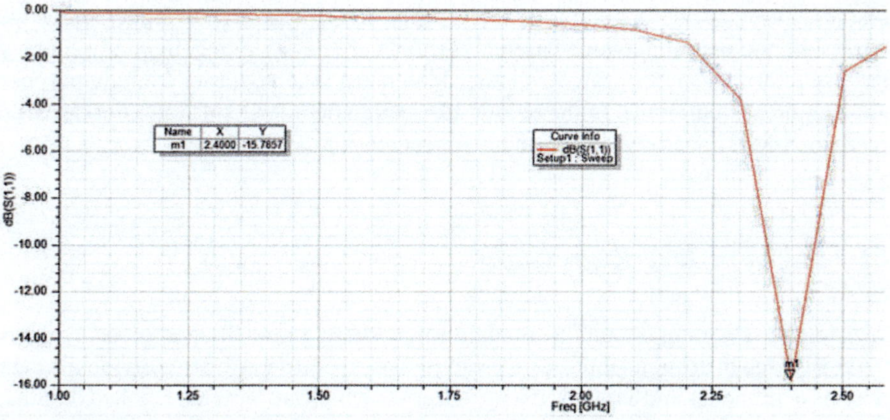

Fig. 2 Return loss of microstrip antenna at 2.4 GHz

Now, the prediction of return loss, gain, and any other necessary parameter is done in less time with efficiency.

3 Machine Learning Algorithms

A machine learning algorithm is a step of a sequence of instructions on a specified dataset. A dataset is a data table with varied combinations of data to be trained. The dataset should have information in it, just to be trained because unnecessary data can lead you to low efficacy. The provided dataset is then imposed into the model with a suitable algorithm. Here in this paper, the antenna parameters dataset is made, which is highly numeric with discrete values. A regression model is a model that works well with discrete values. The role of the algorithm is vital because in machine learning outcomes with labels are obtained when we provide properly pre-processed features. Here, we have taken features in the dataset such as length and width of a patch of divergent ranges, including its theoretical resonant frequency, and labelled outcomes were kept as practical resonant frequency, impedance value and return loss. Algorithms such as linear regression, ridge regressor which are referred to as L2 regularization, decision tree regressor and support vector machine were used to train the optimized valued dataset. Linear regression and ridge regression algorithms are capable of predicting frequency and input impedance with low data such as hundred entries with average 90% accuracy but that less data entries are not capable of predicting return loss because return loss has too much variance in its value as compared to each length and width's return loss. By implementation, we have justified that linear regression and ridge regression does not provide good results with an average accuracy of 89% for resonant frequency on a large dataset and on estimating return loss and input impedance with accuracy less than 65% on a large dataset, which is quite low. Thus, other algorithms such as linear and ridge regression and neural network-based algorithms are not able to gain good accuracy because return loss has negative values which takes much data for training purpose and complex calculations. In the below sub-section detailed explanation of decision tree regression, linear regression and ridge regression is articulated.

3.1 Decision Tree Regression

Decision tree learning is a unique method for predicting modelling approaches. This method is used in artificial intelligence and statistics. The decision tree divides its data in terms of branches and extends that up to leaflets which are targeted labelled outputs. The decision tree algorithm has a tree-like structure. When we have discrete targeted values then that type of model is known as classification trees. The classification tree model gives output in their labelled nodes. In the case of continuous targeted labelled outputs then that model is known as the regression model. Whenever we are making any model, we have a target to achieve higher accuracy. For regression analysis, we need to have an impurity metric that works for continuous variables. So, for this impurity measurement weighted mean-squared error is implemented for leaflet nodes. The weighted mean-squared error formula is taken as

$$WMSE(t) = \frac{1}{N(t)} \sum_{i \in Dt}^{n} (y(i) - y')^2, \tag{5}$$

where $N(t)$ is the total number of training samples at respected node t, D_t is the subset of training at node t, $y(i)$ is the real value and y' is the predicted value. And mean for the model is calculated as

$$y' = \frac{1}{N(t)} \sum_{i \in Dt}^{n} y(i) \tag{6}$$

This algorithm was tested with three different types of combinations in data. Decision tree regressor is a great and efficient algorithm for predicting practical resonant frequency, return loss and input impedance as shown in the table of comparison of various algorithms.

3.2 Linear Regression

Linear regression is part of a supervised learning algorithm. It is the relationship between the input variables and output variables with the linear equation which fits the model. If there is more than one variable for input, it is known as multiple regression. Here, we have taken two features as input into the model that is length and width of the patch. In this, linear equation allocates weights, also referred to as coefficients, which are added by itself to each parameters data entry. A supplementary coefficient, which is also inferred as intercept coefficient or bias, is provided for more space of freedom to thoroughly correlate with the linear model. By means of this linear equation, the model tries to best fit the line. The equation for the linear model is given as

$$Y = a + bX \tag{7}$$

Here, *a* is the intercept at the y-axis and *b* is the slope. *X* is an independent variable and *Y* is the dependent variable. Formulas of *a* and *b* for the calculation of the output variables w.r.t. the input variables are

$$a = \frac{(\Sigma y)\left(\Sigma x^2\right) - (\Sigma x)(\Sigma xy)}{n\left(\Sigma x^2\right) - (\Sigma x)^2} \tag{8}$$

$$b = \frac{n(\Sigma xy) - (\Sigma y)(\Sigma x)}{n\left(\Sigma x^2\right) - (\Sigma x)^2} \tag{9}$$

The equations would try to properly fit with the model and the remaining disturbance calculation is done using error root mean square methodology. Generally, R2 squared error formula is used for error calculation and that formula also tries to increase the prediction rate for the model. The cost function is calculated for each model to take out an estimate of how properly the model is fitting or not. It is calculated with the predicted output value and actual value by using root mean squared error. As we get the minimum value of cost function better will be the prediction rate. Simultaneous updates of weights and intercepts with arbitrary values are made to get a perfect fitting. This algorithm fits well with smaller dataset and low variance. This model was justified on three different combinations as shown in Table 1.

3.3 Ridge Regression

Ridge regression is a type of regularization which is also known as L2 regularization. It is a shrinkage method. It adds penalty to the training parameters and thus obtains the main function. The main purpose during training should be to perfectly assign a penalty to β which helps to minimize the function and tune properly. Algorithm gets tuned with reference to mentioned function $F(\beta)$

$$F(\beta) = ||X\beta - Y||_2^2 + \lambda||\beta||_2^2 \tag{10}$$

where a hyperparameter is denoted by λ, training data is denoted by X and testing data is denoted by Y. When the tuning process is done then for perfect fitting λ is tuned precisely. When hyperparameters are tuned and their values start tending towards zero it is considered good, but the data which have a low rate of moving towards zero is believed to have a higher amount of data in it. Thus, this regression method helps in reducing variance and increasing the prediction rate. But this algorithm does not work well for a large dataset. This model was justified on three different combinations as shown in Table 1.

4 Implementation

A rectangular microstrip patch antenna is designed using HFSS. Optimized parameters were supplied to the training model with radiating patch's length (lp) and width (wp). After training the model, it gives us resonant frequency (fr), return loss (lr) and input impedance (Zi) (Fig. 3).

In the training model, algorithms get trained with the supplied features. In this paper, we have analysed three different algorithms with different size of the dataset. We have split the dataset into 80% training data and 20% testing data. With the help of the python programming language, the model was prepared for training and testing. Table 1 shows the comparison of every algorithm with the desired outcomes. In Table 1, the accuracy of algorithms is compared with two sizes of data, that is, with 200 combinations and 1000 combinations. Linear regression helps us in estimating good values of resonant frequency and input impedance with less combinations but when the size of data gets raised then it stops showing much accuracy. Ridge regression is giving a similar performance as linear regression. Both these algorithms do not fit well with a large dataset as well as when negative valued labelled outputs are provided with normalized data, then also it does not work well and both give accuracy of less than 50% which is inaccurate. Decision tree algorithm is showing good fitting on small and large dataset both.

A common approach for optimization of antenna parameters is the neural network, and its extended architecture-based algorithms with a specified application are prevailing. Neural network-based architecture is able to efficiently predict the frequency and input impedance but not the value of the return loss of the designed antenna accurately. As the decision tree algorithm shows a good outcome, its actual

Fig. 3 Training model of the patch antenna in proper sequence

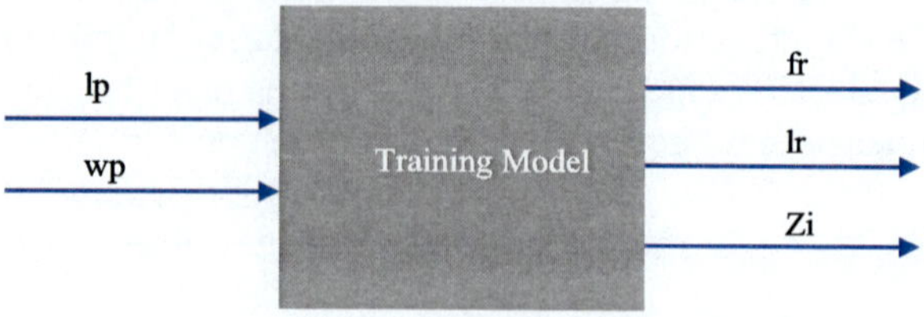

and predicted results for all three outcomes for 1000 combinations-based datasets are described in Table 2.

5 Conclusion

The amalgamation of an antenna and artificial intelligence is imperative in this era to advance technology. Therefore, its parameter optimization concept is prevailing. In this paper, a novice implementation with ridge, linear and decision tree is analysed, which is uncommon as discussed in surveyed literature. The decision tree algorithm has a high rate of prediction. An approach in this paper is to justify that whenever we want less combination-based data then all these algorithms can be preferred and when we want to have more combinations for prediction then the decision tree regressor should be given preference. This is the generalized approach and can be applied to any type of antenna with its corresponding dataset. For estimating return loss only the decision tree algorithm can be used in any antenna. In future, other complex-structured antennas can be implemented with this prepared model. Even return loss and resonant frequency can be achieved with ease by using this model.

References

1. D. Ustun, A. Toktas, A. Akdagli, Deep neural network–based soft computing the resonant frequency of E–shaped patch antennas. Int. J. Electron. Commun. https://doi.org/10.1016/J.AEUE.2019.02.011
2. T. Padmavathy, D. Bhargava, P. Venkatesh, N. Sivakumar, Design and development of microstrip patch antenna with circular and rectangular slot for structural health monitoring. Pers. Ubiquitous Comput. Springer (2018). https://doi.org/10.1007/s00779-018-1130-9
3. A. Kayabasi, A. Toktas, E. Yigit, K. Sabanci, Triangular quad-port multi-polarized UWB MIMO antenna with enhanced isolation using neutralization ring. Int. J. Electron. Commun. **85**, 47–53 (2018). https://doi.org/10.1016/J.AEUE.2017.12.027
4. D. Ustun, A. Akdagli, Design of band–notched UWB antenna using a hybrid optimization based on ABC and DE algorithms. Int. J. Electron. Commun. **87**, 10–21 (2018). https://doi.org/10.1016/J.AEUE.2018.02.001
5. Y. Sharma, H.H. Zhang, H. Xin, Machine learning techniques for optimizing design of double T-shaped monopole antenna. IEEE Trans. Antennas Propag. (2019). https://doi.org/10.1109/TAP.2020.2966051
6. A. Kayabasi, MLP and KNN algorithm model applications for determining the operating frequency of A-shaped patch antennas. Int. J. Intell. Syst. Appl. Eng. Adv. Technol. Sci. (2017). ISSN: 2147-6799. https://doi.org/10.18201/ijisae.2017531432
7. A. Ali, A. Toktas, A. Kayabasi, I. Develi, An application of ANN to compute resonant frequency of E-shaped compact microstrip antennas. J. Electr. Eng. **64**(5), 317–322 (2013). https://doi.org/10.2478/jee-2013-0046
8. S. Gangwar, R. Gangwar, B. Kanaujia, Paras, Resonant frequency on circular microstrip antenna using artificial neural networks. Indian J. Radio Space Phys. **37**, 204–208 (2008)
9. S. Jain, A. Patnaik, S. Sinha, Design of custom-made stacked patch antennas: a machine learning approach. Int. J. Mach. Learn. Cybersecur. **4**, 189–194 (2013). https://doi.org/10.1007/s13042-012-0084-x

10. S. Jain, Bandwidth enhancement of patch antennas using neural network dependent modified optimizer. Int. J. Microw. Wirel. Technol. (2015). https://doi.org/10.1017/S1759078715000616
11. M. Raya, S. Pal, K. Ali, Design of inset fed rectangular shaped microstrip patch antenna using deep neural networks, in *International Conference on Computer and Information Technology* (2019)
12. A. Gehani, D. Pujara, D. Adhyaru, Analysis of an elliptical patch antenna using artificial neural networks, in *IEEE Asia-Pacific Conference on Antennas and Propagation* (2012)
13. C.A. Balanis, *Antenna Theory—Analysis and Design*

Leveraging Transfer Learning for Binary Classification of Images with CNN

Mayank Jha, Om Amrit, Harsh Jain, Mayank Sharma, and Achal Kaushik

Abstract Machine learning and deep learning approaches are widely used for various real-world applications. For the conventional machine learning methodologies, datasets from the same domain are generally used, viz., the training and test data for keeping the same input function space and the same data distribution characteristics. However, this approach is not very effective in scenarios where the cost of training data is high, or they are hard to access or gather. Hence, there is a requirement for trained high-performance learners on the data that is easily accessible in different domains. This technique is called learning transfer, relaxing the hypothesis of an independent and identically distributed dataset for test and training data and leading us to solve the problem of inadequate training data. Transfer of knowledge improves learning performance by drastically avoiding costly data labelling efforts. This work is the study and investigation of modern CNN architecture models used through transfer learning and checking the accuracy and efficiency with new image dataset. We compared the most widely used CNN transfer learning models available and observed that results were quite comparable with decreasing complexity of the latest highly dense neural network models. We also drew a comparison of models in terms of both training and validation accuracy based on model parameters and the layers present in the model.

Keywords Transfer learning · Deep learning · Neural networks · ImageNet · VGGnet · ResNet · DenseNet

1 Introduction

From the beginning of deep learning, there has been a re-vamping interest in neural networks in a various range of applications [1, 2]. Deep learning has solved problems that were considered impossible, but deep neural networks require large clusters of compute servers, vast volumes of training data, and considerable time to train

M. Jha · O. Amrit · H. Jain · M. Sharma · A. Kaushik (✉)
Bhagwan Parshuram Institute of Technology, GGSIPU, New Delhi, India

F. Thakkar et al. (eds.), *Proceedings of the International e-Conference on Intelligent Systems and Signal Processing*, Advances in Intelligent Systems and Computing 1370,
https://doi.org/10.1007/978-981-16-2123-9_53

the network. The deep neural networks connect several layers of network neurons, represented by tens of millions of weights. These weights (typically real values) are fine-tuned during the training process and applied to inputs to feed forward to an output classification.

The fundamental idea of transfer learning is to begin with a deep learning network that is pre-initialized by training of a similar kind of problem [3]. Within a short period of training the knowledge to be applied is transferred for the new, but related problem domain. But there are various options, feature transfer and fine-tuning (which learn on the similitude of the problems at hand), in addition to freezing certain layers of the network and retraining others can be used. The extensive idea is to transfer the knowledge that the model has learned earlier from a task having a considerable amount of available labelled training data to a new task that doesn't have enough data.

Usually, transfer learning is utilized in computer vision and natural language processing (NLP) tasks, like sentiment analysis and image classification. Due to the demand for a large amount of computational power and data [3], computer vision is present in every part of our lives, with significant relevance in medical science, object recognition, drones, mapping and self-driving. It is fundamental to several applications, including visual recognition tasks such as classification of images, localization and detection. Current developments in deep learning technologies have greatly enhanced the efficiency of state-of-the-art image recognition systems, starting with shallow neural networks into very deep networks [4, 5]. For image classification tasks, the latest models are capable of beating a human accuracy, and people are working on it for more improvement, which may lead us to some better developments shortly.

The outline of the paper is as follows: Sect. 2 briefs the transfer learning with some applications. Section 3 discusses the related works on the development of modern transfer learning by surveying the literature. Section 4 details the dataset used in the model. The quantification of various models and their comparison is made in Sect. 5 with the concluding remarks.

2 Transfer Learning

Training data and test data distinguish traditional machine learning with the same input characteristic and the data distribution. The output of a predictive learner will decrease if the data transmission between training data and test data varies. For all the cases, where training dataset characteristic mismatches with the test dataset or when a limited supply of target training data is available, there required a trained high-performance learner on the data that is easily accessible in different domains. This is why transfer learning is used when the data is unique, expensive to obtain, marked or inaccessible [2].

Large data repositories have recently become more common, making moving learning solutions an appealing approach using existing datasets that are not the same as a target domain but connected to the same destination. Transfer learning was successfully put into various machine learning applications, including classification of sentiments, image classification, classification of human activity, classification of multi-language text, and classification of software defects [3]. Instead of starting the learning process from scratch, we use the patterns learned earlier from the same domain to solve a similar problem. For example, through computer vision, neural networks generally try to detect edges in the earlier layers, middle layer shapes, and certain task-specific features in the final layers. We just retrain the later layers with transfer learning and use the early and middle layers. It lets us use the designated data from the role it was initially trained on.

2.1 *Transfer Learning with Linguistic Data*

Nowadays, it is prevalent to use transfer learning in text-based NLP problems. In these kinds of issues, we use word embedding, which maps words into a high-dimensional continuous vector space. In this process, the words with synonymous meaning have a similar vector representation. There are several effective algorithms for learning such distributed projections of words, and the release of such pre-trained models training on a huge corpus of word docs under a restrictive license is common among research organizations. Some famous pre-trained word embeddings are Word2Vec, GloVe, FastText, BERT, etc. [6–9].

2.2 *Transfer Learning with Image Data*

Transfer learning on image data, taken as input for predictive modelling issues, is also popular. This task of prediction may take as input photographs or video data. For a broad and demanding task of image classification such as ImageNet [10], a 1000-class object classification competition, people commonly use a pre-trained deep learning model for these types of problems. Different research organizations design the models and release their final model under a permissive reuse license [11]. These models implemented for new models taking input as image data. Examples of modern CNN architecture models are: visual geometry group (VGG) model, GoogLeNet model (inception), ResNet model, DenseNet model, etc. [12–15].

2.3 *Transfer Learning with CNN*

Training the entire convolution network from scratch (with random initialization) is not preferred in practice in most cases, as having a dataset of appropriate size is relatively rare. Alternatively, use either the ConvNet as initialization or as a fixed extraction feature after pre-train of a ConvNet on the massive dataset, e.g. ImageNet having millions of images in different categories [10]. The three significant scenarios of transfer learning task are:

Using ConvNet as a fixed feature extractor

Use an ImageNet pre-trained ConvNet [10], remove the final fully connected layer, and instead use the remaining ConvNet as a fixed extractor to be utilized for the new dataset.

Fine-tuning the ConvNet

This technique is to replace and retrain the classifier in the new dataset on top of the ConvNet, also by fine-tuning the weights through the backpropagation of the pre-trained network. There is a possibility to fine-tune all the ConvNet layers or fix some earlier layers and only fine-tune the network's higher-level portion to some extent. It may be helpful for many tasks to consider that earlier features of a ConvNet include more generic features. Still, later ConvNet layers are increasingly unique to the class features of the original dataset.

Using pre-trained models

Since ConvNets training on ImageNet today takes 2–3 weeks over several GPUs, we usually see people releasing their final ConvNet checkpoints to help those who can use the networks for fine-tuning [14].

2.4 When and How to Fine Tune

ConvNet apps are more prevalent in the early stages, and in later stages, more original-data-specific. There are several factors that can be used to choose a particular type of transfer learning. Similarity and the size of the original dataset are the two most important factors to be incorporated.

1. The new dataset is small and similar to the original dataset—Due to over-fitting issues and small data size, fine-tuning the ConvNet is not a practical idea. And since there is a high similarity with the original data, ConvNet higher-level features are considered to be important.
2. The new dataset is large but similar to the original dataset—With a large set of data, the chances of over-fitting are reduced if fine-tuning is required through the entire network.
3. The new dataset is small and dissimilar from the original dataset—The data is small; a linear classifier will probably be the best choice. Training the classifier at the top of the network contains more dataset-specific features, since the dataset is quite different. As an alternative, train the SVM classifier from activations somewhere earlier in the network.

4. The new dataset is large and dissimilar from the original dataset—It is presumed that data is sufficient to train a ConvNet from scratch for a very large dataset. The large data adds firmness to fine tune the model over the entire network.

3 Reviewing Development of Modern Transfer Learning

Transfer learning initially used a transition algorithm based on discrimination. Later on, that idea was extended in cognitive science by reusing the conversion of neural networks. After which the area had progressed to incorporate multi-task learning, creating and building on these advances the modern transfer learning models.

3.1 Visual Geometry Group

Introduction

VGGNet showed that for good performance, the network's depth is a crucial component. Their best network is made up of 16 CONV /FC layers that conduct only 3 × convolutions and 2 × 2 pooling across the network with a homogeneous architecture. It is expensive, difficult to calculate and use much more space and parameters [14].

VGG Architecture

In this architecture, the image is moved via a stack of convolutionary layers as shown in Fig. 1 using a low receptive field filter: 3 × 3 and 1 × 1 convolution filters are one of the setups which can be viewed as a linear input channel transformation. The convolution stage is set at 1; the convolution layer's input spatial padding is such that after convolution the spatial resolution is retained, i.e. the padding is 1 for 3 × 3 convolution layers. Spatial pooling is done by five layers of max-pooling that adopt multiple layers of convolution (all layers of convolution do not accompany max-pooling) [16].

Max-pooling is achieved through a 2 × 2 window with step 2. Three fully connected (FC) layers implement a stack of convolution layers (with a similar depth in different architectures): the first two have 4096 channels each, the third performs ILSVRC 1000-way classification, i.e. one channel for each and thus has 1000 channels. In all networks, fully linked layers are set up in the same way. All hidden layers are equipped with nonlinearity (ReLU) rectification [16–18].

One key thing was keeping a very small filter; only 3 × 3 convolution all the way, which is the smallest convolution filter size that is looking at a little bit of the neighbouring pixels. And they just maintained this very simple 3 × 3 convolution structure with the periodic pooling throughout the network. The idea of using smaller filters is beneficial because using this, we have fewer parameters. Stack more of them instead of having larger filters. Stacking three 3 × 3 convolutions on top of others

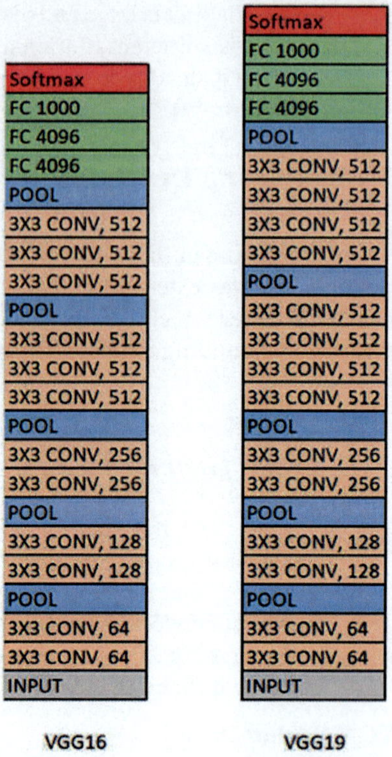

Fig. 1 Architecture diagram of VGG

has the same effective receptive field as if we only have one 7×7 convolution. So, having fewer parameters are computationally efficient [14].

3.2 *GoogLeNet (Inception)*

Introduction

Convolutional network (ConvNet) is from Google, and the main objective is to create an inception module with a reduced number of parameters. This work uses average pooling at the top of the ConvNet instead of fully connected layers, removing many unconnected parameters [7, 15].

GoogLeNet Architecture

The model consists of an inception cell (Fig. 2) in which the results are aggregated after a series of convolutions are performed at different scales. They apply several different types of parallel filter operations on top of the same input coming into the same layer. To save computation, the depth of the input channel is reduced by 1×1

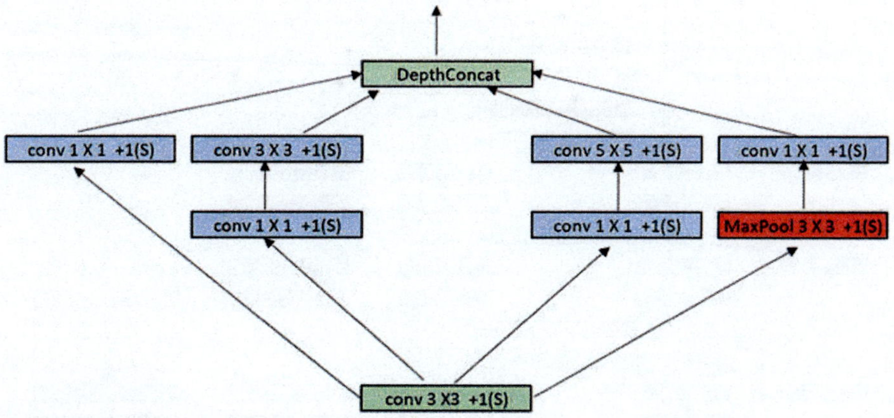

Fig. 2 Inception module

convolutions. We learn a set of 1 × 1, 3 × 3, and 5 × 5 filters for each cell that can learn to extract features from the input at various scales.

Max-pooling is also used to maintain the length, although with the same padding so that the output can be concatenated properly. Also, all the zero-padding are done to maintain the spatial dimensions, and that way this filter concatenation depth-wise can be done. And the key insights used by GoogleNet were using bottleneck layers and designing those feature maps to lower dimensions before convolutional operations or expensive layers. 1 × 1 convolution is taking input volume; it is performing a dot product at each spatial location. It preserves the spatial dimension, but it reduces the depth by projecting our input depth to a lower dimension. It's a linear combination of input feature maps. Two auxiliary outputs are added across the network to improve overall network performance. These additional auxiliary classification outputs get more gradient training injected at the earlier layers, and so more helpful signal flowing in the intermediate layers, as shown in Fig. 3. Adding auxiliary outputs mainly benefited the model's end performance by converging at a slightly better value, and it has a regularizing effect on the network [15].

3.3 ResNet

Introduction

The residual network uses heavy batch standardization and has a special skip connection. In this architecture, the fully connected layers are absent at the end of the network. In the deep neural networks, this was a revolution, a hugely deeper model at 152 layers. ResNets are state-of-the-art models of the convolutional neural network and enabled much deeper networks to be developed. The input representation within the deeper networks is generally accepted for learning more complex

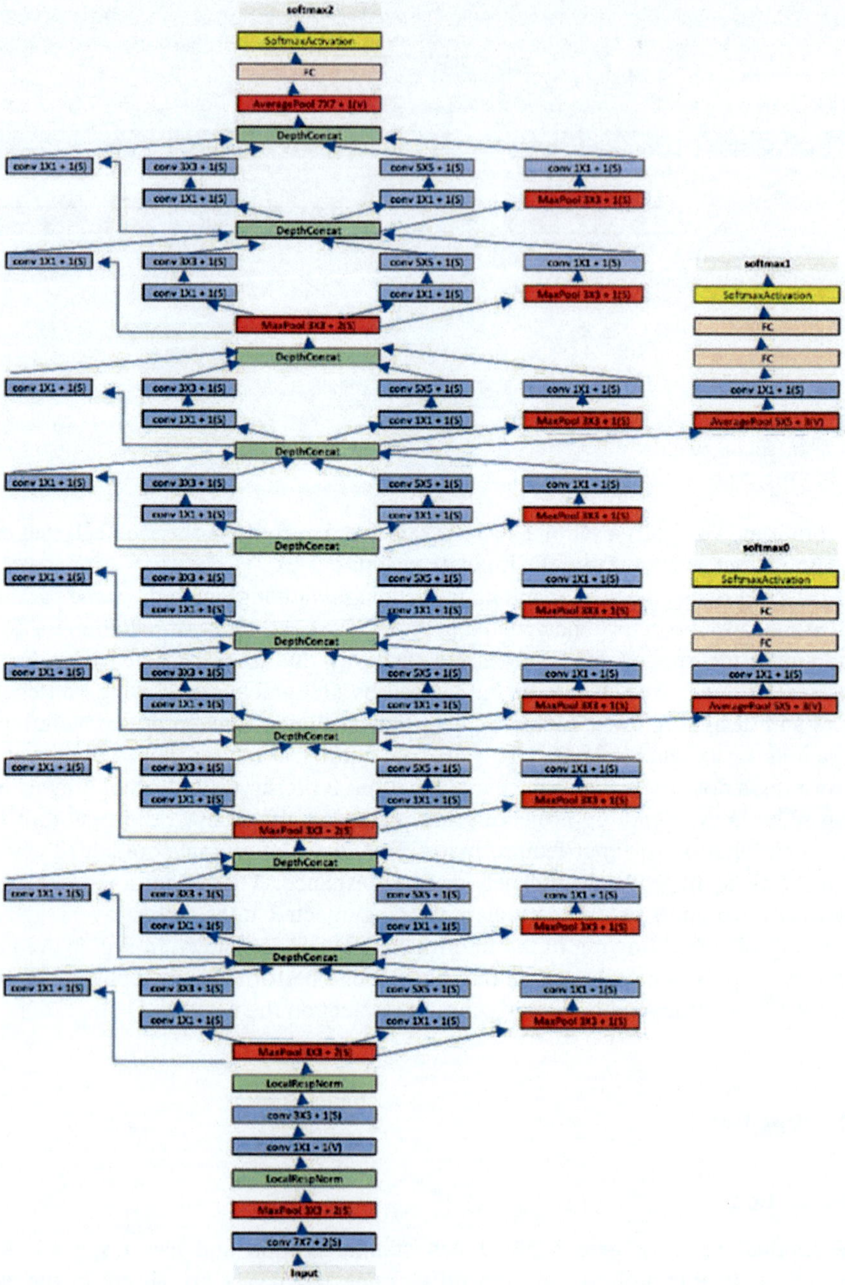

Fig. 3 Inception architecture diagram

functions with improved outcome. More recent developments have also come from tweaking the original architecture. Although several researchers noted that the ultimate effect on the final performance of adding more layers was negative, the problem of disappearing/exploding gradients and hampering convergence from the beginning was solved by standardized initialization and intermediate layers of normalization, enabling networks with tens of layers to begin converging with backpropagation for stochastic gradient descent (SGD) [1, 5, 12].

ResNet Architecture

The deeper networks begin to converge with increased network size; accuracy is saturated and then quickly degrades. The deterioration or loss (of the training accuracy) shows that not all systems can be fairly easily configured. By inserting residual blocks into which intermediate block layers learn a residual input function, the authors suggest a degradation problem solution. Every few stacked layers precisely match a residual mapping but for identity mapping transfer the residual at zero instead of mapping through a stack of nonlinear layers [1, 5].

In this architecture (Fig. 5), the plain baselines are mainly inspired by the VGG net theory. The convolution layers mostly have 3×3 filters and obey two simple design rules: (i) Based on the output feature size the layers have the same number of filters and (ii) to maintain the time complexity per layer the feature size is inversely proportional to the number of filters. With a step of 2, we perform down-sampling straight through convolutionary layers [19].

The network ends with an average global pooling layer and a 1000-way softmax layer which is completely connected. The total weighted layer count is 34. We insert shortcut connections based on the simple network above, which turns the network into a residual version. Two alternatives are considered when the dimensions increase: (A) The workaround only performs identity mapping, padding additional zero entries to increase the dimensions. This method does not add any further parameters; (B) the projection workaround is used to suit the dimensions (done by 11 convolutions). For both choices, when the shortcuts go through two-size function maps, they are executed in two stages [19].

This residual function can be considered to be a step of refinement to map input feature to high-quality features. If there is no need for refinement, the intermediate layers can learn to adjust their weights slowly to zero so that the residual block (Fig. 4) is a function of identity [12].

The total top five error was 3.6% for classification, and this is better than human performance in the ImageNet. There was also a human metric that came from [20] that basically did this task, and the accuracy was around 5.1%, so it was able to do better than the human [12] (Fig. 5).

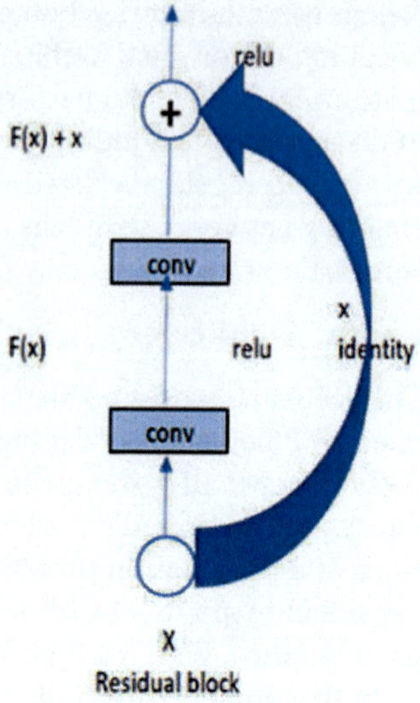

Fig. 4 Residual block

3.4 DenseNet

Introduction

Densely connected convolutional networks (DenseNet) is a CNN architecture that reached much better results on classification datasets using fewer parameters. It is a new use of residual with deeper than the usual networks and still be easy to optimize [13].

DenseNet Architecture

The concept of a dense convolution network is simple: referencing feature maps from any of the previous layers that can be utilized. Therefore, the feature maps of each layer are successively concatenated to the next input layer that came inside a dense block (three-layer model, as shown in Fig. 6). It enables the successive network layers to use features directly from previous layers, allowing features to be utilized again.

The authors [15, 21] state, "concatenating feature-maps learned by different layers increases variation in the input of subsequent layers and improves efficiency." Since the feature maps could be accessed from any previous layer directly, it is found that even with a limited output channel depth of 12 filters per layer the required parameters are significantly reduced. DenseNets are found to achieve better performance over ResNets models with lesser complexity. The researchers mimicked the general ResNet design architecture for most of the paper's experiments, by interchanging the dense block as the recurrent structure [13].

The main takeaway from this is alleviating a vanishing gradient problem because all of these have very dense connections. It strengthens feature propagation and also

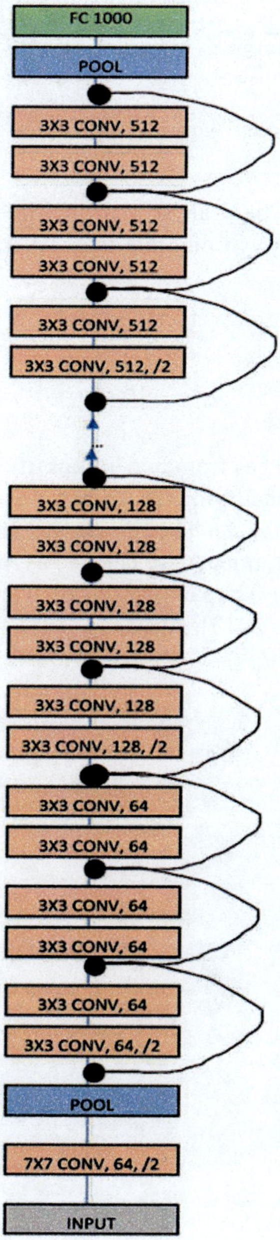

Fig. 5 Residual network architecture

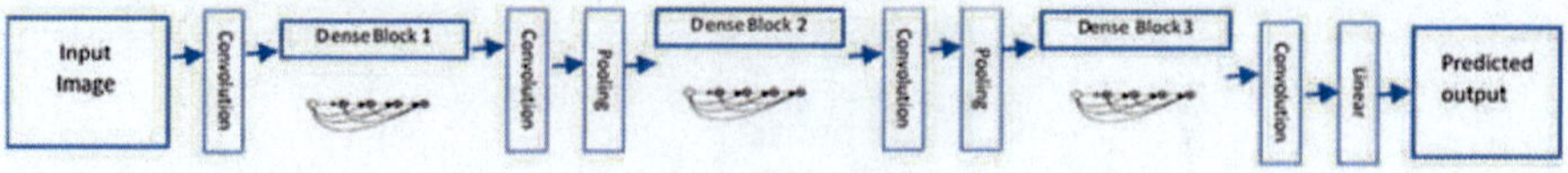

Fig. 6 Deep DenseNet with three dense blocks

encourages future use because there are so many of these connections in each feature map that it is learning to input in multiple later layers and being used numerous times.

4 Dataset

The dataset contains 5226 images from two groups of validated chest X-ray images identified and analyzed in "Identifying Image-based Deep Learning Medical Diagnoses and Treatable Diseases" [22]. The images are divided up into a training and an independent patient test set. Images are labelled as (disease)-(randomized patient ID)-(patient's picture number). The sample of normal and pneumonia-affected chest X-ray images is shown in Fig. 7 [13].

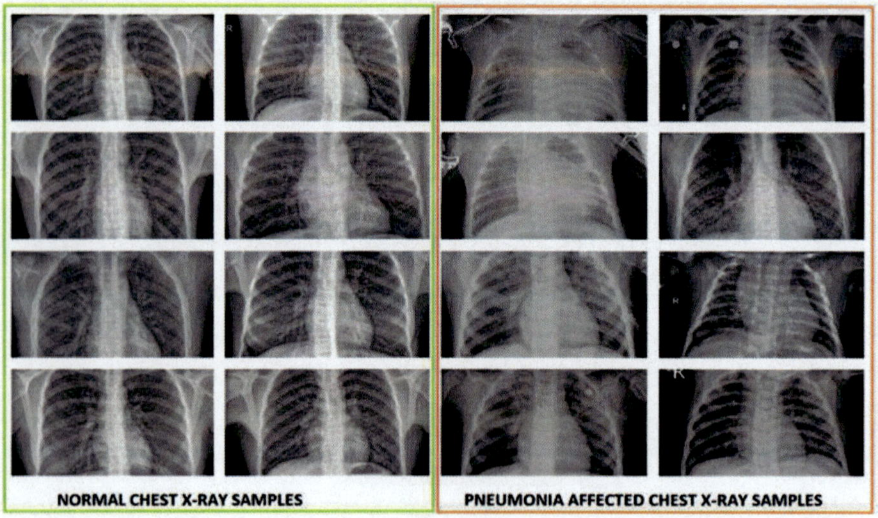

Fig. 7 Dataset samples

5 Results and Future Scope

We compared and evaluated four different CNN transfer learning models as referred to in Table 1, namely VGG, Inception, ResNet and DenseNet over different layers of architecture on pre-trained models which were initially trained over ImageNet dataset [10]. We found that there was a minute difference between the validation accuracy of deep neural network models of DenseNet201 and ResNet152 which is approximately equal to 0.62% and achieving a training accuracy difference of 0.27%. In both cases, the DenseNet model outperformed ResNet while using almost less than three times and the same parameters ResNet utilized in achieving so.

Comparing the traditional models like VGG and Inception we found that the lesser layer VGG16 model gave almost comparable results to the ResNet101 layer model which came as a moment of surprise with almost equal validation accuracy and training accuracy of an approximate difference of 0.08%. Comparing other classification models there was no such model that outperformed DensNet201 on the chest X-ray dataset, with validation accuracy of around 94.84% using such a small dataset. Comparison of different models based on validation and training accuracies is as shown in Fig. 8.

We also found the data to be biased in some manner as it also gave higher validation accuracy than the training in some epochs run during the testing. The comparison of 10 different transfer learning models that were pre-trained on Imagenet [10] weights and applied on a classification model gives us a clear insight that results were quite comparable with decreasing complexity of the latest highly dense neural network models, and one can choose models from this inference. The comparison based on model parameters and validation accuracy can be seen in Fig. 9; also a comparison of depth of model, i.e. layers in model versus validation accuracy is as shown in Fig. 10. Further work can be done in the areas of comparison of these models with a less biased dataset or implementing the same comparison for the detection of pneumonia.

Table 1 Comparison table

Model	Training Acc.	Validation Acc.	Parameters	Depth
VGG16	74.09	62.19	139,001,898	23
VGG19	96.76	90.94	144,311,594	26
InceptionV3	98.44	91.09	23,851,784	159
InceptionResNetV2	98.34	93.75	55,873,736	572
ResNet50	97.9	91.41	25,583,592	50
ResNet101	74.21	62.19	45,246,186	101
ResNet152	99.1	94.22	60,912,874	152
DenseNet121	96.55	88.91	8,621,210	121
DenseNet169	74.17	62.19	14,793,834	169
DenseNet201	99.37	94.84	20,658,202	201

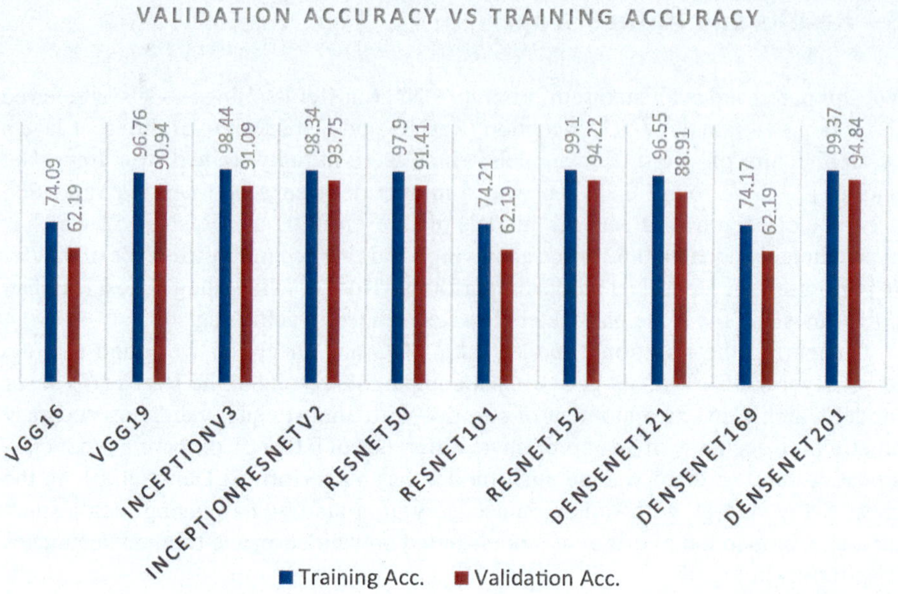

Fig. 8 Validation accuracy versus training accuracy plot

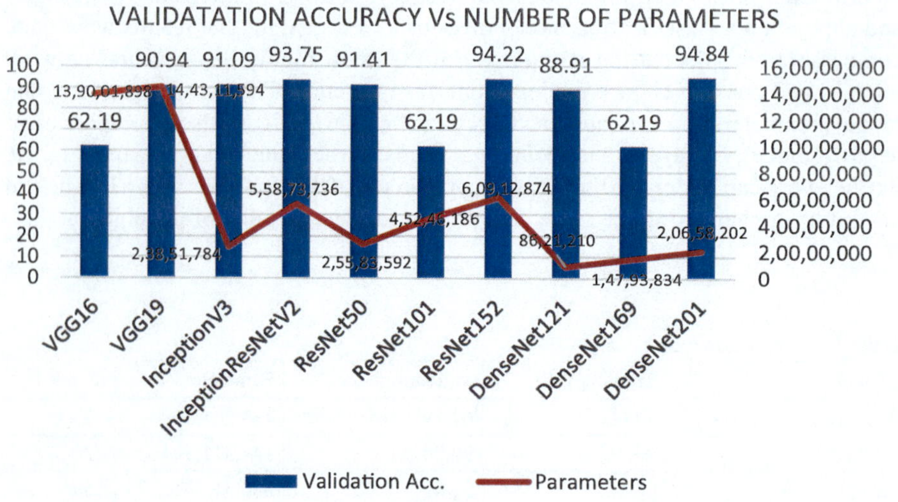

Fig. 9 Validation accuracy versus model parameters plot

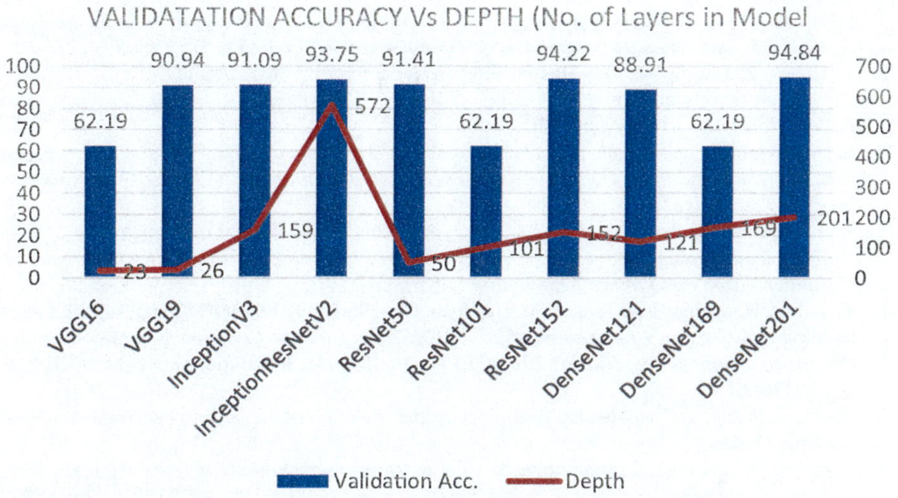

Fig. 10 Validation accuracy versus depth of model plot

References

1. V. Sze, Y.H. Chen, T.J. Yang, J.S. Emer, Efficient processing of deep neural networks: a tutorial and survey. Proc. IEEE **105**(12), 2295–2329 (2017). https://doi.org/10.1109/JPROC.2017.2761740
2. K. Weiss, T.M. Khoshgoftaar, D.D. Wang, *A Survey of Transfer Learning*, vol. 3, no. 1 (Springer International Publishing, 2016)
3. S.J. Pan, Q. Yang, A survey on transfer learning. IEEE Trans. Knowl. Data Eng. **22**(10), 1345–1359 (2010). https://doi.org/10.1109/TKDE.2009.191
4. C. Angermueller, T. Pärnamaa, L. Parts, O. Stegle, Deep learning for computational biology. Mol. Syst. Biol. **12**(7), 878 (2016). https://doi.org/10.15252/msb.20156651
5. C. Zhang, P. Patras, H. Haddadi, Deep learning in mobile and wireless networking: a survey. IEEE Commun. Surv. Tutor. **21**(3), 2224–2287 (2019). https://doi.org/10.1109/COMST.2019.2904897
6. J. Devlin, M.-W. Chang, K. Lee, K. Toutanova, BERT: pre-training of deep bidirectional transformers for language understanding. (2018)
7. J. Zhang, C. Lu, X. Li, H.J. Kim, J. Wang, A full convolutional network based on DenseNet for remote sensing scene classification. Math. Biosci. Eng. **16**(5), 3345–3367 (2019). https://doi.org/10.3934/mbe.2019167
8. C.D. Pennington, J. Socher, R. Manning, Glove: global vectors for word representation, in *Proceedings of the 2014 conference on empirical methods in natural language processing (EMNLP)* (2014), pp. 1532–1543
9. L.Y. Wei, M. Levoy, Fast texture synthesis using tree-structured vector quantization, in *Proceedings of ACM SIGGRAPH Conference on Computer Graphics* (2000), pp. 479–488. https://doi.org/10.1145/344779.345009
10. I. Sutskever, G.E. Hinton, A. Krizhevsky, Imagenet classification with deep convolutional neural networks, in *Advances in Neural Information Processing Systems*, ed. by F. Pereira, C.J.C. Burges, L. Bottou, K.Q. Weinberger (2012), pp. 1097–1105.
11. A. Guadamuz, Open science: open source licences in scientific research. SSRN Electron. J. **7**, 321–366 (2011). https://doi.org/10.2139/ssrn.764064

12. J. Sun, K. He, X. Zhang, S. Ren, Deep residual learning for image recognition, in *Proceedings of the IEEE conference on computer vision and pattern recognition* (2016), pp. 770–778
13. G. Huang, Z. Liu, L. Van Der Maaten, K.Q. Weinbergerocsari, S.S. Stone, Densely connected convolutional networks, in *Proceedings of the IEEE Conference on Computer Vision and Pattern Recognition* (2017), pp. 4700–4708
14. K. Simonyan, A. Zisserman, Very deep convolutional networks for large-scale image recognition, in *3rd International Conference on Learning Representations, ICLR 2015—Conference Track Proceedings* (2015), pp. 1–14
15. A. Rabinovich, C. Szegedy, W. Liu, Y. Jia, P. Sermanet, S. Reed, D. Anguelov, D. Erhan, V. Vanhoucke, Going deeper with convolutions, in *Proceedings of the IEEE Conference on Computer Vision and Pattern Recognition* (2015), pp. 1–9
16. M. Awais, H. Muller, T.B. Tang, F. Meriaudeau, Classification of SD-OCT images using a deep learning approach, in *Proceedings of 2017 IEEE International Conference on Signal Image Processing Applications, ICSIPA 2017* (2017), pp. 489–492. https://doi.org/10.1109/ICSIPA.2017.8120661
17. B. Xu, N. Wang, T. Chen, M. Li, Empirical evaluation of rectified activations in convolutional network. (2015)
18. N. Cohen, A. Shashua, Convolutional rectifier networks as generalized tensor decompositions, in *33rd International Conference on Machine Learning, ICML 2016*, vol. 2 (2016), pp. 1500–1519
19. E.R.S.D. Rezende, G.C.S. Ruppert, T. Carvalho, Detecting computer generated images with deep convolutional neural networks, in *Proceedings—30th Conference on Graphics, Patterns and Images, SIBGRAPI 2017* (2017), pp. 71–78. https://doi.org/10.1109/SIBGRAPI.2017.16
20. O. Russakovsky et al., ImageNet large scale visual recognition challenge. Int. J. Comput. Vis. **115**(3), 211–252 (2015). https://doi.org/10.1007/s11263-015-0816-y
21. C. Szegedy, V. Vanhoucke, S. Ioffe, J. Shlens, Z. Wojna, Rethinking the inception architecture for computer vision, in *Proceedings of IEEE Computer Society Conference on Computer Vision and Pattern Recognition*, vol. 2016 (2016), pp. 2818–2826. https://doi.org/10.1109/CVPR.2016.308
22. M. Goldbaum, D. Kermany, K. Zhang, Labeled optical coherence tomography (OCT) and chest X-ray images for classification. Mendeley data 2 (2018). https://doi.org/10.17632/rscbjbr9sj.3

Big Data, IoT, and Cloud Computing

Shortest Pathfinder for Air Traffic Network: A Graph-Based Analysis

Piyushi Jain, Drashti Patel, and Jai Prakash Verma

Abstract Air Traffic Management is the dynamic, integrated management of air traffic. There has been a dramatic increase in people who are preferring airways over the traditional transportation system as it saves time. The growth will generate countless new routes and require hundreds of new airports and thousands of new planes and pilots. Most of the traffic management is done throughout the flight to help the pilot with ongoing traffic above there. Due to this rapid growth of passengers and the increasing environmental concerns, there is a need to find the shortest and efficient path for any flight considering distance and time. Using graph theory's Dijkstra algorithm, the shortest point between a source and a distance could be calculated. The shortest path based on parameters including distance and time has been implemented with Neo4j and Apache Spark and hence a comparative study between the two software has also been done. The model was trained using a dataset of air routes of Oceania and was able to predict the best route in terms of distance and time separately. Neo4j provides visual results in the form of a graph while spark provides the shortest distance and route followed in the path. An optimal path that is derived using the shortest distance on the cost of time or vice versa is not only the findings of the system. An equation could be derived that would balance both distance and time. That equation can also be extended to take care of parameters like traffic, weather conditions, and country borders or restricted areas. ...

Keywords Graph theory · Dijkstra algorithm · Air traffic management · Graph analytic · Neo4j · Apache spark

---

P. Jain · D. Patel · J. P. Verma (✉)
Department of Computer Science and Engineering, Institute of Technology, Nirma University, Ahmadabad, India
e-mail: jaiprakash.verma@nirmauni.ac.in

P. Jain
e-mail: 17bit081@nirmauni.ac.in

D. Patel
e-mail: 17bce075@nirmauni.ac.in

F. Thakkar et al. (eds.), *Proceedings of the International e-Conference on Intelligent Systems and Signal Processing*, Advances in Intelligent Systems and Computing 1370,
https://doi.org/10.1007/978-981-16-2123-9_54

1 Introduction

Air traffic control includes services provided by ground-based controllers who continuously interact with the pilots to ensure the safe, orderly, and expeditious flow of traffic on the airways and at airports [15]. They supervise aircraft departing and landing, as well as throughout the flight, by instructing pilots to fly at assigned altitudes and on defined routes [3]. According to statistics, there is an average increase of 7% of air passengers every year [1]. This increase leads to an exponential increase in air traffic leading to significant congestion, flight delays, and pollution [14]. To keep these many numbers of airplanes at safe distances from one another, to direct them during takeoff and landing from airports, and also to guide them about bad weather and to ensure that traffic flows smoothly with minimal delays, there is a need to introduce a NextGen air traffic management and optimized methods [7]. The optimization and automation of air traffic control have been the subject of several studies in the last decades and there is still more to go [21]. The objective of this paper is to determine the shortest path for the aircraft [16]. For this purpose, we will be using graph theory and implement the shortest path algorithm on software namely: Neo4J and Apache Spark.

1.1 Background of Air Traffic Management (ATM)

Over the past few years, air transportation has gained worldwide adoption by millions and has become the most widely used method of conveyance. World's busiest air routes are mainly those that connect inter cities and this perhaps increases the risk of accidents as a very short range is covered during those flights [20]. The growing stress of modern ATM is on interoperable and harmonized systems that allow an aircraft to proceed with the minimum of performance change from one airspace to another [22]. Air Traffic Control (ATC) systems have traditionally been developed by individual states that concentrated on their specifications, creating various levels of service and capacity around the world.

The most basic requirement of air travel is communication between aircraft and ground facilities to maintain safe distances between aircraft and accurate flight paths from source to destination and to provide pilots with current information regarding weather and other critical variables [10]. Back then in 1911, in the earliest air transportation, communication between pilot and ground consisted of mere visual cues from the ground with directional pointers, hilltop bonfires, and airport beacons. The radio was invented in 1985 and its usage in air communication became prevalent from 1917 onwards which brought a major change, i.e., voice communications between pilots and ground controllers. Who knew that was just the beginning of active ATC [26]. By the 1950s, the local airport authorities along with the government had established an extensive system of ground stations, radar arrays, and requirements that aircraft be equipped with standard communications equipment [12]. By the 1960s, we could see ATC similar to what we see now. Since then, many new technologies, such as powerful computers in ATC facilities and collision avoidance systems in aircraft, have made air travel progressively safer and more reliable. But the basic features and procedures of the 1960s ATC system have remained remarkably unchanged [18]. In recent years, there is an urge to provide safer, faster, more reliable, and more fuel-efficient air travel and to keep pace with the increasing volume of air traffic.

1.2 Motivation

The passengers are asked by their airlines to arrive a few hours before the flight is scheduled. As a regular practice, the lowered cost of flight airways is preferred by more travelers [20]. For instance, 5,000 airplanes can be seen in the sky during peak hours. This sums up to more than 50,000 aircraft operating in our skies each day [9]. What prevents these aircraft from colliding with each other? How does air traffic move into and out of an airport or across the country?

Air traffic controllers ensure the movement of the aircraft, direct them during takeoff and landing, and keep them at a safe distance to prevent commercial or private airplanes from colliding and delays. On a global scale, an average of 7% increase can be seen in air traffic every year [14]. This increase in passengers leads to an increase in the number of flights scheduled and ultimately leading to flight delays. Delays are the most common passenger complaint received by the Department of Transportation (DOT), accounting for about 40% [17]. Passenger experience may appear to have little to do with what goes on behind the flight deck door, but that's only true when a flight goes smoothly. To approach such a big issue, there is an urgent need for a NextGen transportation system enabling air traffic controllers to identify the shortest route and pilots to follow that route [19].

1.3 Contribution

In the work presented, we intend to achieve the following objectives:

1. Design and develop a graph-based analysis model of shortest pathfinder for Air Traffic Network.
2. Applying Dijkstra Algorithm using Neo4J and Apache Spark for shortest path generation between the airports.

1.4 Organization

The remaining paper is organized as follows. Section 2 summarizes the related work in the field of cancer prediction. Section 3 provides the proposed approach. Methodology, Execution, and implementation of the proposed approach are included in Sects. 4 and 5. The results, discussions, and conclusions in Sects. 6 and 7, respectively.

2 Related Work

The graph theory can be applied to solve many problems in ATM. It can be applied to find the minimum delay in the scheduling of the aircraft, maximum no. of dependent aircraft in precision arrival schedule, and airspace sectorized problem. It finally suggests that when these theories are applied with machine learning techniques on real data, it will be useful for automatic ATM [7]. Most of the research is going on with machine learning techniques used for predicting the path of the aircraft. The predicted graph is a 2D or 3D representation [6]. The path predicted is usually 4D which means they are considering the weather of the place concerning the time [13].

Noriko Etani et al. [6] developed a model that predicts from historical data that flight will be delayed or not. The prediction of the air route majorly depends on the weather. Even if we have a well-planned route before the departure of any flight, when the flight is about to takeoff or it is in the air, it may have to change its route due to weather conditions. Therefore, the weather is an important factor to consider while predicting the route. The prediction of the flight route was done with and without the weather data and it showed significant change and improvement while using the weather data and it gave better accuracy.

Jacob et al. [5] devised a node-based prediction for air route. This will help the future needs of ATM and also provide alternatives to the current methods. The model is effective with minimal use of data. The accuracy of the weather data in the prediction of the node-based trajectory is very important. Weather data with the high-resolution grid will have better accuracy of the weather at a given point in the grid which improves the predictive model. The model is also able to give the prediction about turbulence avoidance during the flight journey. Ahyan et al. [2] developed an algorithm to find the path of the flight. The 3D flight route prediction was done using the HMM and Viterbi algorithm. This method predicts the states of the flight route one by one and using the transition probabilities, emission probabilities, and initial probabilities.

Bosson et al. [4] proposed a predictive model for ATM. They particularly focused on air routes while landing or the runway path prediction. This path is important as the number of flights is increasing day by day and due to traffic at the airport, they are delayed. The delays in flights cost the airlines so much. This prediction is done using some of the supervised learning.

The prediction of the air route has to precise for it to be used in real life. Therefore, we need to consider every factor, the most important being weather. Weather will be taken at each time interval so the prediction will become a 4D trajectory of the aircraft. Liu et al. [13] developed the model to predict the flight trajectories by using the last minute plan of the air route which they prepare just before the takeoff of the plane. The noise from the relevant trajectories can be removed by using the clustering technique in the preprocessing step. Zhengyi Wang et al. [25] proposed that the flight routes should be clustered with some dependencies and then we can apply differently the algorithms to each cluster. Esther Calvo Fern et al. [8] proposed an approach using reinforcement learning to find the delays and dependencies in flights. This planning is done much before the day of the travel. The interacting flights were plotted which can be used to predict the delays and trajectories.

While considering every factor still there are more delays in the flights and due to which it has to change the route considering other flights and getting a runway for landing. Therefore, it is important to reduce this delay in the flights. When a flight is delayed, it costs much to the flight agency. They will want to operate without any delays. This delay can be predicted and the reason for the delay can be identified so that the next time, the issue is solved and there is no delay. Kalliguddi et al. [11] proposed a model to predict the delay and the root cause of it. The data collected was huge so the accuracy is better.

Considering the graph theory and machine learning, we can predict the shortest route for the aircraft. This may not be used currently but in the future, this may help design the routes more efficiently. Machine learning will be helpful as it will have taken into consideration all the previous routes with other data for predicting better results (please refer Table 1 and Table 2).

3 Proposed Work

Modern air transportation systems involve some of the most complex and challenging multi-disciplinary technical challenges of the age. There is an urgent need to provide safer, faster, more reliable, and more fuel-efficient air travel and to keep pace with the increasing volume of air traffic. In recognition of these challenges, a multifaceted analysis and development efforts are underway to realize the Next-Generation Air Transportation System (NextGen), intended to accommodate the projected growth in the demand for air transportation and to achieve increased efficiency in the effective use of the airspace and Air Traffic Control (ATC) resources. In this research, we would be focusing more on the problem of fuel-efficient air travel. Our approach provides the solution for finding the shortest path for an aircraft traveling from source to destination. The graph-based analysis models are implemented with Neo4J and Apache Spark programming frameworks [23, 24]. Elements of the graph and their meaning in this research work are as follows.

Table 1 Comparative analysis of work in the area of air traffic management

Approach	Year	Objective	Dataset	Methodology	Pros	Cons
Big Data [6]	2019	Finding a relation between weather and flight and developing a predictive model for flight arrival	Flightstats-flight dataset and weather data from three weather spots of Wakkana	SVM, GB, RF, AB	It was found that predictive model worked best with RF with weather data and GB performed the best without weather data	More research on disordered time series is needed to improve the predictive model
Dijkstra and other node-based algorithms [5]	2017	Flight planning: node-based trajectory prediction and turbulence avoidance	All EGLL to KJFK flights from FlightAware using the baseline values of oDir = 4 and ng = 25	NBP, HF	The proposed system is similar to the existing one so it is easy for future research and usage of the same	It only used one of the paths with turbulence, more data are required to train the model more accurately
Machine Learning [11]	2017	Predictive modeling of aircraft flight delay	Bureau of Transportation Statistics (BTS)	DT, RF, MLR	Aircraft delay, carrier delay, weather delay, and NAS delay which have the most effect on on-time flight departure	Weather data can be extracted and used to better develop a predictive model for flight delay
Machine Learning [7]	2017	Air traffic management and control using concepts of graph theory	–	Graph theory	Graph theory applied with adequate knowledge of machine learning tools and data science could benefit ATM	It just gives a theoretical approach which we can follow so the implementation of the same is required

(continued)

Table 1 (continued)

Approach	Year	Objective	Dataset	Methodology	Pros	Cons
Deep learning [13]	2018	Predicting aircraft trajectories: a deep generative convolutional recurrent neural networks approach	TFMS-flight dataset, which gives 4-dimensional information of each flight	CRNN with Long Short-Term Memory (LSTM) encoder and decoder network	Generalization of features related to weather was done properly	Predictive model could be extended to include more useful features
Machine learning [2]	2016	Aircraft trajectory prediction made easy with predictive analytics	Global forecast system-rapid refresh weather model's 3-dimensional network	Hidden Markov Model (HMM)	An approach that would ensure better flow-management, safety, and less fuel consumption was proposed	For acquiring more accuracy, we need higher spatial resolution weather data than 13 km
Machine learning [4]	2018	Supervised learning applied to air traffic trajectory Classification	Track data of a month in Integrated Flight Format (IFF) and air traffic in the region	LR, SVM, KNN, RF, NB, MLP, CNN	For 3 min before landing the accuracy was better than 10 min before landing	More practical approach required to help air traffic controller to help in aircraft trajectories
Machine Learning [25]	2017	Short-term 4D trajectory prediction using machine learning methods	July 2017 records of ADS-B records over an airport in China	Principle Component Analysis (PCA), Neural Network (NN)	Clustering helped achieve better prediction than the model without clustering	Comparison with results from model-based methods, as well as on studying prediction accuracy for other trajectory variables besides ETA
Machine learning [8]	2017	Prediction of aircraft trajectory and balancing capacity-demand	Radar tracks and weather data of Spanish flights	HMM, non-linear regression, kernel-based clustering, RL	Positive results of both trajectory prediction using machine learning and demand capacity balancing problems	More operational features helping in improving the benefits for ATM domain

Table 2 Abbreviations table

Sr. no.	Abbreviations	Full form	Sr. no.	Abbreviations	Full form
1	DT	Decision tree	2	LR	Logistic regression
3	CRNN	Convolution recurrent neural network	4	SVM	Support vector machine
5	GB	Gradient boosting classifier	6	AB	AdaBoost
7	NBP	Node-based path finding (using Dijkstra's algorithm)	8	HF	Heuristic function
9	RF	Random forest	10	MLR	Multi-layer regression
11	KNN	K-nearest neighbors	12	NB	Naive Bayes
13	MLP	Multi-layer perceptron	14	CNN	Convolution neural network

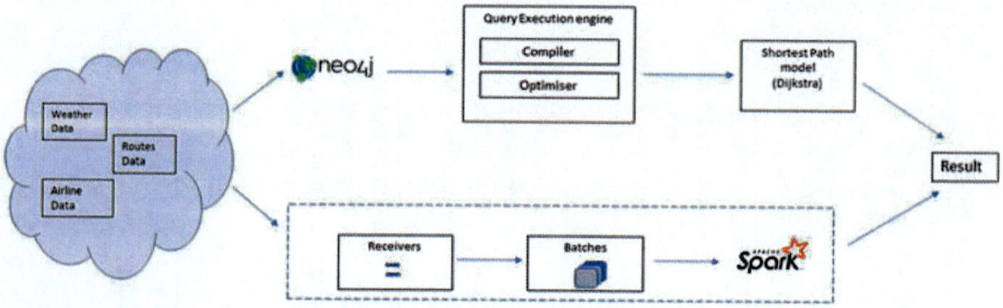

Fig. 1 Proposed system architecture

VERTICES : Airports
VERTEX ATTRIBUTES : Footfall, Terminals, Staff, City, Capacity
EDGES : Airplanes/ Routes
EDGE ATTRIBUTES : Frequency, Passengers, Distance covered, Fuel usage

As shown in Fig. 1, to travel from point A to point B, one can consider parameters like distance and time. Making the route efficient based on one of the parameters at the cost of the other is not going to benefit the air transportation system. The flight plan routing should be decided for the minimum flight time while avoiding bad weather and restricted airspace and making the overall journey fuel-efficient. Air traffic control could change the routes as per to avoid traffic and hazardous collisions of the airspace. Therefore, a model needs to be developed which works on a balanced equation of time and distance, and through that equation, the most desirable air route can be calculated.

4 Methodology and Concepts

Graphs are the most effective way of representing data and obtaining underlying patterns. To understand graphs, we primarily need to learn about graph theory. Airports are represented using nodes of the graph. For instance, Fig. 2a shows the airport POM (Port Moresby Jacksons International Airport) whereas the edges of the graph show the routes of aircraft. Figure 2b shows a route connecting two airports, POM and GKA. The above graph, as shown in Fig. 3, is a part of the huge dataset we have used. The graph is plotted using draw_networkx() function of Neo4J and it depicts the different air routes for different airlines.

As per Algorithm 1, Neo4j is one of the graph platforms where we can visualize and store any type of graph. It is a database management system where you store information in the form of a graph and then you can run queries. In Neo4j, we can create nodes and edges (called as relationships). The edges and nodes can store any type of property and any number of properties. Neo4j has inbuilt Libraries and Plugins which contain graph algorithms and other algorithms which can be applied for graphs. Example query for Neo4j is shown in Fig. 3. After loading the data in Neo4j, we match the origin and destination airport to find the shortest path using the Dijkstra algorithm (Fig. 4).

Algorithm 1: Prediction

Input: Dataset D1, source airport(a), destination airport(y)
Output: Shortest path
Step-1. Create a shortest path tree set (SPTree) which keeps track of vertices included in the minimum path.Initially, the set is empty.
Step-2. Assign 0 to the source node and Infinity to the remaining nodes in the input graph.
Step-3. Add a vertex which is closest to the source vertex and is not present in the SPTree.
Step-4. Update the distance of all the adjacent vertices.

Fig. 2 Fundamentals of graph. **a** Node, **b** Edge

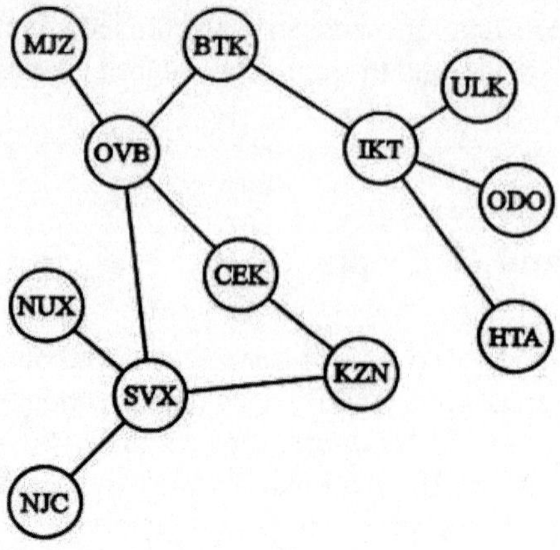

Fig. 3 Graph

```
MATCH (c1:MyNode1 { Name: 'IAD' })
MATCH (c2:MyNode1 { Name: 'TPA' })
CALL apoc.algo.dijkstra(c1, c2, 'TO', 'dist') YIELD path,
weight
RETURN path, weight
```

Fig. 4 Example query

4.1 Data Preprocessing

Data Preprocessing is one of the most important steps of data mining where we prepare the data before mining. Data in the real world is inconsistent, incomplete, and is likely to contain errors, hence there is a need to modify it before processing and applying any of the predictive models. We need to handle null values as they have no significance and cannot be processed with other data types. Some of the most common ways to resolve them are: (1) To ignore the data rows containing them. (2) Using a global constant to fill those places. (3) To fill missing values with the mean value. (4) To replace the null value with the calculated most probable value. We can go with ignoring the data rows only if they are in lesser percentage and if in that row more than one attribute is missing.

4.2 Feature Selection

It is a method of selecting a subset from the relevant set of features for use in model construction. Feature selection includes removing features having missing values and choosing the features which influence the result and have a greater possibility of deviating and giving wrong output if ignored. We used the Correlation Matrix with Heatmap. Heatmap helps us to correlate the data and it shows how much it is related to each other. Using this, we can identify which all features have a greater correlation with the output and choose accordingly. Here, for breast cancer data, we used this technique from which we chose the top 10 features for model 2. While model 1 was kept as it is with all the features. Data is split into two parts: Test set and Training set. The training set is used for training the model and the test set is used to test the trained model. Stratified splitting is done to ensure that each split is similar for some features. In classification, stratified splitting is done so that each subgroup has the same percentage of target class as that of the complete set. After data preprocessing, we will use this data to train our model and apply the Dijkstra algorithm to find the shortest path.

5 Execution and Implementation

The dataset used in this research is taken from Kaggle. It consists of data of 67,664 routes between 3,209 airports on 531 airlines globally. Libraries like GraphFrame which is available with PySpark and networks are used to train the data and predict the shortest path by implementing Dijkstra's shortest path algorithm. The model is implemented on two different platforms: Neo4J and Spark. The two are different in some ways: Neo4J is intuitive whereas Spark is comparatively fast. For the implementation of our project in Neo4j, we have taken only the source and destination airports and the distances or air time between the flights. For calculating the shortest path, the plugin which we will be using is APOC. APOC has already defined functions to call the shortest path algorithm Dijkstra.

6 Results and Discussion

In this research, we focused on the problem of increasing air passengers leading to an exponential increase in air traffic. To approach this problem, we used two software for the model development namely: Neo4j and Apache Spark. We developed an independent working model on both of the software to achieve the best possible results. Neo4j, on one hand, gave us better visuals of the graph and the dataset we were working on. Whereas spark on the other hand supported multiple languages and was fast and easy to use and understand. We implemented the Dijkstra algorithm to determine the shortest path.

The above Figs. 5 and 6 show the output when 'CMA' airport was selected as a source and 'KPR' airport was chosen as the destination. The first path is obtained optimized path according to distance and the second result is obtained when optimization is done based on time.

```
In [15]: shortest_path_distance = nx.dijkstra_path(df, source='CMA', target='KPR', weight='Distance')
         shortest_path_distance

Out[15]: ['CMA', 'SGO', 'TWB', 'SYD', 'LAX', 'ANC', 'ADQ', 'KKB', 'SYB', 'KPR']

In [16]: shortest_path_distance = nx.dijkstra_path(df, source='CMA', target='KRP', weight='AirTime')
         shortest_path_distance

Out[16]: ['CMA', 'SGO', 'TWB', 'SYD', 'PEK', 'CPH', 'KRP']
```

Fig. 5 Shortest path prediction

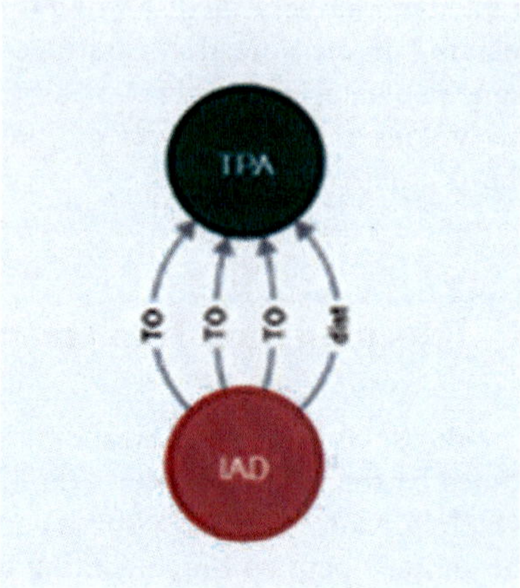

Fig. 6 Results—graph showing all possible paths between two airports

Table 3 All distances between two airports (IAD TPA)

Origin airport	Distance	Destination airport
IAD	888	TPA
IAD	810	TPA
IAD	810	TPA

For 'IAD' airport as a source and 'TPA' as a destination, there are three possible air routes as displayed in Table 3. The possible three distances from IAD to TPA are 888, 810, and 810 as shown in Table 3. The model developed was able to obtain the optimized path with a minimum distance of 810. This is just one of the many applications of Graph Theory. We can apply it to almost any kind of problem and get solutions and visualizations. Graphs provide a convenient way to represent various kinds of mathematical objects. Applications of graph theory are also useful in our day-to-day life.

Graph theory and its applications are beneficial in various fields and everyday life. The shortest path algorithm has been discussed which is very useful when it comes to time and fuel efficiency, which are both vital. We have also discussed the output generated by the two software used in this project. The algorithm can be applied in either way: to search the shortest path by distance or time. For the increasing air population and hence increasing traffic, many other better solutions need to be proposed.

7 Conclusion

In this research paper, a brief introduction of graph theory and its application in air traffic management has been discussed. Graph theory and its applications are beneficial in various fields and everyday life. The shortest path algorithm has been discussed which is very useful when it comes to time and fuel efficiency, which are both vital. We have also discussed the output generated by the two software used in this project. The algorithm can be applied in either way: to search the shortest path by distance or time. For the increasing air population and hence increasing traffic, many other better solutions need to be proposed.

References

1. Air navigation report. https://www.icao.int/airnavigation/documents/. *ICAO* (2016)
2. S, Ayhan, H. Samet, *Aircraft Trajectory Prediction Made Easy with Predictive Analytics* (2016), pp. 21–30
3. J. Bergner, O. Hassa, *Air Traffic Control* (2012), pp. 197–225
4. C. Bosson, T. Nikoleris, *Supervised Learning Applied to Air Traffic Trajectory Classification* (2018)
5. J.C.H. Cheung, Flight planning: node-based trajectory prediction and turbulence avoidance. Meteorol. Appl. **25**(1), 78–85 (2018)
6. N. Etani, Development of a predictive model for on-time arrival flight of airliner by discovering correlation between flight and weather data. J. Big Data **6**(1), 85 (2019)
7. A. Farrahi, A. Goldberg, L. Bagasol, J. Jung, *Applying Graph Theory to Problems in Air Traffic Management* (2017)
8. E.C. Fernández, J.M. Cordero, G.A. Vouros, N. Pelekis, T. Kravaris, H.V. Georgiou, G. Fuchs, E. Casado, P. Costas, S. Ayhan, *Dart: A Machine-Learning Approach to Trajectory Prediction and Demand-Capacity Balancing* (2017)
9. C. Freudenrich, *How Air Traffic Control Works* (2001) p. 456
10. P.P. Jovanis, F.D. Hobbs, *Traffic Control* (Encyclopedia Britannica, inc, 2018). https://www.britannica.com/technology/traffic-control
11. A. Kalliguddi, A. Leboulluec, Predictive modeling of aircraft flight delay. Univ. J. Manage. textbf5, 485–491 (2017)
12. J. Kelvey, *Air and Space Magazine* (2018). https://www.airspacemag.com/flight-today/air-traffic-calming-devices-180967713/
13. Y. Liu, M. Hansen, *Predicting Aircraft Trajectories: A Deep Generative Convolutional Recurrent Neural Networks Approach* (2018). arXiv preprint arXiv:1812.11670
14. Mazareanu, *Global Air Traffic—Annual Growth of Passenger Demand 2006–2020* (2020)
15. S. Mondoloni, N. Rozen, Aircraft trajectory prediction and synchronization for air traffic management applications. Progress Aerospace Sci. **119**, 100640 (2020)
16. V. Rani, Uses of graph theory in day to day life. Int. J. Adv. Res. Sci. Eng. **7** (2018)

17. R.W. Jr. Poole, *A Review of the Small Aircraft Transportation System Concept*, pp. 50–77 (2002)
18. R.W. Jr. Poole, *Hudson Institute Initiative on Future Innovation, Organisation and Innovation in Air Traffic Control* (2013)
19. R.W. Jr. Poole, *Organization and Innovation in Air Traffic Control* (2014)
20. J.-P. Rodrigue, *The Geography of Transport Systems* (5th edn, 2020) p. 456
21. K. Salah-ddine, M. Hafidi, *Optimization and Automation of Air Traffic Control Systems: An Overview* (2018)
22. SESAR Joint Undertaking, *A Proposal for the Future Architecture of the European Airspace* (2019)
23. J.P. Verma, S.H. Mankad, S. Garg, A graph based analysis of user mobility for a smart city project. Commun. Comput. Inform. Sci. Next Gener. Comput. Technol. Comput. Intell. 140–151 (2019)
24. J. Prakash, V. Verma, S.H. Mankad, S. Garg, Geohash tag based mobility detection and prediction for traffic management. SN Appl. Sci. **2**(8) (2020)
25. Z. Wang, M. Liang, D. Delahaye, *Short-Term 4D Trajectory Prediction Using Machine Learning Methods* (2017)
26. S. Yanbin, T. Zhongji, L. Xu, The application of the cognitive radio in the aviation communication spectrum management. Phys. Procedia **25**, 1720–1725 (2012). International Conference on Solid State Devices and Materials Science, April 1–2, 2012, Macao

Power Constrained Performance Evaluation of AODV, OLSR and DSDV Routing Protocols for Vehicular Ad-Hoc Networks

Amit Choksi and Mehul Shah

Abstract VANET is a subclass of MANET in which each vehicle is equipped with on-board sensor units and is compensated for routing by transferring data to other vehicles. With particular properties such as multi-hop communications, high-speed vehicle density, dynamic and heterogeneous networks, bandwidth, security, QoS, and high node energy consumption, the design of VANET routing protocols has become a key intriguing challengeable research topic. Meanwhile many routing protocols inherited from MANET used for VANET and they have been also executed using NS-2. NS-3 has recently gained popularity as a simulator for deploying VANET routing protocols. In comparison to NS-2, NS-3 includes a lot more advanced features. Here, the main goal of this paper is to compare the power constrained performance between routing protocols ADOV, OLSR and DSDV. ADOV is on-demand reactive routing protocol which determines a route to the destination only, when desired source node wants to send a packet to the destination, where packet holds and keeps a table holding information about the destination packet. OLSR and DSDV are table-driven proactive routing protocol mainly developed for MANET which can also be extended for VANET. Finally, these three routing protocols are compared by quantitative metrics like an average throughput, packet delivery ratio and overhead and are evaluated using SUMO and NS-3. Simulation results show that AODV affords a realistic performance rather than OLSR and DSDV and acclimates well for highly dense vehicular networks.

Keywords VANET · AODV · OLSR · DSDV · SUMO and NS-3

A. Choksi (✉)
Birla Vishvakarma Mahavidyalaya Engineering College, V. V. Nagar, Anand, Gujarat, India
e-mail: amit.choksi@bvmengineering.ac.in

M. Shah
G. H. Patel College of Engineering & Technology, V. V. Nagar, Anand, Gujarat, India
e-mail: mehulshah@gcet.ac.in

F. Thakkar et al. (eds.), *Proceedings of the International e-Conference on Intelligent Systems and Signal Processing*, Advances in Intelligent Systems and Computing 1370,
https://doi.org/10.1007/978-981-16-2123-9_55

1 Introduction

As the technology is increasing, the world is expecting to use these fast-developing wireless sensor networks all the way wherever they need. Here, Intelligent Transportation Systems (ITSs) is an evolving solution, which is intended to gather data from vehicles in smart cities. VANET is a subclass of MANETs, which can be defined as a group of vehicles armed with on-board wireless sensors, to gather, compute and exchange the information of safety and infotainment services with other vehicles. Furthermore, vehicles [1] can communicate their sensed information to other vehicles via (1) vehicle-to-vehicle (V2V) ad-hoc communication or else straight to a base station called Roadside Unit (RSU) and (2) vehicle-to-infrastructure (V2I) communication by internet [2], shown in Fig. 1.

Adding more to the characteristics of VANET include high dynamic topology, different communication environment, variable network density, frequently disconnected network and hard delay constraints influence the design of the communication system and its routing protocols [4]. Additionally, VANET allows a large-scale dynamic phenomenon monitored by exploring the various technologies, which have been integrated in smart cars, and fruitful terrain for developing a solution towards the Smart City.

The chief goal of a VANET is to deliver the entertainment and safety services to passengers. Every vehicle is equipped with wireless sensor devices, which can obtain and communicate information or messages like traffic conditions, road maintenance, location tracking, weather information, parking and surrounding fuel stations among

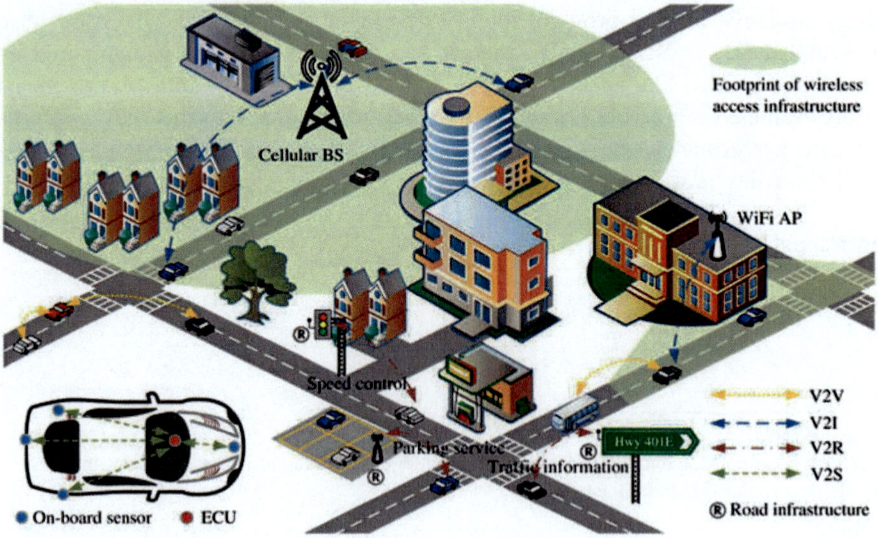

Fig. 1 VANET architecture [3]

other vehicles, which requires a high-energy and fuel consumption that yields to CO_2 emissions accordingly [5].

As such, the key aids of this paper are:

1. To create real-world vehicular mobility (cars, roads, etc.) traces from SUMO (Simulation of Urban MObility) and integrated into NS-3.29 with his features and tools.
2. The evaluation of the AODV, OLSR and DSDV routing protocols in synthetic highway and realistic urban scenarios.
3. Power constrained performance comparison between AODV, OLSR and DSDV routing protocols in terms of quantitative metrics like packet delivery ratio (PDR), average throughput and overhead under highly denser mobility network.

The remaining part of this paper continues as follows: Related literature survey quickly debated in Sect. 2. Section 3 described various VANETs routing protocols briefly. Routing protocols studied for this research work are reviewed and described in Sect. 4. Then the simulation set up and results are analyzed in Sect. 5. Finally, Sect. 6 concludes the paper with challenging future works.

2 Related Works

In this section, the survey has compared and studied the performance of numerous topology-based, broadcast-based, cluster-based, position-based, and infrastructure-based VANET routing protocols, to the best of our knowledge. Yet, the researchers believe the essential of an efficient performance assessment and comparison that help to appraise the performance of various VANET protocols.

In [4], the authors assess PDR, end-to-end delay, throughput and signalling traffic of AODV, DSDV, OLSR, GPCR and GPSR in a realistic urban map of Oujda, Morocco, using node density ranges between 20 and 90 vehicles over an area of 1.7×1.5 km. They appraised OLSR achieves well in case of throughput and PDR, while GPSR and GPCR reveal their best performance in case of end-to-end delay and overhead.

In [6], vehicular mobility traces in crossroad scenario is generated using CAVENET and performance of AODV and OLSR routing protocols studied using NS-3. Authors presented that OLSR attains better than AODV for bigger transmission rates.

In [7], AODV, AOMDV, DSDV and DSR protocols are evaluated using NS-2.34 through different speed of vehicles. Authors demonstrated that DSR does well in case of the throughput and end-to-end delay, but AOMDV and AODV in case of PDR and packet loss.

In [8], authors equaled the performance of DSDV, AODV and OLSR on sub-urban scenario by means of VanetMobiSim and NS-3. They showed that OLSR beats others in case of end-to-end delay, PDR and overhead.

In [9], performance of GPSR and AODV in VANET under various scenarios has been analyzed by using NS-2.35 in combination with VanetMobiSim. The authors showed that AODV and GPSR accomplish better in case of PDR and average end-to-end delay, respectively.

The work in [10] is contrasted GPSR and GPCR, which are position-based routing protocols simulated using VanetMobiSim and NS-2.35. Authors showed that GPCR performs well than GPSR under fluctuating characteristics like the node density, the node speed and the transmission ranges of vehicles.

In [11], authors used SUMO for exhibiting the real-world traffic road map with cars and NS-3 for equating the performance of DSR, AODV and DSDV routing protocols to forward TCP and UDP packets to the destination cars. Simulation results showed that AODV gives a sensible performance for vehicles driving at high velocity 100 km/h.

We demonstrated our research work in this article, which is quite dissimilar from the earlier works discussed in this section. We centred on creating a real-world traffic mobility fact to assess the execution of OLSR, DSDV and AODV using synthetic highway scenario, a realistic road map scenario of Zurich city, Switzerland and Mumbai city, India, from Open Street Map (OSM) [12]. We have extracted a highly denser vehicle mobility, as well traffic files, that have been generated from SUMO-v1.6.0 and taken it as input to NS-3.29 [13] for further assessment of network.

3 VANET Routing Protocols

Due to highly dynamic mobility of vehicles in vehicular ad-hoc networks, designing an efficient routing protocol capable of managing and maintaining effective routing paths among vehicles is a challenging research field. Consequently, numerous routing algorithms have been exhibited, modified and enhanced from algorithms that suggested in the MANETs. These routing algorithms have been proven to get good performance for MANETs, but they might have not able to guarantee the same level of proficiency into a different VANET scenarios yet. Hereafter, more novel concepts and advanced strategies must be developed depending upon VANET applications.

The supreme famous VANET routing protocols, which may be categorized in the following five distinct categories, are based on: (1) Routing Topology, (2) Broadcasting, (3) Clustering, (4) Position and (5) Infrastructure.

3.1 Based on Routing Topology

This group of routing algorithms might be segmented into three types: (1) On-demand or Reactive routing protocol like DSR and AODV. (2) Table-driven or Proactive routing protocols such as DSDV and OLSR and (3) Combination of both Reactive and Proactive called as Hybrid routing protocol alike ZRP–Zone Routing Protocol, etc.

Reactive involves the route from a source node to a destination node, which is discovered only when an interaction is needed between them. In Proactive, every node in a network maintains the information about topology by exchanging the routing table when a source node tries to find a path by sending the packets. Finally, Hybrid is a grouping of both reactive and proactive protocol.

3.2 *Based on Broadcasting*

This type of routing protocol is frequently used to share weather, emergency, traffic, and road condition information among vehicles, as well as to provide basic safety message alerts and advertisements in VANET. When a broadcast routing protocol is used into a VANET, instead of sending the packets individually to other nodes, source node broadcasts the packets to all neighbours in a network for route discovery. Here the only main objective is to escape remaining bandwidth from disinterested node in the network.

Most well identified routing protocols in this category are: (1) DVCAST—Distributed Vehicular broadcast, (2) PBSM—Parameterless Broadcast in Static to highly Mobile, (3) SRB—Secure Ring Broadcasting and (4) EAEP—Edge-Aware Epidemic Protocol.

3.3 *Based on Clustering*

In vehicular networks, the network topology is very often altered and encompasses large zones. So, here, the creation of routing method is a tricky challenge. When the number of vehicles increases in VANET, scalability becomes an important issue. In the absence of any central infrastructure, VANET suffers high packet loss due to a large volume of message dissemination among the vehicles for V2V communication. VANET also suffers from issues such as the hidden terminal problem, high latency for safety message transmission, message security, broadcast storm problem, quality of service (QoS), packet routing, congestion control, and resource management. So, clustering is an important concept in VANET where several vehicles join to form a group based on common features. Also, each cluster has its own cluster-head, which is liable for nodes communication into the cluster and a cluster-gateway, which is responsible for the communication among other clusters in a network. In VANET clustering, cluster head (CH) plays a key role in the formation process of a cluster. A cluster can be created in various ways based on the input metrics. The member vehicle of a cluster is called cluster member (CM). So, CH and CM are participants of clusters. Likewise, the geographic information of vehicles is also needed for clustering by GPS system mounted on vehicles.

Supreme famous routing protocols based on clustering are: (1) CBLR—Cluster Based Location Protocol, (2) CBDRP—Cluster Based Directional Routing Protocol and (3) EDCBRP—Euclidean Distance Cluster Based Routing Protocol.

3.4 *Based on Position*

VANET environment must manage with a huge task such as rapidly moving topology and the high speed of vehicles. Aimed at this reason, other methods to assemble the routes and forward the packets must be set up. Position-based routing protocols were developed since decades using the same exciting concept and with commendable results. In that protocol, geographical locations of vehicles have been used to create routing decisions. For that, data or information is taken from GPS device and other local services like V2I and V2V. Geocast routing is a particular class of position-based routing protocol, in which the total number of vehicles are being split into predefined geographical areas. There are three methods for forwarding packets that can be found here: (1) Excessive forwarding (2) Hierarchical directional flooding and (3) Restricted directional flooding.

Most recognized routing protocols in this category are: (1) GPSR—Greedy Perimeter Stateless Protocol, (2) LABAR—Location Area Based Ad-hoc Routing, (3) GYTAR—GreedY Traffic Aware Routing, (4) ROVER—RObust VEhicular Routing, (5) AMAR—Adaptive Movement Routing and (6) DREAM—Distance Routing Effect Algorithm for Mobility.

3.5 *Based on Infrastructure*

This type of protocols spread on static infrastructure stations to improve routing matters. RAR—Roadside-Aided Routing is the well-known protocol under this category.

4 Studied Routing Protocols

This portion explains routing algorithms like DSDV, AODV and OLSR. AODV is labelled into the topology-based reactive routing, while OLSR and DSDV are under proactive routing [4].

4.1 AODV—Ad-Hoc On-Demand Distance Vector

In AODV [14], when a node wants to send a packet, at that time a route is formed between source to destination. Each node forwards hello packets regularly to discover its neighbour node. Once a source node demands a new route to destination node, which is not neighbour of it, source node sends a route request (RREQ) packet to its neighbours to catch a path. These RREQ packets are also sent by in-between nodes, until complete path to destination node found [5]. Here all in-between nodes update their routing table by counting the latest node pursued by the RREQ packet and adding the hop count value. AODV finds the optimum routing path between source node and destination node by two important packets 1) RREQ - Route request and 2) RREP - Route Reply. As soon as the source node received RREP packets, it starts forwarding application packets to the destination node using path established by in-between nodes from source to destination. Very important notice, here, is that the destination node replies to the first received RREQ packet, but when it receives RREQ with lesser hop count rather than previous one, only then the destination node will produce another RREP to form a shortest path [5]. If there is an error in the current path, AODV is also capable to manage and create new paths and proficient for handling both unicast and multicast routing.

4.2 OLSR—Optimized Link State Routing

OLSR [15] is a proactive table-driven routing protocol, in which each node reserves its routing table and shares or swaps the network information with all nodes of the network frequently. This permits each node to create a universal vision of the network environment. In OLSR, nodes are selected as a MPR—Multi-Point Relays by other neighbours to broadcast this information continually by control messages to decrease the payload from flooding and enhance the transmission method. In that way, a node has reachability to other nodes in the entire network by broadcasting, which is designated as MPR. These MPRs are used in route calculation to form the path from any given node to destination node in the network.

4.3 DSDV—Destination-Sequenced Distance Vector

DSDV [16] is a table-driven proactive routing protocol, routes to all destinations are always easily available at every node. The routing tables are switched between neighbours at regular periods to keep an up-to-date view of the network topology. The tables are also sent if a node sees a significant change in topology. Routing shows different entries holding a sequence number. If a link is present, then these sequence numbers are usually even, otherwise these numbers are odd. The sequence number is created by the destination node and intermediate nodes are updated frequently using this number. Despite the advantages of DSDV, such as the absence of loops, simplicity, and lack of wait time [4] during the path discovery process. Yet the DSDV deteriorate from the overhead because of the regular updates produced by the highly dense networks.

5 Simulation Results

In vehicular networks, node moves with high velocity and arms by good quality embedded resources such as processing unit with high battery power and memory. So, for measuring the performance of the VANET routing algorithms, it might be good to implement routing algorithms and install the communication units on the vehicles driving in the realistic world, which is very expensive in price with human capitals.

To carry out simulation, SUMO-v1.6.0 as road traffic simulator and NS-3.29 as network simulator under the Linux OS Ubuntu 20.04.01 has been used here. The programming language C++ is used in NS-3.29 to test the performance of the AODV, OLSR, and DSDV routing protocols in the following scenarios: (1) a synthetic highway, and (2) a real-world metropolitan map of Zurich City, Switzerland, and Mumbai City, India, both acquired from Open Street Map (OSM) [12].

The performance indices such as packet delivery ratio, overhead and average throughput are evaluated among routing protocols OLSR, DSDV and AODV by focusing on two important simulation parameters: (1) node density and (2) node power.

In the simulation, two ray propagation reflection model is used as it predicts more for long distances. Standard WAVE—Wireless Access in Vehicular Environment protocol IEEE 802.11p, which is the origin of DSRC—Dedicated Short Range Communications has been used for vehicle-to-vehicle communication. The transmission power of all nodes in the network is varied from 7.5 dBm to 20 dBm. The basic safety message (BSM) transmission range is also varied from 50 to 500 m.

5.1 Synthetic Highway

Table 1 shows simulation parameters for synthetic highway scenario using node density of 30, 50 and 70 vehicles along with node power equal to 7.5 dBm over an area of 1500 $\times$ 300 m.

PDR [4] is the ratio of the total number of packets received to packet transmitted, shows the percentage of successful transmissions. The Fig. 2, demonstrates that once the node density increases, PDR decreased in both OLSR and DSDV. But in case of AODV, PDR increased when node density increases.

Table 1 Simulation configuration for synthetic highway

Parameters	Values
No. of nodes	30, 50, 70
Simulation area	1500 m × 300 m
Mobility model	Random way point
Propagation loss model	Two ray ground
Mac standard	802.11P
Bit rate	2 kbps
Packet size	64 bytes
BSM message size	200 bytes
Total simulation time	300 s
Node speed	72 km/h
GPS accuracy	40 ns
Transmission range	50 m, 100 m, ……, 500 m
Node transmission power	7.5 dBm

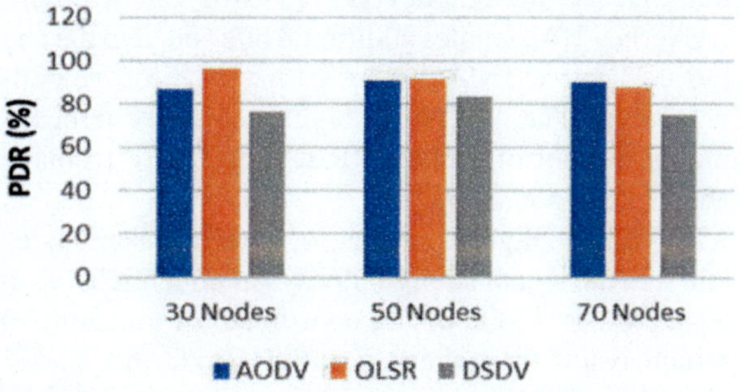

Fig. 2 PDR comparison for synthetic highway scenario

Throughput [4], typically measured in kbps—kilobits per second, denotes the total number of packets effectively transmitted to total simulation time through the network. In the Fig. 3, it has been observed that as node density increases, throughput of all these three protocols gained better. But throughput of OLSR is little improved

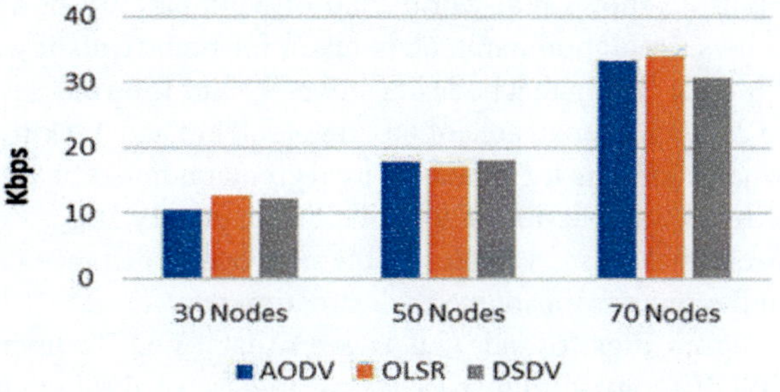

Fig. 3 Average throughput comparison for synthetic highway scenario

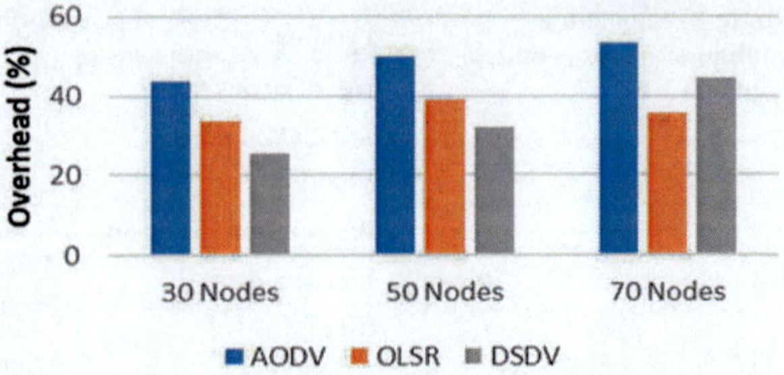

Fig. 4 Overhead comparison for synthetic highway scenario

than AODV and DSDV. This might be elucidated by the point that DSDV has extra network overhead because of broadcasting the entire routing table after fixed interval of time and that affect its throughput, while AODV that floods the network with high control packets might have affected the throughput also. While forwarding a broadcast packet, OLSR employs the MPR method, which 1) reduces the amount of duplicate retransmissions and reduces network overhead by limiting the number of nodes that retransmit a packet from all nodes (normal flooding) to a subset of all nodes. Hence throughput achieves using OLSR little better than AODV and DSDV.

Overhead [4] denotes additional bits added in the packet with the actual message at physical layer and MAC layer, so that the message will be in understandable form for the receiver. The AODV has highest control overhead, because it broadcasts a large number of control packets RREQ and RREP to maintain the route for forwarding data packets in network.

In synthetic highway scenario, we have taken node transmission power 7.5 dBm. Note that the signal strength decays exponentially as the distance between two vehicles increases, so the power consumed for transmitting packets to closer vehicles is ominously less than when transmitting to farther ones. Hence, less node transmission power directly affects the PDR, throughput and overhead of network.

5.2 *Realistic Vehicular Traces of Sub-urban Zurich City, Switzerland and Mumbai City, India*

Figure 5 shows real-world map of Mumbai city downloaded from OSM. Table 2 shows simulation parameters taken for both realistic vehicular traces in Zurich and Mumbai city, in that node density of 99 and 156 vehicles along with node power equal to 20 dBm over an area of 4.6 km $\times$ 3.0 km and 4.0 km $\times$ 3.0 km, respectively. Thus, when node density increases, then greater number of vehicles might have participated in route construction of network. Subsequently, less power is required for transmitting packets to closer vehicles in the network as distance between intermediate vehicles in the route from source to destination decreased.

From Figs. 6 and 7, it is seen that AODV outperforms well than OLSR and DSDV to attain higher PDR; but in case of throughput, OLSR gives slight decent

Fig. 5 Real world map of sub-urban Mumbai City (4.0 km × 3.0 km)

Table 2 Simulation configuration for Zurich and Mumbai city

Parameters	Zurich city	Mumbai city
No. of nodes	99	156
Simulation area	4.6 km × 3.0 km	4.0 km × 3.0 km
Mobility model	Random way point	
Propagation loss model	Two ray ground	
Mac standard	802.11P	
Bit rate	4 kbps	8 kbps
Packet size	512 bytes	1024 bytes
BSM message size	200 bytes	
Total simulation time	200 s	
Node speed	72 km/h	
GPS accuracy	40 ns	
Transmission range	50 m, 100 m, ..., 500 m	
Node transmission power	20 dBm	

performance than others as it uses MPR mechanism in path calculation. Moreover, as AODV floods the network with higher number of RREQ packets to in route discovery process to set up a route from source node to destination node, which results higher network overhead compared to OLSR and DSDV.

Fig. 6 Throughput, overhead and PDR comparison for Zurich City

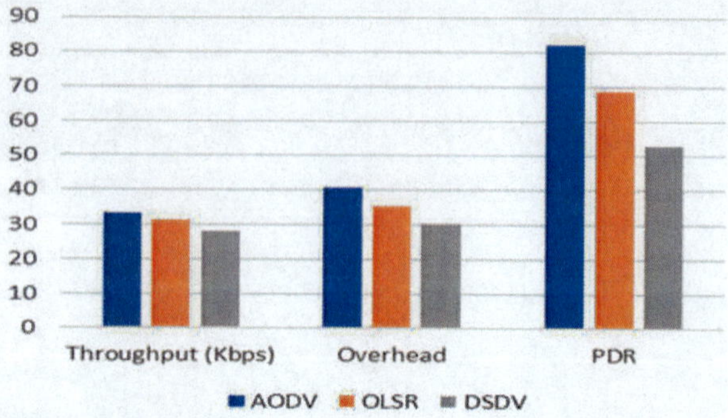

Fig. 7 Throughput, overhead and PDR comparison for Mumbai City

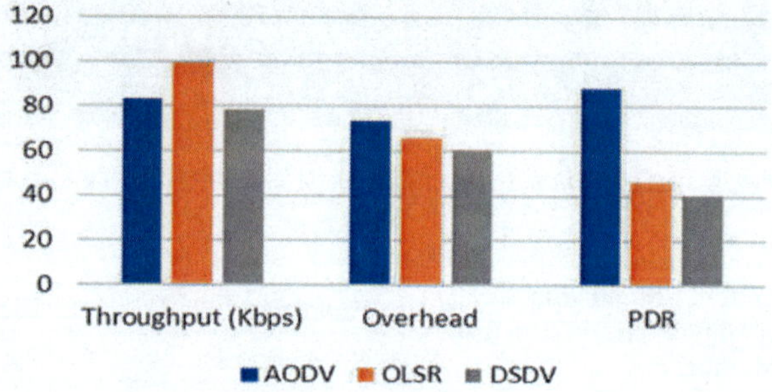

6 Conclusion and Future Work

This work offers the SUMO and NS-3 tools for evaluating and comparing the best known reactive (AODV) and proactive (OLSR and DSDV) routing protocols for VANETs, in order to determine the difficulties and benefits of each routing strategy. Hence, it is concluded from the above research work that the performance of AODV could be considered as the best one for (1) synthetic highway and (2) realistic urban scenarios, in terms of throughput and PDR. While OLSR and DSDV are the finest in case of network overhead. When transferring packets to closer vehicles, the amount of power required is much lower than when broadcasting to faraway vehicles. As a result, a network with low node transmission power has lower PDR and throughput, and high overhead. When node density rises, a larger number of vehicles may have engaged in network route building. As the distance between intermediate vehicles in the route from source to destination reduces, less power is required for transmitting packets to nearby vehicles in the network. As a result, the **PDR, throughput, overhead and life time of a network** are directly influenced by the node transmission power and density. In future, we will propose an energy efficient routing protocol aimed for choosing the efficient and less energy consummated routes among the vehicles in vehicular network and also optimize the performance of the same using intelligent computational techniques like a machine learning and deep learning.

References

1. S. Al-Sultan, M.M. Al-Doori, A.H. Al-Bayatti, H. Zedan, A comprehensive survey on vehicular ad-hoc network. J. Netw. Comput. Appl. **37**, 380–392 (2014)
2. B.T. Sharef, R.A. Alsaqour, M. Ismail, Vehicular communication Ad-hoc routing protocols: a survey. J. Netw. Comput. Appl. **40**(1), 363–396 (2014)
3. https://ece.uwaterloo.ca/~kan.yang/security_bbcr/vanet.htmlL.T.
4. B. Amina, E. Mohamed, Performance evaluation of VANETs routing protocols using SUMO and NS3, in *IEEE 5th International Congress on Information Science and Technology* (2018), pp. 525–530
5. T. Baker et al., GreeAODV: an energy efficient routing protocol for vehicular Ad Hoc networks, in *Intelligent Computing Methodologies. ICIC 2018. Lecture Notes in Computer Science*, vol 10956, ed. by D.S. Huang, M. Gromiha, K. Han, A. Hussain (Springer, Cham, 2018). https://doi.org/10.1007/978-3-319-95957-3_69.
6. E. Spaho, M. Ikeda, L. Barolli, F. Xhafa, M. Younas, M. Takizawa, Performance evaluation of OLSR and AODV protocols in a VANET crossroad scenario, in *2013 IEEE 27th International Conference on Advanced Information Networking and Applications (AINA), Barcelona* (2013), pp. 577–582
7. S. Singh, P. Kumari, S. Agrawal, Comparative analysis of various routing protocols in VANET, in *2015 Fifth International Conference on Advanced Computing & Communication Technologies* (2015), pp. 315–319
8. B.V. Coutinho, E.C.G. Wille, H.I. Del Monego, Performance of routing protocols for VANETs: a realistic analysis format, in *Proceedings of the 9th International Conference on Ubiquitous Information Management and Communication (IMCOM '15)*, vol. 1 (2015), pp. 1–6
9. R. Bala, C.R. Krishna, Scenario based Performance Analysis of AODV and GPSR Routing Protocols in a VANET, in *2015 IEEE International Conference on Computational Intelligence & Communication Technology* (2015), pp. 432–437
10. A. Ashtaiwi, A. Saoud, Performance comparison of position based routing protocols for VANETs, in *2017 International Conference on Internet of Things, Embedded Systems and Communications (IINTEC)* (2017), pp. 78–84
11. S. Kang, Y. Chae, S. Yeon, VANET routing algorithm performance comparison using NS-3 and SUMO, in *2017 4th International Conference on Computer Applications and Information Processing Technology (CAIPT)* (2017), pp. 1–5
12. OpenStreetMap Foundation, OpenStreetMap (2013). [Online]. Available: http://www.openstreetmap.org/.
13. NS-3, NS-3 manual: Release ns-3.29.
14. C.E. Perkins, E.M. Royer, Ad Hoc on-demand distance vector routing, in *Proceeding of IEEE Workshop on Mobile Computing System and Applications (WMCSA)* (1999), pp. 1–11
15. T. Clausen, P. Jacquet, *RFC3626: Optimized Link State Routing Protocol (OLSR)* (RFC Editor, USA, 2003), p. 75.
16. G. He, *Destination Sequenced Distance Vector (DSDV) Protocol.* Networking Laboratory, Helsinki University of Technology, (2002), pp. 1–9.

Remote Monitoring and Predictive Maintenance of Medical Devices

Mohammed Salleh Ahmed Qaid, Anas Mohd Noor, Ahmad Nasrul Norali, Zulkarnay Zakaria, A. Z. Ahmad Firdaus, Asyraf Hakimi Abu Bakar, and Chong Yen Fook

Abstract Equipment maintenance is the primary task performed by mechanical and electrical equipment personnel to ensure the optimum output, good efficacy and long service life. Conventional types of traditional equipment maintenance, such as calendar-based maintenance, are most commonly referred to as preventive maintenance schedules (PPM). PPM is a scheduled event performed by qualified personnel for the systematic analysis, detection and repair of incipient failures. However, scheduled PPM is not effective as it is occurs only biyearly or depending on the requirements or even worse, during the unplanned maintenance which is time-consuming, costly and affect the ongoing process due to the downtime. The improvement of the typical preventive maintenance strategy through incorporating the Internet of Things (IoT) enabled by Industrial Revolution 4.0 is indeed advantageous. The Internet of Things (IoT) is used to facilitate the continuous transmission of data from tracked sensors on the equipment underuse to the person in-charge. As a result, it shows that the maintenance person in-charge will remotely monitor the equipment. Continuous monitoring system will help the maintenance team identify any unusual activity that may cause device failure and predict the equipment downtime.

Keywords Predictive preventive maintenance · Internet of thing · Continuous-time monitoring system

---

M. S. Ahmed Qaid · A. Mohd Noor (✉) · A. N. Norali · Z. Zakaria · A. Z. Ahmad Firdaus · A. H. Abu Bakar · C. Y. Fook
Faculty of Electronic Engineering Technology, Universiti Malaysia Perlis, 02600 Arau, Perlis, Malaysia
e-mail: anasnoor@unimap.edu.my

A. Mohd Noor · A. N. Norali · Z. Zakaria · A. H. Abu Bakar · C. Y. Fook
Sports Engineering Research Centre (SERC), Universiti Malaysia Perlis, 02600 Arau, Perlis, Malaysia

F. Thakkar et al. (eds.), *Proceedings of the International e-Conference on Intelligent Systems and Signal Processing*, Advances in Intelligent Systems and Computing 1370,
https://doi.org/10.1007/978-981-16-2123-9_56

1 Introduction

There are two types of maintenance on the medical system: The Preventive Planned Maintenance (PPM) and the Predictive Maintenance (PdM). PPM is a scheduled event for routine inspection, detection and repair of incipient defects by trained staff before or prior to the occurrence of serious defects [1]. Conventional PPMs are planned based on the risk score of each medical device. The medical device risk rating typically is based on the relative frequency of maintenance.

Meanwhile, PdM is determined on the basis of equipment performance instead of statistics on fact, average or expected [2]. PdM, which has been described as condition-based maintenance, continually tracks the effectiveness of medical equipment during routine operation, suggesting the corrective steps should be taken before failure [3]. The prime objective of the PdM is to be able to predict errors and to perform corrective maintenance before the device has failed. PDM can operate with condition monitoring equipment, where the system tracks the performance of the devices in real time.

The Internet of Things (IoT) is an essential element in the PdM process. One of the key issues with PdM is that it must deal with vast quantities of data and information to store and analyze an intelligent algorithm. IoT-based methods allow large quantities of data to be stored in a safe location and accessed by authorized individuals. It also allows the machine learning algorithm to process the data obtained by predicting the performance of the medical device [4]. The Cloud Gateway offers efficient data transmission and networking for various field gateways.

For example, an infusion pump is one of the key medical devices currently dependent on PPM, which is scheduled once a year. In recent years, many security issues related to infusion pumps have been raised. The efficacy of the device is related to various problems such as alarm failure, battery loss, incorrect drug movement, air bubbles and other problems. There are many cases of failures of infusion pumps, resulting in patient loss and death due to unnecessary infusions according to the Medicines and Health Products Regulatory Agency (MHRA). However, no proven system can track the efficiency of the infusion pump continuously and predict its efficiency prior to its occurrence.

Operating hours can also affect the maintenance time. Present technologies play no part in operating time recording. The highly in-use infusion pump maintains the typical PPM schedule in the same way as the regular operation of infusion pump. However, this planned PPM is not adequate to prevent the infusion pump from malfunctioning since it only happens bi-yearly or depends on the device requirements. The infusion pump may have some undetected issues that are not found in the maintenance process. In this situation, it is necessary for the ongoing monitoring system to continue updating the performance of the device and to make predictions for future maintenance.

The goal of this work is to establish a new framework for continuous-time preventive maintenance for medical devices. The system monitors and improves the medical performance of medical devices and thus predicts future failures before they occur. In addition, a effective real-time monitoring system also forms the main objective of this work. In order to build real-time monitoring to deliver user warnings via email and mobile devices, IoT technologies such as ThingSpeak from MATLAB (Mathworks Inc. USA) and IFTTT (IFTTT Inc. USA) have been used as Thingview app. The maintenance team might also be alerted to any malforming influences even though they are not or around the facility. In addition, the efficient system can increase productivity and performance of the device and the management.

2 Methodology

2.1 Framework

The system's features in the block diagram (see Fig. 1). The device can be applied in this work for any electrical instrument that needs to be tracked. Second, the computer has to connect to the display system, where the monitor system can monitor the current of the device. Data from the system-integrated sensing system will be tracked under various conditions, such as control, idle and off-mode conditions. Both sensing and data analysis can be transmitted to the cloud via IoT technology, where data can be

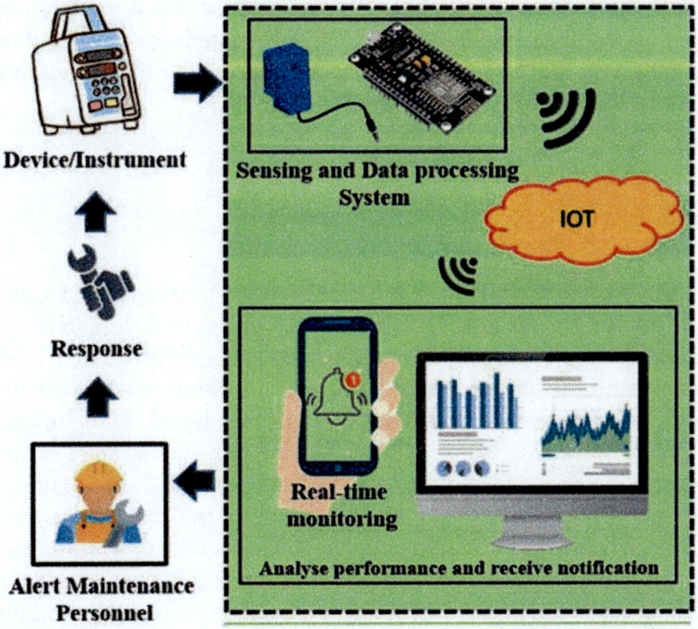

Fig. 1 Block diagram of the system

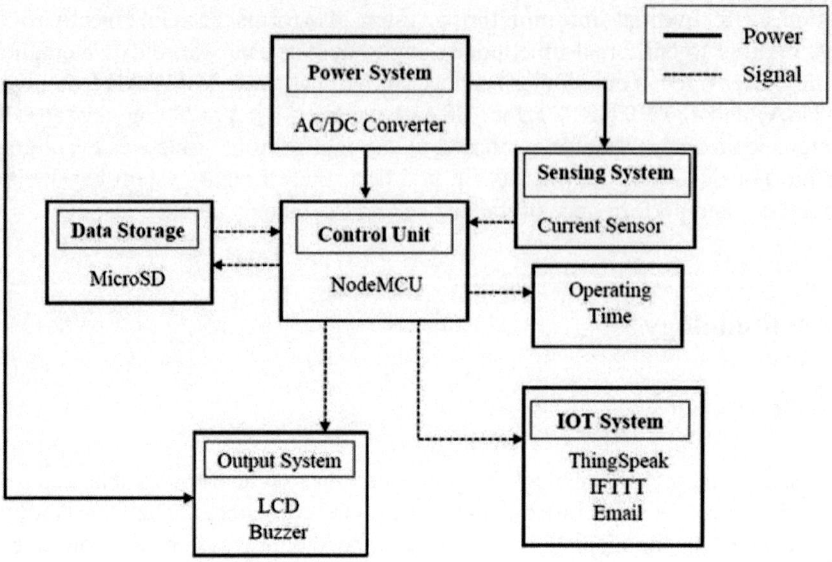

Fig. 2 System layout

tracked and a mobile phone app built via the central network. Once the maintenance staff have been alerted to the system detail, the response will be given to the PPM.

The Continuous-time Predictive Preventive maintenance system layout (see Fig. 2). To power up the control unit, sensing system and the output system, the system relies on the AC as a power system.

2.2 Hardware

The system needs the controller to instruct the memory arithmetic and logic unit and the input and output devices how to respond to the instructions sent to the processor. The controller device in this project is the NodeMCU (ESP8266 ESP-12E. China) where it informs all the other components how to respond to the instructions. The performance of the current transformer (CT) sensor is based on the magnetic field principle. The CT sensor has clapped onto a live or natural cable, and the current flow induces a magnetic field in the middle of the primary winding. The secondary winding consists of several fine wires, which are stored within the shell of the transformer. The current also allows the current to circulate in the secondary winding circuit. The current is equal to the current flowing in the main winding in the secondary winding. The NodeMCU reads the data of the Current Transformer (CT) sensor, the data of the sensor range from zero to a single voltage only and the controller system can calculate the current value in the ampere and the amount of power in the watt based

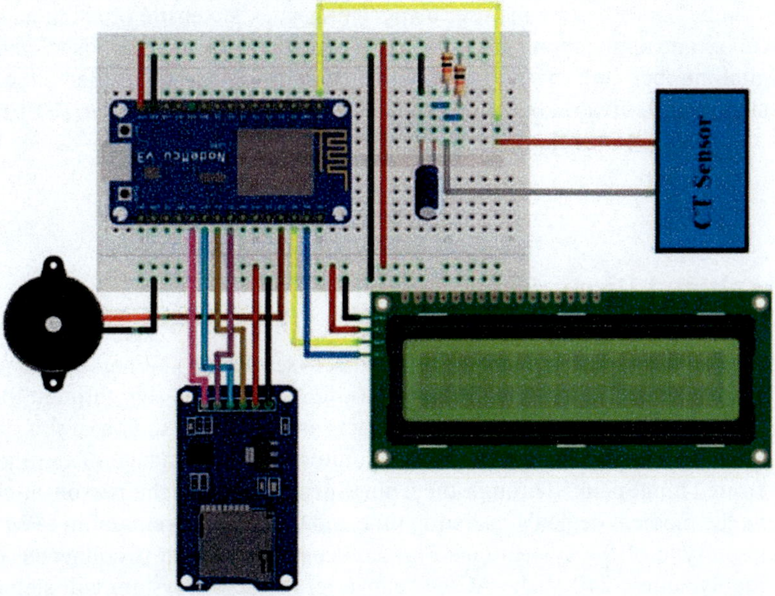

Fig. 3 Schematic diagram

on the reading of the output voltage. The CT sensor used a voltage divider, as the input voltage of the CT sensor did not exceed 2.5 V [5].

All components are supplied with power from the AC/DC converter, which is considered to be the primary source of the system. The LCD is attached to the controller. When the buzz is needed, the buzzer is connected to the NodeMCU to the buzz. The adapter of the SD card is connected to the controller device known as the system memory, making it possible for the system to work offline. Once the computer is connected to the Internet coverage area, all information will be sent to the IoT cloud portal. The IFTTT app has been developed, including running time, remaining time, current number and power number. All ties are connected (see Fig. 3).

2.3 Software

The Arduino IDE is a software programme that writes programming code and uploads the code part of the board. All coding for this project is performed using the Arduino IDE, which allows communication between the administration and the system; calibration is also possible if the computer has a different current rate for each mode of operation.

The ThingSpeak platform is a software used for tracking data in this project. A special channel for each device that needs to be regulated should be created. The system can be controlled by the user using Thingspeak's website or using an established cell phone application (ThingSpeak Viewer). The data can be visualized both as a digital number and as a graph. Multiple devices may be controlled simultaneously on the same network, but with a separate channel for another user. IFTTT shall alert maintenance staff if the system reaches the running time or if the device needs a PPM.

3 Result and Discussion

The total system of development of this project (see Fig. 4). When the abnormal conditions of medical device happens, the notification will be sent through the cell phone using IFTTT application. The laptop acts as the central station of this system since the charge person can observe and monitor the performance of the medical device from ThingSpeak. Through the ThingSpeak platform, the person-in-charge can know the medical device's operating time and the device's remaining PPM time.

The prototype of the system (see Fig. 5). When the system is connected to the power supply source 240 V, the AC/DC converter inside the system will step down the voltage to 5 V, which will turn on the whole system. Then, the system will be waiting for an assigned medical device to be plugged in, and at the same time, LCD shows "Plug in the Device".

The prototype of the hardware system consists of the sensing and data processing system with the user-interface system. The CT sensor YHDC-SCT-013–030 is the sensing element in this project. NodeMCU acts as the control unit of the system responsible for the data collection and IoT system. The circuitry of the hardware system (see Fig. 6).

There is a microSD card module which used for storage of data when there is no WiFi connection. Once the NodeMCU is connected to WiFi, the parameters that are stored inside the microSD will be updated to ThingSpeak and IFTTT apps. The LCD

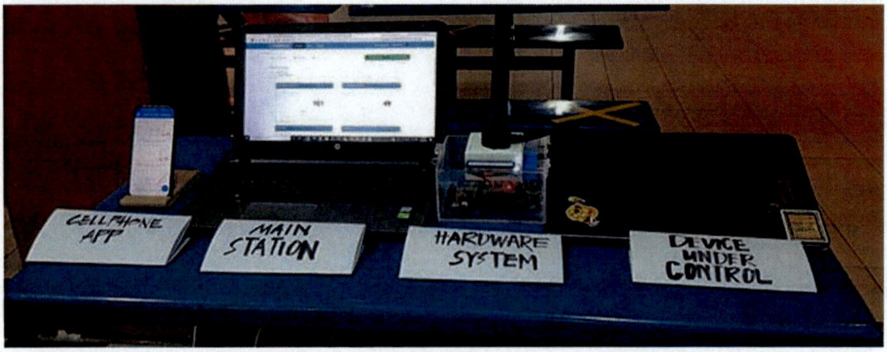

Fig. 4 Developed system

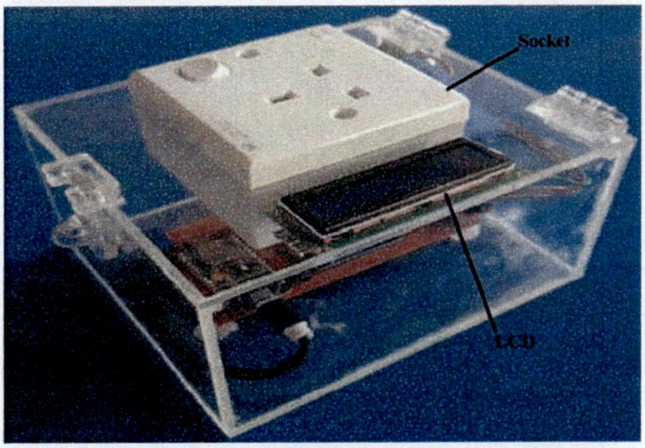

Fig. 5 Developed hardware system

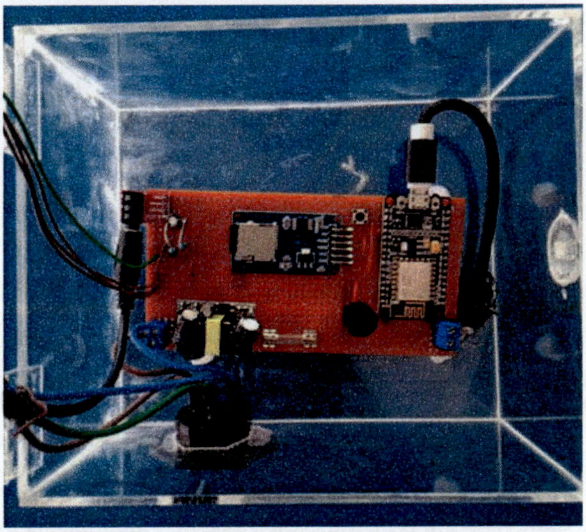

Fig. 6 Circuitry of the system

beside the plug is used to display the system notification to the end-user. If there is no medical device plug on it, the LCD shows "Plug-In The Device" when a medical device plugs on it, it will show "Device In Use." If there is any abnormal condition or over usage of the device, it shows "Device Needs Maintenance."

The sensor placement of the hardware system (see Fig. 7). It can be seen that the current sensor is located beside the plug. It is used to detect medical device underuse.

The data can be visualized by the ThingView app that is downloaded on the cellphone. The graphs will represent each parameter. At the beginning of the system, the charts will display the accumulated value for the medical device's power and current. The operating time and remaining time of the medical device in digital form (see Fig. 8).

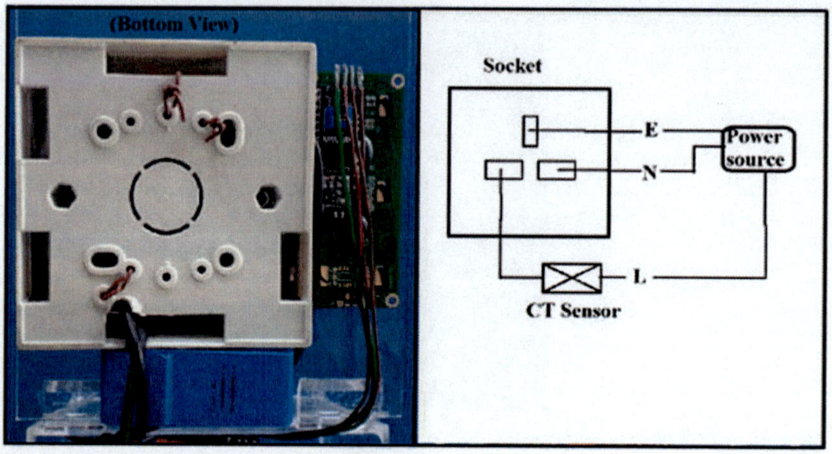

Fig. 7 Sensor placement

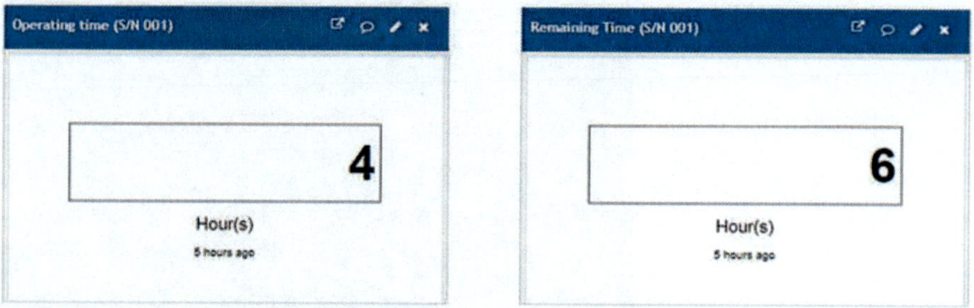

Fig. 8 Operating time and Remaining Time

In this case, it's clear that a specific device's operating time is shown as 4 h as a demo and the remaining time is 6 h, which means that the device can operate for ten hours before undergoing maintenance. The remaining time should be set by the medical person-in-charge based on the studies that have been done previously to identify the operation of the device before getting maintenance. After that, the person-in-charge can adjust the remaining time based on the continuous-time predictive preventive maintenance system. Each device in different departments doesn't have the same operation time. As instance, some devices having an extensive usage or overload operations over others. In that case, this continuous-time predictive preventive maintenance system provides an accurate reading and monitoring for such a particular device.

The device's operating time concerning the date can be displayed in graphical form (see Fig. 9).

Also, the power usage of the medical device per day can be easily visualized by the person-in-charge through ThingSpeak. The graph of date versus power (see Fig. 10).

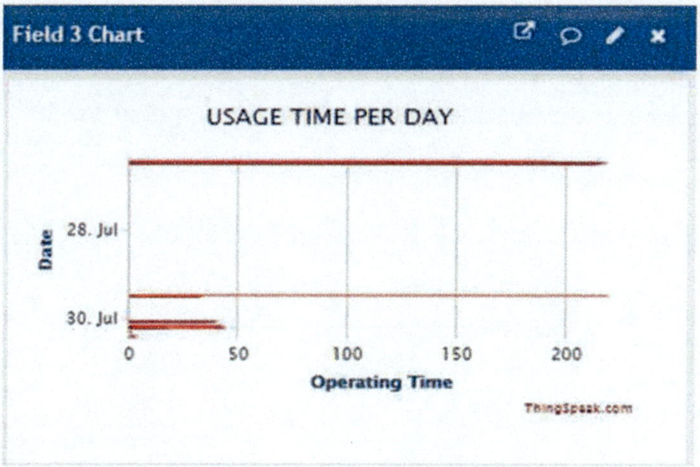

Fig. 9 Operating time

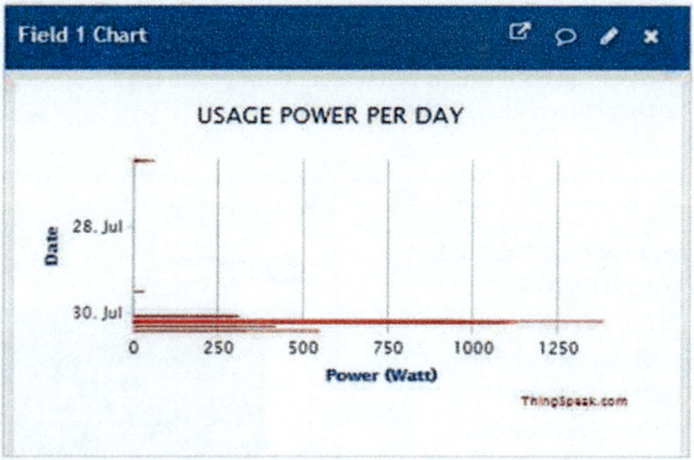

Fig. 10 Usage of power per day

The graph shows a very high usage before 30 Jul after that reset to the operating time results in zero operating time.

The person-in-charge also can visualize the usage of the current of the medical device per day through the graph (see Fig. 11). This will make the device under monitoring easier such the exact date of the abnormal condition occured. This result helps in monitoring the consumption of power and the current drawn for such a particular device. The person-in-charge will be able to know which device is more valuable and which device is more economic in terms of power and current consumption.

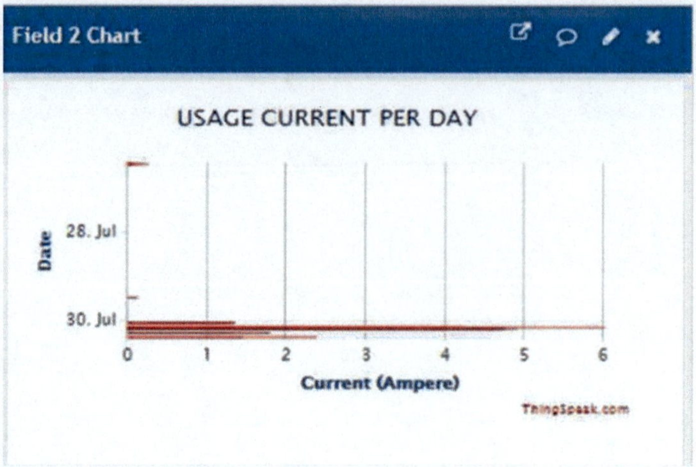

Fig. 11 Usage of current per day

When the operating time of the medical device reached the thresholding value which means that the remaining time of PPM reached zero or less than zero hours. The alert messages and notification will be sent through the IFTTT app and email to the person-in-charge (see Fig. 12).

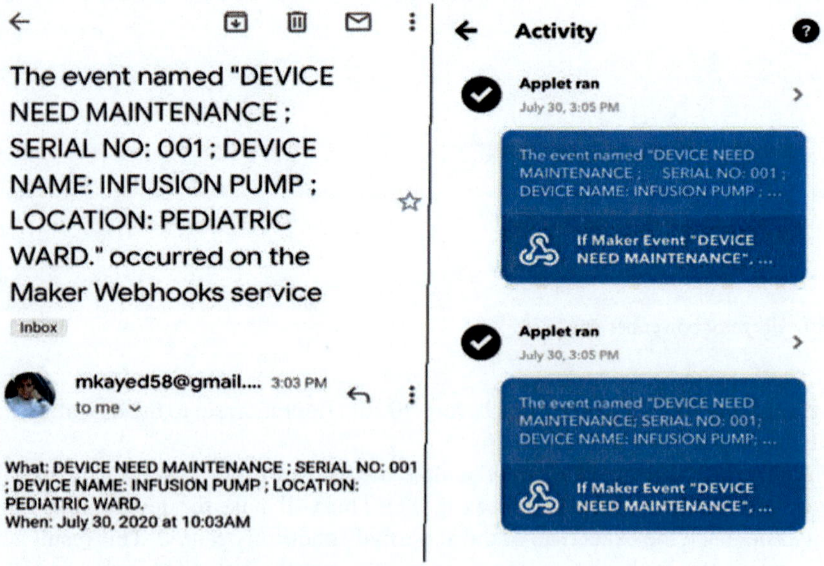

Fig. 12 Alert messages from email and IFTTT app

The alert notification will be forwarded to the maintenance team for the IFTTT app and the email, including the infusion pump's serial number and location. These notifications will enable and help the responsible person find a medical device for which PPM is required.

When the remaining time for PPM is 0, the buzzer will be activated. It will alert the hospital staff and LCD will show message "Unit Needs Maintenance". Once the person-in-charge has done PPM on the medical device, the medical device's operating period and remaining PPM period will reset to its original value.

4 Conclusion

This project aims to develop the traditional preventive maintenance framework for medical devices in hospitals and electrical devices in the industry by introducing the Internet of Things (IoT). The main objective would be to develop a continuous monitoring system to monitor electrical and medical devices' performance remotely. The ongoing data analysis of the system's output will help the maintenance team detect an unexpected event that could cause device failure and forecast device downtime. The IoT system was also used to wirelessly send data about current usage by the device to the maintenance team to reduce the workload of the maintenance team. Multi-sensors and integration can still be improved in this design project, for example, calendar-based maintenance with usage-based maintenance. Multi-sensors other than the current sensor, such as the flow rate sensor and bubble trap detector, are suggested. Perhaps a better use of the user interface can be created to collect and observe other data than use power and time, particularly for various medical devices' performance monitoring.

Acknowledgements The authors would also like to thank UEM Edgenta Sdn. Bhd as well as staff from Hospital Tuanku Fauziah (HTF) for their help and assistant.

References

1. G. Aiello et al., A decision support system to assure high-performance maintenance service. J. Qual. Maint. Eng. (2020).
2. S. Selcuk, Predictive maintenance, its implementation and latest trends. Proc. Inst. Mech. Eng. Part B J. Eng. Manuf. **231**(9), 670–1679 (2017)
3. S. Coban, M.O. Gokalp, E. Gokalp, P.E. Eren, A. Kocyigit, Predictive maintenance in healthcare services with big data technologies, in *Proceedings of the 2018 IEEE 11th International Conference Services Oriented Computing and Applications (SOCA)* (2019), pp. 93–98.
4. J. Maktoubian, K. Ansari, An IoT architecture for preventive maintenance of medical devices in healthcare organizations. Health Technol. (Berl) **9**(3), 233–243 (2019)
5. R.W.B. Sc, *Yhdc SCT-013–000 Current Transformer. A Report on the properties of the Yhdc Current Transformer and Its Suitability for Use with the OpenEnergy Nonitor*, vol. 6 (2014), pp. 1–10

A Review of Approaches to Energy Aware Multi-hop Routing for Lifetime Enhancement in Wireless Sensor Networks

M. Abdul Jawad and Farida Khurshid

Abstract Enhancing the network lifetime is one of the most challenging tasks in research involving wireless sensor networks. The phenomenon of the 'hotspot issue' that is the uneven energy depletion of sensor nodes becomes more prominent as the number of sensor nodes increase. These sensor nodes disperse their energy quickly and finally die down. Since immediate battery replacement is not possible in operational environments, researchers have suggested various alternative methods to address the issue. This paper brings out a comprehensive and exhaustive literature review discussing methods and techniques used in sensor networks outlining the improvements achieved so far. Moreover, many theoretical and numerical calculations have been illustrated to substantiate the practicability of these various methods.

Keywords Wireless networks · EASR · MERR · Base station · Relay nodes · Energy cost · EEBR · DEBR

1 Introduction

Recent technological advances in micro-electronics have made it possible to manufacture small and low-cost sensors that are both technically and economically feasible. A Wireless Sensor Network (WSN) is an internetwork of hundreds or even thousands of tiny nodes known as Sensors. Wireless Sensor Network (WSN) forms a subset of Ad hoc networks. Wireless sensor networks have many restrictions compared to ad hoc networks in terms of its sensor nodes' capability of memory storage, processing and available energy source. Wireless sensor networks are generally assumed to be energy restrained because sensor nodes operate with small capacity DC source or may be placed in such a way that replacement of its energy source is not possible. A WSN consists of spatially distributed autonomous sensor nodes which

M. A. Jawad (✉) · F. Khurshid
NIT Srinagar, Srinagar, India

F. Thakkar et al. (eds.), *Proceedings of the International e-Conference on Intelligent Systems and Signal Processing*, Advances in Intelligent Systems and Computing 1370,
https://doi.org/10.1007/978-981-16-2123-9_57

cooperatively monitor physical or environmental conditions. These nodes communicate wirelessly and often self-organize after being deployed in an ad hoc fashion. Source nodes transmit their data to destination nodes through intermediate nodes. Whereas destination node is connected to a central gateway (also known as base station). The central gateway provides a connection to the wired world where the data can be collected, processed and analyzed. In WSNs a node depletes its energy mainly in processing and communicating with adjacent nodes. This energy dissipation follows first order radio model, an accepted model for radio communications. To address this issue, hop-by-hop data transfer is used by nodes to deliver data to the destination. The hop-by-hop communication increases the burden on sensor nodes that are close to the base station as all transmission takes place through these nodes only (Hotspot Effect). In addition to it, a lot of energy is consumed in various states that are useless from application point of view like idle listening, overhearing, interference and collision.

The wireless sensor networks find a lot of applications in diverse fields ranging from environmental monitoring [1], through weather monitoring [2], military surveillance [3], smart grid [4], traffic monitoring [5], forest fire detection [7] to habitat monitoring [6] and the like. Such network systems particularly those located in remote areas encounter problems of coverage holes, communication holes etc. on account of battery exhaustion. Replacement of batteries or re-energizing them by any means is not possible because of their remote location. So, in order to combat these problems, a few WSN system designs are planning efficient strategies to maximize the residual energy levels in sensor nodes. One such strategy is having a duty cycle scheduling in sensor networks that allows some of the sensor nodes to enter the sleep state in order to moderate energy consumption in wireless sensor network. Also, efficient energy algorithms are designed aimed at diminishing the battery depletion of each sensor node. This is done by aggregating various sensory information into a single data/transmission unit that reduces the network traffic thereby extending the lifetime of the wireless sensor network [34].

2 Related Work

The highly application specific nature of WSNs together with their inbuilt constraints make them quite different from general wireless networks. Consequently, WSNs pose different research challenges. Network based routing in WSNs can be divided into flat-based routing, hierarchical-based routing, and location-based routing depending on the network structure. In flat-based routing, all nodes are typically assigned equal roles or functionality. In hierarchical-based routing, however, nodes will play different roles in the network. In location-based routing, sensor nodes' positions are exploited to route data in the network. Next section discusses in detail the various energy aware multi-hop routing protocols specified for wireless sensor networks over time, followed by a brief survey of mobility and sleep scheduling aided protocols that accrue a promising result in improving the network lifetime.

2.1 Energy Aware Multi-hop Routing in Sensor Networks

In multi-Hop Routing a sensor node sends data to the sink node via multiple intermediate nodes. In the process of data transmission all the nodes involved in data transmission dissipate some part of their energy based on First-Order Radio Model. For the radio hardware, the transmitter dissipates energy to run the transmitter radio electronics and power amplifier, and the receiver dissipates energy to run the receiver radio electronics. $E_T(k, d)$ denotes the energy required by source sensor to transmit data and ER(k) is the energy required to receive the sensed data. The transmitter energy $E_T(k, d)$ of any sensor node will comprise of first energy required to drive the transmitter circuitry—$k * E_{elec}$, second reliably transferring data to receiver node—$k* \in_{amp} *d^n$ and for receiving node the energy required $E_R(k)$ will simply be equal to $k * E_{elec}$. So total energy dissipation in wireless sensor network will be:

$$E_T(k, d) = k * E_{elec} + k* \in_{amp} *d^n \tag{1}$$

$$E_R(k) = k * E_{elec} \tag{2}$$

For simulation purposes, most researchers used radio energy dissipation $E_{elec} = 50nJ/bit$ to run transmitter or receiver circuitry and $\in_{amp}= 100pJ/bit/m^2$ to run the transmitter amplifier.

Chang et al. [8] proposed a Maximum Residual Energy Path (MREP) algorithm in order to maximize the network lifetime of WSNs. The main idea is to route the data through paths with maximum residual energy for balanced energy consumption. The authors define link cost C_{ij} which is the function of node's remaining energy E_i and required transmission energy. There are two ways to calculate this link cost:

$$C_{ij} = \frac{1}{(E_i - e_{ij})} \; and \; C_{ij} = \frac{e_{ij}}{E_i} \tag{3}$$

The Shortest path from node i to destination is obtained using Bellman Ford algorithm. The main issue with this scheme is that it creates large coverage holes due to data being frequently transmitted over minimum residual paths.

Huang et al. [9] proposed an energy-aware, load balanced routing scheme known as Maximum Capacity Path Scheme. In MCP, the sensor network is first constructed as a layered network. Each sensor node selects the shortest path with maximum capacity to sink. To understand the concept of MCP routing scheme, consider a layered network as illustrated in Fig. 1.

From any node in the network, there exist multiple shortest paths to the sink. The weight associated with each node represents its available energy. Suppose node 'e' wants to relay the data to the sink, there are three routing paths available: e → c → a → s, e → c → b → s, and e → d → b → s. If the node 'e' selects 'd' as the forward relay node because 'd' has maximum available energy, then the node 'b' will run out of its energy rapidly. To avoid this, MCP scheme ensures data to be routed

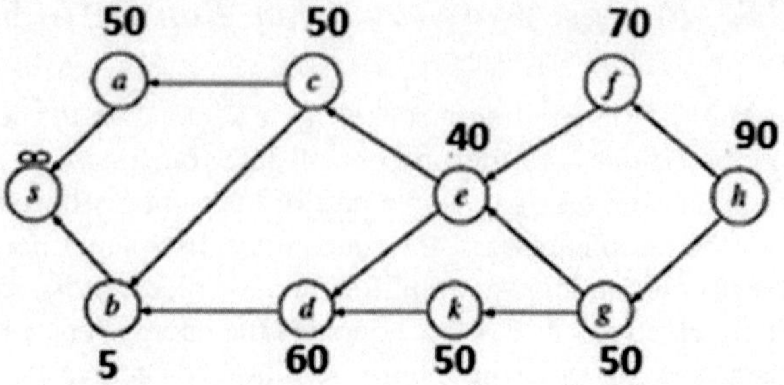

Fig. 1 Path selection in a layered network

through the path with maximum capacity. So, according to MCP scheme node ‘e’ relays its data along the path e → c → a → s. This MCP provides a balanced energy dissipation among sensor nodes, however, in some instances a node having energy less than a particular threshold may still fall in a path having more average residual energy. This can increase chances of coverage holes.

Zimmerling et al. [10] presented MERR (Minimum Energy Relay Routing), a novel approach to energy-efficient data routing to a single control center in a linear sensor topology. Nodes in a sensor network are said to communicate with the base-station in two ways: Direct transmission and Multi-hop routing. With direct transmission each sensor node sends its data directly to the base station. The nodes that are present far away from the base station will quickly drain their batteries. If nodes are closer to the base station, the direct transmission is preferred. In multi-hop routing intermediate nodes are used to relay the data to the base station. Multi-hop routing is preferred for long-distance transmissions. In linear sensor networks, for efficient data transmission from node to base station located at a distance D is performed by using optimal hop number. This optimal hop number, $K_{opt} = \frac{D}{d_{char}} or \frac{D}{d_{char}}$ where d_{char} is the characteristic distance given by

$$d_{char} = \sqrt[\gamma]{\frac{\alpha}{\varepsilon(\gamma - 1)}} \tag{4}$$

The characteristic distance is constant and is considered as an optimal transmission range where energy needed for routing is minimum. Moreover, the total relaying energy can be minimized if the distance between neighboring hops is made equal to D/K_{opt}. This means $(K_{opt} - 1)$ relays can be placed at constant intervals of D/K_{opt}. In MERR the sensed data is routed to the base station using intermediate nodes as relays which are spaced at distances equal to characteristic distance. The operation of MERR is illustrated in Fig. 2.

In Fig. 2, it is clearly illustrated that nodes do not choose their direct downstream neighbor as next relay instead a node is chosen with distance equal to the characteristic instance. The main problem with this approach is that, with every data transmission some nodes need to be identified so as to drop them between successive relays for finding optimal path. This adds more computation overhead.

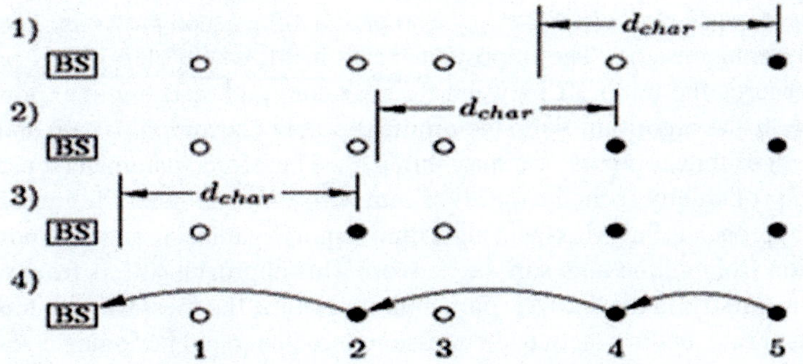

Fig. 2 MERR information relay [10]

Toh et al. in [11] presented a power-aware routing scheme to satisfy two main constraints, i.e., even distribution of the power consumption and minimization of the overall transmission power. In Conditional Max-Min Battery Capacity Routing (CMMBCR) scheme, battery capacity is used as a route selection metric instead of cost function. A threshold value γ (ranging between 0 and 100) is specified and all the possible routes between source and destination with nodes having energy greater than this threshold value are selected. The route with minimum transmission power is selected. The performance of CMMBCR depends on the value γ. This algorithm provides an optimal energy dissipation but the operation is very complex and time consuming.

Traditionally, route computation algorithms are classified as *online* and *offline* routing algorithms. Online routing algorithms carry out route computation at the time of transmission, while offline computation is used on a slower time scale for more efficient utilization of network resources. Li et al. [12] presented an online power-aware routing in large wireless ad hoc networks namely Max-Min zP_{min} algorithm. There are two complementary levels at which the power of a node can be optimized: (i) minimizing power during the idle time, and (ii) minimizing power consumption during communication. However, the authors focused mainly on how to minimize the power consumption during communication. A number of local metrics like minimizing the ratio of cost/packet and minimizing max. node cost have been devised. But these metrics focus on individual nodes rather than the whole system. Hence it is required to have a global metric to maximize the lifetime of the network. Max-min zP_{min} algorithm optimizes both the criteria i.e. selecting the path with the minimum power consumption (P_{min}) and the path that maximizes the minimal residual power in the nodes of the network. The proposed algorithm routes messages along the path with the maximal-minimal fraction of remaining power after the message is transmitted. This path is known as Max-Min path. The algorithm restricts the power consumption for sending the message to zP_{min}, where z is a parameter whose value is greater or equal to 1, and maximizes the minimal residual power fraction by adaptively choosing the parameter z. The algorithm finds the path with least power consumption, P_{min} in the graph G (V, E) using the Dijkstra's algorithm.

If the power consumption is > zP_{min} or no path is found then previous shortest path is used for transmission. The important factor in Max-Min algorithmis parameter z that measures the trade off between the max-min path and minimal power path. When z = 1, the algorithm finds the minimal power consumption path and when z = ∞ the algorithm computes the max-min path. Therefore , parameter z needs to be computed proficiently such that the algorithm will lead to a longer lifetime. The issue with this approach is that Max-Min algorithm requires gathering accurate power level information from all the nodes in the network.This approach suffers from two main limitations. Firstly, in the analysis part, authors assume the messages are transmitted at a constant rate, but in practice, the messages are generated periodically. Secondly, the competitive bound is dependent on values that aren't included as part of input to the algorithm.

Kar et al. in [13] defined the network lifetime as the measure of time before the first message loss is noticed. Thus, the objective is to maximize the total number of messages successfully sent across the network without worrying about future message arrivals. The authors proposed an online routing algorithm known as CMAX. If l_k, s_k, and d_k denote the length, the source, and the destination nodes,, respectively, of the message *k*. Also, if $E_i(k)$ denotes the residual energy of node *i* at the time when message *k* is generated. The algorithm defines a quantity α_i as the *energy utilization* of node *i* and $\alpha_i(k)$ as the fraction of the energy of node *i* that is used at the time of the arrival of message *k*, where

$$\alpha_i(k) = 1 - \frac{E_i(k)}{E_i} \tag{5}$$

The energy consumed for transmitting a unit message along the link (i, j) is represented by e_{ij}. The proposed algorithm initially creates a reduced network by eliminating all links (i, j) for which $e_{ij} > \frac{E_i(k)}{l_k}$. The weights w_{ij} are associated with each link where $w_{ij} = e_{ij}\left(\lambda^{\alpha_i(k)} - 1\right)$ and the shortest path from S_k to d_k is found in the reduced graph. Let γ_k be the length of shortest path found. If $\gamma_k \leq \sigma$, route the message along the shortest path, otherwise reject it. λ and σ are arbitrarily chosen constants. The main advantage with their approach is, the algorithm tries to avoid links which require very high energy for transmission and also avoids the nodes where the residual energy fraction is low.

Park et al. [14] proposed an online heuristic for maximum lifetime routing in WSN. This heuristic approach finds a path for each routing request $r_i = (s_i, t_i)$, where s_i is the source and t_i is the destination. This objective is achieved by using a two-step algorithm for each routing request. In the first step, all the edges are pruned from the graph G for which $ce(u) < w(u, v)$, where $ce(u)$ represents the current energy in sensor 'u' and $w(u, v)$ represents the energy required to do single-hop transmission from sensor node 'u' to 'v'. Let the resulting graph be G′ (V, E′). A minimum energy path $P_i^{‘}$ is determined from $s_i tot_i$ in the pruned graph using Dijkstra's algorithm. The residual energy for the edge (u, v) on the path $P_i^{‘}$ is computed which is equal to $re(u) = ce(u) - w(u, v)$. If the minimum residual energy be denoted by $minRE$, then a graph G″ (V, E″) is obtained by removing all the edges from G′ with $ce(u) - w(u, v) < minRE$. This pruning prevents depletion of low energy nodes.

The second step of the algorithm finds the path to be used for routing. In the graph G″, weights $w(u, v)$ are assigned to each edge $(u, v) \in E''$. Let $eMin(u) = \min\{w(u, v)\}$ be the energy needed by sensor 'u' to transmit message to its nearest neighbor G′. The algorithm defines the function α(u) as $\alpha(u) = \frac{minRE}{ce(u)}$. For each edge (u, v) of E″ compute the weights $w''(u, v)$ given by:

$$(u, v) = (w(u, v) + \rho(u, v))\left(\lambda^{\alpha(u)} - 1\right) \tag{6}$$

where $\rho(u, v)$ is a function defined as

$$\rho(u, v) = \begin{cases} 0 & if ce(u) - w(u, v) > eMin(u) \\ c & otherwise, \end{cases} \tag{7}$$

c and λ are non negative constants and algorithmic parameters. The weighing function $w''(u, v)$ assigns a high weight to edges that can lower sensors' residual energy. If P_i'' be shortest path from source to destination, then this path is used to relay data to destination.,Since the algorithm performs two shortest path computations for each request contrary to CMAX and Max-Min zPmin algorithms, which perform O(log n) shortest path computaions per route request, this makes the approach computationally more expensive.

A good energy aware routing technique maintains the balance between choosing a path with maximal residual energy and minimal energy consumption. The residual energy along a path corresponds to concave metric referred as *Metric 1*, while as energy consumed along a path corresponds to additive metric referred as *Metric 2*. Mohanoor et al. [15] proposed three polynomial time combinatorial techniques that provide a good balance between *Metric 1* and *Metric 2*. The weight of vertex in energy graph represents the residual energy available at node and weight on edge (u, v) represents the amount of energy required by node u (resp. v) to transmit one unit of data to node v (resp. u). Shortest Widest Path maximizes metric 1 and minimizes metric 2. Shortest Widest Path is a two phased routing algorithm that is applied on Energy Graph. Given source node 's' and destination node 't' the goal is to find a suitable path between s and t as described in Fig. 3. In phase 1, a variant of Dijkstra's algorithm known as Max Residual Energy is applied on the Energy Graph. This returns a path whose residual energy will be maximum in the network, say the residual energy of the path is '*R*'. There can be other paths between s and t whose residual energy is equal to '*R*'. In phase 2, a path with lowest energy consumption is chosen from set of all paths with a residual energy of '*R*'. let E′ be the set of edges in EG whose residual energy is less than R. Shortest Width Constrained Path is the derivative of shortest widest path. First prune all the edges in the graph whose residual energy level is below a cut off value. The algorithm does not use the path with maximum residual energy but a path with slightly lesser residual energy is chosen that provides the benefit of consuming lesser energy along the path. A constraint

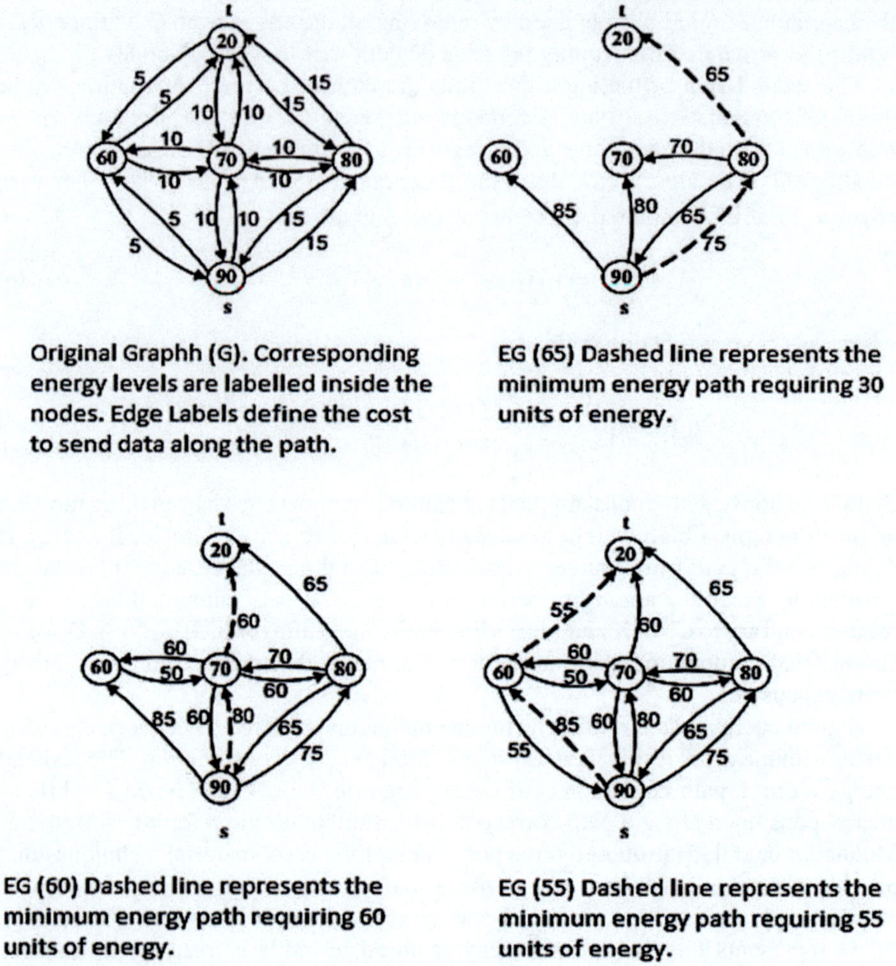

Fig. 3 Energy Graph for finding minimum energy path [15]

is placed on the width (residual energy) to be a certain fraction of the maximum possible residual energy for the given source destination pair.

Zytoune et al. [16] proposed a uniform balancing energy routing protocol (UBERP), a method that combines both transmission energy reduction and the distribution of this energy over the entire network. Before each round of data transmission, the sink calculates the energy state of whole network, thereby updating the information on each node. This procedure starts with the active nodes in the network sending a message frame containing the sensed data and nodes' remaining battery energy. The base station then calculates a dynamic threshold to determine the set of nodes that are allowed to act as routers. Then, the processing center compares the energy of each node to the threshold, if it is less than this value, the node is not authorized to participate to act as relay node. Then all nodes receive the Base Station Broadcasting Message signaling the dis-allowed nodes for the path. For each allowed path nodes, every node is added to distance matrix. Using the Dijkstra's algorithm, it determines the routes that can be used to transmit. The message is transmitted and nodes' path energy is updated accordingly. The nodes with energy less than the threshold value do not act as routers making the energy depletion of the all nodes in the network decrease uniformly. This way, the lifetime of network increases. If energy level of nodes between a sending node and base station is lower than the average, then the node tries to reach the base station directly. If the node is not in the visibility of sink node, it chooses a path with minimal transmission energy using the metric

$$C_{ij} = \frac{e_{ij}}{E_i} \tag{8}$$

where e_{ij} is energy consumed when a packet is transmitted between node i and node j. E_i is the residual energy at node i and C_{ij} is the cost of sending data over a link from node i and node j. The authors also presented the decentralized Uniform Balancing Energy Routing protocol based on local network energy average estimation. The major issue with this scheme is broadcasting message to the non-participating nodes, this also consumes good energy though not necessary.

Ok et al. [17] presented a Distributed Energy Balanced routing for WSNs and introduce a new metric, Energy Cost, to balance the energy sufficiency and energy efficiency. Energy sufficiency depends on the available energy, and energy efficiency depends on the required energy. By using a composite of both quantities, a good path that achieves energy balance can be found. The authors define a composite measure -energy cost (EC_{ij}) for a transmission from node i to j as

$$EC_{ij} = \frac{Required\, energy\, from\, node\, i\, to\, j}{Available\, energy\, at\, node\, i} \tag{9}$$

The main idea of the proposed algorithm is to use a path with minimum *EC*. A sensor node sends data to the base station either directly or to one of its neighbors (N_i). For evaluation, the sensor node i calculates the total energy required to reach the base station via neighboring nodes. The *total energy cost* ((TEC_{ik}) of a neighboring node k at sensor i is simply the sum of the energy costs from node i to k and from node k to the base station. The sensor node i selects best candidate—node *K*, for direct communication with the base station $K = Arg \min_{j \varepsilon Ni+\{i\}} \left(TEC_{ij}\right)$. If the sensor node i itself is the best candidate, it sends data directly to the base station and hence completes the routing process. Otherwise, it forwards the data to the best candidate among its neighboring nodes and that node then repeats the same routing process. This process continues until a node selects itself as the best candidate and sends directly to the base station. Each sensor keeps a small EC table. The EC table contains node identification number, minimum transmission power to the base station, and available energy for each neighboring node. This concept of relay and direct communication is a good idea in delaying the creation of coverage holes in the network. However, too many information exchanges for updating the EC table regularly can reduce the overall network lifetime particularly when nodes have less residual energy.

Cost based routing has been studied thoroughly because of its distributed nature and good Energy performance. A well-designed cost function will lead to good energy efficiency and hence prolonged network lifetime. Liu et al. [18] proposed two energy aware cost-based routing algorithms named Exponential and Sine Cost Function based Routing (ESCFR) and Double Cost Function based Routing (DCFR). The performance of ESCFR is almost comparable to DEBR, however, ESCFR does not always eliminate energy imbalance between established routing paths. The schemes discussed in the literature so far take into account only the end-to-end energy consumption and residual energy of a node, thereby achieving sub-optimal balanced energy consumption. DCFR additionally includes energy consumption rate in its cost function. The results demonstrated that the cost functions of ESCFR and DCFR are more effective than DEBR, Minimum Transmission Energy (MTE), Direct Communication (DC), and hence provide more prolonged network lifetime.

Deepika et al. [19] proposed a distributed routing algorithm called EEBR (Energy Efficient and Balanced Routing) that considers both energy consumption as well as energy balancing. The proposed algorithm uses energy density of sensor nodes to balance energy consumption and distance between the nodes. The scheme operates in two phases: Bootstrapping and Distributed Routing. In the beginning, all nodes undergo bootstrapping process in which a hop-packet indicates the hop value is that flooded by BS, with initial value set to 1. After receiving the packet, a sensor node increments it by one and transmits it to their neighbor sensor nodes. The process continues until all the sensor nodes are aware of their hop values. The hop value of sensor node s_i is denoted by $H_c(s_i)$ and denotes the number of next hops required to reach to the BS. In the distributed routing phase, sensor nodes select next-hop relay nodes in such a way that it requires less energy and also take care of energy balancing. To achieve this, the sensor node s_j chooses next hop as relay node s_r if $H_c(s_j) > H_c(s_r)$ based on the cost function. The cost function is $cost(s_j, s_r)$ which indicates cost of selecting s_r as next-hop relay node by s_j and is equal to

$$\cos t(s_j, s_r) = \frac{E_{density}(s_r) \times E_{residual}(s_r)}{dis(s_j, s_r)} \tag{10}$$

where $E_{density}(s_r)$ represents the energy density of the sensor node, $E_{residual}(s_r)$ represents the residual energy of the sensor node and $dis(s_j, s_r)$ represents the distance between the two nodes. The sensor node s_j selects the s_r as relay node if it has maximum cost value. The sensor nodes which are within communication range of the BS will not execute the data routing algorithm. They send their data directly to the base station. Other sensor nodes forward the data to the next-hop relay node using a non-persistent CSMA/CA MAC protocol. On comparison, EEBR provides prolonged network lifetime as compared to DEBR and ESCFR with lesser algorithm execution time. The classification of energy aware routing protocols in

Table 1 Classification of energy aware protocols based on deployment and mode of routing

Scheme	Deployment	Complexity	Cost function routing	Online algorithm
MREP [8]	Distributed	Less	Yes	Yes
MCP [9]	Centralized	Less	Yes	No
MERR [10]	Centralized	Moderate	No	No
CMMBCR [11]	Distributed	High	No	Yes
Max-Min zPmin [12]	Distributed	High	Yes	No
CMAX [13]	Centralized	Moderate	Yes	No
OML [14]	Centralized	Less	Yes	No
SWP [15]	Centralized	High	No	No
UBERP [16]	Centralized	High	No	Yes
DEBR [17]	Distributed	Moderate	Yes	Yes
ESCFR [18]	Distributed	Moderate	Yes	Yes
DCFR [18]	Distributed	Moderate	Yes	Yes
EEBR [19]	Distributed	Less	Yes	Yes

context of their deployment, algorithmic complexity, and the nature of their operation is summarized in Table 1. Similarly, the various optimizations performed and the constraint functions used by these protocols is summarized in Table 2.

2.2 *Sink Mobility and Sleep Scheduling Aware Routing*

In a wireless sensor network, the sensor nodes must transmit/relay the information sensed and which reaches base station only when the sensor nodes are in mutual transmission range (also termed as coverage). Deng et al. [20] studied the coverage issues faced by sensor nodes when deployed for agricultural applications. It is assumed that each node has the ability to perform different sensing tasks and the objective is to ensure the reliable coverage throughout the operation of sensor network. Chen et al. [21] devised a novel approach that permits the activation of a small subset of sensors. This method uses traffic balancing so as to provide all reliable coverage of sensing area and maintain all time connectivity to the base station. Similarly, a reliable traffic flow-aware approach was proposed by wang et al. [22], which balanced the traffic flow and minimized the energy dissipation across the network.

Routing has an important role in minimizing the energy dissipation in a network. To aid routing, the transmission of messages must be uniform across the network since using the same path over and again will only result in depletion of battery of nodes, resulting in reduction of network lifetime. Hsu et al. [23] proposed a joint routing and sleep scheduling of sensors aimed at balancing the energy dissipation in the network. The idle sensors are allowed to go into sleep mode. The authors implemented an organized layout of sensors that jointly optimized sleep state and opportunistic

Table 2 Optimizations and constraint functions of energy aware routing protocols

Scheme	Optimization function(s)	Constraint function(s)	Remarks
MREP [8]	Network Lifetime, Balanced Energy	Maximum residual energy path, Least cost path	Sub-optimal results are obtained
MCP [9]	Network Lifetime, Balanced Energy	Residual Energy, shortest path	Optimal for initial stages of operation only. Later stages show connectivity issues
MERR [10]	Network Lifetime, Characteristic Distance	Hop number, transmission range	Only sub-optimal results obtained. Subset of nodes don't participate in data transmission process
CMMBCR [11]	Network Lifetime, Energy Dissipation	Transmission range, route selection metric	Optimal energy dissipation and the procedure is highly complex and time consuming
Max-Min zPmin [12]	Ratio of cost/packet, node cost, Network Lifetime	Idle time, Energy consumption	Some information travelling along sub-optimal is suppressed. Improved network lifetime
CMAX [13]	Energy utilization, Network Lifetime	Shortest path, residual energy, transmission energy	Improved network lifetime is achieved using greedy approach
OML [14]	Transmission energy dissipation, Energy Efficiency.	Residual energy, link cost, shortest path	Heuristic approach prevents formation of coverage holes. Optimal energy efficient paths are obtained
SWP [15]	Balanced energy dissipation, Energy consumption	Residual energy, link energy cost	Sub-optimal network lifetime results show incurrence of connectivity issues at later stages of operation
UBERP [16]	Balanced energy dissipation, Network Lifetime	Transmission energy, threshold node energy	Near optimal network lifetime and efficient path selection provide better reliability

(continued)

Table 2 (continued)

Scheme	Optimization function(s)	Constraint function(s)	Remarks
DEBR [17]	Balanced energy dissipation, end-to end connectivity	Energy cost, residual energy	Operation of data transmission is time consuming. Marginal increase of network lifetime
ESCFR [18]	Cost function, Network Lifetime	Residual energy of node	An imbalance of energy among routing paths result in only an incremental improvement in network lifetime
DCFR [18]	End-to-end energy consumption, Cost function	Energy consumption rate	Effective cost function-based routing makes network work for longer duration
EEBR [19]	Energy dissipation, Network Lifetime	Energy density of nodes, distance between node and sink	The network converges in very less time and optimal energy dissipation is achieved

routing. The optimized sleep state improved the energy efficiency and their optimized opportunistic routing balanced the traffic thereby increased the network lifetime.

Lu et al. [24] proposed a sleep scheduling approach for maximizing the network lifetime. The algorithm turns on a group of sensors that sense specific target spots. These active sensors relay the sensed data over multiple hops, finally reaching the base station. Zhu et al. [25] extended the concept of sleep scheduling and incorporated a mobile cloud for providing reliability to the information sensed by sensors thereby improving the network lifetime. This novel scheme WSN-MCC (wireless sensor network—mobile cloud computing) operates in two phases. In the first phase, TPSDT (Time and priority-based selective data transmission) allows a gateway to selectively transmit data according to the time and priority features of data requested by a user. In the second phase, PSS (Priority Sleep Scheduling) allows a subset of nodes to go into sleep state in order to increase the residual energy of nodes.

The data reaching base station usually routes through specific set of nodes only. These nodes are mostly located around the base station and are known as Hotspot nodes. The consequence is quick battery drainage of hotspot nodes resulting in reduction of network lifetime and creation of coverage holes. This phenomenon is known as Hotspot issue. To circumvent the hotspot issue, sensor mobility or sink mobility is provided to the network. As a result, the traffic is balanced uniformly across the network. Wang et al. [26] proposed a relocation scheme for mobile sink, for maximizing the network lifetime. The proposed scheme used residual energy of sensor nodes for dynamically adjusting the transmission range of these nodes. Also, the sink node is allowed to move at some predefined points based on maximum residual energy of sensor node. Tashtarian et al. [27] implemented sink mobility in context to event-driven applications. The scheme associated an optimal single-hop link where the sink node captures the number of specific event occurrences from a set of sensors till a deadline is over. An alternate shortest path [28] approach uses a maximum capacity path to send data to the base station. Once any node drains the battery below a minimum threshold instead of triggering a sink relocation, alternate possible paths are used for relaying data to sink. Finally, the sink relocates when no alternate paths are available.

Asmaa et al. [29] proposed a hybrid approach combining sink-improved energy efficient-based routing protocol with the direct transmission (DT) protocol. This restricts the movement of sink so as to reduce the data loss that takes place when sink moves from one location to next. Moreover, mobile sink must spend minimum time known as sojourn time at the designated sojourn locations. The results show an improvement in energy efficiency and network lifetime when compared with both MIEEPB (Mobile Sink improved energy efficient PEGASIS based routing) and DT protocols. Zone-based sink mobility (ZBSM) studied by Prasanth et al. [30] deployed the network into a number of zones optimally placing the sink using fuzzy logic. The optimal sink position mitigates any network partition problem that may arise. The results provide an evidence of better performance than stationary and random mobile sink deployments. Kamran et al. [31] devised a framework for joint optimization of sink mobility and data aggregation with pattern matching for minimizing the node propagation delay, maximizing throughput, and minimizing the energy consumption. The framework simulates the network with mobile sink and dense deployment of nodes with heterogenous transmission ranges. This dense deployment and varying transmission ranges resulted in lesser transmissions, better residual energies, and improved throughput.

Advertising the position information of mobile sinks becomes an additional overhead in terms of energy being wasted. Energy Efficient Secured Ring Routing (E^2SR^2) [34] aims to mitigate this overhead. E^2SR^2 deploys rechargeable sensors and uses MCP as underlying routing protocol for balancing traffic load. Further it uses 2ACK that provides an efficient method for detecting routing misbehavior, therefore, enhancing the network security. The simulation results show a better performance than rail road ad ring routing. Virtual grid based hierarchical routing [35] suits best for delay bound applications. The authors used virtual grid that conservatively selects a mobile sink's route taking into account hop-count and data aggregation rates of individual nodes. Data aggregation helps to reduce traffic load and improve throughput. The results confirm better performance in end-to-end delay and network lifetime.

Biswa et al. [32] used particle swarm optimization (PSO) algorithm together with an efficient clustering and sink mobility (PSO-ECSM) for optimization of sink placement and clustering head selection. The PSO-ECSM tries to optimize the values of residual energy, node degree, average energy, and energy conservation rate. Statistical results demonstrate an improvement in network lifetime and throughput by a good percentage compared to the schemes already discussed. The prominent optimization functions, constraint functions, and the findings of various techniques that contributed to improvement of network lifetime are characterized in Table 3. These techniques are based on either Node mobility or Sleep Scheduling methods.

3 Conclusion

Due to their widespread applications, wireless sensor network has remained the focus of attention of researchers over the last decade. Enhancing the battery lifetime has been the main issue in WSN. In such networks, sensor nodes are deployed particularly in areas where recharging facilities are not available. Naturally, more stress in researches has been in designing techniques saving energy consumption, thereby improving network lifetime. Many technological advances have been made in this field, but the issue is still not over.

In this paper, a comprehensive review of the state-of-the-art approaches formulated over time for improving the network lifetime in sensor networks has been presented. For comparison and evaluation, we considered the generic network parameters which are very essential in context to the operation of a sensor network. In the initial part of our work, we evaluated the energy aware routing protocols based on parameters like energy dissipation, transmission range, transmission cost, residual energy of nodes, the approach of algorithm which can be either centralized or distributed and the algorithm computing complexity. In the later part, sink mobility and sleep scheduling aided protocols are evaluated on the basis of parameters like end-to-end connectivity, optimal sink position, energy consumption, load/traffic balancing, and throughout. The tables have ample information that represent various optimizations performed. The tables reproduced also outlines the constraints used to maximize or minimize the required objectives. The techniques discussed in the text and the innovations proposed indeed may result in optimization of a number of battery-life related issues in WSN, but a lot more is yet to be done to completely resolve the problem.

Table 3 Mobility and Sleep scheduling aided techniques for improving network lifetime

Author(s)	Optimization function(s)	Constraint function(s)	Remarks
Deng et al. [20]	Network Connectivity, Network Lifetime	Reliable coverage, Node activity scheduling	Coverage hole issue is solved to a larger extent using both centralized and heuristic algorithm
Chen et al. [21]	Node residual energy	Traffic balancing, Any-time connectivity	Sub-optimal solution obtained for traffic balancing using heuristic coverage control
Wang et al. [22]	Network Lifetime	Balanced energy dissipation, traffic flow, maximized coverage	Combination of shortest path and low-energy adaptive routing obtains sub-optimal solution
Hsu et al. [23]	Network Lifetime, Residual Energy	Sleep Scheduling, Balanced traffic and reliable data transmission	Combination of sleep-awake scheduling and opportunistic routing obtains sub-optimal solution
Zhu et al. [25]	Data reliability, Residual energy, Network Lifetime	Sleep scheduling, Time and priority-based data transmission	Nearly optimal solution is obtained less coverage hole incurrence
Lu et al. [24]	End-to-End connectivity, Network Lifetime	Multi-hop routing, sleep scheduling.	Extended connectivity issue is solved in polynomial time using approximation algorithm
Tashtarian et al. [27]	Network Lifetime	Sink mobility, continuous optimal trajectory	Approximation algorithm computes the continuous optimal trajectory and provides near optimal solution
Wang et al. [26]	Network Lifetime	Sink mobility, Sink relocation, residual battery, adaptive transmission range	Energy aware sink relocation using maximum capacity path obtains the sub-optimal solution
Kamran et al. [31]	Node propagation delay, Network Lifetime	Maximum throughput, Data aggregation	Varying transmission ranges resulted in better network lifetime with increased throughput. Optimal performance for delay intolerant applications

(continued)

Table 3 (continued)

Author(s)	Optimization function(s)	Constraint function(s)	Remarks
Biswa et al. [32]	Residual energy, throughput, Network Lifetime	Sink mobility, energy consumption rate	The sink placement using PSO provided optimal solution for throughput and network lifetime
Rajanpreet et al. [35]	End-to-End delay, network lifetime	Sink mobility, data aggregation	Virtual grid provides near optimal solution for delay bound applications
Mitra et al. [34]	Traffic load, security	Routing misbehavior, energy dissipation	2ACK provides a better performance and improved security

References

1. M. Srbinovska, C. Gavrovski, V. Dimcev, A. Krkoleva, Borozan: Environmental parameters monitoring in precision agriculture using wireless sensor networks. J. Clean. Prod. **88**, 297–307 (2015)
2. O. Amale, R. Patil, IOT Based rainfall monitoring system using WSN enabled architecture, in *Proceedings of the 2019 3rd International Conference on Computing Methodologies & Communication (ICCMC), Erode, India*, 27–29 March 2019, pp. 789–791 (2019)
3. K. Ghosh, S. Neogy, P.K. Das, Mehta: Intrusion detection at international borders and large military barracks with multi-sink wireless sensor networks: an energy efficient solution. Wirel. Pers. Commun. **2018**(98), 1083–1101 (2018)
4. R. Ullah, Y. Faheem, B. Kim, Energy and congestion-aware routing metric for smart grid AMI networks in smart city. IEEE Access **2017**(5), 13799–13810 (2017)
5. Z. Hao, Design of WSN traffic forecasting system with delayed self-sensing. Int. J. Comput. Appl. **2020**(42), 30–35 (2020)
6. A. Boulmaiz, D. Messadeg, N. Doghmane, A. Taleb-Ahmed, Robust acoustic bird recognition for habitat monitoring with wireless sensor networks. Int. J. Speech Technol. **2016**(19), 631–645 (2016)
7. A. Molina-Pico, D. Cuesta-Frau, A. Araujo, J. Alejandre, A. Rozas, Forest monitoring and wildland early fire detection by a hierarchical wireless sensor network. J. Sens, 1–8 (2016)
8. J.H. Chang, L. Tassiulas, Maximum lifetime routing in wireless sensor networks, in *Proceedings of the Advanced Telecommunications and Information Distribution Research Program (ATIRP'2000)*, vol. 12, no. 4, College Park, MD, pp. 609–619 (2000)
9. S.C. Huang, R.H. Jan, Energy-aware load balanced routing schemes for sensor networks, in *Proceedings of the 10th International Conference on Parallel Distribution Systems,* July 2004, pp. 419–425 (2004)
10. M. Zimmerling, W. Dargie, J.M. Reason, Energy-efficient routing in linear wireless sensor networks, in *IEEE Conference on Mobile Adhoc and Sensor Networks* (2007)
11. C.K. Toh, H. Cobb, D.A. Scott, Performance evaluation of battery-life aware routing schemes for wireless ad hoc networks, in *IEEE International Conference on Communications (ICC)*, vol. 9, pp. 2824–2829 (2001)
12. Q. Li, J. Aslam, D. Rus, Online power-aware routing in wireless Adhoc networks, in *Proceedings of the 7th Annual International Conference on Mobile Computing and Networking (MobiCom)*, pp. 97–107 (2001)

13. K. Kar, M. Kodialam, T.V. Lakshman, L. Tassiulas, Routing for network capacity maximization in energy-constrained ad-hoc networks, in *International Conference on Computer Communications (INFOCOM)*,pp. 673–681 (2003)
14. J. Park, S. Sahni, An online heuristic for maximum lifetime routing in wireless sensor networks. IEEE Trans. Comput. **55**(8), 1048–1056 (2006)
15. A.B. Mohanoor, S. Radhakrishnan, V. Sarangan, Online energy aware routing in wireless networks, in *Ad Hoc Networks*, vol. 7, no. 5, pp. 918–931 (2009)
16. O. Zytoune, M. El-aroussi, D. Aboutajdine, A uniform balancing energy routing protocol for wireless sensor networks. Wirel. Pers. Commun. **55**(2), 147–161 (2009)
17. C.S. Ok, S. Lee, P. Mitra, S. Kumara, Distributed energy balanced routing for wireless sensor networks. Comput. Ind. Eng. **57**(1), 125–135 (2009)
18. A. Liu, J. Ren, X. Li, Z. Chen, X.S. Shen, Design principles and improvement of cost function-based energy aware routing algorithms for wireless sensor networks. Comput. Netw. **56**(7), 1951–1967 (2012)
19. S. Deepika, P. Kuila, P.K. Jana, A distributed energy efficient & energy balanced routing algorithm for wireless sensor network, in *International Conference on Advances in Computing, Communications and Informatics* (2014)
20. X. Deng, B. Wang, W. Liu, L. Yang, Sensor scheduling for multi-modal confident information coverage in sensor networks. IEEE Trans. Parallel Distrib. Syst. **26**(3), 902–913 (2015)
21. C.P. Chen, S. Mukhopadhyay, C.L. Chuang, M.Y. Liu, J.A. Jiang, Efficient coverage and connectivity preservation with load balance for wireless sensor networks. IEEE Sens. J. **15**(1), 48–62 (2015)
22. D. Wang, E. Liu, Z. Zhang, R. Wang, S. Zhao, X. Huang, F. Liu, A flow-weighted scale-free topology for wireless sensor networks. IEEE Commun. Lett. **19**(2), 235–238 (2015)
23. C.C. Hsu, M.S. Kuo, S.C. Wang, C.F. Chou, Joint design of asynchronous sleep-wake scheduling and opportunistic routing in wireless sensor networks. IEEE Trans. Comput. **63**(7), 1840–1846 (2014)
24. Z. Lu, W.W. Li, M. Pan, Maximum lifetime scheduling for target coverage and data collection in wireless sensor networks. IEEE Trans. Veh. Technol. **64**(2), 714–727 (2015)
25. C. Zhu, Z. Sheng, C.M. Victor, L.S. Leung, T.Y. Laurence, Towards offering more useful data reliability to mobile cloud from wireless sensor network. IEEE Trans. Emerg. Top. Comput. (2014)
26. C.F. Wang, J.D. Shih, B.H. Pan, T.Y. Wu, A network lifetime enhancement method for sink relocation and its analysis in wireless sensor networks. IEEE Sens. J. **14**(6), 1932–1943 (2014)
27. F. Tashtarian, Y.M. Hossein, K. Sohraby, S. Effati, On maximizing the lifetime of wireless sensor networks in event-driven applications with mobile sinks. IEEE Trans. Veh. Technol. **64**(7), 3177–3189 (2015)
28. M.A. Jawad, F. Mir, Network Lifetime enhancement wireless sensor networks using secure alternate path, in *Proceedings of the 2017 International Conference On Wireless Communications, Signal Processing And Networking. WiSPNET*, pp. 414–419 (2018)
29. A. Rady, M. Shokair, S.M. El-Rabaie1, W. Saad, A. Benaya, Energy -efficient routing protocol based on sink mobility for wireless sensor networks. IET Wirel. Sens. Syst. **9**(6), 405–415 (2019)
30. A. Prasanth, S. Pavalarajan, Zone-based mobility in wireless sensor networks. Sensor Review. **39**(6), 874–880 (2019)
31. K. Latif, N. Javaid, I. Ullah, Z. Kaleem, Z. Abbas Malik, L.D. Nguyen, DIEER: delay-intolerant energy-efficient routing with sink mobility in underwater wireless sensor networks. Sensors **20**, 3467. https://doi.org/10.3390/s20123467 (2020)
32. B. Bhushan, G. Sahoo, E^2SR^2: an acknowledgement-based mobile sink routing protocol with rechargeable sensors for wireless sensor networks. Wirel. Netw. **25**, 2697–2721. https://doi.org/10.1007/s11276-019-01988-7 (2019)
33. B.M. Sahoo, T. Amgoth, H.M. Pandey, Particle swarm optimiz ation based energy efficient clustering and sink mobility in heterogeneous wireless sensor network. Ad Hoc Netw. **106**, 2020,102237 (2020)

34. R. Mitra, S. Sharma, Proactive data routing using controlled mobility of a mobile sink in wireless sensor networks. Comput. Electr. Eng. **2018**(70), 21–36 (2018)
35. R. Bhatti, G. Kaur, Virtual Grid based energy efficient mobile sink routing algorithm for WSN, in *International Conference on Intelligent Systems and Control (ISCO)* pp. 30–33 (2017)

Industrial Automation, Robotics, and Interdisciplinary Areas

Sensor-Based Smart Glove for Rehabilitation of Paralysis Patients

U. Snekhalatha, Haritha Nair, and Nehaa Pravin

Abstract This work aims at representing the design of a system which helps the robotic arm controlling through a sensory glove which can be used for the purpose of rehabilitation for the disabled. The smart glove consists of the controller component that features sensors, which include the Inertial Measurement Units (IMUs) and a sensor board (MPU6050). We have only tested the smart glove with MPU6050IMU, which is a sensory IMU board. We have also used Encoders for the measurement of the rotational movement of the hand. The result showed that the sensor glove successfully showed the finger and the wrist flexion movements. However, the glove requires further changes due to a few drawbacks which includes accuracy. Nonetheless, an improved glove design could even help us increase its applications like the manipulation of the robotic hand to mimic the actual hand movements, or for gesture recognition for the disabled or as a diagnostic tool for finger movements.

Keywords Rehabilitation · Sensors · Smart glove

1 Introduction

The primitive design of the glove-based systems in order to track the motion of hand with the help of mechanical based sensors [1, 2]. Klahr et al. [2] came up with a glove system to measure the bending angles of the fingers precisely based on a wire-driven mechanism. Feisel and Rosa [3] used the term "hands-on mindson" that suggested the relation between the objective of the practical and the comprehensive domains. Thomas de Fanti & Daniel Sandin developed the very first specimen of a glove system in 1977, which was named as the Sayre Glove [4]. It was only soon after the marketing and advertising of the VPL Data-Glove [5] and Mattel's Power Glove [6], that the research based on glove-based system became extremely popularized. And so, Flex sensors started being used for the measurement of the general flexion of the

U. Snekhalatha · H. Nair · N. Pravin (✉)
Biomedical Engineering Department, College of Engineering and Technology, SRMIST, Kattankulathur, Chennai, Tamil Nadu, India
e-mail: snehalau@srmist.edu.in

F. Thakkar et al. (eds.), *Proceedings of the International e-Conference on Intelligent Systems and Signal Processing*, Advances in Intelligent Systems and Computing 1370,
https://doi.org/10.1007/978-981-16-2123-9_58

fingers, as most of the glove- based system during that time featured flex sensors. Power Glove was then used to pacify video games. Following this progression, various glove-based systems with different applications have emerged till date. The numerous applications of the glove-based system include gaming, rehabilitation, gesture recognition, etc. Oz and Leu [7] found yet another application, which is the sign language identification with the help of neural networks. Borghetti et al. [8] work included a feedback system along with the flex sensors for the purpose of rehabilitation. Popescu et al. [9] became the first person to develop VR-based rehabilitation system mainly for domestic purpose. It made use of the Master II [10] glove. For a much better understanding of the hand glove design, the authors prefer to study the works carried out by Heo et al. [11], and Maciejasz et al. [12].

Our design currently stresses on the application of a sensory glove for rehabilitative purposes. The 5DT data glove helped in developing the first software iteration [13]. In the previous years, Aubin et al. designed wearable hand devices [14], and the hand exoskeleton (wearable). Cempini et al. [15] demonstrated the exoskeletal model based on the thumb contact (opposition) and grasping of different objects. Also, a number of smart gloves have been designed for the purpose of rehabilitation for the disabled. Some of them used flex sensors, while others used IMU sensors, along with servo motors and Arduino. A lot of previously developed smart gloves had the drawback of limited degrees of freedom, so we tried building a glove with MPU6050 and encoder along with Arduino and a Bluetooth module, which could transmit the data wirelessly to the arm controller which could mimic the actual hand movement.

2 Materials and Methods

2.1 Proposed Architecture

The proposed system consists of a sensor-based glove system. This sensor glove consists of IMU sensors and encoders as well as one MPU6050 sensor board. These sensors provide the accurate values for fingers and wrist flexion, thus allowing us to calculate the angle of flexion of the fingers and the wrist. Keeping this into consideration, the IMU sensors and the encoders were put together in the sensor glove to measure the flexion of each of the five fingers as well as the rotation of the wrist. Along with it, one MPU6050IMU sensor board, which is a 6 DOF IMU sensor, i.e., it provides 6 output values, 3 from the accelerometer, and 3 from the gyroscope, were placed on the glove. The used IMU was a MPU6050 model. The IMU used in the system is for the measurement of the hand's rotation along with the flexion movements. All the sensors, encoders, and the sensor board were interfaced with an Arduino, along with the encoder (a sensor which is used to measure rotation).

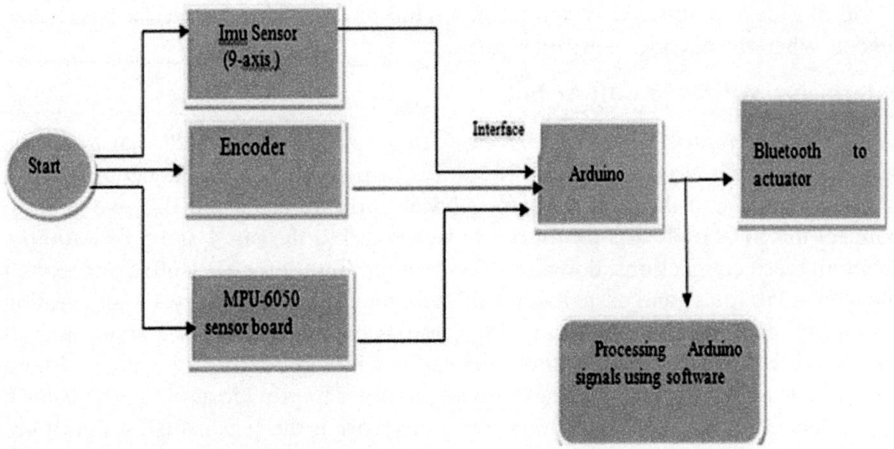

Fig. 1 System architecture of proposed work

2.2 Proposed Method

The IMU sensor measures the movements of finger in rotation and flexion. The sensors (both IMU and Encoders) are interfaced with an Arduino UNO microcontroller. Sensor angles were transferred to the microcontroller from the sensors. These sensor angles were transformed into voltage values using the Arduino library folder. We have also used MPU6050 and encoders and interfaced both the sensor board and the sensor to the Arduino UNO. The Arduino is, in turn, connected to the Bluetooth module. The MPU uses I2C protocol in order to communicate with the Arduino UNO. Bluetooth module was interfaced with the Arduino microcontroller for conducting a wireless transmission. The graphs for the finger movements (from the gyroscope and the accelerometer) can be computed using processing. The graph consists of the accelerometer and the gyroscope values represented in the x and the y-axis.

The system Architecture shown above is a representation of the work flow of the system (Fig. 1). It gives a flow representation of the different components being used in this work, i.e., MPU 6050 sensor board, imu sensors, encoders, and their interference to Arduino, which is, in turn, connected to the arm actuator.

2.3 Hardware Connections

Interfacing encoder to Arduino

Connect the v + pin of the encoder to the 5v of the Arduino and connect ground of the encoder and the Arduino and then connect the clock and DT pins to digital pins

2 and 3. Finally, connect SW pin to the digital pin 4. Then, we need a sketch that detects when the encoder is being rotated.

Interfacing MPU6050 with Arduino

MPU uses I2C protocol to communicate with the Arduino. The MPU has a module pin of 5v which is connected to the Arduino's 5v pin. After that, connect the ground of the Arduino and the sensor board. Now in order to setup the I2C, we need to connect the SDA and SCL pin of the sensor board to the pin 4 and 5 of Arduino. After all these connections, download the Arduino library for MPU6050 and extract the MPU6050 folder and paste it in the folder named Arduino library. Finally, testing is done by sending a DMP programming demo, in order to get the yaw, pitch, and roll values. MPU has an in built motion processor, which processes the values of both the accelerometer, as well as the gyroscope, in order to provide accurate 3D values.

The basic approach or algorithm used in this work is the sensor-fusion algorithm. This algorithm combines the data from various sensors and synthesizes those data, and thereby helps in reducing the uncertainty involved in the machine perception in order to give out the most accurate positions of objects. This sensor-fusion algorithm takes the data from multiple sensor's which might be unreliable if considered individually, and converts them into accurate truth.

3 Result and Discussions

Robotic applications demand the use of sensors with high accuracy and repeatability, and previous versions of sensor-based smart gloves consisted of sensors like flex sensors, pressure sensors, position sensors, along with servo motors, and RF module or Bluetooth. IMU sensors are considered as the most practical sensors while designing gloves, as they have the advantage of being light and can precisely capture even the slightest finger motion by applying an algorithm to the primitive values, the ROM(range of motion) can be determined. Kortier et al. [16] and Moreira et al. [17] proposed the use of IMU sensors, but they had the drawback of using wired transmission. Choi et al. proposed a 9-axis sensor, but it conducted only static validation [18]. So later, a data glove with 9-axis IMU sensors were developed and used in data gloves. In this work, we have made use of sensors like Encoders and MPU6050 sensor board along with Arduino and Bluetooth, which makes our work different from the previously developed versions of the smart gloves. Figure 2a depicts the developed prototype of sensor-based glove. Figure 2b represents the wearing glove in subject's hand. Figure 3 displays the graph illustrating the gyrometer response to finger movements. As a future process, we could also consider the addition of new components to the glove which would allow us to access more DOF, which, in turn, could allow the user to produce movements similar to a human hand exhibiting similar functions as well. Finally, the designing of this glove-based system has allowed us to understand more about the mechanics of a human hand, along with the number of applications that this work could find.

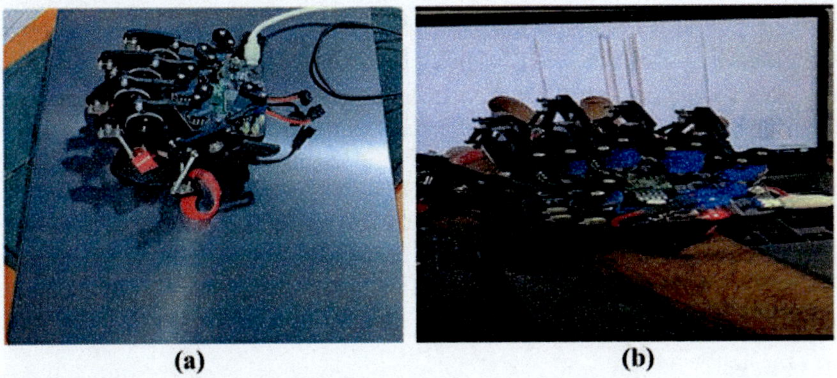

Fig. 2 **a** Prototype of sensor-based glove. **b** Wearing glove in hand

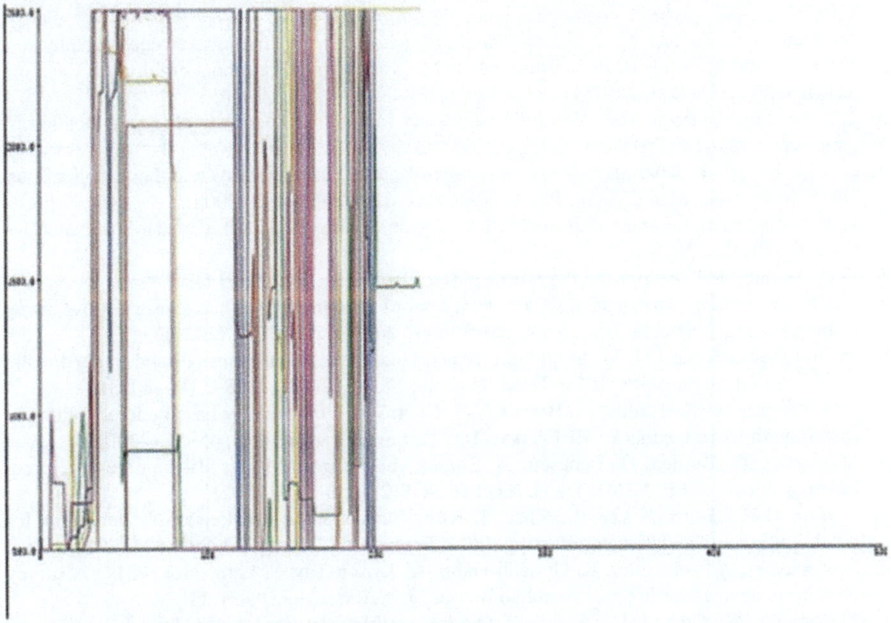

Fig. 3 Graph showing the response to increased finger movements

The use of Imu sensors and the sensor board along with the sensor-fusion algorithm gives us the accurate determination of the position and angle of the sensors due to the flexion/extension of the arm thereby giving out the hand movements.

4 Conclusion

In the proposed work, we have developed a smart glove which could be used along with an actuator component to mimic actual hand movement or can be used along with a receiver component for gesture recognition. This work has enabled us to move from wired connections to wireless technology. We can conclude that the designed glove gives the accurate values for the hand and wrist flexion movement. We could also conclude that our proposed glove can find wider applications.

References

1. D. Pusca, R.J. Bowers, D.O. Northwood, Hands-on experiences in engineering classes: the need, the implementation and the results. World Trans. Eng. Technol. Educ. **15**(1), 12–18 (2017)
2. D. Klahr, L.M. Triona, C. Williams, Hands on what? The relative effectiveness of physical versus virtual materials in an engineering design work by middle school children. J. Res. Sci. Teach. **44**(1), 183–203 (2007)
3. L.D. Feisel, A.J. Rosa, The role of the laboratory in undergraduate engineering education. J. Eng. Educ. 121–130 (2005)
4. L. Dipietro, A.M. Sabatini, P. Dario, A survey of glove-based systems and their applications. IEEE Trans. Syst. Man Cybern. Part C Appl. Rev. **38**(4), 461–482 (2008)
5. G.T. Zimmerman, J. Lanier, C. Blanchard, S. Bryson, Y. Harvill (1987) A hand gesture interface device. VPL Res. Inc.
6. D.L. Gardner, Inside story on: the power glove. Des. News **45**(23), 63 (1989)
7. C. Oz, M.C. Leu, American Sign Language word recognition with a sensory glove using artificial neural networks. Eng. Appl. Artif. Intell. **24**(7), 1204–1213 (2011)
8. M. Borghetti, E. Sardini, M. Serpelloni, Sensorized glove for measuring hand finger flexion for rehabilitation purposes. IEEE Trans. Instrum. Meas. **62**(12), 3308–3314 (2013)
9. V.G. Popescu, G.C. Burdea, M. Bouzit, V.R. Hentz, A virtual-reality-based telerehabilitation system with force feedback. IEEE Trans. Inf. Technol. Biomed. **4**(1), 45–51 (2000)
10. M. Bouzit, G. Burdea, G. Popescu, R. Boian, The Rutgers Master II—new design force-feedback glove. IEEE/ASME Trans. Mechatron. **7**(2), 256–263 (2002)
11. P. Heo, G.M. Gu, S.-J. Lee, K. Rhee, J. Kim, Current hand exoskeleton technologies for rehabilitation and assistive engineering. Int. J. Precis. Eng. Manuf. **13**, 807–824 (2012)
12. P. Maciejasz, J. Eschweiler, K. Gerlach-Hahn, A. Jansen-Toy, S. Leonhardt (2014) A survey on robotic devices for upper limb rehabilitation. J. Neuroeng. Rehabil. **11**
13. J. Connolly, K. Curran, J. Condell, P. Gardiner, Wearable rehab technology for automatic measurement of patients with arthritis, in *Sensors (Peterborough)* (2011), pp. 508–509
14. P. Aubin, H. Sallum, C. Walsh, A. Correia, L. Stirling, A pediatric robotic thumb exoskeleton for at-home rehabilitation, in *IEEE International Conference on Rehabilitation Robotics*, ICORR (2013)
15. M. Cempini, M. Cortese, N. Vitiello, A powered finger—wearable thumb—hand exoskeleton with self-aligning joint axes. IEEE/ASME Trans. Mechatron. (2014)
16. H.G. Kortier, V.I. Sluiter, D. Roetenberg, P.H. Veltink, Assessment of hand kinematics using inertial and magnetic sensors. J. Neuroeng. Rehabil. **11**, 1–14 (2014)
17. A.H.J. Moreira, S. Queirós, J. Fonseca, P.L. Rodrigues, N.F. Rodrigues, J.L. Vilaça, Real-time hand tracking for rehabilitation and character animation, in *Proceedings of the IEEE 3nd International Conference on Serious Games and Applications for Health (SeGAH)*. (Rio de Janeiro, Brazil, 2014)

18. Y. Choi, K. Yoo, S.J. Kang, B. Seo, S.K. Kim, Development of a low-cost wearable sensing glove with multiple inertial sensors and a light and fast orientation estimation algorithm. J. Supercomput. **72**, 1–14 (2016)

Wireless Robot Control Using Wrist Movements from Surface Electromyogram Signal

Ng Chooi Rou, Chong Yen Fook, Lim Chee Chin, Zulkarnay Zakaria, Asyraf Hakimi Abu Bakar, Ahmad Nasrul Norali, and Anas Mohd Noor

Abstract Nowadays, the rates of disability are increasing in part due to aging populations and an increase in chronic health conditions. Human-Computer Interaction (HCI) system become vital for helping the temporary and permanently disabled patients. In this framework, sEMG signals of five motions which are neutral, left, right, up and down will be performed and recorded from subjects by using ADInstruments PowerLab 4/25T device. Then, the signal will be filtered with 4th order band-pass Butterworth filter and an 50 Hz notch filter used to remove the noise of the power line. Empirical Mode Decomposition (EMD) method and wavelet packet decomposition method were implemented for extracting the 11 features. Classifiers used are probabilistic neural network (PNN) and generalized regression neural network (GRNN) in order to analyze the performance of controlling the wireless patient remote system prototype via wrist sEMG signal. From the results obtained, PNN is the best classifier with the highest recognition rate of 87.0% by using the 10% of testing data and 90% of training data compared with GRNN which is 86.25%. After that, the graphical user interface (GUI) will be connected with the Blynk app to show the IoT platforms that allows patients for controlling the robotic system.

Keywords EMG · Human-computer interaction (HCI) · Internet of things · Blynk · EMD · PNN · GRNN

N. C. Rou · C. Y. Fook (✉) · L. C. Chin · Z. Zakaria · A. H. A. Bakar · A. N. Norali · A. M. Noor
Faculty of Electronic Engineering Technology, Universiti Malaysia Perlis, 02600 Arau, Perlis, Malaysia
e-mail: yenfook@unimap.edu.my

C. Y. Fook · L. C. Chin · Z. Zakaria · A. H. A. Bakar · A. N. Norali · A. M. Noor
Sports Engineering Research Centre (SERC), Universiti Malaysia Perlis, 02600 Arau, Perlis, Malaysia

F. Thakkar et al. (eds.), *Proceedings of the International e-Conference on Intelligent Systems and Signal Processing*, Advances in Intelligent Systems and Computing 1370,
https://doi.org/10.1007/978-981-16-2123-9_59

1 Introduction

Nowadays, the insights have been demonstrated that over a billion people are assessed to live with some type of disability comparing to about 15% of the total populace. At that point, the general population whose 15 years and more seasoned that between 110 million (2.2%) and 190 million (3.8%) have noteworthy challenges in their functioning [1]. Human-Computer Interaction (HCI) system is turned out to be a more imperative and critical part of our daily lives because it is a method that very valuable for temporary and permanently disabled patients through surface electromyogram signals. Through the movement of the wrist and the muscle signal from the forearm, the physically disabled patient can do things more easily. There are five specific movements such as neutral, flexion, extension, adduction and abduction.

In this paper, it is proposed to develop sEMG signal based Human Machine Interface for physically disabled persons and patients. The suitable data acquisition module is to be identified and the data acquisition protocol is to be developed to record sEMG signals [2]. Data collection will be taken by AD Instruments to capture the muscle activity of both Extensor Digitorum (ED) and Flexor Carpi Ulnaris (FCU) muscles. Feature extraction and classification algorithms will be developed using artificial intelligence techniques and digital signal processing with MATLAB software. The proposed project can control wireless the movement of a robotic system through the sEMG signal.

2 Methodology

In Fig. 1, it showed an overview of the workflow for this project from data acquisition until the robotic system prototype to design and develop a wireless patient remote system.

2.1 Data Acquisition

In this paper, sEMG signals of five wrist motions are recorded from 20 healthy subjects (10 male and 10 female, age 22–24 years old). For sEMG signal acquisition, ED and FCU muscles are used to record and collect data via the movements of the

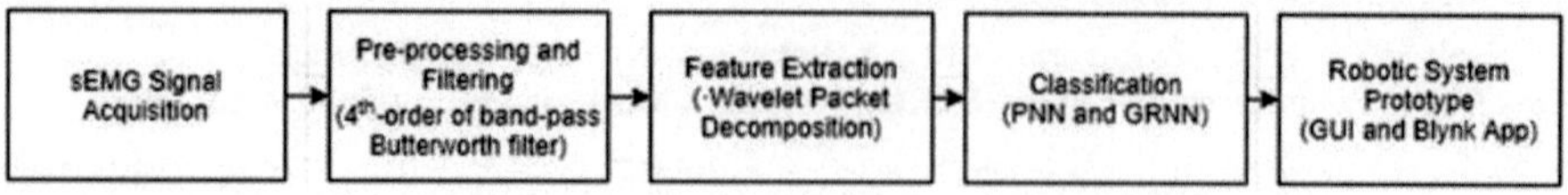

Fig. 1 Block diagram of the proposed system

wrist. The sEMG signals are recorded by utilizing the ADInstruments PowerLab 4/25T device. Five surface electrodes which are two-channel pairs and a reference electrode are used for each subject. Each subject will have adhered one electrode pairs to the upper forearm and the other electrode pairs will be adhered to the lower forearm [3, 4]. Besides that, sEMG signals are recorded with a sampling frequency 1000 Hz and an amplitude of 500 μV. The high-pass and low-pass of LabChart7 will be set at 0.3 and 500 Hz.

2.2 *Data Pre-processing*

All of the raw data obtained were recorded and brought into MATLAB software for analysis of feature extractions and classification processes. The raw data of sEMG signals were filtering with 4th order band-pass Butterworth filter. It contained a high-pass filter 10 Hz cut-off frequency and a low-pass filter 500 Hz cut-off frequency [5, 6]. This process will be helped in removing various unwanted noises and maintain the relevant information data of 5 motions. Moreover, an 50 Hz notch filter was also used to remove the noise of the power line.

2.3 *Feature Extraction*

Feature extraction is an essential step for pattern recognition. It can be partitioned into three classes such as frequency domain, time domain and time-frequency domain. In the project, wavelet packet decomposition has been used to extract features. The chosen features will be used which were frequency median (FMD), median amplitude spectrum (MFMD), modified mean absolute value 1 (MMAV1), modified mean absolute value 2 (MMAV2), simple square integral (SSI), waveform length (WL), root mean square (RMS), mean absolute value (MAV), variance (VAR), power spectrum and standard deviation [7, 8].

Wavelet Packet Decomposition. Wavelet packet decomposition is a wavelet transform where the discrete-time signal is passed through more filters than the discrete wavelet transform (DWT). Wavelet packet decomposition helps to divide the high frequency side into smaller frequency bands, which is not possible with general discrete wavelet transform. In this project, the sampling frequency used 1000 Hz and frequency bands of each level have been decomposed into three levels as shown in Table 1. This three-level decomposition will decompose the 11 features from every eight coefficients.

Table 1 Wavelet packet decomposition levels

Wavelet packet decomposition level	Frequency band (Hz)
1	0–250, 250–500
2	0–125, 125–250, 250–375, 375–500
3	0–62.5, 62.5–125, 125–187.5, 187.5–250, 250–312.5, 312.5–375, 375–437.5, 437.5–500

2.4 Classification

Probabilistic Neural Network (PNN). PNN is an artificial neural network that used to estimate the probability density function of a given data set. It consists of four layers which are the input layer, pattern layer, summation layer and output. When there is an input, the first layer will calculate the distance from the input vector to a training input vector. This will generate a vector where the element represents how close the input is to the training input. The second layer sums the contributions of each type of input and uses its net output as the probability vector. Lastly, complete transfer function of output on the second layer selects the maximum of these probabilities and produces 1 (positive recognition) for the class and 0 (negative recognition) for the non-target class. Moreover, the performance of the PNN classifier is dependent on the spread factor (σ). Based on this project, the σ value varies that used in PNN is between 0.01 and 0.10 in steps of 0.01 [9].

General Regression Neural Network (GRNN). GRNN is a single-pass neural network that uses a Gaussian activation function in the hidden layer. It consists four layers which are input, hidden, summation, and division layers. Based on the paper, the σ value is varied between 0.01 and 0.10 in steps of 0.01. Lastly, all the feature extraction and classifications are progressed under MATLAB software [9, 10].

2.5 Robotic System Prototype

A robotic system prototype will be developed by using graphical user interface (GUI) and Blynk app. GUI is a part of MATLAB software used to interact with an electronic device using icons or visual indicators, rather than just textual interaction through the command line. Blynk is an apps that performing an IoT platform to connect devices to the cloud or design apps to control them [11]. In this paper, the accuracy of classification was recorded and observed as result and used it for connecting and controlling the Blynk app wireless via a NodeMCU.

3 Results and Discussion

3.1 Result of Pre-processing

The results of data acquisition depend on wrist movement from the surface. LabChart7 software was used to develop a signal processing and record the data of signal. The raw EMG data was band-pass filtered, extracted during the signal processing process. The high-pass frequency and low-pass frequency were set as 0.3 and 500 Hz in LabChart7 software. The signal ED and FCU muscles will be selected and record the data as a notepad via five wrist motions which were up, down, neutral, left and right. Figures 2 and 3 shown the sEMG signals of the ED and FCU muscles during filtering.

3.2 Result of Wavelet Packet Decomposition

After filtering processing, the signal of ED and FCU muscles will be undergo wavelet packet decomposition with three levels. This three levels can decompose different

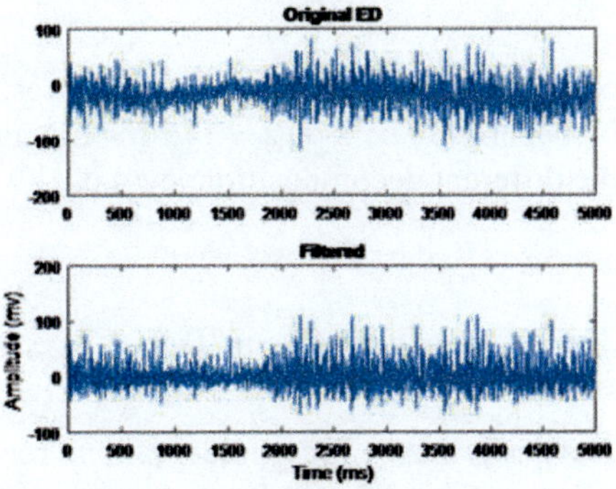

Fig. 2 Original and filtered sEMG signals of ED muscle

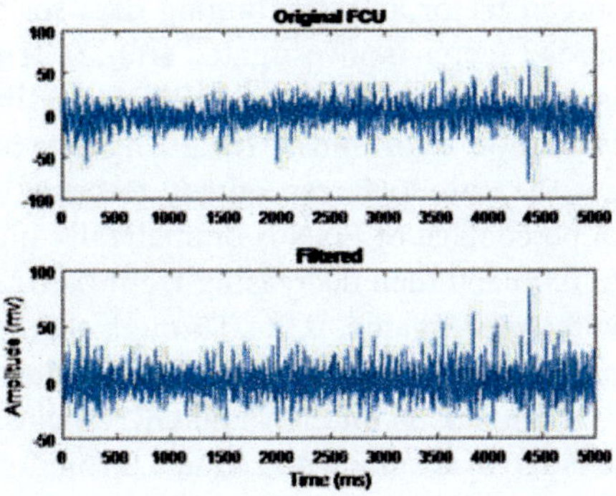

Fig. 3 Original and filtered sEMG signals of FCU muscle

Fig. 4 Wavelet packet decomposition of ED muscle with 3 levels

Fig. 5 Wavelet packet decomposition of FCU muscle with 3 levels

eight coefficients for extracting features. Hence, the total number of 11 features will be become $11 \times 8 \times 2 = 176$ for ED and FCU muscles. Figures 4 and 5 show that the different decomposition level of ED and FCU muscles.

3.3 Comparison on Two Classifiers, PNN and GRNN

The performance for wireless patient remote system by using PNN and GRNN classifiers will be affected by the value of spread factor. In this section, the analysis of spread factor and the training data set and testing data set will be discussed. The spread factor is investigated from 0.01 to 0.1. The performance of remote system is evaluated by PNN and GRNN classifiers and validated by using 10 k-fold cross validation algorithms. The testing data and training data ratio from the dataset is set as 10% and 90% respectively to carry out the both tests. From the Fig. 6, the rate of percentage of PNN is dramatically increasing while spread factor vary from 0.01 to 0.06 and then decreasing from 0.06 to 0.10. The highest percentage rate is 87% with spread value, 0.06. Through the figure below, it can be seen that the highest percentage is 86.25% when GRNN classifier is employed with 0.04 spread factor.

Comparison on PNN and GRNN classifiers has been carried out according to the average of accuracy percentage. From figure below, it can be seen that PNN classifier

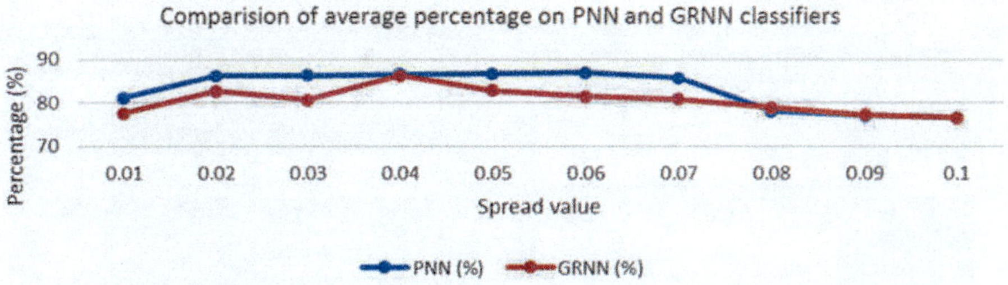

Fig. 6 Comparison of average percentage on PNN and GRNN classifiers

got the highest average of accuracy percentage, 87.00% and GRNN classifier has the lowest average of accuracy percentage, 76.55%. By reviewing previous researches, the classifiers used in this project are different from previous researches. Hence, there is no direct comparison between this project and previous researches. Most of the researchers did their studies on the other classification methods. However, Hariharan et al. [3] tested their data by using LDA, PNN and GRNN and obtained average accuracy 99%.

3.4 *GUI for Wireless Patient Remote System*

In this project, GUI was constructed to show the wireless patient remote system by using MATLAB software. The instructions to operate the system are stated in the GUI in the start button as shown in Fig. 7. For Fig. 8, it show the main page of GUI after pressed the start button from front page. After pressing the load button, the chosen file will be appeared and start to choose the motions.

The different motions have different instruction, for example, the left and right signal will be shown to move right and left to choose the application on Blynk app, up signal used to enter or start the application, down signal used to start the alarm system and neutral signal is for resting. Besides that, original signal of ED and FCU muscles and wavelet packet decomposition level signal with different eight coefficients will be shown in GUI. Then, the features of signal will be carried out to classify by using PNN classifier and continuously used it to control action on Blynk app. To obtain the

Fig. 7 Main page of GUI

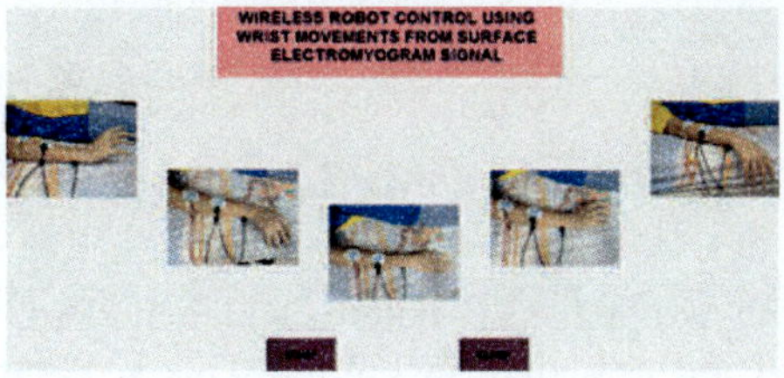

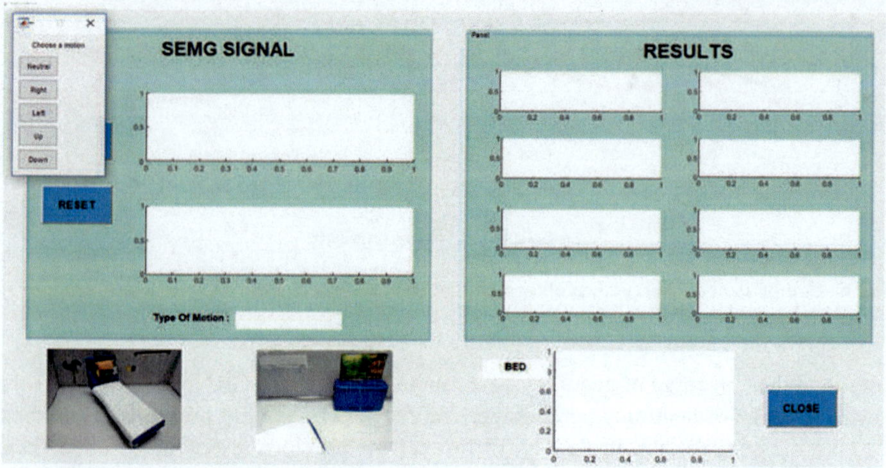

Fig. 8 Choose a motion file

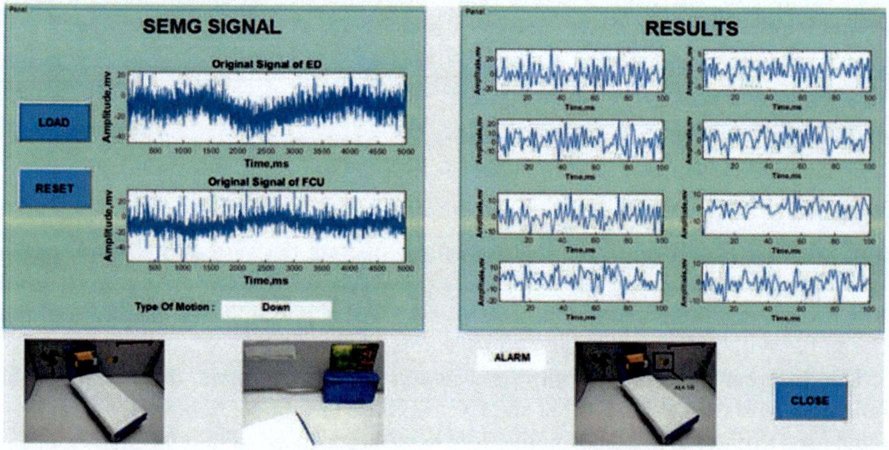

Fig. 9 GUI page for down motion signal

other file of wrist motions from other individual, press reset and the whole signal that shown will be cleared. To close the GUI system, press close button and the system will be shut down as shown in Fig. 9.

3.5 *Blynk App and Prototype*

Blynk app has six buttons which are bed, fan, light, alarm, air conditioner and television as shown in Fig. 10. This six buttons connected with the six applications

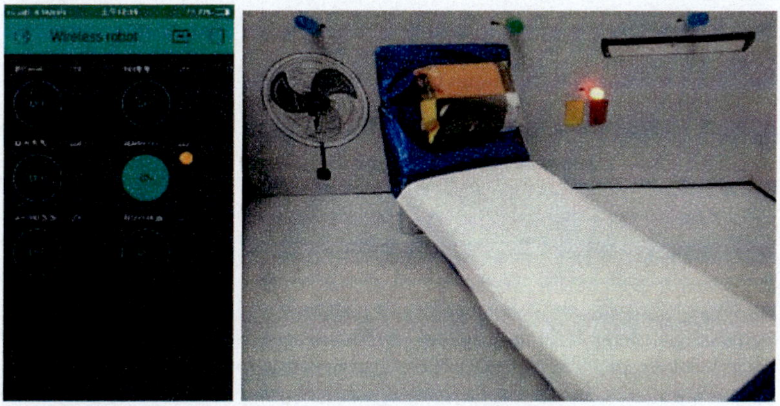

Fig. 10 Alarm ON

synchronous. Besides that, Blynk app can be connected and started the whole six application on once. It will very simple and helpful to the disable patients to on the application. This system like a remote control for them.

4 Conclusion

In this paper, based on the process of develop the protocol for collecting wrist sEMG signal, extract features for classification of wrist motions and develop a GUI and Blynk app, the main objectives were achieved and a wireless patient remote system prototype was successfully developed. From the project, sEMG signals are subjected to feature extraction using wavelet packet decomposition with three levels. 11 features were classified by PNN and GRNN classifiers. But, from the analysis of results, PNN has performed well and obtained the highest recognition rate of 5 motions with 87.0% compared to GRNN. From the experimental results, it can be concluded that PNN is the best classifier for this system. Then, system will be constructed with PNN classifier to classify the motions of wrist of patient and used the signal of motions to control the Blynk app on mobile via GUI of MATLAB.

References

1. C.A. Riddle, Disability and health, Hum. Rights Disabil. Capab. 11–30 (2016)
2. Y. Kamei, S. Okada, Classification of forearm and finger motions using electromyogram and arm-shape-changes, in *Proceedings of Annual International Conference of the IEEE Engineering in Medicine and Biology Society, EMBS*, vol. 2016 (2016), pp. 5680–5683

3. M. Hariharan, C.Y. Fook, R. Sindhu, B. Ilias, S. Yaacob, A comparative study of wavelet families for classification of wrist motions (2012)
4. K. Soo Kim, H. Ho Choi, C. Soo Moon, C. Woong Mun, Comparison of k-nearest neighbor, quadratic discriminant and linear discriminant analysis in classification of electromyogram signals based on the wrist-motion directions (2011)
5. N. Nazmi, M.A.A. Rahman, S.I. Yamamoto, S.A. Ahmad, H. Zamzuri, S.A. Mazlan, A review of classification techniques of EMG signals during isotonic and isometric contractions. Sensors (Switzerland) **16**(8), 1–28 (2016)
6. O.S. Powar, K. Chemmangat, S. Figarado, A novel pre-processing procedure for enhanced feature extraction and characterization of electromyogram signals. Biomed. Signal Process. Control **42**, 277–286 (2018). Apr.
7. S. Thongpanja, A. Phinyomark, P. Phukpattaranont, C. Limsakul, Mean and median frequency of EMG signal to determine muscle force based on time dependent power spectrum. Elektron. ir Elektrotechnika **19**(3), 51–56 (2013)
8. A. Gupta, T. Sayed, R. Garg, R. Shreyam, EMG signal analysis of healthy and neuropathic individuals. IOP Conf. Ser. Mater. Sci. Eng. **225**, 012128 (2017)
9. C.Y. Fook, M. Hariharan, S. Yaacob, A.H. Adom, Malay speech recognition in normal and noise condition, in *2012 IEEE 8th International Colloquium on Signal Processing and its Applications*, Melaka (2012), pp. 409–412
10. J. Qi, G. Jiang, G. Li et al., Surface EMG hand gesture recognition system based on PCA and GRNN. Neural Comput. Appl. **32**, 6343–6351 (2020)
11. M.Y. Loh, C.C. Lim, C.Y. Fook, D. Maslia, N.B. Shafriza, S.C. Lim, Internet of things (IoT) fall detection using wearable sensor. IOP Conf. Ser. J. Phys. **1372**, 012048 (2019)

COVID-19 in Malaysia: A Correlation and Regression Analysis on the Imported Cases and Local Transmission

Chow-Khuen Chan, Ho-Kin Tang, Sim-Kuan Goh, and Masni Azian Akiah

Abstract COVID-19 pandemic has been seen to be uncontrollable with some countries suffering several surging waves. This pandemic will continue to threaten many nations in the foreseeable future, particularly the health government, public health systems, socio-economy, and national economy. Different countries use dissimilar containing measures in combating the pandemic. In Malaysia, lockdown has been implemented in different phases, namely, Movement Control Order (MCO), Conditional Movement Control Order (CMCO), and Recovery Movement Control Order (RMCO). Nevertheless, the border control has been loosely adhered with the spiking of cases that leads to several clusters, namely, Sivagangga and Benteng LD clusters. Therefore, the consequences of imported cases in local transmission cannot be neglected. The aim of this study is to apprehend the correlation and relationship between the imported and local occurrences in every MCO phase. Through the correlation analysis, the imported cases have greatly affected the local transmission at the onset of the MCO. From the regression analysis, results showed a positive relationship of the local and imported cases in the MCO phases. Thence, the returnees to Malaysia should be strictly observed and quarantined in avoiding the surging of the COVID-19 cases. The proposed analysis in this work will help us to comprehend the dynamics of COVID-19 in Malaysia, which helps to verify the effectiveness of MCO.

Keywords Regression · Correlation · COVID-19 · Pandemic · Transmission

C.-K. Chan (✉)
Universiti Malaya, Kuala Lumpur, Malaysia
e-mail: ckchan@um.edu.my

H.-K. Tang
Shenzhen JL Computational Science and Applied Research Institute, Beijing, China

S.-K. Goh
Nanyang Technological University, Singapore, Singapore

M. A. Akiah
Universiti Teknikal Malaysia Melaka, Melaka, Malaysia

F. Thakkar et al. (eds.), *Proceedings of the International e-Conference on Intelligent Systems and Signal Processing*, Advances in Intelligent Systems and Computing 1370,
https://doi.org/10.1007/978-981-16-2123-9_60

1 Introduction

COVID-19 pandemic has been seen to be uncontrollable in some countries. It is prominent that this pandemic disease will continue to threaten many countries in the foreseeable future. The rising of this pandemic has posed a serious threat to public health [1], public health systems [2] and the national economy. Various countries have implemented dissimilar approaches and standard operating procedures (SOPs) to contain the virus, which ultimately flatten the curve and reduce the reproduction number. In Malaysia, lockdown has been implemented in different phases to curb the disease. However, the border control has loosely adhered as there was a sudden spike in cases reported on September 20, 2020. Many clusters have been resurging due to the imported cases that elicit to the local transmission. This can be seen in Sivagangga Cluster from India (July 28, 2020) [3] and Benteng LD cluster from foreign nationals of Indonesia and Philippines [4]. Hence, the effect of imported cases in local transmission cannot be neglected. In this study, we ought to understand the correlation and linear relationship of the imported cases and local occurrences. To our knowledge, there has been no analytical studies on this objective. Identification of the correlation and relationship of the imported and local cases will provide a useful strategy for better COVID-19 containing measures.

2 Methodology

In this study, COVID-19 data was gathered from the Malaysia COVID-19 Healthy Ministry website (http://covid-19.moh.gov.my/terkini). In Malaysia, Movement Control Order (MCO) had been implemented by the government to lockdown the nation in several phases. Hence, the daily number of imported cases and local occurrences was collected for the following MCO phases: (i) Movement Control Order (MCO) ranges from March 18, 2020 to May 12, 2020; (ii) Conditional Movement Control Order (CMCO) spans from May 13, 2020 to June 7, 2020; and (iii) Recovery Movement Control Order ranges from June 8, 2020 to October 8, 2020; (iv) partial CMCO from October 9, 2020 due to the third wave. The SOPs were strict during MCO but gradually relaxed till RMCO. The data was then analyzed for the imported and local cases for all the MCO phases. Thereafter, the correlation coefficients between the imported and local cases for every MCO phases were observed using Pearson, Kendall, and Spearman rank correlations. The correlation is essential in determining whether there is correlation between imported cases and local occurrences based on week of the MCO phases. Thereafter, the linear relationship was computed to attain the relationship between imported cases and local occurrences. A designated Python programming was developed to compute the correlation and relationship.

3 Results

The distribution of imported and local cases in Malaysia is shown in Fig. 1. The number of cases was segregated into different MCO phases. It can be seen that the imported and local cases were declining from MCO to RMCO phases. Nonetheless, the COVID-19 cases showed signs of escalating in the third wave starting from September 20, 2020.

Correlation analysis was performed on the imported cases and local occurrences using Pearson, Kendall, and Spearman rank correlations (Fig. 2). Additionally, the correlation coefficients were computed according to weeks for MCO-CMCO and RMCO. MCO, and CMCO were consolidated in the analysis as the period for CMCO was merely 4 weeks. From the analysis, there was a positive correlation in the first week of MCO-CMCO followed by a negative correlation in the subsequent weeks. In RMCO, the first 2 weeks reported a positive correlation while the last week reported a negative correlation.

Linear regression analysis for the number of local cases against the number of import cases was performed for MCO-CMCO and RMCO phases (Fig. 3). From the period (22/4 to 7/6), there was significant evidence that the imported cases had a

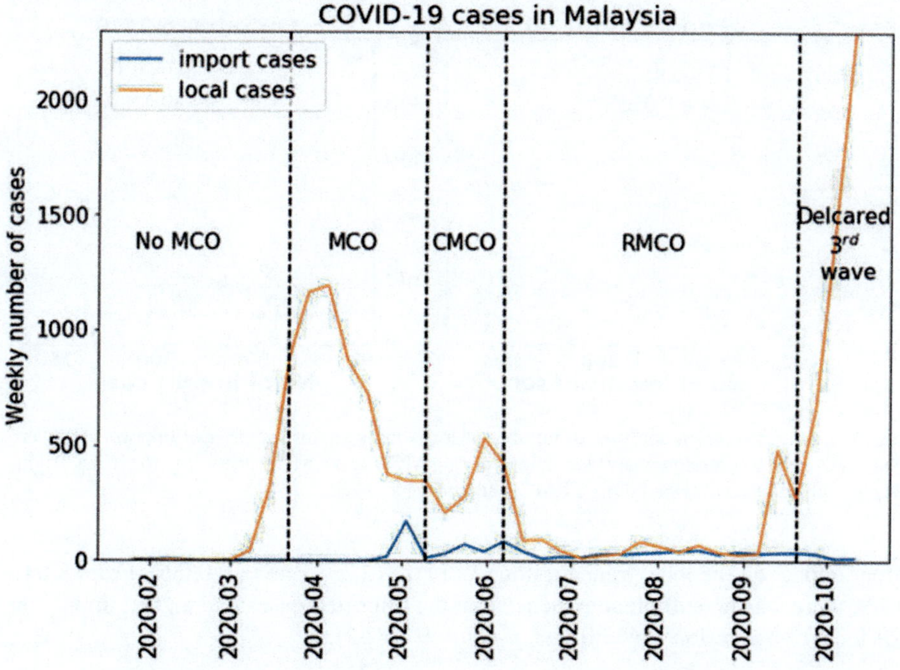

Fig. 1 Distribution of the imported and local cases in Malaysia according to different MCO phases (MCO: Movement Control Order; CMCO: Conditional Movement Control Order; and RMCO: Recovery Movement Control Order)

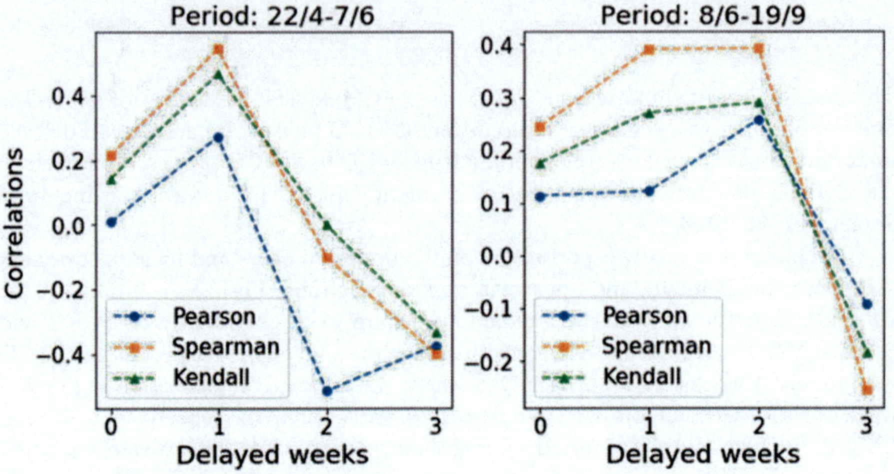

Fig. 2 The correlation coefficients between the imported and local cases were observed using Pearson, Kendall, and Spearman rank correlations. Left: Correlation coefficients obtained during MCO and CMCO according to weeks. Right: Correlation coefficients obtained during RMCO according to weeks

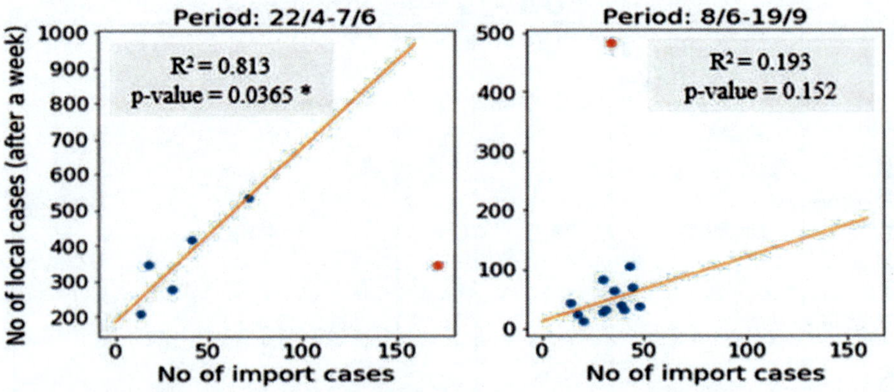

Fig. 3 Linear regression analyses of number of local cases against number of import cases were observed. Left: Regression analysis obtained during MCO and CMCO according to weeks. Right: Regression analysis obtained during RMCO according to weeks

great impact on the local transmission ($R^2 = 0.813$, p-value $= 0.0365$). From 8/6 to 19/9, there was no sufficient evidence that the imported cases had a great impact on the local transmission ($R^2 = 0.193$, p-value $= 0.152$).

4 Discussion

The data reported in this study is referring to the second pandemic wave in Malaysia. Rigorous study on the first wave was not performed in view of the lack of suitable data and the uncertainty of the parameters, namely, imported cases and local transmission. Additionally, the number of cases was relatively low in the first wave compared to the second one.

From the findings obtained through correlation analysis, imported cases have been unveiled to have a great impact on the local inhabitants during the onset of the movement control order phases in Malaysia. Nevertheless, the imported cases may have a lower impact on the local transmission due to strict SOP measures that have been taken by the Malaysian authorities. Collective efforts by the government and the society in obeying the measures set by the authorities may contribute to the negative correlation. The results from the analysis may be essential during the beginning of the wave. The time within the wave might not be greatly affected by the imported cases as most of the incidents are due to local transmission. This may reveal that there is a positive result of the authorities in monitoring the imported cases through measures, namely, quarantine and Home Surveillance Order.

The imported cases pose a direct relationship with the local occurrences, which can be seen in MCO-CMCO. As such, the relaxation of the border control should be well observed, particularly when the number of incidents is low.

5 Conclusion

The correlation and linear regression approaches can be seen to unfold the trend of the imported cases in local transmission incidents in Malaysia for different MCO phases. With the findings obtained from this study, it is expected that more emphasis should be given on the imported cases, particularly on the inhabitants who have returned to Malaysia during the pandemic period. Furthermore, the findings will assist the Malaysian authorities in structuring proper measures for the imported cases adherence to MCO. The future direction of this study is to investigate on the correlation and linear regression trends in the third outbreak wave in Malaysia, together with the reproduction ratio from the pandemic modeling curve using Susceptible-Exposed-Infectious-Removed (SEIR) model.

References

1. L. Heng, L. Shang-Ming, Y. Xiao-Hua, T. Shi-Lin, T. Chao-Ke, Coronavirus disease 2019 (COVID-19): current status and future perspectives. Int. J. Antimicrob. Agents **55**, 1–9 (2020)
2. M. Simsek, B. Kantarci, Artificial intelligence-empowered mobilization of assessments in COVID-19-like pandemics: a case study for early flattening of the curve. Int. J. Environ. Res. Public Health **17**, 1–17 (2020)
3. M. V. Ang, *COVID-19 Cluster in Kedah Possibly Caused By Super-Spreader Strain Imported From Egypt* (2020). Available: https://says.com/my/news/covid-19-cluster-in-kedah-possibly-super-spreader-strain-imported-from-egypt
4. CodeBlue, *Malaysia's Daily Covid-19 Cases Surge to 100, Most From Benteng LD Cluster* (2020). Available: https://codeblue.galencentre.org/2020/09/08/malaysias-daily-covid-19-cases-surge-to-100-most-from-benteng-ld-cluster/

AROGYAKAVACHAM—Automatic Hand Sanitizer Dispenser with Temperature Measurement

Harshil Sathwara, Parth Vaghela, and Smita Joshi

Abstract The prevention steps taken in coronavirus disease outbreak of using the mask and frequent cleaning of hands have become common in our daily routine along with social distancing. In schools, colleges, malls, and restaurants, additionally, temperature is also measured with IR temperature sensors. The device named AROGYAKAVACHAM—an automatic hand sanitizer dispenser with contactless temperature measurement device was designed, developed, and tested for a small group of Registrar Office of Anand Agriculture University and G. H. Patel College of Engineering and Technology, Gujarat, India. It consists of 12 V, 1 A adapter, voltage regulator LM7805, two Arduino Uno, one ultrasonic sensor, MLX90614 infrared temperature sensor, motor driver L298, sanitizer container, DC submersible pump, display with I2C module, and a buzzer. Initially, the device was powered by rechargeable DC battery which was charged by the solar panel. For cutting down its cost and make the device more compact, it was connected to AC supply through a 12 V, 1 A adapter. Additionally, ambient temperature was also measured along with respondent temperature. It was in good agreement with the actual ambient temperature. Although there was some difference in temperature of the respondent when compared with MLX90614 readings and temperature gun readings. This developed device will be a boon to the society as it reduces the risk of transmission of severe acute respiratory syndrome coronavirus 2.

Keywords Arduino · Coronavirus · Contactless sanitization · Infrared temperature sensor · Ultrasonic sensor

H. Sathwara (✉) · P. Vaghela
Department of Electrical Engineering, G.H. Patel College of Engineering and Technology, Vallabh Vidhyanagar, Anand, India

S. Joshi
Department of Applied Science and Humanities, G.H. Patel College of Engineering and Technology, Vallabh Vidhyanagar, Anand, India

F. Thakkar et al. (eds.), *Proceedings of the International e-Conference on Intelligent Systems and Signal Processing*, Advances in Intelligent Systems and Computing 1370,
https://doi.org/10.1007/978-981-16-2123-9_61

1 Introduction

More priority has been given to hygiene as a prevention measure to prevent the transmission of the disease, as suggested by the World Health Organization (WHO), owing to the global severe acute respiratory syndrome coronavirus 2 (SARS-CoV-2) epidemics [1]. Although the use of soap and water is the preferred way to clean one's hand, another option has recently become common. When one is on the run, or don't have access to soap and water, hand sanitizers are the safest choices [2]. According to the Centre for Disease Control and Prevention (CDC), one should use a hand sanitizer that has at least 60% alcohol to effectively eliminate bacteria from one's hands [3]. To use sanitizer effectively, one needs to ensure that hands are free from any visible germ or dirt before one applies the product on palm [4]. Apart from this temperature screening has also been brought to implementation. A thermometer is designed that uses infrared sensors to detect temperatures without bringing into contact by [5]. Infrared thermometers are largely based on concept of blackbody radiations to measure the infrared radiation wavelength of human body, preceded by human body temperature measurement. Infrared sensors are used to absorb human body infrared radiation without any emission which uses passive non-contact methods and can prevent cross infection of human body effectively and hence does not harm human body [6, 7]. As per WHO guidelines, the temperature below or at 37.3 °C is acceptable or say temperature above 37.4 °C should be the threshold temperature [8]. A person who may be sick has a body temperature then tested for a double search using a tympanic infrared thermometer. The body's temperature measurements are used by two forms of infrared thermometers: tympanic and forehead [9]. The coronavirus pandemic has contributed to a radical change in the trajectory of consumption, which would have a significant effect on the future scenario [10]. Health maintenance and control goods, as well as advances and development in the health sector, are commonly seen in all sectors. One such sub-sector consists of hand sanitizer dispensers, temperature screening systems, sanitizer spray tunnels, etc. [11]. However, it has been found that most hand sanitizer dispensers are situated at the entrance-exit point of different public meeting places, at workplaces, in specific schools, universities, workplaces, etc. [12]. It has to be used manually, either by hands or using feet. This can increase the contact as well as the risk of getting infections [13]. Apart from these, it has been observed that there are specific jobs given to 2–3 people for measuring temperature, noting down the readings and name of the person. Hence, hereby the solution to the abovementioned problems are explained.

1.1 Literature Review

WHO has called for safe hand washing and sanitizing procedures to be maintained, but the biggest concern is how one does it, that is, through physical touch. Touching containers with contaminated hands or hand sanitizers will spread the virus to the next individual. Various models are circulating across the web for automatic hand sanitizer dispenser; one of them is using metal–oxide–semiconductor field-effect transistor (MOSFET) and IR sensor [14] to reduce the cost [15]. However, the main disadvantage of these categories (without a microcontroller) is that spilling cannot be avoided. Another model available in the market is simply using IR sensor, DC pump, and wood for the compartment [16]. This DIY is simple to build and can only be a suitable choice for indoor use, as in the case of outdoor environments, this device would automatically dispense the sanitizer if someone cuts off the sunlight falling on the sensor. However, in [17], it has been demonstrated another way of building an automatic sanitizer dispenser by replacing an IR sensor with an ultrasonic sensor for hand detection, which reduces the abovementioned drawback and can be used both indoors and outdoors. In the design and installation of the automated hand sanitizing system with door controller using ATMEGA328P, the ultrasonic sensor, the electromagnetic lock, and other simple components are seen, thus replacing the security guard stationed at the gates and doors, which performs the opening-closing of the gates and hand sanitizer for people in person. Apart from automatic hand sanitizer dispenser, there is non-contact infrared temperature available made using ATMEGA328P and Mlx90614 used for measuring body temperature, surface temperature, heat ventilation, and much more [17]. However, this allows a person to carry a temperature gun that places a person in close contact with others, and this can raise the likelihood of illness spreading. The issue can then be overcome by removing the person-to-person contact defined by the KIOSK Information System, USA by developing and installing a facial recognition system that monitors the temperature [18]. NIRAMAI, India has also developed a temperature screening method based on Artificial Intelligence, which monitors fever and COVID-19-based respiratory symptoms. They also mounted a high-resolution thermal camera device that monitors temperature and other main Covid-19 symptoms and, in case of any identification, a local alarm is raised [19]. The radio-frequency identification-based method has also been developed where the name and temperature are entered in the Excel sheets while the individual uses RFID for attendance at school, colleges, offices, etc. [20]. There is a system developed by SIFSOF, USA which combines both a contactless temperature control system and alarm system with an automatic hand-held dispenser [21]. However, it costs US 585$ as of October 7, 2020 which could be very expensive and could not be afforded by everyone. Also, a high-precision real-time temperature control approach based on distance correction was proposed and a device architecture was adopted to satisfy the need for real-time tracking of the human body surface temperature for optical recovery therapy [22].

2 Methodology

2.1 Circuit Explanation

As shown in Fig. 1, 12 V, 1 A adapter has been used for the system. However, since the system requires only 5 V for the components used in the system, LM-7805 has been used for voltage regulation. Mlx90614 takes the regulated voltage as its input and the ground is connected to negative of the battery. Further SCL-SDA connections are made with Arduino UNO. LCD display with I2C and buzzer are also connected with the same Arduino UNO. For accurate measurement, one has to place hand within the range of 1 cm above the temperature sensor. The display will show both respondent temperature and ambient temperature. As the threshold temperature has been kept to be of 37.4 °C, the buzzer will alert when the temperature will exceed above the threshold. Further, the regulated voltage is provided to the ultrasonic sensor and L298 motor driver. Whenever the hand is in range of up to 8 cm, the ultrasonic sensor will be activated and will give the signal to another Arduino UNO used as shown in the above figure. This will further provide the signal to L298 motor driver to run the pump for 0.6 s and will later cool down for 2 s and then the process will be repeated. During the cool-down period, no sanitizer will be dispensed even if the hands are in proximity. Also one can use solar panel when there is power cut-off by simply connecting LM 7812 voltage regulator. The design of the system in form of schematic diagram is shown in Fig. 2.

Figure 3a gives an explanation of the working procedure of temperature measurement using MLX90614. When the hand is within 1 cm range, as mentioned above

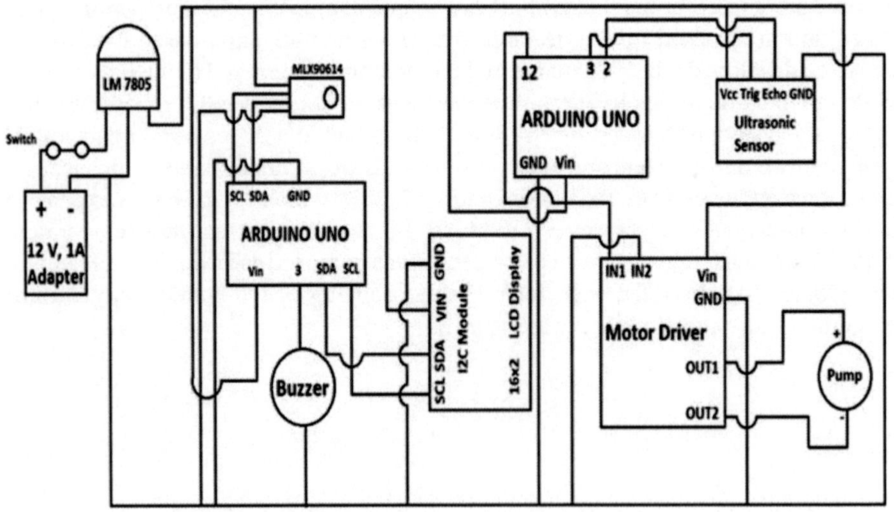

Fig. 1 AROGYAKAVACHAM circuit diagram

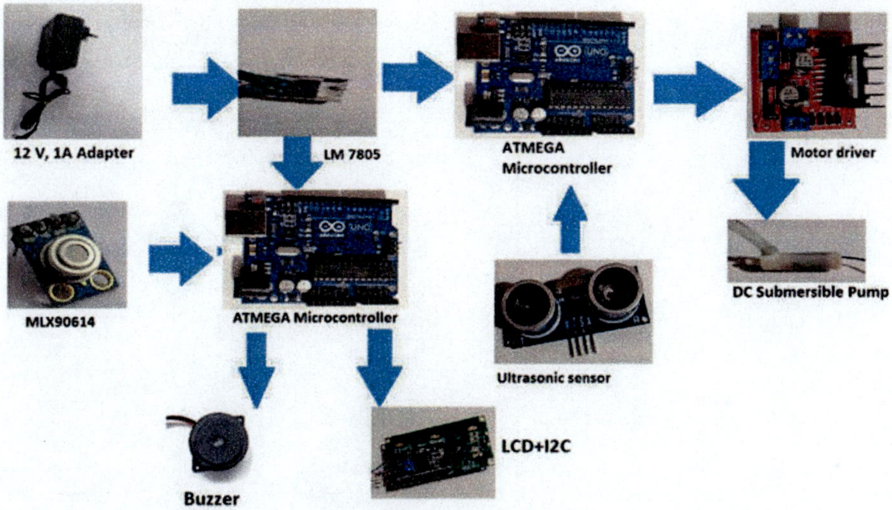

Fig. 2 AROGYAKAVACHAM schematic diagram

for accurate measurement, both ambient temperature and respondent temperature are displayed in LCD. Also in case, the temperature goes above the threshold, the buzzer will give alert or else the procedure is repeated. Similarly, Fig. 3b gives an explanation of automatic hand sanitizer dispenser. When the hand is in the proximity of the ultrasonic sensor, about 0.83 ml of sanitizer will be dispensed for 600 ms and then it will cool down for about 2 s. After that, the procedure will be repeated.

2.2 Hardware Prototype Development

The development of contactless hand sanitizer dispenser with temperature measurement is as shown in Fig. 4. PVC foam sheet of 5 mm thickness was used to make the outer box of the device. On the front side of the device, there is a dispenser outlet, ultrasonic sensor, slit to indicate the level of sanitizer. On the top section, there are IR temperature sensor and inlet pipe have been fixed to refill the sanitizer. In the left side of the device, power pin and on/off switch has been fixed. On the right side of the device, a USB port has been fixed to connect it with LCD display. Figures 4 and 5 show the hardware prototype and frame dimensions prepared using AutoCAD software with 3D View of the system.

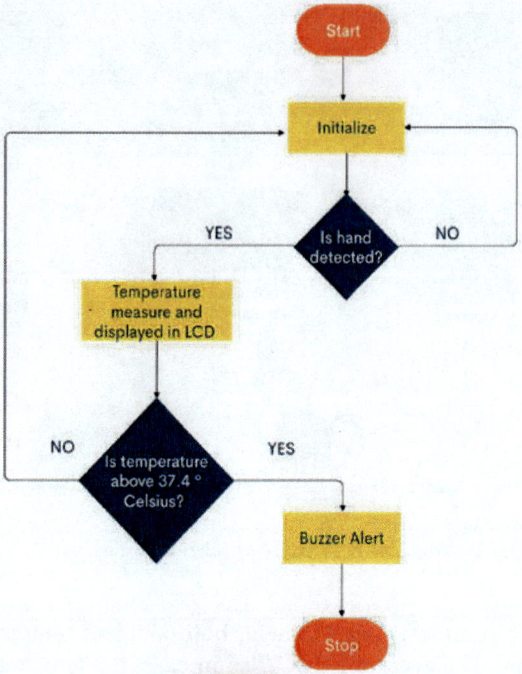

(a)

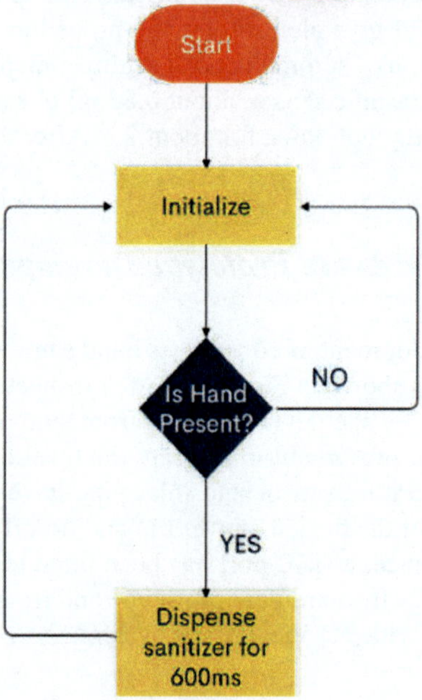

(b)

Fig. 3 **a** and **b** AROGYAKAVACHAM flowchart

Fig. 4 AROGYAKAVACHAM hardware prototype

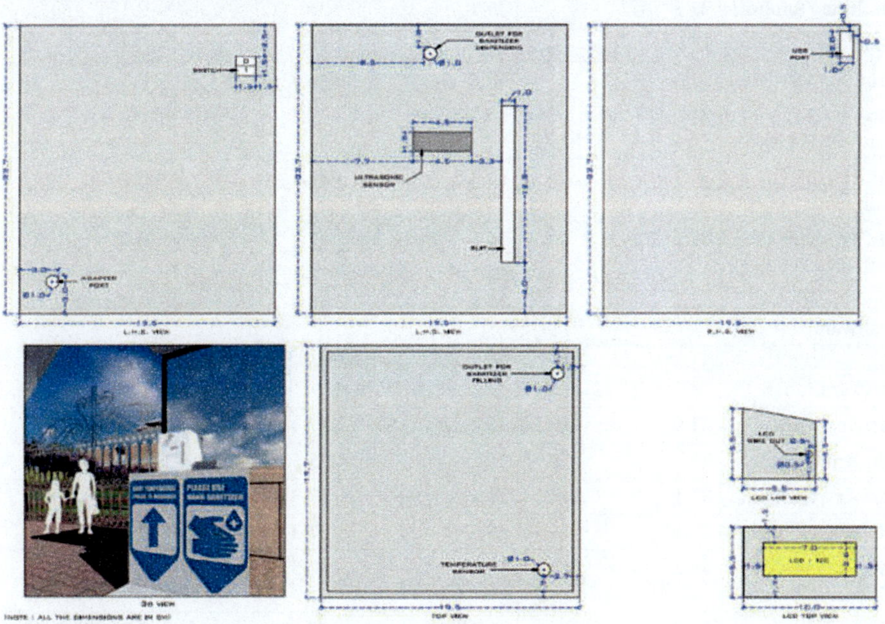

Fig. 5 AROGYAKAVACHAM frame dimension

3 Results and Discussions

The test was conducted on December 8, 2020 at 15:30 h at G. H. Patel College of Engineering and Technology and at 16:30 h in the Office of Registrar of Anand Agricultural University, Anand, Gujarat. Data of 20 respondents have been collected and the graph has been plotted as shown in Table 1 and Fig. 6. There are errors while comparing the reading of MLX90614, temperature gun, and clinical thermometer. Here, the readings measured by MLX90614 and clinical thermometer are found to be similar but there are errors when compared with a temperature gun. The difference

Table 1 Reading of MLX90614 and temperature gun of 20 respondents

Name	Ambient temp (°C)	MLX90614 (°C)	Temp gun (°C)	Difference (°C)
Prof. Mukesh Joshi	31	36.3	36	−0.3
Dr. Vinod Patel	31.2	35.9	36	0.1
Mr. NiravMehta	31	37	36.5	−0.5
Dr. KetanTamboli	31.3	36.1	36	−0.1
Dr. Heena Raj	31.2	36.8	36.4	−0.4
Prof. Sameer Patel	31	35.8	36.1	0.3
Prof. PartheshMankodi	31	36.5	37.1	0.6
Mr. Harshil Sathwara	31.7	37.4	37.1	−0.3
Mr. Prashant Sathwara	31.6	36.5	37.2	0.7
Mrs. Avani Sathwara	31.2	36.8	36.4	−0.4
Mr. DevkarnRabari	31.8	36.4	36.7	0.3
Mr. SudeepBagul	31.9	36.4	37.1	0.7
Mr. Amit Patel	32	35	35.5	0.5
Mr. ArvindDabhi	32.1	37	37.3	0.3
Mr. Ronak Patel	32.1	36.5	37.3	0.8
Mr. MukeshRathod	32.2	37	37.3	0.3
Mr. GanshyamValani	32.8	36.7	37	0.3
Mr. MansingChavda	32.9	36.8	37	0.2
Mr. Nirav Patel	33	36.2	37	0.8
Mr. Bharat Satasiya	31.9	36	36.6	0.6

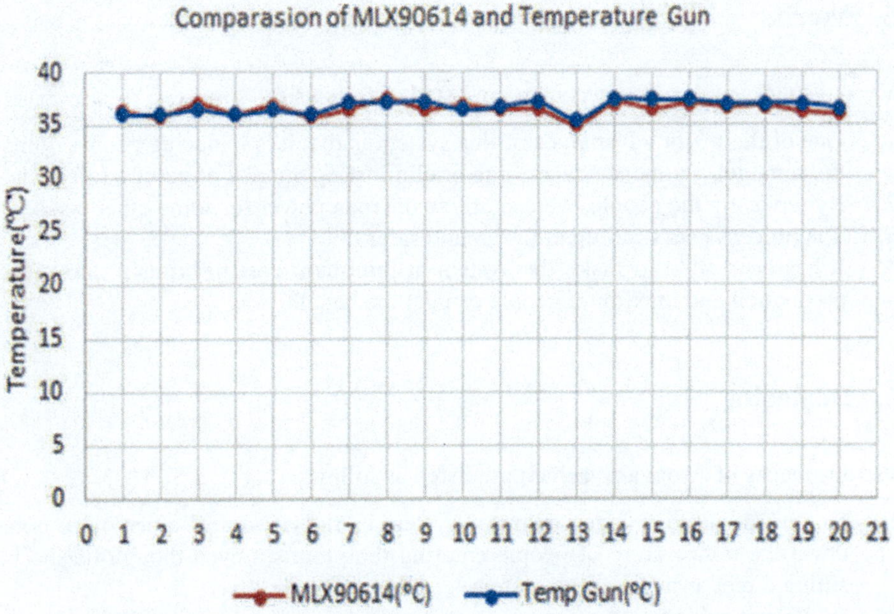

Fig. 6 Graph shows comparison between temperature gun and MLX90614

is calculated between temperature gun and mlx90614. The mean of the difference is found to be 0.225 and hence the system is accurate enough. The variations are mainly due to interference light, measuring distance, ambient temperature, time, and measuring objects. However, large errors can be avoided by keeping the measuring distance in the range of 2–4 cm, under a suitable environment and with the hands should be kept steady while measuring the temperature [23]. In reality, a shift incident angle occurs between the infrared sensor and the measured object for real infrared measurements due to the actual measuring condition and cost limitation which is explained in [24].

4 Applications

Various applications of Arogyakavacham are discussed as follows:

1. This setup can be replaced with the person/s doing the task of temperature measurement, entering data, and advising people to properly sanitize the hands.
2. Can be used mostly at the entrance of offices, restaurants, University campus, shopping malls, hospitals, etc.

5 Merits

Various merits of Arogyakavacham are listed as follows:

1. One of the major advantages of this system is that it is replacing people doing the same job and hence decreasing or eliminating the risk of getting infected.
2. By replacing the people, we can utilize the man power for some other tasks.
3. The process gets simplified, faster, and safer.
4. It has been tried to make the system at minimum cost to bring it into mass production and maximum people can get the benefit.

6 Demerits

Various merits of Arogyakavacham are listed as follows:

1. One of the disadvantages with this system is that it requires a person to note down the temperature of people entering the premises even though he/she is sitting 6 feet away from the system.
2. While operating using solar panel, the system may not function during lower solar insolation.
3. Temperature sensor will not provide accurate reading if the hands are not within the mentioned range and unsteady.

7 Conclusion and Future Scope

In this paper, it has been discussed about the automatic sanitizer dispenser with contactless temperature reader usage to avoid the transmission of SARS-Cov-2. This system replaces the people doing the task of measuring and noting down of the temperature and hence decreasing their chance of getting infected by the virus. Also, the process becomes faster by replacing the sanitizer pump which was operated manually by hand or feet and hence, in turn, decreasing the risk of getting infected by touching the same sanitizer pump repetitively. It has been found that around 0.83 ml of sanitizer dispenses which is found to be sufficient. Also, the temperature sensor gives both readings of ambient temperature and respondent temperature. The errors in the reading, the reason behind it, and ways to solve it are discussed in the results section. Apart from these, LCD display is separated from the device using 6 feet USB cable, so that the person noting down the temperature remains 6 feet away and the social distancing norms are followed which in turn reduces the risk of getting infected by a virus. Future scope or modification to be done in the device is discussed as follows:

- Introduction of automatic registering of names of people with their temperature [25]. Hence, replacing the person who is sitting to register the temperature and his getting risks of getting infected during this job are eliminated.
- Introduction of system which feeds attendance and date-time directly into the computer at the time when the person is measuring the temperature.
- Ultrasonic sensor can be introduced in the system and programmed so that when the hand is within the mentioned range of MLX90614 as detected by ultrasonic sensor, the temperature sensor will be activated and the readings will be taken and hence accuracy can be increased [26].

Acknowledgements The authors wish to thank Dr. H B Soni, Principal, GCET and Dr. Ritesh Patel, HOD of Electrical Department. The authors also like to thank Dr. Subrota Hati, Assistant Professor at AAU, Anand, Gujarat for their very generous and valuable help in paper correction as well as writing the paper as per manuscript. The authors also wish to thank Ar. Prem Sathwara for helping with the CAD model sketch and Er. Krushi Modi and Er. Ishan Bhatt for helping with the testing of the MLX90614 temperature sensor and IR sensor. Also, the author wishes to thank Office of the Registrar of Anand Agricultural University, Gujarat for their cooperation in taking temperature sensor data.

References

1. WHO publication on water, sanitation, hygiene, and waste management for SARS-CoV-2, the virus that causes COVID-19. https://www.who.int/publications/i/item/water-sanitation-hygiene-and-waste-management-for-covid-19. Accessed 10 Dec 2020
2. J.L.J. Jing, T. Pei Yi, R.J.C. Bose, J.R. McCarthy, N. Tharmalingam, T. Madheswaran, Hand sanitizers: a review on formulation aspects, adverse effects, and regulations. Int. J. Environ. Res. Public Health **17**(9), 3326 (2020). https://doi.org/10.3390/ijerph17093326
3. Centre for Disease Control and Prevention (CDC) community guideline. https://www.cdc.gov/handwashing/hand-sanitizer-use.html. Accessed 10 Dec 2020
4. Verywell health. https://www.verywellhealth.com/should-you-use-hand-sanitizer-770727. Accessed 10 Dec 2020
5. J. Zhang, Development of a non-contact infrared thermometer, in *International Conference Advanced Engineering and Technology Research*, vol. 153, ed. by J.X.N. Wang, H. Zhou. (Atlantis Press, Shenyang, China, 2018), pp. 308–312. https://doi.org/10.2991/aetr-17.2018.59
6. Y. Luo, Design of data acquisition system based on LabVIEW for infrared thermometer. Adv. Mater. Res. **569**, 808–813 (2012)
7. N. Li, Design of infrared temperature measurement system based on MLX90615 and MSP430. J. Sens. Syst. **9**, 115–117 (2011)
8. Temperature screening guidelines. shorturl.at/vQWY3. Accessed 10 Dec 2020
9. H. Chen, A. Chen, C. Chen, Investigation of the impact of infrared sensors on core body temperature monitoring by comparing measurement sites. Sensors **20**(10), 1–17 (2020). https://doi.org/10.3390/s20102885
10. M. Chaudhary, P.R. Sodani, S. Das, Effect of COVID-19 on economy in India: some reflections for policy and programme. J. Health Manag. **22**(2), 169–180 (2020)
11. The global hand sanitizer market growth over 600% due to impact COVID—19 in 2019. https://www.prnewswire.com/news-releases/the-global-hand-sanitizer-market-to-grow-over-600-due-to-impact-covid-19-in-2019---market-research-by-arizton-301041017.html. Accessed 10 Dec 2020

12. The importance of hand sanitizer placement in workplace. https://info.debgroup.com/blog/the-importance-of-hand-sanitizer-placement-in-the-workplace. Accessed 10 Dec 2020
13. E. Enerst, J. Wantimba, K. Zaina, Design and implementation of a smart hand sanitizer dispenser with door controller using ATMEGA328P. Int. J. Eng. Inf. Syst. **4**(9), 114–119 (2020)
14. P. Ajmera, A review paper on infrared sensor. Int. J. Eng. Res. Technol. (IJERT) VIMPACT **5**(23), 1–3(2017)
15. Automatic hand sanitizer dispenser (No Arduino). https://www.hackster.io/taifur/automatic-hand-sanitizer-dispenser-no-arduino-ce852c. Accessed 10 Dec 2020
16. TPC automation for indoor usage. https://tpcautomation.com/how-make-hand-sanitizer-dispenser/. Accessed 10 Dec 2020
17. Make a non-contact infrared thermometer with IR temperature sensor. https://circuitdigest.com/microcontroller-projects/ir-thermometer-using-arduino-and-ir-temperature-sensor. Accessed 10 Dec 2020
18. Temperature check-in KIOSK overview. shorturl.at/anEO0. Accessed 10 Dec 2020
19. NIRAMAI contactless COVID fever screening. shorturl.at/buxA3. Last accessed 10 Dec 2020
20. RFID based contactless body temperature screening using arduino and MLX90614 IR temperature sensor. shorturl.at/dfxT9. Last accessed 10 Dec 2020
21. SIFSOF temperature detector and automatic disinfectant dispenser. https://www.sifsof.com/product/temperature-detector-with-automatic-disinfectant-dispenser-sifcleantemp-1-4/. Last accessed 10 Dec 2020
22. J. Chen. J. Wang, T. Shen, D. Xiong, G. Liquan, High precision infrared temperature measurement system based on distance compensation, in *ITM Web of Conferences 2017*, vol. 12, ed. by L. Yongjun, Z. Lei, C. Hong. (EDP Sciences, China, 2017), pp. 1–5
23. X. Luo, The calibration method of infrared thermometers for measurement of human temperature. Measure. Test. Technol. **7**, 10–11 (2013)
24. S. Wei, W. Qin, L. Han et al., The research on compensation algorithm of infrared temperature measurement based on intelligent sensors. Cluster Comput. **22**, 6091–6100 (2019). https://doi.org/10.1007/s10586-018-1828-5
25. S. Patel, R. Kumar, Face recognition based smart attendance system using IOT. Int. J. Comput. Sci. Eng. **6**(5), 871–877 (2018)
26. M. Gan, J. Sy, E. Calibara, A. Comendador, W. Degife, A. Sisay, Non-contact temperature reader with sanitizer dispenser (NCTRSD). Int. J. Sci. Res. Publ. **10**(9), 583 (2020)

Author Index

F. Thakkar et al. (eds.), *Proceedings of the International e-Conference on Intelligent Systems and Signal Processing*, Advances in Intelligent Systems and Computing 1370,
https://doi.org/10.1007/978-981-16-2123-9